FUNDAMENTALS OF ELECTRICAL ENGINEERING

FUNDAMENTALS OF ELECTRICAL ENGINEERING

By

Dr. Yaduvir Singh
Ph.D., M.B.A., M.E., B.E.
Professor and Head
Department of Electrical Engineering
Harcourt Butler Technological Institute,
Kanpur (U.P.)

Mandhir Verma
Ph.D., M.E., B.Tech.*
Associate Professor
Department of Electrical and Electronics Engineering,
Accurate Institute of Management and Technology,
Greater Noida, (U.P.)

UNIVERSITY SCIENCE PRESS
(An Imprint of Laxmi Publications Pvt. Ltd.)
An ISO 9001:2008 Company

BENGALURU • CHENNAI • COCHIN • GUWAHATI • HYDERABAD
JALANDHAR • KOLKATA • LUCKNOW • MUMBAI • RANCHI • NEW DELHI
BOSTON (USA) • ACCRA (GHANA) • NAIROBI (KENYA)

FUNDAMENTALS OF ELECTRICAL ENGINEERING

Printed and bound in India
Typeset at Shubham Composer, Delhi
First Edition : 2010; Second Edition : 2016
ISBN 978-93-80386-76-8

PUBLISHED IN INDIA BY

UNIVERSITY SCIENCE PRESS
(An Imprint of Laxmi Publications Pvt. Ltd.)

An ISO 9001:2008 Company
113, GOLDEN HOUSE, DARYAGANJ,
NEW DELHI - 110002, INDIA
Telephone : 91-11-4353 2500, 4353 2501
Fax : 91-11-2325 2572, 4353 2528
www.laxmipublications.com info@laxmipublications.com

Branches

Bengaluru	080-26 75 69 30	
Chennai	044-24 34 47 26,	24 35 95 07
Cochin	0484-237 70 04,	405 13 03
Guwahati	0361-254 36 69,	251 38 81
Hyderabad	040-27 55 53 83,	27 55 53 93
Jalandhar	0181-222 12 72	
Kolkata	033-22 27 43 84	
Lucknow	0522-220 99 16	
Mumbai	022-24 91 54 15,	24 92 78 69
Ranchi	0651-220 44 64	

C—
Printed at: Repro Knowledgecast Limited, Thane

Dedicated

to

Our Parents

Contents

PREFACE TO THE SECOND EDITION

"Books tap the wisdom of our species — the greatest minds, the best teachers — from all over the world and from all our history. And they're patient."

—Carl Sagan

These lines exactly portray the zeal of the authors behind the creation of this marvel entitled, "Fundamentals of Electrical Engineering". This is a perfect textbook having utmost reader friendly approach. It retains all the modern concepts. Effective teaching of basics and fundamentals of electrical engineering had always been a challenging task. This book presents electrical engineering subject in a terse and simplified manner so that it is easy to comprehend and logically tractable.

This is second edition presented to our readers with more solved problems, new methodologies to solve conventional problems and new chapters.

Our aim in writing this book is to bridge the gap between specialist textbooks (which are pitched at a level too academic for an average student) and the more prosaic 'handbooks', which are full of useful detail but provide little opportunity for the development of any real insight or understanding.

The book has evolved from long teaching experiences in several technical institutions. The subject matter in each chapter is presented in a way that enables students to graduate from very basic simple problems to the complex problems in electrical engineering easily.

However, we lay no claim to the original research in writing this work.

Chapter 1 and chapter 2 present DC circuit concepts and their analyses with different methods and approaches. These chapters provide all basic and necessary concepts of DC circuits as developed over the years. Chapters 3, 4 and 5 discuss AC circuit fundamentals, their steady state analysis and concept of resonance in series as well as in parallel AC circuits. Chapter 6 deals with basic concepts of magnetic circuits – which later become foundation stones for study of transformers.

Chapter 7 deals with electrical measuring instruments. The basic principles and working of most of the electrical measuring instruments is dealt with precise ideas in this chapter.

Chapter 8 discusses single phase transformers the most basic structural blocks involved in the electrical engineering. A transformer is a static electromagnetic energy conversion device, which transfers electrical energy from one electrical circuit to another electrical circuit, without any change in supply frequency.

A big part of electrical power is used to drive machines in industry and homes. Useful gadgets like fans, tape-recorders, mixers, etc., all run with the help of a *motor* fitted inside them. Basically, these motors take energy from electrical power mains and convert it into mechanical energy which is used to turn the blades of the fan or juicers. Reverse conversion takes place in a machine called *generator*. Motors and generators are electromechanical energy conversion (EMEC) devices or machines. Chapter 9 presents generalize theory of electrical machines in which basic principles of rotating machines are discussed followed by chapter 10 through 14 related to rotating machines. DC machines with both DC motor and DC generator are discussed in chapter 10 and chapter 11. The three phase induction and single phase induction motors are described in chapters 12 and 13. Chapter 14 deals with three phase synchronous machines in which both synchronous generator (commonalty known as alternator) and synchronous motor are discussed.

Chapter 15 is concerned with three phase AC circuits. It gives insight of connections of three phases and provides a brief knowledge of how windings are placed in rotating electrical machines.

In the end, chapter 16 and chapter 17 deals with power system basics and introduction to earthing and electrical safety, which is an important basic aspect for any stream engineer for planning and executing minor or major electrical power dependent projects. Grounding will help protect the person working on the system, the system itself, and any appliances and equipment that are connected to the system.

To provide motivation, numerous examples and numerical problems have been discussed along with the text and also at end of the each chapter.

—Authors

PREFACE TO THE FIRST EDITION

"Books tap the wisdom of our species — the greatest minds, the best teachers — from all over the world and from all our history. And they're patient."

—Carl Sagan

These lines exactly portray the zeal of the authors behind the creation of this marvel entitled, "Fundamentals of Electrical Engineering". This is a perfect textbook having utmost reader friendly approach. It retains all the modern concepts. Effective teaching of basics and fundamentals of electrical engineering had always been a challenging task. This book presents electrical engineering subject in a terse and simplified manner so that it is easy to comprehend and logically tractable.

Our aim in writing this book is to bridge the gap between specialist textbooks (which are pitched at a level too academic for an average student) and the more prosaic 'handbooks', which are full of useful detail but provide little opportunity for the development of any real insight or understanding.

The book has evolved from long teaching experiences in several technical institutions. The subject matter in each chapter is presented in a way that enables students to graduate from very basic simple problems to the complex problems in electrical engineering easily.

However, we lay no claim to the original research in writing this work.

Chapter 1 and chapter 2 present DC circuit concepts and their analyses with different methods and approaches. These chapters provide all basic and necessary concepts of DC circuits as developed over the years. Chapters 3, 4 and 5 discuss AC circuit fundamentals, their steady state analysis and concept of resonance in series as well as in parallel AC circuits. Chapter 6 deals with basic concepts of magnetic circuits— which later become foundation stones for study of transformers. Chapter 7 deals with electrical measuring instruments. The basic principles and working of most of the electrical measuring instruments is dealt with precise ideas in this chapter.

Chapter 8 discusses single phase transformers the most basic structural blocks involved in the electrical engineering. A transformer is a static electromagnetic energy conversion device, which transfers electrical energy from one electrical circuit to another electrical circuit, without any change in supply frequency.

Chapter 9 presents generalized theory of electrical machines which is followed by chapter 10 through 14 related to rotating machines. DC machines with both DC motor and DC generator are discussed in chapter 10 and chapter 11. The three phase induction and single phase induction motors are described in chapters 12 and 13. Chapter 14 deals with three phase synchronous machines in which both synchronous generator (commonalty known as alternator) and synchronous motor are discussed.

Chapter 15 is concerned with three phase AC circuits. It gives insight of connections of three phases and provides a brief knowledge of how windings are placed in rotating electrical machines. Finally, Chapter 16 deals with power system and its concepts.

To provide motivation, numerous examples and numerical problems have been discussed along with the text and also at end of the each chapter.

—Authors

Acknowledgement

We are always be thankful to Almighty without whose blessings this work will not have come to its present shape.

In the course of writing this book many people have supported us in countless different ways. We express our deepest sense of gratitude to all of them, whether mentioned here or not. We greatly acknowledge the subject rendered by our buddies, our young and old friends from IIT Roorkee, IIT Kanpur, IIT Delhi, IIT Kharagpur, NIT Bhopal, NIT Allahabad, NIT Hamirpur, NIT Jalandhar, IET Lucknow, HBTI Kanpur, MMMEC Gorakhpur, KNIT Sultanpur, BIET Jhansi, G B Pant University, Pantnagar, Indian Railways, BSNL, NTPC, GAIL, SAIL, CSIR, DRDO, BARC, TIFR, RITES, Future First, D.E. Shaw, TCS, Infosys, Samsung India, Samtel, ICICI etc.

This book is a tribute to our parents and our family members. We couldn't have done this without you all. We express our sincere thanks to all those who have directly or indirectly helped and encouraged in producing this work.

—Authors

1

DC Circuit Concepts

1.1 INTRODUCTION

A *circuit* is an unbroken loop of conductive material that allows electrons to flow through itself continuously without beginning or end.

One might have been wondering how electrons can continuously flow in a uniform direction through wires without the benefit of these hypothetical electron sources and destinations. In order for the source-and-destination scheme to work, both would have to have an infinite capacity for electrons in order to sustain a continuous flow.

Using the marble-and-tube analogy, the marble source and marble destination buckets would have to be infinitely large to contain enough marble capacity for a "flow" of marbles to be sustained.

The answer to this paradox is found in the concept of a *circuit*: **a never-ending looped pathway for electrons.** If we take a wire, or many wires joined end-to-end, and loop it around so that it forms a continuous pathway, we have the means to support a uniform flow of electrons without having to resort to infinite sources and destinations.

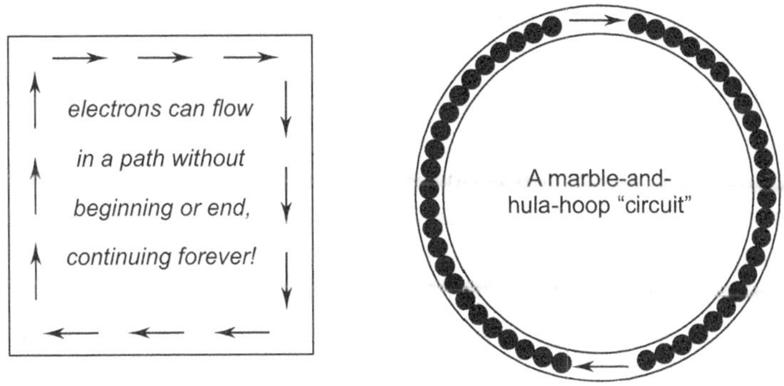

Fig. 1.1

Each electron advancing clockwise in this circuit (Fig. 1.1) pushes on the one in front of it, which pushes on the one in front of it, and so on, and so on, just like a hula-hoop filled with marbles.

Now, we have the capability of supporting a continuous flow of electrons indefinitely without the need for infinite electron supplies and dumps. All we need to maintain this flow is a continuous means of motivation for those electrons.

It must be realized that continuity is just as important in a circuit as it is in a straight piece of wire. Just as in the example with the straight piece of wire between the electron source and destination, any break in this circuit will prevent electrons from flowing through it.

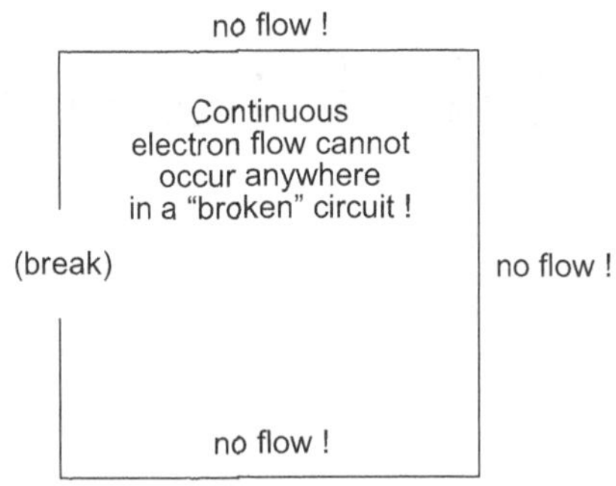

Fig. 1.2

An important principle to realize here is that ***it doesn't matter where the break occurs.*** Any discontinuity in the circuit will prevent electron flow throughout the entire circuit. ***Unless*** there is a continuous, unbroken loop of conductive material for electrons to flow through, a sustained flow simply cannot be maintained.

1.2 ELEMENTARY CONCEPTS AND DEFINITIONS

First we will discuss the elementary concepts and definitions related to electrical networks and circuits for a better insight.

Circuit

A circuit is a closed conducting path through which an electric current either flows or is intended to flow.

(*i*) ***Linear Circuit:*** Linear circuit is one whose parameters are constant, *i.e.,* they do not change with voltage and current.

(*ii*) ***Non-linear Circuit:*** Non-linear circuit is that circuit whose parameters changes with voltage or current.

(*iii*) ***Bilateral Circuit:*** Bilateral circuit is one whose properties or characteristics are the same in the either direction. For example, a circuit consists of R, L or C.

(*iv*) ***Unilateral Circuit:*** Unilateral circuit is that circuit whose properties or characteristics change with the direction of its operation. For example, circuit containing diodes.

(*v*) ***Passive Circuit:*** Passive circuit is one which has no source of e.m.f. in it.

(*vi*) ***Active Circuit:*** Active circuit is one which contains one or more than one source of e.m.f. in it.

Parameters

The various elements of a circuit are called its parameter like resistance, inductance and capacitance.

(*i*) ***Lumped Parameters***

Lumped parameters are those parameters which can be separate out from a given circuit. For example, an FM Radio circuit, we can separate out the resistors, capacitors etc.

(*ii*) ***Distributed Parameters***

Those parameters which are internal property of a circuit and which can't be separated out from the circuit are called as Distributed parameters. For example, resistance, inductance and capacitance of a 'Transmission line'.

(*iii*) ***Linear Parameters***

Linear parameters are those which follow Ohm's law. For example, resistance.

(*iv*) ***Non-Linear Parameters***

Non-linear parameters are those which do not follow Ohm's law. For example, diode.

Node

Node is a junction in a circuit where more than two branches are connected together or a node can also be defined as that point in a circuit, where the current divides in different branches.

Branch

Branch is that part of a network which lies between two junctions or nodes.

Loop

Loop is closed path in a circuit in which no elements of the node is encountered more than once.

Mesh

Mesh is a loop that contains no other loop in it.

In Fig. 1.3(*a*), I_1 and I_2 represent mesh (as well as loop also) and the bigger dotted path I_3 shown in Fig. 1.3(*b*) represents loop only. There are two nodes in the circuit represent by '1' and '2'. Sometimes this bigger dotted path [Fig. 1.3(*b*)] is also known as ***super mesh***.

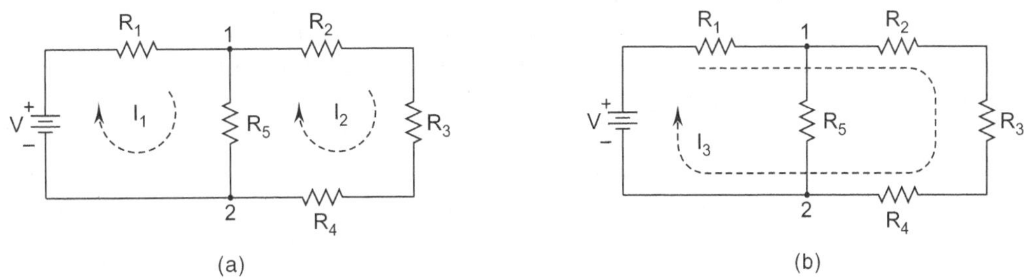

Fig. 1.3

Voltage

Voltage is the electrical force that causes current to flow in a circuit. It is measured in Volts (V).

Current

Current is the movement of electrical charge–the flow of electrons through the electronic circuit. The direction of a current is opposite to electrons flow direction. Current is measured in Amperes (A).

Power

Power and *Energy* are two terms that are often misused. Energy can be thought of as the ability to do work, and it has units such as joules. Power, on the other hand, is the *rate* at which energy is generated or used, and therefore it has rate units such as joules/s. There is often confusion about the units for electrical power and energy. Electrical power is measured in watts, which is a rate (1 J/s = 1 watt), so electrical energy is watts multiplied by time, for example, watt-hours. Be careful not to say "watts per hour," which is incorrect.

When a battery delivers current to a load, power is generated by the battery and is dissipated by the load. The expression for instantaneous power supplied, or consumed, by a component of a circuit is given as

$$\boxed{P = VI} \qquad \ldots(1.1)$$

Above equation tells us that the power supplied at any instant by a source, or consumed by a load, is given by the current through the component times the voltage across the component. When current is given in amperes, and voltage in volts, the units of power are *Watts* (W). Thus, a 12 V battery delivering 10 A to a load is supplying 120 W of power.

Energy

Since power is the rate at which work is being done, and energy is the total amount of work done, energy is just the integral of power.

$$\boxed{W = \int P\,dt} \qquad \ldots(1.2)$$

In an electrical circuit, energy can be expressed in terms of joules (J), where 1 watt-second = 1 joule. In the electric power industry the units of electrical energy are more often given in watt-hours, or for larger quantities kilowatt-hours (kWh) or Megawatt-hours (MWh). Thus, for example, a 100 W computer that is operated for 10 hours will consume 1000 Wh, or 1 kWh of energy.

1.3 OHM'S LAW

An excellent analogy for the simplest of electrical circuits is the water in a hose connected to a pressure valve. Think of the electrons in the copper wire as the water in the hose, the pressure valve as the applied voltage, and the size of the hose as the factor that determines the resistance. If the pressure valve is closed, the water simply sits in the hose without motion, much like the electrons in a conductor without an applied voltage. When we open the pressure valve, water will flow through the hose much like the electrons in a copper wire when the voltage is applied.

In other words, the absence of the 'pressure' in one case and the voltage in the other will simply result in a system without motion or reaction. The rate at which the water will flow in the hose is a function of the size of the hose. A hose with a very small diameter will limit the rate at which water can flow through the hose, just as a copper wire with a small diameter will have a high resistance and will limit the current.

In summary, therefore, the absence of an applied 'pressure' such as voltage in an electric circuit will result in no reaction in the system and no current in the electric circuit. Current is a reaction to the applied voltage and not the factor that gets the system in motion. To continue with the analogy, the more the pressure at the valve, the more the rate of water flow through the hose, just as applying a higher voltage to the same circuit will result in a higher current. Substituting the above discussed analogy for an electrical circuit, results in

$$\text{Current} = \frac{\text{Potential difference}}{\text{Resistance}} \qquad \ldots(1.3)$$

$$I = \frac{V}{R} \text{ (Ampere)} \qquad \ldots(1.4)$$

Equation (1.4) is known as Ohm's law in honor of Georg Simon Ohm. The law clearly reveals that for a fixed resistance, the greater the voltage across a resistor, the more the current, and the more the resistance for the same voltage, the less the current. In other words, the current is proportional to the applied voltage and inversely proportional to the resistance.

Note in Fig. 1.4 that the voltage source 'pressures' current in a direction that passes from the negative to the positive terminal of the battery. This will always be the case for single-source circuits. The effect of more than one source in the network will be examined in the chapters to follow.

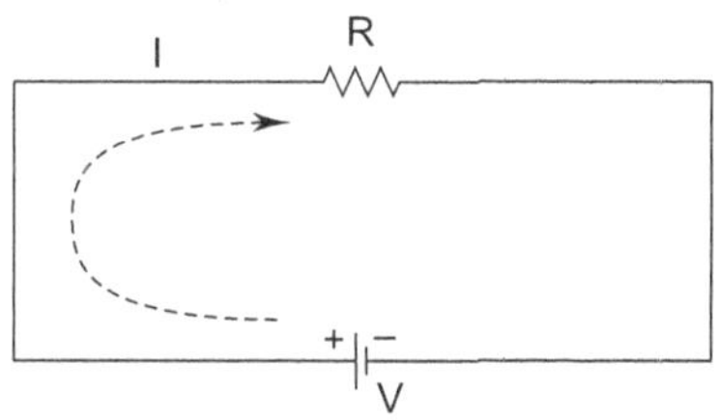

Fig. 1.4

1.4 VOLTAGE AND CURRENT SOURCES

To supply electrical power to the circuit or load, source is required. The source is either DC (Direct Current) or AC (Alternating Current). Alternating current source (AC) will be discussed later on the upcoming chapters. Here we are concern only with direct current source.

Basically there is no practical difference between a voltage source and a current source.

A source [Fig. 1.5(*a*)] which has very ***low*** internal resistance as compared to load resistance is considered as ***Voltage source.***

A source [Fig. 1.5(*b*)] which has very ***high*** internal resistance as compared to load resistance is considered as ***Current source.***

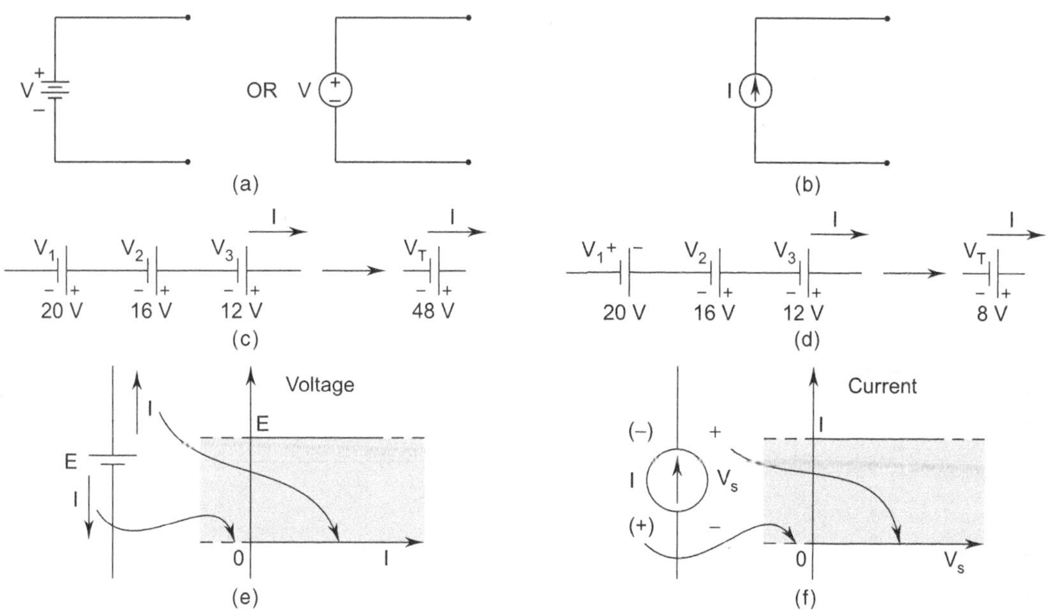

Fig. 1.5

Voltage sources can be connected in series, as shown in Figs. 1.5 (*c*) and (*d*), to increase or decrease the total voltage applied to a system. The net voltage is determined simply by summing the sources with the same polarity and subtracting the total of the sources with the opposite 'pressure'.

The net polarity is the polarity of the larger sum. In Fig. 1.5 (*c*), for example, the sources are all 'pressuring' current to the right, so the net voltage is

$$V_T = V_1 + V_2 + V_3 = 20\text{ V} + 16\text{ V} + 12\text{ V} = 48\text{ V}$$

as shown in the figure. In Fig. 1.5 (*d*), however, the greater 'pressure' is to the left, with a net voltage of

$$V_T = V_2 + V_3 + V_1 = 16\text{ V} + 12\text{ V} - 20\text{ V} = 8\text{ V}$$

and the polarity shown in the figure.

The current source is often referred to as the dual of the voltage source. A battery supplies a fixed voltage, and the source current can vary; but the current source supplies a fixed current to the branch in which it is located, while its terminal voltage may vary as determined by the network to which it is applied. Note from the above that duality simply implies an interchange of current and voltage to distinguish the characteristics of one source from the other.

For further comparison, the terminal characteristics of an ideal DC voltage and current source are presented in Fig. 1.5 (*e*), ideal implying perfect sources, or no internal losses sensitive to the demand from the applied load. Note that for the voltage source, the terminal voltage is fixed at E volts independent of the direction of the current I. The direction and magnitude of I will be determined by the network to which the supply is connected.

The characteristics of the ideal current source, shown in Fig. 1.5 (*f*), reveal that the magnitude of the supply current is independent of the polarity of the voltage across the source. The polarity and magnitude of the source voltage vs will be determined by the network to which the source is connected.

For all one-voltage-source networks the current will have the direction indicated to the right of the battery and for all single-current-source networks, it will have the polarity indicated to the right of the current source.

1.5 SOURCE TRANSFORMATION

In a circuit, a given voltage source in series with a resistance can be converted into an equivalent current source with a parallel resistance. The magnitude of current source will be

$$I_1 = \frac{V_1}{R_1} \qquad \text{...(1.5)}$$

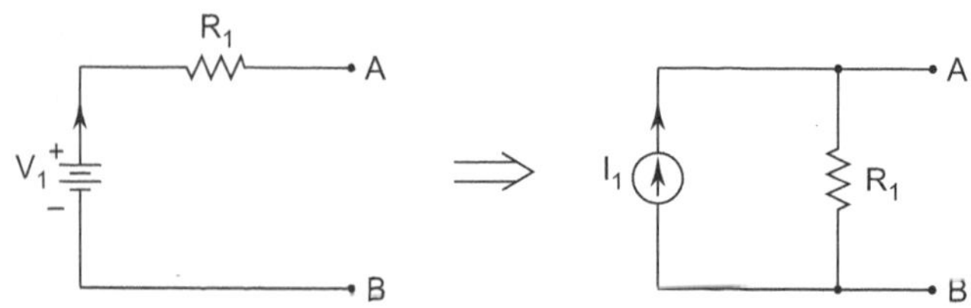

Fig. 1.6

It should be noted that the direction of the current of the current source should be same to that of the current coming out of the voltage source.

The same resistance is connected in parallel to the current source which was in series with the voltage source (Fig. 1.6).

Similarly a given current source with a parallel resistance can be converted into or replace by an equivalent voltage source in series with the resistance.

The magnitude of voltage source will be

$$V_2 = I_2R_2 \qquad ...(1.6)$$

The same resistance is connecting in series with the voltage source which was in parallel to the current source (Fig. 1.7).

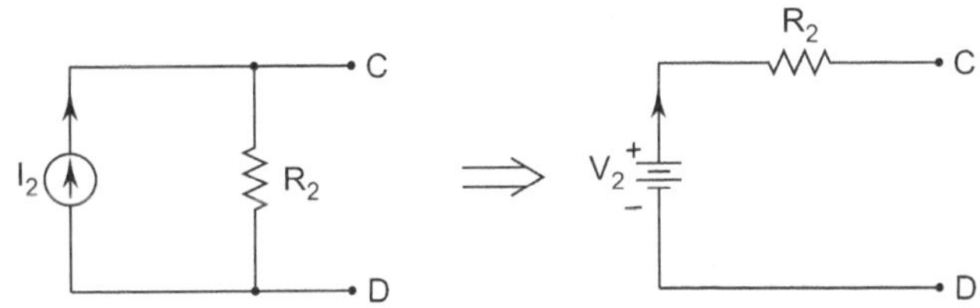

Fig. 1.7

1.6 BASIC ELEMENTS

The two circuit elements, current source and voltage source are basically active elements. Along with these two basic active elements in a circuit we have passive elements also. These passive elements mainly comprises of—Resistor, Inductor and Capacitor. Most of the electrical circuits are formed using these three basic passive elements in combination. Next we will discuss these elements in detail.

1.7 THE RESISTANCE PARAMETER

Resistance is one of the three basic parameters of electrical circuit. In general resistance can be described as, *"that property of a circuit element which offers opposition to the flow of current and while doing so it converts electrical energy into heat energy."*

According to Ohm's law we have

$$R = \frac{V}{I} \qquad ...(1.7)$$

The graphical representation of Ohm's law is shown in Fig. 1.9.

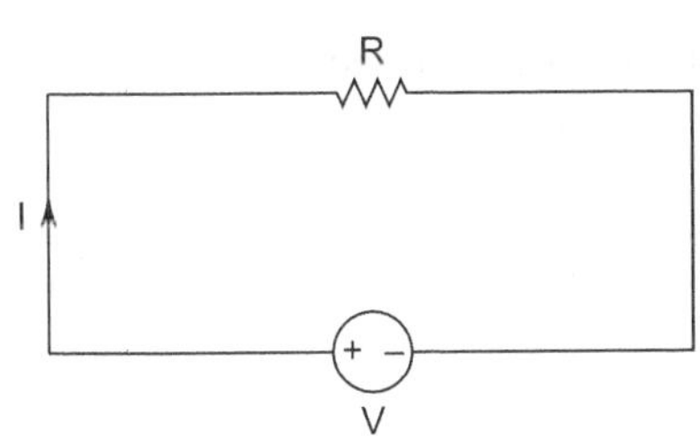

Fig. 1.8

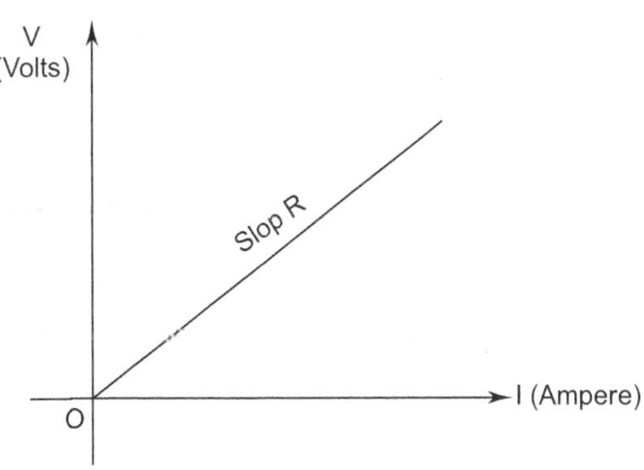

Fig. 1.9

In practical, the resistance of most metallic conductors varies with temperature.

Specifically, if the resistance of a metal at temperature T_1 is R_1, then for normal range of temperatures, the resistance, at temperature T_2 is given by

$$\boxed{R_2 = R_1 [1 + \alpha_1 (T_2 - T_1)]} \quad \text{...(1.8)}$$

where, α_1 is the temperature coefficient of resistance at temperature T_1.

Ohm's also discovered that the resistance of a conductor of uniform dimensions was directly dependent upon the length (L) and inversely proportional to the cross-section area (A).

$$\boxed{R = \rho \frac{L}{A}} \quad \text{...(1.9)}$$

where ρ is the resistivity of the material in Ω m.

1.7.1 Series Combination of Resistor

Circuit elements that carry the same current are said to be in series.

Let us consider three resistors R_1, R_2 and R_3 in series as shown in Fig. 1.10.

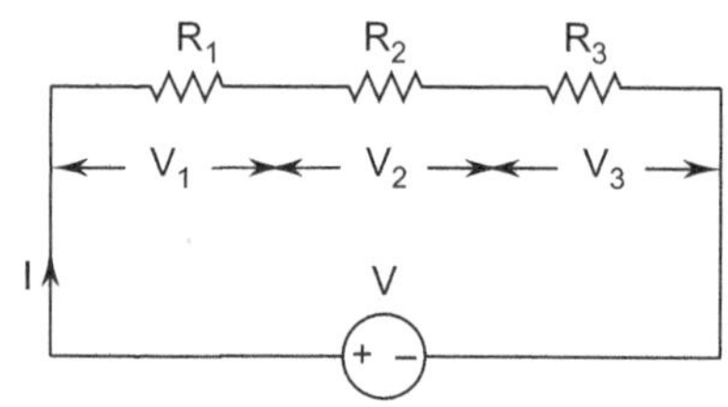

Fig. 1.10

By Kirchhoff's voltage law*

$$V = V_1 + V_2 + V_3 \quad \text{...(1.10)}$$

where V_1 is the voltage across R_1

V_2 is the voltage across R_2

V_3 is the voltage across R_3.

Applying Ohm's law to equation (1.10) we have

$$V = IR_1 + IR_2 + IR_3 \quad \text{...(1.11)}$$

or $$V = I(R_1 + R_2 + R_3) \quad \text{...(1.12)}$$

or $$V = IR_S \quad \text{...(1.13)}$$

where $$R_S = R_1 + R_2 + R_3 \quad \text{...(1.14)}$$

= Equivalent resistance of the series connected resistors.

1.7.2 Parallel Combination of Resistor

Let us consider two resistors R_1 and R_2 connected in parallel as shown in Fig. 1.11.

---

*Kirchhoff's Laws will be discussed in detail in next chapter.

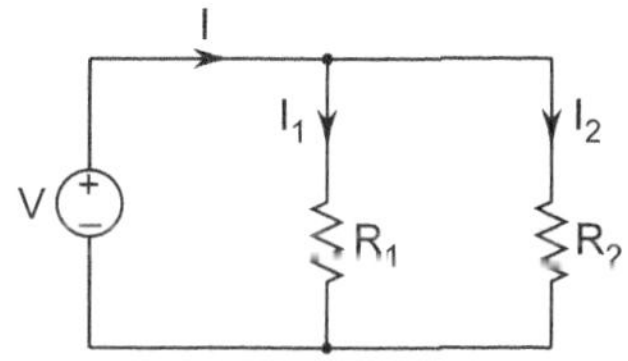

Fig. 1.11

As both the resistors are in parallel, hence the potential difference across both the resistors will be same. Applying Kirchhoff's current law to the circuit shown in Fig. 1.11.

$$I = I_1 + I_2 \quad \text{...(1.15)}$$

or
$$I = \frac{V}{R_1} + \frac{V}{R_2} \quad \text{...(1.16)}$$

or
$$I = \left(\frac{1}{R_1} + \frac{1}{R_2}\right)V \quad \text{...(1.17)}$$

or
$$I = \frac{V}{R_P} \quad \text{...(1.18)}$$

where
$$\frac{1}{R_P} = \frac{1}{R_1} + \frac{1}{R_2} \quad \text{...(1.19)}$$

= Equivalent resistance of two parallel connected resistors.

Equation (1.19) can also be written as

$$R_P = \frac{R_1 R_2}{R_1 + R_2} \quad \text{...(1.20)}$$

Problem 1.1. *A DC circuit comprises two resistors; resistor R_1 of value 25 Ω and resistor R_2 of unknown value connected in parallel, together with a third resistor R_3 of value of 5 Ω connected in series with the parallel branch. Find the voltage to be applied across the whole circuit and the value of resistor R_2 if the potential difference across R_3 is 90 V and the total power consumed is 4,320 W.*

Solution. The circuit diagram is shown in Fig. 1.12

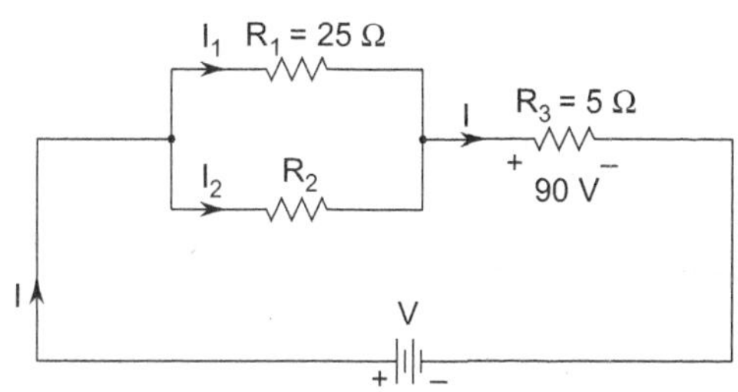

Fig. 1.12

Total current coming out from the voltage source

I = Current flowing through resistor R_3

$$I = \frac{\text{Voltage drop across resistor } R_3}{\text{Value of resistor } R_3} = \frac{90}{5} = \mathbf{18\ V}$$

Voltage applied across the whole circuit

$$V = \frac{\text{Total power consumed}}{\text{Current through the circuit}} = \frac{4320}{18} = \mathbf{240\ V} \qquad \textbf{Ans.}$$

Voltage drop across parallel combination

$$= V - 90 = 240 - 90 = \mathbf{150\ V}$$

Current through resistor, $R_1 = 25\ \Omega$ is

$$I_1 = \frac{150}{25} = \mathbf{6\ A}$$

Current through unknown resistor, R_2

$$I_2 = 18 - 6 = \mathbf{12\ A}$$

Unknown resistor, R_2

$$R_2 = \frac{150}{12} = \mathbf{12.5\ \Omega} \qquad \textbf{Ans.}$$

Problem 1.2. *Find the equivalent resistance between terminals A and B of the circuit shown in Fig. 1.13.*

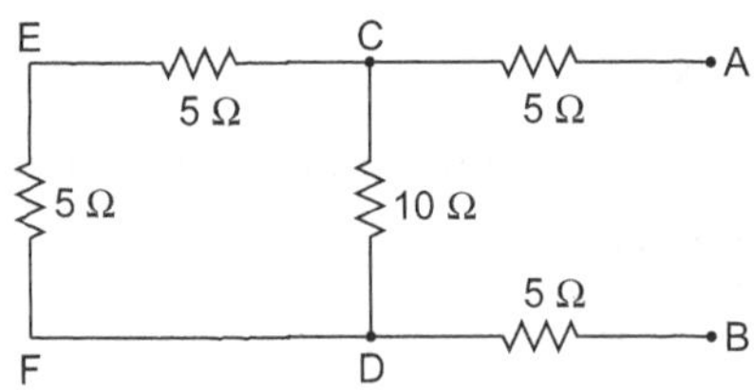

Fig. 1.13

Solution. Observing Fig. 1.13, we find that the 5 Ω resistance connected between E and C points and 5 Ω resistance connected between E and F points are in series.

$\therefore \qquad R_{FEC} = 5 + 5 = 10\ \Omega$

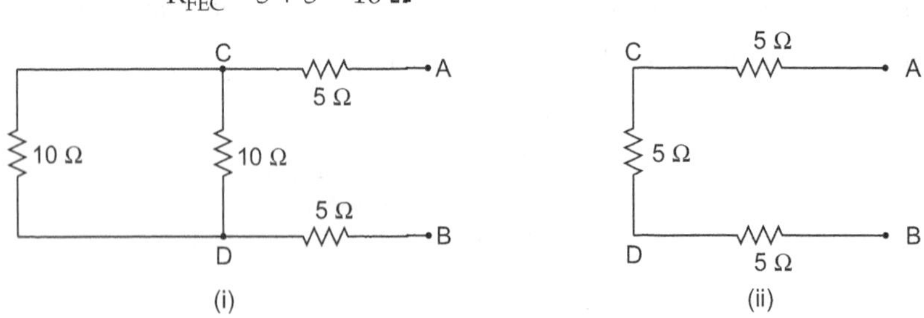

Fig. 1.14

$$R_{CD} = (10 \,||\, 10) = \frac{10 \times 10}{10 + 10} = \frac{100}{20}$$

$$R_{CD} = 5\ \Omega$$

$$R_{AB} = 5 + 5 + 5$$

$$R_{AB} = \mathbf{15\ \Omega}$$ **Ans.**

Problem 1.3. *Find the equivalent resistance between terminals A and B of the circuit shown in Fig. 1.15.*

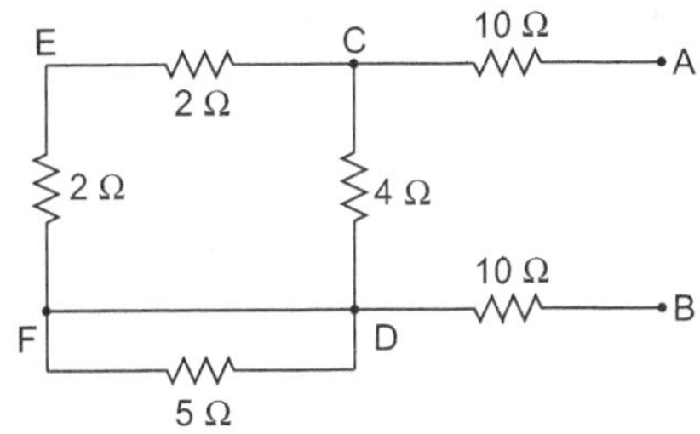

Fig. 1.15

Solution. When, between any two nodes there are more than one branch, connecting those two nodes and there is any one branch in which there is no element present then it is a short-circuit and those two nodes are considered as one node.

Other resistances connected between those two nodes are of no use and should be removed.

Fig. 1.15 can be redrawn as shown in Fig. 1.16 (*i*)

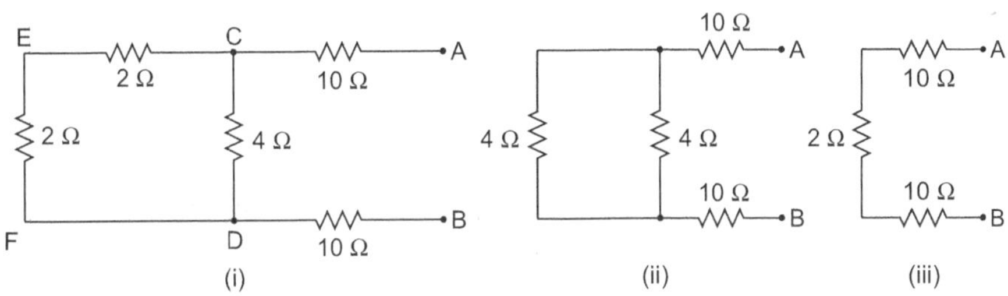

Fig. 1.16

$$R_{FEC} = 2 + 2 = 4\ \Omega$$

$$R_{CD} = (4 \parallel 4) = \frac{4 \times 4}{4 + 4} = 2\ \Omega$$

$$R_{AB} = 10 + 2 + 10$$

$$R_{AB} = \mathbf{22\ \Omega}$$ **Ans.**

Problem 1.4. *Find the equivalent resistance between terminals A and B of the circuit shown in Fig. 1.17.*

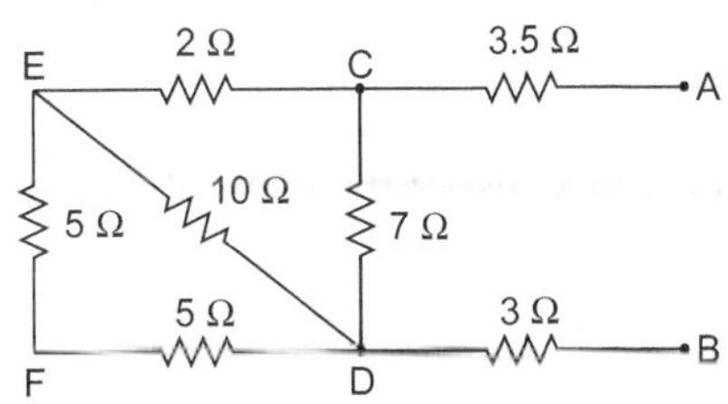

Fig. 1.17

Solution. The resistances connected between E and F points and F and D points are in series. Hence,

$$R_{EFD} = 5 + 5 = 10\ \Omega$$
$$R_{ED} = (10 \parallel 10) = 5\ \Omega$$

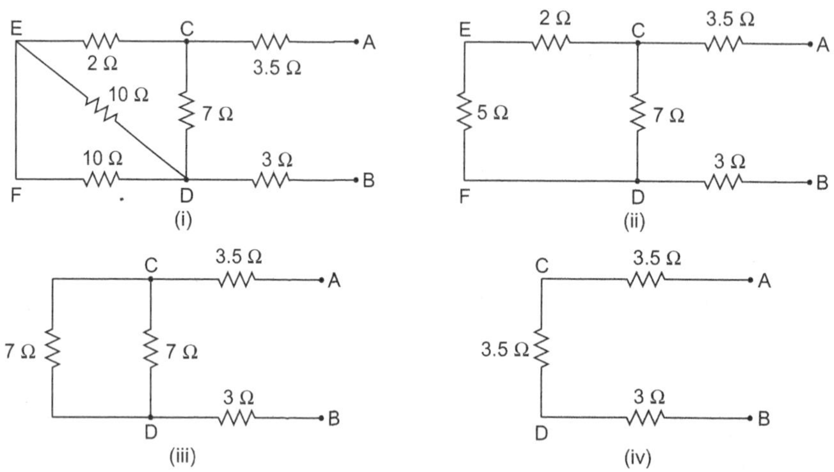

Fig. 1.18

$$R_{FEC} = 5 + 2 = 7\ \Omega$$
$$R_{CD} = (7 \parallel 7) = 3.5\ \Omega$$
$$R_{AB} = 3.5 + 3.5 + 3 = \mathbf{10\ \Omega}$$ **Ans.**

Problem 1.5. *A 100 V, 60 Watt bulb is to be operated from a 220 V supply. What resistance must be connected in series with the bulb to glow normally?*

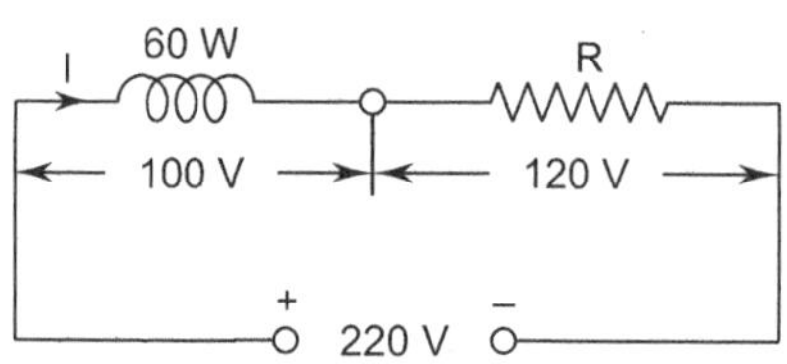

Fig. 1.19

Solution. Rated power of lamp, P = 60 W

Rated voltage of lamp, V = 100 V

Current drawn by the lamp*, when operated on rated voltage, *i.e.*,

$$\text{Rated current, } I = \frac{P}{V} = \frac{60}{100} = 0.6\ \text{A}$$

Lamp will operate normally on 220 V also if the current flowing through the lamp remains the rated current, *i.e.*, 0.6 A.

Let the resistance connected in series with the lamp to make it glow normally on 220 V be of R ohms, as shown in Fig. 1.19.

Now since the resistance R is in series with the lamp, the same current with flow through the resistance R, as in the lamp, *i.e.*, 0.6 A.

---

* A bulb or a lamp is consider as pure resistive load.

Voltage drop across the series resistance =

$$IR = \text{Supply voltage} - \text{rated voltage of the lamp}$$
$$= 220 - 100 = 120 \text{ V}$$

or $$IR = 120 \text{ V}$$

or $$R = \frac{120}{I} = \frac{120}{0.6} = \mathbf{200\ \Omega}$$ **Ans.**

1.8 THE INDUCTANCE PARAMETER

According to Faraday's law of electromagnetic induction,

"A changing current produced an e.m.f. of self-induction in a coil of wire which varies directly with rate of change of current with respect to time". Mathematically

$$e = -L\frac{di}{dt} \qquad \ldots(1.21)$$

where L is a proportionality factor called as the coefficient of Self-Inductance. ***Inductance*** can be defined as ***"that property of a circuit element by which energy is capable of being stored in a magnetic flux field".***

Here, negative sign in the e.m.f. expression indicates that the induced e.m.f. is in opposite direction to the applied voltage.

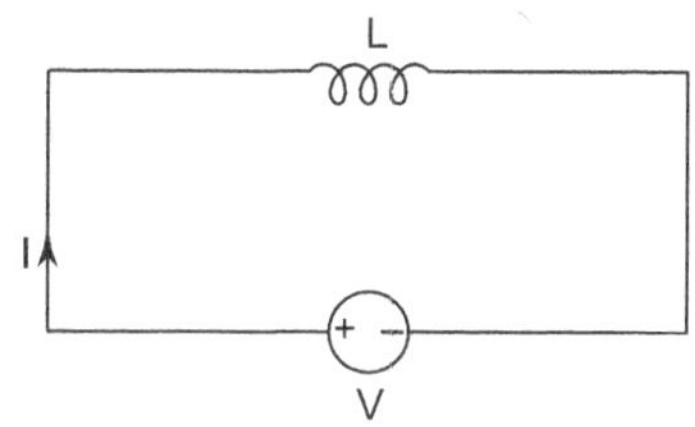

Fig. 1.20

One of the distinguishing factors of inductance is that it makes itself felt in a circuit only when, there is a changing current (For example, in case of an alternating current). If we are applying direct current (DC) to a circuit containing inductor only, then in that case the inductor will simply behave as a short-circuit.

In ideal case, an inductor is considered as resistance free. Although in practical it must contain the resistance of the wire out of which the inductor will be formed.

Equation (1.21) is plotted in the graph of Fig. 1.21.

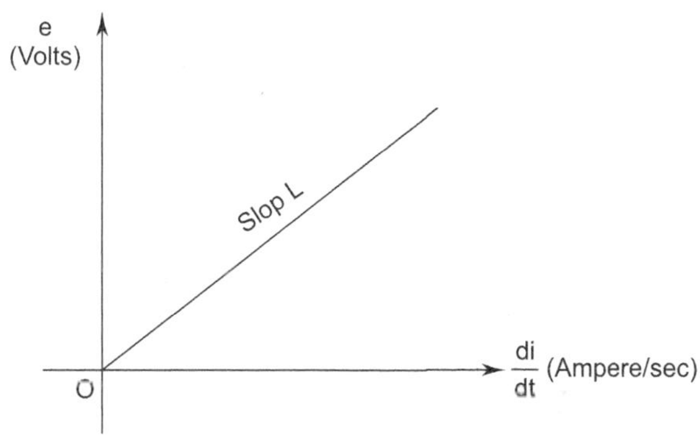

Fig. 1.21

From equation (1.21) we can find the expression for current also.

$$e = -L\frac{di}{dt} \quad \text{[Eqn. (1.21)]}$$

(where negative sign in the expression indicates that the e.m.f. induced is in opposite direction to the applied voltage.)

or $$di = \frac{1}{L}e.dt \quad ...(1.22)$$

or $$i = \frac{1}{L}\int_0^t v dt + i(0) \quad ...(1.23)$$

If the initial current in the inductor is zero, then

$$i = \frac{1}{L}\int_0^t v dt \quad ...(1.24)$$

1.8.1 Series Combination of Inductor

Let us consider two inductors L_1 and L_2 connected in series as shown in Fig. 1.22.

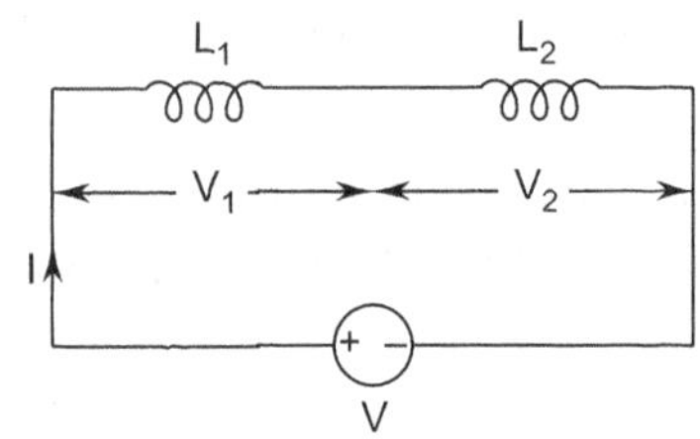

Fig. 1.22

By Kirchhoff's voltage law

$$V = V_1 + V_2 \quad ...(1.25)$$

From equation (1.21), V_1 and V_2 can be expressed as

$$V_1 = L_1\frac{dI}{dt} \quad ...(1.26)$$

and $$V_2 = L_2\frac{dI}{dt} \quad ...(1.27)$$

From equations (1.26) and (1.27), equation (1.25) will become

$$V = L_1\frac{dI}{dt} + L_2\frac{dI}{dt} \quad ...(1.28)$$

or $$V = (L_1 + L_2)\frac{dI}{dt} \quad ...(1.29)$$

or $$V = L_S\frac{dI}{dt} \quad ...(1.30)$$

where $$L_S = L_1 + L_2 \quad ...(1.31)$$

= Equivalent inductance of two series connected inductors.

It is clear from equation (1.31) that when two or more **inductors are connected in *series***, then they can be treated in the same fashion as **resistances connected in *series***.

1.8.2 Parallel Combination of Inductor

Let us consider two inductors L_1 and L_2 are connected in parallel as shown in Fig. 1.23.

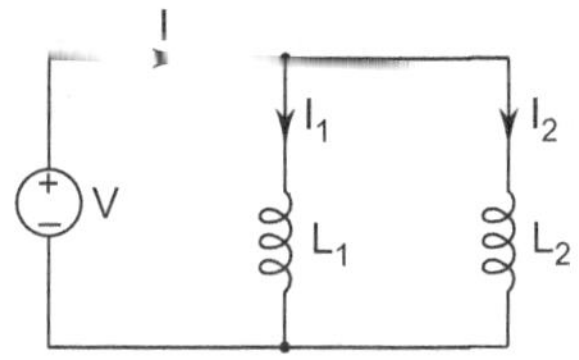

Fig. 1.23

As both the inductors are in parallel, hence the potential difference across both the inductors will be same. Applying Kirchhoff's current law to the circuit shown in Fig. 1.23,

$$I = I_1 + I_2 \qquad \ldots(1.32)$$

Assuming zero initial current through each inductor, from equation (1.24), I_1 and I_2 can be expressed as

$$I_1 = \frac{1}{L_1}\int_0^t V\,dt \qquad \ldots(1.33)$$

$$I_2 = \frac{1}{L_2}\int_0^t V\,dt \qquad \ldots(1.34)$$

From equations (1.33) and (1.34), equation (1.32) will become

$$I = \frac{1}{L_1}\int_0^t V\,dt + \frac{1}{L_2}\int_0^t V\,dt \qquad \ldots(1.35)$$

or

$$I = \left(\frac{1}{L_1} + \frac{1}{L_2}\right)\int_0^t V\,dt \qquad \ldots(1.36)$$

or

$$I = \frac{1}{L_P}\int_0^t V\,dt \qquad \ldots(1.37)$$

where

$$\frac{1}{L_P} = \frac{1}{L_1} + \frac{1}{L_2} \qquad \ldots(1.38)$$

= Equivalent inductance of two parallel connected inductors.

Equation (1.38) can also be written as

$$L_P = \frac{L_1 L_2}{L_1 + L_2} \qquad \ldots(1.39)$$

Equation (1.39) indicates that when two or more **inductors are connected in *parallel*,** then they can be treated in the same fashion as **resistances connected in *parallel*.**

1.9 THE CAPACITANCE PARAMETER

The third basic parameter of electrical circuits is capacitance. A capacitor is an energy storing element in which energy is stored in the form of electrical field. The charge stored between two plates of the capacitor is directly proportional to the potential difference across the two plates, *i.e.*,

$$Q \propto V \quad \text{...(1.40)}$$

or

$$Q = CV \quad \text{...(1.41)}$$

where Q is the charge stored between two plates.

V is the potential difference across two plates.

C is the proportionality factor known as capacitance.

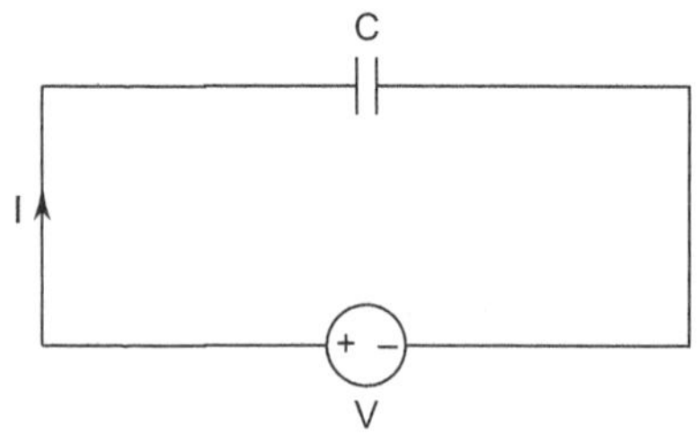

Fig. 1.24

As in general the current is expressed as

$$I = \frac{dQ}{dt} = C\frac{dV}{dt} \qquad \text{[From equation (1.41)]} \quad \text{...(1.42)}$$

or

$$dV = \frac{1}{C} I dt \quad \text{...(1.43)}$$

$$V = \frac{1}{C}\int_0^t I dt + V(0) \quad \text{...(1.44)}$$

When there is no initial voltage across the capacitor, in that case

$$V = \frac{1}{C}\int_0^t I dt \quad \text{...(1.45)}$$

The graphical representation of equation (1.45) is shown in Fig. 1.25.

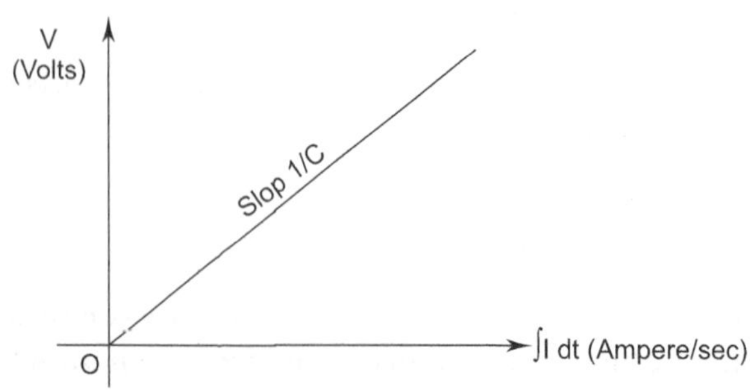

Fig. 1.25

1.9.1 Series Combination of Capacitor

Let us consider a series combination of two capacitors C_1 and C_2 connected as shown in Fig. 1.26.

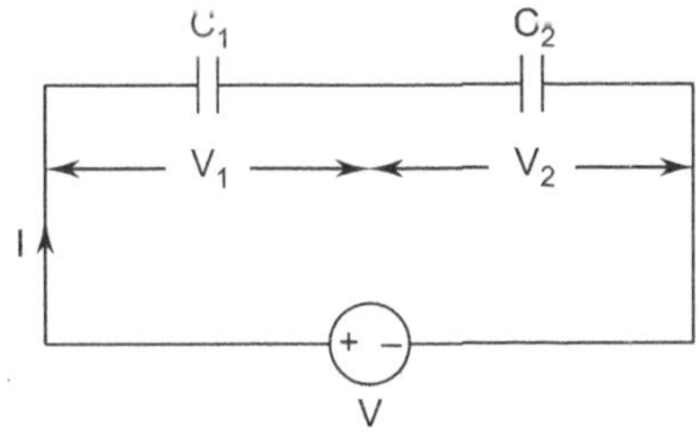

Fig. 1.26

When a voltage is applied across the combination, the capacitor will have equal displacement of charge. But the potential difference across their terminal will be different (provided C_1 is not equal to C_2).

However by Kirchhoff's voltage law the sum of these two voltages must add up to the source voltage. Thus,

$$V = V_1 + V_2 \quad \text{...(1.46)}$$

For zero initial capacitor voltage, from equation (1.45) we have

$$V_1 = \frac{1}{C_1}\int_0^t I\,dt \quad \text{...(1.47)}$$

and

$$V_2 = \frac{1}{C_2}\int_0^t I\,dt \quad \text{...(1.48)}$$

Inserting equations (1.47) and (1.48) in equation (1.46) we have

$$V = \frac{1}{C_1}\int_0^t I\,dt + \frac{1}{C_2}\int_0^t I\,dt \quad \text{...(1.49)}$$

or

$$V = \left(\frac{1}{C_1} + \frac{1}{C_2}\right)\int_0^t I\,dt \quad \text{...(1.50)}$$

or

$$V = \frac{1}{C_S}\int_0^t I\,dt \quad \text{...(1.51)}$$

where

$$\frac{1}{C_S} = \frac{1}{C_1} + \frac{1}{C_2} \quad \text{...(1.52)}$$

= Equivalent capacitance of two series connected capacitors.

Equation (1.52) can also be written as

$$C_S = \frac{C_1 C_2}{C_1 + C_2} \quad \text{...(1.53)}$$

It is clear from equation (1.53) that when two or more capacitors are connected in series then they can be treated as resistances connected in parallel. Hence the equivalent **capacitance** in ***series*** circuit can be calculated in the same manner as equivalent **resistance** is calculated in case of ***parallel*** circuit.

1.9.2 Parallel Combination of Capacitor

Let us consider two capacitors C_1 and C_2 are connected in parallel as shown in Fig. 1.27.

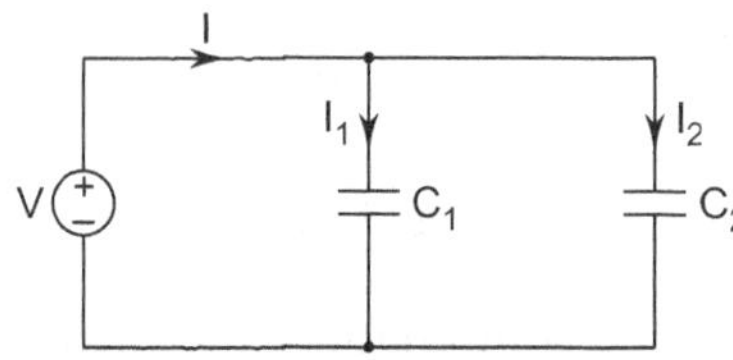

Fig. 1.27

As both the capacitors are connected in parallel, hence the potential difference across both the capacitors will be same. Applying Kirchhoff's current law,

$$I = I_1 + I_2 \qquad ...(1.54)$$

But from equation (1.42)

$$I_1 = C_1 \frac{dV}{dt} \qquad ...(1.55)$$

and

$$I_2 = C_2 \frac{dV}{dt} \qquad ...(1.56)$$

From equations (1.55) and (1.56), equation (1.54) will be

$$I = C_1 \frac{dV}{dt} + C_2 \frac{dV}{dt} \qquad ...(1.57)$$

$$I = (C_1 + C_2)\frac{dV}{dt} \qquad ...(1.58)$$

or

$$I = C_P \frac{dV}{dt} \qquad ...(1.59)$$

where

$$C_P = C_1 + C_2 \qquad ...(1.60)$$

= Equivalent capacitance of two parallel connected capacitors.

It can be concluded from expression (1.60) that **capacitance in *parallel*** can be treated in same manner as **resistance in *series*.**

1.10 CURRENT DIVIDER RULE

This rule can be applied to a circuit where we have to find current in two parallel branches. For example, let us consider a circuit shown in Fig. 1.28. Total current I will be

$$I = \frac{V}{R} \qquad ...(1.61)$$

where $$R = \frac{R_1 \times R_2}{R_1 + R_2} \quad \text{...(1.62)}$$

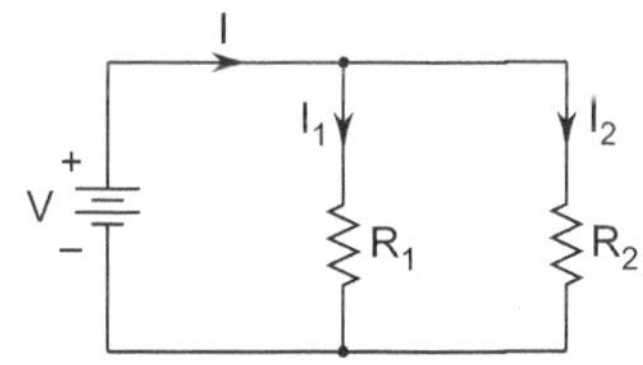

Fig. 1.28

According to current divider rule, branch current is given by

$$\boxed{\text{Branch Current} = \frac{\text{Total current} \times \text{Resistance of opposite branch}}{\text{Sum of resistances of both branches}}} \quad \text{...(1.63)}$$

Current I_1 through resistance R_1 is given by

$$I_1 = \frac{I \times R_2}{R_1 + R_2} \quad \text{...(1.64)}$$

Similarly current I_2 through resistance R_2 is given by

$$I_2 = \frac{I \times R_1}{R_1 + R_2} \quad \text{...(1.65)}$$

Problem 1.6. *Using current divider rule find current I_1 and I_2 in circuit shown in Fig. 1.29.*

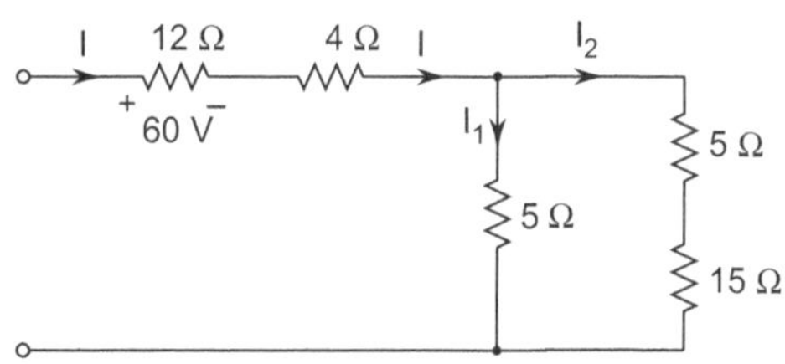

Fig. 1.29

Solution. Voltage drop across 12 Ω resistance is 60 volts hence current through 12 Ω resistance (or total current) will be

$$I = \frac{60}{12} = 5 \text{ A}$$

Hence, $$I_1 = \frac{5 \times 20}{5 + 20} = \mathbf{4\ A} \quad \textbf{Ans.}$$

$$I_2 = \frac{5 \times 5}{5 + 20} = \mathbf{1\ A} \quad \textbf{Ans.}$$

1.11 VOLTAGE DIVIDER RULE

This rule can be applied to a circuit where we have to find voltage across two resistances connected in series. For example, consider a circuit shown in Fig. 1.30.

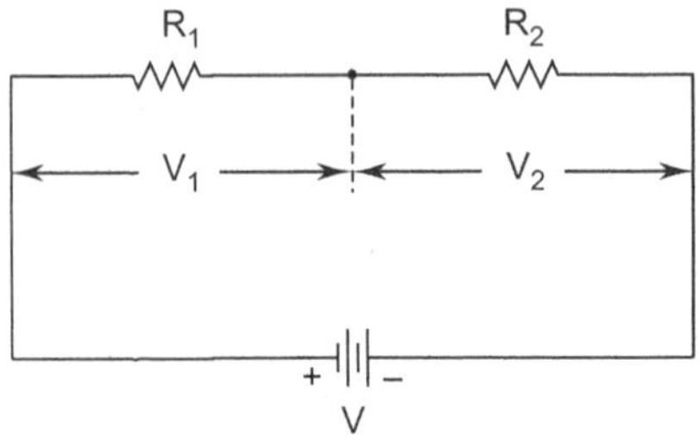

Fig. 1.30

The voltage across any of the resistance connected in series is given by

$$\boxed{\text{Voltage} = \frac{\text{Total voltage} \times \text{Resistance across which the voltage is to be calculated}}{\text{Total resistance of branch}}} \quad ...(1.66)$$

The voltage V_1 across resistance R_1 is given by

$$V_1 = \frac{V \times R_1}{R_1 + R_2} \quad ...(1.67)$$

Similarly, the voltage V_2 across resistance R_2 is given by

$$V_2 = \frac{V \times R_2}{R_1 + R_2} \quad ...(1.68)$$

SOLVED PROBLEMS

Problem 1.7. *Using voltage divider rule find the voltage drop across the 10 ohms resistance in the circuit shown in Fig. 1.31.*

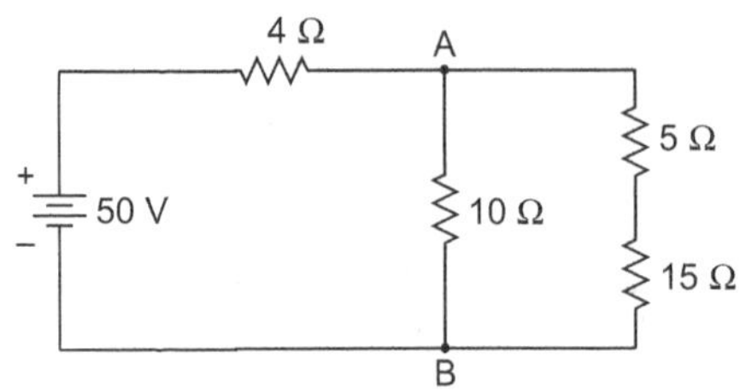

Fig. 1.31

Solution. The voltage drop across 10 Ω resistance will be same and equal to voltage drop across (5 + 15) Ω resistance.

Total resistance between AB is = [10 || (5 + 15)]

$$= \frac{10 \times 20}{10 + 20} = \frac{200}{30} = \frac{20}{3}$$

$$= 6.67\ \Omega.$$

Now the 50 volt will divide across 4 Ω resistance and this total resistance of 6.67 Ω between A and B.

Hence, voltage across AB or 10 Ω resistance will be

$$V_{AB} = \frac{50 \times 6.67}{4 + 6.67} = 31.26 \text{ V}$$

$$V_{AB} = \mathbf{31.26 \text{ V.}}$$ **Ans.**

Problem 1.8. *Two resistances of 10 Ω and 20 Ω respectively are connected in parallel. These two parallel resistances are further connected in series with a resistance of 15 Ω.*

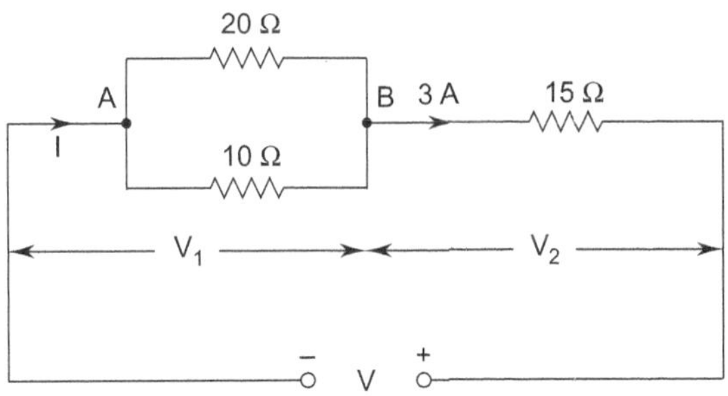

Fig. 1.32

If current through 15 Ω resistance is 3 A, find

(a) The current through 10 Ω resistance.

(b) The current through 20 Ω resistance.

(c) Voltage across the whole circuit.

(d) Total power consumed.

Solution. If a current of 3 A is flowing through 15 Ω resistance, it means it is total current of the circuit, *i.e.*,

$$I = 3 \text{ A}$$

(*a*) Let current through 10 Ω resistance be I_1

$$I_1 = \frac{3 \times 20}{20 + 10} = \mathbf{2 \text{ A}}$$ **Ans.**

(*b*) Let current through 20 Ω resistance be I_2

$$I_2 = \frac{3 \times 10}{20 + 10} = \mathbf{1 \text{ A}}$$ **Ans.**

(*c*) Voltage drop across branch AB is

$$V_1 = 20 \times 1 = 10 \times 2 = 20 \text{ V}$$

Voltage drop across BC branch will be

$$V_2 = 3 \times 15 = 45 \text{ V}$$

and voltage across the whole circuit will be

$$V = V_1 + V_2 = 20 + 45 = \mathbf{65 \text{ V}}$$ **Ans.**

(*d*) Total power consumed

$$P = VI = 65 \times 3$$

$$P = \mathbf{195 \text{ W}}$$ **Ans.**

Problem 1.9. *A coil has a resistance of 18 Ω when its mean temperature is 20°C and of 20 Ω when its mean temperature is 50°C. Find its mean temperature rise when its resistance is 21 Ω and the surrounding temperature is 15°C.*

Solution. Let R_0 be the resistance of the coil and α_0 its temperature coefficient at 0°C.

Then,
$$18 = R_0 (1 + \alpha_0 \times 20) \quad \ldots(i)$$
and
$$20 = R_0 (1 + 50\alpha_0) \quad \ldots(ii)$$

Dividing equation (*ii*) by (*i*), we get

$$\frac{20}{18} = \frac{1 + 50\,\alpha_0}{1 + 20\,\alpha_0}$$

or
$$\alpha_0 = \frac{1}{250} \text{ per °C}$$

If t°C is the temperature of the coil when its resistance is 21 Ω, then

$$21 = R_0 \left(1 + \frac{t}{250}\right) \quad \ldots(iii)$$

Dividing equation (*iii*) by equation, (*i*) we have

$$\frac{21}{18} = \frac{R_0 \left(1 + \frac{t}{250}\right)}{R_0 (1 + 20\,\alpha_0)}$$

or
$$t = 65°C$$

Temperature rise $= 65 - 15 =$ **50°C.** **Ans.**

Problem 1.10. *The equivalent resistance of four resistors connected in parallel is 30 Ω. The current flowing through them are 4 A, 2 A, 1.33 A and 0.66 A. Determine the value of each resistor.*

Solution. The circuit diagram is shown in Fig. 1.33.

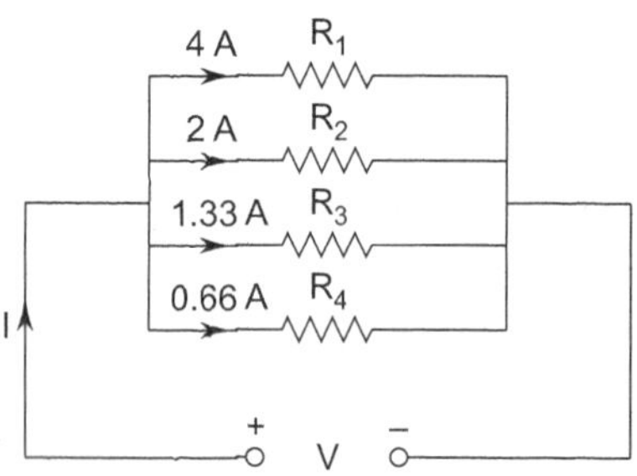

Fig. 1.33

Total current from the source

$$I = I_1 + I_2 + I_3 + I_4$$
$$= 4 + 2 + 1.33 + 0.66 = 8 \text{ A}$$

Equivalent resistance of the network is

$$R = 30\,\Omega$$

Total voltage across the circuit is

$$V = I \cdot R = 8 \times 30 = 240 \text{ V}$$

$$R_1 = \frac{V}{I_1} = \frac{240}{4} = \mathbf{60\ \Omega}$$ **Ans.**

$$R_2 = \frac{V}{I_2} = \frac{240}{2} = \mathbf{120\ \Omega}$$ **Ans.**

$$R_3 = \frac{V}{I_3} = \frac{240}{1.33} = \mathbf{180\ \Omega}$$ **Ans.**

$$R_4 = \frac{V}{I_4} = \frac{240}{0.66} = \mathbf{360\ \Omega.}$$ **Ans.**

Problem 1.11. *In the given circuit shown in Fig. 1.34, find the resistance between the points B and C.*

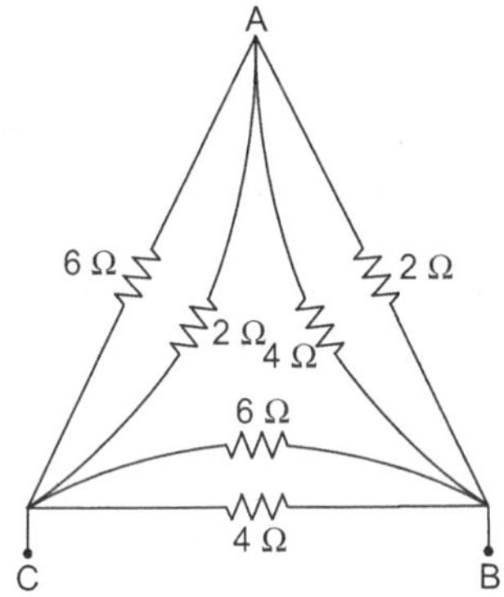

Fig. 1.34

Solution. Resistance between points C and A will be

$$R_{CA} = 6 \parallel 2 = \frac{6 \times 2}{6+2}$$

$$= 1.5\ \Omega$$

Resistance between points A and B will be

$$R_{AB} = 4 \parallel 2 = \frac{4 \times 2}{4+2}$$

$$= 1.33\ \Omega$$

Now the circuit will become as shown in Fig. 1.35 (*a*) and (*b*)

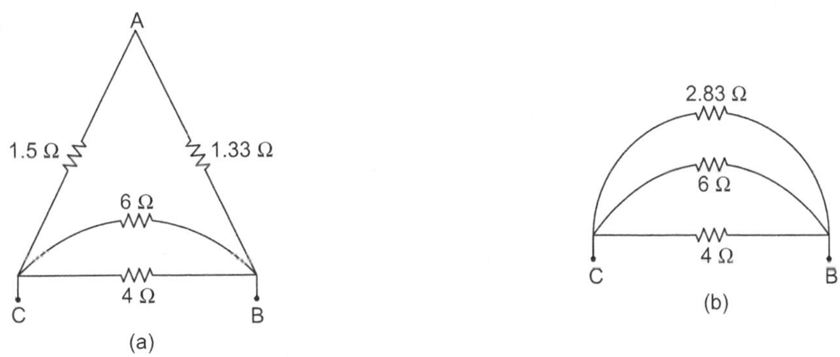

Fig. 1.35

Total resistance between points B and C will be

$$R_{BC} = 4 \,||\, 6 \,||\, 2.83$$

$$= \left(\frac{4 \times 6}{4 + 6}\right) || \, 2.83 = 2.4 \,||\, 2.83$$

$$= \frac{2.4 \times 2.83}{2.4 + 2.83} = \frac{6.792}{5.23}$$

$$R_{BC} = \mathbf{1.3\ \Omega}$$ **Ans.**

EXERCISE

1. A lamp has been designed to consume 60 Ω when it is connected to a 12 V power source. What is the resistance of the filament, and what amount of current will flow? If the actual voltage is only 11 V, how much energy would it consume over a 100 h period?
2. How is energy related to power? State the SI units of work, power and energy. What are the practical units of electrical energy and mechanical power?
3. What are the limitations of Ohms law?
4. On what factors do the resistance of a conductor depends?
5. What is meant by conductivity and how is it related with resistivity?
6. On what factors do the force experienced by a current carrying conductor placed in a magnetic field depends?
7. Define the following terms:
   (*i*) Circuits (*ii*) Parameters
   (*iii*) Node (*iv*) Branch.
8. What is the difference between a mesh and loop?

□ □ □

2

DC CIRCUIT ANALYSIS

2.1 INTRODUCTION

Generally speaking, *network analysis* is any structured technique used to mathematically analyze a circuit. Quite often the engineer will encounter circuits containing multiple sources of power or component configurations which challenge simplification by series/parallel analysis techniques. In those cases, he or she will be forced to use other means. This chapter presents a few techniques useful in analyzing such complex circuits.

Complex DC networks consist of more than one active elements and resistances in series-parallel combination.

While analyzing a particular circuit, the various responses of current and voltage are to be determined.

This analysis is carried out by writing the governing equations of the given circuit based on some well-known laws and network theorems and then solving these equations for the unknown.

For achieving this, there are certain laws and network theorems which simplify the circuit for these analysis. In this chapter we will study those related laws and theorems.

2.2 KIRCHHOFF'S LAWS

Kirchhoff discover two basic laws concerning to networks, one commonly known as Kirchhoff's Current Law (KCL) and the second one is known as Kirchhoff's Voltage Law (KVL).

These laws are very useful in determining the current following in the various branches and equivalent resistance of a complicated network.

2.2.1 Kirchhoff's Current Law (KCL)

It states that ***"In any electrical network, the algebraic sum* of the currents meeting at a point (or junction) is zero".***

* The algebric sum is the sum which takes into account the polarities of the currents, voltage drops and e.m.fs sources.

Or

"The total current leaving a junction is equal to the total current entering in that junction."

Considering Fig. 2.1, according to KCL, we have

$$\sum I = 0 \quad ...(2.1)$$

or
$$I_1 + I_2 = I_3 + I_4 + I_5 \quad ...(2.2)$$

or
$$I_1 + I_2 - I_3 - I_4 - I_5 = 0 \quad ...(2.3)$$

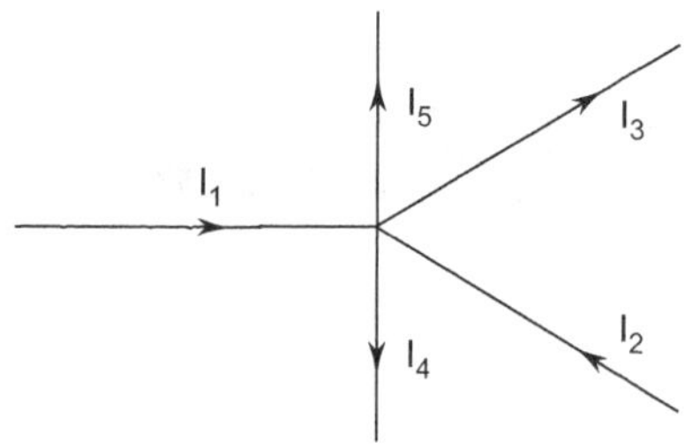

Fig. 2.1

Some schematic examples are shown in Fig. 2.2, to understand Kirchhoff's current law in a better way.

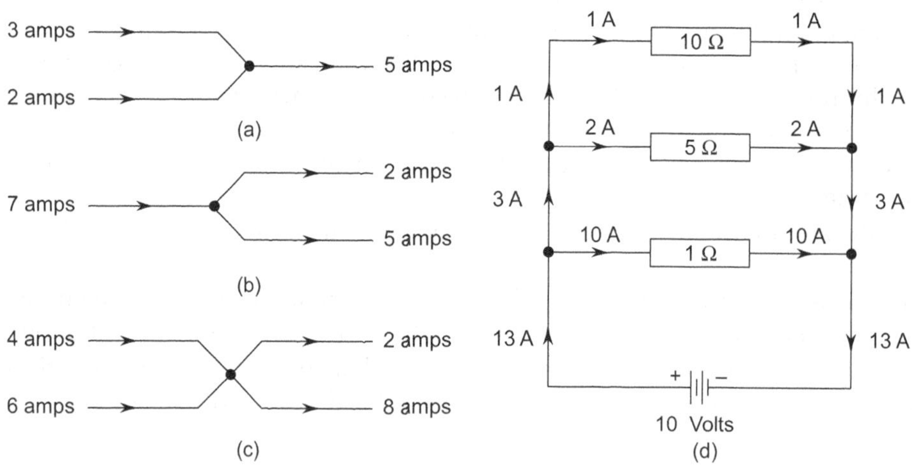

Fig. 2.2

Problem 2.1. *A node of a circuit is shown in Fig. 2.3 with current direction arrows chosen arbitrarily. Having picked those directions, $i_1 = -5$ A, $i_2 = 3$ A, and $i_3 = -1$ A. Write an expression for Kirchhoff's current law and solve for i_4.*

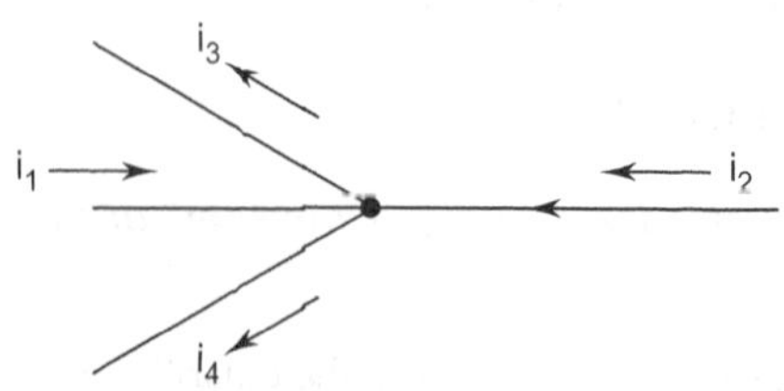

Fig. 2.3

Solution. By Kirchhoff's current law,

$$i_1 + i_2 = i_3 + i_4$$

$$-5 + 3 = -1 + i_4$$

So that $$i_4 = -1 \text{ A}$$

That is, i_4 is actually 1 A flowing into the node. Note that i_2, i_3, and i_4 are all entering the node, and i_1 is the only current that is leaving the node.

2.2.2 Kirchhoff's Voltage Law (KVL)

It states that *"the algebraic sum of products of current and resistance (i.e., voltage drops) in each of the conductors in any closed path (or mesh) in a network plus the algebraic sum of the e.m.f. sources in that path is zero"*

$$\Sigma \text{IR} + \Sigma \text{e.m.f.} = 0 \quad \text{...(2.4)}$$

For example, let us consider a simple circuit shown in Fig. 2.4. Applying Kirchhoff's voltage law, we have

$$\text{I R}_1 + \text{I R}_2 = \text{V} \quad \text{...(2.5)}$$

or $$\text{V} - \text{I R}_1 - \text{I R}_2 = 0 \quad \text{...(2.6)}$$

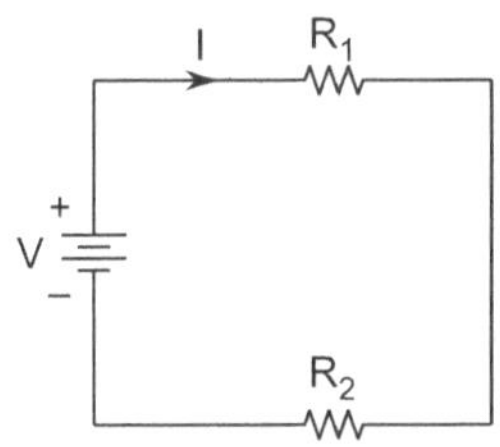

Fig. 2.4

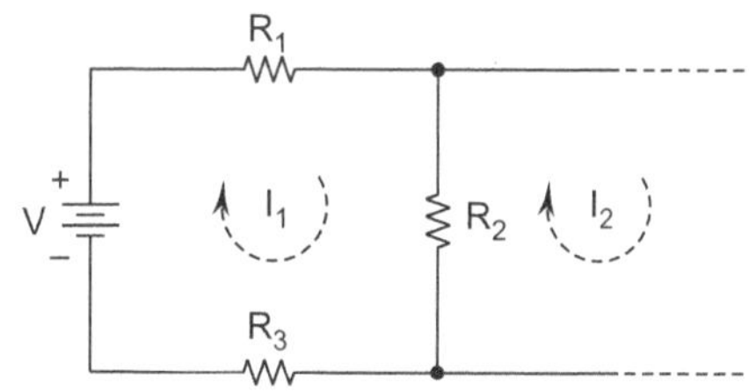

Fig. 2.5

Now let us consider a part of an electrical circuit shown in Fig. 2.5.

For loop 1

Current through R_1 and R_3 will be I_1

Current through R_2 will be $(I_1 - I_2)$

Therefore, $$\text{V} = I_1 R_1 + I_1 R_3 + (I_1 - I_2) R_2 \quad \text{...(2.7)}$$

or $$\text{V} = I_1 (R_1 + R_2 + R_3) - I_2 R_2 \quad \text{...(2.8)}$$

2.3 SOLVING EQUATIONS WITH THREE UNKNOWNS PARAMETERS

A lot of numerical problems arrive where three unknowns or three equations will be encounter. In that case one can use the following method for solving those three equations. This method is commonly known as ***Cramer's Rule***. Let the three simultaneous equations be as under:

$$ax + by + cz = d \quad \text{...(2.9)}$$

$$ex + fy + gz = h \quad \text{...(2.10)}$$

$$jx + ky + lz = m \quad \text{...(2.11)}$$

The above equations can be put in matrix form as under:

$$\begin{bmatrix} a & b & c \\ e & f & g \\ j & k & l \end{bmatrix}\begin{bmatrix} x \\ y \\ z \end{bmatrix} = \begin{bmatrix} d \\ h \\ m \end{bmatrix} \quad \ldots(2.12)$$

We have to find the value of x, y and z (or there may be some other constants). We will solve these equations with Cramer's rule.

The common determinant is given by

$$\Delta = \begin{bmatrix} a & b & c \\ e & f & g \\ j & k & l \end{bmatrix} \quad \ldots(2.13)$$

$$= a(fl - gk) - e(bl - ck) + j(bg - cf) \quad \ldots(2.14)$$

The determinant of x can be found by replacing coefficients of x in the original matrix by the constants.

$$\Delta_1 = \begin{bmatrix} d & b & c \\ h & f & g \\ m & k & l \end{bmatrix} \quad \ldots(2.15)$$

$$= d(fl - gk) - h(bl - ck) + m(bg - cf) \quad \ldots(2.16)$$

Similarly the determinant of y is given by

$$\Delta_2 = \begin{bmatrix} a & d & c \\ e & h & g \\ j & m & l \end{bmatrix} \quad \ldots(2.17)$$

$$= a(hl - mg) - e(dl - mc) + j(dg - hc) \quad \ldots(2.18)$$

And the determinant of x is given by

$$\Delta_3 = \begin{bmatrix} a & b & d \\ e & f & h \\ j & k & m \end{bmatrix} \quad \ldots(2.19)$$

$$= a(fm - hk) - e(bm - dk) + j(bh - df) \quad \ldots(2.20)$$

As per Cramer's rule

$$x = \frac{\Delta_1}{\Delta} \quad \ldots(2.21)$$

$$y = \frac{\Delta_2}{\Delta} \quad \ldots(2.22)$$

$$z = \frac{\Delta_3}{\Delta} \quad \ldots(2.23)$$

2.4 MAXWELL'S MESH/LOOP CURRENT METHOD

In Maxwell's mesh/loop current method, for solving complex electrical networks, we employ a system of loop or mesh current instead of branch current. Here, the current in different meshes are assigned continuous path so that they do not split at a junction into branch currents*.

It is based up Kirchhoff's Voltage Law (KVL).

For solving a circuit with Maxwell's mesh/loop current method, one can follow the following procedure describe below.

For describing the procedure let us consider an electrical network shown in Fig. 2.6.

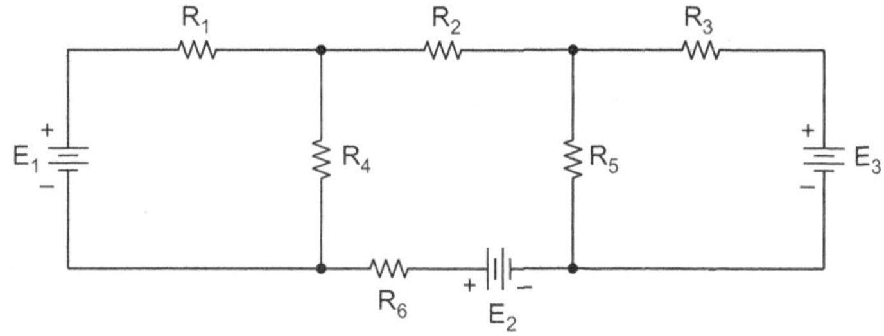

Fig. 2.6

Step 1. Decide number of meshes in the given circuit.

(In above case numbers of meshes are 3)

Step 2. Mark mesh current in each mesh with every current having clockwise direction (shown in Fig. 2.7). [The mesh current direction can be taken in any direction (*i.e.*, in clockwise or in anti-clockwise). The polarity of the mesh currents obtained will automatically verify whether the direction of the currents taken is right or wrong].

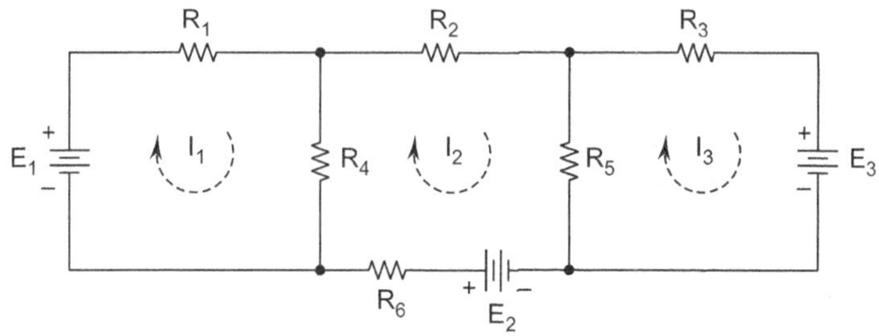

Fig. 2.7

Step 3. Consider each mesh at a time and apply KVL to it. First put an equal to sign (=). Write the e.m.f. sources on left hand side and all voltage drops on right hand side.

Step 4. The polarity of the voltage source can be determined by direction of its current. If the current coming out from the e.m.f. source is in same direction in which the mesh current is

*While solving a DC electrical circuit, with Maxwell's mesh/loop current method, one can assume that the current consider in the loop is similar to the continuous movement of a current in a closed room in one direction. And branch current can be considered as movement of current inside the four walls of that room.

moving, then consider that e.m.f. source as positive otherwise consider it as negative. (For example, in mesh 1, the current coming out from e.m.f. source E_1 will be in the same direction as that of mesh current I_1, so it will be considered as positive, but in case of mesh 3, the current coming out from e.m.f. source E_3 is in opposite direction as that of mesh current I_3, so this e.m.f. will be considered as negative.)

Step 5. If a branch is common between two meshes (for example, in mesh 1 and mesh 2 the branch containing resistance R_4 is common in both meshes) then consider that mesh current greater which is being considered for writing KVL equation.

Step 6. Simplify the equations and solve them simultaneously or by Cramer's rule.

Step 7. If after finding the values of I_1, I_2 and I_3 any of the result comes out to be negative, then it means the direction of that current is in anti-clockwise direction.

For circuit shown in Fig. 2.7, three mesh equations will be

For Mesh 1

$$E_1 = R_1 I_1 + R_4 (I_1 - I_2) \qquad \ldots(2.24)$$

Note that current I_1 is taken as greater than current I_2 while writing equation for mesh 1

For Mesh 2

$$E_2 = R_2 I_2 + R_5 (I_2 - I_3) + R_6 I_2 + R_4 (I_2 - I_1) \qquad \ldots(2.25)$$

Note that current I_2 is taken as greater than current I_1 and I_3 while writing equation for mesh 2.

For Mesh 3

$$- E_3 = R_3 I_3 + R_5 (I_3 - I_2) \qquad \ldots(2.26)$$

Note that current I_3 is taken as greater than current I_2 while writing equation for mesh 3.

Problem 2.2. *Find the current in 2 ohms resistance of the circuit shown in Fig. 2.8 using mesh (or loop) analysis.*

Solution.

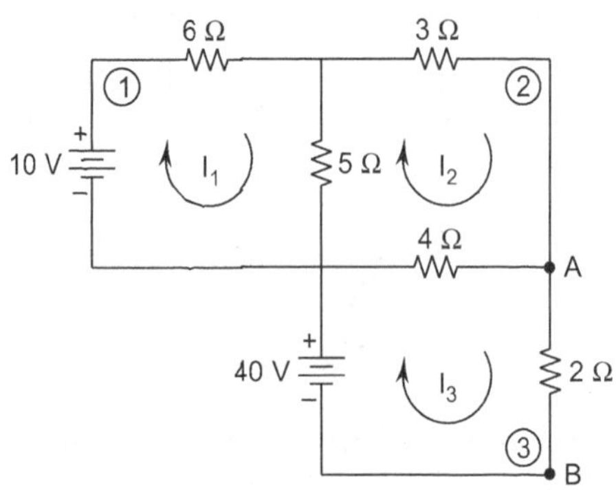

Fig. 2.8

For Mesh 1, Applying KVL, we have

$$10 = 6 I_1 + 5 (I_1 - I_2)$$

or

$$10 = 11 I_1 - 5 I_2 + 0 I_3 \qquad \ldots(i)$$

For Mesh 2, Applying KVL, we have

$$0 = 3\,I_2 + 4\,(I_2 - I_3) + 5\,(I_2 - I_1)$$

or $$0 = 5\,I_1 - 12\,I_2 + 4\,I_3 \quad \ldots(ii)$$

For Mesh 3, Applying KVL, we have

$$40 = 2\,I_3 + 4\,(I_3 - I_2)$$

or $$40 = 0\,I_1 - 4\,I_2 + 6\,I_3$$

or $$20 = 0\,I_1 - 2\,I_2 + 3\,I_3 \quad \ldots(iii)$$

Solving equations (*i*), (*ii*) and (*iii*) by Cramer rule, we have

$$\begin{bmatrix} 11 & -5 & 0 \\ 5 & -12 & 4 \\ 0 & -2 & 3 \end{bmatrix}\begin{bmatrix} I_1 \\ I_2 \\ I_3 \end{bmatrix} = \begin{bmatrix} 10 \\ 0 \\ 20 \end{bmatrix}$$

$$\Delta = 11\,(-36 + 8) + 5\,(15 - 0)$$

$$= 11\,(-28) + 75 = -308 + 75$$

$$\Delta = -233$$

$$\Delta_1 = \begin{bmatrix} 10 & -5 & 0 \\ 0 & -12 & 4 \\ 20 & -2 & 3 \end{bmatrix}$$

$$= 10\,(-36 + 8) + 5\,(0 - 80) = -280 - 400$$

$$\Delta_1 = -680$$

$$\Delta_2 = \begin{bmatrix} 11 & 10 & 0 \\ 5 & 0 & 4 \\ 0 & 20 & 3 \end{bmatrix}$$

$$= 11\,(0 - 80) - 10\,(15 - 0) = -880 - 150$$

$$\Delta_2 = -1030$$

$$\Delta_3 = \begin{bmatrix} 11 & -5 & 10 \\ 5 & -12 & 0 \\ 0 & -2 & 20 \end{bmatrix}$$

$$= 11\,(-240 + 0) - 5\,(-1000 + 20) = -2640 + 4900$$

$$\Delta_3 = 2260$$

$$I_1 = \frac{\Delta_1}{\Delta} = \frac{-680}{-233} = 2.92 \text{ A}$$

$$I_2 = \frac{\Delta_2}{\Delta} = \frac{-1030}{-233} = 4.42 \text{ A}$$

$$I_3 = \frac{\Delta_3}{\Delta} = \frac{2260}{-233} = -9.70\text{ A}$$

Hence, the current in 2 Ω resistance will be

$$I = I_3 = -\mathbf{9.7\ A}$$ **Ans.**

Negative sign in the answer indicates that the current is moving in the upward direction in 2 ohm resistance.

2.5 NODAL ANALYSIS

With the help of Nodal Analysis, unknown voltages at different nodes can be determined.

It is based upon Kirchhoff's current law (KCL). The number of equations formed due to nodal analysis is either equal or less than the number of equation formed due mesh analysis. In actual, the number of equations formed due to nodal analysis is equal to

$$N - 1 \qquad \text{...(2.27)}$$

where, N is number of nodes.

For solving a circuit with nodal analysis, one can follow the following procedure describe below. For describing the procedure let us consider an electrical network shown in Fig. 2.9.

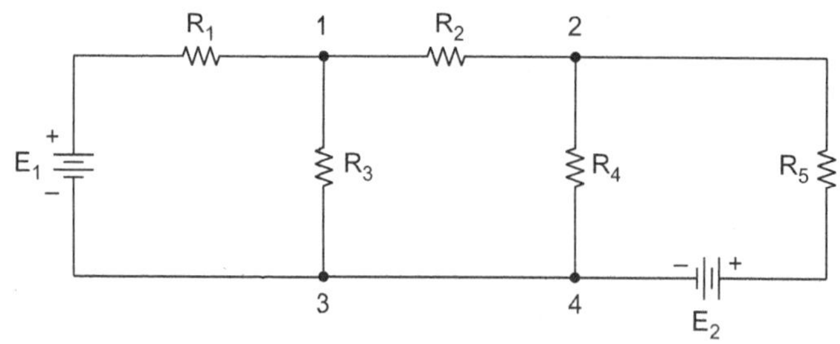

Fig. 2.9

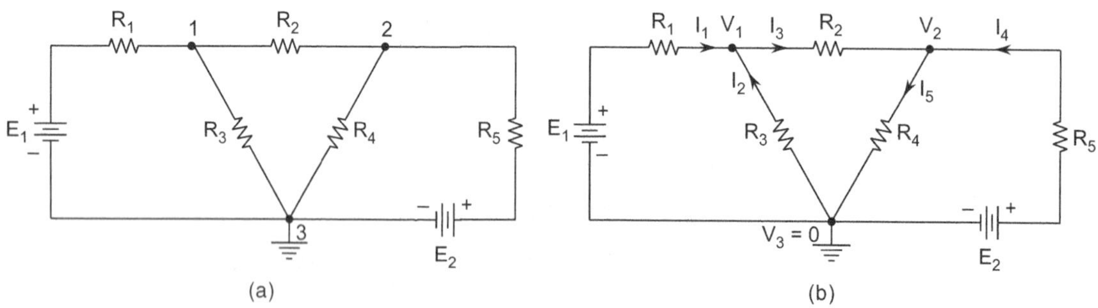

Fig. 2.10

Step 1. Decide number of nodes in the given circuit. If there is no element connected in a particular branch between two nodes then those two nodes can be combined into one node. For example, in the circuit of Fig. 2.9, there is no element between node 3 and node 4. So these two nodes can be combines into one node as shown in Fig. 2.10.

Step 2. Mark any one node as reference node (reference node is also known as datum node or ground node). Take potential of reference node as zero. In above case node 3 is taken as reference node.

Step 3. Mark rest of the nodes as 1, 2, 3, ... etc., and takes potential on those nodes as V_1, V_2, V_3, ... etc.

Step 4. Mark currents in all branches and name them as I_1, I_2, I_3, ... etc.

The direction of the branch currents can be taken in any direction. The polarity of branch currents obtained in the final answer will automatically verify whether the direction of the currents taken is right or wrong.

Step 5. Consider each node at a time. Apply KCL on that node. For example, at node 1 the KCL will be [Fig. 2.10 (*b*)]

$$I_1 + I_2 = I_3 \qquad \text{...(2.28)}$$

Step 6. Find the value of these branch currents in term of node potentials and branch resistances.

A branch is that part of network which lies between two nodes. Hence branch current will start from any one of those two nodes and end or terminate into another node. First consider potential of that node from which the branch current starts, for example, consider a single branch (Fig. 2.11) of a network.

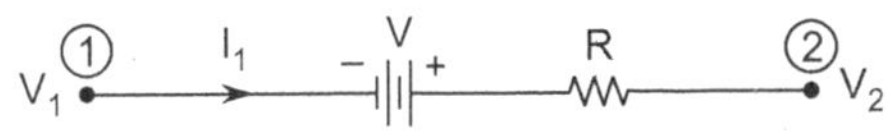

Fig. 2.11

The branch current I_1 starts from node '1' and terminals or ends at node '2'. Also the current coming out from the battery source 'V' (connected in the branch) is in same direction in which the branch current 'I' (denoted by us) is moving.

The branch current I_1 is given by

$$I_1 = \frac{V_1 + V - V_2}{R} \qquad \text{...(2.29)}$$

If the polarity of the e.m.f. source connected in the branch is in opposite direction (as shown in Fig. 2.12), then

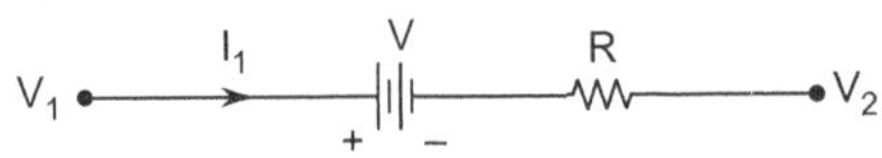

Fig. 2.12

The branch current I_1 will be

$$I_1 = \frac{V_1 - V - V_2}{R} \qquad \text{...(2.30)}$$

and if starting node is reference node (as shown in Fig. 2.13), then

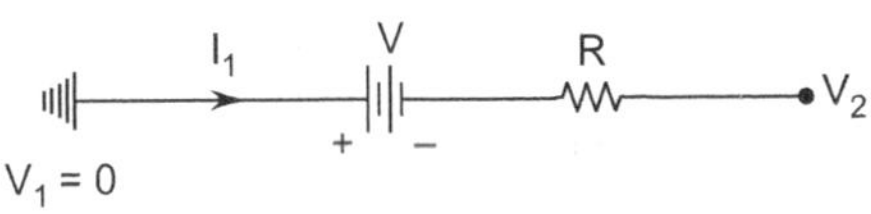

Fig. 2.13

The branch current I_1 will be

$$I_1 = \frac{0 - V - V_2}{R} \qquad \text{...(2.31)}$$

Step 7. In the same way, write value of each branch current and put them in the KCL equation of that node.

Step 8. Then consider next node and follow same procedure as described above (from step 5 to step 7).

Step 9. After obtaining equations for all the nodes, solve them simultaneously to find the numerical value of each node potential.

Step 10. Each branch current can be calculated easily if all node potentials are known as these branch currents are already expressed and calculated in terms of node potential and branch resistances.

Problem 2.3. *Find current I in AB branch of circuit shown in Fig. 2.14 using nodal analysis.*

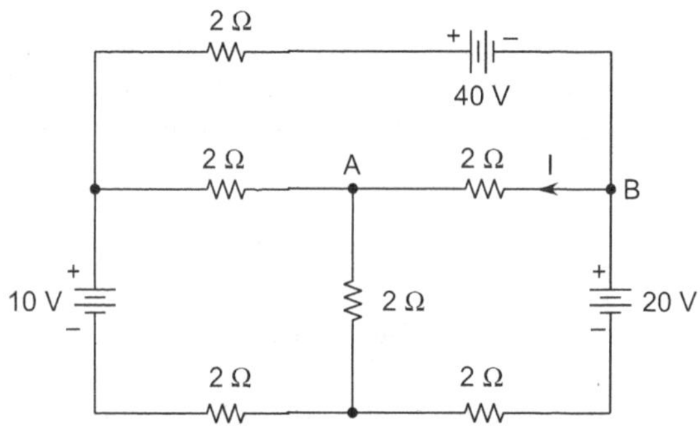

Fig. 2.14

Solution. First, the branch currents and node voltages are assigned as shown in Fig. 2.15.

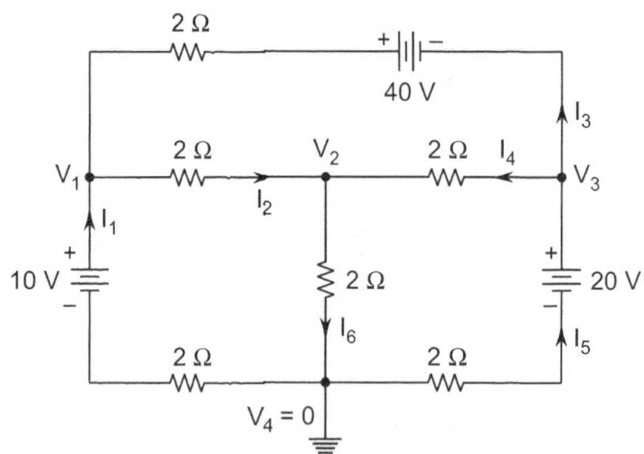

Fig. 2.15

At Node I, Applying KCL, we have

$$I_1 + I_3 = I_2$$

or
$$\frac{0+10-V_1}{2} + \frac{V_3+40-V_1}{2} = \frac{V_1-V_2}{2}$$

or
$$10 - V_1 + V_3 + 40 - V_1 = V_1 - V_2$$

or
$$3V_1 - V_2 - V_3 = 50 \qquad ...(i)$$

At Node II, Applying KCL, we have

$$I_2 + I_4 = I_6$$

$$\frac{V_1 - V_2}{2} + \frac{V_3 - V_2}{2} = \frac{V_2 - 0}{2}$$

$$V_1 - V_2 + V_3 - V_2 = V_2$$

$$V_1 - 3V_2 + V_3 = 0 \quad \text{...}(ii)$$

At Node III, applying KCL, we have

$$I_5 = I_3 + I_4$$

$$\frac{0 + 20 - V_3}{2} = \frac{V_3 + 40 - V_1}{2} + \frac{V_3 - V_2}{2}$$

or $$20 - V_3 = V_3 + 40 - V_1 + V_3 - V_2$$

or $$V_1 + V_2 - 3V_3 = 20 \quad \text{...}(iii)$$

Solving equations (*i*), (*ii*) and (*iii*) by Cramer rule, we have

$$\begin{bmatrix} 3 & -1 & -1 \\ 1 & -3 & 1 \\ 1 & 1 & -3 \end{bmatrix} \begin{bmatrix} V_1 \\ V_2 \\ V_3 \end{bmatrix} = \begin{bmatrix} 50 \\ 0 \\ 20 \end{bmatrix}$$

$$\Delta = \begin{bmatrix} 3 & -1 & -1 \\ 1 & -3 & 1 \\ 1 & 1 & -3 \end{bmatrix}$$

$$\Delta = 3\,(9 - 1) + 1\,(-3 - 1) - 1\,(1 + 3) = 24 - 4 - 4$$

$$\Delta = 16.$$

$$\Delta_1 = \begin{bmatrix} 50 & -1 & -1 \\ 0 & -3 & 1 \\ 20 & 1 & -3 \end{bmatrix}$$

$$= 50\,(9 - 1) + 20\,(-1 - 3) = 400 - 80$$

$$\Delta_1 = 320$$

$$\Delta_2 = \begin{bmatrix} 3 & 50 & -1 \\ 1 & 0 & 1 \\ 1 & 20 & -3 \end{bmatrix}$$

$$= 3\,(0 - 20) - 1\,(-150 + 20) + 1\,(50 - 0) = -60 + 130 + 50$$

$$\Delta_2 = 120$$

$$\Delta_3 = \begin{bmatrix} 3 & -1 & 50 \\ 1 & -3 & 0 \\ 1 & 1 & 20 \end{bmatrix}$$

$$= 3\,(-60 - 0) - 1\,(-20 - 50) + 1\,(0 + 150) = -180 + 70 + 150$$

$$\Delta_3 = 40.$$

Now, $$V_1 = \frac{\Delta_1}{\Delta} = \frac{320}{16} = 20\text{ V}$$

$$V_2 = \frac{\Delta_2}{\Delta} = \frac{120}{16} = 7.5\text{ V}$$

$$V_3 = \frac{\Delta_3}{\Delta} = \frac{40}{16} = 2.5 \text{ V}.$$

Now, current in branch AB is

$$I = I_4 = \frac{V_3 - V_2}{2}$$

$$= \frac{2.5 - 7.5}{2} = \frac{-5}{2}$$

$$I = -\ 2.5 \text{ A}. \qquad \textbf{Ans.}$$

Negative sign in the answer shows that the direction of current I is from A to B.

Problem 2.4. *Frame and solve node equations of network shown in Fig. 2.16. Hence find total power consumed by passive elements of the network.*

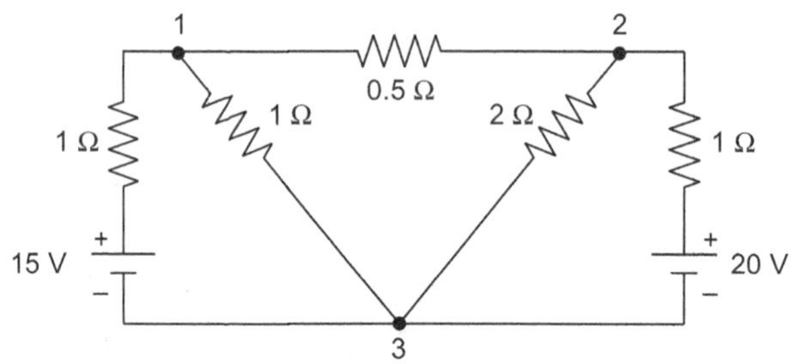

Fig. 2.16

Solution. Number of nodes in the given network are 3. Considering 3rd node as reference, or ground or datum node and taking its potential as zero, we have

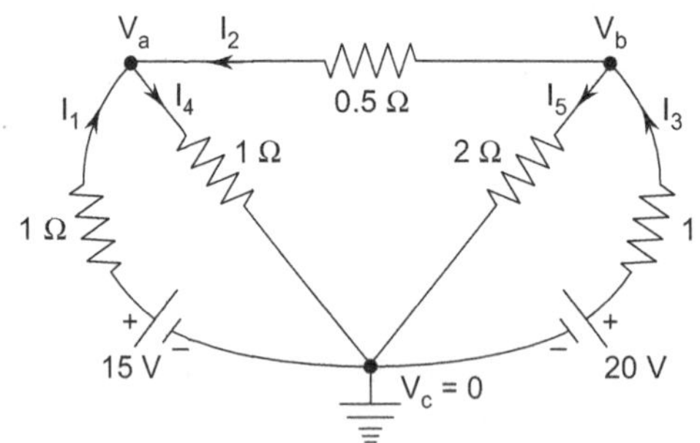

Fig. 2.17

Branch currents with directions and node potentials are marked as shown in Fig. 2.17.

Applying KCL on Node I

$$I_1 + I_2 = I_4$$

$$\frac{0 + 15 - V_a}{1} + \frac{V_b - V_a}{0.5} = \frac{V_a - 0}{1}$$

or simplifying above equation, we get

$$4\,V_a - 2\,V_b = 15 \qquad ..(i)$$

Applying KCL on Node II

$$I_3 = I_2 + I_5$$

$$\frac{0 + 20 - V_b}{1} = \frac{V_b - V_a}{0.5} + \frac{V_b - 0}{2}$$

Simplifying above equation, we get

$$4\,V_a - 7\,V_b = -\,40 \qquad ..(ii)$$

Solving equations (*i*) and (*ii*) simultaneously, we get

$$V_a = \frac{37}{4} = \mathbf{9.25\ V} \qquad \textbf{Ans.}$$

$$V_b = \mathbf{11\ V} \qquad \textbf{Ans.}$$

Now, $$I_1 = \frac{15 - V_a}{1} = 15 - 9.25$$

or $$I_1 = \mathbf{5.75\ A} \qquad \textbf{Ans.}$$

and $$I_2 = \frac{V_b - V_a}{0.5} = \frac{11 - 9.25}{0.5}$$

or $$I_2 = \mathbf{3.5\ A} \qquad \textbf{Ans.}$$

Similarly, $$I_3 = \mathbf{9\ A} \qquad \textbf{Ans.}$$

$$I_4 = \mathbf{9.25\ A} \qquad \textbf{Ans.}$$

$$I_5 = \mathbf{5.5\ A} \qquad \textbf{Ans.}$$

The passive elements of the network are its five resistances.

Total power consumed by them is

$$= 5.75^2 \times 1 + 3.5^2 \times 0.5 + 9^2 \times 1 + 9.25^2 \times 1 + 5.5^2 \times 2$$

$$P = \mathbf{266.25\ Watt.} \qquad \textbf{Ans.}$$

Problem 2.5. *By using mesh analysis, find the current I_4 and I_5 in the circuit shown in Fig. 2.18.*

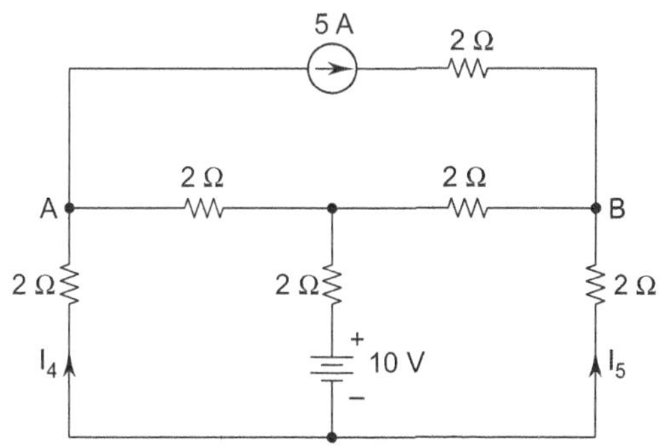

Fig. 2.18

Solution. *This is a typical case when a current source is*

(*i*) *in series with a resistance,*

(*ii*) *in a independent branch.*

This current source in AB branch cannot be converted into voltage source. And hence mesh analysis cannot be applied to that mesh, in which the branch containing current source, is present.

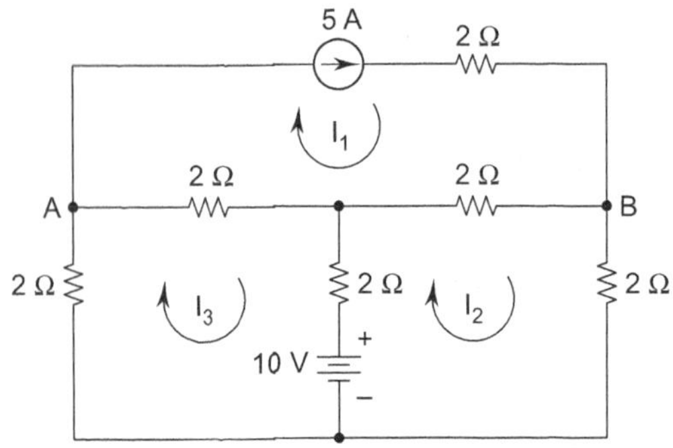

Fig. 2.19

For Mesh I,

$$I_1 = 5\text{ A} \qquad ...(i)$$

For Mesh II,

$$10 = 2I_2 + 2(I_2 - I_3) + 2(I_2 - I_1)$$

$$10 = 2I_2 + 2(I_2 - I_3) + 2(I_2 - 5) \qquad \text{[From equation (i)]}$$

or $$10 = 6I_2 - 2I_3 - 10$$

or $$10 = 3I_2 - I_3 \qquad ...(ii)$$

In Mesh III,

$$-10 = 2I_3 + 2(I_3 - I_1) + 2(I_3 - I_2)$$

$$-10 = 2I_3 + 2(I_3 - 5) + 2(I_3 - I_2)$$

$$0 = 6I_3 - 2I_2$$

or $$I_2 = 3I_3 \qquad ...(iii)$$

Putting equation (*iii*) in equation (*ii*)

$$10 = 3(3I_3) - I_3$$

or $$10 = 9I_3 - I_3$$

or $$I_3 = \frac{10}{8} = 1.25\text{ A}$$

and from equation (*iii*),

$$I_2 = 3 \times 1.25 = 3.75\text{ A}$$

Now, current I_4 is

$$I_4 = I_3 = \mathbf{1.25\ A} \qquad \textbf{Ans.}$$

and current I_5 is

$$I_5 = -I_2 = \mathbf{-3.75\ A.} \qquad \textbf{Ans.}$$

Problem 2.6. *By using mesh analysis, find current in 1 Ω resistance of circuit shown in Fig. 2.20.*

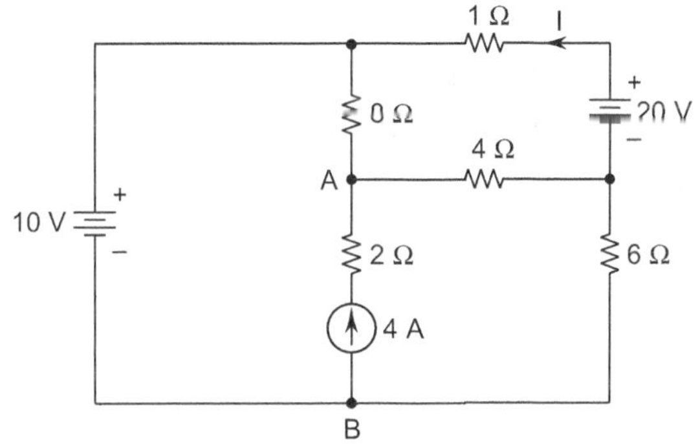

Fig. 2.20

Solution. ***This is a typical case when a current source is***

(*i*) ***in series with a resistance,***

(*ii*) ***in a common branch between two meshes.***

This current source in AB branch cannot be converted into voltage source. And hence mesh analysis cannot be applied to those meshes, in which, current source is present in any branch.

In this case, we have to use ***loop or super-mesh***

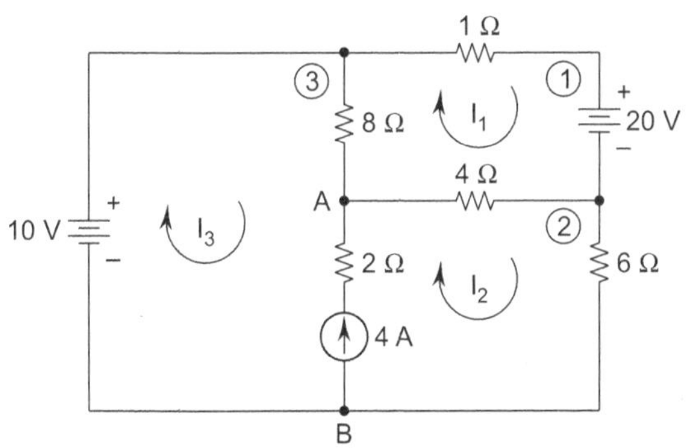

Fig. 2.21

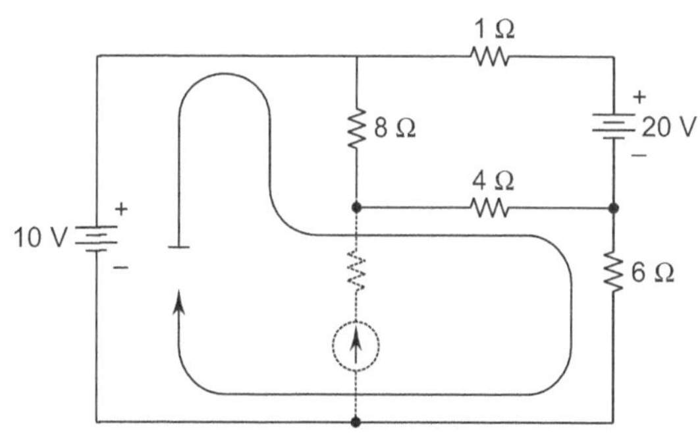

Fig. 2.22

From Fig. 2.21

In Mesh I

$$-20 = I_1 + 4(I_1 - I_2) + 8(I_1 - I_3)$$

or $$-20 = 13I_1 - 4I_2 - 8I_3 \quad ...(i)$$

In AB branch,

The direction of 4 A current source is upward.

The direction of I_2 current is upward.

The direction of I_3 current is downward.

Hence, $$I_2 - I_3 = 4 \quad ...(ii)$$

In the loop (or super mesh) shown in Fig. 2.22

$$10 = 8(I_3 - I_1) + 4(I_2 - I_1) + 6I_2$$

or $$10 = -12I_1 + 10I_2 + 8I_3$$

or $$5 = -6I_1 + 5I_2 + 4I_3 \quad \text{...}(iii)$$

Solving equations (*i*), (*ii*) and (*iii*) by Cramer method.

$$\begin{bmatrix} -13 & 4 & 8 \\ -6 & 5 & 4 \\ 0 & 1 & -1 \end{bmatrix}\begin{bmatrix} I_1 \\ I_2 \\ I_3 \end{bmatrix} = \begin{bmatrix} 20 \\ 5 \\ 4 \end{bmatrix}$$

$$\Delta = \begin{bmatrix} -13 & 4 & 8 \\ -6 & 5 & 4 \\ 0 & 1 & -1 \end{bmatrix}$$

$$\Delta = -1(-52 + 48) - 1(65 + 24)$$

$$\Delta = -37$$

$$\Delta_1 = \begin{bmatrix} 20 & 4 & 8 \\ 5 & 5 & 4 \\ 4 & 1 & -1 \end{bmatrix}$$

$$= 20(-5 - 4) - 4(-5 - 16) + 8(5 - 20)$$

$$\Delta_1 = 216$$

$$\Delta_2 = \begin{bmatrix} -13 & 20 & 8 \\ -6 & 5 & 4 \\ 0 & 4 & -1 \end{bmatrix}$$

$$= -13(-5 - 16) - 20(6 - 0) + 8(-24 - 0)$$

$$\Delta_2 = -39$$

$$\Delta_3 = \begin{bmatrix} -13 & 4 & 20 \\ -6 & 5 & 5 \\ 0 & 1 & 4 \end{bmatrix}$$

$$= -13(20 - 5) - 4(-24 - 0) + 20(-6 - 0)$$

$$\Delta_3 = -219.$$

Now, $$I_1 = \frac{\Delta_1}{\Delta} = \frac{216}{-37} = -5.84 \text{ A}$$

$$I_2 = \frac{\Delta_2}{\Delta} = \frac{-39}{-37} = 1.05 \text{ A}$$

$$I_3 = \frac{\Delta_3}{\Delta} = \frac{-219}{-37} = 5.92 \text{ A}$$

Hence, the current in 1 Ω resistance will be

$$I = I_1 = -5.84 \text{ A.}$$ **Ans.**

Problem 2.7. *Using nodal method analysis, find current in AB branch of circuit shown in Fig. 2.23.*

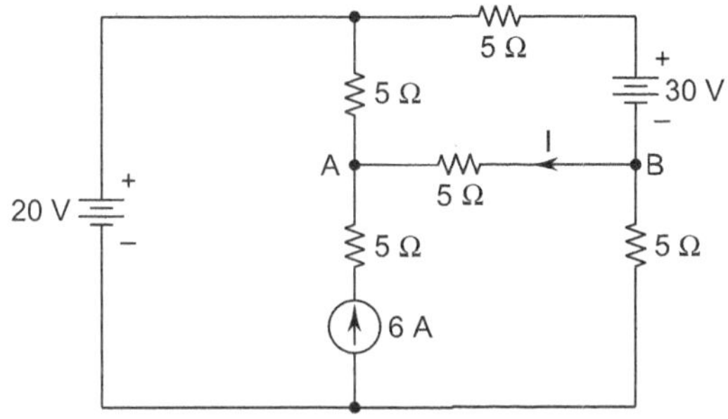

Fig. 2.23

Solution. ***This is a typical case, when a voltage source is***

(i) ***alone in a branch,***

(ii) ***a current source is in series with a resistance.***

This voltage source can't be converted into a current source. In such cases, take that node as datum node, which is connected with negative terminal of that voltage source, which is alone in the branch.

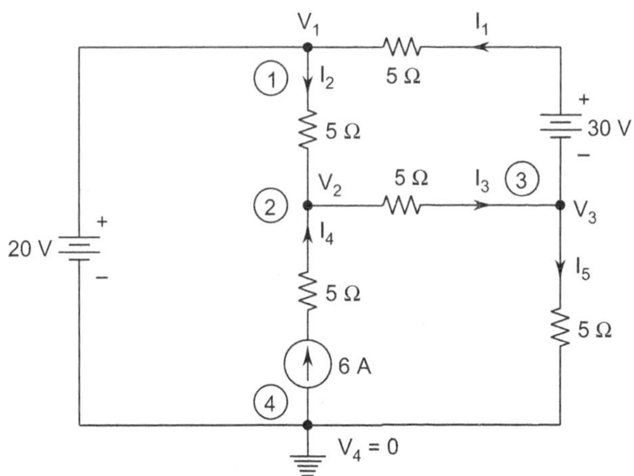

Fig. 2.24

In this case, $V_4 = 0$ ...(*i*)

and hence $V_1 = 20$ V ...(*ii*)

Nodal analysis will not be applicable at Node 1.

Taking branch currents as shown in Fig. 2.24

At Node 2, Applying KCL

$$I_2 + I_4 = I_3$$

or $$\frac{V_1 - V_2}{5} + 6 = \frac{V_2 - V_3}{5}$$

or $$\frac{20 - V_2}{5} + 6 = \frac{V_2 - V_3}{5}$$ [From equation (*ii*)]

or $$20 - V_2 + 30 = V_2 - V_3$$

$$2V_2 - V_3 = 50 \quad ...(iii)$$

At Node 3, Applying KCL

$$I_3 = I_1 + I_5$$

$$\frac{V_2 - V_3}{5} = \frac{V_3 + 30 - V_1}{5} + \frac{V_3 - 0}{5}$$

or $$V_2 - V_3 = V_3 + 30 - 20 + V_3$$

or $$V_2 - 3V_3 = 10 \quad ...(iv)$$

Solving equations (*iii*) and (*iv*) simultaneously, we get

$$V_2 = 28 \text{ V}$$

$$V_3 = 6 \text{ V}$$

Hence, current in AB branch I will be

$$I = -I_3 = -\left(\frac{V_2 - V_3}{5}\right) = -\left(\frac{28 - 6}{5}\right)$$

$$I = \mathbf{4.4\ A.}$$ **Ans.**

2.6 THEVENIN'S THEOREM

Statement

"Any linear bilateral active network can be replaced (across a given load resistance R_L) by a Thevenin's equivalent voltage source V_{TH} in series with a Thevenin's equivalent resistance R_{TH}."

The Thevenin's equivalent voltage source V_{TH} is obtained by removing the load resistance R_L and by finding the potential difference across the open terminals of R_L.

The Thevenin's equivalent resistance R_{TH} can be obtained by finding total internal resistance of the circuit as viewed back into the open circuit terminals of R_L.

While finding R_{TH},

- All voltage sources are replaced by their *internal resistance*, if any.
- If voltage sources do not have their internal resistances, then they are replaced by a *short-circuit*.
- All current sources are replaced by *open circuit*.

Procedure/Proof

Let us consider a simple circuit shown in Fig. 2.25 for describing the procedure.

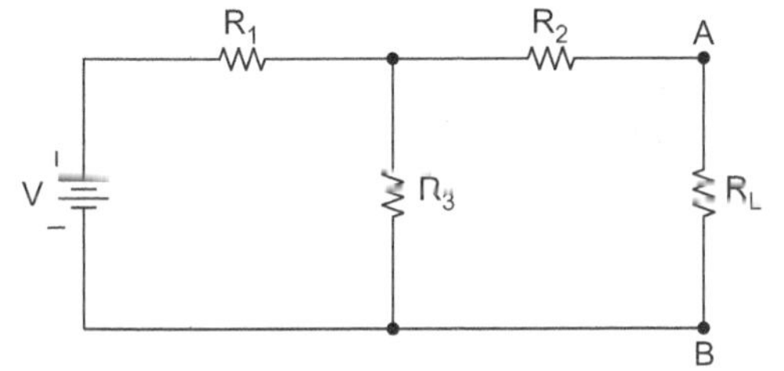

Fig. 2.25

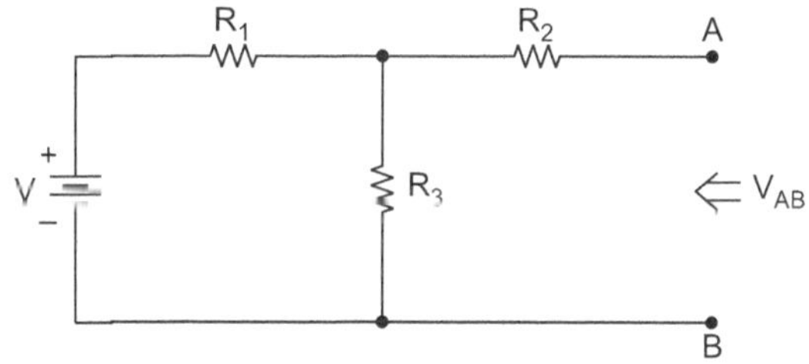

Fig. 2.26

Suppose that we have to find Thevenin's equivalent circuit across the given two terminals AB. For finding Thevenin's equivalent circuit, follow the following procedure step-by-step.

Step 1. Remove load resistance across, which we have to find the Thevenin's equivalent circuit or through which we have to find the current with the help of Thevenin's theorem (Fig. 2.26).

Step 2. Find the open circuit voltage V_{AB} or V_{OC}, which appear across the two terminals A and B. It is called Thevenin's Voltage V_{TH}. This V_{TH} can be calculated by moving from terminal B to A (or from A to B) and following the shortest path.

- If there is any voltage source in the path and the current coming out from the source is in same direction in which you are moving then consider it as positive or otherwise negative.
- If there is any resistance in between and no current is flowing through it, then ignore it and move further.
- If there is any resistance in between and there is a current flowing through it then takes the product of current and resistance. Follow the following polarity rule.

{(– +) (– +) (– +) (– +)} **Addition**

{(+ –) (+ –) (+ –) (+ –)} **Subtraction**

Step 3. Compute the resistance of whole network as looked from these two terminals (Fig. 2.27) after all voltage sources have been removed leaving behind their internal resistance (if any) and the current source have been removed by open circuit. This resistance is called Thevenin's equivalent resistance R_{TH}.

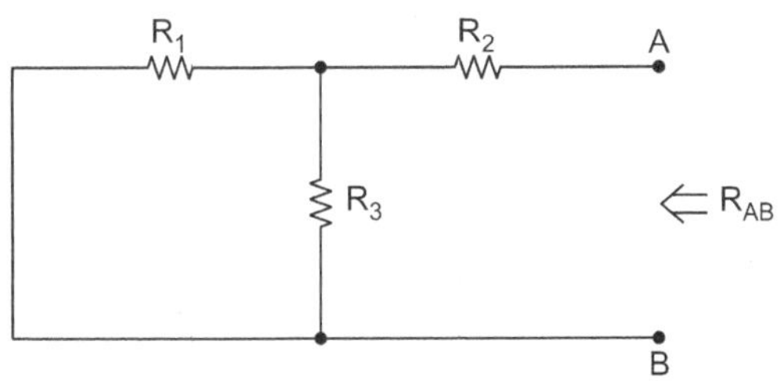

Fig. 2.27

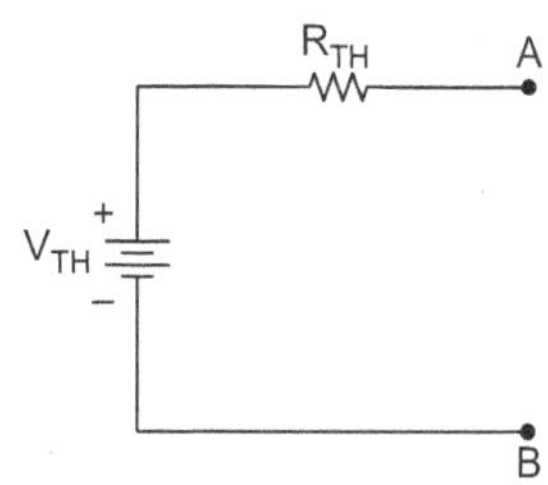

Fig. 2.28

Step 4. Replace entire network by a single Thevenin's equivalent voltage source V_{TH} in series with a Thevenin's equivalent resistance R_{TH} (Fig. 2.28).

Step 5. Connect load resistance R_L back to its terminals from where it was previously removed (Fig. 2.29) and find the current through the circuit as

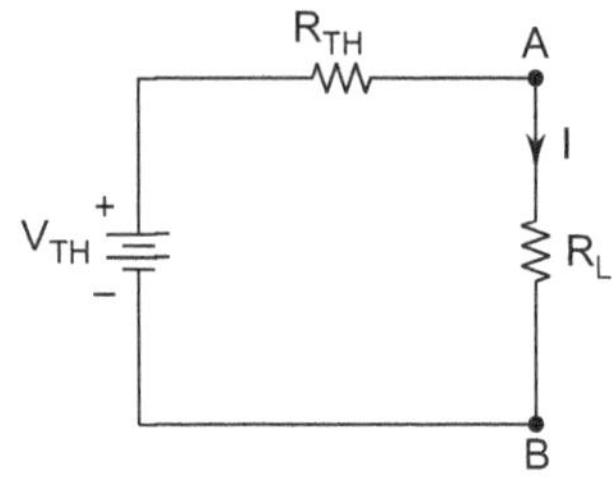

Fig. 2.29

$$I = \frac{V_{TH}}{R_{TH} + R_L}.$$

Problem 2.8. *Replace network of Fig. 2.30 across AB by its Thevenin's equivalent circuit and hence determine the current I through load resistance R_L.*

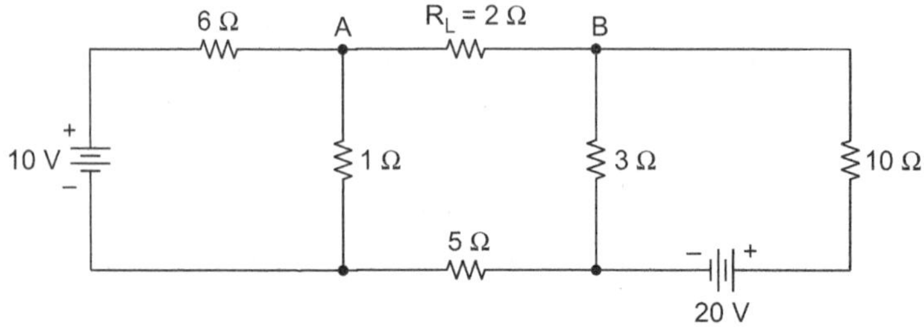

Fig. 2.30

Solution. After removing the load resistance, $R_L = 2\ \Omega$ from the terminals AB, we have

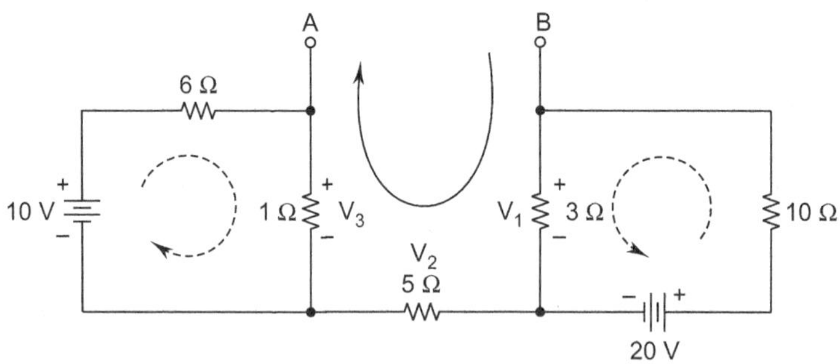

Fig. 2.31

While moving from B to A we will encounter with three parameters:

(*i*) 3 Ω resistance across which the voltage drop occur due to 20 volt battery source only.

$$V_1 = \frac{20 \times 3}{10+3} = \frac{60}{13} = 4.62 \text{ V (Negative)}$$

(*ii*) 5 Ω resistance across which no voltage drop occur because no current will flow through it due to any source hence

$$V_2 = 0.$$

(*iii*) 1 Ω resistance across which the voltage drop occur due to 10 volts battery source only hence

$$V_3 = \frac{10 \times 1}{6+1} = \frac{10}{7} = 1.43 \text{ V (Positive)}$$

Hence, $V_{TH} = V_{AB} = -V_1 + V_2 + V_3 = -4.62 + 0 + 1.43$

$V_{TH} = V_{AB} = -\mathbf{3.19\ V.}$ **Ans.**

For finding R_{TH}, all voltage sources are removed by short-circuit (Fig. 2.32).

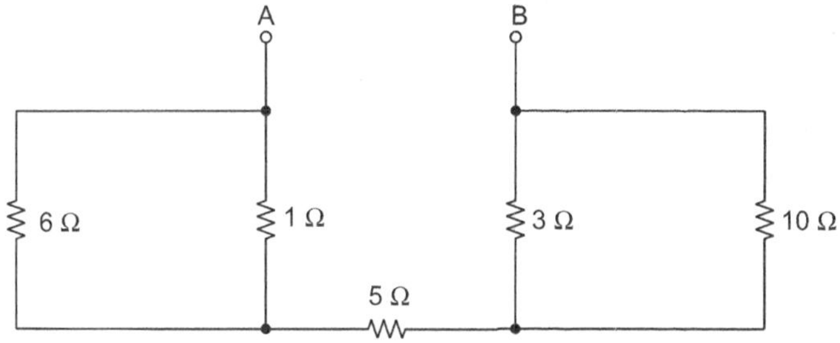

Fig. 2.32

$$R_{AB} = (6 \parallel 1) + 5 + (3 \parallel 10)$$

$$= \frac{6 \times 1}{6+1} + 5 + \frac{3 \times 10}{3+10} = 0.857 + 5 + \frac{30}{13}$$

$R_{TH} = R_{AB} = \mathbf{8.16\ \Omega.}$ **Ans.**

Hence the Thevenin's equivalent circuit will be

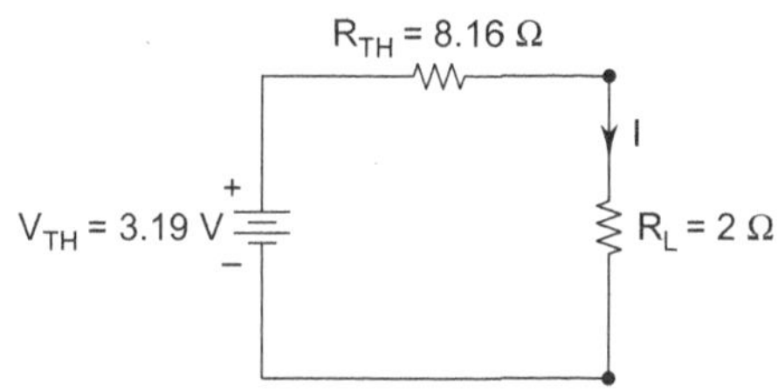

Fig. 2.33

Current I through R_L will be

$$I = \frac{V_{TH}}{R_{TH} + R_L} = \frac{3.19}{8.16 + 2}$$

$I = \mathbf{0.314\ A.}$ **Ans.**

Problem 2.9. *Replace the network of Fig. 2.34 across AB by its Thevenin's equivalent circuit and hence determine the current I.*

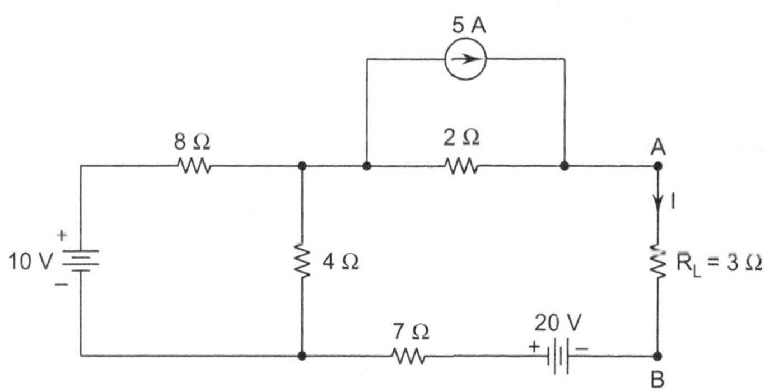

Fig. 2.34

Solution. While taking shortest path from B to A, we will encounter 4 elements.

(*i*) 20 V source

Hence, $V_1 = 20\text{ V (Positive)}$

(*ii*) 7 Ω resistance across which no voltage drop appears because there is no current flowing through it. Hence,

$$V_2 = 0\text{ V}.$$

(*iii*) 4 Ω resistance in branch CD, through which current will flow due to 10 volt source. Hence the voltage drop across 4 Ω resistance will be

$$V_3 = \frac{10 \times 4}{8+4} = \frac{40}{12}$$

$$V_3 = 3.33\text{ V (Positive)}$$

(*iv*) 2 Ω resistance between C and A through which current will flow only due to 5 Ampere current source. Hence, voltage drop across 2 Ω resistance will be

$$V_4 = 2 \times 5 = 10\text{ V (Positive)}$$

Hence, $V_{TH} = V_{AB} = V_1 + V_2 + V_3 + V_4$

$$= 20 + 0 + 3.33 + 10$$

$$V_{TH} = \mathbf{33.33\ V.}$$ **Ans.**

For finding R_{TH}, all voltage sources are removed by short-circuit and all current sources are removed by open circuit.

$$R_{AB} = R_{TH} = 2 + (8 \parallel 4) + 7 = 9 + \frac{8 \times 4}{8+4} = 9 + \frac{32}{12}$$

$$R_{TH} = \mathbf{11.67\ \Omega.}$$ **Ans.**

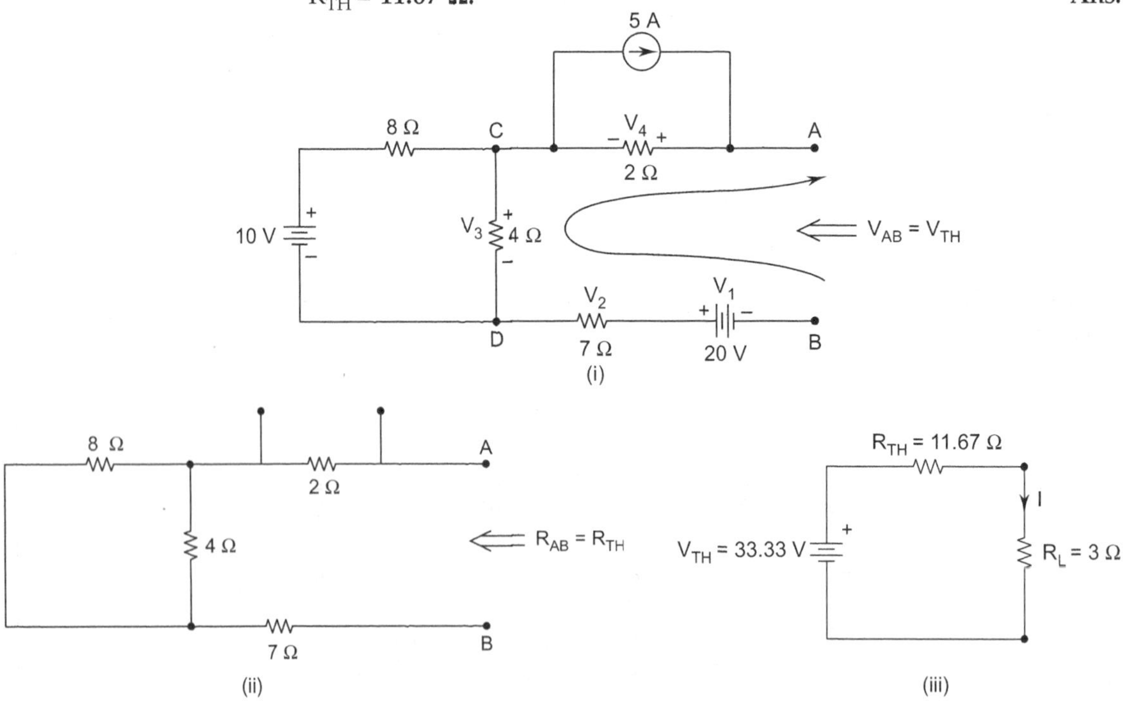

Fig. 2.35

$$I = \frac{V_{TH}}{R_{TH} + R_L} = \frac{33.33}{11.67 + 3}$$

$I = \mathbf{2.272\ A.}$ **Ans.**

2.7 NORTON'S THEOREM

Norton's theorem is an alternate to Thevenin's theorem. Infect it is dual of Thevenin's theorem.

Statement

"Any linear bilateral active network containing active sources and resistances can be replaced (across a given load resistance R_L) by a Norton's equivalent constant current source I_N in parallel with Norton's equivalent resistance R_N."

The Norton's equivalent constant current source I_N is obtained by removing the load resistance R_L, short-circuiting those terminals and finding the current through the short-circuited terminals of R_L.

The Norton's equivalent resistance R_N can be obtained by finding total internal resistance of the circuit as viewed back into the open circuit terminals of R_L.

While finding R_N,

- All voltage source replaced by their *internal resistance,* if any.
- If voltage sources do not have their internal resistances then they are replaced by a *short-circuit.*
- All current sources are replaced by *open circuit.*

Procedure/Proof

Let us again consider a simple circuit shown in Fig. 2.36 for describing the procedure.

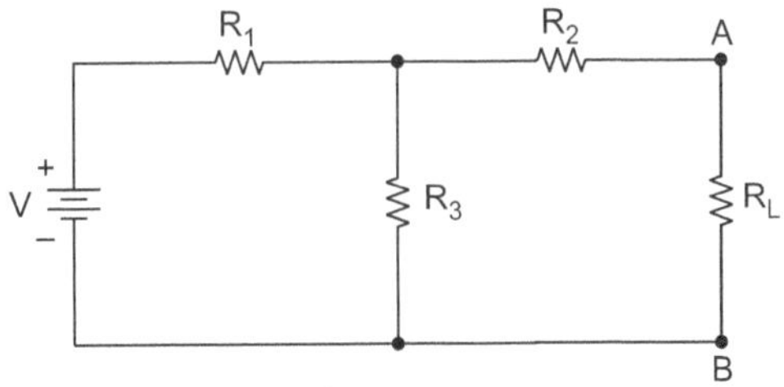

Fig. 2.36

Suppose that we have to find Norton's equivalent circuit across given two terminals AB. For finding this Norton's equivalent circuit follow the following procedure step-by-step.

Step 1. Remove load resistance across the given two terminals and short two given terminals (Fig. 2.37).

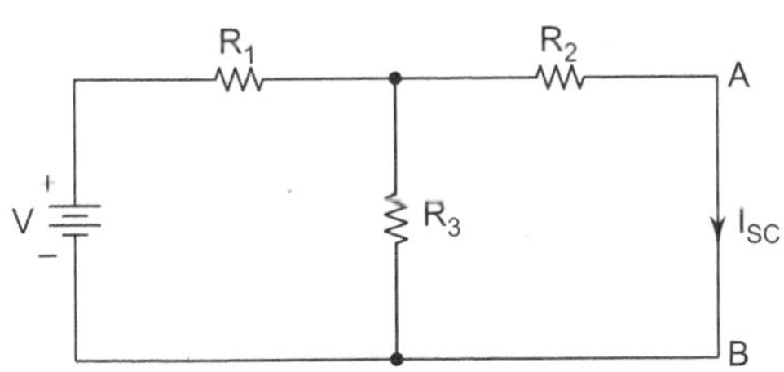

Fig. 2.37

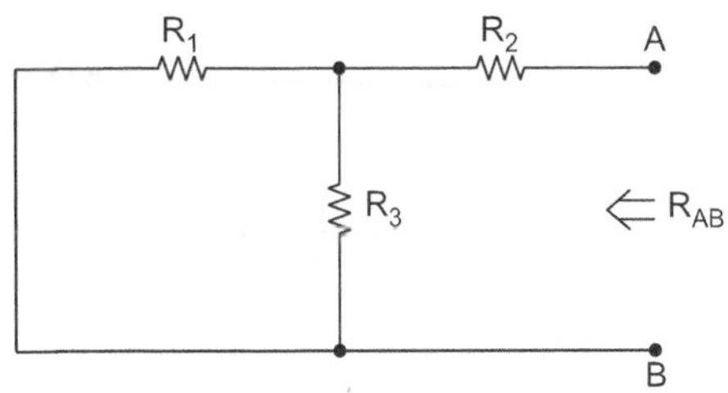

Fig. 2.38

Step 2. Compute the short-circuit current I_{SC} using whatever method one prefer.

Step 3. Compute the resistance of the whole network as looked from these two terminals (Fig. 2.38) after all voltage sources have been removed leaving behind their internal resistance if any and the current source have been removed by open circuit. This resistance is called Norton's equivalent resistance R_N.

Step 4. Replace the entire network by a single Norton's equivalent current source I_N or I_{SC} in parallel with a Norton's equivalent resistance R_N (Fig. 2.39).

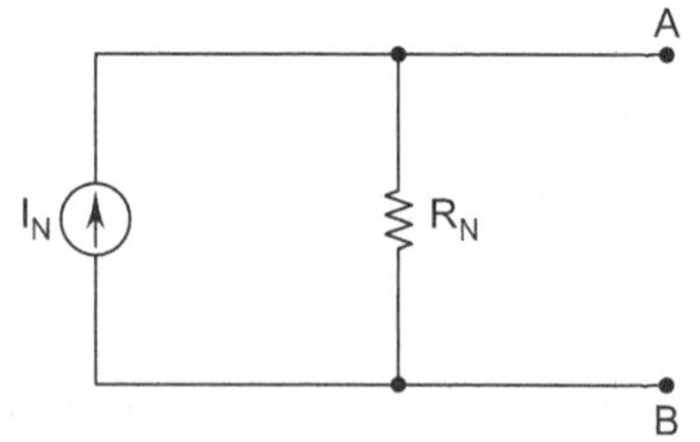

Fig. 2.39

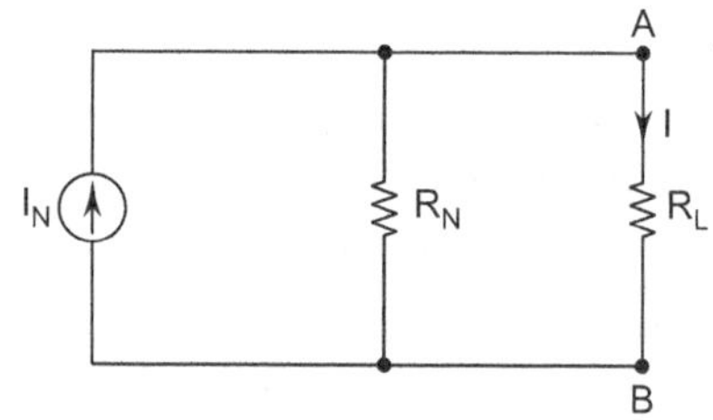

Fig. 2.40

Step 5. Connect the load resistance R_L back to its terminals from where it was previously removed (Fig. 2.40) and find the current through the circuit as (by current divider rule)

$$I = \frac{I_N \times R_N}{R_N + R_L} \qquad \text{...(2.32)}$$

Problem 2.10. *Replace the network of Fig. 2.41 across AB by its Norton's equivalent circuit and hence determine current I.*

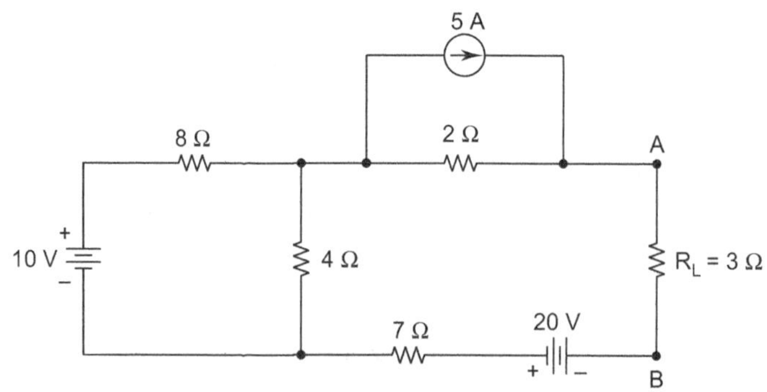

Fig. 2.41

Solution. For finding short-circuit current or Norton's equivalent current source magnitude, load resistance is removed from the given terminals AB and 5 Ampere current source is converted into voltage source. The terminal AB are short-circuited.

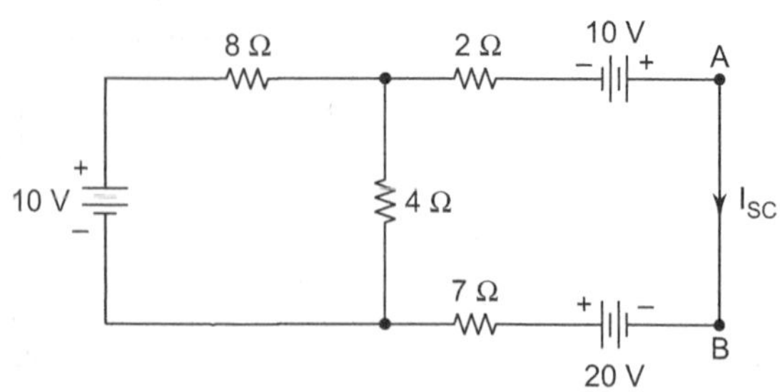

Fig. 2.42

When the current source is converted into voltage source, then the voltage source magnitude will be = $5 \times 2 = 10$ V.

Let us apply nodal analysis to find current I_{SC}.

The circuit shown in Fig. 2.42 can be made more simplified as shown in Fig. 2.43.

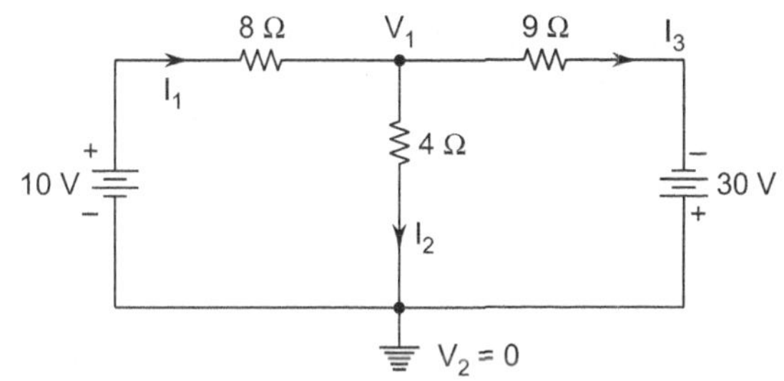

Fig. 2.43

Applying KCL at Node I, we have

$$I_1 = I_2 + I_3$$

or $$\frac{10 - V_1}{8} = \frac{V_1}{4} + \frac{V_1 + 30}{9}$$

or $$9(10 - V_1) = 18V_1 + 8(V_1 + 30)$$

or $$90 - 9V_1 = 18V_1 + 8V_1 + 240$$

$$-35V_1 = 150$$

$$V_1 = -\frac{150}{35} = -4.29 \text{ V}$$

Hence the short-circuit current will be

$$I_{SC} = I_3 = \frac{V_1 + 30}{9}$$

$$= \frac{-4.29 + 30}{9}$$

or $$I_N = I_{SC} = \mathbf{2.86\ A.}$$ **Ans.**

Equivalent resistance will be

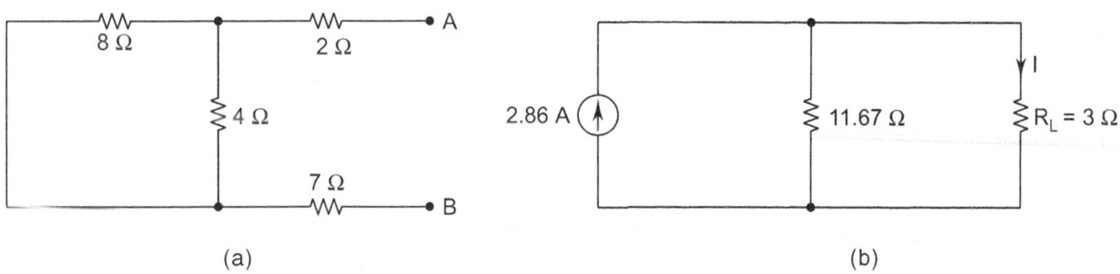

Fig. 2.44

$$R_{AB} = 2 + (8 \| 4) + 7$$

or $$R_N \approx R_{AB} = \mathbf{11.67\ \Omega}.$$ **Ans.**

$$I = \frac{2.86 \times 11.67}{11.67 \times 3}$$

$$I = \mathbf{2.275\ A}.$$ **Ans.**

Problem 2.11. *By using Norton's theorem, find current through the 8 Ω resistance of Fig. 2.45 (a). Also verify the result by Thevenin's theorem.*

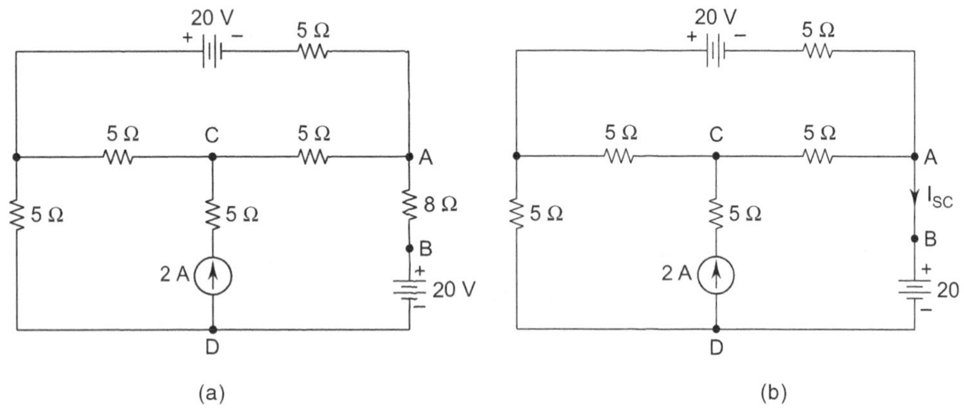

Fig. 2.45

Solution. Removing the load resistance (8 Ω) from terminals A and B and short-circuit terminals A and B, we have circuit as shown in Fig. 2.45(*b*).

We will solve the circuit with mesh analysis. Mesh currents are shown in Fig. 2.46(*a*).

As there is a current source in series with a resistance hence we cannot write mesh equation for mesh-II and mesh-III. Consider a super mesh as shown in Fig. 2.46(*b*).

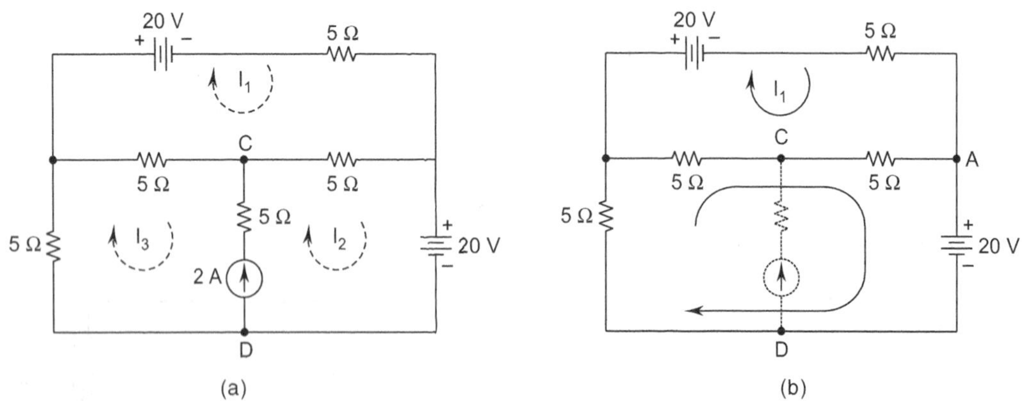

Fig. 2.46

First considering mesh - I, we have [From Fig. 2.46 (*a*)]

$$-20 = 5I_1 + 5(I_1 - I_2) + 5(I_1 - I_3)$$

$$-20 = 15I_1 - 5I_2 - 5I_3$$

$$-4 = 3I_1 - I_2 - I_3 \quad \text{...(i)}$$

In Branch CD, [From Fig. 2.46(*b*)]

$$I_2 - I_3 = 2 \qquad ...(ii)$$

Considering super mesh, we have [From Fig. 2.46(*b*)]

$$-20 = 5I_3 + 5(I_3 - I_1) + 5(I_2 - I_1)$$

$$-20 = -10I_1 + 5I_2 + 10I_3$$

or

$$4 = 2I_1 - I_2 - 2I_3 \qquad ...(iii)$$

Solving equations (*i*), (*ii*) and (*iii*) by Cramer rule, we have

$$\begin{bmatrix} 3 & -1 & -1 \\ 0 & 1 & -1 \\ 2 & -1 & -2 \end{bmatrix} \begin{bmatrix} I_1 \\ I_2 \\ I_3 \end{bmatrix} = \begin{bmatrix} -4 \\ 2 \\ 4 \end{bmatrix}$$

$$\Delta = 3(-2-1) + 2(1+1)$$

$$= 3(-3) + 2(2)$$

$$\Delta = -5$$

$$\Delta_1 = \begin{bmatrix} -4 & -1 & -1 \\ 2 & 1 & -1 \\ 4 & -1 & -2 \end{bmatrix}$$

$$= -4(-2-1) + 1(-4+4) - 1(-2-4) = 12 + 0 + 6$$

$$\Delta_1 = 18$$

$$\Delta_2 = \begin{bmatrix} 3 & -4 & -1 \\ 0 & 2 & -1 \\ 2 & 4 & -2 \end{bmatrix}$$

$$= 3(-4+4) + 2(4+2) = 0 + 2(6)$$

$$\Delta_2 = 12$$

$$\Delta_3 = \begin{bmatrix} 3 & -1 & -4 \\ 0 & 1 & 2 \\ 2 & -1 & 4 \end{bmatrix}$$

$$= 3(4+2) + 2(-2+4) = 18 + 4$$

$$\Delta_3 = 22$$

$$I_1 = \frac{\Delta_1}{\Delta} = \frac{18}{-5} = -3.6 \text{ A}$$

$$I_2 = \frac{\Delta_2}{\Delta} = \frac{12}{-5} = -2.4 \text{ A}$$

$$I_3 = \frac{\Delta_3}{\Delta} = \frac{22}{-5} = -4.4 \text{ A}$$

Hence, $I_{SC} = I_2 = \mathbf{-2.4\ A.}$ **Ans.**

Equivalence resistance

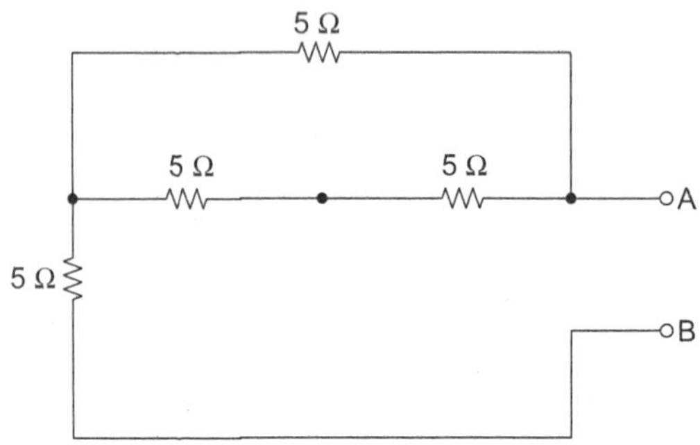

Fig. 2.47

$$R_N = R_{AB} = (5 \parallel 10) + 5 = \frac{5 \times 10}{5 + 10} + 5 = 8.33\ \Omega$$

Current through load resistance, $R_L = 8\ \Omega$ will be

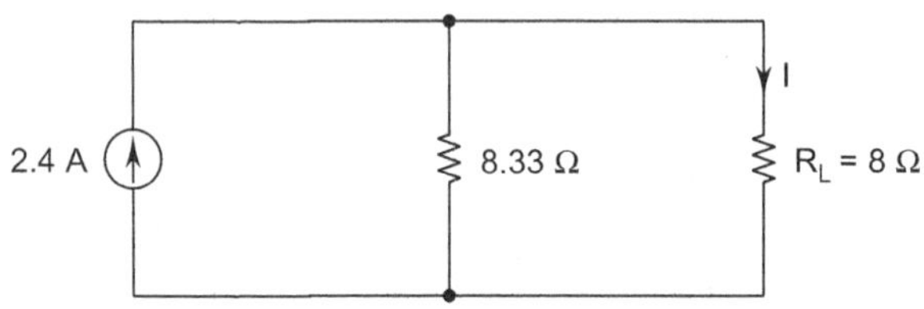

Fig. 2.48

$$I = \frac{2.4 \times 8.33}{8 + 8.33} = \mathbf{1.22\ A}$$ **Ans.**

Verification of result by Thevenin's Theorem

First remove load reistance $R_L = 8\ \Omega$ (Fig. 2.49) and find open circuit voltage across terminals A and B.

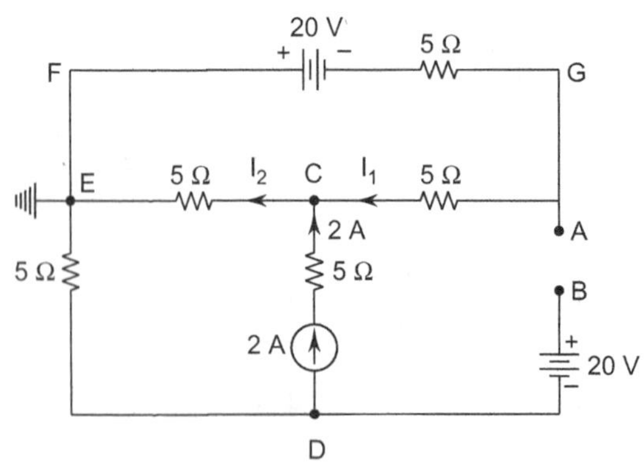

Fig. 2.49

To find V_{AB} or V_{TH}, let us apply Nodal analysis to the circuit shown in Fig. 2.49, in which direction of branch currents and node potentials are shown.

At Node C, applying KCL, we have

$$2 = I_1 + I_2$$

or

$$2 = \frac{V_C + 20}{10} + \frac{V_C}{5}$$

or

$$V_C = 0$$

$\therefore$

$$I_1 = \frac{V_C + 20}{10} = \frac{0 + 20}{10} = 2\text{ A}$$

and

$$I_2 = \frac{V_C}{2} = 0\text{ A}$$

Now, to find V_{TH}, we cannot follow the path BDCA because 2 A current source lies in this path (Fig. 2.50).

Also note that we cannot follow the path BDECA, because after doing Nodal analysis it is clear that no current is flowing in EC branch.

So to find V_{TH}, we have to follow the path BDEFGA.

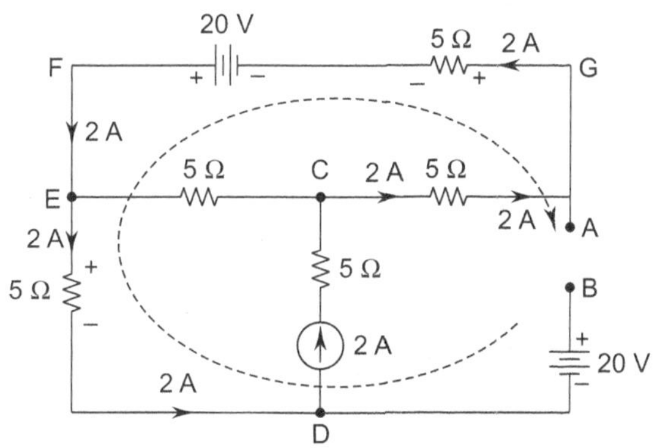

Fig. 2.50

While moving from B to A (following the path BDEFGA),

We will encounter with the following elements:

1. 20 V source (between B and D)

   It is in opposite direction as compare to our direction, hence it is taken as negative

   $$V_1 = -20\text{ V}$$

2. 5 Ω resistance (between D and E)

   The current through this 5 Ω resistance is 2 A and polarities are shown in Fig. 2.50. Voltage drop will be

   $$V_2 = 5 \times 2 = 10\text{ V (Positive)}$$

3. 20 V source (between F and G)

   It is in opposite direction as compare to our direction, hence it is taken as negative

   $$V_3 = -20\text{ V}$$

4. 5 Ω resistance (between F and G)

The current through 5 Ω resistance is 2 A and hence the voltage drop will be

$V_4 = 5 \times 2 = 10$ V (Positive)

Hence, $$V_{TH} = V_1 + V_2 + V_3 + V_4$$

or $$V_{TH} = -20 + 10 - 20 + 10 = -20 \text{ V}$$

We have already calculated R_N. Hence,

$$R_{TH} = R_N = 8.33\ \Omega$$

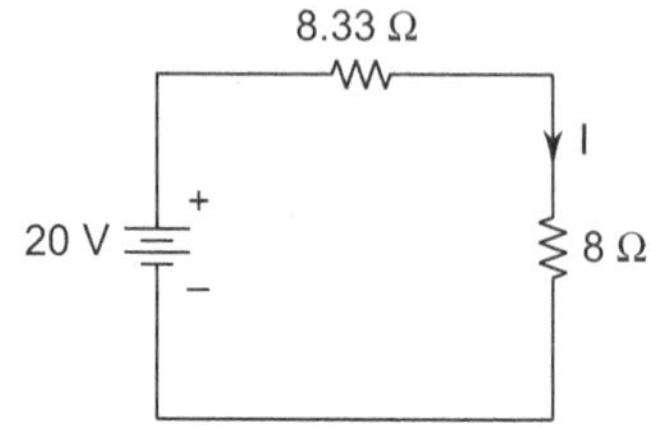

Fig. 2.51

Current through 8 Ω resistance will be

$$I = \frac{V_{TH}}{R_{TH} + R_L} = \frac{20}{8.33 + 8}$$

or $$I = \mathbf{1.22\ A.}$$ **Ans.**

The current through 8 Ω resistance as calculated with Norton's theorem is same as calculated from Thevenin's theorem.

Problem 2.12. *By using Norton's theorem, find current in load resistance R_L of Fig. 2.52.*

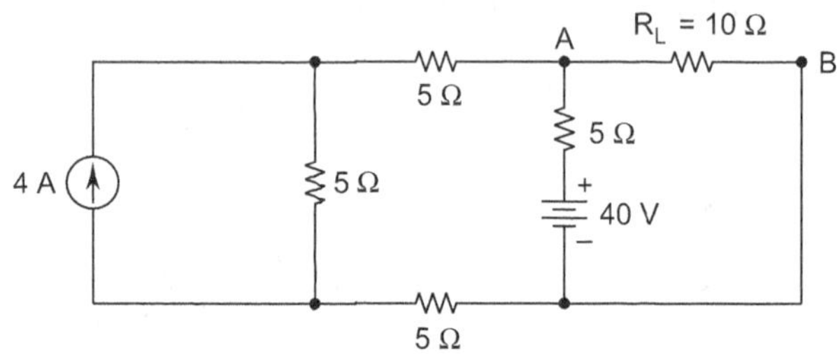

Fig. 2.52

Solution. Removing the load resistance and short-circuiting terminals A and B.

Converting 4 A current source into voltage source, we have

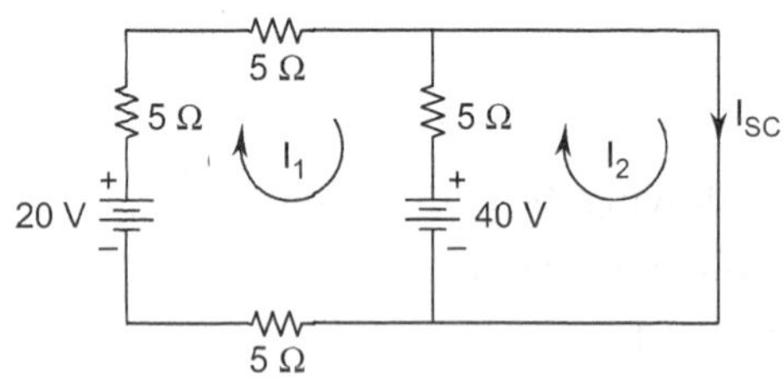

Fig. 2.53

Solving the circuit by mesh analysis.

In Mesh I,

$$20 - 40 = 5I_1 + 5I_1 + 5\,(I_1 - I_2) + 5I_1$$
$$-20 = 20I_1 - 5I_2$$
$$-4 = 4I_1 - I_2 \qquad ...(i)$$

In Mesh II,

$$40 = 5(I_2 - I_1)$$

$$40 = 5I_2 - 5I_1$$

$$8 = -I_1 + I_2 \qquad (ii)$$

Solving equations (*i*) and (*ii*) simultaneously, we get

$$I_1 = 1.33 \text{ A}$$

and $$I_2 = 9.33 \text{ A}$$

Hence, $I_{SC} = I_2 = \mathbf{9.33\ A.}$ **Ans.**

Equivalent resistance will be

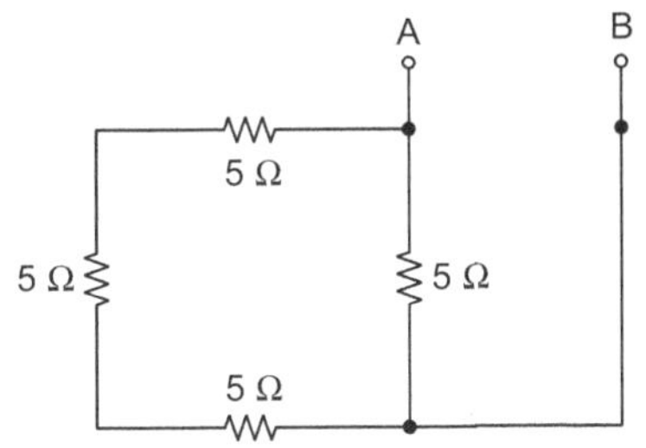

Fig. 2.54

$$R_{AB} = (15 \,||\, 5)$$

or $R_{AB} = R_N = \mathbf{3.75\ \Omega}$ **Ans.**

Current through load resistance will be

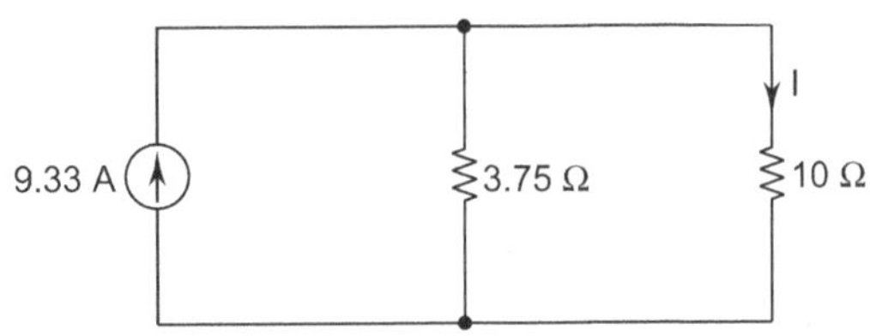

Fig. 2.55

$$I = \frac{9.33 \times 3.75}{3.75 + 10}$$

or $I = \mathbf{2.55\ A.}$ **Ans.**

Problem 2.13. *By using Norton's theorem, find current in load resistance R_L of Fig. 2.56.*

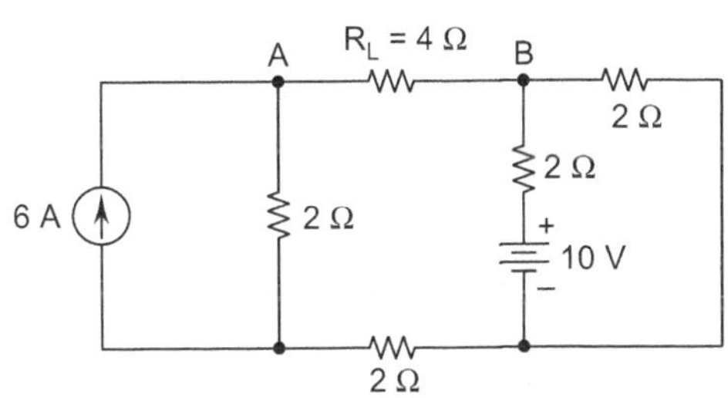

Fig. 2.56

Solution. For finding Norton's equivalent current source, Load Resistance R_L is removed and AB terminals are short-circuited.

The current source of 6 A is converted into voltage source whose value is

$$= 6 \times 2 = 12 \text{ V}.$$

Solving the circuit with nodal analysis, we have

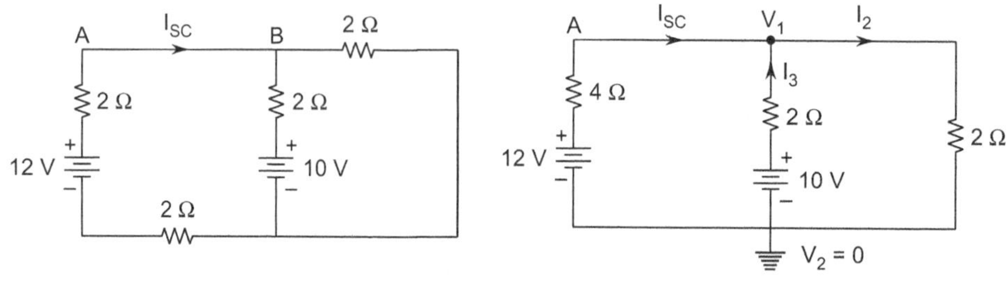

Fig. 2.57

Applying KCL at Node 1, we have

$$I_{SC} + I_3 = I_2$$

or
$$\frac{12 - V_1}{4} + \frac{10 - V_1}{2} = \frac{V_1}{2}$$

or
$$V_1 = 6.4 \text{ V}$$

Short-circuit current will be

$$I_{SC} = \frac{12 - V_1}{4} = \frac{12 - 6.4}{4}$$

or
$$I_N = I_{SC} = \mathbf{1.4 \ A.}$$ **Ans.**

Norton's equivalent resistance will be

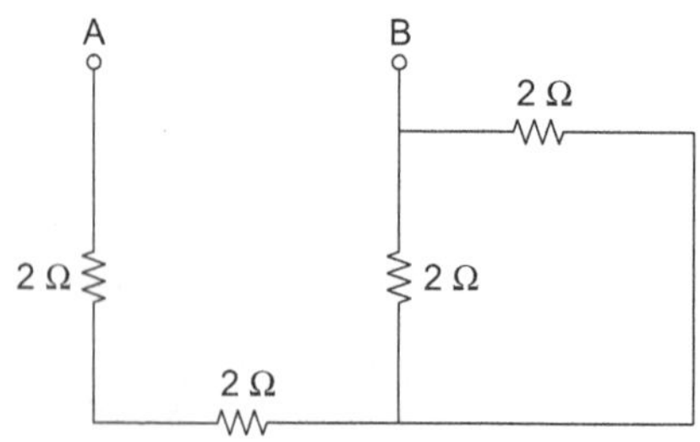

Fig. 2.58

$$R_{AB} = R_N = 2 + 2 + (2 \,||\, 2)$$

$$R_N = \mathbf{5 \ \Omega}$$ **Ans.**

Current through load resistance will be

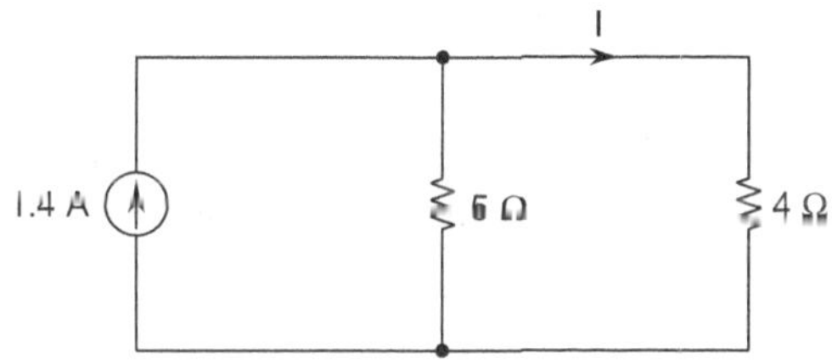

Fig. 2.59

$$I = \frac{1.4 \times 5}{5 + 4} = 0.78 \text{ A.}$$ **Ans.**

2.8 SUPERPOSITION THEOREM

Statement

"In a network of linear resistances containing more than one source of e.m.f. the current in any branch (or point) will be equal to algebraic sum of all those currents which would flow through that branch (or at that point) if each source of e.m.f. is considered separately and all the other sources of e.m.f. are replaced for time being by their internal resistances if any".

Procedure/Proof

Let us consider a circuit shown in Fig. 2.60 for describing the procedure.

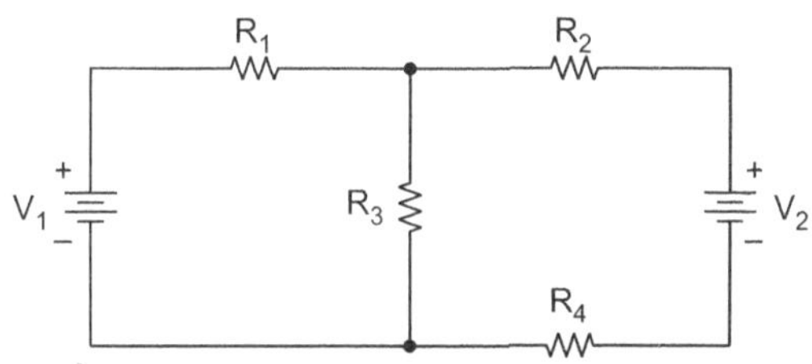

Fig. 2.60

Step 1. Each source is considered separately.

Step 2. While considering a single source, other sources are treated as follows:

- If there are voltage sources, then they are replaced by their *internal resistances* if any.
- If there are voltage sources without their internal resistances, then they are replaced by *short-circuit.*
- If there are current sources, then they are replaced by *open circuit.*

Step 3. The current is calculated due to that individual source in the desired branch (or through the point). (Consider that current as I′)

Step 4. Then next source is considered and steps 2 and 3 are repeated. (Consider so obtained currents as I″, I‴,...).

Step 5. After considering all sources individually and obtaining currents, total current through that branch (or point) will be algebraic sum of those currents.

$$I = I' + I'' + I''' + \ldots \qquad \ldots(2.33)$$

where **I′, I″, I‴** etc., are currents obtained due to each source and **I** is total current through that branch (or point).

Step 6. Note that if any current obtained due to individual source is in opposite direction to the main current, then, while summing up, that current will be considered as negative.

Problem 2.14. *Using super position theorem, find current in AB branch in circuit shown in Fig. 2.61.*

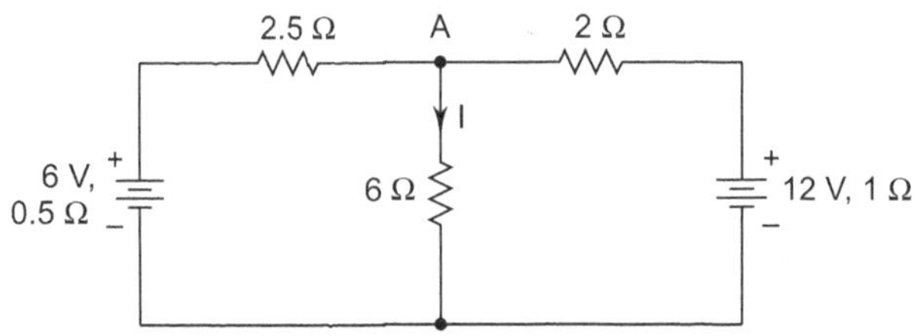

Fig. 2.61

Solution. First consider 6 Volt source and short 12 Volt source with its internal resistance of 1 Ω. Consider AB branch current as I′ (Fig. 2.62) and compute value of that current.

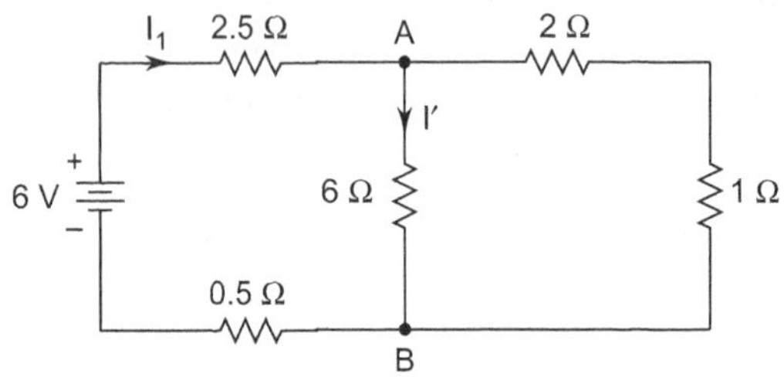

Fig. 2.62

Total resistance of the circuit as seen from 6 V source will be

$$R = 2.5 + 0.5 + (3 \,||\, 6)$$

$$R = 5\ \Omega$$

Total current from 6 Volts battery will be

$$I_1 = \frac{6}{5}\ \Omega$$

Applying current divider rule, current in AB branch due to 6 V source will be

$$I' = \frac{\frac{6}{5} \times 3}{6+3} = 0.4$$

$$I' = 0.4\ \text{A}$$

Now consider 12 Volt source and short 6 Volt source with its internal resistance (*i.e.*, 0.5 Ω). Consider AB branch current as I″ (Fig. 2.63) and compute value of that current.

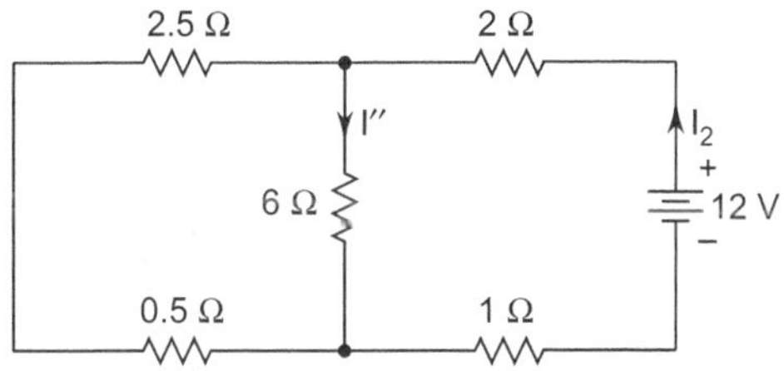

Fig. 2.63

Total resistance of the circuit as seen from 12 V source will be

$$R = 2 + 1 + (3 \parallel 6)$$

or $$R = 5\ \Omega$$

Total current from 12 volts battery will be

$$I_2 = \frac{12}{5}\ \Omega$$

Applying current divider rule, current in AB branch due to 12 V source will be

$$I'' = \frac{\frac{12}{5} \times 3}{6+3} = 0.8$$

or $$I'' = 0.8\text{ A}$$

According to super position theorem the algebraic sum of these currents due to individual sources will give total current in that branch when both the sources acted in the circuit simultaneously *i.e.*,

$$I = I' + I''$$

or $$I = 0.4 + 0.8$$

or $$I = \mathbf{1.2\ A}$$ **Ans.**

Note that both currents (I′ and I″) are in the same direction as that of main current (I) *i.e.*, in downward direction. Hence both are considered as positive.

2.9 MAXIMUM POWER TRANSFER THEOREM

Statement

"The maximum amount of power will be dissipated by a load resistance if it is equal to Thevenin's or Norton's resistance of the network supplying power".

Or

"For a linear network with fixed source impedance, the maximum power is delivered from the source to the load when load impedance is complex conjugate of the source impedance",

i.e., $$R_L = R_{TH} \quad \ldots(2.34)$$

Procedure/Proof

Consider Thevenin's equivalent circuit shown in Fig. 2.64.

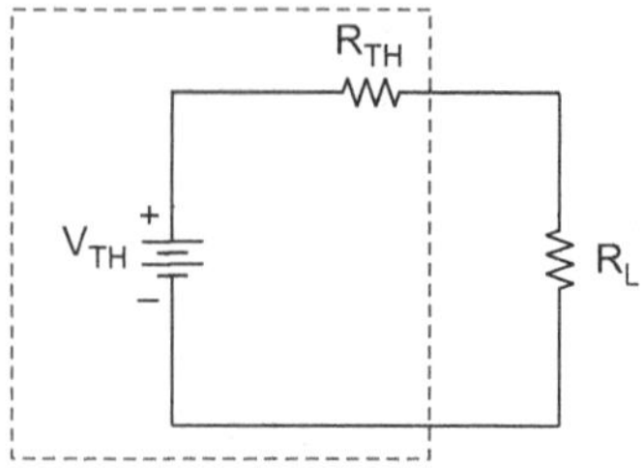

Fig. 2.64

Circuit current is $$I = \frac{V_{TH}}{R_{TH} + R_L} \qquad ...(2.35)$$

Power consumed by the load is $$P = I^2 R_L \qquad ...(2.36)$$

From equation (2.35) and equation (2.36) will be

$$P = \left(\frac{V_{TH}}{R_{TH} + R_L}\right)^2 R_L \qquad ...(2.37)$$

Now, for power to be maximum, $$\frac{dP}{dR_L} = 0 \qquad ...(2.38)$$

or $$V_{TH}^2 \left[\frac{1}{(R_L + R_{TH})^2} - \frac{-2R_L}{(R_L + R_{TH})^3}\right] = 0 \qquad ...(2.39)$$

or $$\frac{1}{(R_L + R_{TH})^2} - \frac{-2R_L}{(R_L + R_{TH})^3} = 0 \qquad ...(2.40)$$

or $$1 = \frac{2R_L}{(R_L + R_{TH})} \qquad ...(2.41)$$

or $$2R_L = R_L + R_{TH} \qquad ...(2.42)$$

or $$\boxed{R_L = R_{TH}} \qquad ...(2.43)$$

From equation (2.43) maximum power [equation (2.37)] will be

$$P_m = \left(\frac{V_{TH}}{R_{TH} + R_L}\right)^2 R_L = \frac{V_{TH}^2 R_L}{R_L + R_L} \qquad ...(2.44)$$

or $$\boxed{P_m = \frac{V_{TH}^2}{4R_L} = \frac{V_{TH}^2}{4R_{TH}}} \qquad ...(2.45)$$

Problem 2.15. *Assuming maximum power transfer from source to load, find value of load resistance R_L in Fig. 2.65 and also, find the value of maximum power.*

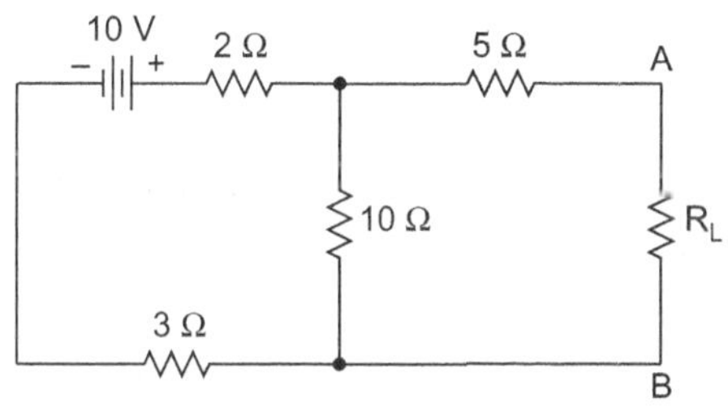

Fig. 2.65

Solution. For finding V_{TH}, load resistance R_L is removed and V_{AB} is calculated (Fig. 2.66).

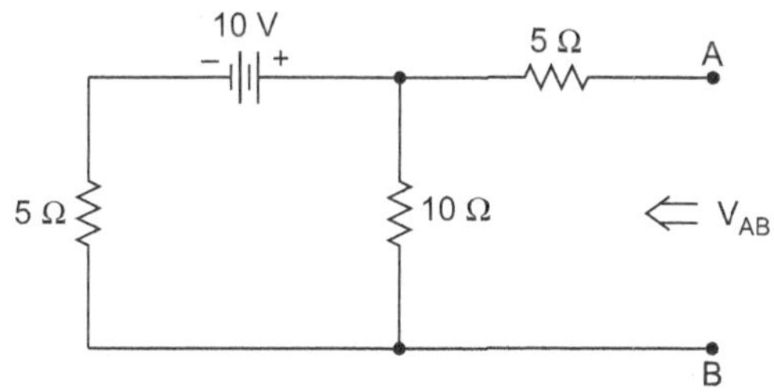

Fig. 2.66

$$V_{AB} = \text{Voltage across } 10\ \Omega \text{ resistance}$$

or
$$V_{TH} = V_{AB} = \frac{10 \times 10}{10 + 5} = 6.67 \text{ V}$$

For finding R_{TH}, voltage source is removed by short-circuit (Fig. 2.67)

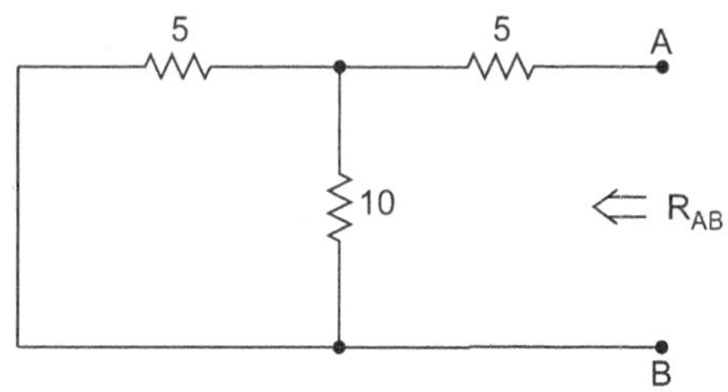

Fig. 2.67

$$R_{AB} = (5 \,||\, 10) + 5 = \frac{5 \times 10}{5 + 10} + 5$$

$$R_{TH} = R_{AB} = 8.33\ \Omega$$

For maximum power to be transferred from source to load

$R_L = R_{TH} = \mathbf{8.33\ \Omega.}$ **Ans.**

Maximum power will be

$$P_M = \frac{V_{TH}^2}{4\,R_L} = \frac{(6.67)^2}{4 \times 8.33}$$

$P_M = \mathbf{65.37\ Watt.}$ **Ans.**

2.10 RECIPROCITY THEOREM

Statement

"In any linear bilateral active network, if a source of e.m.f 'E' in any one branch produces a current 'I' in any other branch, then the same e.m.f. source 'E' acting in second branch would produce same current 'I' in first branch".

In other words, it simply means that 'E' and 'I' are mutually transferable.

Procedure/Proof

Let us consider a circuit shown in Fig. 2.68 for describing the procedure.

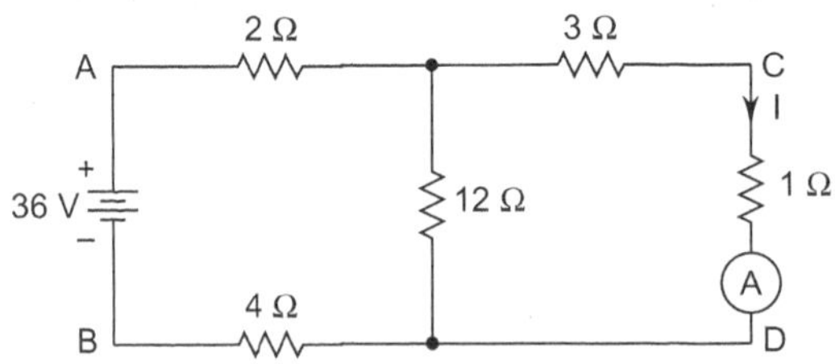

Fig. 2.68

Step 1. Initially the voltage source is in branch AB. The ammeter is connected in branch CD in which current is I.

Step 2. Current I can be calculated by mesh or nodal analysis or by simply current divider rule.

Total resistance $R = 2 + (12 \parallel 4) + 4$

$$R = 6 + \frac{12 \times 4}{12 + 4}$$

$$R = 9\,\Omega$$

Battery current $= \dfrac{36}{9} = 4\text{ A}$

Current in branch CD, by current divider rule, will be

$$I = \frac{4 \times 12}{12 + 3 + 1} = 3\text{ A}$$

Step 3. Now the voltage source is transferred to branch CD (Fig. 2.69). The current is now calculated due to that individual source in the AB branch. The ammeter is connected in branch AB.

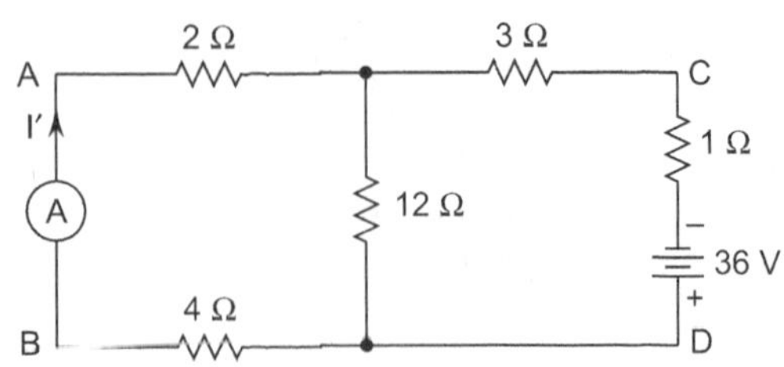

Fig. 2.69

Note. *It should be noted that while transferring the e.m.f. source from one branch to other, the polarity of the e.m.f. source should be in the same direction in which current was flowing in that branch previously.*

Step 4. Current I′ can be calculated by mesh or nodal analysis or by simply current divider rule.

Total resistance, $R = 1 + 3 + (6 \,||\, 12)$

$$R = 4 + \frac{12 \times 6}{12 + 6}$$

$$R = 8\,\Omega$$

Battery current $= \frac{36}{8} = 4.5\text{ A}$

Current in branch CD

$$I' = \frac{4.5 \times 12}{12 + 6} = 3\text{ A}$$

Step 5. $I = I' = 3\text{ A}$

Step 6.

Transfer resistance, $\frac{E}{I} = \frac{36}{3} = 12\,\Omega$

2.11 MILLMAN'S THEOREM

This theorem can be stated either in terms of voltage sources or current sources or both.

As applicable to Voltage Sources

It is used to find the common voltage across any two terminals of a network which contains a number of parallel voltage sources as shown in Fig. 2.70.

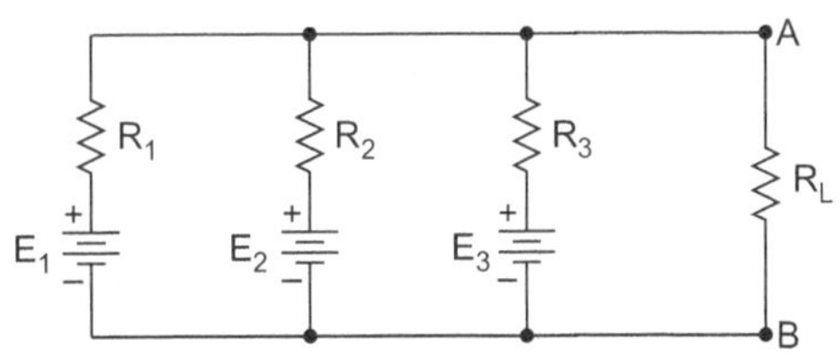

Fig. 2.70

The common voltage V_{AB} which appear across the output terminal A and B is given by

$$V_{AB} = \frac{\frac{E_1}{R_1} + \frac{E_2}{R_2} + \frac{E_3}{R_3}}{\frac{1}{R_1} + \frac{1}{R_2} + \frac{1}{R_3}} \qquad \text{...(2.46)}$$

This voltage V_{AB} represents the Thevenin's equivalent voltage.

And the total equivalent resistance will be

$$R_{eq} = R_{TH} = R_{AB} = (R_1 \,||\, R_2 \,||\, R_3) \qquad \text{...(2.47)}$$

As applicable to Current Sources

It is used to find total current through any two terminals of a network which contains a number of parallel current sources as shown in Fig. 2.71.

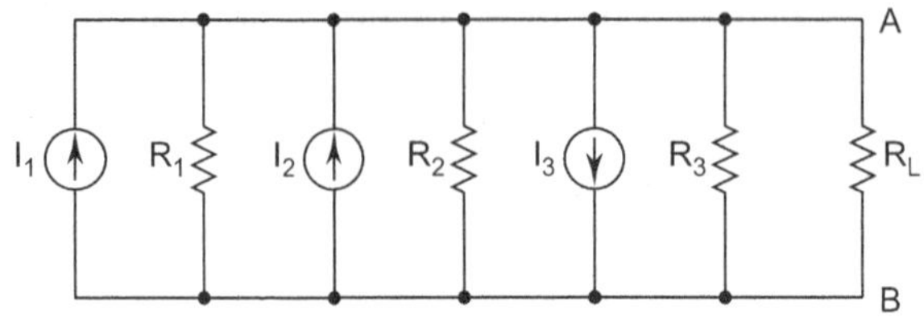

Fig. 2.71

The equivalent current source will be algebraic sum of all current sources

$$I = I_1 + I_2 + (-I_3) \quad \ldots(2.48)$$

Here I_3 is negative because of its opposite direction as compare to other current sources.

Total equivalent resistance will be

$$R_N = R_{eq} = R_{TH} = R_{AB} = (R_1 \parallel R_2 \parallel R_3) \quad \ldots(2.49)$$

The reduced circuit or equivalent circuit is shown in Fig. 2.72.

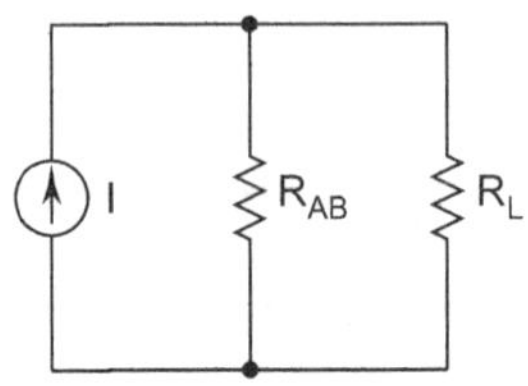

Fig. 2.72

Problem 2.16. *Find the current (Fig. 2.73) through the load resistance R_L = 5 ohms with the help of Millman's theorem.*

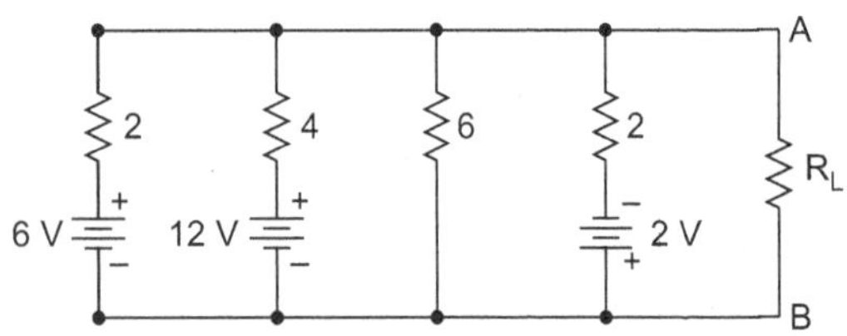

Fig. 2.73

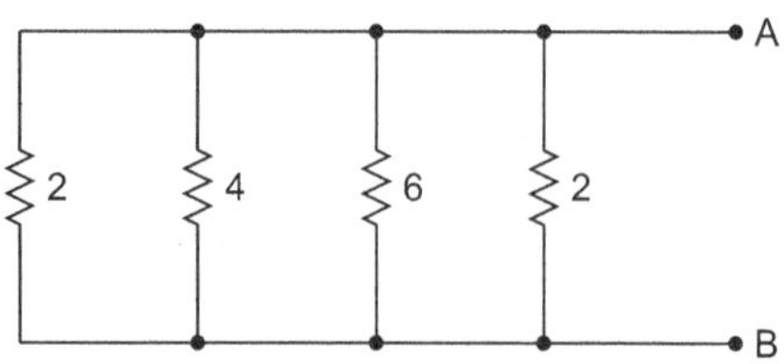

Fig. 2.74

Solution. The common voltage V_{AB} which appear across the output terminals A and B is given by

$$V_{AB} = \frac{\frac{6}{2} + \frac{12}{4} + \frac{0}{6} + \left(-\frac{2}{2}\right)}{\frac{1}{2} + \frac{1}{4} + \frac{1}{6} + \frac{1}{2}} = 3.529 \text{ V}$$

And the total equivalent resistance (Fig. 2.74) will be

$$R_{eq} = R_{TH} = R_{AB} = (2 \parallel 4 \parallel 6 \parallel 2) = 0.706\ \Omega$$

Hence the current through the load resistance R_L will be

$$I = \frac{V_{AB}}{R_{AB} + R_L} = \frac{3.529}{0.706 + 5}$$

$$I = \mathbf{0.618\ A}$$ **Ans.**

Problem 2.17. *Find current (Fig. 2.75) through load resistance R_L = 8 ohms with the help of Millman's theorem.*

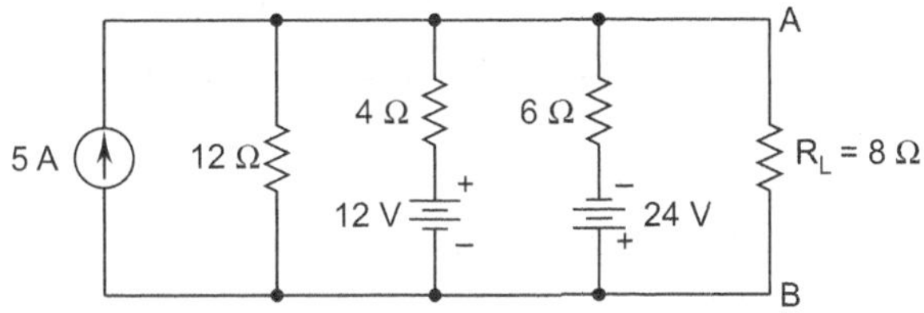

Fig. 2.75

Solution. In circuit diagram shown in Fig. 2.75, the voltage sources can be converted into current source as shown in Fig. 2.76.

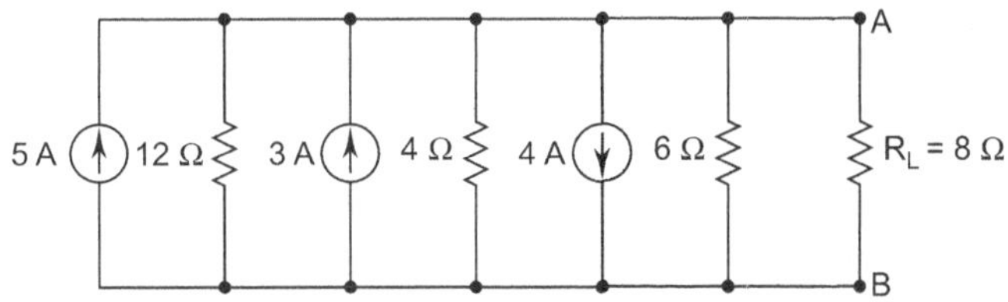

Fig. 2.76

The equivalent current source will be algebraic sum of all the current sources

$$I = I_1 + I_2 + (-I_3) = 5 + 3 - 4 = 4 \text{ A}$$

Here I_3 is negative because of its opposite direction as compare to other current sources.

Total equivalent resistance will be

$$R_{eq} = R_{TH} = R_{AB} = (12 \,||\, 4 \,||\, 6) = 2\,\Omega$$

Reduced circuit or equivalent circuit is shown in Fig. 2.77.

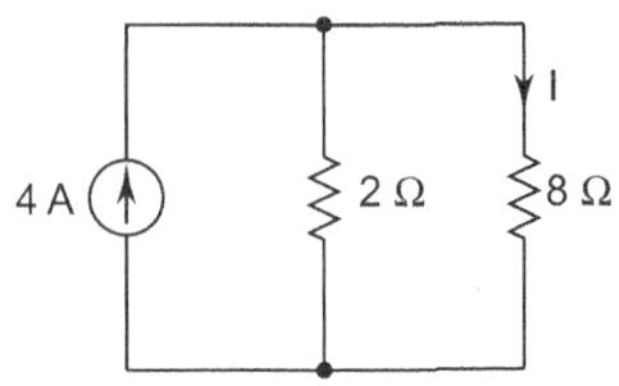

Fig. 2.77

Current through the load resistance will be

$$I = \frac{4 \times 2}{2 + 8} = \frac{8}{10} = \mathbf{0.8\ A}$$ **Ans.**

2.12 DELTA-STAR TRANSFORMATION

Whenever we have to find equivalent resistance between any two terminals of a network in which every node has 3 branches (***not more or less than 3***), delta-star or star-delta transformation is very helpful. Without changing the equivalent resistance between the two terminals, the resistance connected in delta (Δ) configuration can be replaced by or converted into star (Y) configuration.

Let us first consider delta-to-star transformation. The delta-to-star transformation is shown in Fig. 2.78.

- Consider Fig. 2.78(*a*) in which delta connection is shown.
- The terminals are named 1, 2 and 3 and while moving in clockwise direction, the resistances are named according to their presence in their respective terminals.
- Before converting a delta network into star network, first remove those resistances which are to be converted into star configuration and keep the rest of the terminals as it is [Fig. 2.78 (*b*)].
- Now take a neutral point (N) as shown in Fig. 2.78(*c*).
- Connect the star equivalent resistances as shown in Fig. 2.78(*d*).

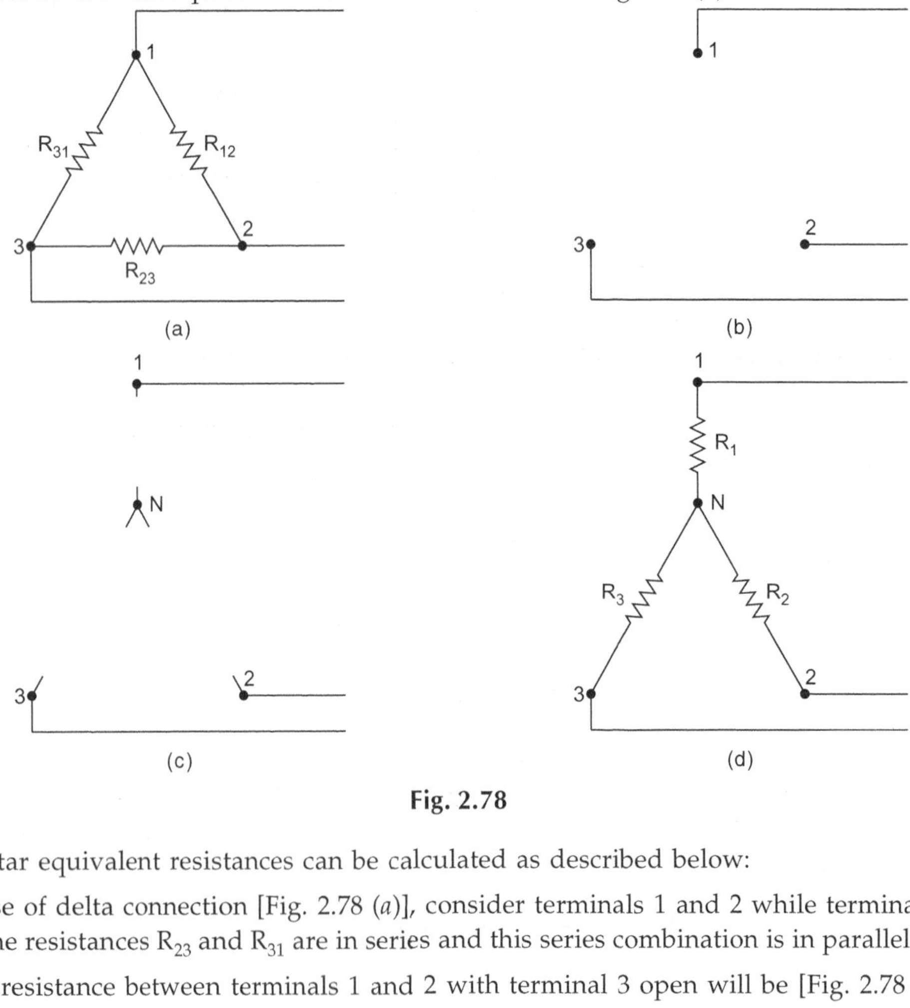

Fig. 2.78

The star equivalent resistances can be calculated as described below:

In case of delta connection [Fig. 2.78 (*a*)], consider terminals 1 and 2 while terminal 3 is left open. The resistances R_{23} and R_{31} are in series and this series combination is in parallel with R_{12}.

Thus resistance between terminals 1 and 2 with terminal 3 open will be [Fig. 2.78 (*a*)]

$$= (R_{23} + R_{31}) \,||\, R_{12} = \frac{R_{12}\left(R_{23} + R_{31}\right)}{R_{12} + R_{23} + R_{31}} \qquad \text{...(2.50)}$$

And resistance between terminals 1 and 2 with terminal 3 open *in case of star connection* [Fig. 2.78(*d*)] will be

$$= R_1 + R_2 \qquad \text{...(2.51)}$$

For these two networks to be electrically equivalent, equation (2.50) must be equal to equation (2.51)

$$R_1 + R_2 = \frac{R_{12}\left(R_{23} + R_{31}\right)}{R_{12} + R_{23} + R_{31}} \qquad \text{...(2.52)}$$

Similarly, resistance between terminals 2 and 3 with terminal 1 open in both cases will be

$$R_2 + R_3 = \frac{R_{23}\left(R_{12} + R_{31}\right)}{R_{12} + R_{23} + R_{31}} \qquad \text{...(2.53)}$$

And resistance between terminal 3 and 1 with terminal 2 open in both cases will be

$$R_3 + R_1 = \frac{R_{31}\left(R_{12} + R_{23}\right)}{R_{12} + R_{23} + R_{31}} \qquad \text{...(2.54)}$$

Subtracting equation (2.54) from equation (2.52)

$$(R_1 + R_2) - (R_3 + R_1) = \frac{\left(R_{12}R_{23} + R_{12}R_{31}\right) - \left(R_{31}R_{12} + R_{31}R_{23}\right)}{R_{12} + R_{23} + R_{31}} \qquad \text{...(2.55)}$$

or $$R_2 - R_3 = \frac{\left(R_{12}R_{23} - R_{31}R_{23}\right)}{R_{12} + R_{23} + R_{31}} \qquad \text{...(2.56)}$$

or $$R_2 - R_3 = \frac{R_{23}\left(R_{12} - R_{31}\right)}{R_{12} + R_{23} + R_{31}} \qquad \text{...(2.57)}$$

Adding equations (2.53) and (2.57) and dividing by 2, we get

$$(R_2 + R_3) + (R_2 - R_3) = \frac{R_{23}\left(R_{12} + R_{31}\right)}{R_{12} + R_{23} + R_{31}} + \frac{R_{23}\left(R_{12} - R_{31}\right)}{R_{12} + R_{23} + R_{31}} \qquad \text{...(2.58)}$$

or $$2R_2 = \frac{2R_{23}R_{12}}{R_{12} + R_{23} + R_{31}} \qquad \text{...(2.59)}$$

or $$\boxed{R_2 = \frac{R_{12}R_{23}}{R_{12} + R_{23} + R_{31}}} \qquad \text{...(2.60)}$$

Similarly $$\boxed{R_1 = \frac{R_{12}R_{31}}{R_{12} + R_{23} + R_{31}}} \qquad \text{...(2.61)}$$

and $$\boxed{R_3 = \frac{R_{23}R_{31}}{R_{12} + R_{23} + R_{31}}} \qquad \text{...(2.62)}$$

2.13 STAR-DELTA TRANSFORMATION

Now consider star-to-delta transformation. The star-to-delta transformation is shown in Fig. 2.79.

- Consider Fig. 2.79(*a*) in which star connection is shown.
- The terminals are named 1, 2 and 3 and while moving in clockwise direction, the resistances are named according to their presence in between their respective terminals and neutral point (N).
- Before converting a star network into delta network, first remove those resistances and keep the rest of the terminals as it is [Fig. 2.79 (*b*)]
- Now remove the neutral point (N) as shown in Fig. 2.79(*c*).
- Connect the delta equivalent resistances as shown in Fig. 2.79(*d*).

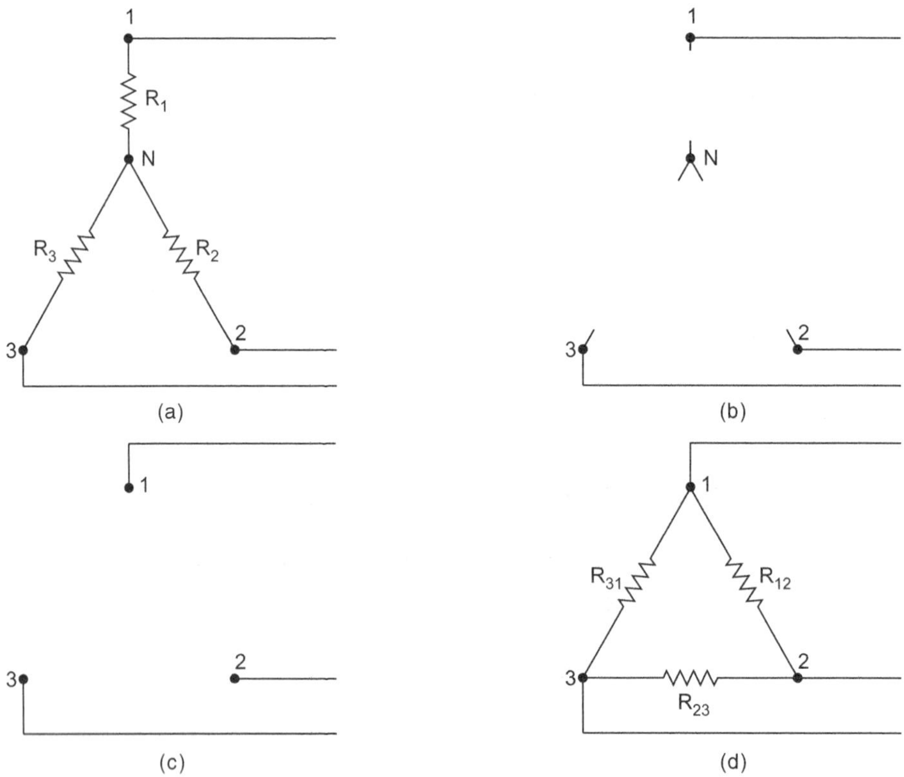

Fig. 2.79

The star equivalent resistances can be calculated as described below. Basic equations guiding this conversion will remain same, *i.e.*, equations (2.53), (2.54) and (2.55). Writing equations (2.60), (2.61) and (2.62) again we have

$$R_1 = \frac{R_{12}R_{31}}{R_{12} + R_{23} + R_{31}} \qquad \text{[Eqn. (2.61)]}$$

$$R_2 = \frac{R_{12}R_{23}}{R_{12} + R_{23} + R_{31}} \qquad \text{[Eqn. (2.60)]}$$

$$R_3 = \frac{R_{23}R_{31}}{R_{12} + R_{23} + R_{31}} \qquad \text{[Eqn. (2.62)]}$$

Now dividing equation (2.62) by equation (2.61), we get

$$\frac{R_3}{R_1} = \frac{R_{23}}{R_{12}} \qquad ...(2.63)$$

or $$R_{23} = R_{12} \times \frac{R_3}{R_1} \qquad ...(2.64)$$

Dividing equation (2.60) by equation (2.61)

$$\frac{R_2}{R_1} = \frac{R_{23}}{R_{31}} \qquad ...(2.65)$$

or $$R_{31} = \frac{R_1}{R_2} \times R_{23} \qquad ...(2.66)$$

Substitute the value of R_{23} and R_{31} from equations (2.64) and (2.66) in equation (2.61), we get

$$R_1\left[1 + \frac{R_3}{R_1} + \frac{R_3}{R_2}\right]R_{12} = R_{31} \,.\, R_{12} \qquad ...(2.67)$$

or $$R_{31} = \frac{R_1R_2 + R_2R_3 + R_3R_1}{R_2} \qquad ...(2.68)$$

or $$R_{31} = \frac{R_3 + R_1 + R_3R_1}{R_2} \qquad ...(2.69)$$

or $$\boxed{R_{31} = R_3 + R_1 + \frac{R_3R_1}{R_2}} \qquad ...(2.70)$$

Similarly $$R_{12} = \frac{R_1 + R_2 + R_1R_2}{R_3} \qquad ...(2.71)$$

or $$\boxed{R_{12} = R_1 + R_2 + \frac{R_1R_2}{R_3}} \qquad ...(2.72)$$

and $$R_{23} = \frac{R_2 + R_3 + R_2R_3}{R_1} \qquad ...(2.73)$$

or $$\boxed{R_{23} = R_2 + R_3 + \frac{R_2R_3}{R_1}} \qquad ...(2.74)$$

SOLVED PROBLEMS

Problem 2.18. *Find the total current from the 10 Volt source (Fig. 2.80) with the help of Δ-Y conversion.*

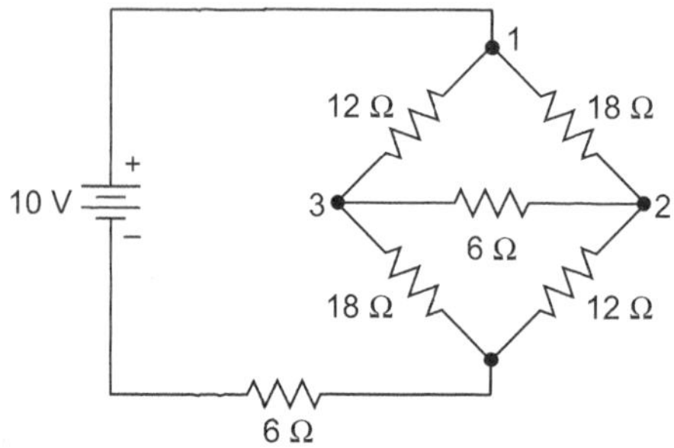

Fig. 2.80

Solution. If we were to treat resistors R_{12}, R_{23}, and R_{31} as being connected in a Δ configuration and generate an equivalent Y network to replace them, we could turn this bridge circuit into a (simpler) series/parallel combination circuit.

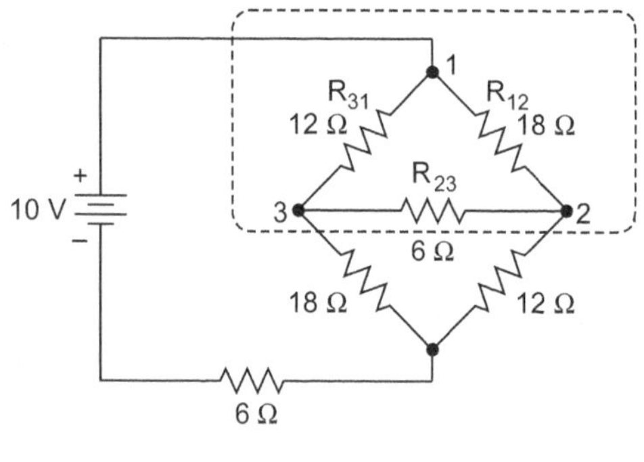

Fig. 2.81

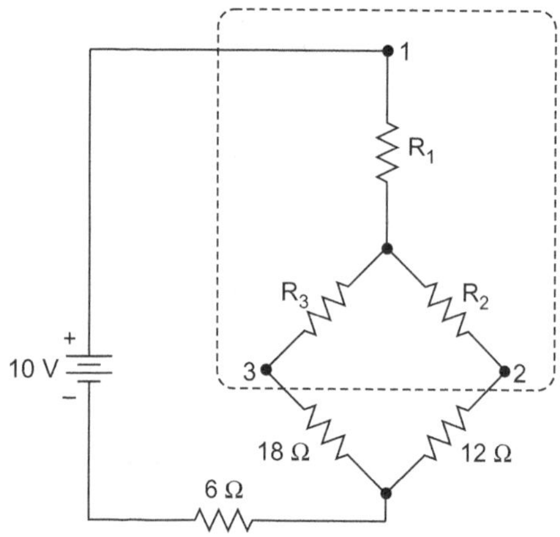

Fig. 2.82

After the Δ-Y conversion the circuit is shown in Fig. 2.82. The resistances of star connection R_1, R_2, and R_3 are given as

$$R_1 = \frac{18 \times 12}{18 + 6 + 12} = \frac{216}{36} = 6\ \Omega$$

$$R_2 = \frac{18 \times 6}{18 + 6 + 12} = \frac{108}{36} = 3\ \Omega$$

$$R_3 = \frac{6 \times 12}{18 + 6 + 12} = \frac{72}{36} = 2\ \Omega$$

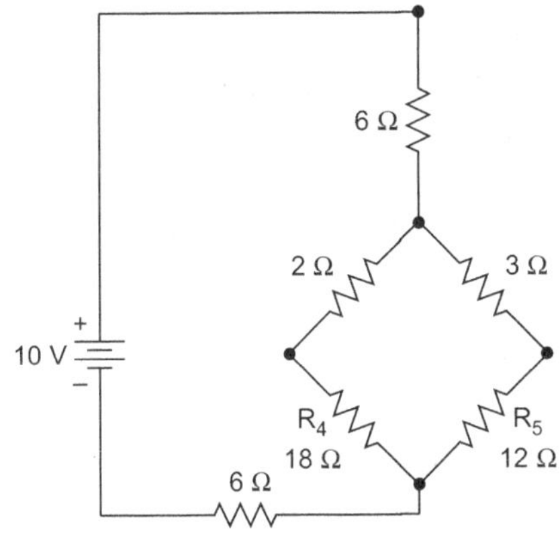

Fig. 2.83

Resistors R_4 and R_5, of course, remain the same at 18 Ω and 12 Ω, respectively. Analyzing the circuit now as a series/parallel combination, total resistance of the circuit will be

$$R = (2 + 18) \parallel (3 + 12) + 6 + 6$$

$$R = \frac{20 \times 15}{20 + 15} + 12 = 20.57\ \Omega$$

Total current will be $$I = \frac{10}{20.57} = \mathbf{0.486\ A}$$ **Ans.**

Problem 2.19. *Find the current I in AB branch of the given circuit with the help of both mesh and nodal analysis.*

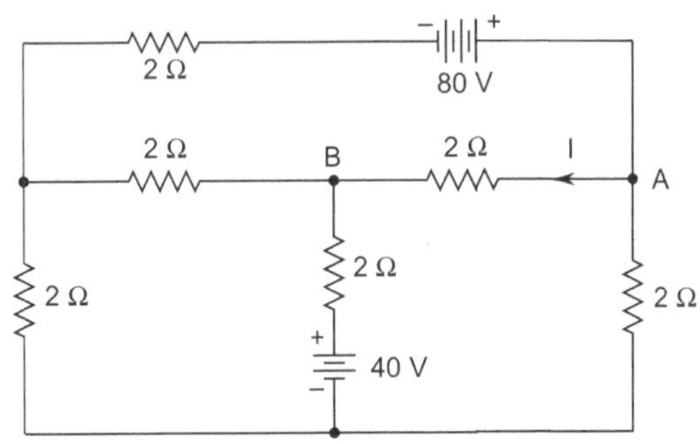

Fig. 2.84

Solution. ***Using mesh analysis:***

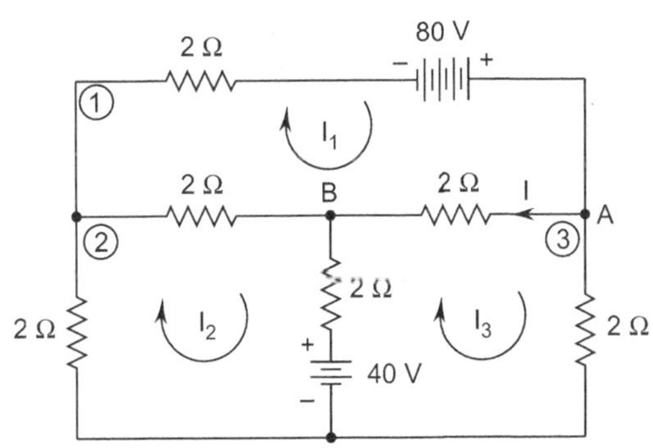

Fig. 2.85

In Mesh – I, applying KVL we have

$$80 = 2(I_1 - I_3) + 2(I_1 - I_2) + 2I_1$$

or $$40 = 3I_1 - I_2 - I_3 \quad ...(i)$$

In Mesh – II, applying KVL we have

$$-40 = 2I_2 + 2(I_2 - I_1) + 2(I_2 - I_3)$$

or $$20 = I_1 - 3I_2 + I_3 \quad ...(ii)$$

In Mesh – III, applying KVL we have

$$40 = 2(I_3 - I_1) + 2(I_3 - I_2) + 2I_3$$

or $$20 = -I_1 - I_2 + 3I_3 \quad ...(iii)$$

Solving equations (*i*), (*ii*) and (*iii*) by Cramer rule

$$\begin{bmatrix} 3 & -1 & -1 \\ 1 & -3 & 1 \\ -1 & -1 & 3 \end{bmatrix} \begin{bmatrix} I_1 \\ I_2 \\ I_3 \end{bmatrix} = \begin{bmatrix} 40 \\ 20 \\ 20 \end{bmatrix}$$

$$\Delta = \begin{bmatrix} 3 & -1 & -1 \\ 1 & -3 & 1 \\ -1 & -1 & 3 \end{bmatrix}$$

$$= 3(-9 + 1) + (3 + 1) - (-1 - 3)$$

$$\Delta = -16$$

$$\Delta_1 = \begin{bmatrix} 40 & -1 & -1 \\ 20 & -3 & 1 \\ 20 & -1 & 3 \end{bmatrix}$$

$$= 40(-9 + 1) - 20(-3 - 1) + 20(-1 - 3)$$

$$\Delta_1 = -320$$

$$\Delta_2 = -\begin{bmatrix} 3 & 40 & -1 \\ 1 & 20 & 1 \\ -1 & 20 & 3 \end{bmatrix}$$

$$= 3(60 - 20) - (120 + 20) - (40 + 20)$$

$$\Delta_2 = -80$$

$$\Delta_3 = -\begin{bmatrix} 3 & -1 & 40 \\ 1 & -3 & 20 \\ -1 & -1 & 20 \end{bmatrix}$$

$$= 3(-60 + 20) - (-20 + 40) - (-20 + 120)$$

$$\Delta_3 = -240$$

Hence the mesh currents will be

$$I_1 = \frac{\Delta_1}{\Delta} = \frac{-320}{-16} = 20 \text{ A}$$

$$I_2 = \frac{\Delta_2}{\Delta} = \frac{-80}{-16} = 5\text{ A}$$

$$I_3 = \frac{\Delta_3}{\Delta} = \frac{-240}{-16} = 15\text{ A}$$

Current in AB branch will be

$$I = I_1 - I_3 = 20 - 15$$

or $$I = \mathbf{5\ A}$$ **Ans.**

Using Nodal analysis:

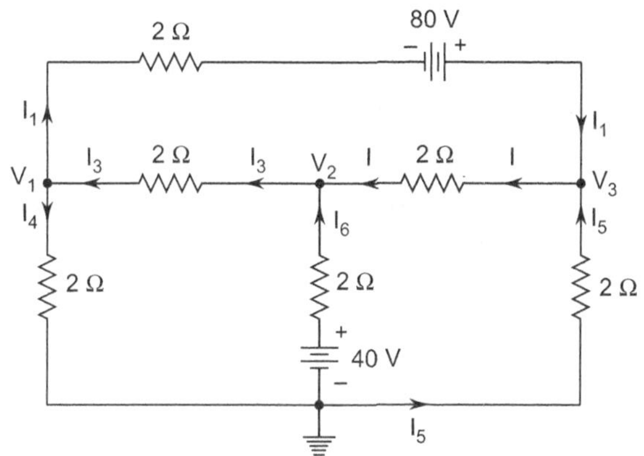

Fig. 2.86

At Node 1 (having potential V_1) applying KCL, we have

$$I_3 = I_1 + I_4$$

$$\frac{V_2 - V_1}{2} = \frac{V_1 + 80 - V_3}{2} + \frac{V_1}{2}$$

or $$-80 = 3V_1 - V_2 - V_3 \qquad ...(i)$$

At Node 2 (having potential V_2) applying KCL, we have

$$I_3 = I + I_6$$

$$\frac{V_2 - V_1}{2} = \frac{V_3 - V_2}{2} + \frac{40 - V_2}{2}$$

or $$40 = -V_1 + 3V_2 - V_3 \qquad ...(ii)$$

At Node 3 (having potential V_3) applying KCL, we have

$$I = I_1 + I_5$$

$$\frac{V_3 - V_2}{2} = \frac{V_1 + 80 - V_3}{2} + \frac{-V_3}{2}$$

or $$80 = -V_1 - V_2 + 3V_3 \qquad ...(iii)$$

Solving equations (*i*), (*ii*) and (*iii*) by Cramer rule, we have

$$3V_1 - V_2 - V_3 = -80 \qquad ...(i)$$

$$-V_1 + 3V_2 - V_3 = 40 \qquad ...(ii)$$

$$-V_1 - V_2 + 3V_3 = 80 \qquad ...(iii)$$

$$\begin{bmatrix} 3 & -1 & -1 \\ -1 & 3 & -1 \\ -1 & -1 & 3 \end{bmatrix} \begin{bmatrix} V_1 \\ V_2 \\ V_3 \end{bmatrix} = \begin{bmatrix} -80 \\ 40 \\ 80 \end{bmatrix}$$

$$\Delta = 3(9-1) + 1(-3-1) - 1(1+3)$$

$$\Delta = 24 - 4 - 4 = 16$$

$$\Delta_1 = \begin{bmatrix} -80 & -1 & -1 \\ 40 & 3 & -1 \\ 80 & -1 & 3 \end{bmatrix}$$

$$= -80(9-1) + 1(120+80) - 1(-40-240)$$

$$= -640 + 200 + 280$$

$$\Delta_1 = -160$$

$$\Delta_2 = \begin{bmatrix} 3 & -80 & -1 \\ -1 & 40 & -1 \\ -1 & 80 & 3 \end{bmatrix}$$

$$= 3(120+80) + 80(-3-1) - 1(-80+40)$$

$$= 600 - 320 + 40$$

$$\Delta_2 = 320$$

$$\Delta_3 = \begin{bmatrix} 3 & -1 & -80 \\ -1 & 3 & 40 \\ -1 & -1 & 80 \end{bmatrix}$$

$$= 3(240+40) + 1(-80+40) - 80(1+3)$$

$$= 840 - 40 - 320$$

$$\Delta_3 = 480$$

$$V_1 = \frac{\Delta_1}{\Delta} = \frac{-160}{16} = -10 \text{ V}$$

$$V_2 = \frac{\Delta_2}{\Delta} = \frac{320}{16} = 20 \text{ V}$$

$$V_3 = \frac{\Delta_3}{\Delta} = \frac{480}{16} = 30 \text{ V}$$

$$I = \frac{V_3 - V_2}{2} = \frac{30-20}{2} = \mathbf{5\ A}$$ **Ans.**

Problem 2.20. *Find the current I in the circuit given of Fig. 2.87.*

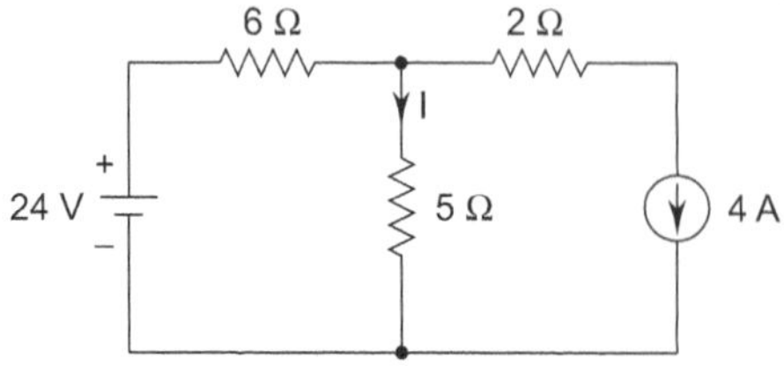

Fig. 2.87

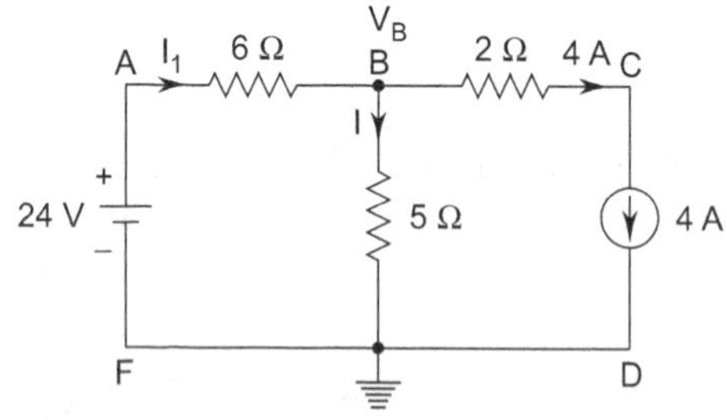

Fig. 2.88

Solution. Applying Kirchhoff's current law at junction B we have

Current supplied by 24 V battery, $I_1 = (I + 4)$ amperes

Applying Kirchhoff's voltage law to mesh ABEFA we have

$$6\,I_1 + 5\,I = 24$$

or $$6\,(I + 4) + 5\,I = 24$$

or $$11\,I + 24 = 24$$

or $$I = \mathbf{0}$$ **Ans.**

Alternative Method:

Applying Nodal analysis

$$I_1 = I + 4$$

$$\frac{24 - V_B}{6} = \frac{V_B}{5} + 4$$

or $$V_B = 0$$

∴ $$I = \mathbf{0}$$ **Ans.**

Problem 2.21. *Fig. 2.89 shows two batteries connected in parallel each represented by an e.m.f along with its internal resistance. A load resistance of 6 Ω is connected across the end of the batteries. Calculate the current through each battery and the load.*

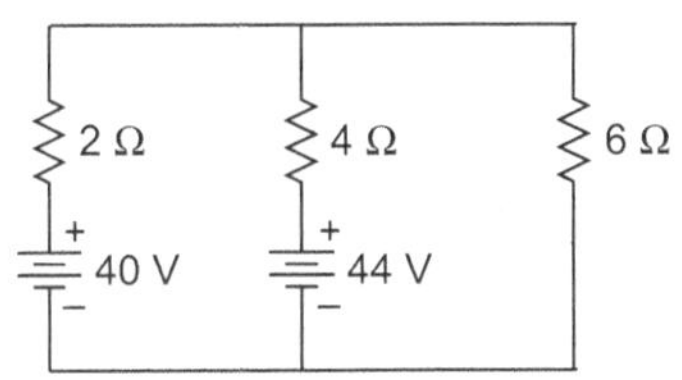

Fig. 2.89

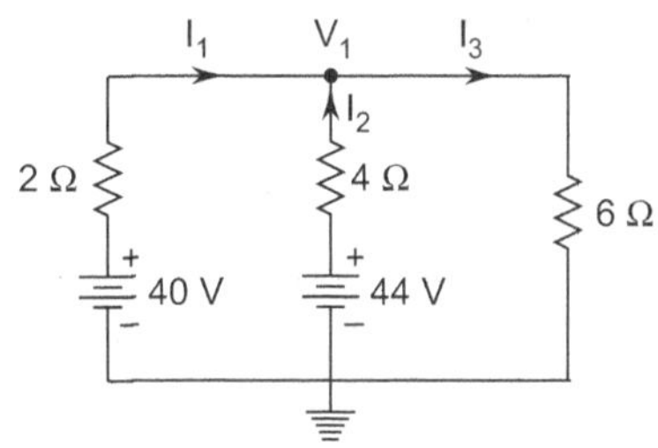

Fig. 2.90

Solution. Let us solve this circuit of Fig. 2.89 with the help of nodal analysis. The branch currents and node potentials are shown in Fig. 2.90.

At Node 1, applying KCL we have

$$I_1 + I_2 = I_3$$

or $$\frac{40 - V_1}{2} + \frac{44 - V_1}{4} = \frac{V_1}{6}$$

or $$240 - 6V_1 + 132 - 3V_1 = 2V_1$$

or $$V_1 = 33.82\ \text{V}$$

Hence the current through each battery and load resistance will be

$$I_1 = \frac{40 - V_1}{2} = \frac{40 - 33.82}{2} = \mathbf{3.09\ A}$$ **Ans.**

$$I_2 = \frac{44 - V_1}{4} = \frac{44 - 33.82}{4} = \mathbf{2.55\ A}$$ **Ans.**

$$I_3 = \frac{V_1}{6} = \frac{33.82}{6} = \mathbf{5.64\ A}$$ **Ans.**

Problem 2.22. *Calculate the current through the galvanometer in the following bridge circuit (Fig. 2.91).*

Solution. Assume current distribution in the bridge network as shown in Fig. 2.92.

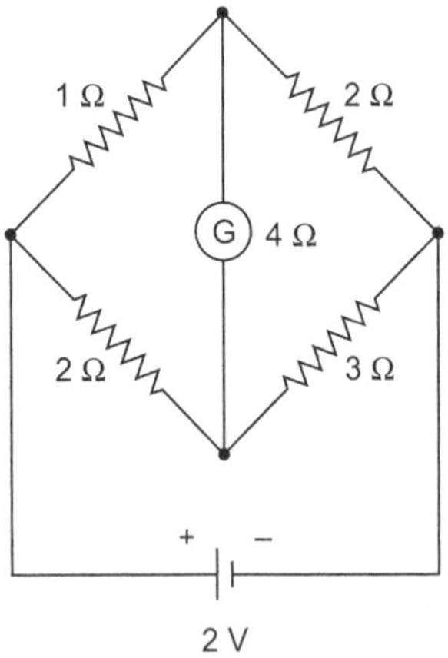

Fig. 2.91

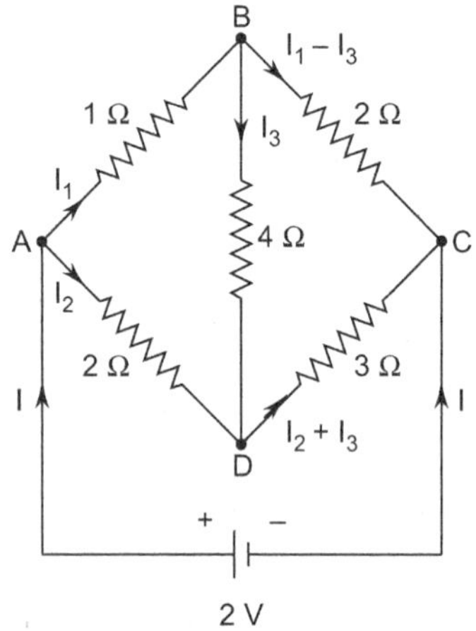

Fig. 2.92

Applying Kirchhoff's second law to meshes ABDA, BCDB and ABCA we have

$$I_1 + 4\,I_3 - 2\,I_2 = 0 \qquad ...(i)$$

$$2\,(I_1 - I_3) - 3\,(I_2 + I_3) - 4\,I_3 = 0$$

or $$2\,I_1 - 3\,I_2 - 9\,I_3 = 0 \qquad ...(ii)$$

and $$I_1 + 2\,(I_1 - I_3) = 2$$

or $$3\,I_1 - 2\,I_3 = 2 \qquad ...(iii)$$

Solving equations (*i*), (*ii*) and (*iii*), we get $I_3 = \frac{1}{44}$ A ; $I_2 = \frac{17}{44}$ A and $I_1 = \frac{30}{44}$ A

So current through galvanometer $= I_3 = \mathbf{\frac{1}{44}\ A}$ **Ans.**

Problem 2.23. *For the circuit of Fig. 2.93, find I using superposition theorem.*

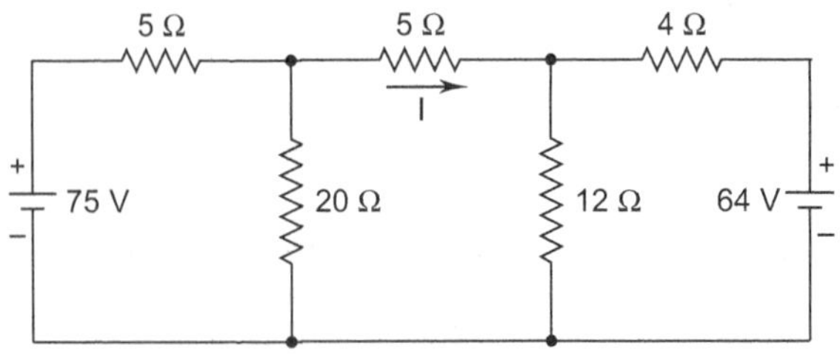

Fig. 2.93

Solution. First consider 75 Volts source and replacing 64 Volts source by a short-circuit as shown in Fig. 2.94 (*a*). Let the current through BC branch be denoted by I′:

From series-parallel circuit law equivalent resistance of the circuit shown in Fig. 2.94 (*a*) is given as

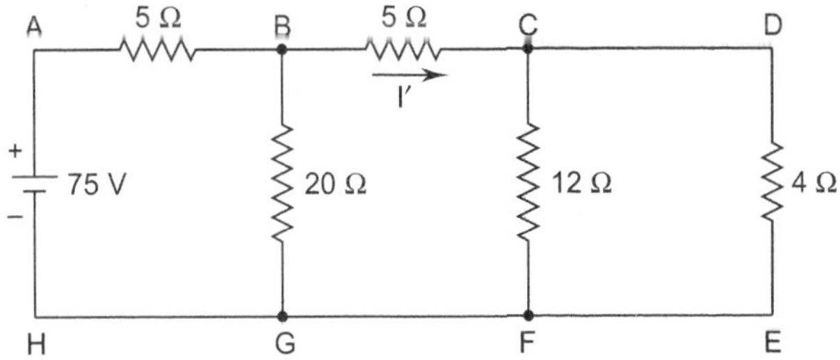

Fig. 2.94 (a)

$$R'_{eq} = R_{AB} + R_{BG} \,||\, [R_{BC} + R_{CF} \,||\, R_{DE}]$$
$$= 5 + 20 \,||\, [5 + 12 \,||\, 4]$$
$$= 5 + 20 \,||\, \left[5 + \frac{12 \times 4}{12 + 4}\right]$$
$$= 5 + 20 \,||\, (5 + 3)$$
$$= 5 + 20 \,||\, 8 = 5 + \frac{20 \times 8}{20 + 8} = \frac{75}{7}\ \Omega$$

Current through branch AB, $I_{AB} = \dfrac{V}{R'_{eq}} = \dfrac{75}{75/7} = 7\ \text{A}$

Current through branch BC, $I' = I_{AB} \times \dfrac{R_{BG}}{R_{BG} + R_{BF}} = 7 \times \dfrac{20}{20 + 8} = 5$ A from B toward C

Replacing 75 V source by a short-circuit, the circuit is reduced to a circuit shown in Fig. 2.94(*b*). Let current through branch BC be denoted by I″.

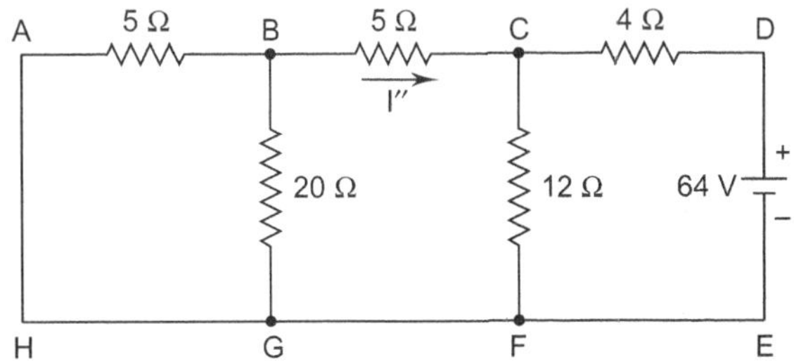

Fig. 2.94 (b)

Equivalent resistance of the circuit,

$$R''_{eq} = R_{DC} + R_{CF} \,||\, [R_{CB} + R_{BG} \,||\, R_{BA}]$$
$$= 4 + 12 \,||\, [5 + 20 \,||\, 5] = 4 + 12 \,||\, \left[5 + \frac{20 \times 5}{20 + 5}\right]$$
$$= 4 + 12 \,||\, (5 + 4) = 4 + \frac{12 \times 9}{12 + 9} = \frac{64}{7}\ \Omega$$

Current through branch DC, $I_{DC} = \dfrac{64}{64/7} = 7\text{ A}$

Current in branch CB, $I'' = I_{DC} \times \dfrac{R_{CF}}{R_{CF} + R_{BAG}} = 7 \times \dfrac{12}{12+9} = 4\text{ A from C toward B}$

Hence current, $I = \text{Current in branch BC} = I' - I'' = 5 - 4 = \mathbf{1\ A.}$ **Ans.**

Problem 2.24. *Using superposition theorem, determine currents in all the resistors of the following network (Fig. 2.95).*

Solution. Replacing 10 V source by a short-circuit and utilizing 2 Ampere source, the circuit is reduced to a simple circuit as shown in Fig. 2.96 (*a*).

The current of 2 A is divided into resistances of 5 Ω and 5 Ω equally as shown in Fig. 2.96 (*a*). No current will flow through 10 Ω resistance as there is a short-circuit across the 10 Ω resistance. The current will pass through the short-circuit path rather than through 10 Ω resistance (as current always follows shortest path). Replacing 2 A source by an open-circuit, the circuit is reduced to a simple circuit as shown in Fig. 2.96 (*b*).

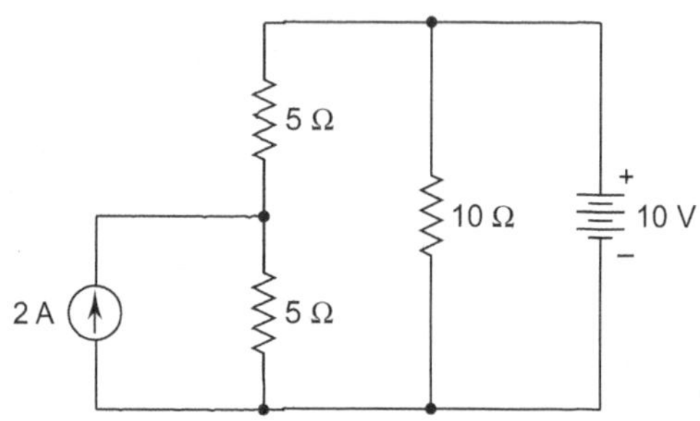

Fig. 2.95

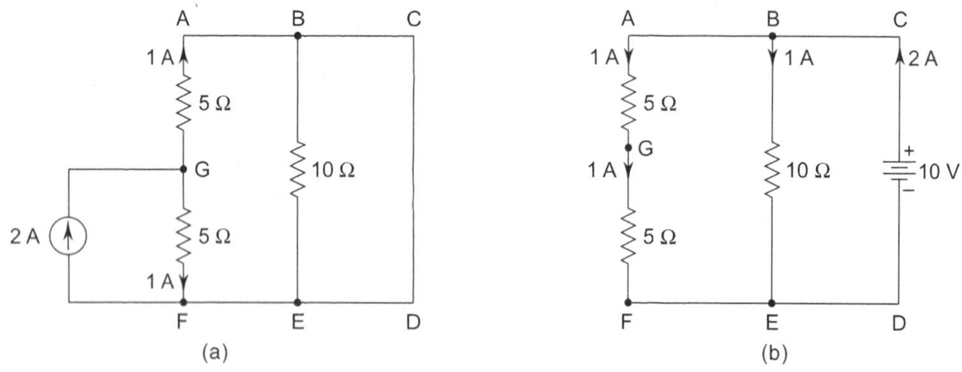

Fig. 2.96

Equivalent resistance of the circuit, $R = (5 + 5) \parallel 10 = \dfrac{10 \times 10}{10 + 10} = 5\ \Omega$

Current supplied by the source, $I = \dfrac{10}{5} = 2\text{ A}$

Current of 2 A is divided between resistance arms BE and AGF equally because their resistances are equal, so current distribution will be as shown in Fig. 2.96 (*b*). From Figs. 2.96 (*a*) and (*b*), we have

Current in 10 Ω resistance (branch BE) = 0 + 1 = **1 A** **Ans.**

Current in 5 Ω resistance (branch AG) = 1 – 1 = **0 A** **Ans.**

Current in 5 Ω resistance (branch GF) = 1 + 1 = **2 A** **Ans.**

Note. *When 2 A current source is utilized and 10 V voltage source is short circuited, then the path BCDE is a short-circuit. In such a case, 10 Ω resistance have no role left in the circuit because current always follows shortest path.*

Problem 2.25. *In the circuit shown in Fig. 2.97, find the current through the 6 Ω resistor using superposition theorem.*

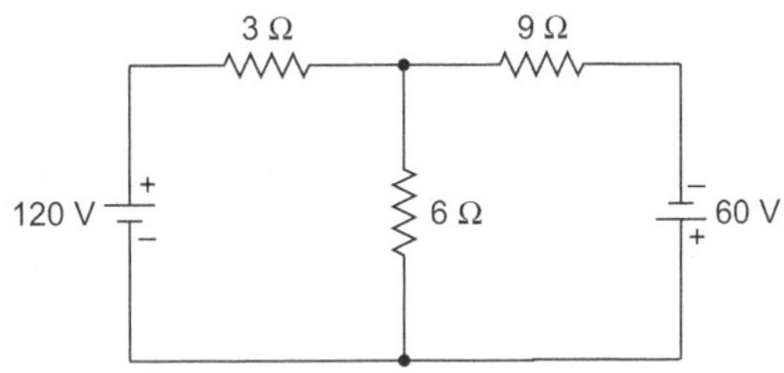

Fig. 2.97

Solution. Replacing 60 V source by a short-circuit, the circuit is reduced to a simple circuit shown in Fig. 2.98 (*a*).

Equivalent resistance of the circuit shown in Fig. 2.98 (*a*),

$$R' = R_{AB} + R_{BC} \,||\, R_{BD}$$
$$= 3 + 6 \,||\, 9 = 3 + 3.6 = 6.6\ \Omega$$

Current supplied by 120 V source

$$I_1 = \frac{120}{R'} = \frac{120}{6.6} = \frac{200}{11}\ \text{A}$$

Current in resistance, R_{BD},

$$I' = I_1 \times \frac{R_{BC}}{R_{BC} + R_{CD}} \qquad \text{(By current division rule)}$$

$$= \frac{200}{11} \times \frac{9}{9+6} = \frac{120}{11}\ \text{A} \qquad \text{(From B to D)}$$

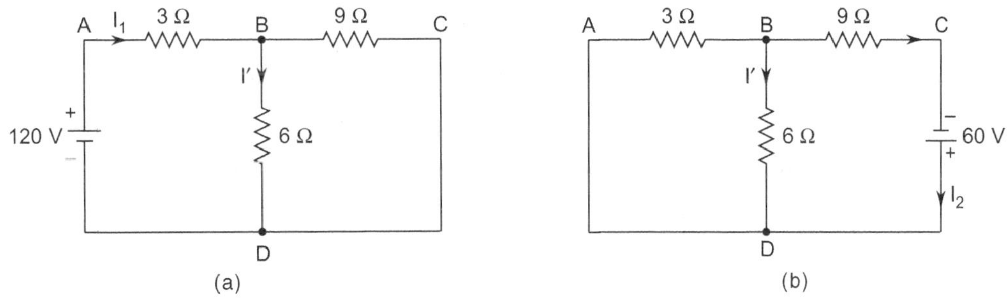

Fig. 2.98

Similarly on replacement of 120 V source by a short-circuit, the given circuit is reduced to a simple circuit shown in Fig. 2.98 (*b*).

Equivalent resistance of this circuit is

$$R'' = R_{CB} + R_{BD} \,||\, R_{BA} = 9 + 6 \,||\, 3 = 9 + 2 = 11\ \Omega$$

Current supplied by 60 V source,

$$I_2 = \frac{60}{11}\ A$$

Current in resistor BD,

$$I'' = I_2 \times \frac{R_{BA}}{R_{BA} + R_{CB}}$$

$$= \frac{60}{11} \times \frac{3}{3+6} = \frac{20}{11}\ A \qquad \text{(From D to B)}$$

Total current, through 6 Ω resistor is equal to the algebraic sum of currents I′ and I″

$$I = I' + I'' = \frac{120}{11} + \frac{-20}{11} = \mathbf{\frac{100}{11}\ A.} \qquad \textbf{Ans.}$$

Problem 2.26. *Using superposition theorem, find the current I in the following circuit shown in Fig. 2.99.*

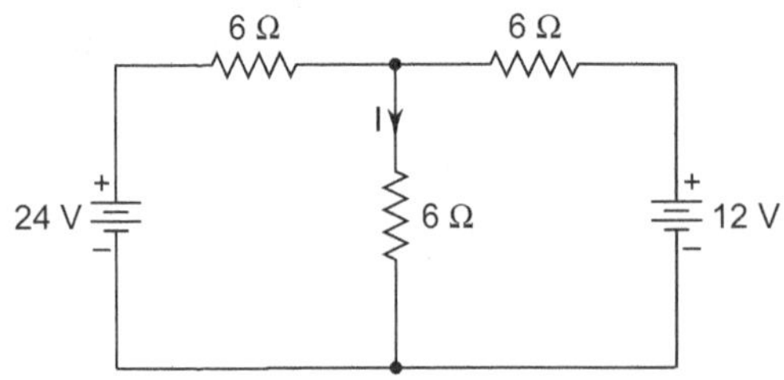

Fig. 2.99

Solution. Replacing 12 V source by a short-circuit, the circuit is reduced to a simple circuit shown in Fig. 2.100 (*a*). Equivalent resistance of this circuit will be

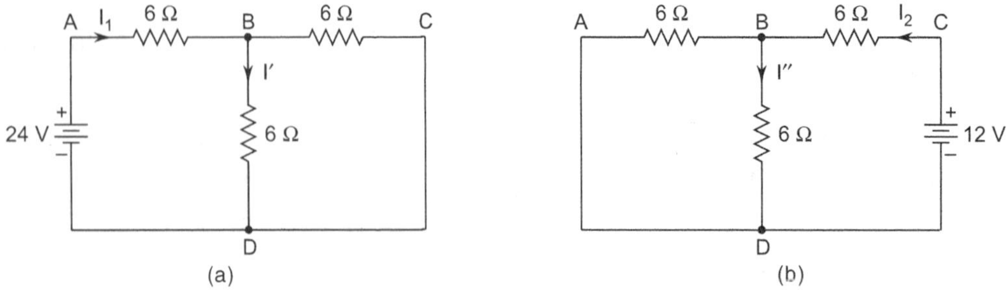

Fig. 2.100

$$R' = R_{AB} + R_{BC} \,||\, R_{BD} = 6 + 6 \,||\, 6 = 6 + \frac{6}{2} = 9\ \Omega$$

Current supplied by 24 V source, $I_1 = \frac{24}{R'} = \frac{24}{9} = \frac{8}{3}$ A

Current in resistance R_{BD}, $I' = I_1 \times \frac{R_{BC}}{R_{BC} + R_{BD}} = \frac{8}{3} \times \frac{6}{6+6} = \frac{4}{3}$ A.

Similarly on replacement of 24 V source by a short-circuit, the given circuit is reduced to a simple circuit shown in Fig. 2.100 (*b*). Equivalent resistance of this circuit will be

$$R'' = R_{BC} + R_{BD} \parallel R_{BA} = 6 + 6 \parallel 6 = 6 + \frac{6}{2} = 9\ \Omega$$

Current supplied by 12 V source, $I_2 = \frac{12}{R'} = \frac{12}{9} = \frac{4}{3}$ A

Current in resistor R_{BD}, $I'' = I_2 \times \frac{R_{AB}}{R_{AB} + R_{BD}} = \frac{4}{3} \times \frac{6}{6+6} = \frac{2}{3}$ A

Total current through 6 Ω resistor connected between terminals B-D, when both sources are present,

$$I = I' + I'' = \frac{4}{3} + \frac{2}{3} = \mathbf{2\ A.}$$ **Ans.**

Problem 2.27. *By superposition theorem, find the current in the resistance, R (Fig. 2.101).*

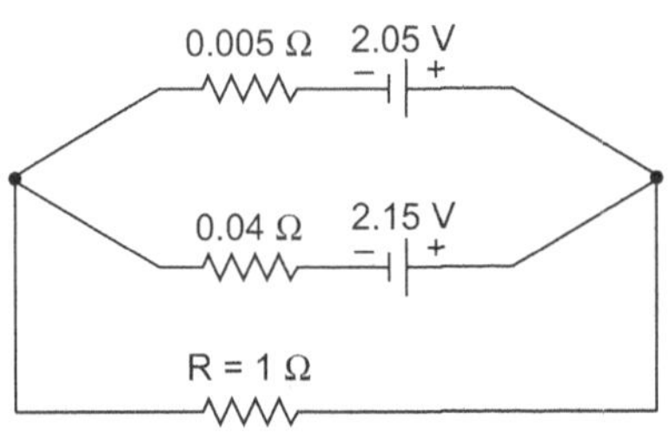

Fig. 2.101

Solution. Taking $E_2 = 0$, the circuit is reduced to a simple circuit shown in Fig. 2.102(*a*) in which resistances of 0.04 Ω and 1 Ω are in parallel across points A and C and combination acts in series with resistor R_1 of 0.05 Ω. Equivalent resistance of the circuit shown in Fig. 2.102(*a*).

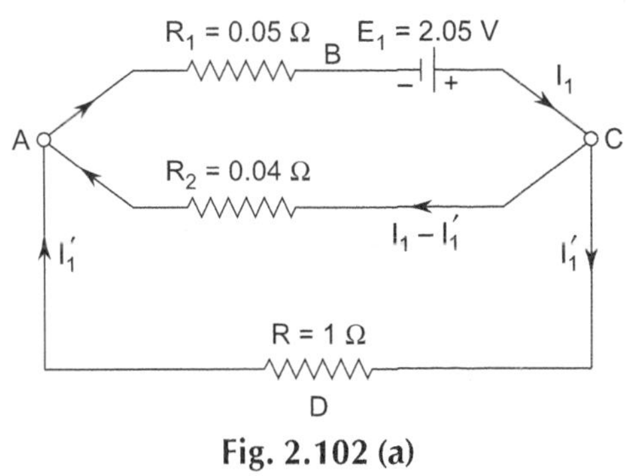

Fig. 2.102 (a)

$$R' = R_1 + \frac{R_2 \times R}{R_2 + R} = 0.05 + \frac{0.04 \times 1}{0.04 + 1} = 0.0885\ \Omega$$

Current supplied by battery E_1, $I_1 = \frac{E_1}{R'} = \frac{2.05}{0.0885} = 23.2$ A

Current in 1 Ω resistor, $I_1' = I_1 \times \frac{R_2}{R + R_2}$

$$= \frac{23.2 \times 0.04}{1.04} = 0.892 \text{ A from C to A.}$$

Taking $E_1 = 0$, the circuit is reduced to the simple circuit in Fig. 2.102 (*b*) in which resistors of 0.05 Ω and 1 Ω are in parallel across points A and C and the combination is in series with the resistor R_2 of 0.04 Ω. Equivalent resistance of the circuit shown in Fig. 2.102 (*b*).

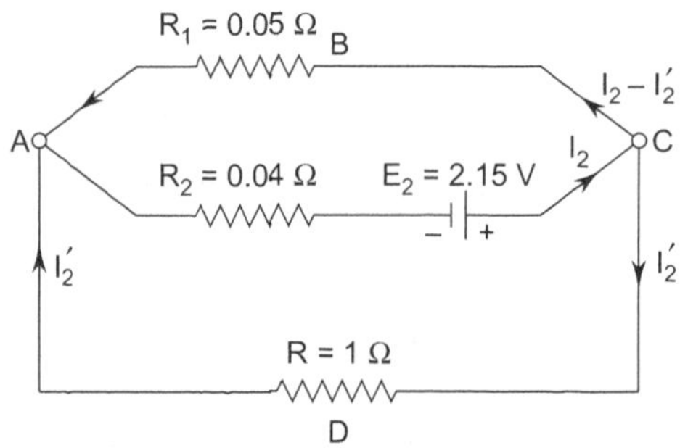

Fig. 2.102 (b)

$$R'' = R_2 + \frac{R_1 \cdot R}{R_1 + R} = 0.04 + \frac{0.05 \times 1}{0.05 + 1} = 0.0876\ \Omega$$

Current supplied by battery E_2, $I_2 = \frac{E_2}{R''} = \frac{2.15}{0.0876} = 24.54$ A

Current in 1 Ω resistor, $I_2' = I_2 \times \frac{R_1}{R + R_1} = \frac{24.54 \times 0.05}{1.05} = 1.169$ A from C to A

Total current through 1 Ω resistor when both batteries are present

$= I_1' + I_2' = 0.892 + 1.169 =$ **2.06 A.** **Ans.**

Problem 2.28. *Out of mesh and nodal analysis, which method will require least number of equations to solve the network in Fig. 2.103. Write the equations in matrix form following the method requiring least effort.*

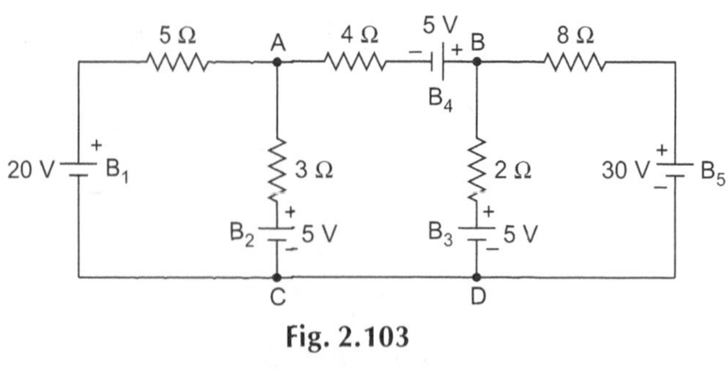

Fig. 2.103

Solution. In the given circuit, there are three meshes. Hence to solve the circuit with the help of mesh analysis, we require ***three*** equations. There are three nodes in the given circuit. (As there are no elements between C and D nodes, hence these two nodes are consider as one node). So to solve the circuit with nodal analysis, we require only ***two*** equations. For solving the circuit with nodal analysis, consider Fig. 2.104.

Applying KCL at node-A, we have

$$I_1 + I_2 + I_3 = 0$$

or $$\frac{V_A - 20}{5} + \frac{V_A - 5}{3} + \frac{V_A + 5 - V_B}{4} = 0$$

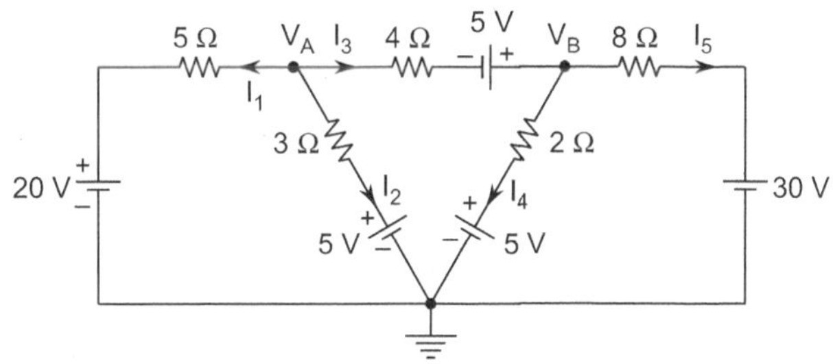

Fig. 2.104

or $$47\,V_A - 15\,V_B = 265 \quad ...(i)$$

At node-B applying KCL, we have

$$I_3 = I_4 + I_5$$

or $$\frac{V_A + 5 - V_B}{4} = \frac{V_B - 5}{2} + \frac{V_B - 30}{8}$$

or $$2V_A - 7V_B = -60 \quad ...(ii)$$

The above equations in matrix form are written as

$$\begin{bmatrix} 47 & -15 \\ 2 & -7 \end{bmatrix} \begin{bmatrix} V_A \\ V_B \end{bmatrix} = \begin{bmatrix} 265 \\ -60 \end{bmatrix}$$ **Ans.**

Problem 2.29. *Calculate currents in all the resistances of the circuit shown in Fig. 2.105 using node anlaysis method.*

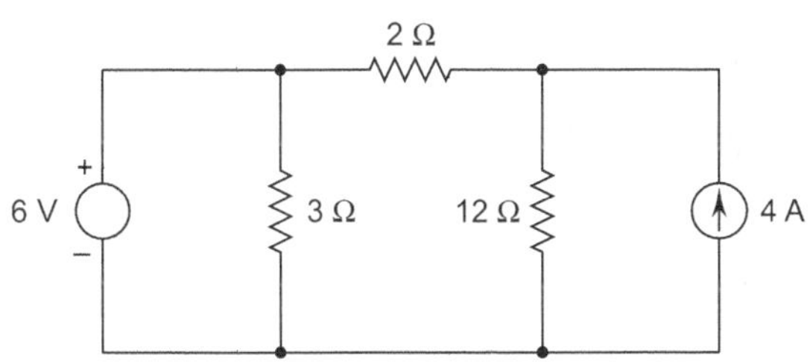

Fig. 2.105

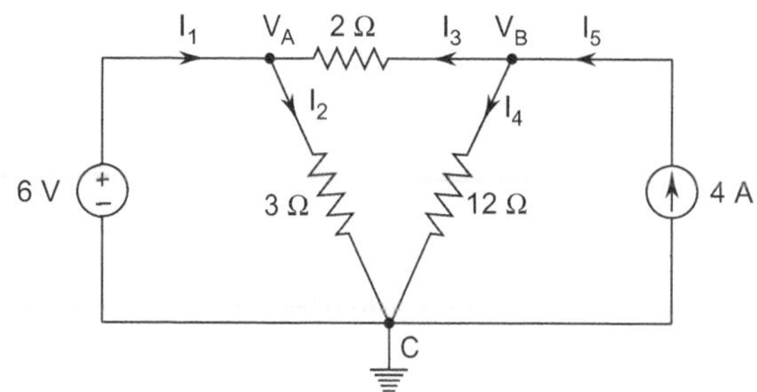

Fig. 2.106

Solution. Let the given circuit be redrawn with terminal C as the reference node and arbitrarily assumed distribution of currents, as shown in Fig. 2.106. The node A will have a potential of 6 V and the current shown I_5 will have value of 4 Ampere.

Applying Kirchhoff's current law to node B we have

$$I_4 = I_3 + I_5$$

or $$\frac{V_B}{12} = \frac{V_A - V_B}{2} + 4$$

or $$\frac{V_B}{12} = \frac{6 - V_B}{2} + 4 \qquad (\because \; V_A = 6\,V)$$

or $$V_B = 36 - 6\,V_B + 48$$

or $$7\,V_B = 84$$

or $$V_B = 12\text{ V}$$

So we have

Current in 3 Ω resistance, $$I_2 = \frac{V_A}{3} = \frac{6}{3} = \mathbf{2\ A} \qquad \textbf{Ans.}$$

Current in 2 Ω resistance, $$I_3 = \frac{V_A - V_B}{2} = \frac{6 - 12}{2} = \mathbf{-3\ A} \qquad \textbf{Ans.}$$

i.e., 3 A from node B to node A

Current in 12 Ω resistance, $$I_4 = \frac{V_B}{12} = \frac{12}{12} = \mathbf{1\ A} \qquad \textbf{Ans.}$$

Problem 2.30. *For the circuit shown in Fig. 2.107, find voltages of nodes B and C, and determine current in 8 Ω resistor.*

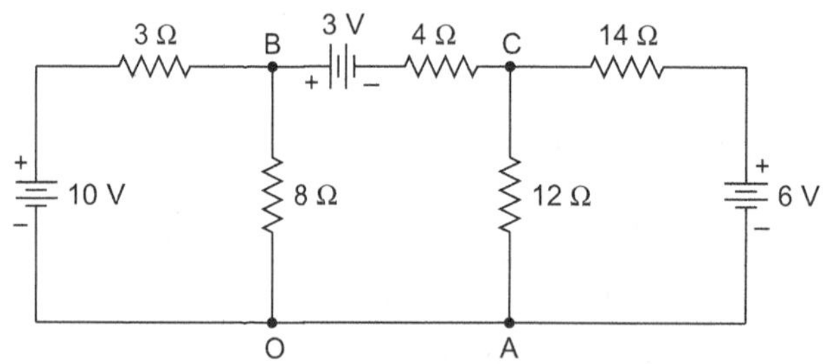

Fig. 2.107

Solution. Let the circuit be redrawn with terminal O as the reference node and arbitrarily assumed distribution of current as shown in Fig. 2.108.

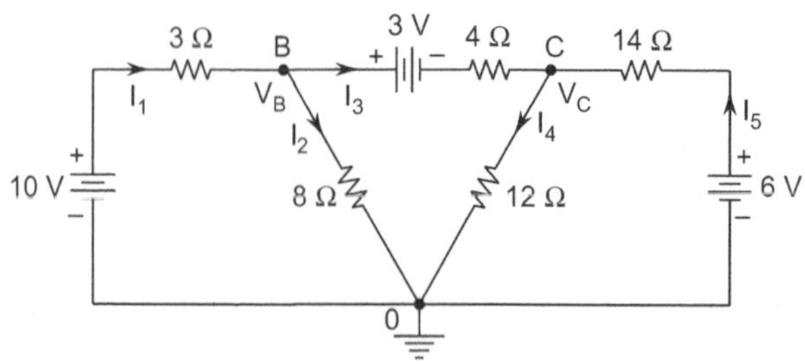

Fig. 2.108

Applying KCL to nodes B and C respectively we have

$$I_1 = I_2 + I_3$$

or $$\frac{10 - V_B}{3} = \frac{V_B}{8} + \frac{V_B - V_C - 3}{4} \quad ...(i)$$

and $$I_4 = I_5 + I_3$$

or $$\frac{V_C}{12} = \frac{6 - V_C}{14} + \frac{V_B - V_C - 3}{4} \quad ...(ii)$$

or $$17\,V_B - 6\,V_C = 98 \quad ...(iii)$$

and $$21\,V_B - 34\,V_C = 27 \quad ...(iv)$$

Solving above equations (*iii*) and (*iv*) we have

Voltage of node B, $V_B = \mathbf{7.0133\ V}$ **Ans.**

Voltage of node C, $V_C = \mathbf{3.5376\ V}$ **Ans.**

Current in 8 Ω resistor $= I_2 = \frac{V_B}{8} = \frac{7.0133}{8} = \mathbf{0.87\ A}$ **Ans.**

Problem 2.31. *Use nodal anlaysis to find the currents in various resistors of the circuit shown in Fig. 2.109.*

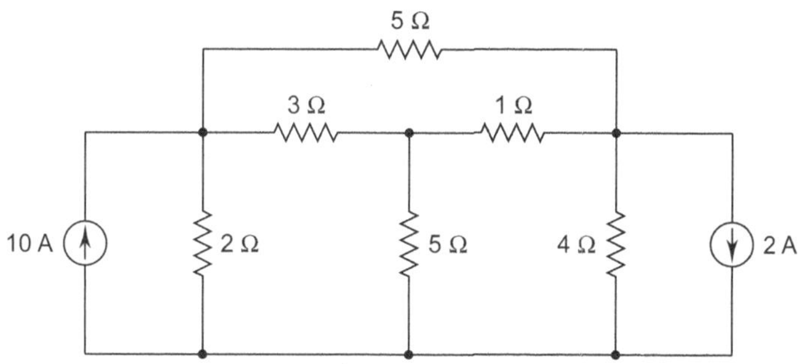

Fig. 2.109

Solution. Let the circuit be redrawn with its different nodes marked A, B, C and D, the last one being taken as reference or datum node (Fig. 2.110). Applying KCL to nodes A, B and C respectively we have

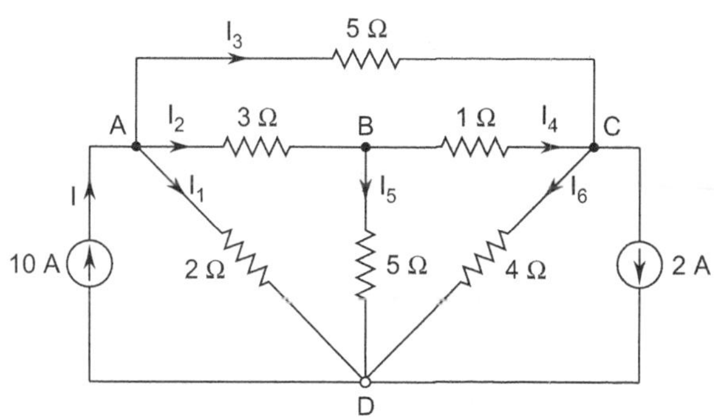

Fig. 2.110

$$I_1 + I_2 + I_3 = I$$

or $$\frac{V_A}{2} + \frac{V_A - V_B}{3} + \frac{V_A - V_C}{5} = 10$$

or $$15V_A + 10V_A - 10V_B + 6V_A - 6V_C = 300$$

or $$31V_A - 10V_B - 6V_C = 300 \quad ...(i)$$

$$I_2 = I_4 + I_5$$

or $$\frac{V_A - V_B}{3} = \frac{V_B - V_C}{1} + \frac{V_B}{5}$$

or $$5V_A - 5V_B = 15V_B - 15V_C + 3V_B$$

or $$5V_A - 23V_B + 15V_C = 0 \quad ...(ii)$$

and $$I_3 + I_4 - I_6 = 2$$

or $$\frac{V_A - V_C}{5} + \frac{V_B - V_C}{1} - \frac{V_C}{4} = 2$$

or $$4V_A - 4V_C + 20V_B - 20V_C - 5V_C = 40$$

or $$4V_A + 20V_B - 29V_C = 40 \quad ...(iii)$$

Solving equations (*i*), (*ii*) and (*iii*) we have

$$V_A = \frac{6.572}{545} \text{ A}$$

$$V_B = \frac{556}{109} \text{ V}$$

$$V_C = \frac{2.072}{545} \text{ V}$$

Thus, current $$I_1 = \frac{V_A}{2} = \frac{6,572}{2 \times 545} = \mathbf{6.029\ A}$$ **Ans.**

Current $$I_2 = \frac{V_A - V_B}{3} = \frac{6,572 - 2,780}{3 \times 545} = \mathbf{2.319\ A}$$ **Ans.**

Current, $$I_3 = \frac{V_A - V_C}{5} = \frac{6,572 - 2,072}{5 \times 545} = \mathbf{1.651\ A}$$ **Ans.**

Current, $$I_4 = \frac{V_B - V_C}{1} = \frac{556}{109} - \frac{2,072}{545} = \mathbf{1.299\ A}$$ **Ans.**

Current, $$I_5 = \frac{V_B}{5} = \frac{556}{109 \times 5} = \mathbf{1.02\ A}$$ **Ans.**

Current, $$I_6 = \frac{V_C}{4} = \frac{2,072}{4 \times 545} = \mathbf{0.95\ A}$$ **Ans.**

Problem 2.32. *Find the Thevenin's equivalent of the network (Fig. 2.111) across terminals a and b.*

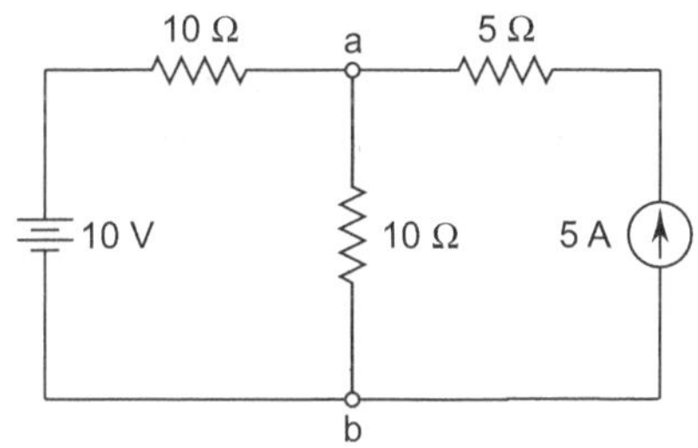

Fig. 2.111

Solution. By opening 5 A current source and short-circuiting 10 V voltage source [Fig. 2.112 (*a*)] we have

$$R_{Th} = \mathbf{10\ \Omega} \qquad \textbf{Ans.}$$

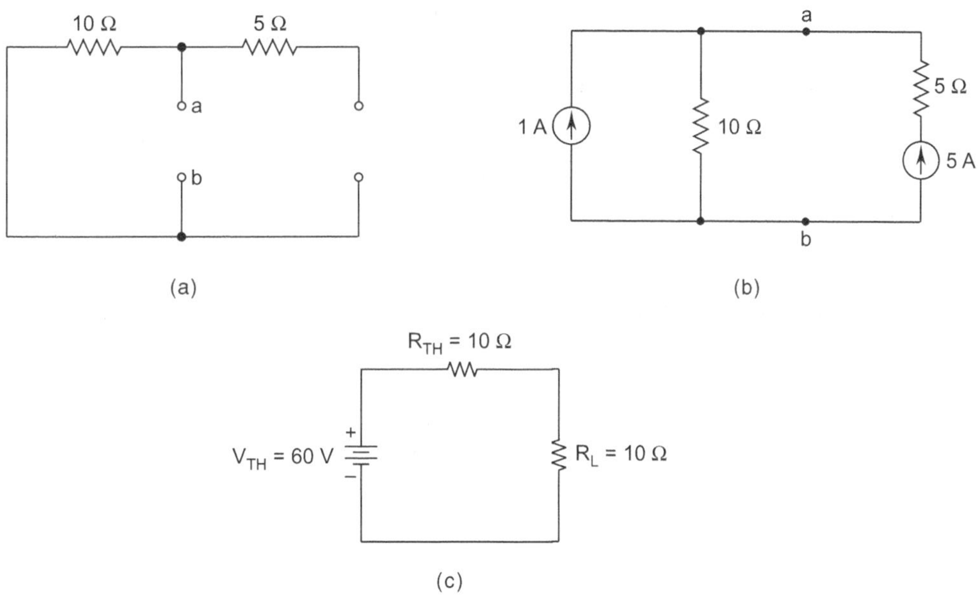

Fig. 2.112

Let us convert 10 V voltage source with its series resistance of 10 Ω into an equivalent current source of 1 A with a parallel resistance of 10 Ω as shown in Fig. 2.112 (*b*).

From the circuit it is obvious that 1 + 5 = 6 A current flows through 10 Ω resistor, thereby producing a voltage drop of 6 × 10 = 60 V. Hence

$$V_{TH} = \mathbf{60\ V} \qquad \textbf{Ans.}$$

So Thevenin equivalent of the given network is equivalent to a voltage source of 60 V with 10 Ω resistance in series with it [Fig. 2.112 (*c*)].

Problem 2.33. *Calculate current in a 1,000 Ω resistor connected between terminals A and B, as shown in Fig. 2.113 with the help of Thevenin's theorem.*

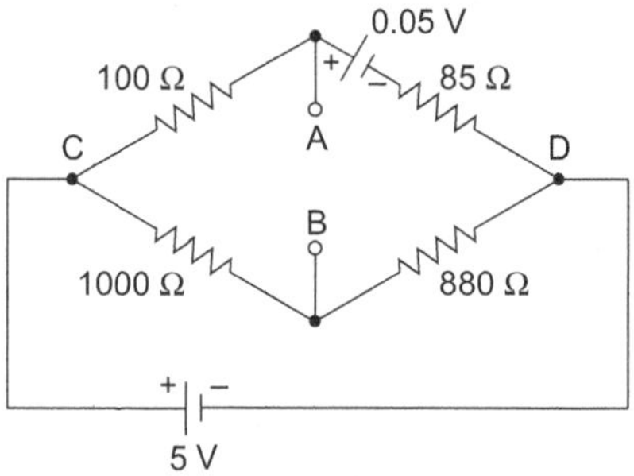

Fig. 2.113

Solution. In the circuit shown in Fig. 2.113,

Potential of point B w.r.t. terminal D

$$V_{BD} = \frac{5 \times 880}{1,000 + 880} = 2.340426 \text{ V}$$

Potential of point A w.r.t. terminal D

$$V_{AD} = \frac{5 - 0.05}{100 + 85} \times 85 + 0.05 = 2.324324 \text{ V}$$

PD between terminals B and A,

$$V_T = 2.340426 - 2.324324 = 0.0161 \text{ V}$$

After removing batteries from the circuit, equivalent resistance of the network with reference to terminals A and B (Fig. 2.114).

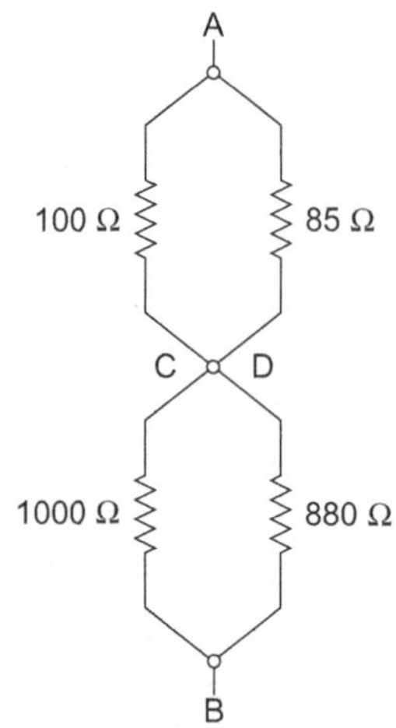

Fig. 2.114

$$R_T = 100 \parallel 85 + 1,000 \parallel 880$$

$$= \frac{1700}{17 + 20} + \frac{22,000}{22 + 25} = 514 \ \Omega$$

Current in 1,000 Ω resistor connected between terminals A and B,

$$I = \frac{V_T}{R_T + R_L} = \frac{0.0161}{1,000 + 514}$$

$$= \mathbf{10.625\ \mu A\ (from\ terminals\ B\ to\ A)}$$ **Ans.**

Problem 2.34. *Determine current in 10 Ω resistance using Norton's theorem in the network shown in Fig. 2.115.*

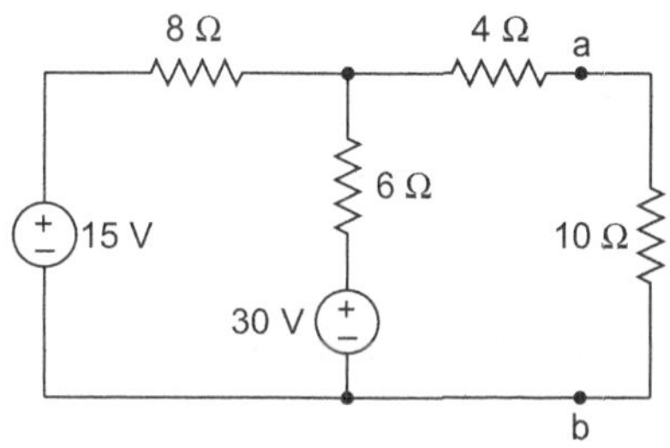

Fig. 2.115

Solution. Norton's equivalent resistance of given network (*i.e.*, equivalent resistance of the network after terminals *a* and *b* are open-circuited and the two batteries are short-circuited) as viewed from terminals *a* and *b* shown in Fig. 2.116 (*a*).

$$R_N = 4 + (6 \,||\, 8) = 4 + \frac{6 \times 8}{6 + 8} = \frac{52}{7} = \mathbf{7.43\ \Omega}$$ **Ans.**

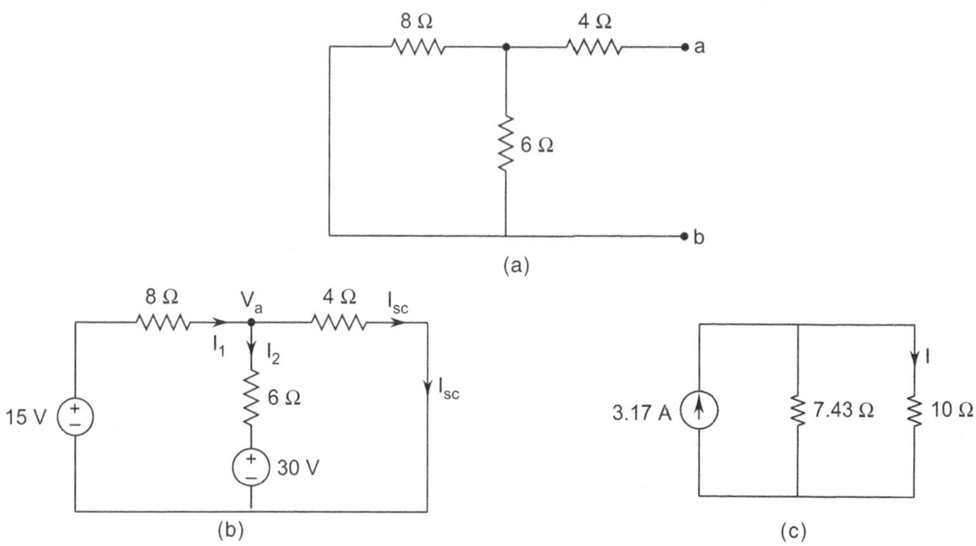

Fig. 2.116

For determining short-circuit current I_{SC} *i.e.*, current in zero resistance connected across terminals *a* and *b* [10 Ω resistance is replaced by a zero resistance as shown in Fig. 2.116 (*b*)] let us apply Nodel analysis to Fig. 2.116 (*b*). Applying KCL at Node *a*, we have

$$I_1 = I_2 + I_{SC}$$

or

$$\frac{15 - V_a}{8} = \frac{V_a - 30}{6} + \frac{V_a}{4}$$

or $$45 - 3V_a = 4V_a - 120 + 6V_a$$

or $$V_a = 12.69 \text{ V}$$

Hence the short-circuit current or Norton's equivalent source magnitude will be

$$I_N = I_{SC} = \frac{V_a}{4} = \frac{12.69}{4} = \mathbf{3.17\ A}$$ **Ans.**

The Norton's equivalent circuit is shown in Fig. 2.116 (*c*)

Current through 10 Ω resistance will be

$$I = \frac{3.17 \times 7.43}{7.43 + 10} = \mathbf{1.35\ A}$$ **Ans.**

Problem 2.35. *Find the Norton's equivalent circuit as seen by R_L in the circuit of Fig. 2.117.*

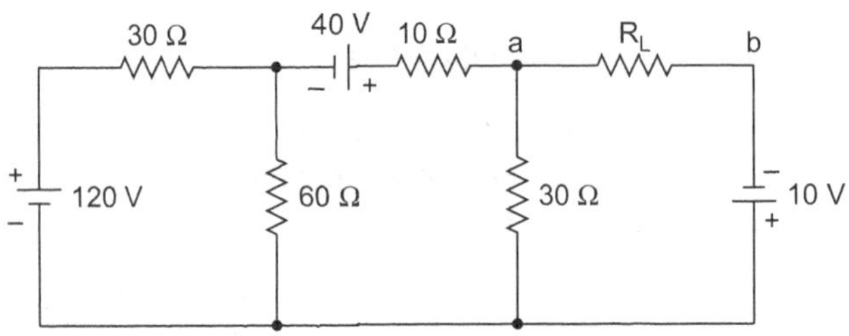

Fig. 2.117

Solution. Equivalent resistance of the circuit when viewed from terminals *ab* with voltage sources short-circuited (Fig. 2.118)

$$R_N = 30 \parallel (10 + 30 \parallel 60)$$

$$= 30 \parallel \left(10 + \frac{30 \times 60}{30 + 60}\right) = 30 \parallel 30 = \mathbf{15\ \Omega}$$ **Ans.**

For determination of short-circuit current I_{SC} *i.e.*, current in zero resistance connected between terminals *a* and *b*, R_L is replaced by zero resistance as shown in Fig. 2.119.

Let the node *j* be the reference node and let the arbitrary distribution of current as shown in Fig. 2.119.

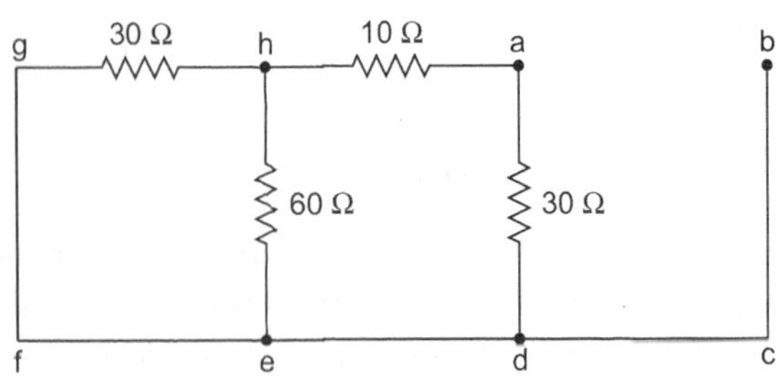

Fig. 2.118

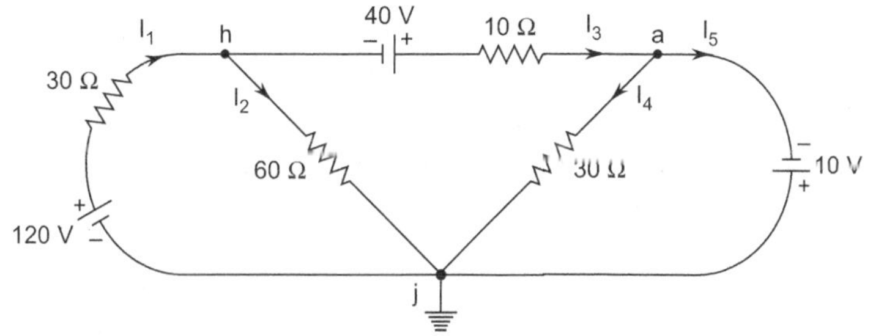

Fig. 2.119

Applying KCL to node ***h*** we have $I_1 = I_2 + I_3$

or $$\frac{120 - V_h}{30} = \frac{V_h}{60} + \frac{V_h + 40 - V_a}{10}$$

or $$2(120 - V_h) = V_h + 6(V_h + 40 + 10) \qquad [\because \; V_a = -10 \text{ V}]$$

or $$V_h = \frac{-20}{3} \text{ V}$$

Applying KCL to node ***a*** we have $I_3 = I_4 + I_5$

or $$I_5 = I_3 - I_4 = \frac{V_h + 40 - V_a}{10} - \frac{V_a}{30} = \frac{\frac{-20}{3} + 40 - (-10)}{10} - \frac{(-10)}{30} = \frac{14}{3} \text{A}$$

and short-circuit current $$I_{SC} = I_5 = \frac{14}{3} = \mathbf{4.67\ A}$$ **Ans.**

Thus Norton's equivalent circuit is a current source of 4.67 A with a resistance of 15 Ω across it as shown in Fig. 2.120.

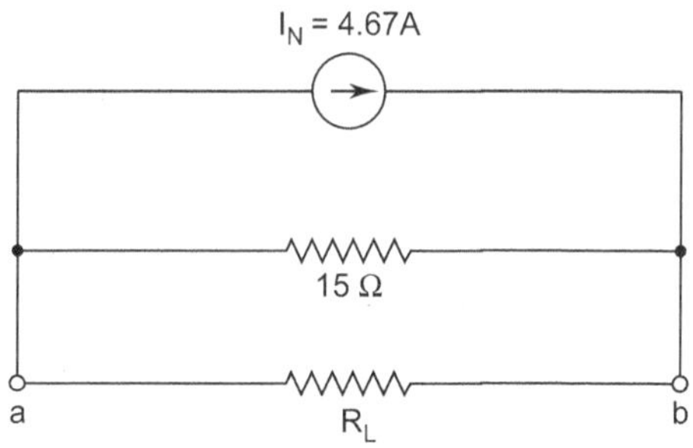

Fig. 2.120

Problem 2.36. *Find the value of resistance 'R' to have maximum power transfer in the circuit as shown in Fig. 2.121. Also obtain the amount of maximum power.*

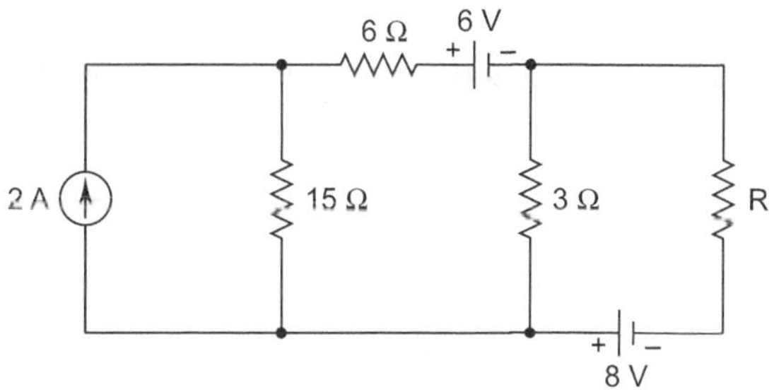

Fig. 2.121

Solution. Let current source of 2 A shunted by 15 Ω resistor be converted into its equivalent voltage source of 2 × 15 = 30 V connected in series with a resistance of 15 Ω and load resistor R be removed. The circuit is shown in Fig. 2.122 (*a*).

The equivalent resistance of the network with voltage sources short-circuited when viewed from terminals A-B [Fig. 2.122 (*b*)],

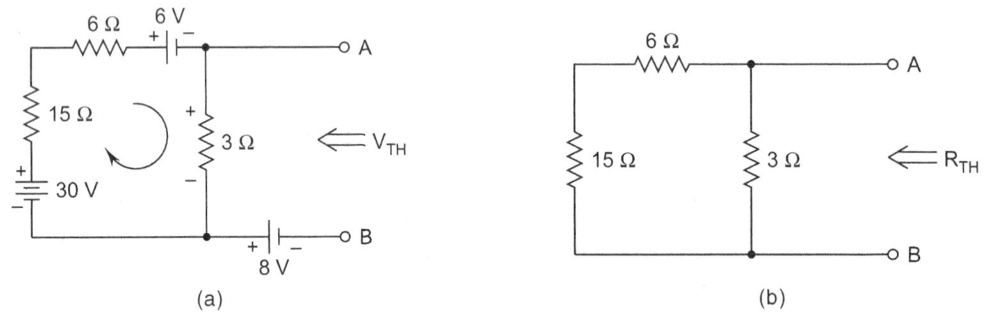

Fig. 2.122

$$R_{TH} = 3 \parallel (6+15) = 3 \parallel 21 = \frac{1}{\frac{1}{3}+\frac{1}{21}} = \frac{21}{8} = 2.625\ \Omega$$

Current flowing through a resistance of 3 Ω in circuit shown in Fig. 2.122 (*a*).

$$I = \frac{30-6}{15+6+3} = 1\ \text{A}.$$

Open-circuit voltage, V_{TH} = Voltage across 3 Ω resistor + 8 V

$$= 1 \times 3 + 8 = 11\ \text{V}$$

According to maximum power transfer theorem, R will absorb maximum power when it is equal to R_{TH} *i.e.*,

$$R = R_{TH} = R_L = \mathbf{2.625\ \Omega}$$ **Ans.**

Maximum power transferred,

$$P_{max} = \frac{V_{TH}^2}{4R_L} = \frac{11^2}{4 \times 2.625} = \mathbf{11.524\ W}$$ **Ans.**

Problem 2.37. *Find the value of R in the circuit of Fig. 2.123 such that the maximum power transfer takes place.*

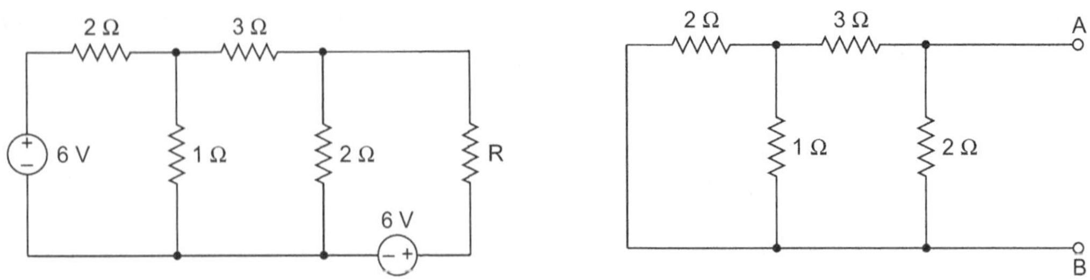

Fig. 2.123 **Fig. 2.124**

Solution. The equivalent resistance of the network when viewed from terminals A and B after removing resistor R from the circuit and short-circuiting the voltage (Fig. 2.124) sources, we have

$$R_{TH} = 2 \parallel [3 + 2 \parallel 1]$$

$$= 2 \parallel \left[3 + \frac{2}{3}\right]$$

$$= \left(2 \parallel \frac{11}{3}\right) = \frac{22}{17}\ \Omega$$

Maximum power will be transferred to the resistor R when it is equal to Thevenin's equivalent resistance R_T

i.e., $$R = R_{TH} = \frac{22}{17} = \mathbf{1.29\ \Omega}$$ **Ans.**

Problem 2.38. *Find the resistance between AB of the circuit shown in Fig. 2.125. Use Y-Δ (star-delta) transformation.*

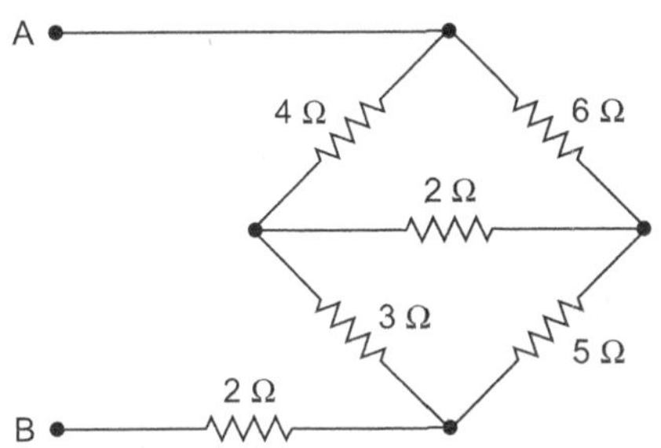

Fig. 2.125

Solution. As shown in Fig. 2.126 (*a*), delta ACD has been reduced to equivalent star

$$R_A = \frac{R_{AC} \cdot R_{AD}}{R_{AC} + R_{CD} + R_{DA}} = \frac{6 \times 4}{6 + 2 + 4} = 2\ \Omega$$

Similarly, $$R_C = \frac{6 \times 2}{12} = 1\ \Omega$$

and $$R_D = \frac{4 \times 2}{12} = \frac{2}{3}\ \Omega$$

Hence the given circuit is reduced to the circuit shown in Fig. 2.126(*c*). As seen, there are two parallel paths between terminals N and B′, one of resistance $1 + 5 = 6\ \Omega$ and the other of resistance

$$3 + \frac{2}{3} = \frac{11}{3}\ \Omega.$$

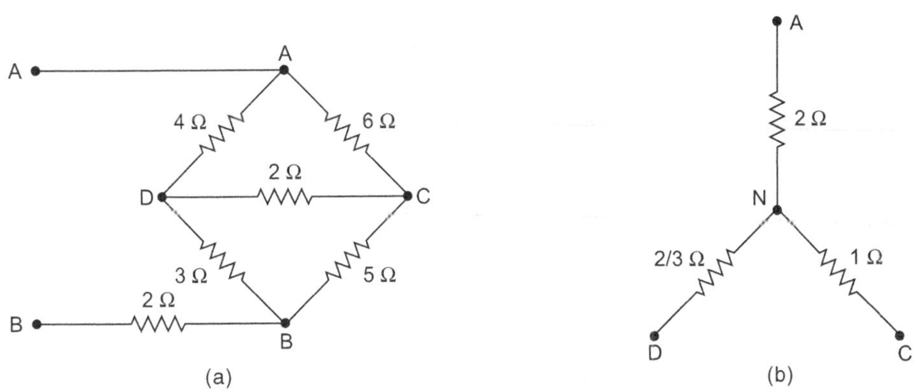

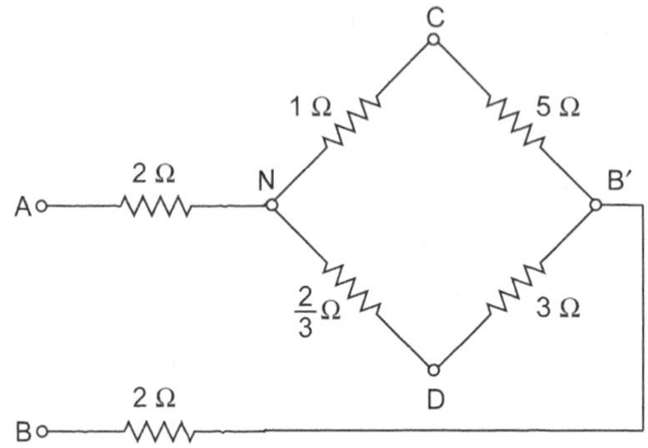

(c)

Fig. 2.126

The equivalent resistance of parallel combination (Between N and B′)

$$= \frac{1}{\frac{1}{6} + \frac{3}{11}} = \frac{66}{29}\ \Omega$$

So total resistance of the network between terminals A and B

$$= \frac{66}{29} + 2 + 2 = \mathbf{6.276\ \Omega}$$ **Ans.**

Problem 2.39. *Using delta to star transformation determine the resistance between terminals a-b and the total power drawn from the supply in the circuit shown in Fig. 2.127.*

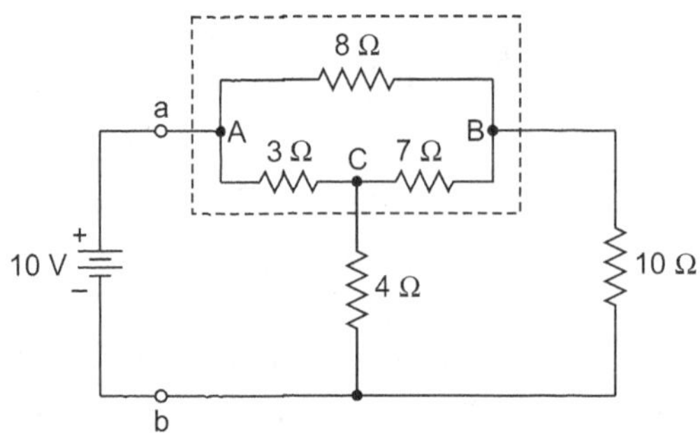

Fig. 2.127

Solution. By transforming delta formed by resistors of 8 Ω, 7 Ω and 3 Ω (Fig. 2.128) into star, we have

$$R_A = \frac{R_{AC} \times R_{AB}}{R_{AC} + R_{AB} + R_{BC}} = \frac{3 \times 8}{3 + 8 + 7} = \frac{4}{3}\ \Omega$$

Similarly $$R_B = \frac{8 \times 7}{18} = \frac{28}{9}\ \Omega$$

and $$R_C = \frac{3 \times 7}{18} = \frac{7}{6}\ \Omega$$

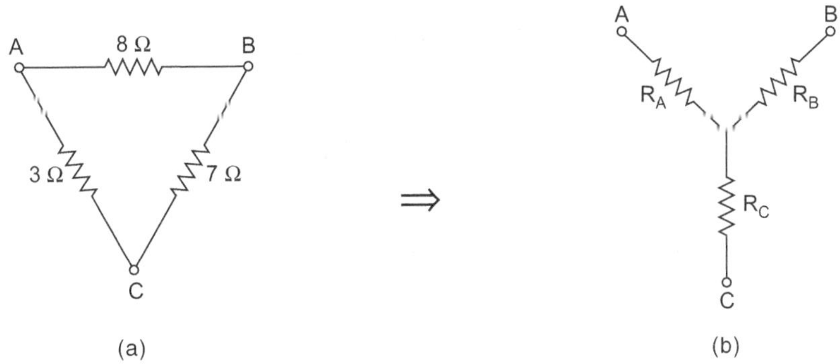

Fig. 2.128

Now, the given circuit is reduced to that as shown in Fig. 2.129.

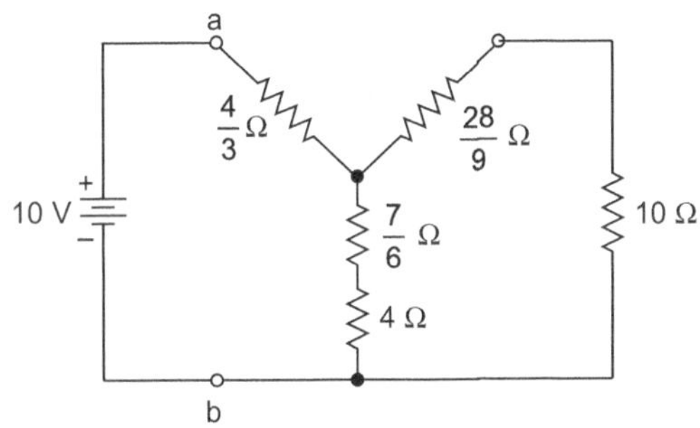

Fig. 2.129

Resistance between terminals a and b,

$$R = \frac{4}{3} + \left(\frac{7}{6} + 4\right) \| \left(\frac{28}{9} + 10\right) = \frac{4}{3} + \left(\frac{31}{7} \| \frac{118}{9}\right)$$

$$= 1.3333 + 3.7062 = \mathbf{5.04\ \Omega} \qquad \textbf{Ans.}$$

Total power drawn,

$$P = \frac{V^2}{R} = \frac{10^2}{5.04} = \mathbf{19.84\ Watts} \qquad \textbf{Ans.}$$

Problem 2.40. *Determine the current i_x and i_y in the following network shown in Fig. 2.130.*

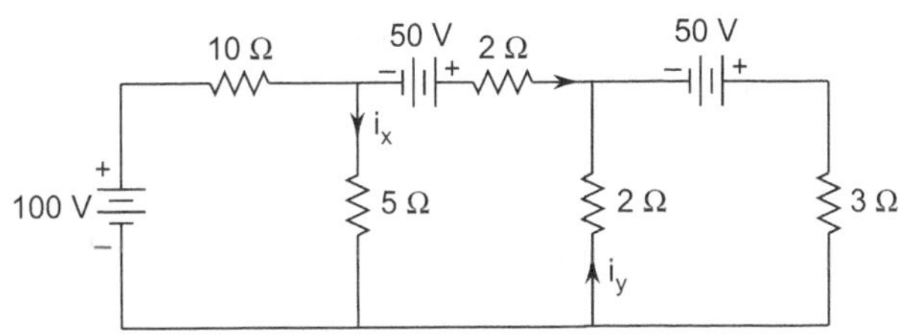

Fig. 2.130

Solution. The two arms in which the currents are to be calculated can be redrawn as shown in Fig. 2.131. Using nodal analysis

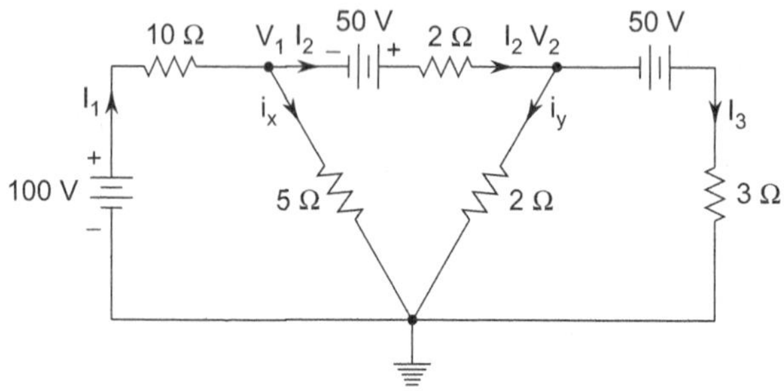

Fig. 2.131

At node (1) applying KCL

$$I_1 = I_2 + i_x$$

$$\frac{100 - V_1}{10} = \frac{V_1 + 50 - V_2}{2} + \frac{V_1}{5}$$

or
$$100 - V_1 = 5\,V_1 + 250 - 5\,V_2$$

$$-8\,V_1 + 5\,V_2 = 150 \qquad ...(i)$$

At node (2) applying KCL

$$I_3 = I_2 + i_y$$

$$\frac{V_2 + 50}{3} = \frac{V_1 + 50 - V_2}{2} - \frac{V_2}{2}$$

or
$$3\,V_1 - 8\,V_2 = -50 \qquad ...(ii)$$

Solving equations (*i*) and (*ii*) simultaneously we get

$$V_1 = -18.11\text{ V}$$

$$V_2 = 1.02\text{ V}$$

Hence the currents in the branches will be

$$i_x = \frac{V_1}{5} = \frac{-18.11}{5} = \mathbf{-3.62\ A} \qquad \textbf{Ans.}$$

$$i_y = \frac{-V_2}{2} = \frac{-1.02}{2} = \mathbf{-0.51\ A} \qquad \textbf{Ans.}$$

Problem 2.41. *Replace the network of following Fig. 2.132 to the left of terminals AB by its Thevenin's equivalent circuit. Hence determine I.*

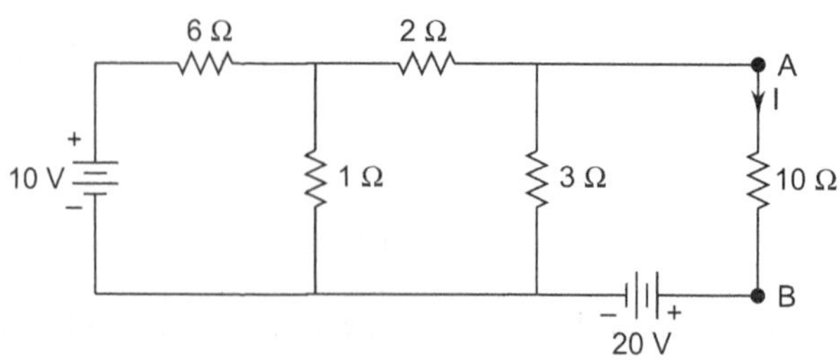

Fig. 2.132

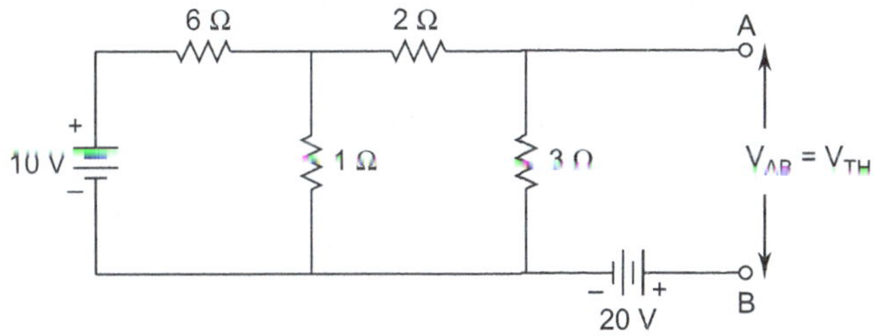

Fig. 2.133

Solution. To find V_{TH}, R_L is removed (Fig. 2.133)

Moving from B to A we will encounter with 2 elements

1. 20 V source (considered as negative)
2. 3 ohms resistance

Current through this 3 ohms resistance will be

Total resistance as seen from the 10 volt source

$$R = [(5 || 1) + 6] = 6.83\ \Omega$$

Total current
$$I = \frac{10}{6.83} = 1.46\ \text{A}$$

Current through 3 ohm resistance will be

$$I_1 = \frac{1.46 \times 1}{1 + 2 + 3} = 0.244\ \text{A}$$

Voltage drop across 3 ohm resistance will be

$$V = 3 \times 0.244 = 1.032\ \text{V (taken as positive)}$$

Hence V_{TH} will be

$$V_{TH} = -20 + 1.032 = \mathbf{-18.968\ V}$$ **Ans.**

To find the R_{TH} voltage sources are short-circuited (Fig. 2.134)

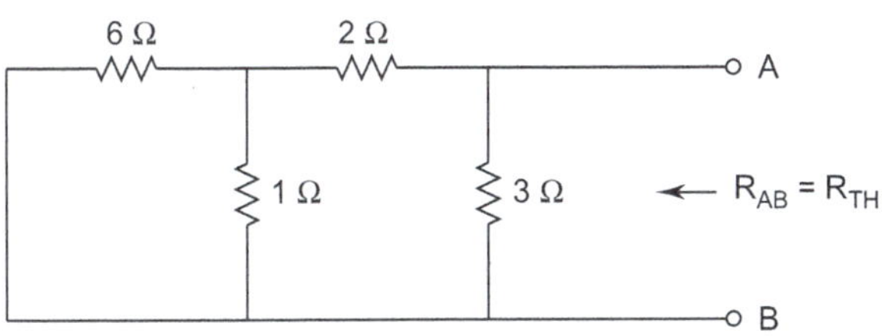

Fig. 2.134

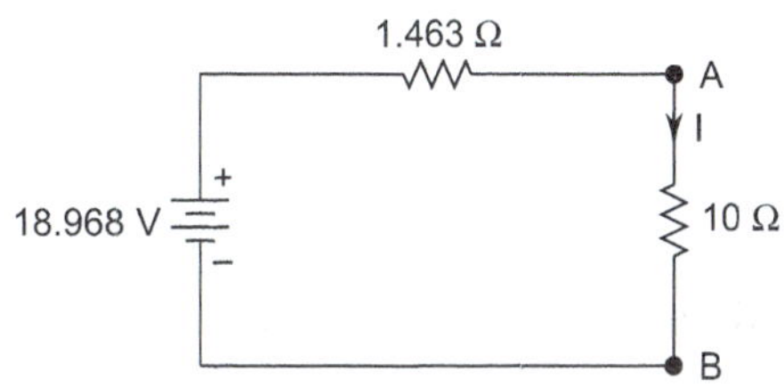

Fig. 2.135

$$R_{TH} = [(6 || 1) + 2] || [3] = \mathbf{1.463\ \Omega}$$ **Ans.**

Hence the Thevenin's equivalent circuit will be as shown in Fig. 2.135

And the current through the load resistance will be

$$I_1 = \frac{18.968}{1.463 + 10} = \mathbf{1.655\ A}$$ **Ans.**

Problem 2.42. *Find the current in and voltage across, the 2Ω resistance in the following Fig. 2.136.*

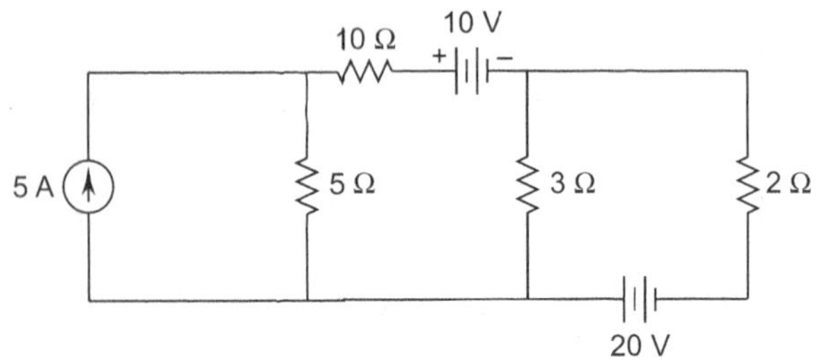

Fig. 2.136

Solution. The two arms in which the currents are to be calculated can be redrawn as shown in Fig. 2.137. Using nodal analysis

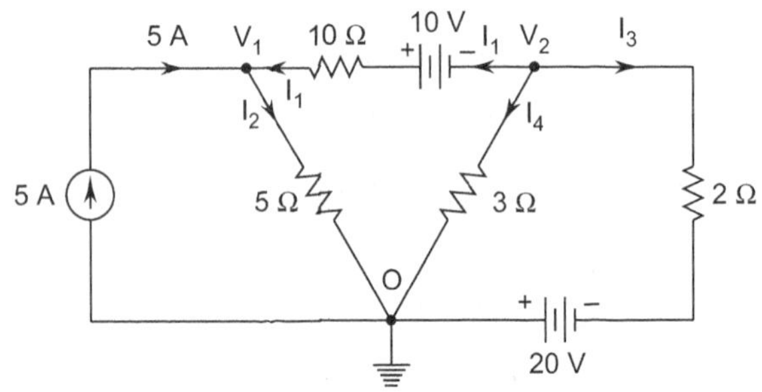

Fig. 2.137

At node (1) applying KCL

$$I_2 = I_1 + 5$$

or $$\frac{V_1}{5} = \frac{V_2 + 10 - V_1}{10} + 5$$

or $$3\,V_1 - V_2 = 60 \qquad \text{...}(i)$$

At node (2) applying KCL

$$I_4 = I_1 + I_3$$

or $$-\frac{V_2}{3} = \frac{V_2 + 10 - V_1}{10} + \frac{V_2 + 20}{2}$$

or $$3\,V_1 - 28\,V_2 = 330 \qquad \text{...}(ii)$$

Solving equations (*i*) and (*ii*) simultaneously, we get

$$V_1 = 16.67 \text{ V}$$

$$V_2 = -10 \text{ V}$$

Hence the current in the 2 ohms resistance I_3 will be

$$I_3 = \frac{V_1 + 20}{2} = \frac{-10 + 20}{2} = \mathbf{5\ A} \qquad \textbf{Ans.}$$

Voltage drop across 2 ohms resistance will be

$$V = 2 \times 5 = \mathbf{10\ V} \qquad \textbf{Ans.}$$

Problem 2.43. *Using Norton's theorem determines the current in 10 Ω resistance, in the circuit shown in Fig. 2.138.*

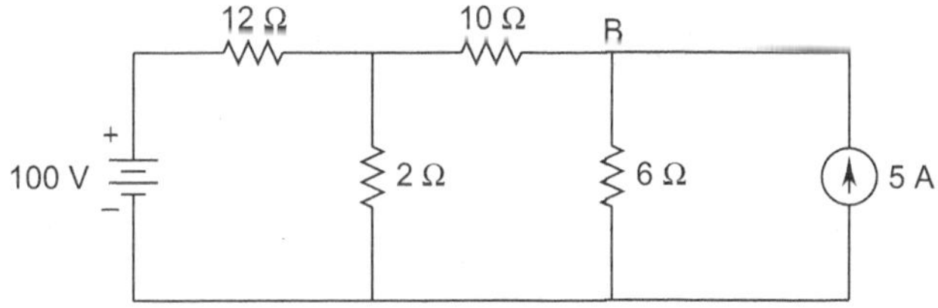

Fig. 2.138

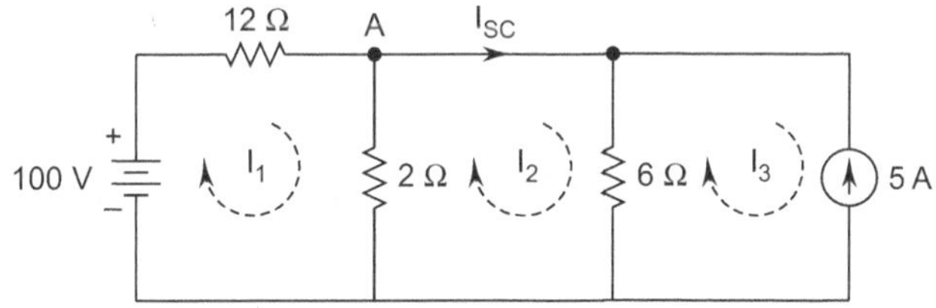

Fig. 2.139

Solution. To find the Norton's equivalent current source the load resistance terminals are short-circuited (Fig. 2.139).

Now applying mesh analysis to the network, in mesh (1) applying the KVL, we have

$$100 = 12I_1 + 2(I_1 - I_2)$$

or $$7I_1 - I_2 = 50 \quad ...(i)$$

In mesh (2) applying the KVL, we have

$$0 = 2(I_2 - I_1) + 6(I_2 - I_3)$$

or $$2I_2 - 2I_1 + 6I_2 - 6 \times (-5) = 0$$

or $$I_1 - 4I_2 = 15 \quad ...(ii)$$

In mesh (3), we have

$$I_3 = -5 \text{ A} \quad ...(iii)$$

Solving equations (*i*) and (*ii*) simultaneously, we get

$$I_1 = 6.85 \text{ A}$$

$$I_2 = -2.0375 \text{ A}$$

The short-circuit current I_{SC} or Norton's equivalent current source will be

$$I_N = I_{SC} = I_2 = -2.0375 \text{ A}$$

To find the Norton's equivalent resistance, all the voltage sources are short-circuited and the all the current sources are open circuited as shown in Fig. 2.140

$$R_N = R_{TH} = R_{AB} = [(12 \parallel 2) + 6]$$

or $$R_N = 7.714 \ \Omega$$

Hence the Norton's equivalent circuit will be as shown in Fig. 2.141

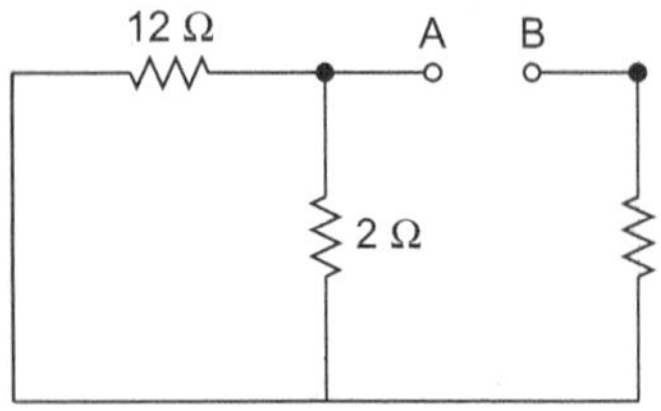

Fig. 2.140

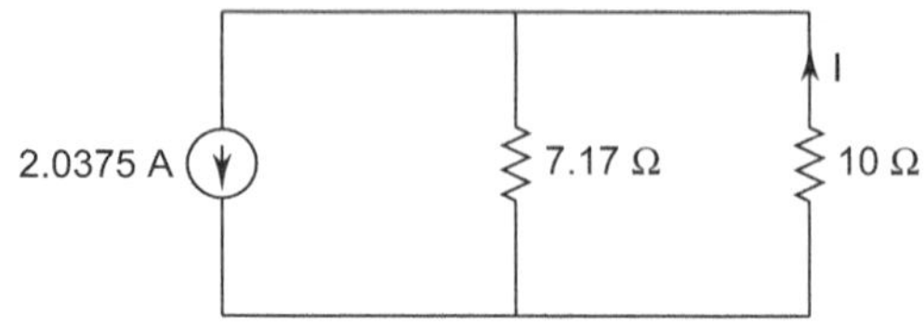

Fig. 2.141

The current through the load resistance will be

$$I = \frac{2.0375 \times 7.714}{7.714 + 10} = \mathbf{0.887\ A}$$ **Ans.**

Problem 2.44. *Using superposition thermo determine I in the network of Fig. 2.142*

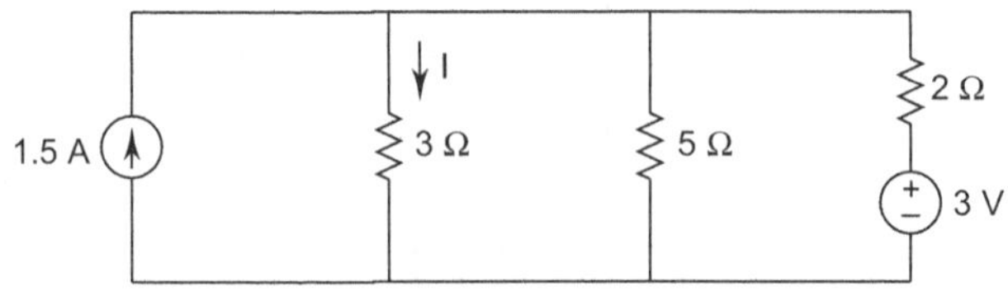

Fig. 2.142

Solution. First consider the current source of 1.5 A

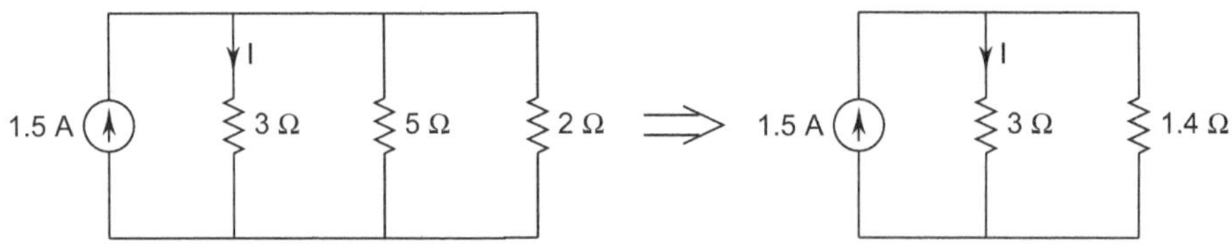

Fig. 2.143

$$I' = \frac{1.5 \times 1.43}{3 + 1.43} = 0.484\ \text{A}$$

Now, consider the voltage source of 3 Volts

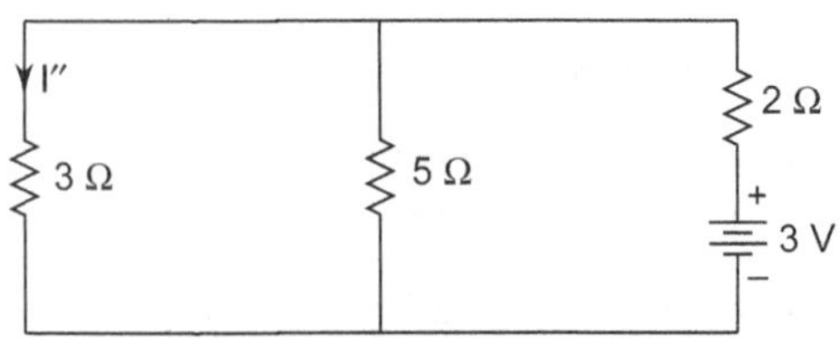

Fig. 2.144

Total resistance of the circuit as seen from the voltage source

$$R = [(3 \parallel 5) + 2] = 3.875\ \Omega$$

Total current coming from voltage source

$$i = \frac{V}{R} = \frac{3}{3.875} = 0.774 \text{ A}$$

And the current through the 3 ohms resistance is

$$I' = \frac{0.774 \times 5}{3+5} = 0.484 \text{ A}$$

Hence total current due to both the sources will be

$$I = I' + I'' = 0.484 + 0.484 = \mathbf{0.968\ A}$$ **Ans.**

Problem 2.45. *Obtain the Norton's equivalent circuit at terminals AB of the network given in Fig. 2.145.*

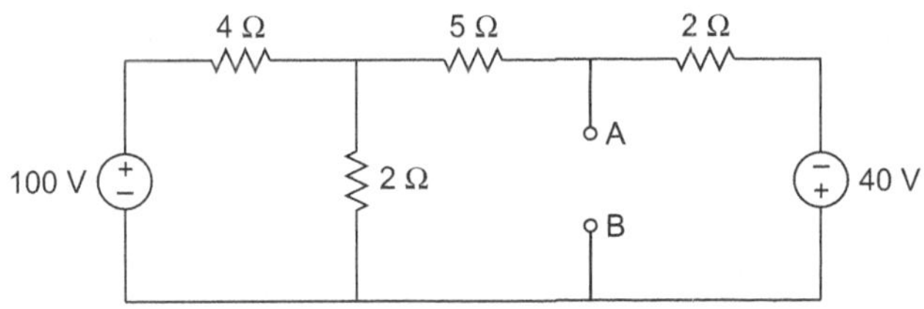

Fig. 2.145

Solution.

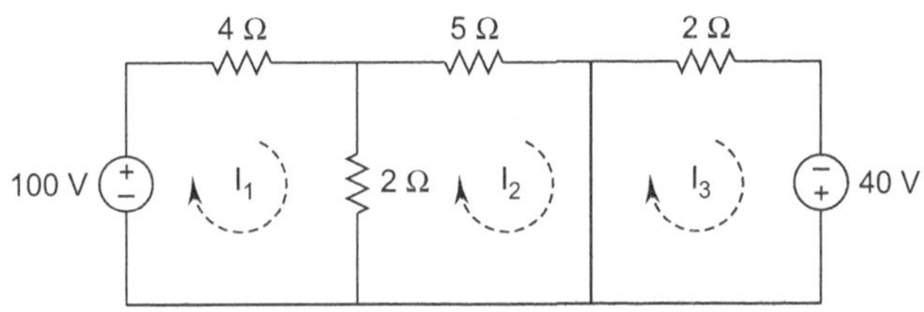

Fig. 2.146

To find the Norton's equivalent current source the load resistance terminals are short-circuited (Fig. 2.146).

Now, applying mesh analysis to the network, in mesh (1) applying the KVL, we have

$$100 = 6I_1 - 2I_2$$

or $$3I_1 - I_2 = 50 \quad ...(i)$$

In mesh (2) applying the KVL, we have

$$0 = 7I_2 - 2I_1$$

or $$I_1 = \frac{7}{2} I_2 \quad ...(ii)$$

In mesh (3) applying the KVL, we have

$$40 = 2I_3$$

or $$I_3 = 20 \text{ A} \quad ...(iii)$$

Solving equations (*i*) and (*ii*) simultaneously, we get

$$I_1 = 38.80 \text{ A}$$

$$I_2 = 5.26 \text{ A}$$

The short-circuit current I_{SC} or Norton's equivalent current source will be

$$I_N = I_{SC} = I_3 - I_2 = 20 - 5.26 = \mathbf{14.74\ A}$$ **Ans.**

To find the Norton's equivalent resistance, all the voltage sources are short-circuited and the all the current sources are open circuited as shown in Fig. 2.147

$$R_N = [(4 \parallel 2) + 5] \parallel [2]$$

or $$R_N = \mathbf{1.52\ \Omega}$$ **Ans.**

Hence the Norton's equivalent circuit will be as shown in Fig. 2.148

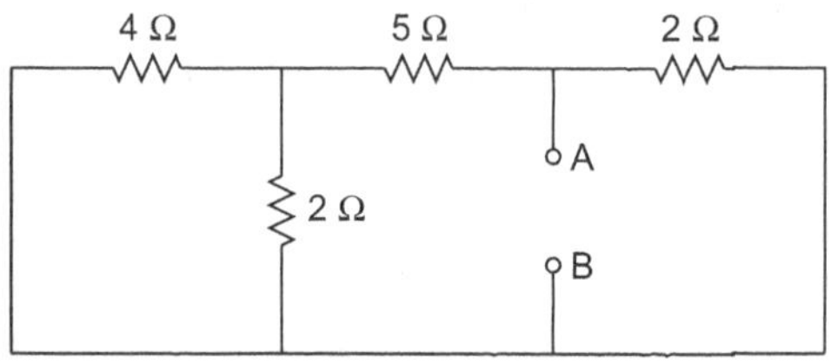

Fig. 2.147

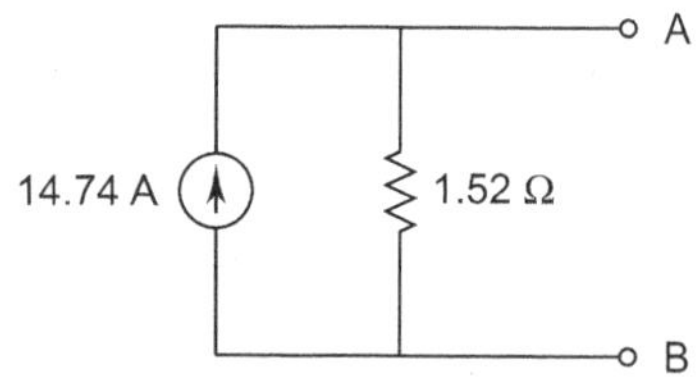

Fig. 2.148

Problem 2.46. *For the circuit shown in Fig. 2.149, find the maximum power that can be supplied to the load.*

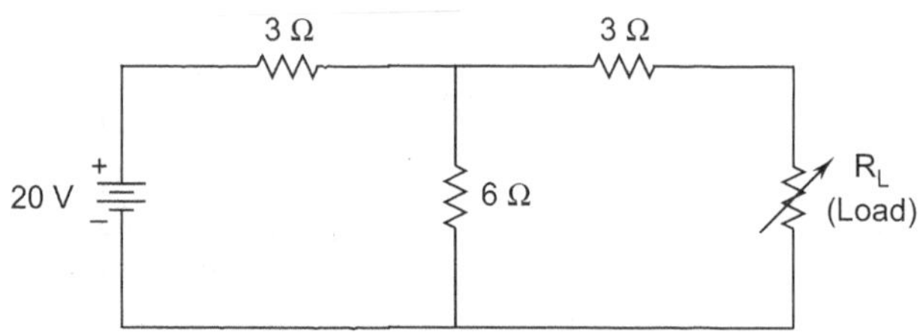

Fig. 2.149

Solution. For maximum power to be transferred from the source to load,

$$R_{TH} = R_L$$

For finding R_{TH}, consider Fig. 2.150

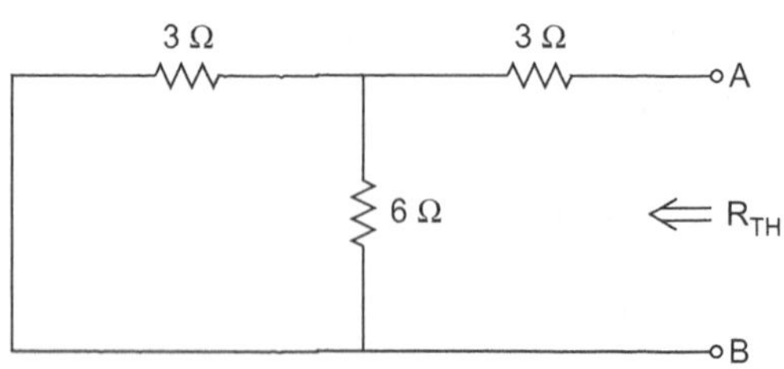

Fig. 2.150

$$R_{TH} = (3 \parallel 6) + 3$$

$$= \frac{3 \times 6}{3 + 6} + 3 = 5\ \Omega$$

For finding V_{TH}, consider Fig. 2.151.

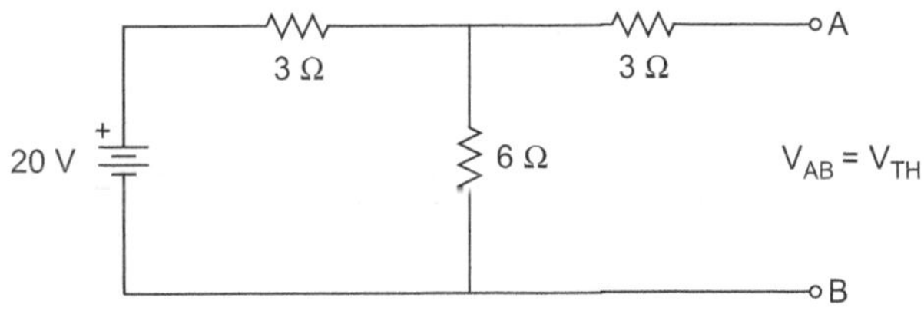

Fig. 2.151

The voltage V_{AB} (or V_{TH}) will be equal to the voltage across 6 Ω resistance only, *i.e.*,

$$V_{TH} = \frac{20 \times 6}{3 + 6} = 13.33 \text{ V (by voltage divider rule)}$$

$$\therefore \quad \text{Maximum Power, } P_m = \frac{V_{TH}^2}{4 R_L} = \frac{(13.33)^2}{4 \times 5}$$

or $$P_m = \mathbf{8.89\ \Omega}$$ **Ans.**

Problem 2.47. *Determine current in 4 Ω resistance using Thevenin's theorem in the circuit shown in Fig. 2.152.*

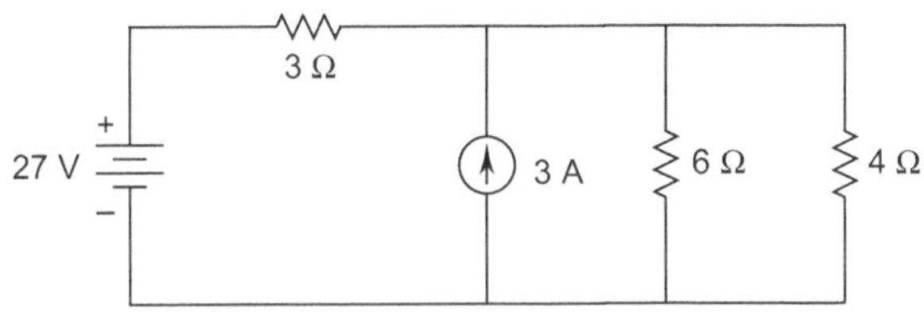

Fig. 2.152

Solution. For finding R_{TH}, consider the circuit shown in Fig. 2.153.

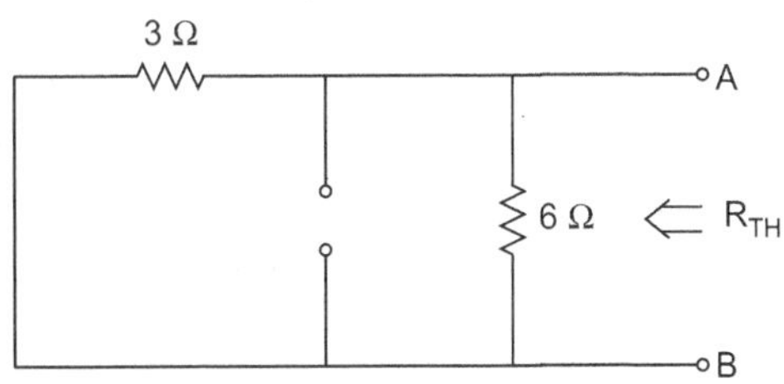

Fig. 2.153

$$R_{TH} = 6 \parallel 3 = \frac{6 \times 3}{6 + 3}$$

$$R_{TH} = 2\ \Omega$$

For finding V_{TH}, consider the circuit shown in Fig. 2.154 in which the 3 A current source is converted into voltage source.

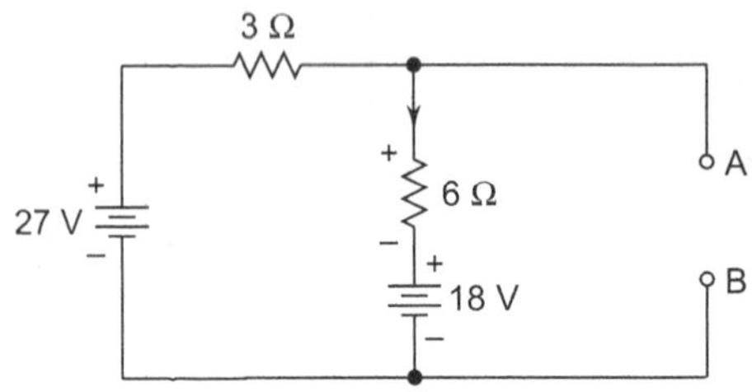

Fig. 2.154

While moving from B to A we will encounter with two elements.

(*i*) 18 V voltage source

$$V_1 = 18\text{ V (positive)}$$

(*ii*) 6 Ω resistance

voltage drop across 6 Ω resistance will be

$$V_2 = \frac{(27-18)\times 6}{3+6} = 6\text{ V (negative)}$$

$$\therefore \qquad V_{TH} = 18 - 6 = 12\text{ Volts}$$

Current through the 4 Ω resistance will be

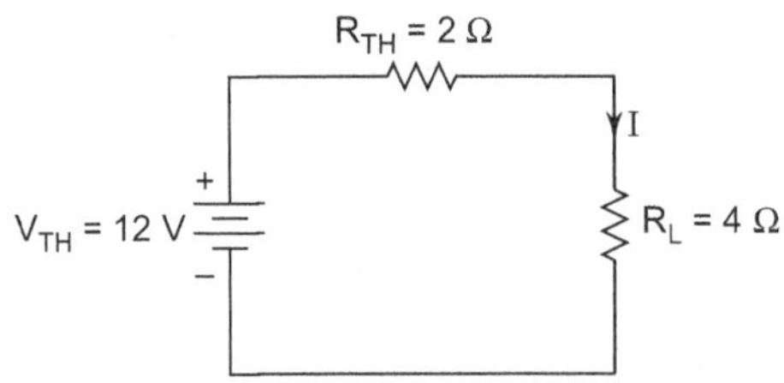

Fig. 2.155

$$I = \frac{12}{2+4} = \frac{12}{6} = \mathbf{2\ A}$$ **Ans.**

Problem 2.48. *Find voltage V_1 across 6 Ω resistance in the following circuit using Loop analysis method.*

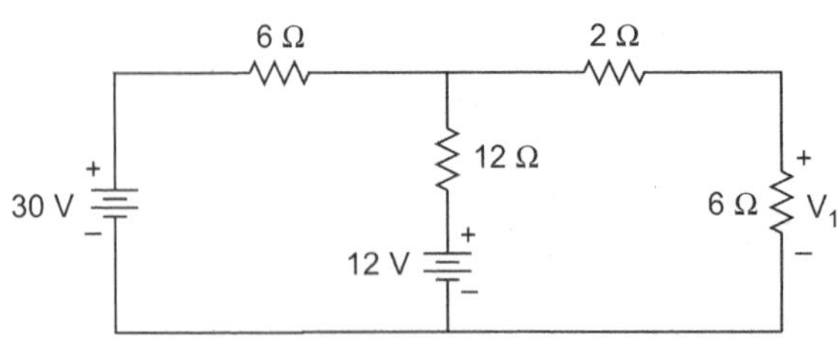

Fig. 2.156

Solution. The given circuit Fig. 2.156 has two meshes. The mesh currents are shown in Fig. 2.157.

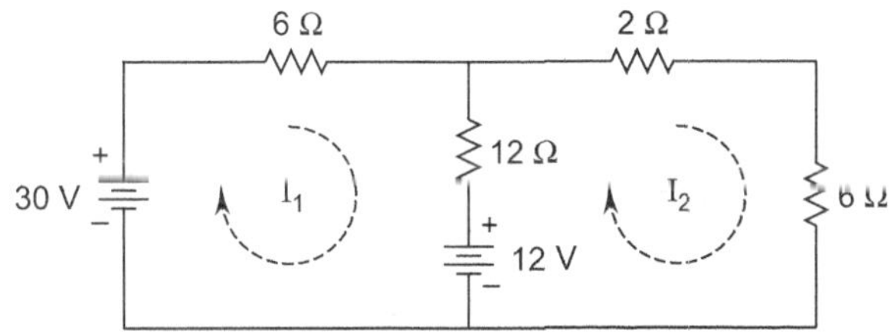

Fig. 2.157

For mesh (1), applying KVL, we have

$$30 - 12 = 6\,I_1 + 12\,(I_1 - I_2)$$

or $$18 = 18\,I_1 - 12\,I_2$$

or $$3 = 3\,I_1 - 2\,I_2 \qquad ...(i)$$

For mesh (2), applying KVL, we have

$$12 = 2\,I_2 + 6\,I_2 + 12\,(I_2 - I_1)$$

or $$12 = -12\,I_1 + 20\,I_2$$

or $$3 = -3\,I_1 + 5\,I_2 \qquad ...(ii)$$

Solving equation (*i*) and (*ii*), we get

$$I_2 = 2\text{ A}$$

$$I_1 = 7/3\text{ A}$$

The current through the 6 Ω resistance (across which, we have to find the voltage) is I_2. Hence

$$V_1 = I_2 \times 6$$

or $$V_1 = 2 \times 6 = \mathbf{12\ V}$$ **Ans.**

Problem 2.49. *Using Super position theorem, find the current in 20 Ω resistor of the circuit shown in Fig. 2.158.*

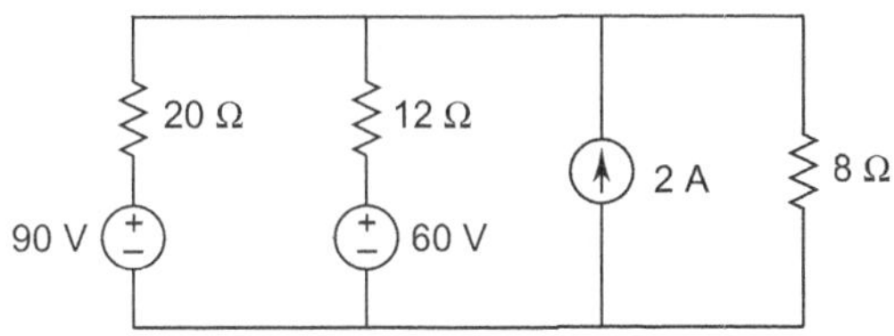

Fig. 2.158

Solution.

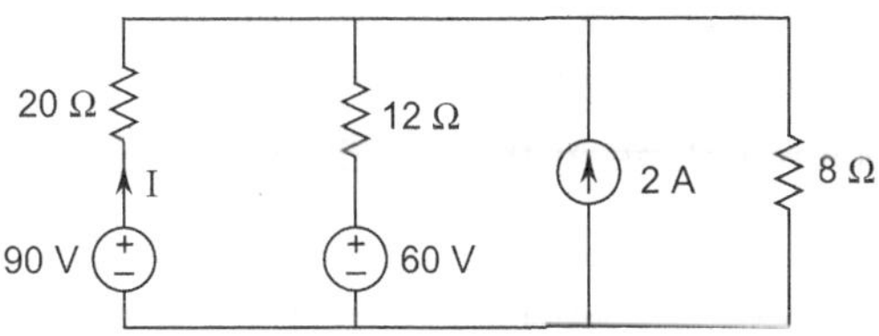

Fig. 2.159

First, consider 90 V source, removing all other sources (voltage source by short-circuit and current source by open circuit) as shown in Fig. 2.160.

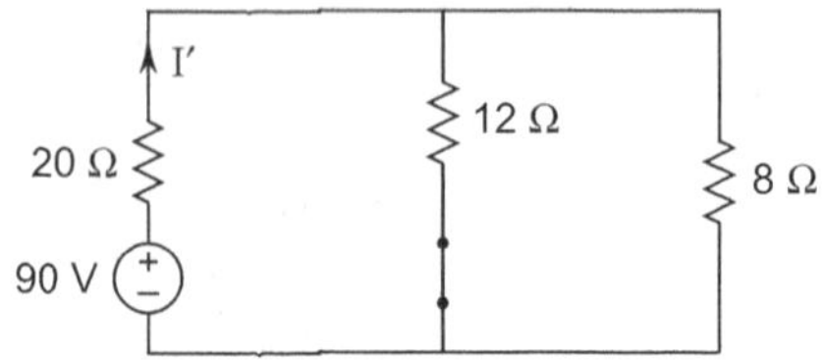

Fig. 2.160

Let the current due to 90 V source in 20 Ω resistance be I′ (in upward direction)

Total resistance of the circuit will be

$$R = (12 \,||\, 8) + 20$$

$$= \frac{12 \times 8}{12 + 8} + 20 = 24.8\ \Omega$$

$$I' = \frac{90}{24.8} = 3.63\ A$$

Next, consider 60 V source only, as shown in Fig. 2.161

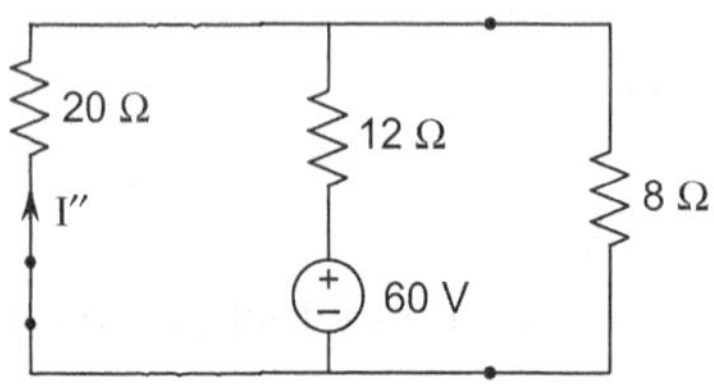

Fig. 2.161

Let the current through 20 Ω resistance due to 60 V source be I″ (in upward direction)

Total resistance of the circuit

$$R = (20 \,||\, 8) + 12$$

$$= 17.71\ \Omega$$

Total current through 60 V source will be

$$I = \frac{60}{17.71} = 3.39\ A$$

By current divider rule

$$I'' = \frac{I \times 8}{20 + 8} = \frac{3.39 \times 8}{20 + 8}$$

$$I'' = 0.97\ A \text{ (in downward direction)}$$

or $$I'' = -0.97\ A \text{ (in upward direction)}$$

Now, consider the 2 A source alone in the circuit as shown in Fig. 2.162

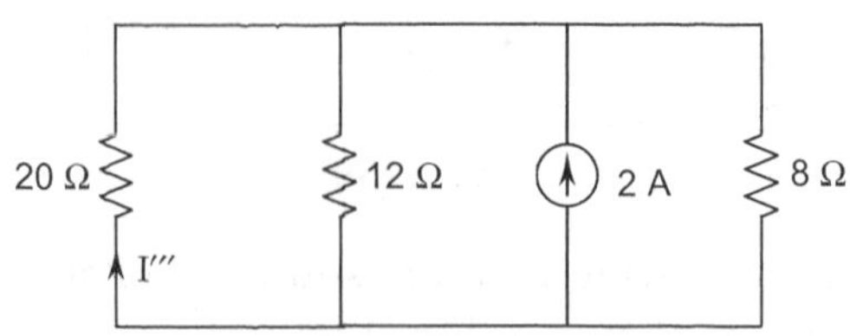

Fig. 2.162

Let the current through the 20 Ω resistance be I‴ in the upward direction.

12 Ω and 8 Ω resistances are in parallel and these two resistances can be combined into single resistance as shown in Fig. 2.163.

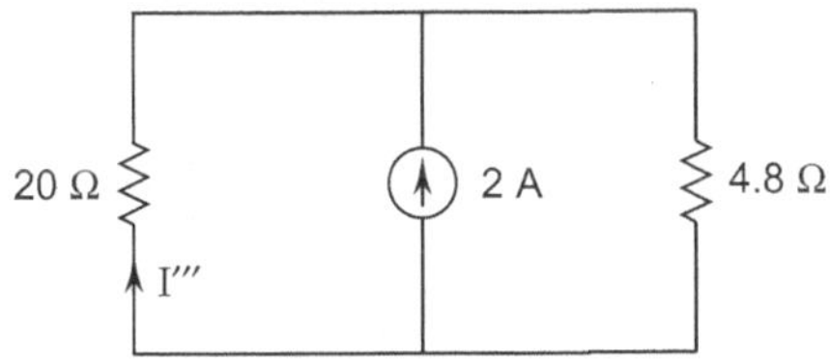

Fig. 2.163

$$12 \parallel 8 = \frac{12 \times 8}{12 + 8} = 4.8\ \Omega$$

Now, the current through 20 Ω resistance, according to current divider rule, will be

$$I''' = \frac{2 \times 4.8}{20 + 4.8} = 0.39\ \text{A (in downward direction)}$$

or $$I''' = -0.39\ \text{A (in upward direction)}$$

Hence the total current through the 20 Ω resistance will be the algebraic sum of all the above currents calculated while considering each source at a time, *i.e.*,

$$I = I' + I'' + I'''$$
$$= 3.63 - 0.97 - 0.39$$
$$I = \mathbf{2.27\ A}\ \text{(in upward direction)}$$ **Ans.**

Problem 2.50. *Find the current delivered by the voltage source of 140 V.*

Solution.

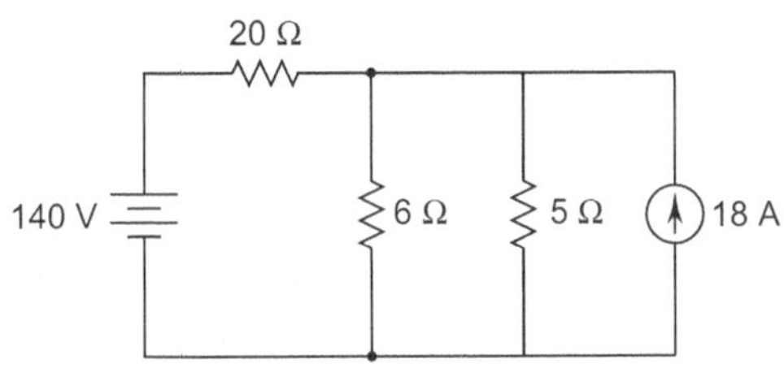

Fig. 2.164

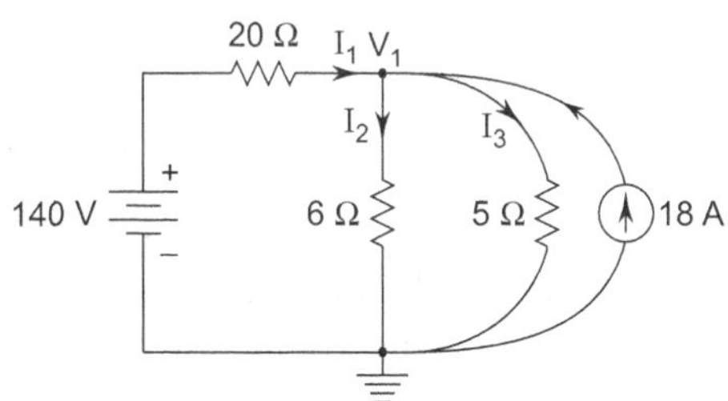

Fig. 2.165

The circuit shown in Fig 2.164 can be redrawn as shown in Fig. 2.165. Applying Nodal analysis, we have

At Node 1, having potential V_1 by KCL

$$18 + I_1 = I_2 + I_3$$

or $$18 + \frac{140 - V_1}{20} = \frac{V_1}{6} + \frac{V_1}{5}$$

or $$V_1 = 60\ \text{V}$$

∴ The current delivered by the voltage source of 140 V will be

$$I_1 = \frac{140 - V_1}{20} = \frac{140 - 60}{20}$$

or $\quad I_1 = \mathbf{4\ A}$ **Ans.**

Problem 2.51. *For the circuit shown in Fig. 2.166, find the value of R_L for which power dissipated in it is maximum and also calculate this maximum power.*

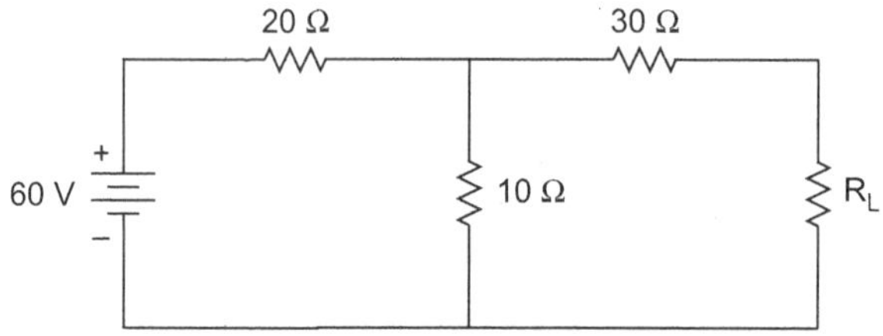

Fig. 2.166

Solution. For maximum power transfer from source to load

$$R_L = R_{TH}.$$

To find R_{TH}, remove R_L, remove the voltage source by short-circuit

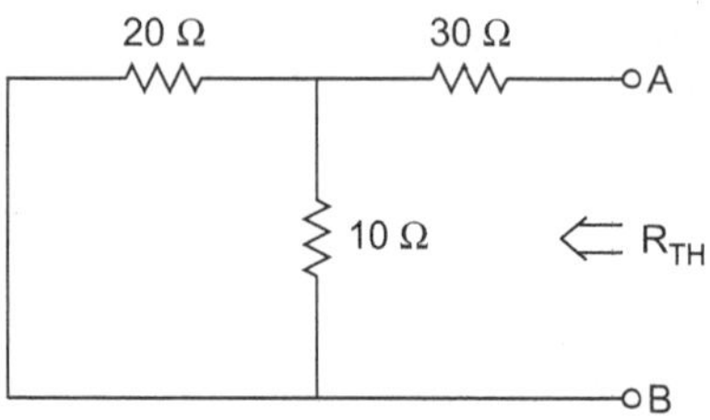

Fig. 2.167

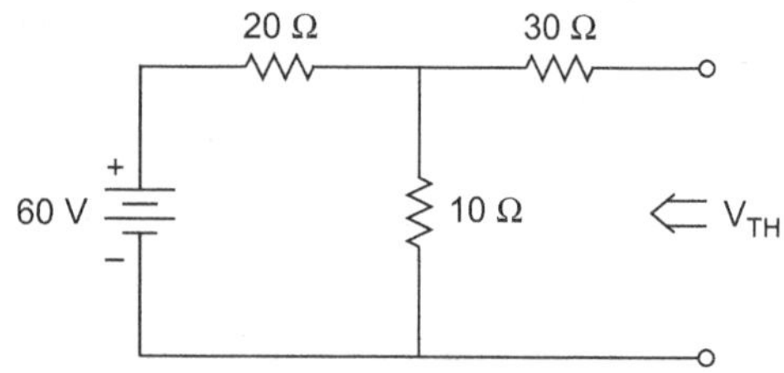

Fig. 2.168

$$R_{TH} = (20 \parallel 10) + 30$$

or
$$R_{TH} = \frac{20 \times 10}{20 + 10} + 30 = 36.67\ \Omega$$

$\therefore \quad R_L = R_{TH} = \mathbf{36.67\ \Omega}$ **Ans.**

To find V_{TH} consider Fig. 2.168. It is clear from Fig. 2.168 that the Thevenin's equivalent voltage V_{TH} will be equal to the voltage drop across 10 Ω resistance alone (as there is no current flowing in 30 Ω resistance)

Hence by voltage divider rule

$$V_{TH} = \frac{60 \times 10}{20 + 10} = 20\ V$$

The maximum power transferred will be

$$P_m = \frac{V_{TH}^2}{4 R_L} = \frac{(20)^2}{4 \times 36.67}$$

$P_m = \mathbf{2.73\ \Omega}$ **Ans.**

Problem 2.52. *Fig. 2.169 shows a number of resistances connected in star and delta. Using star/delta conversion method, calculate the equivalent resistance between B and C.*

Solution.

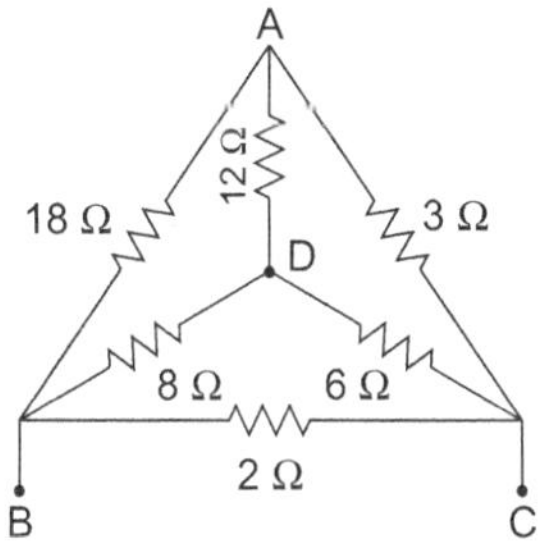

Fig. 2.169

⇒

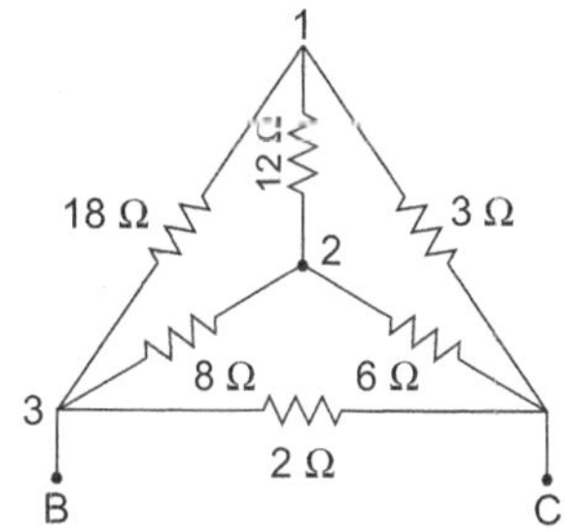

Fig. 2.170

Consider the delta formed by ADBA. Replacing this delta by star, we have

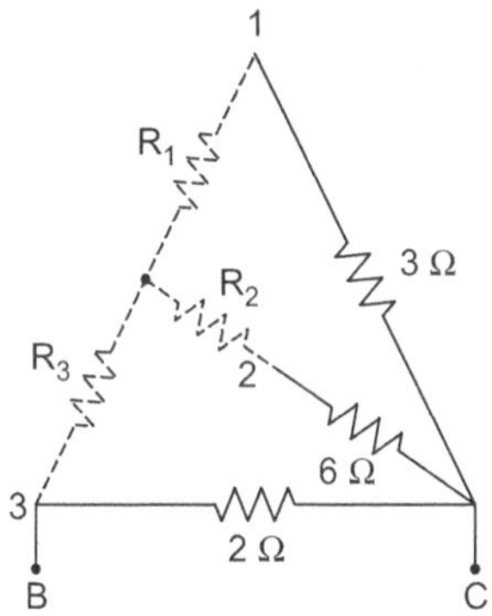

Fig. 2.171

⇒

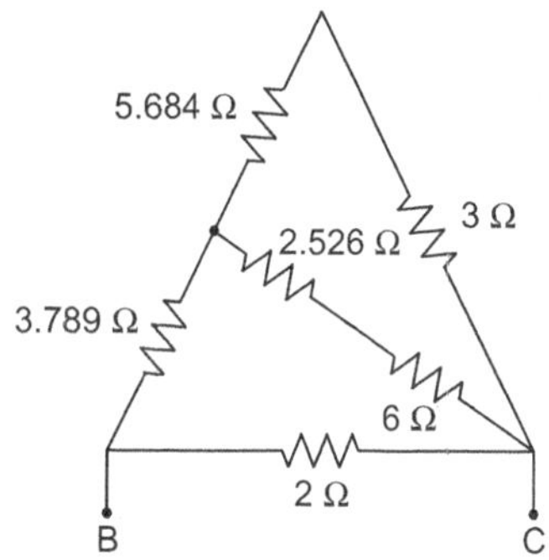

Fig. 2.172

$$R_1 = \frac{R_{12} \times R_{31}}{R_{12} + R_{23} + R_{31}} = \frac{18 \times 12}{12 + 8 + 18} = 5.684\,\Omega$$

$$R_2 = \frac{R_{23} \times R_{12}}{R_{12} + R_{23} + R_{31}} = \frac{12 \times 8}{12 + 8 + 18} = 2.526\,\Omega$$

$$R_3 = \frac{R_{31} \times R_{23}}{R_{12} + R_{23} + R_{31}} = \frac{8 \times 18}{12 + 8 + 18} = 3.789\,\Omega$$

Fig. 2.172 can further be reduced as shown in Fig. 2.173 in which series resistances are combined to form one resistance.

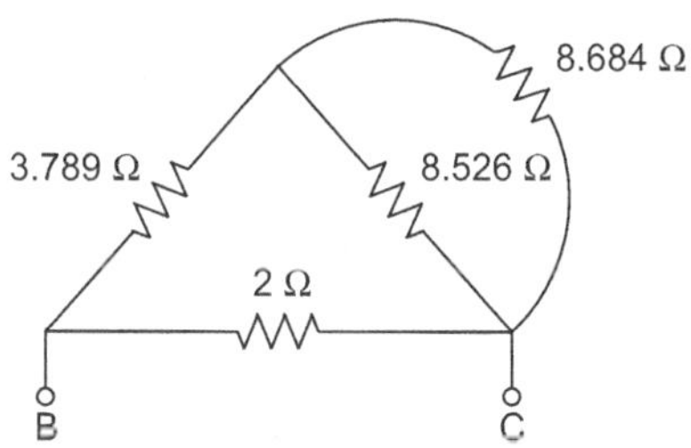

Fig. 2.173

⇒

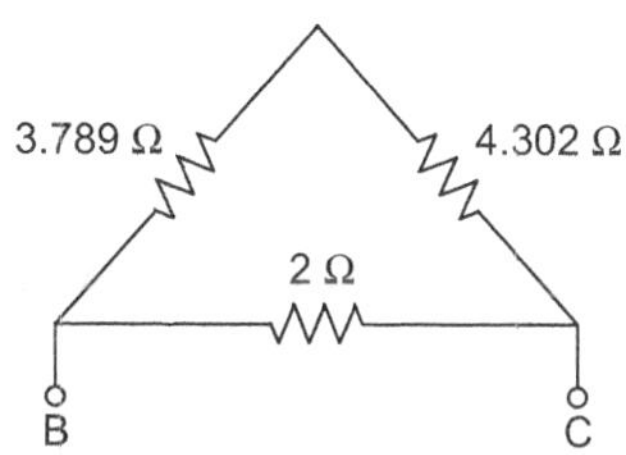

Fig 2.174

Resistance between B and C will be [Fig. 2.174]

$$R_{BC} = [3.789 + 4.302] \,||\, [2]$$

or $$R_{BC} = \frac{8.091 \times 2}{8.091 + 2} = \mathbf{1.604\ \Omega}$$ **Ans.**

Problem 2.53. *Solve for I_1 and I_2 of the network in Fig. 2.175 by nodal analysis.*

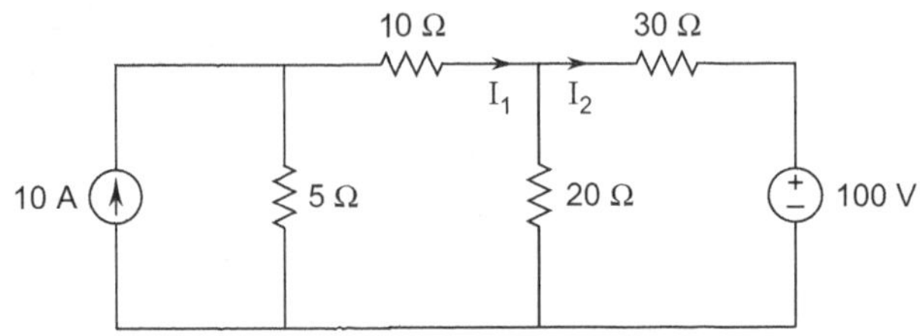

Fig. 2.175

Solution. In the given network, there are four nodes as shown in Fig. 2.175. These four nodes can be reduced to three nodes as shown in Fig. 2.176.

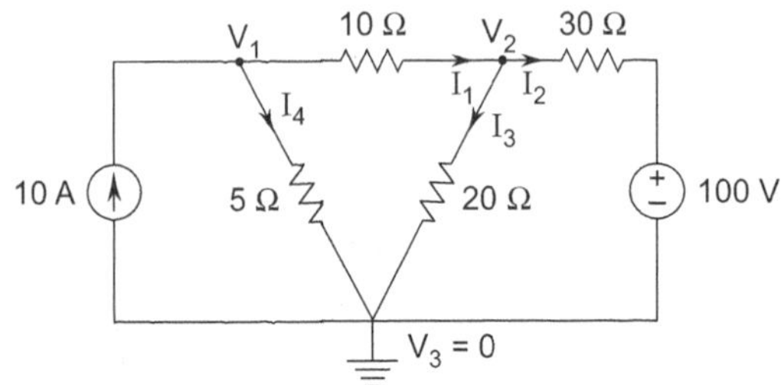

Fig. 2.176

The branch currents are shown in Fig. 2.176

Let the potential at node 1 be V_1

and the potential at node 2 be V_2

Third node is assumed as reference node, hence $V_3 = 0$

Applying Nodal Analysis,

At Node 1 by KCL

$$10 = I_4 + I_1$$

or $$10 = \frac{V_1}{5} + \frac{V_1 - V_2}{10}$$

or $$100 = 2V_1 + V_1 - V_2$$

or $$100 = 3V_1 - V_2 \quad \ldots(i)$$

At Node 2, by KCL

$$I_1 = I_2 + I_3$$

or $$\frac{V_1 - V_2}{10} = \frac{V_2 - 100}{30} + \frac{V_2}{20}$$

or $$6V_1 - 6V_2 = 2V_2 - 200 + 3V_2$$

or $$200 = -6\,V_1 + 11\,V_2 \quad \ldots(ii)$$

Solving equations (*i*) and (*ii*) simultaneously, we get

$$V_1 = 48.15\text{ V}$$

$$V_2 = 44.44\text{ V}$$

∴ The branch currents I_1 and I_2 will be

$$I_1 = \frac{V_1 - V_2}{10} = \frac{48.15 - 44.44}{10}$$

or $I_1 = \mathbf{0.371\ A}$ **Ans.**

or $$I_2 = \frac{V_2 - 100}{30} = \frac{44.44 - 100}{30}$$

or $I_2 = \mathbf{-1.852\ A}$ **Ans.**

Problem 2.54. *Apply Norton's theorem to find the Norton's equivalent circuit as seen by R_L in the circuit shown in Fig. 2.177.*

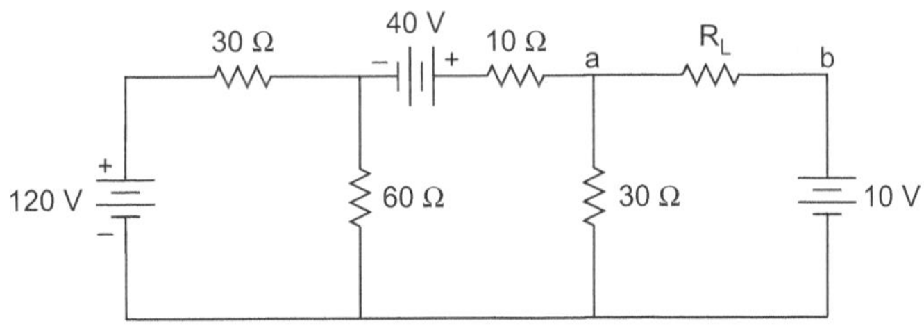

Fig. 2.177

Solution. To find Norton's equivalent resistance R_N (or R_{TH}), remove load resistance.

Remove all voltage sources by short-circuit. Find the resistance across the two terminal *a* and *b* as shown in Fig. 2.178.

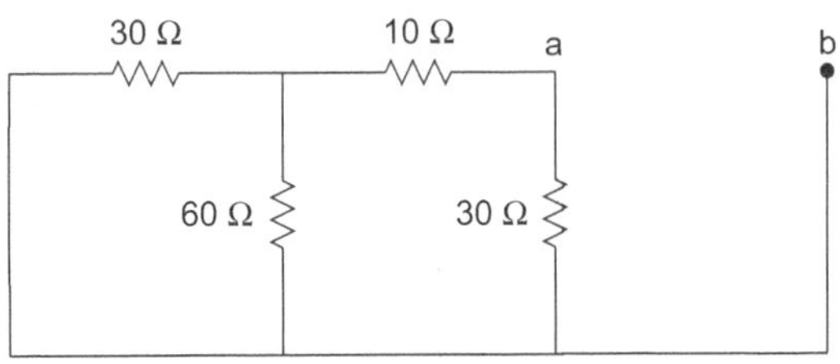

Fig. 2.178

$$R_{ab} = R_N = R_{TH} = [(30 \parallel 60) + 10] \parallel [30]$$

$$= \left[\frac{30 \times 60}{30 + 60} + 10\right] \parallel [30]$$

$$= 30 \parallel 30 = \frac{30 \times 30}{30 + 30} = 15\ \Omega$$

$R_N = \mathbf{15\ \Omega}$ **Ans.**

To find the Norton's equivalent current source, I_N

Remove the load resistance

Short-circuit the two terminals *a* and *b*

Find the short-circuit through branch *ab*.

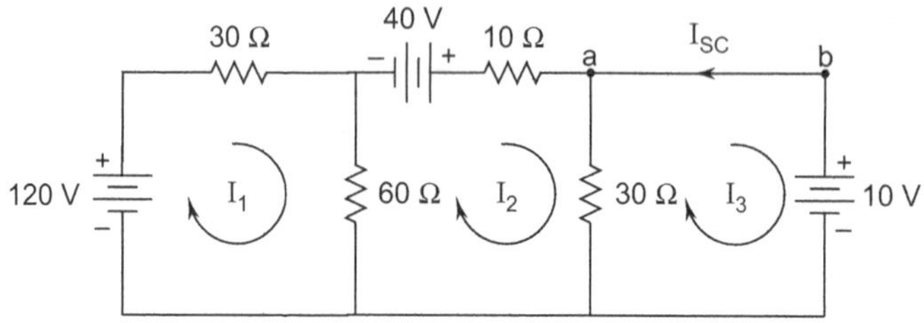

Fig. 2.179

Applying mesh analysis in Fig. 2.179

In mesh 1

$$120 = 90\, I_1 - 60\, I_2$$

or $$4 = 3I_1 - 2\, I_2 + 0\, I_3 \quad \text{...(1)}$$

In mesh 2

$$40 = -\,60\, I_1 + 100\, I_2 - 30\, I_3$$

or $$4 = -\,6\, I_1 + 10\, I_2 - 3\, I_3 \quad \text{...(2)}$$

In mesh 3

$$-\,10 = -\,30\, I_2 + 30\, I_3$$

or $$1 = 0\, I_1 + 3\, I_2 - 3\, I_3 \quad \text{...(3)}$$

Solving equations (1) (2) and (3) by Cramer rule

$$\begin{bmatrix} 3 & -2 & 0 \\ -6 & 10 & -3 \\ 0 & 3 & -3 \end{bmatrix} \begin{bmatrix} I_1 \\ I_2 \\ I_3 \end{bmatrix} = \begin{bmatrix} 4 \\ 4 \\ 1 \end{bmatrix}$$

$$\Delta = 3(-\,30 + 9) + 2\,(18)$$

$$\Delta = -\,27$$

$$\Delta_3 = \begin{bmatrix} 3 & -2 & 4 \\ -6 & 10 & 4 \\ 0 & 3 & 1 \end{bmatrix}$$

$$= 3\,(10 - 12) + 6\,(-\,2 - 12) = -\,6 - 84$$

$$\Delta_3 = -\,90$$

∴ $$I_3 = \frac{\Delta 3}{\Delta} = \frac{-\,90}{-\,27} = 3.33 \text{ A}$$

or $$I_{SC} = -\,I_3 = -\,3.33 \text{ A}$$

∴ $I_N = \mathbf{-\,3.33\ A}$ **Ans.**

Therefore, the Norton's equivalent circuit will be as shown in Fig. 2.180.

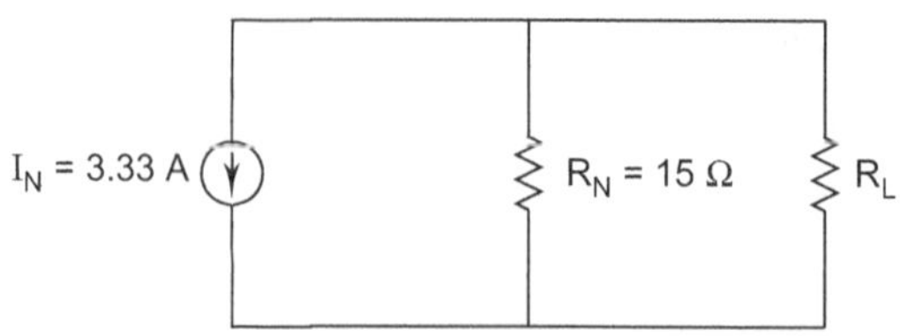

Fig. 2.180

Problem 2.55. *Use Thevenin's theorem to replace the three loop circuit of Fig. 2.181 by a single–loop equivalent circuit in which the identity of R_L is present.*

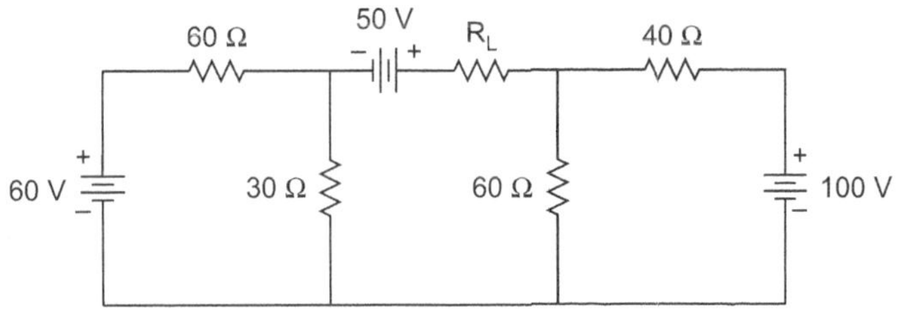

Fig. 2.181

Solution. To find R_{TH}, remove all the voltage sources by short-circuit, as shown in Fig. 2.182.

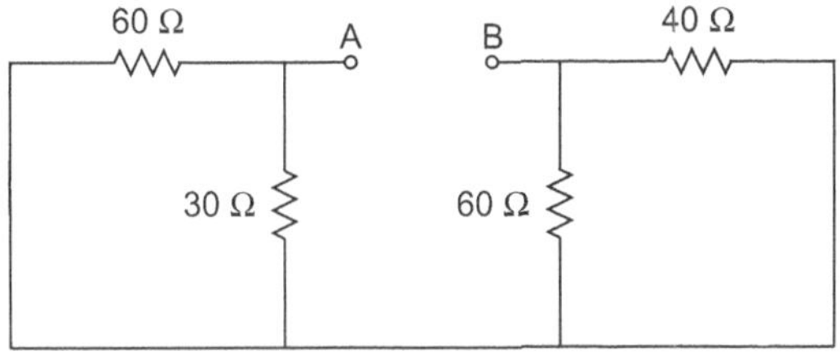

Fig. 2.182

$$R_{AB} = R_{TH} = (60 \,||\, 30) + (40 \,||\, 60)$$

$$= \frac{60 \times 30}{60 + 30} + \frac{40 \times 60}{40 + 60}$$

$$= 20 + 24 = \mathbf{44\ \Omega}$$ **Ans.**

To find V_{TH}, removing load resistance and find the voltage drop across open terminals A and B.

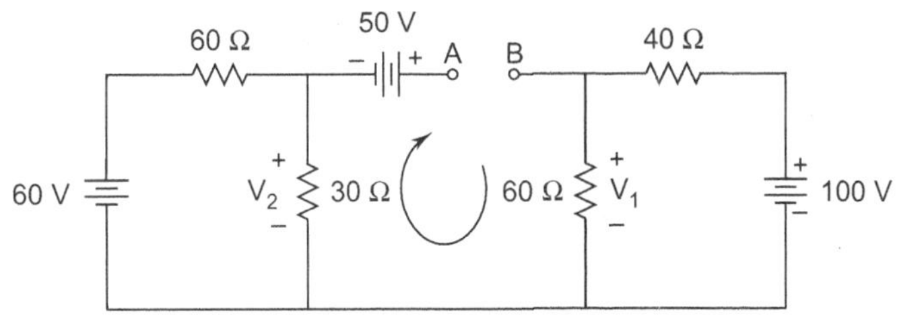

Fig. 2.183

While moving from B and A, we encounter with three elements:

(*i*) *60 Ω resistance*: Voltage drop across 60 Ω resistance is only due to 100 volt source alone. By voltage divider rule

$$V_1 = \frac{100 \times 60}{40 + 60} = 60 \text{ V}$$

or $V_1 = \mathbf{60\ V}$ (negative) **Ans.**

(*ii*) *30 Ω resistance:* Voltage drop across 30 Ω resistance is due to 60 volt source alone.

By voltage divider rule

$$V_2 = \frac{60 \times 30}{60 + 30} = 20 \text{ V}$$

or $V_2 = \mathbf{20\ V}$ (positive) **Ans.**

(*iii*) *50 volt source:* 50 volt source polarity is in the same direction in which we are moving (*i.e.* from B to A)

Therefore, $V_3 = 50$ V (positive)

Hence,

$$V_{TH} = -V_1 + V_2 + V_3$$
$$= -60 + 20 + 50$$

or $V_{TH} = \mathbf{10\ V}$ **Ans.**

Hence, the thevenin's equivalent circuit will be as shown in Fig. 2.184.

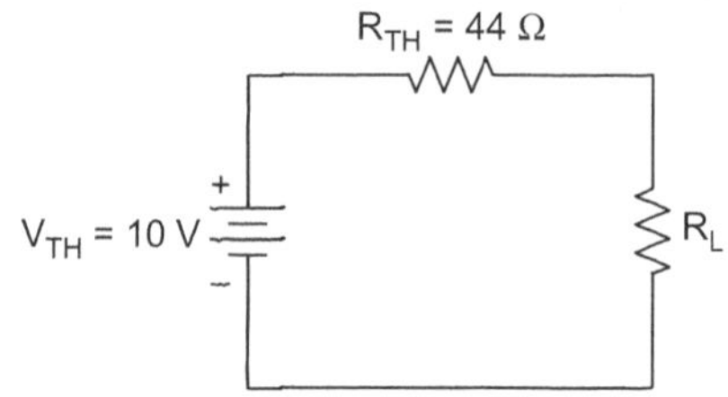

Fig. 2.184

EXERCISE

1. By super position theorem, find the current in R of Fig. 2.185.
2. In Fig. 2.186, find the Thevenin's equivalent voltage source. Also give the value of load resistance to be connected across terminals A and B to give maximum power delivered to it.
3. Find the current drawn I from the source voltage of 100 V in the circuit of Fig. 2.187.

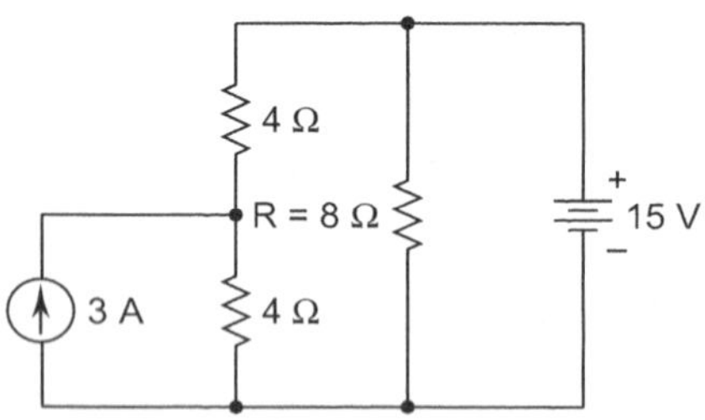

Fig. 2.185

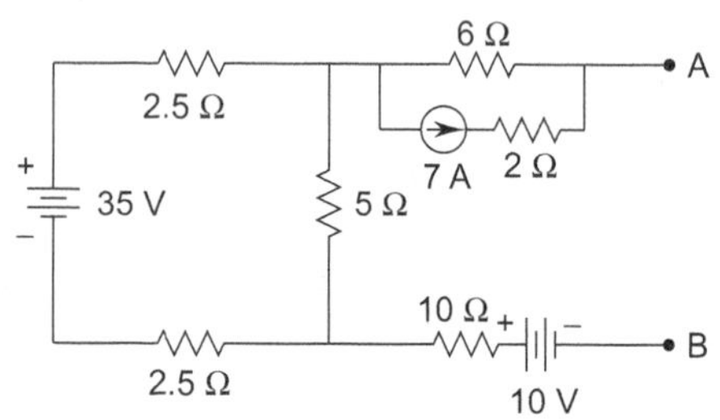

Fig. 2.186

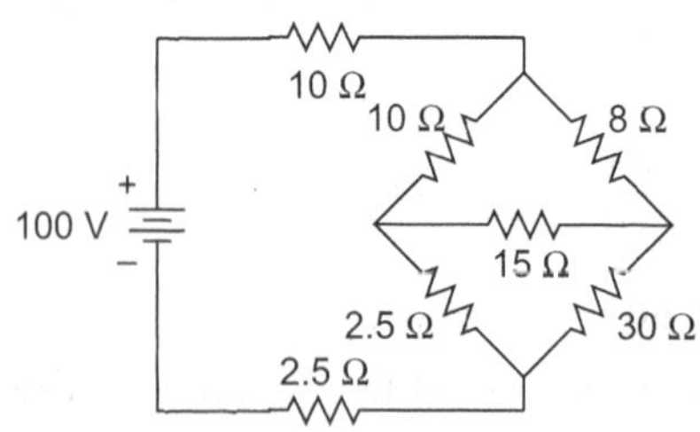

Fig. 2.187

4. In Fig. 2.188 determine the value and direction of current in BD using Thevenin's theorem.
5. Out of Mesh and Nodal Analysis, which method will require least number of equations to solve network shown in Fig. 2.189. Write the equation in matrix form for both the methods.
6. Using loop current (mesh current) method, finds the current I_1 and I_2 in Fig. 2.190.

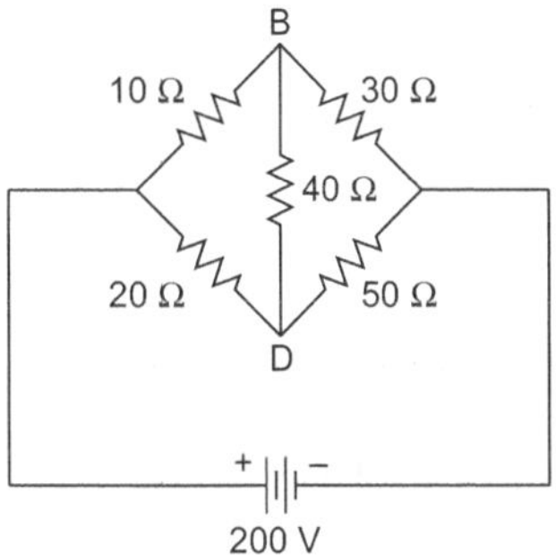

Fig. 2.188

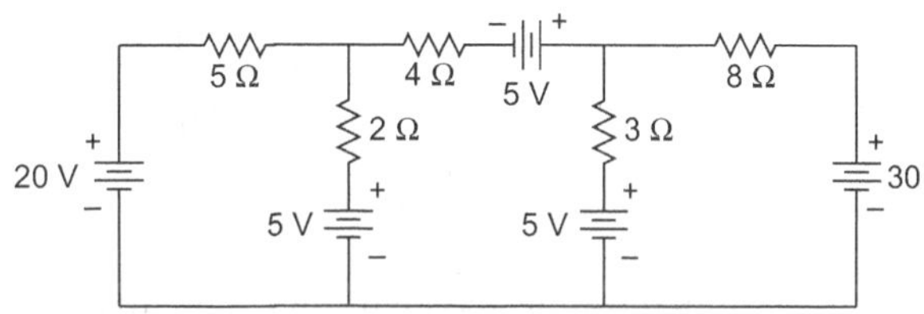

Fig. 2.189

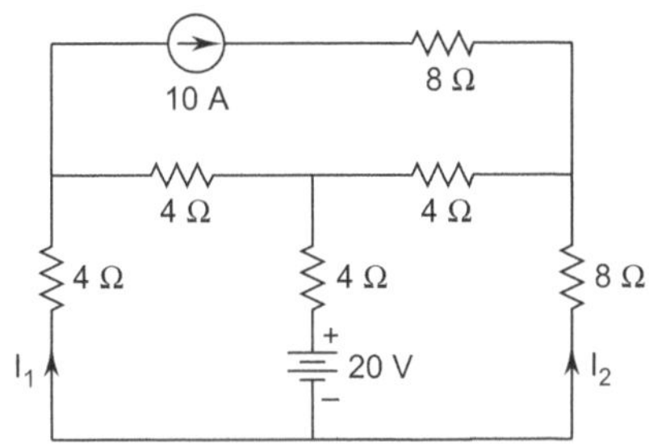

Fig. 2.190

7. Use nodal method analysis to find the current in 1 ohm resistance of circuit shown in Fig. 2.191.
8. Find Thevenin's equivalent circuit across AB shown in Fig. 2.192.
9. For the circuit shown in Fig. 2.193 calculate the current in 8 ohms resistor by using Norton's theorem. Also verify the result by using Thevenin's theorem.

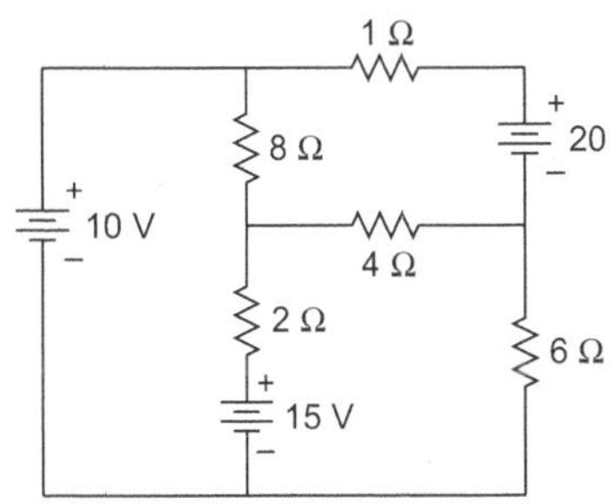

Fig. 2.191

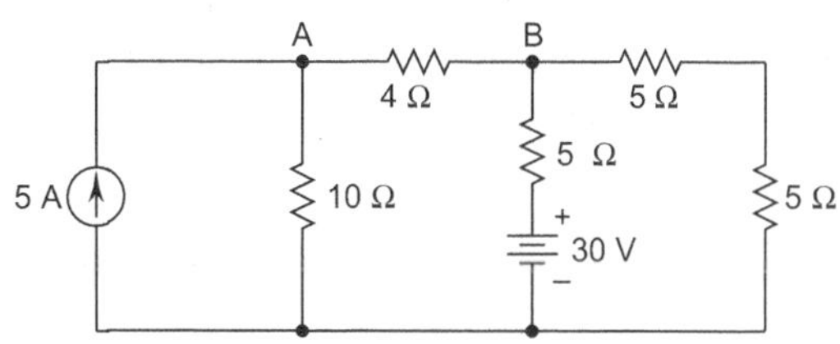

Fig. 2.192

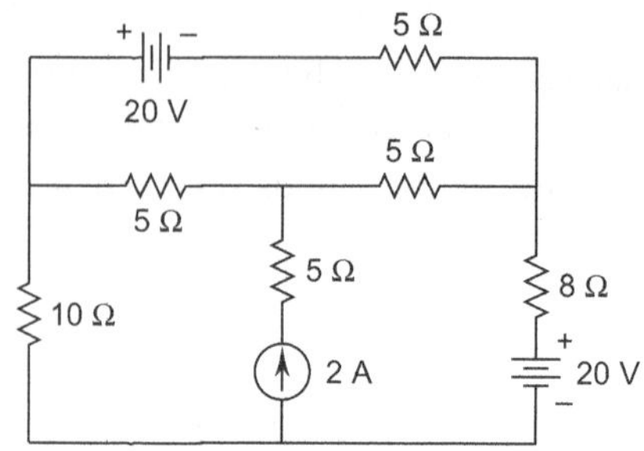

Fig. 2.193

10. Using loop analysis (mesh) determines the current through 5 ohms resistor in the network shown in Fig. 2.194.
11. By using superposition theorem, find the current in the resistances R of Fig. 2.195.
12. By source transformation as well as by mesh current method determine current I in the circuit shown in Fig. 2.196.

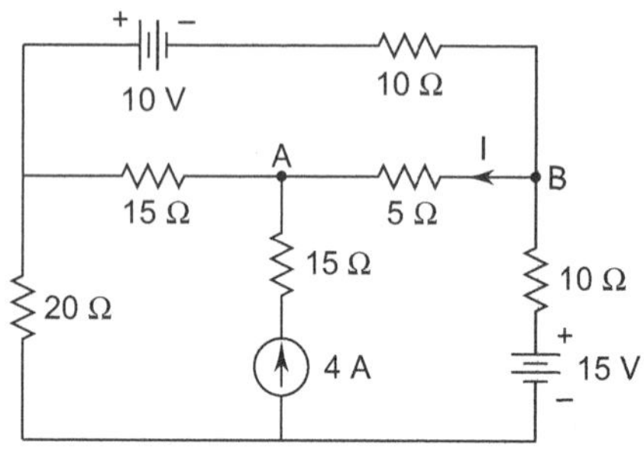

Fig. 2.194

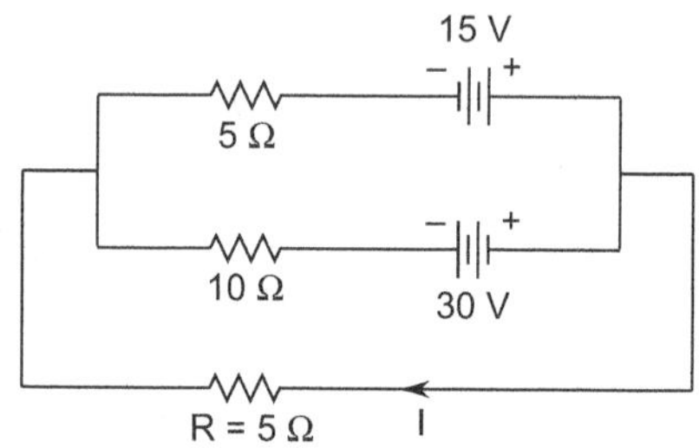

Fig. 2.195

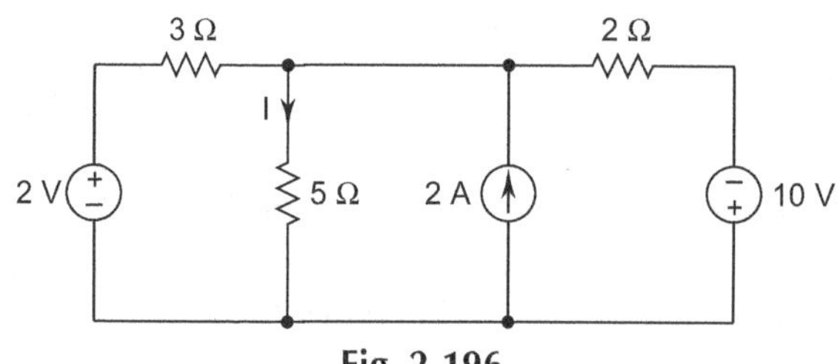

Fig. 2.196

13. By using superposition theorem, find the current I of Fig. 2.197.
14. What is the criterion for star-delta transformation? Develop the relationship for (*i*) Delta-to-Star, and (*ii*) Star-to-Delta transformation.

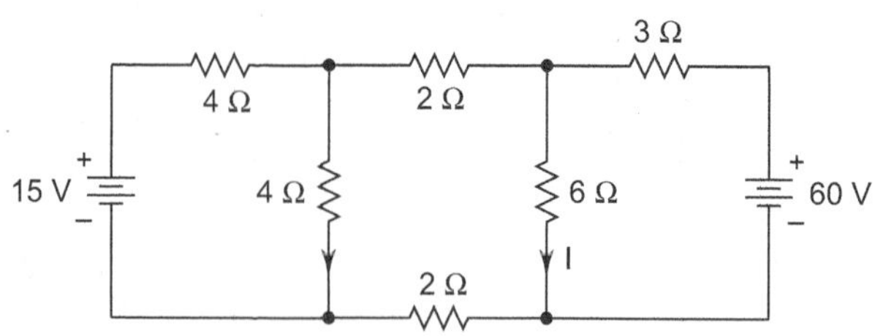

Fig. 2.197

□ □ □

3

AC CIRCUITS CONCEPTS

3.1 INTRODUCTION

The electrical quantities which vary or changes continuously with time are known as Alternating quantities. Alternating voltage and current are some of those quantities.

The magnitude of these alternating quantities changes continuously with time and the direction also vary or changes after definite time period. These alternating quantities may have square waveform, triangular waveform, sawtooth waveform, sinusoidal waveform etc.

Here we are concern mainly with sinusoidal waveforms. These alternating sinusoidal quantities will be applied to electrical circuits having different configuration of components and their steady state behaviour will be studied.

3.2 REPRESENTATION OF SINUSOIDAL ALTERNATING VOLTAGE AND CURRENT

When an alternator produces AC voltage, the voltage switches polarity overtime, but does so in a very particular manner. When graphed overtime, the "wave" traced by this voltage of alternating polarity from an alternator takes on a distinct shape, known as a sine wave (Fig. 3.1).

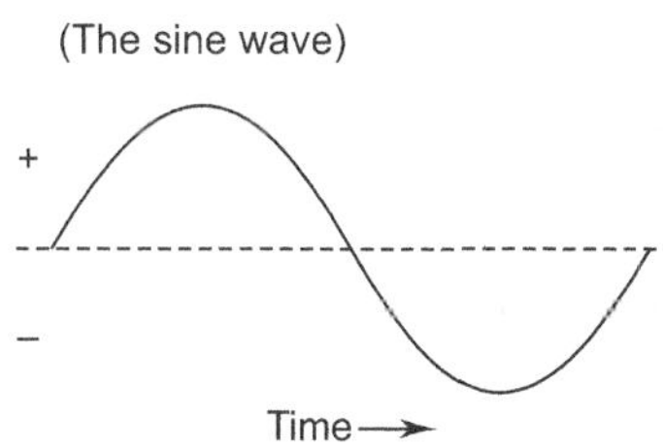

Fig. 3.1

3.2.1 Graph of AC Voltage Overtime (The Sine Wave)

In the voltage plot from an electromechanical alternator, the change from one polarity to the other is a smooth one, the voltage level changing most rapidly at the zero ('crossover') point and most slowly at its peak. If we were to graph the trigonometric function of 'sine' over a horizontal range of 0 to 360 degrees, we would find the exact same pattern as in Table 3.1.

Table 3.1

Angle (°)	sin (angle)	Wave	Angle (°)	sin (angle)	Wave
0	0.0000	zero	180	0.0000	zero
15	0.2588	+	195	– 0.2588	–
30	0.5000	+	210	– 0.5000	–
45	0.7071	+	225	– 0.7071	–
60	0.8660	+	240	– 0.8660	–
75	0.9659	+	255	– 0.9659	–
90	1.0000	+ peak	270	– 1.0000	– peak
105	0.9659	+	285	– 0.9659	–
120	0.8660	+	300	– 0.8660	–
135	0.7071	+	315	– 0.7071	–
150	0.5000	+	330	– 0.5000	–
165	0.2588	+	345	– 0.2588	–
180	0.0000	zero	360	0.0000	zero

The reason why an electromechanical alternator outputs sine wave AC is due to the physics of its operation. The voltage produced by the stationary coils by the motion of the rotating magnet is proportional to the rate at which the magnetic flux is changing perpendicular to the coils (Faraday's law of electromagnetic induction). That rate is greatest when the magnet poles are closest to the coils and least when the magnet poles are furthest away from the coils. Mathematically, the rate of magnetic flux change due to a rotating magnet follows that of a sine function, so the voltage produced by the coils follows that same function.

A sinusoidal alternating voltage changes polarity after every half cycle (Fig. 3.2). It can be represented by equation

$$v = V_m \sin \omega t \qquad \ldots(3.1)$$

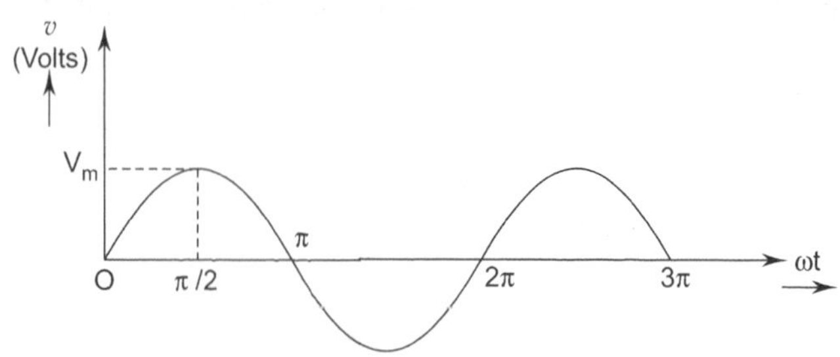

Fig. 3.2

where, v = Instantaneous value of alternating voltage,

V_m = Maximum value of alternating voltage,

ω = Angular velocity = $2\pi f$, and

f = Frequency in **Hz**.

Similarly a sinusoidal alternating current is also cyclic in nature. After every half cycle the direction of flow of current changes. It can be represented by equation

$$i = I_m \sin \omega t \quad \ldots(3.2)$$

where, i = Instantaneous value of alternating current,

I_m = Maximum value of alternating current.

While electromechanical alternators and many other physical phenomena naturally produce sine waves, this is not the only kind of alternating wave in existence. Other 'waveforms' of AC are commonly produced within electronic circuitry. Here are a few sample waveforms and their common designations in Fig. 3.3.

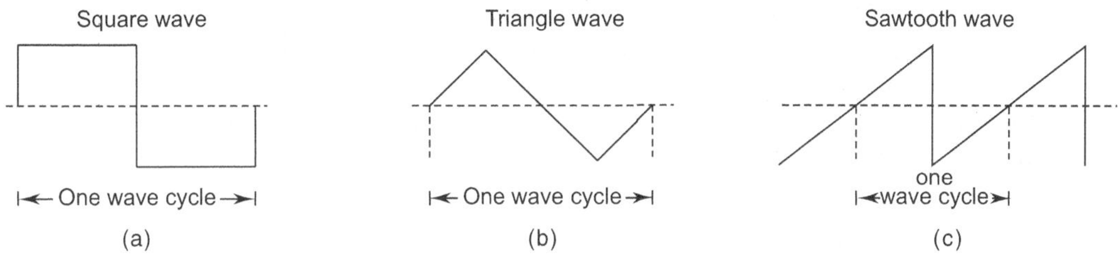

Fig. 3.3

Apart from these waveforms, these are also other kind of waveforms in existence. These are simply a few that are common enough to have been given distinct names. Even in circuits that are supposed to manifest 'pure' sine, square, triangle, or sawtooth voltage/current waveforms, the real-life result is often a distorted version of the intended wave shape. Some waveforms are so complex that they defy classification as a particular 'type' (including waveforms associated with many kinds of musical instruments). Generally speaking, any wave shape bearing close resemblance to a perfect sine wave is termed *sinusoidal*, anything different being labelled as *non-sinusoidal*. Being that the waveform of an AC voltage or current is crucial to its impact in a circuit, we need to be aware of the fact that AC waves come in a variety of shapes.

3.3 GENERATION OF SINUSOIDAL ALTERNATING VOLTAGE AND CURRENT

According to Faraday's law of electromagnetic induction when a coil rotates in a magnetic field, an e.m.f. is induced in it, which is given by

$$e = -N\frac{d\Phi}{dt} \quad \ldots(3.3)$$

An alternating voltage may be generated:

(*a*) By rotating a coil at constant angular velocity in a uniform magnetic field as shown in Fig. 3.4.

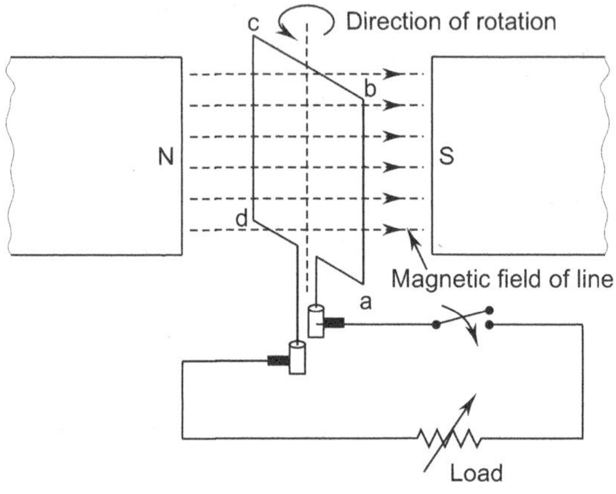

Fig. 3.4

(*b*) By rotating a magnetic field at a constant angular velocity within a stationary coil as shown in Fig. 3.5. If a machine is constructed to rotate a magnetic field around a set of stationary wire coils with the turning of a shaft, AC voltage will be produced across the wire coils as that shaft is rotated, in accordance with Faraday's law of electromagnetic induction. This is the basic operating principle of an AC generator, also known as analternator (Fig. 3.5).

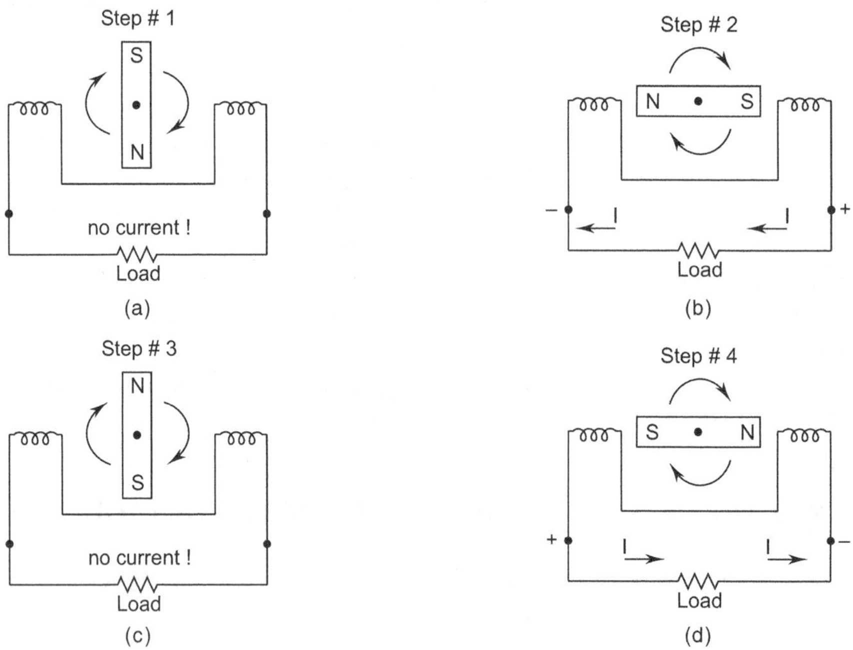

Fig. 3.5

Let us consider the first case (Fig. 3.4) in which a coil is rotated in the stationary magnetic field.

As the coil is rotated in anticlockwise direction, it cut the lines of magnetic field and intern e.m.f. is generated in the coil.

When the coil (abcd) is parallel to the magnetic field minimum magnetic field lines are cut, and hence no e.m.f. is induced in the coil. As the coil moves in anticlockwise direction, the angle between the coil and the magnetic field lines increases. Hence flux cut will be more and more e.m.f. is induced in the coil. When the coil is perpendicular to the magnetic field, maximum lines are cut and hence maximum e.m.f is generated.

Let the generating system have the following specifications:

Number of turns of the coil = N

Area of the coil = A

Flux density of the magnetic field = B

Angular speed of the coil = ω

The flux linked with the coil = Φ

The flux linked with the coil can be expressed as

$$\Phi = \vec{B} . \vec{A} = BA \cos\theta \qquad ...(3.4)$$

As $\theta = \omega t$, equation (3.4) will become

$$\Phi = BA \cos \omega t \qquad ...(3.5)$$

From equation (3.5), equation (3.3) will become

$$e = -N \frac{d(BA \cos \omega t)}{dt} \qquad ...(3.6)$$

or

$$e = \omega NBA \sin \omega t \qquad ...(3.7)$$

The generated e.m.f. will be maximum when

$$\sin \omega t = 1 \qquad ...(3.8)$$

And that maximum e.m.f. is given by

$$E_m = \omega NBA \qquad ...(3.9)$$

From equation (3.9), equation (3.7) will become

$$e = E_m \sin \omega t \qquad ...(3.10)$$

This is an equation of alternating sinusoidal induced voltage.

where e = Instantaneous value of induced voltage.

E_m = Maximum value of induced voltage.

Problem 3.1. *An alternating current is given by i = 20 sin 600t Amperes. Find the*

(i) frequency

(ii) peak value of current and

(iii) time taken from t = 0 for the current to reach a value of 10 A.

Solution. (*i*) Frequency, $f = \frac{\omega}{2\pi} = \frac{600}{2\pi}$ = **99.5 Hz** **Ans.**

(*ii*) Peak value of current, I_{max} = **20 A** **Ans.**

(*iii*) Time taken from t = 0 for the current to reach a value of 10 A,

$$t = \frac{1}{600} \sin^{-1} \frac{10}{20} = \frac{\pi}{6 \times 600} \text{ s} = \frac{\pi}{3600} \text{ s}$$

t = **0.873 ms** **Ans.**

3.4 AVERAGE VALUE OF ALTERNATING QUANTITIES

Mathematically, average value is given by

$$\text{Average value} = \frac{\text{Area under the curve}}{\text{Base}} \quad \ldots(3.11)$$

The average value of an alternating quantity (*whether it is sinusoidal or square or triangular*) if calculated over a complete cycle will be zero because it has identical positive and negative half cycles. Hence the average value of an alternating quantity is calculated over a half cycle.

The sinusoidal alternating current is given by

$$i = I_m \sin \omega t \quad \text{[Eqn. (3.2)]}$$

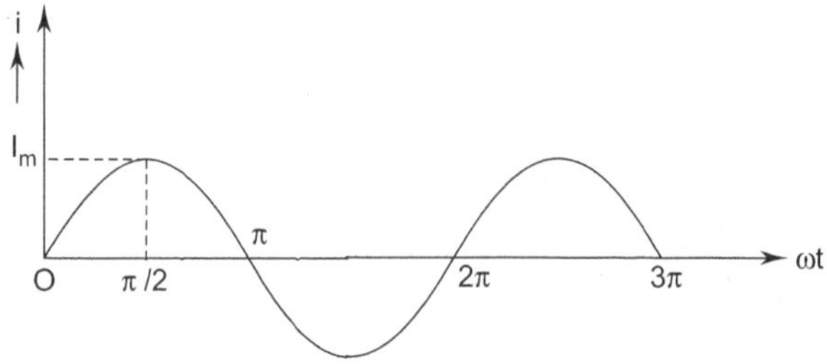

Fig. 3.6

For an alternating sinusoidal quantity

$$\text{Average value } (I_{av}) = \frac{\text{Area of one alternation}}{\text{Base length of one alteration}} \quad \ldots(3.12)$$

or
$$I_{av} = \frac{1}{\pi}\left[\int_0^{\pi} i\, d(\omega t)\right] \quad \ldots(3.13)$$

or
$$= \frac{1}{\pi}\left[\int_0^{\pi} I_m \sin(\omega t)\, d(\omega t)\right] \text{ [From equation (3.2)]} \quad \ldots(3.14)$$

or
$$I_{av} = \frac{I_m}{\pi}\left|-\cos \omega t\right|_0^{\pi}$$

$$= \frac{I_m}{\pi}\left(-\cos 180° + \cos 0°\right)$$

or
$$\boxed{I_{av} = \frac{2I_m}{\pi} = 0.637\, I_m} \quad \ldots(3.15)$$

Similarly for sinusoidal alternating voltage the average value is given by

$$\boxed{V_{av} = \frac{2V_m}{\pi} = 0.637\, V_m} \quad \ldots(3.16)$$

Problem 3.2. *A sinusoidally varying alternating current of frequency 60 Hz has a maximum value of 15 amperes.*

(i) Write down equation for instantaneous value of current.

(ii) Find value of current after 1/200 second.

(iii) Find time taken to reach 10 amperes for the first time.

(iv) Find its average value.

Solution. Maximum value of current, $I_m = 15$ A

Frequency of alternating current, $f = 60$ Hz

(*i*) The equation for instantaneous value of sinusoidal alternating current with θ as zero when time is zero, is given as

$$i = I_m \sin 2\pi ft$$
$$= 15 \sin 2\pi \times 60t$$
$$i = \mathbf{15 \sin 120\, \pi t\ A} \qquad \textbf{Ans.}$$

(*ii*) Substituting $t = \frac{1}{200}$ second in the expression $i = 15 \sin 120\, \pi t$, we have value of current after $\frac{1}{200}$ second

$$= 15 \sin 120\, \pi \times \frac{1}{200} = 15 \sin 0.6\, \pi$$

or $$i = \mathbf{14.266\ A} \qquad \textbf{Ans.}$$

(*iii*) Let the instantaneous value of current be 10 A for the first time t seconds after the instant the current is zero and becoming positive

$$10 = 15 \sin 120\, \pi t$$

or $$t = \frac{1}{120\pi} \sin^{-1} \frac{10}{15} = \frac{1}{120\pi} \sin^{-1} 0.6667 = \mathbf{0.001936\ s.} \qquad \textbf{Ans.}$$

(*iv*) Average value, $$I_{av} = \frac{2}{\pi} I_m = \frac{2}{\pi} \times 15 = \mathbf{9.55\ A} \qquad \textbf{Ans.}$$

3.5 ROOT MEAN SQUARE (R.M.S) OR EFFECTIVE VALUE

The effective or r.m.s value of an alternating current is *"**that steady current (DC) which when flowing through a given resistance for a given time produces the same amount of heat as produced by the alternating current (AC) when flowing through the same resistance for same time**"*.

For example, when we say that the r.m.s or effective value of an alternating current is 5 A, it means that the alternating current will do work or produce heat at the same rate as 5 A direct current under similar conditions.

The r.m.s value of non-sinusoidal quantities can be calculated by Graphical method

$$I_{rms} = \sqrt{\frac{i_1^2 + i_2^2 + i_3^2 + \ldots + i_n^2}{n}} \qquad \ldots(3.17)$$

The r.m.s value of sinusoidal quantities **over a complete cycle** can be calculated by Integral method

$$I_{rms} = \left[\frac{1}{2\pi} \int_0^{2\pi} i^2 d(\omega t) \right]^{1/2} \qquad \ldots(3.18)$$

From equation (3.2),

$$i = I_m \sin \omega t \qquad \text{[Eqn. (3.2)]}$$

$$I_{rms} = \left[\frac{1}{2\pi}\int_0^{2\pi} (I_m \sin \omega t)^2 \, d(\omega t)\right]^{1/2} \qquad ...(3.19)$$

or

$$= \left[\frac{1}{2\pi}\int_0^{2\pi} I_m^2 \sin^2 \omega t \, d(wt)\right]^{1/2} \qquad ...(3.20)$$

or

$$= \left[\frac{I_m^2}{2\pi}\int_0^{2\pi} \frac{(1-\cos 2\omega t)}{2} d(\omega t)\right]^{1/2} \qquad ...(3.21)$$

or

$$= \left[\frac{I_m^2}{4\pi}\left| \omega t - \frac{\sin 2\,\omega t}{2}\right|_0^{2\pi}\right]^{1/2} \qquad ...(3.22)$$

or

$$= \left[\frac{I_m^2}{4\pi}\left(2\pi - \frac{\sin 4\pi}{2}\right) - (0-0)\right]^{1/2} \qquad ...(3.23)$$

or

$$\boxed{I_{rms} = \frac{I_m}{\sqrt{2}}} \qquad ...(3.24)$$

Similarly for sinusoidal alternating voltage the r.m.s value is given by

$$\boxed{V_{rms} = \frac{V_m}{\sqrt{2}}} \qquad ...(3.25)$$

3.6 FORM FACTOR

Form factor of an alternating quantity is defined as the ratio of r.m.s value of an alternating quantity to the maximum value of that alternating quantity.

$$\boxed{K_f = \frac{\text{r.m.s value}}{\text{Average value}}} \qquad ...(3.26)$$

For a ***sinusoidal alternating current***

$$K_f = \frac{0.707 I_m}{0.637 I_m} \qquad ...(3.27)$$

or

$$\boxed{K_f = 1.11} \qquad ...(3.28)$$

From the definition of form factor it is clear that form factor enable us to find the r.m.s value of an sinusoidal alternating quantity if average value of an sinusoidal alternating quantity is given or vice versa.

It is to noted that the constant value of the form factor ($K_f = 1.11$) is for sinusoidal alternating quantity only and not for any other alternating quantity.

3.7 PEAK FACTOR

Peak factor is also known as crest factor or amplitude factor. It is defined as the ratio of maximum value of an alternating quantity to the r.m.s value of that alternating quantity

$$\boxed{K_p = \frac{\text{Maximum value}}{\text{r.m.s value}}} \qquad \ldots(3.29)$$

For a ***sinusoidal alternating current***

$$K_p = \frac{I_m}{0.707 I_m} \qquad \ldots(3.30)$$

or

$$\boxed{K_p = 1.414} \qquad \ldots(3.31)$$

It is to noted that the constant value of the Peak factor ($K_p = 1.414$) is for sinusoidal alternating quantity only and not for any other alternating quantity. Notations of these factors may vary from book to book.

The peak factor is of much greater importance because it indicates the maximum voltage being applied to the various parts of the apparatus. For instance when an alternating voltage is applied across a cable, the breakdown of the insulation will depend upon the maximum voltage. The insulation must be able to withstand the maximum value of the voltage rather than the r.m.s value.

Problem 3.3. *Compute the average value, r.m.s (or effective) value, form factor and peak factor for the square wave shape shown in Fig. 3.7.*

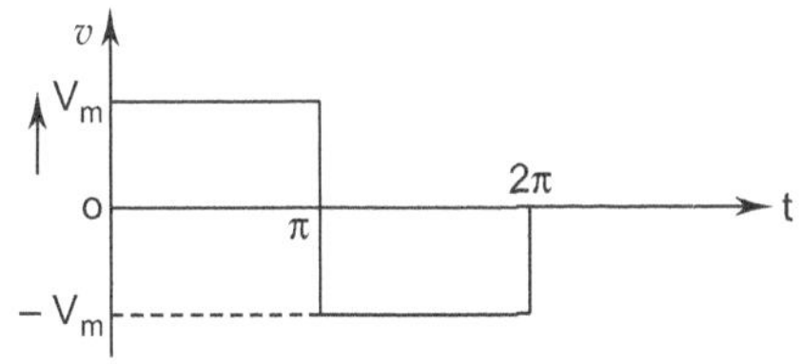

Fig. 3.7

Solution. (*i*) Average Value

For an alternating quantity average value is given by

$$V_{av} = \frac{1}{T}\left[\int_0^T v\,dt\right]$$

$$= \frac{1}{\pi}\left[\int_0^\pi V_m\,dt\right]$$

$$V_{av} = V_m\frac{(\pi - 0)}{\pi} = \mathbf{V_m} \qquad \textbf{Ans.}$$

i.e., the average value of an square waveform is equal to the maximum value.

(*ii*) RMS Value

$$V_{rms} = \left[\frac{1}{T}\int_0^T v^2\, dt\right]^{1/2}$$

$$V_{rms} = \left[\frac{1}{\pi}\int_0^{\pi} V_m^2\, dt\right]^{1/2}$$

$$V_{rms} = \left[V_m^2\,\frac{(\pi - 0)}{\pi}\right]^{1/2} = \mathbf{V}_m \qquad \textbf{Ans.}$$

i.e., the r.m.s value of an square waveform is equal to the maximum value.

(*iii*) Form Factor

Form factor of an alternating quantity is given by

$$K_f = \frac{\text{r.m.s value}}{\text{Average value}}$$

For the square waveform

$$K_f = \frac{V_m}{V_m} = \mathbf{1} \qquad \textbf{Ans.}$$

(*iv*) Peak Factor

Peak factor of an alternating quantity is given by

$$K_p = \frac{\text{Maximum value}}{\text{r.m.s value}}$$

For the square waveform

$$K_p = \frac{V_m}{V_m} = \mathbf{1} \qquad \textbf{Ans.}$$

Problem 3.4. *Compute average value, r.m.s (or effective) value, form factor and peak factor for the half wave rectified sinusoidal alternating voltage shown in Fig. 3.8.*

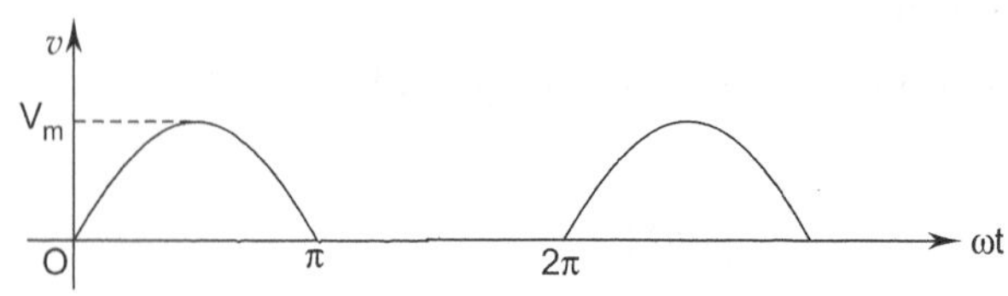

Fig. 3.8

Solution. (*i*) Average Value

$$V_{av} = \frac{1}{T}\left[\int_0^T v\, d\omega t\right]$$

$$= \frac{1}{2\pi}\left[\int_0^{\pi} V_m \sin\omega t\, d\omega t + \int_{\pi}^{2\pi} 0\, d\omega t\right] = \frac{V_m}{2\pi}\Big|-\cos\omega t\Big|_0^{\pi}$$

$$V_{av} = \frac{V_m}{\pi} = \mathbf{0.318\ V_m} \qquad \textbf{Ans.}$$

(*ii*) RMS Value

$$V_{rms} = \left[\frac{1}{T}\int_0^{T} v^2\, d\omega t\right]^{1/2} = \left[\frac{1}{2\pi}\int_0^{\pi} V_m^2 \sin^2 \omega t\, d\omega t\right]^{1/2}$$

$$= \left[\frac{V_m^2}{2\pi}\int_0^{\pi}\left(1-\cos^2\omega t\right) d\omega t\right]^{1/2} = \left[\frac{V_m^2}{2\pi}\left|\omega t - \frac{\sin^2\omega t}{2}\right|_0^{\pi}\right]^{1/2}$$

$$V_{rms} = \frac{V_m}{2} = \mathbf{0.5\ V_m} \qquad \textbf{Ans.}$$

(*iii*) Form Factor

Form factor for the half wave rectified waveform is

$$K_f = \frac{\text{r.m.s. value}}{\text{Average value}} = \frac{0.5\ V_m}{0.318\ V_m} = \mathbf{1.572} \qquad \textbf{Ans.}$$

(*iv*) Peak Factor

Peak factor for the half wave rectified waveform is

$$K_p = \frac{\text{Maximum value}}{\text{r.m.s value}} = \frac{V_m}{0.5\ V_m} = \mathbf{2} \qquad \textbf{Ans.}$$

Problem 3.5. *For the triangular wave shape shown in Fig. 3.9 (a), find the average value, r.m.s (or effective) value, form factor and peak factor.*

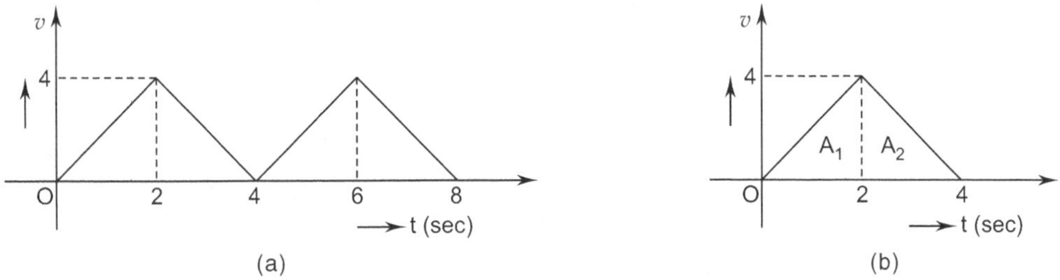

Fig. 3.9

Solution. (*i*) Average Value

The triangular wave shape as shown in Fig. 3.9 has the time period of 4 seconds (as for finding average value we always consider half wave). The average value can be calculated as

$$V_{av} = \frac{1}{T}\left[\int_0^T V_m\,dt\right] = \frac{1}{4}\left[\int_0^2 v_1(t)dt + \int_2^4 v_2(t)dt\right]$$

where $v_1(t)$ is given as

$$y = mx$$

or $v_1(t) = 2t$ (2 is the slop of the line)

Hence area A_1 is calculated as [Fig. 3.9(*b*)]

$$A_1 = \left[\int_0^2 2t\,dt\right] = \left[\frac{2t^2}{2}\right]_0^2 = 4$$

And $v_2(t)$ is given as

$$y = mx + c$$

$v_2(t) = -2t + c$ (– 2 is the slop of the line)

At $t = 4$, $v_2(t) = 0$

Therefore, $c = 8$

And $v_2(t) = -2t + 8$

Hence area A_2 is calculated as

$$A_2 = \left[\int_2^4 (-2t + 8)\,dt\right] = \left[-\frac{2t^2}{2} + 8t\right]_2^4 = 4$$

$$V_{av} = \frac{1}{4}[A_1 + A_2] = \frac{1}{4}[4 + 4] = \mathbf{2\ V}$$ **Ans.**

In general the average value of a triangular wave is

$$\boxed{V_{av} = \frac{V_m}{2} = 0.5\,V_m}$$

Note. *Remember that for a symmetrical wave shape $A_1 = A_2$.*

(*ii*) RMS Value

$$V_{rms} = \left[\frac{1}{4}\left\{\int_0^2 v_1^2(t)dt + \int_2^4 v_2^2(t)dt\right\}\right]^{1/2} = \left[\frac{1}{4}\left\{\int_0^2 (2t)^2 dt + \int_2^4 (-2t+8)^2\,dt\right\}\right]^{1/2}$$

or

$$= \left[\frac{1}{4}\left\{\int_0^2 4t^2 dt + \int_2^4 \left(4t^2 + 64 - 32t\right)dt\right\}\right]^{1/2}$$

or

$$= \left[\frac{1}{4}\left\{\left(\frac{4t^3}{3}\right)_0^2 + \left(\frac{4t^3}{3} + 64t - \frac{32t^2}{2}\right)_2^4\right\}\right]^{1/2}$$

or $V_{rms} = \mathbf{2.30\ V}$ **Ans.**

In general the r.m.s value of a triangular wave is

$$V_{rms} = \frac{V_m}{\sqrt{3}} = 0.5773\ V_m$$

(iii) Form Factor

Form factor for the triangular wave form is

$$K_f = \frac{\text{r.m.s value}}{\text{Average value}} = \frac{\frac{V_m}{\sqrt{3}}}{\frac{V_m}{2}} = \mathbf{1.155} \qquad \textbf{Ans.}$$

(iv) Peak Factor

Peak factor for the triangular waveform is

$$K_p = \frac{\text{Maximum value}}{\text{r.m.s value}} = \frac{V_m}{\frac{V_m}{\sqrt{3}}} = \sqrt{3} \qquad \textbf{Ans.}$$

Problem 3.6. *Compute the average value, r.m.s (or effective) value, form factor and peak factor for the full wave rectified sinusoidal alternating voltage shown in Fig. 3.10.*

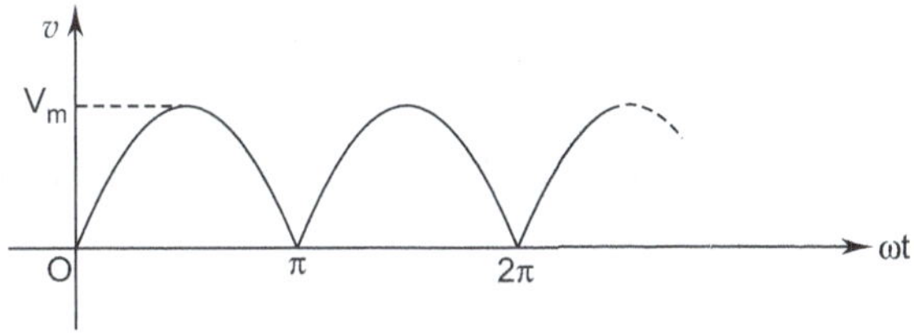

Fig. 3.10

Solution. (i) Average Value

$$V_{av} = \frac{1}{T}\left[\int_0^T v\, d\omega t\right] = \frac{1}{\pi}\left[\int_0^\pi V_m \sin \omega t\, d\omega t\right]$$

$$= \frac{V_m}{\pi}\left|-\cos \omega t\right|_0^\pi$$

$$V_{av} = \frac{2V_m}{\pi} = \mathbf{0.637\ V_m} \qquad \textbf{Ans.}$$

(ii) RMS Value

$$V_{rms} = \left[\frac{1}{T}\int_0^T v^2\, d\omega t\right]^{1/2} = \left[\frac{1}{\pi}\int_0^\pi V_m^2 \sin^2 \omega t\, d\omega t\right]^{1/2}$$

$$= \left[\frac{V_m^2}{\pi}\int_0^\pi \left(1 - \cos^2 \omega t\right) d\omega t\right]^{1/2} = \left[\frac{V_m^2}{\pi}\left|\omega t - \frac{\sin^2 \omega t}{2}\right|_0^\pi\right]^{1/2}$$

$$V_{rms} = \frac{V_m}{\sqrt{2}} = \mathbf{0.707\ V_m} \qquad \textbf{Ans.}$$

(*iii*) Form Factor

Form factor for the full wave rectified sinusoidal alternating voltage waveform is

$$K_f = \frac{\text{r.m.s value}}{\text{Average value}} = \frac{0.707\, V_m}{0.637\, V_m} = \mathbf{1.11} \qquad \textbf{Ans.}$$

(*iv*) Peak Factor

Peak factor for full wave rectified sinusoidal alternating voltage waveform is

$$K_p = \frac{\text{Maximum value}}{\text{r.m.s value}} = \frac{V_m}{0.707\, V_m} = \mathbf{1.414} \qquad \textbf{Ans.}$$

3.8 POWER IN AC CIRCUITS

Suppose that the voltage and current in a circuit are out of phase and the phase difference between them be angle ϕ.

Apparent Power

The product of the applied voltage and total current is known as apparent power. It is also known as real power.

$$\textbf{Apparent Power} = \textbf{VI} \qquad \text{Volt Ampere (VA)} \qquad ...(3.32)$$

Active Power

The power actually used in the circuit (or dissipated in the form of heat) is known as active power. It is also known as true power.

$$\textbf{Active Power} = \textbf{VI}\cos\phi \qquad \text{Watt (W)} \qquad ...(3.33)$$

Reactive Power

Power which is returned to the source by reactive components (for *e.g.*, inductor, capacitor etc.) in the circuit is called reactive power.

$$\textbf{Reactive Power} = \textbf{VI}\sin\phi \qquad \text{Volt Ampere Reactive (VAR)} \qquad ...(3.34)$$

3.9 COMPLEX NOTATION

All the phasors related to electrical quantities can be resolve into two components namely horizontal component and vertical component. Accordingly, these phasors can be expressed in four forms:

- Polar Form
- Rectangular Form
- Trigonometric Form
- Exponential Form

Out of these four types, two are the most important and commonly used forms to express the phasors. They are polar form and rectangular form. These will be discussed in detail in the next articles.

3.9.1 Polar Form

The voltage and current phasor need not always be along X or Y-axis as indicated in Fig. 3.11.

Fig. 3.11

They may be at an inclined direction also (as in case of RL, RC and RLC series circuits which will be discussed in the next chapter). Let us consider four different types of voltage equations.

$$v_1 = V_1 \sin \omega t \qquad \text{where } \theta_1 = 0° \qquad ...(3.35)$$

$$v_2 = V_2 \sin (\omega t + 45°) \qquad \text{where } \theta_2 < 90° \qquad ...(3.36)$$

$$v_3 = V_3 \sin (\omega t + 90°) \qquad \text{where } \theta_3 = 90° \qquad ...(3.37)$$

$$v_4 = V_4 \sin (\omega t - 60°) \qquad \text{where } \theta_4 \text{ is negative} \qquad ...(3.38)$$

In Fig. 3.12 these four phasors are represented.

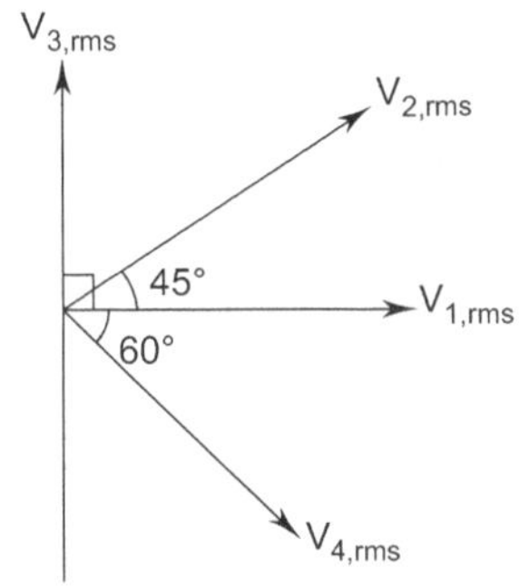

Fig. 3.12

Note. *While drawing the phasors, their r.m.s values are consider to represent the voltages.*

These equations in polar form can be represented as

$$v_1 = |V_1| \angle 0 \qquad ...(3.39)$$

$$v_2 = |V_2| \angle 45° \qquad ...(3.40)$$

$$v_3 = |V_3| \angle 90° \qquad ...(3.41)$$

$$v_4 = |V_4| \angle -60° \qquad ...(3.42)$$

3.9.2 Rectangular Form

Any inclined phasor can be resolved into two components, one along the horizontal or X-axis and the other along the vertical or Y-axis.

Phasor v_1 equation (3.39) is represented in polar form by its magnitude and direction as

$$v_1 = |V_1|\angle 0 \qquad \text{[Eqn. (3.39)]}$$

Horizontal component of v_1

$$v_{1h} = |V_1|\cos(0°) = |V_1| \qquad \ldots(3.43)$$

Vertical component of v_1

$$v_{1r} = |V_1|\sin(0°) = 0 \qquad \ldots(3.44)$$

It has horizontal component alone. It is along positive X-axis, and is considered as the reference phasor.

Phasor v_2 equation (3.40) is in the I quadrant, leading v_1 by an angle 45° and can be represented in polar form as

$$v_2 = |V_2|\angle 45° \qquad \text{[Eqn. (3.40)]}$$

Horizontal component of v_2

$$v_{2h} = |V_2|\cos(45°) \qquad \ldots(3.45)$$

Vertical component of v_2

$$v_{2r} = |V_2|\sin(45°) \qquad \ldots(3.46)$$

Phasor v_3 equation (3.41) is leading v_1 by an angle 90° and can be represented in polar form as

$$v_3 = |V_3|\angle 90° \qquad \text{[Eqn. (3.41)]}$$

Horizontal component of v_3

$$v_{3h} = |V_3|\cos(90°) = 0 \qquad \ldots(3.47)$$

Vertical component of v_3

$$v_{3r} = |V_3|\cos(90°) = |V_3| \qquad \ldots(3.48)$$

Phasor v_3 is along positive Y-axis. It has only a vertical component.

Phasor v_4 equation (3.42) is an inclined phasor shown in IV quadrant lagging reference phasor v_1 by an angle 60°. It can be represented in polar form as

$$v_4 = |V_4|\angle -60° \qquad \text{[Eqn. (3.42)]}$$

Horizontal component of v_4

$$v_{4h} = |V_4|\cos(-60°) = |V_4|\cos(60°) \qquad \ldots(3.49)$$

Vertical component of v_4

$$v_{4r} = |V_4|\sin(-60°) = -|V_4|\sin(60°) \qquad \ldots(3.50)$$

Therefore, in general, any phasor is said to have:

- Either a horizontal component as v_1 or
- Vertical component as v_3 or
- Both horizontal and vertical components as phasors v_2 and v_4

To distinguish a vertical component from horizontal component the complex operator 'j' is introduced.

The operator 'j' is added along with the vertical component of the phasor. Its value is

$$j = 1\angle +90 = \sqrt{-1} \quad \text{...(3.51)}$$

$$j^2 = 1\angle +180 = -1 \quad \text{...(3.52)}$$

$$j^3 = 1\angle +270 = 1\angle -90 = -j \quad \text{...(3.53)}$$

$$j^4 = 1\angle +360 = 1\angle 0 = 1 \quad \text{...(3.54)}$$

$$\frac{1}{j} = 1\angle -90 = -j \quad \text{...(3.55)}$$

With the complex operator 'j', the four phasors v_1, v_2, v_3 and v_4 shown in Fig. 3.12 can be represented in a clear complex form *i.e.*, rectangular form as

$$v_1 = |V_1|\angle 0 = |V_1|(\cos 0 + j\sin 0) = |V_1|(1 + j0) \quad \text{...(3.56)}$$

$$v_2 = |V_2|\angle 45° = |V_2|(\cos 45 + j\sin 45) = |V_2|\left(\frac{1}{\sqrt{2}} + j\frac{1}{\sqrt{2}}\right) \quad \text{...(3.57)}$$

$$v_3 = |V_3|\angle 90° = |V_3|(\cos 90 + j\sin 90) = |V_3|(0 + j) \quad \text{...(3.58)}$$

$$v_4 = |V_4|\angle -60° = |V_4|(\cos(-60) + j\sin(-60)) \quad \text{...(3.59)}$$

or

$$v_4 = |V_4|\left(\frac{1}{2} - j\frac{\sqrt{3}}{2}\right) \quad \text{...(3.60)}$$

Conversion facility from the polar form to the complex (or rectangular) form and vice versa are available in most of the calculators so that one need not always multiply the modulus value with sine and cosine functions.

Once the phasors are represented in the complex form, then their addition, subtraction, multiplication and division are made as simple as with normal numbers.

The only change is that the calculator mode of operation has to be changed to the complex mode and each phasor is to be provided with two values within a set of brackets, an active or real value and another reactive or imaginary value along with the complex operator 'j' or sometimes represented as 'i'.

3.10 RESULTANT OF ALTERNATING QUANTITIES

When a number of similar alternating quantities are acting simultaneously in an AC circuit, then their combined effect or the resultant can be calculated by the following methods:

(*i*) Trigonometric method

(*ii*) Polar method

(*iii*) Graphical method

3.10.1 Trigonometric Method

Let the following alternating current are acting simultaneously in an AC circuit

$$i_1 = 20 \sin \omega t \qquad \text{where } \theta_1 = 0° \qquad ...(3.61)$$

$$i_2 = 15 \sin (\omega t + 45°) \qquad \text{where } \theta_2 < 90° \qquad ...(3.62)$$

$$i_3 = 25 \sin (\omega t + 135°) \qquad \text{where } \theta_3 > 90° \qquad ...(3.63)$$

$$i_4 = 10 \sin (\omega t - 60°) \qquad \text{where } \theta_4 \textit{ is negative} \qquad ...(3.64)$$

$$i_5 = 20 \cos (\omega t + 45°) \qquad \text{where } \theta_5 \textit{ is angle of cosine} \qquad ...(3.65)$$

Step 1. First step to find the resultant of an alternating quantity is that if any equation is given in cosine form convert it into sine form. For example in the above equations, equation (3.65) is in cosine form.

Now when we convert any cosine equation into sine equation, there is a problem with the polarity sign. Let us draw the cosine equation and sine equation:

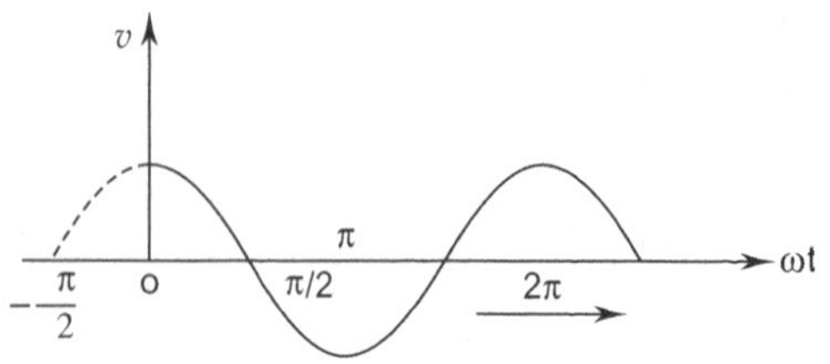

Fig. 3.13 *Cosine Equation*

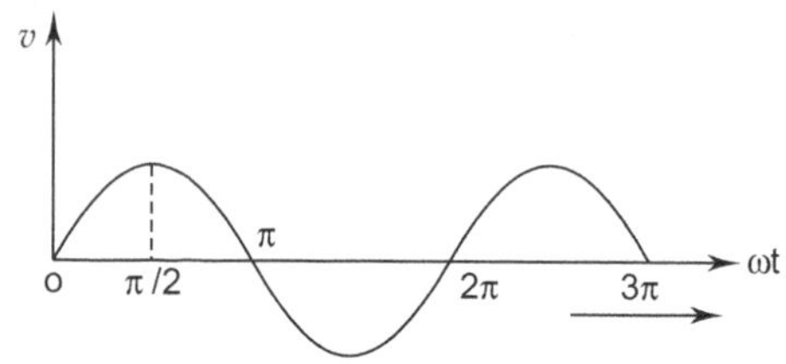

Fig. 3.14 *Sine Equation*

Sine equation will start from zero position or from the origin (as sin 0 = 0) and cosine equation will have a peak at zero position (as cos 0 = 1). It means that a cosine equation peak appear first as compare to sine equation's peak. That is the cosine equation is in leading position to that of the sine equation. Hence, when we draw both the waves [(Fig. 3.13) and (Fig. 3.14)] , it is clear that, the sine wave lags the cosine wave by an angle of 90°. Hence while conveting the cosine equation into sine equation, 90° (or $\frac{\pi}{2}$) must be added to the angle of sine equation to make both equation's angle equal.

i.e., if we consider a cosine equation as

$$v = V_m \cos \omega t \qquad ...(3.66)$$

then its sine conversion will be

$$v = V_m \sin (\omega t + \frac{\pi}{2}) \qquad ...(3.67)$$

And to remember

$$\cos \theta = \sin (\theta + 90°) \qquad ...(3.68)$$

Now changing equation (3.65) from cosine to sine, we have

$$i_5 = 20 \cos (\omega t + 45°) \qquad \text{[Eqn. (3.65)]}$$

or

$$i_5 = 20 \sin (\omega t + 45° + 90°) \qquad ...(3.69)$$

or

$$i_5 = 20 \sin (\omega t + 135°) \qquad ...(3.70)$$

Note. *If in case, any equation has negative magnitude, then to change it from negative value to positive value, that equation phasor is roated by 180°. While doing so, the magnitude becomes positive and the phase angle 'ϕ' becomes ϕ + 180°.*

Step 2. Draw the phasor diagram of all the above equation, as shown in Fig. 3.15.

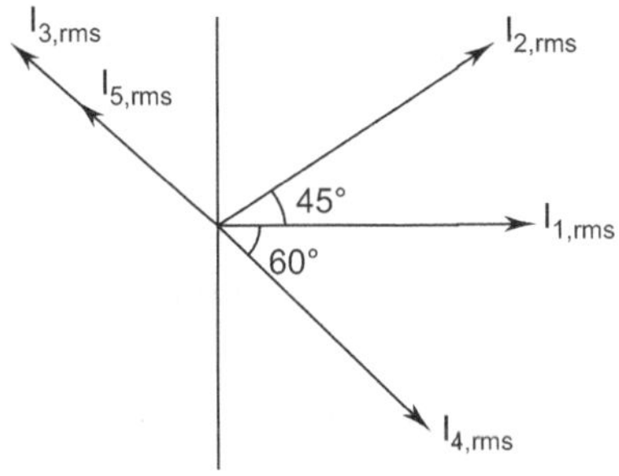

Fig. 3.15

Resolve all the phasors into their horizontal (X) and vertical (Y) components

Step 3. Find X-component

$$X = I_1 \cos \theta_1 + I_2 \cos \theta_2 + I_3 \cos \theta_3 + I_4 \cos \theta_4 + I_5 \cos \theta_5 \qquad \ldots(3.71)$$

or $$X = 20 \cos (0°) + 15 \cos (45°) + 25 \cos (135°) + 10 \cos (-60°) + 20 \cos (135°) \qquad \ldots(3.72)$$

$$= 20 + 10.61 - 17.68 + 5 - 14.14$$

$$X = 3.79 \qquad \ldots(3.73)$$

Step 4. Find Y-component

$$Y = I_1 \sin \theta_1 + I_2 \sin \theta_2 + I_3 \sin \theta_3 + I_4 \sin \theta_4 + I_5 \sin \theta_5 \qquad \ldots(3.74)$$

or $$Y = 20 \sin (0°) + 15 \sin (45°) + 25 \sin (135°) + 10 \sin (-60°) + 20 \sin (135°) \qquad \ldots(3.75)$$

or $$Y = 0 + 10.61 + 17.68 - 8.66 + 14.14$$

$$Y = 33.77 \qquad \ldots(3.76)$$

Step 5. Find magnitude (or maximum value) of the resulant phasor

$$I_m = \sqrt{X^2 + Y^2} = \sqrt{(3.79)^2 + (33.77)^2}$$

or $$I_m = 33.98 \text{ A} \qquad \ldots(3.77)$$

Step 6. Find phase angle of the resultant phasor

$$\phi = \tan^{-1} \frac{Y}{X} = \tan^{-1} \frac{33.77}{3.79} \qquad \ldots(3.78)$$

or $$\phi = 83.6° \qquad \ldots(3.79)$$

Step 7. The resultant of all the above current phasors will be

$$i = I_m \sin (\omega t \pm \phi) \qquad \ldots(3.80)$$

From equations (3.77) and (3.79), equation (3.80) will become

$$i = 33.98 \sin (\omega t + 83.6°) \qquad \ldots(3.81)$$

Note. *For subtraction of vectors:*

Suppose we have to find

$$e_1 + e_2 - e_3 + e_4 \quad \text{...(3.82)}$$

i.e., we have to add e_1, e_2 *and* e_4 *and we have to subtract* e_3 *from their sum.*

In such case, after drawing the phasor diagram, draw another phasor diagram in which invert or shift the phasor (which is to be subtracted) by 180° and take its magnitude negative.

3.10.2 Polar Method

Considering the same equations of trigonometric method:

$i_1 = 20 \sin \omega t$ where $\theta_1 = 0°$ [Eqn. (3.61)]

$i_2 = 15 \sin (\omega t + 45°)$ where $\theta_2 < 90°$ [Eqn. (3.62)]

$i_3 = 25 \sin (\omega t + 135°)$ where $\theta_3 > 90°$ [Eqn.(3.63)]

$i_4 = 10 \sin (\omega t - 60°)$ where θ_4 is negative [Eqn. (3.64)]

$i_5 = 20 \cos (\omega t + 45°)$ where θ_5 is angle of cosine [Eqn. (3.65)]

Step 1. First if any equation is given in cosine form convert it into sine form. For example in the above equations, equation (3.65) is in cosine form

$$i_5 = 20 \cos (\omega t + 45°) \quad \text{[Eqn. (3.65)]}$$

$$= 20 \sin (\omega t + 45° + 90°)$$

or $$i_5 = 20 \sin (\omega t + 135°) \quad \text{...(3.83)}$$

Step 2. Now the above alternating quantities can be converted into polar form and then in rectangular form as

$$i_1 = 20 \angle 0° \quad \text{...(3.84)}$$

$$= 20 (\cos 0° + j \sin 0°)$$

or $$i_1 = 20 + j\,0 \quad \text{...(3.85)}$$

$$i_2 = 15 \angle 45° \quad \text{...(3.86)}$$

$$= 15 (\cos 45° + j \sin 45°)$$

or $$i_2 = 10.61 + j\,10.61 \quad \text{...(3.87)}$$

$$i_3 = 25 \angle 135° \quad \text{...(3.88)}$$

$$= 25 (\cos 135° + j \sin 135°)$$

or $$i_3 = -17.68 + j\,17.68 \quad \text{...(3.89)}$$

$$i_4 = 10 \angle -60° \quad \text{...(3.90)}$$

$$= 10 (\cos (-60°) + j \sin (-60°))$$

$$i_4 = 5 - j\,8.66 \quad \text{...(3.91)}$$

or $$i_5 = 20 \angle 135° \quad \text{...(3.92)}$$

$$= 20 (\cos 135° + j \sin 135°)$$

or $$i_5 = -14.14 + j\,14.14 \quad \text{...(3.93)}$$

Step 3. The resultant of above alternating quantities is

$$i = i_1 + i_2 + i_3 + i_4 + i_5 \quad \text{...(3.94)}$$

$$= (20 + j\,0) + (10.61 + j\,10.61) + (-17.68 + j\,17.68) + (5 - j\,8.66) + (-14.14 + j\,14.14)$$

or $$i = 3.79 + j\,33.77 \quad \text{...(3.95)}$$

or $$i = 33.98 \angle 83.6° \quad \text{...(3.96)}$$

In sinusoidal form, the equation can be written as

$$i = 33.98 \sin (\omega t + 83.6°) \qquad ...(3.97)$$

3.10.3 Graphical Method

The resultant of two or more alternating sinusoidal quantities acting simultaneously in an AC circuit can also be calculated with the help of graphical method.

Following steps can be followed to solve for the resultant:

- Take the phasor first whose phase angle is zero (or minimum among the given phasors). Consider it as reference phasor and draw it along X-axis. The length of this phasor must be equal to its magnitude (or maximum value).
- Draw rest of the given phasors from the tip of the previous one and draw them at their respective phase angles.
- Joint the end point (or the tip of last phasor) with the origin to complete the graphical representation.
- This closing phasor will be the resultant phasor of the addition of all the phasors.
- Measure its length (which is equal to maximum value) and angle with X-axis in positive direction or anticlockwise direction (which will be equal to its phase angle)

  An example of graphical method is shown in solved problems.

SOLVED PROBLEMS

Problem 3.7. *An alternating voltage is given by v = 141.4 sin 314 t. Find (i) frequency, (ii) r.m.s value, (iii) average value, (iv) the intantaneous value of voltage when 't' is 3 m s, (v) the time taken for the voltage to reach 100 V for the first time after passing through zero value.*

Solution. Instantaneous value of alternating voltage is given by expression

$$v = 141.4 \sin 314\, t \qquad ...(i)$$

(*i*) The frequency is given by the coefficient of time divided by 2π, therefore,

$$\text{frequency}, f = \frac{314}{2\pi} = \mathbf{50\ Hz} \qquad \textbf{Ans.}$$

Peak value of an alternating voltage is given by the coefficient of sine of time angle, therefore,

$$V_{max} = 141.4 \text{ V}$$

(*ii*) r.m.s value,

$$V_{rms} = \frac{V_{max}}{\sqrt{2}} = \frac{141.4}{\sqrt{2}} = \mathbf{100\ V} \qquad \textbf{Ans.}$$

(*iii*) Average value,

$$V_{av} = 0.637 \times V_{max} = 0.637 \times 141.4 = \mathbf{90\ V} \qquad \textbf{Ans.}$$

(*iv*) Instantaneous value of voltage at 3 m s (0.003 s) after the instant the voltage is zero and increasing in positive direction may be obtained by substituting t = 0.003 second in equation (*i*). So, we have

$$v = 141.4 \sin 314 \times 0.003 = 141.4 \sin 0.942 = \mathbf{114.4\ V} \qquad \textbf{Ans.}$$

(*v*) Time taken from $t = 0$ for the voltage to reach 100 V

$$t = \frac{1}{314}\sin^{-1}\frac{100}{141.4} = \frac{\pi/4}{314}\text{ second} = \frac{1}{400}\text{ s} = \mathbf{2.5\ ms} \qquad \textbf{Ans.}$$

Problem 3.8. *An alternating voltage is v = 100 sin 100 t. Find (i) Amplitude, (ii) Time period and frequency, (iii) Angular velocity, (iv) Form factor, (v) Peak factor.*

Solution. Instantaneous value of alternating voltage is given by expression

$$v = 100 \sin 100\, t$$

(*i*) Amplitude of alternating voltage is given by the coefficient of the sine of the time angle, so Amplitude of given wave $= V_{max} = \mathbf{100\ V}$ **Ans.**

(*ii*) Frequency is given by coefficient of time t, divided by 2π

$$\therefore \quad \text{Frequency}, f = \frac{100}{2\pi} = \mathbf{15.9\ Hz} \qquad \textbf{Ans.}$$

$$\text{Time period,} \quad T = \frac{1}{f} = \frac{1}{15.9} = \mathbf{0.063\ second\ or\ 63\ m\ s.} \qquad \textbf{Ans.}$$

(*iii*) Angular velocity, $\omega = 2\pi f = 2\pi \times 15.9 = \mathbf{100\ radians/second.}$ **Ans.**

$$\text{r.m.s value, } V_{rms} = \frac{V_{max}}{\sqrt{2}} = \frac{100}{\sqrt{2}} = 70.71\text{ V}$$

$$\text{average value, } V_{av} = 0.637\, V_{max} = 0.637 \times 100 = 63.7\text{ V}$$

(*iv*) Form factor, $K_f = \dfrac{V_{rms}}{V_{av}} = \dfrac{70.71}{63.7} = \mathbf{1.11.}$ **Ans.**

(*v*) Peak factor, $K_p = \dfrac{V_{max}}{V_{rms}} = \dfrac{100}{70.71} = \mathbf{1.4142.}$ **Ans.**

Problem 3.9. *The equation of an alternating current i = 42.42 sin 628 t. Determine (i) its maximum value, (ii) frequency, (iii) r.m.s value, (iv) average value, (v) form factor.*

Solution. Instantaneous value of alternating current is given by the expression

$$i = 42.42 \sin 628\, t$$

(*i*) Maximum value of an alternating current is given by the coefficient of the sine of the time angle

i.e., $I_{max} = \mathbf{42.42\ A}$ **Ans.**

(*ii*) Frequency, $F = \dfrac{628}{2\pi} = \mathbf{100\ Hz}$ **Ans.**

(*iii*) r.m.s value,

$$I_{rms} = \frac{I_{max}}{\sqrt{2}} = \frac{42.42}{\sqrt{2}} = \mathbf{30\ A} \qquad \textbf{Ans.}$$

(*iv*) Average value,

$$I_{av} = 0.637 \times I_{max} = 0.637 \times 42.42 = \mathbf{27\ A} \qquad \textbf{Ans.}$$

(*v*) Form factor $K_f = \dfrac{I_{rms}}{I_{av}} = \dfrac{30}{27} = \mathbf{1.11.}$ **Ans.**

Problem 3.10. *Two AC voltages are represented by*

$$v_1(t) = 30 \sin(314\,t + 45°)$$
$$v_2(t) = 60 \sin(314\,t + 60°)$$

Calculate the resultant voltage $v(t)$ and express it in the form $v(t) = V_m \sin(314\,t \pm \phi)$

Solution.
$$v_1(t) = 30 \sin(314\,t + 45°)$$
$$v_2(t) = 60 \sin(314\,t + 60°)$$

Resolving voltages $V_{1\,max}$ and $V_{2\,max}$ along X-axis and Y-axis, we have

$$\text{X-components} = 30 \cos 45° + 60 \cos 60° = 30 \times \frac{1}{\sqrt{2}} + 60 \times \frac{1}{2} = 51.21 \text{ V}$$

$$\text{Y-component} = 30 \sin 45° + 60 \sin 60° = 30 \times \frac{1}{\sqrt{2}} + 60 \times \frac{\sqrt{3}}{2} = 73.17 \text{ V}$$

Maximum value of resultant voltage,

$$V_{max} = \sqrt{(\text{X-components})^2 + (\text{Y-components})^2}$$
$$= \sqrt{(51.21)^2 + (73.17)^2} = 89.31 \text{ V}$$

Phase angle, $$\phi = \tan^{-1}\frac{\text{Y-component}}{\text{X-component}} = \tan^{-1}\frac{73.17}{51.21} = 55°$$

So expression for resultant voltage is

$$v(t) = V_{max} \sin(\omega t + \phi) = \mathbf{89.31 \sin(314\,t + 55°)}$$ **Ans.**

Problem 3.11. *Find the average value for the waveform shown in Fig. 3.16.*

Solution. Maximum value of alternating current, $I_{max} = 10$ A

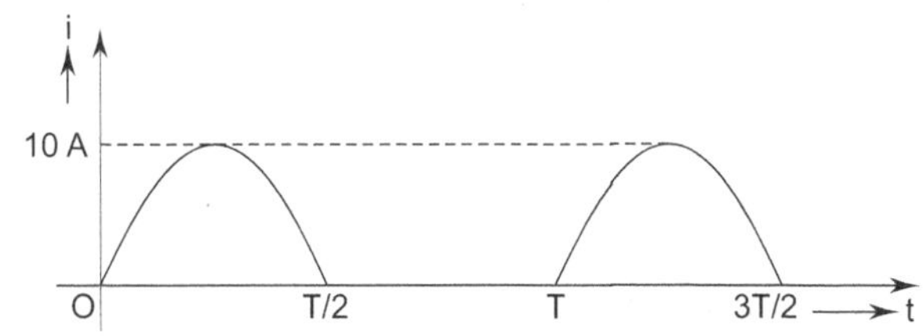

Fig. 3.16

Average value, $$I_{av} = \frac{I_{max}}{\pi} = \frac{10}{\pi} = \mathbf{3.183 \text{ A}}$$ **Ans.**

Problem 3.12. *Draw a phasor diagram showing the following voltages:*

$$v_1 = 100 \sin 500t \qquad v_2 = 200 \sin\left(500t + \frac{\pi}{3}\right)$$

$$v_3 = -50 \cos 500t \qquad v_4 = 150 \sin\left(500t - \frac{\pi}{4}\right)$$

Find r.m.s value of resultant voltage.

Solution. $v_1 = 100 \sin 500\, t$

$$v_2 = 200 \sin\left(500\, t + \frac{\pi}{3}\right)$$

$$v_3 = -50 \cos 500t = -50 \sin\left(500\, t + \frac{\pi}{2}\right) = 50 \sin\left(500\, t - \frac{\pi}{2}\right)$$

$$v_4 = 150 \sin\left(500\, t - \frac{\pi}{4}\right)$$

The phasor diagram is shown in Fig. 3.17.

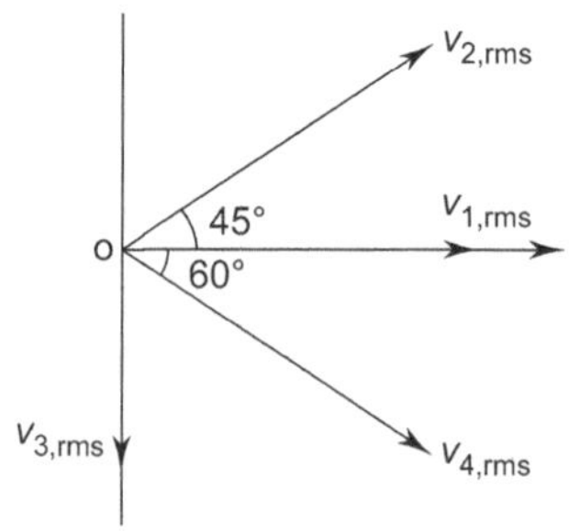

Fig. 3.17

Resolving voltages v_1, v_2, v_3 and v_4 along X-axis and Y-axis, we have

Algebraic sum of X-components

$$= 100 \cos 0° + 200 \cos \frac{\pi}{3} + 50 \cos\left(-\frac{\pi}{2}\right) + 150 \cos\left(-\frac{\pi}{4}\right)$$

$$= 100 + 200 \times 0.5 + 0 + \frac{150}{\sqrt{2}} = 306 \text{ V}$$

Algebraic sum of Y-components

$$= 100 \sin 0° + 200 \sin \frac{\pi}{3} + 50 \sin\left(-\frac{\pi}{2}\right) + 150 \sin\left(-\frac{\pi}{4}\right)$$

$$= 0 + 200 \times 0.866 - 50 - 106 = 17.2 \text{ V}$$

Maximum value of resultant voltage, $V_{max} = \sqrt{(306)^2 + (17.2)^2} = 306.5 \text{ V}$

r.m.s value of resultant voltage, $V_{rms} = \frac{V_{max}}{\sqrt{2}} = \frac{306.5}{\sqrt{2}} = \mathbf{216.72\ V}$ **Ans.**

Problem 3.13. *A sinusoidally varying alternating current of frequency 60 Hz has a maximum value of 15 amperes.*

(i) Write down the equation for instantaneous value,

(ii) Find the value of current after 1/200 second,

(iii) Find the time taken to reach 10 amperes for the first time, and

(iv) Find its average value.

Solution. Maximum value of current, $I_{max} = 15$ A

Frequency of alternating current, $f = 60$ Hz

(*i*) The equation for instantaneous value of sinusoidal alternating current is given as

$$I = I_{max} \sin 2\pi ft = 15 \sin 2\pi \times 60t = \mathbf{15 \sin 120\pi t.} \quad \textbf{Ans.}$$

(*ii*) Substituting $t = \frac{1}{200}$ second in the expression I = 15 sin $120\pi t$ we will get the value of current after $\frac{1}{200}$ second.

$$I = 15 \sin 120\pi \times \frac{1}{200} = 15 \sin 0.6\pi = 15 \times 0.951 = \mathbf{14.266\ A} \quad \textbf{Ans.}$$

(*iii*) Let the instantaneous value of current be 10 A for the first time t seconds after the instant the current is zero and becoming positive

$$\therefore \quad 10 = 15 \sin 120\pi t$$

or
$$t = \frac{1}{120\pi} \sin^{-1} \frac{10}{15} = \frac{1}{120\pi} \sin^{-1} 0.6667 = \mathbf{0.001936\ second} \quad \textbf{Ans.}$$

(*iv*) Average value,
$$I_{av} = \frac{2}{\pi} I_{max} = \frac{2}{\pi} \times 15 = \mathbf{9.55\ A} \quad \textbf{Ans.}$$

Problem 3.14. *A sinusoidal current having r.m.s value of 8 ∠ 0° A is added to another sinusoidal current of r.m.s value 6 ∠ 90° A. What will be the r.m.s value of the resultant current?*

Solution.

Given
$$i_1 = 8 \angle 0° = 8 + j0 \text{ A}$$
$$i_2 = 6 \angle 90° = 0 + j6 \text{ A}$$

Resultant current will be

$$i = i_1 + i_2$$
$$= 8 + j0 + 0 + j6 = 8 + j6$$

or
$$i = \mathbf{10 \angle 36.87°\ A} \quad \textbf{Ans.}$$

Problem 3.15. *For two phasors* $A = a_1 + jb_1$ *and* $B = a_2 - jb_2$, *obtain their multiplication and division using polar form of representation.*

Solution.

Given
$$A = a_1 + j\,b_1 \text{ (Rectangular form)}$$
$$B = a_2 - j\,b_2 \text{ (Rectangular form)}$$

In polar form
$$A = \sqrt{a_1^2 + b_1^2} \angle \tan^{-1} \frac{b_1}{a_1}$$

$$B = \sqrt{a_2^2 + b_2^2} \angle -\tan^{-1} \frac{b_2}{a_2}$$

(*i*) Multiplication

$$A \times B = \left(\sqrt{a_1^2 + b_1^2} \angle \tan^{-1} \frac{b_1}{a_1}\right) \times \left(\sqrt{a_2^2 + b_2^2} \angle -\tan^{-1} \frac{b_2}{a_2}\right)$$

$$= \left(\sqrt{a_1^2 + b_1^2}\right)\left(\sqrt{a_2^2 + b_2^2}\right) \angle \left(\tan^{-1} \frac{b_1}{a_1} - \tan^{-1} \frac{b_2}{a_2}\right) \quad \textbf{Ans.}$$

(*ii*) Division

$$\frac{A}{B} = \frac{\sqrt{a_1^2 + b_1^2} \angle \tan^{-1}\frac{b_1}{a_1}}{\sqrt{a_2^2 + b_2^2 \angle - \tan^{-1}\frac{b_2}{a_2}}}$$

$$= \frac{\sqrt{a_1^2 + b_1^2}}{\sqrt{a_2^2 + b_2^2}} \angle\left(\tan^{-1}\frac{b_1}{a_1} + \tan^{-1}\frac{b_2}{a_2}\right) \quad \textbf{Ans.}$$

Problem 3.16. *Three voltages are represented by the following equations acting together in an AC circuit.*

$$e_1 = 15 \sin \omega t$$

$$e_2 = 5 \sin\left(\omega t + \frac{\pi}{6}\right)$$

$$e_3 = 10 \cos \omega t$$

(*a*) *Represent these voltages by phasor diagram.*

(*b*) *Calculate the expression for the resultant voltage.*

(*c*) *Check the result so obtained graphically.*

Solution. Given

$$e_1 = 15 \sin (\omega t)$$

$$e_2 = 5 \sin\left(\omega t + \frac{\pi}{6}\right)$$

$$e_3 = 10 \cos (\omega t) = 10 \sin (\omega t + \pi/2)$$

Phasor diagram for the above voltage will be as shown in Fig. 3.18.

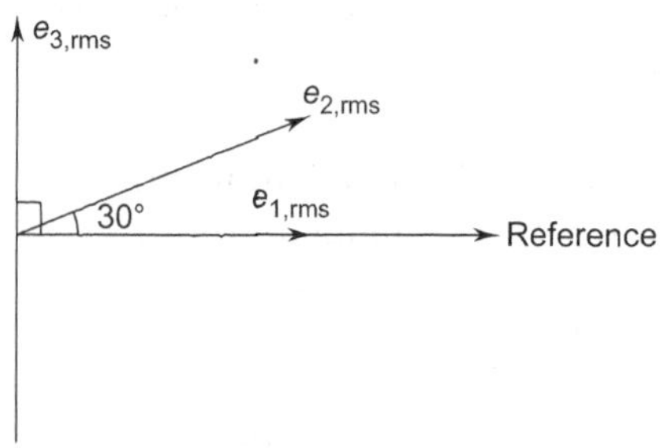

Fig. 3.18

The given equations in polar and rectangular form can be expressed as:

$$e_1 = 15 \angle 0° = 15 + j\,0$$

$$e_2 = 5 \angle 30° = 4.33 + j\,2.5$$

$$e_3 = 10 \angle 90° = 0 + j\,10$$

Resultant will be

$$e = e_1 + e_2 + e_3$$
$$= 15 + 4.33 + j\,2.5 + j\,10$$
$$= 19.33 + j\,12.5$$
$$= 23.02 \angle 32.89°$$
$$\therefore \quad e = 23.02 \sin(\omega t + 32.89°)$$

Verification of Result by Graphical Method:

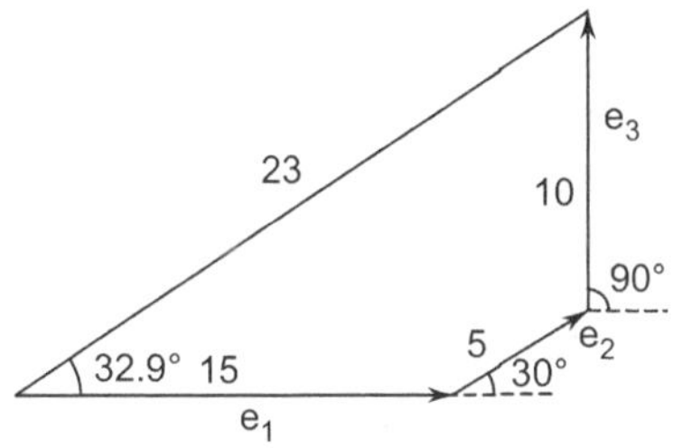

Fig. 3.19

Note. Steps to find the resultant by graphical method:

1. Take that phasor first whose phase angle is zero (or least). Consider it as reference phasor and draw it along x-axis.
2. The length of this reference phasor should be equal to its maximum value on the scale.
3. Draw rest of the given phasors from the tip of one another at their given respective phase angles.
4. Join the end point with the origin to complete the graphical representation.
5. This closing phasor will be the resultant phasor of the addition.
6. Measure its length (which is equal to maximum value) and angle with x-axis in positive direction (which will be equal to its phase angle).

Problem 3.17. *A voltage source of e(t) = 141 sin 377t V is applied to two parallel branches. The time expression for the current in the first branch is*

$$i_1(t) = 7.07 \sin\left(\omega t - \frac{\pi}{3}\right)$$

In the second branch it is

$$i_2(t) = 10 \sin\left(\omega t + \frac{\pi}{6}\right)$$

Compute the total power supplied by the source.

Solution. Given

$$e(t) = 141 \sin 377\,t$$

$$i_1(t) = 7.07 \sin\left(\omega t - \frac{\pi}{3}\right)$$

$$i_2(t) = 10 \sin\left(\omega t + \frac{\pi}{6}\right)$$

Total power supplied by the source will be

$$P = VI \cos \phi$$

where V is the rms value of applied voltage.

I is the rms value of total current in the circuit.

ϕ is the angle between V and I

To find the total current, we have to find the resultant of the two branch currents.

$$X = 7.07 \cos (-60°) + 10 \cos (30°)$$
$$= 3.535 + 8.66$$
$$= 12.195$$
$$Y = 7.07 \sin (-60°) + 10 \sin (30°)$$
$$= -6.123 + 5$$
$$= -1.123$$

$$I_m = \sqrt{X^2 + Y^2} = \sqrt{(12.195)^2 + (1.123)^2}$$
$$I_m = 12.247 \text{ A}$$

$$\phi = \tan^{-1} \frac{Y}{X} = \tan^{-1} \frac{-1.123}{12.195}$$
$$\phi = -5.261°$$

$$V_{rms} = \frac{V_m}{\sqrt{2}} = \frac{141}{\sqrt{2}} \approx 100 \text{ V}$$

$$I_{rms} = \frac{V_m}{\sqrt{2}} = \frac{12.247}{\sqrt{2}} \approx 8.7 \text{ A}$$

$\therefore$ $$P = VI \cos \phi$$
$$P = 100 \times 8.7 \times \cos (5.261) = \mathbf{866.34 \text{ W}}$$ **Ans.**

Problem 3.18. *Convert the following to polar form:*

(a) $3 + j4$ *(b)* 6

(c) $j8$ *(d)* $5 - j5$

(e) $-2 + j2$.

Solution.

(a) $$3 + j4 = r \angle \theta$$

$$r = \sqrt{(3^2 + 4^2)} = 5$$

$$\tan \theta = 4/3,$$
$$\theta = \tan^{-1} 4/3 = 53.1$$

$\therefore$ $$3 + j4 = \mathbf{5 \angle 53.1°}$$ **Ans.**

(b) 6 can be written as $6 + j0$

$$6 + j0 = r \angle \theta$$

$$r = \sqrt{(6^2 + 0^2)} = \mathbf{6}$$ **Ans.**

$$\theta = \tan^{-1} 0/6 = 0°$$

$\therefore$ $$6 = 6 + j0 = \mathbf{6 \angle 0°}$$ **Ans.**

(*c*) $j8$ can be written as $0 + j8$

$$r = \sqrt{(0^2 + 8^2)} = 8$$

$$\theta = \tan^{-1} 8/0 = 90°$$

$\therefore$ $$j8 = \mathbf{8 \angle 90°}$$ **Ans.**

(*d*) $$5 - j5 = r \angle \theta$$

$$r = \sqrt{(5^2 + 5^2)} = 7.07$$

$$\tan\theta = \frac{-5}{5} \tan^{-1}(-45°)$$

$$\theta = -45°$$

$\therefore$ $$5 - j5 = \mathbf{7.07 \angle -45°}$$ **Ans.**

(*e*) $$-2 + j2 = r \angle \theta$$

$$r = \sqrt{(2^2 + 2^2)} = 2.828$$

$$\tan\theta = \frac{2}{-2} = -1 = \tan(-45°)$$

$\therefore$ $$-2 + j2 = \mathbf{2.828 \angle -45°}$$ **Ans.**

EXERCISE

1. An alternating current is represented by $i = 400 \sin\left(157t + \frac{\pi}{6}\right)$.

Determine: (*i*) Peak value; (*ii*) Average value; (*iii*) *rms* value; (*iv*) Frequency; (*v*) Time period; (*vi*) Form factor; (*vii*) Peak factor, and; (*viii*) Phase angle of the current represented.

2. Calculate the rms value, the form factor and the peak factor of a periodic voltage having the following values for equal time intervals, changing suddenly from one value to the next:

0, 5, 10, 20, 50, 60, 50, 20, 10, 5, 0 – 5, – 10 V etc.

What would be the rms value of the sine wave having the same peak value ?

3. Three sinusoidal voltages acting in series are given by

$$V_1 = 10 \sin 440t$$

$$V_2 = 10\sqrt{2} \sin(440t - 45°)$$

$$V_3 = 20 \cos 440t$$

Determine:

(*i*) an expression for the resultant voltage,

(*ii*) the frequency and rms value of the resultant voltage.

4. Find the average value, *rms*, Form Factor and Peak Factor of the waveform shown in Fig. 3.20 to Fig. 3.22.

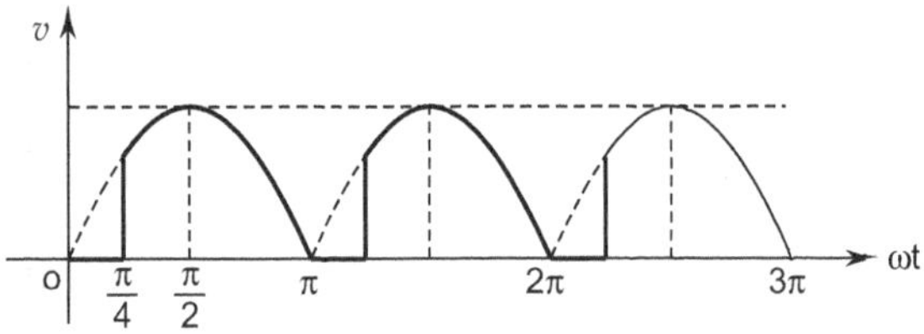

Fig. 3.20

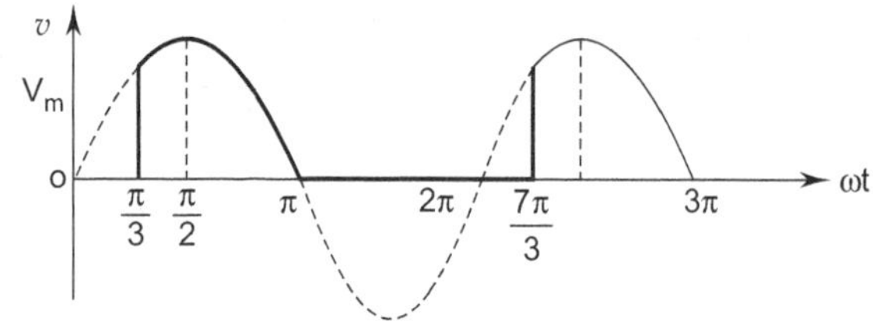

Fig. 3.21

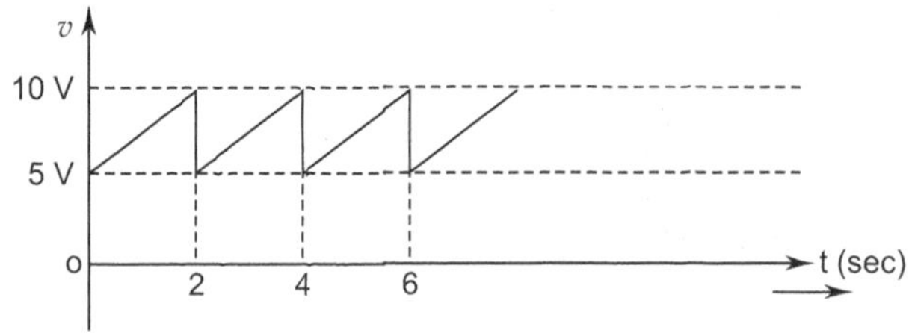

Fig. 3.22

□ □ □

4

STEADY STATE ANALYSIS OF AC CIRCUITS

4.1 INTRODUCTION

Single-phase AC circuits may contain resistors, inductors and capacitors alone or may have combination of these three. Steady state analysis of AC circuit is study of relationship between current and voltage, when we apply alternating sinusoidal voltage to AC circuit containing combinations of all the three above mentioned elements.

To fix direction of different phasors, initially one of the phasor which is a common quantity is considered as reference phasor and may be drawn along positive X-axis.

For example, in a series circuit, the current which passes through different elements in the circuit is same and may be considered as reference phasor.

On the other hand, in a parallel circuit, supply voltage which is common for all parallel paths may be considered as reference phasor. Other phasors are drawn at suitable phase angles counted from the reference phasor in clockwise or anticlockwise direction.

Counting phase angle in a clockwise direction corresponds to a lagging direction and in an anticlockwise direction corresponds to a leading direction. The terms lagging and leading have simple dictionary meanings. A lagging phasor is one which falls back in action with respect to the reference phasor and its direction is fixed by counting phase angle in clockwise direction.

Contrary to this, a leading phasor is one which comes ahead in action with respect to the reference phasor and its direction is fixed by counting phase angle in anticlockwise direction.

The principle is with reference to normal clock as shown in Fig. 4.1. Suppose 3 o'clock is chosen as reference time, then, as 5 o'clock occurs only after 2 hr from 3 o'clock, 5 o'clock is said to lag 3 o'clock by 2 hr. The position of 5 o'clock is situated at an angle of 60° moving clockwise from 3 o'clock position.

Similarly as 12 o'clock occurs ahead of 3 o'clock by 3 hr, one can say 12 o'clock leads 3 o'clock by 3 hr. The 12 o'clock position is traced by moving anticlockwise from 3 o' clock through **90°**. Hence, a phasor whose action is ahead of reference phasor is said to be a leading phasor and its direction is fixed by counting the phase angle in anticlockwise direction.

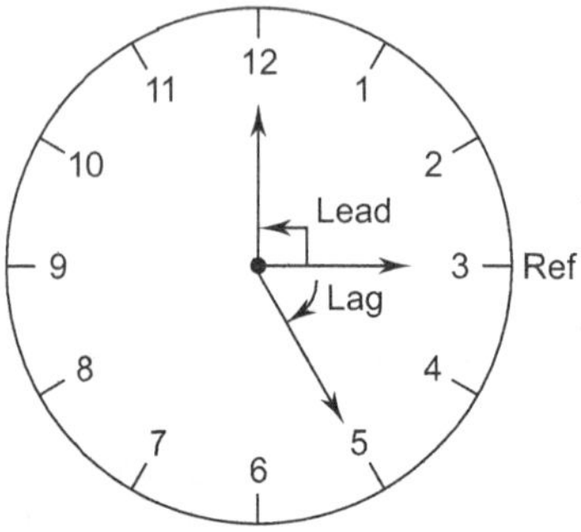

Fig. 4.1

In this chapter we will discuss about behaviour of the components (resistor, inductor and capacitor) when alternating current is applied across them.

In later sections, steady state analysis of their combined behaviour will also be studied.

4.2 AC CIRCUIT CONTAINING PURE RESISTANCE ONLY

Consider a pure resistor 'R' connected across an alternating voltage source, as shown in Fig. 4.2.

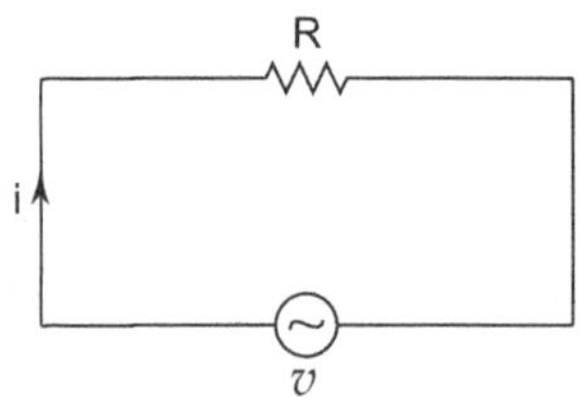

Fig. 4.2

Let the applied voltage be sinusoidal, given by

$$\boldsymbol{v = V_m \sin \omega t} \qquad \text{...(4.1)}$$

According to Ohm's law,

or $$i = \frac{v}{R} = \frac{V_m}{R} \sin \omega t \quad \text{[From equation (4.1)]} \qquad \text{...(4.2)}$$

or $$\boldsymbol{i = I_m \sin \omega t} \qquad \text{...(4.3)}$$

where $$I_m = \frac{V_m}{R} \qquad \text{...(4.4)}$$

= Maximum current when $\sin \omega t = 1$ or $\omega t = 90°$.

From equations (4.1) and (4.3) we can easily conclude that voltage and current are in phase in case of pure resistance.

Phasor diagram and waveform of voltage and current are shown in Fig. 4.3(*a*) and (*b*) respectively.

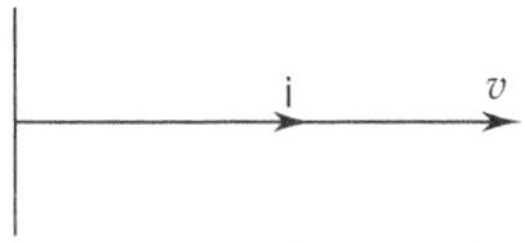

(a) Phasor diagram

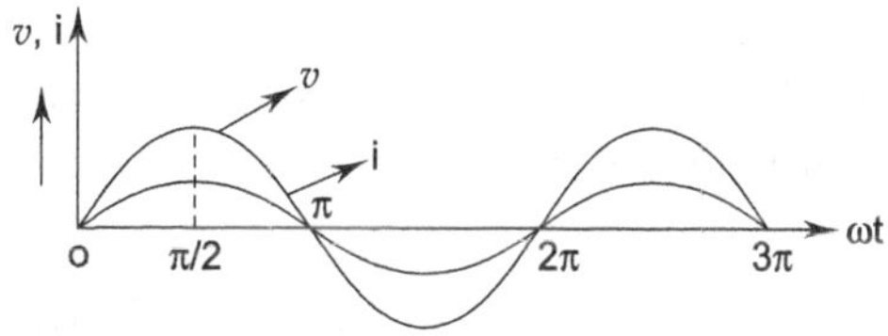

(b) Waveforms

Fig. 4.3

Hence from the phasor diagram:

Phase angle between i and v is $\phi = 0°$ ...(4.5)

Power factor of circuit will be $\cos\phi = \cos 0° = 1$ ...(4.6)

Impedance of the circuit $Z = R$ ...(4.7)

Instantaneous power is given by

$$P_i = vi \quad ...(4.8)$$

or $$P_i = V_m \sin \omega t \times I_m \sin \omega t \quad ...(4.9)$$

$$P_i = V_m I_m \sin^2 \omega t \quad ...(4.10)$$

Average power is given by

$$P_{av} = \frac{1}{\pi}\left[\int_0^{\pi} P_i \, d(\omega t)\right] \quad ...(4.11)$$

Note. *In case of sinusoidal alternating quantity the average power, similar to the average value of the alternating voltage, is considered for half cycle only.*

$$P_a = \frac{1}{\pi}\left[\int_0^{\pi} V_m I_m \sin^2 \omega t \, d(\omega t)\right] \quad ...(4.12)$$

or $$= \left[\frac{V_m I_m}{\pi}\int_0^{\pi}\frac{(1-\cos 2\omega t)}{2} d\omega t\right] \quad ...(4.13)$$

or $$= \left[\frac{V_m I_m}{\pi}\left|\frac{\omega t}{2} - \frac{\sin 2\omega t}{2}\right|_0^{\pi}\right] \quad ...(4.14)$$

or $$= \frac{V_m I_m}{2} = \frac{V_m}{\sqrt{2}} \cdot \frac{I_m}{\sqrt{2}} \quad ...(4.15)$$

or $$P_a = V_{rms} I_{rms} \quad ...(4.16)$$

or $$\mathbf{P_a = V\,I} \quad ...(4.17)$$

where V and I are rms values

Also, $$P_a = \frac{V^2}{R} = I^2 R \quad ...(4.18)$$

Conclusion

In case of AC circuit containing pure resistance only, current is in phase with applied voltage and power factor of the circuit is unity.

4.3 AC CIRCUIT CONTAINING PURE INDUCTANCE ONLY

Consider a pure inductor 'L' connected across an alternating voltage source, as shown in Fig. 4.4.

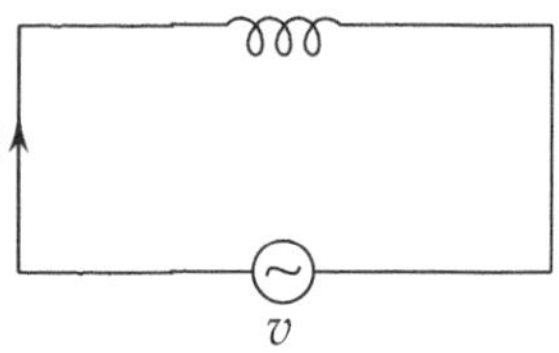

Fig. 4.4

Let the applied voltage be

$$v = \mathbf{V}_m \sin \omega t \qquad \text{[Eqn. (4.1)]}$$

By pure inductive coil we mean a coil that has no ohmic resistance and hence no I^2R loss.

Whenever an alternating voltage is applied to a pure inductive coil, a back e.m.f is produced due to the self-inductance of the coil.

This back e.m.f will oppose rise or fall of current through the coil.

As there is no ohmic voltage drop, the applied voltage has to overcome this self-induced e.m.f only. So if back e.m.f is '*e*' then

$$e = L\frac{di}{dt} = v \qquad \dots(4.19)$$

or

$$v = L\frac{di}{dt} \qquad \dots(4.20)$$

or

$$L\frac{di}{dt} = V_m \sin \omega t \qquad \text{[From equation (4.1)]} \qquad \dots(4.21)$$

or

$$di = \frac{V_m}{L} \sin \omega t \, dt \qquad \dots(4.22)$$

Integrating both sides, we get

$$i = \frac{-V_m}{\omega L} \cos \omega t \qquad \dots(4.23)$$

or

$$i = I_m \sin\left(\omega t - \frac{\pi}{2}\right) \qquad \dots(4.24)$$

(As $\cos \omega t = \sin(\omega t - 90°)$)

where

$$I_m = \frac{V_m}{\omega L} = \frac{V_m}{X_L} \qquad \dots(4.25)$$

And X_L is known as ***Inductive Reactance*** and

$$\omega = 2\pi f$$

$$f = \text{frequency in Hz}$$

From equations (4.1) and (4.24), it is clear that in case of pure inductance, current lags behind the voltage by an angle 90°. The phasor diagram and waveform for the current and voltage are shown in Fig. 4.5(*a*) and (*b*) respectively.

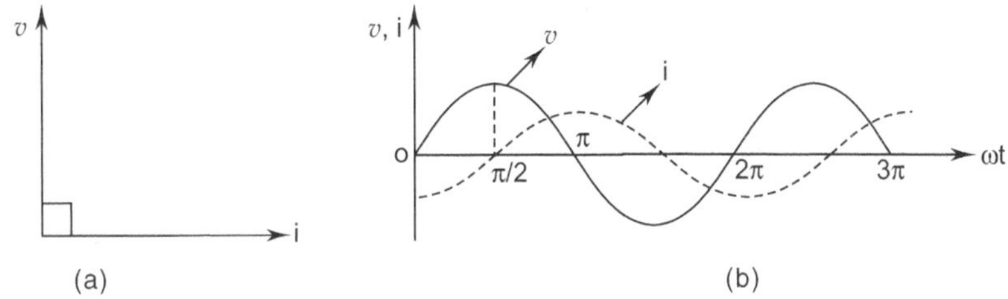

Fig. 4.5

Instantaneous power

$$P_i = vi \quad \text{...(4.26)}$$

Average power, P_a = Average of P_i over half cycle

$$= \frac{1}{\pi}\left[\int_0^{\pi} P_i \, d(\omega t)\right] \quad \text{...(4.27)}$$

$$= \frac{1}{\pi}\left[\int_0^{\pi} V_m \sin\omega t \times I_m \sin\left(\omega t - \frac{\pi}{2}\right) d(\omega t)\right] \quad \text{...(4.28)}$$

$$P_a = 0 \quad \text{...(4.29)}$$

Conclusion

In case of AC circuit containing pure inductance only, the current lags behind the applied voltage by an angle $\pi/2$ or 90° and power factor of circuit is lagging.

It is clear from equation (4.29) that in case of a pure inductor average power consumed is zero. Actually in case of a pure inductor, whatever the power is stored in the inductor in first quarter cycle that is returned back to the source in the next quarter cycle.

4.4 AC CIRCUIT CONTAINING PURE CAPACITANCE ONLY

Consider a pure capacitor 'C' connected across an alternating voltage source, as shown in Fig. 4.6.

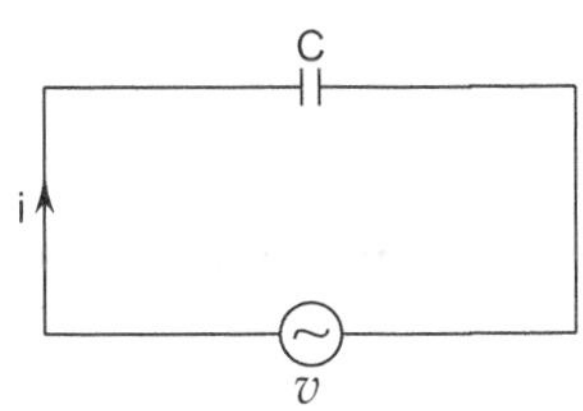

Fig. 4.6

Let the applied voltage be

$$v = V_m \sin \omega t \qquad \text{[Eqn. (4.1)]}$$

In case of a capacitor the charge and current relation is given as

$$i = \frac{dq}{dt} = \frac{d(cv)}{dt} = c\frac{dv}{dt} \quad (\text{as } q = cv) \qquad ...(4.30)$$

$$i = c\frac{d(V_m \sin \omega t)}{dt} \quad \text{[From equation (4.1)]} \qquad ...(4.31)$$

$$= V_m C\omega \sin\left(\omega t + \frac{\pi}{2}\right) \qquad ...(4.32)$$

or

$$i = \frac{V_m}{X_C} \sin\left(\omega t + \frac{\pi}{2}\right) \qquad ...(4.33)$$

Current will be maximum when $\sin\left(\omega t + \frac{\pi}{2}\right) = 1$

And this maximum value will be

$$I_m = \frac{V_m}{X_C} \qquad ...(4.34)$$

The instantaneous current equation will be

$$i = I_m \sin\left(\omega t + \frac{\pi}{2}\right) \qquad ...(4.35)$$

The phasor diagram and waveform for the current and voltage are shown in Fig. 4.7 (*a*) and (*b*) respectively.

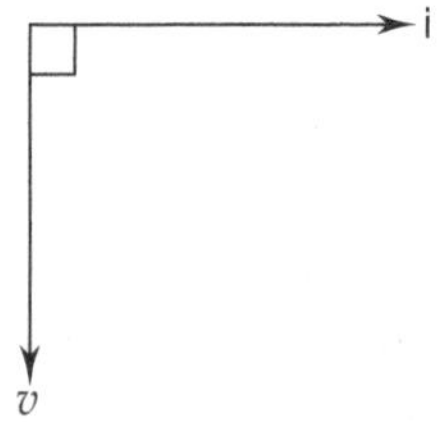

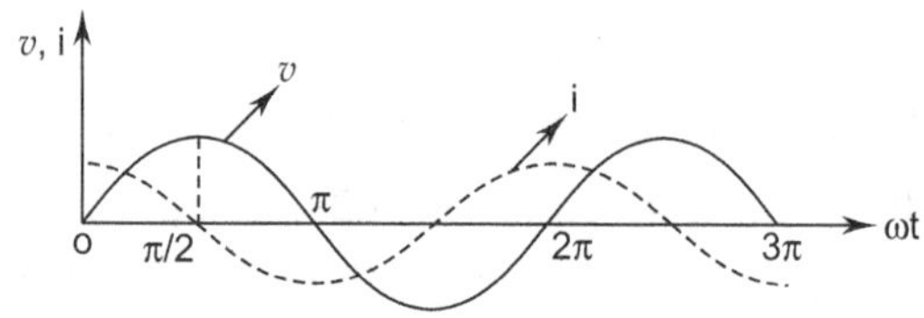

Fig. 4.7

where

$$X_C = \frac{1}{\omega C} = \frac{1}{2\pi f C} \qquad ...(4.36)$$

X_C is known as ***Capacitance Reactance.***

Instantaneous power, $P_i = vi$ ...(4.37)

Average power, P_a – Average of P_i over half cycle

$$= \frac{1}{\pi}\left[\int_0^{\pi} P_i \, d(\omega t)\right] \qquad ...(4.38)$$

$$= \frac{1}{\pi}\left[\int_0^{\pi} V_m \sin\omega t \times I_m \sin\left(\omega t + \frac{\pi}{2}\right) d(\omega t)\right] \quad ...(4.39)$$

$$P_a = 0 \quad ...(4.40)$$

Conclusion

In case of AC circuit containing pure capacitor only, the current leads the applied voltage by an angle $\pi/2$ or 90° and power factor of the circuit is leading.

It is clear from equation (4.40) that in case of a pure capacitor, average power consumed is zero. Similar to an inductor, in case of a pure capacitor also, whatever the power is stored in the capacitor in the first quarter cycle that is returned back to the source in the next quarter cycle.

4.5 AC CIRCUIT CONTAINING RESISTANCE AND INDUCTANCE (RL CIRCUIT)

Consider a pure inductor 'L' and a pure resistor 'R' connected across an alternating voltage source, as shown in Fig. 4.8.

Let the applied voltage be

$$v = V_m \sin \omega t \quad \text{[Eqn. (4.1)]}$$

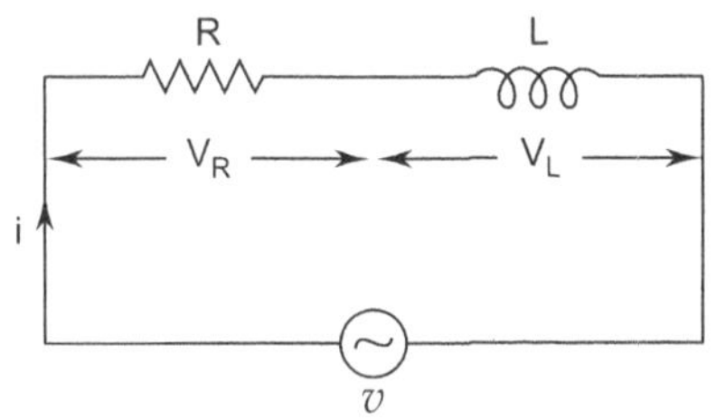

Fig. 4.8

Phasor diagrams related to circuit shown in Fig. 4.8 are shown in Fig. 4.9.

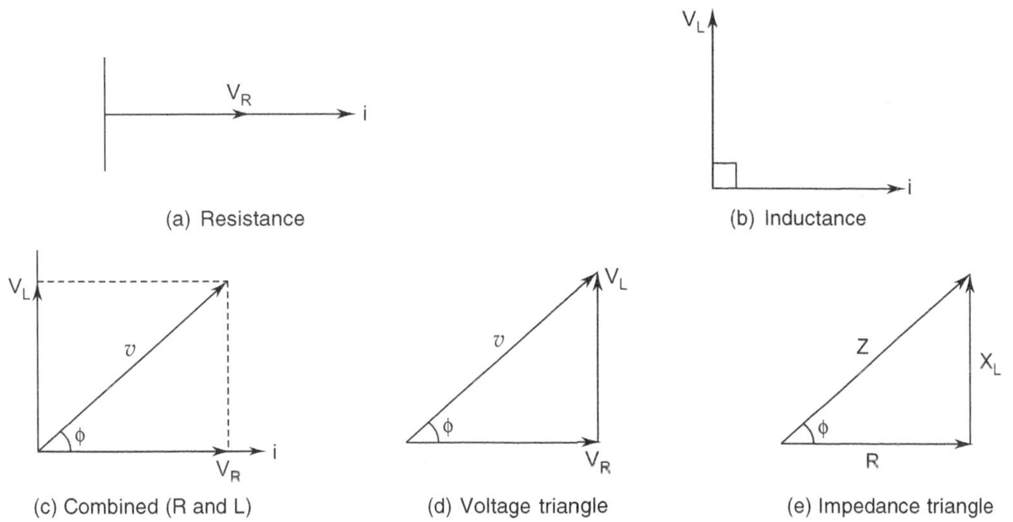

Fig. 4.9

From voltage triangle, shown in Fig. 4.9(*d*),

$$v = \sqrt{V_R^2 + V_L^2} \quad ...(4.41)$$

or
$$v = \sqrt{(iR)^2 + (iX_L)^2} = i\sqrt{R^2 + X_L{}^2} \quad ...(4.42)$$

$$v = iZ \quad ...(4.43)$$

where
$$Z = \sqrt{R^2 + X_L^2} \quad ...(4.44)$$

= Impedance of the circuit

or
$$i = \frac{v}{Z} \quad ...(4.45)$$

From the combined phasor diagram of R and L, power factor of the circuit is

$$\cos\phi = \frac{V_R}{V} = \frac{iR}{iZ} \quad ...(4.46)$$

$$\cos\phi = \frac{R}{Z} \quad ...(4.47)$$

In an R-L circuit, it is clear form voltage (or impedance) triangle that current lags behind applied voltage by an angle ϕ. Hence from equations (4.45) and (4.47), equation for instantaneous current will be

$$\boldsymbol{i = I_m \sin(\omega t - \phi)} \quad ...(4.48)$$

Instantaneous power, $P_i = v.i$ ...(4.49)

Active power, $P_a = VI\cos\phi$ (Watts) ...(4.50)

Reactive power, $P_R = VI\sin\phi$ (VAR) ...(4.51)

Apparent power, $P_{AP} = VI$ (VA) ...(4.52)

where, V and I are rms values.

Conclusion

In case of AC circuit containing resistance and inductance, the current lags applied voltage by an angle ϕ (< 90°) and power factor of the circuit is lagging.

It is clear from equations (4.49) to (4.52) that in case of a RL circuit, due to presence of resistance all the powers exits in the circuit.

4.6 AC CIRCUIT CONTAINING RESISTANCE AND CAPACITANCE (RC CIRCUIT)

Consider a pure capacitor 'C' and a pure resistor 'R' connected across an alternating voltage source, as shown in Fig. 4.10.

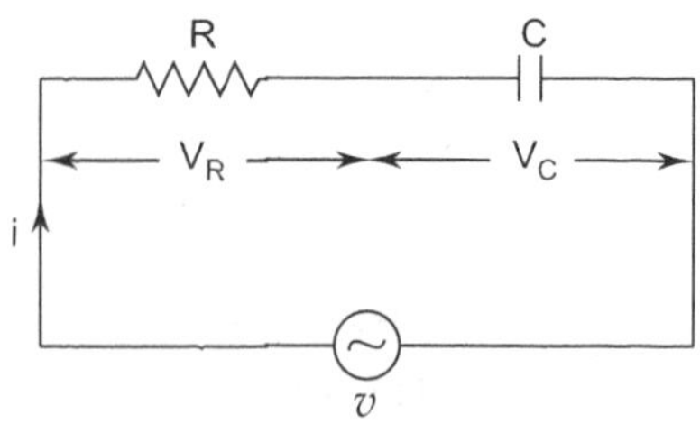

Fig. 4.10

Let the applied voltage be

$$v = V_m \sin \omega t \quad \text{[Eqn. (4.1)]}$$

Phasor diagram related to circuit shown in Fig. 4.10 are shown in Fig 4.11

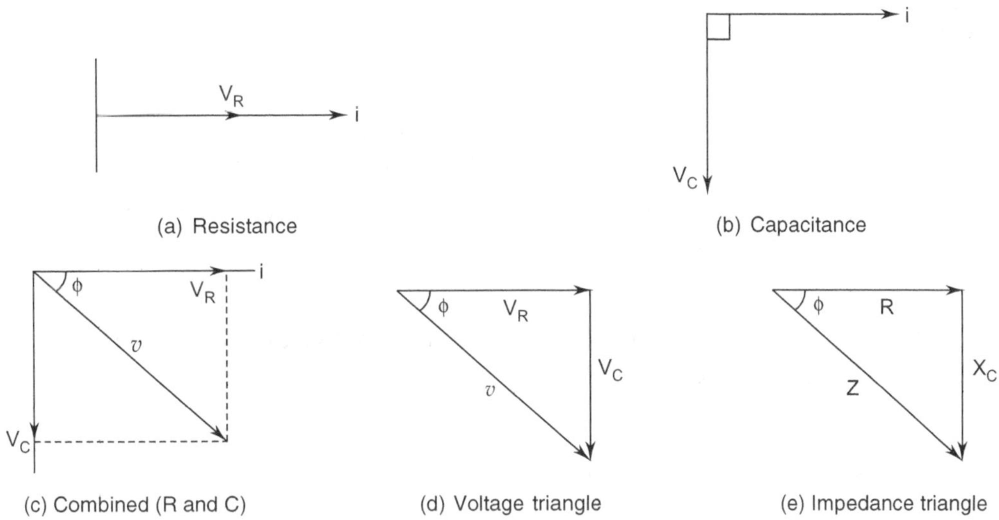

Fig. 4.11

From voltage triangle shown in Fig. 4.11(*d*)

$$v = \sqrt{V_R^2 + V_C^2} \quad \text{...(4.53)}$$

or $$v = \sqrt{(iR)^2 + (iX_C)^2} = i\sqrt{R^2 + {X_C}^2} \quad \text{...(4.54)}$$

or $$v = iZ \quad \text{...(4.55)}$$

where $$Z = \sqrt{R^2 + X_C^2} \quad \text{...(4.56)}$$

= Impedance of the circuit

or $$i = \frac{v}{Z} \quad \text{...(4.57)}$$

From combined phasor diagram of R and C, the power factor of the circuit is

$$\cos \phi = \frac{V_R}{V} = \frac{iR}{iZ} \quad \text{...(4.58)}$$

$$\cos \phi = \frac{R}{Z} \quad \text{...(4.59)}$$

In an R-C circuit, it is clear form voltage (or impedance) triangle that **current leads** applied voltage by an angle ϕ. Hence from equations (4.57) and (4.59), equation for instantaneous current will be

$$i = I_m \sin (\omega t + \phi) \quad \text{...(4.60)}$$

Instantaneous power, $P_i = v.i$ ...(4.61)

Active power, $P_a = VI \cos \phi$ (Watts) ...(4.62)

Reactive power, $P_R = VI \sin \phi$ (VAR) ...(4.63)

Apparent power, $P_{AP} = VI$ (VA) ...(4.64)

where, V and I are rms values.

Conclusion

In case of AC circuit containing resistance and capacitance, the current leads the applied voltage by an angle ϕ (< 90°) and power factor of the circuit is leading.

It is clear from equations (4.62) to (4.64) that in case of a RL circuit, due to presence of resistance all the power exits in the circuit.

4.7 AC CIRCUIT CONTAINING RESISTANCE, CAPACITANCE AND INDUCTANCE (RLC CIRCUIT)

Consider resistor 'R', a pure capacitor 'C' and a pure inductor 'L' connected in series across an alternating voltage source, as shown in Fig. 4.12.

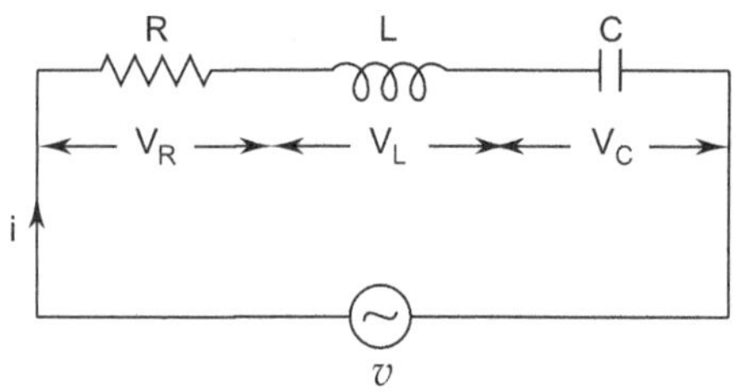

Fig. 4.12

Let the applied voltage be

$$v = V_m \sin \omega t \quad \text{[Eqn. (4.1)]}$$

Phasor diagram related to circuit shown in Fig. 4.12 are shown in Fig. 4.13.

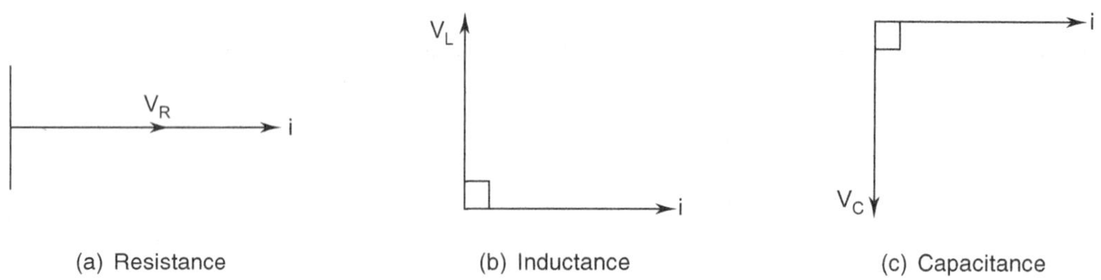

Fig. 4.13

And let $X_L > X_C$ then the overall phasor diagram will be

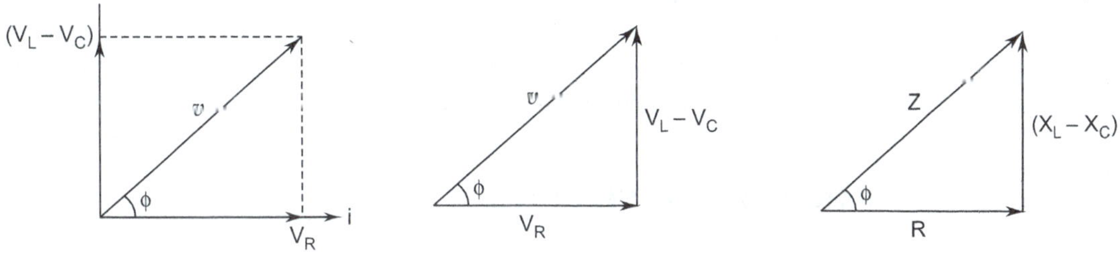

(a) Combined (b) Voltage triangle (c) Impedance triangle

Fig. 4.14

From the voltage triangle shown in Fig. 4.14 (*b*)

$$v = \sqrt{V_R^2 + (V_L - V_C)^2} \quad \text{...(4.65)}$$

or

$$v = i\sqrt{R^2 + (X_L - X_C)^2} \quad \text{...(4.66)}$$

or

$$v = iZ \quad \text{...(4.67)}$$

where

$$Z = \sqrt{R^2 + (X_L - X_C)^2} \quad \text{...(4.68)}$$

= Impedance of the circuit

or

$$i = \frac{v}{Z} \quad \text{...(4.69)}$$

From the combined phasor diagram, the power factor of the circuit is

$$\cos\phi = \frac{V_R}{V} = \frac{iR}{iZ} \quad \text{...(4.70)}$$

$$\cos\phi = \frac{R}{Z} \quad \text{...(4.71)}$$

In an R-L-C circuit, it is clear form the voltage (or impedance) triangle that

If $X_L > X_C$

Then the circuit is overall inductive and hence **current lags behind** the applied voltage by an angle ϕ.

If $X_L < X_C$

Then the circuit is overall capacitive and hence **current leads** the applied voltage by an angle ϕ.

Hence from equations (4.69) and (4.71), equation for instantaneous current will be

$$\boldsymbol{i = I_m \sin(\omega t \pm \phi)} \quad \text{...(4.72)}$$

Positive sign is to be used when current leads *i.e.,* $X_C > X_L$

Negative sign is to be used when current lays *i.e.,* $X_L > X_C$

Instantaneous power, $P_i = v.i$ ...(4.73)

Active power, $P_a = VI\cos\phi$ (Watts) ...(4.74)

Reactive power, $P_R = VI\sin\phi$ (VAR) ...(4.75)

Apparent power, $P_{AP} = VI$ (VA) ...(4.76)

where, V and I are rms values.

Conclusion

In case of AC circuit containing resistance, inductance and capacitance, nature of current and power factor (leading or lagging) depends upon magnitude of inductive reactance and capacitive reactance. If overall circuit is inductive then current lags behind voltage by an angle $\phi(\phi < 90°)$ and if overall circuit is capacitive then current leads voltage by an anlge ϕ $(\phi < 90°)$.

4.8 ANALYSIS OF PARALLEL AC CIRCUITS

When impedances are joined in parallel, there are three methods available to solve such circuits.

(*a*) Phasor Method

(*b*) Admittance Method

(*c*) Complex or Phasor Algebra.

4.8.1 Phasor Method

Let us consider a parallel RLC circuit having two branches in parallel as shown in Fig. 4.15.

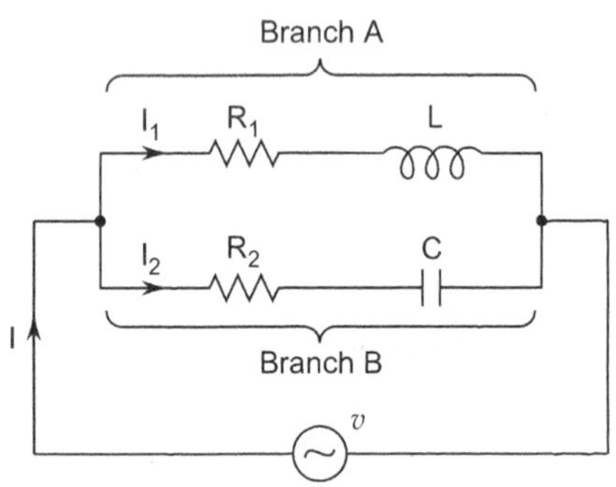

Fig. 4.15

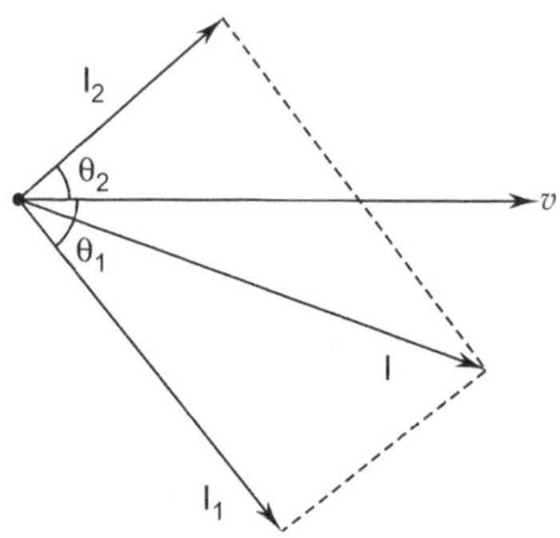

Fig. 4.16

The voltage across two parallel branches is same but current through them are different.

For branch A

In Branch A Current I_1 lags behind the applied voltage, V by θ_1 (Fig. 4.16), as it is an inductive branch.

$$Z_1 = \sqrt{R_1^2 + X_L^2} \quad \ldots(4.77)$$

$$I_1 = \frac{V}{Z_1} \quad \ldots(4.78)$$

$$\cos\theta_1 = \frac{R_1}{Z_1} \quad \ldots(4.79)$$

For branch B

In Branch B current I_2 leads the applied voltage V by θ_2 (Fig. 4.16), as it is an capacitive branch.

$$Z_2 = \sqrt{R_2^2 + X_C^2} \quad \ldots(4.80)$$

$$I_2 = \frac{V}{Z_2} \qquad \text{...(4.81)}$$

$$\cos \theta_2 = \frac{R_2}{Z_2} \qquad \text{...(4.82)}$$

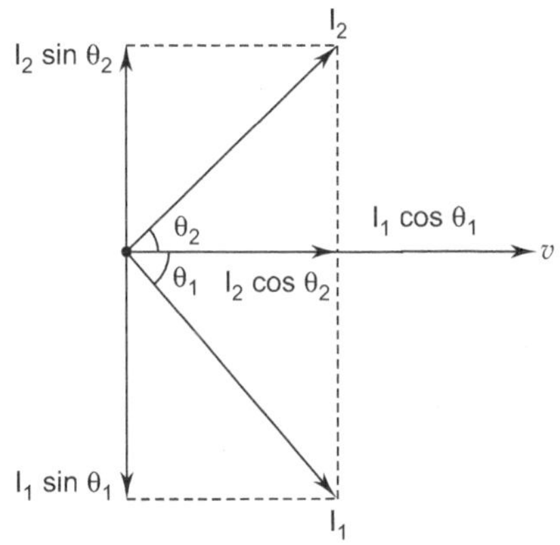

Fig. 4.17

The resultant current, I is vector sum of branch current I_1 and I_2 and can be found by resolving I_1 and I_2 into their X and Y components (or active or reactive components respectively) and then by combining these components (Fig. 4.17).

Sum of active components of I_1 and I_2

$$X = I_1 \cos \theta_1 + I_2 \cos \theta_2 \qquad \text{...(4.83)}$$

Sum of reactive components of I_1 and I_2

$$Y = I_1 \sin \theta_1 + I_2 \sin \theta_2 \qquad \text{...(4.84)}$$

Magnitude of the resultant vector $\quad I_m = \sqrt{X^2 + Y^2} \qquad \text{...(4.85)}$

Phase angle of the resultant vector $\quad \theta = \tan^{-1} \frac{Y}{X} \qquad \text{...(4.86)}$

The resultant of the two branch currents will be

$$i = I_m \sin (\omega t \pm \theta) \qquad \text{...(4.87)}$$

4.8.2 Admittance Method (Y)

Admittance of a circuit is defined as the reciprocal of its impedance.

$$Y = \frac{1}{Z} \text{ (Siemens or mho)} \qquad \text{...(4.88)}$$

Step 1. Find the impedance of each parallel branch *i.e.*, Z_1 and Z_2.

$$\left. \begin{aligned} Z_1 &= \sqrt{R_1^2 + X_L^2} \\ Z_2 &= \sqrt{R_2^2 + X_C^2} \end{aligned} \right\} \qquad \text{...(4.89)}$$

Step 2. Find the admittance of each parallel branch *i.e.*, Y_1 and Y_2.

$$Y_1 = \frac{1}{Z_1} \quad \text{...(4.90)}$$

$$Y_2 = \frac{1}{Z_2} \quad \text{...(4.91)}$$

Step 3. Find the total admittance of each parallel branch *i.e.*, Y_1 and Y_2.

$$Y = Y_1 + Y_2 = \frac{1}{Z_1} + \frac{1}{Z_2} \quad \text{...(4.92)}$$

Step 4. Total current in the circuit, I, will be

$$I = VY \quad \text{...(4.93)}$$

4.8.3 Complex or Phasor Algebra

As voltage across both parallel branches will be same hence branch current will be

$$I_1 = \frac{V}{Z_1} \quad \text{...(4.94)}$$

and

$$I_2 = \frac{V}{Z_2} \quad \text{...(4.95)}$$

Impedance of branch **A** in rectangular form will be

$$Z_1 = R_1 + jX_L \quad \text{...(4.96)}$$

Impedance of branch **A** in rectangular form will be

$$Z_2 = R_2 + (-jX_C) = R_2 - jX_C \quad \text{...(4.97)}$$

$$I = \frac{V}{Z_1} + \frac{V}{Z_2} = V\left[\frac{1}{Z_1} + \frac{1}{Z_2}\right] \quad \text{...(4.98)}$$

4.9 POWER FACTOR

To understand power factor, we will first start with definition of some basic terms which we have already discussed in previous sections.

kW is working power (also called actual power or active power or real power).

It is the power that actually powers the equipment and performs useful work.

kVAR is reactive power.

It is the power that magnetic equipment (transformer, motor and relay) needs to produce the magnetizing flux.

kVA is apparent power.

It is the "vectorial summation" of kVAR and kW.

Power factor is defined by IEEE and IEC as ***"the ratio of true power to apparent power".***

Let's look at a simple analogy in order to better understand these terms.

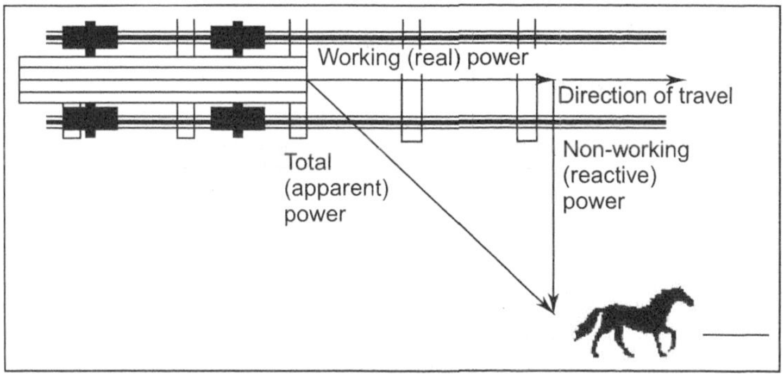

Fig. 4.18

To understand the concept, visualize a horse pulling a railroad car down a railroad track (Fig. 4.18). Because the railroad ties are uneven, the horse must pull the car from the side of the track. The horse is pulling the railroad car at an angle to the direction of the car's travel. The power required to move the car down the track is the working (real) power. The effort of the horse is the total (apparent) power. Because of the angle of the horse's pull, not all of the horse's effort is used to move the car down the track. The car will not move sideways; therefore, the sideways pull of the horse is wasted effort or non-working (reactive) power.

The angle of the horse's pull is related to power factor, which is defined as the ratio of real (working) power to apparent (total) power. If the horse is led closer to the center of the track, the angle of side pull decreases and the real power approaches the value of the apparent power. Therefore, the ratio of real power to apparent power (the power factor) approaches 1. As the power factor approaches 1, the reactive (non-working) power approaches zero (0).

$$\text{Power factor} = \frac{\text{Real power}}{\text{Apparent power}} \qquad \ldots(4.99)$$

Let's look at another analogy.

Mr. Bond here is dragging a heavy load (Fig. 4.19). Bond's working power (or actual power) in the forward direction, where he most wants his load to travel, is kW.

Unfortunately, Bond can't drag his load on a perfect horizontal (he would get a tremendous backache), so his shoulder height adds a little reactive power, or kVAR.

The Apparent Power Bond is dragging, kVA, is this "vectorial summation" of kVAR and kW.

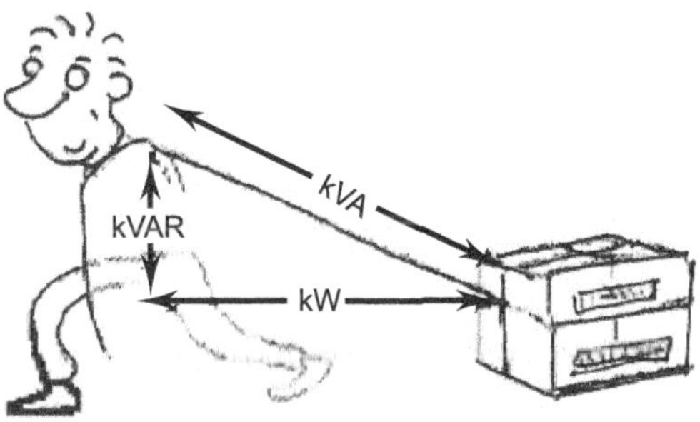

Fig. 4.19

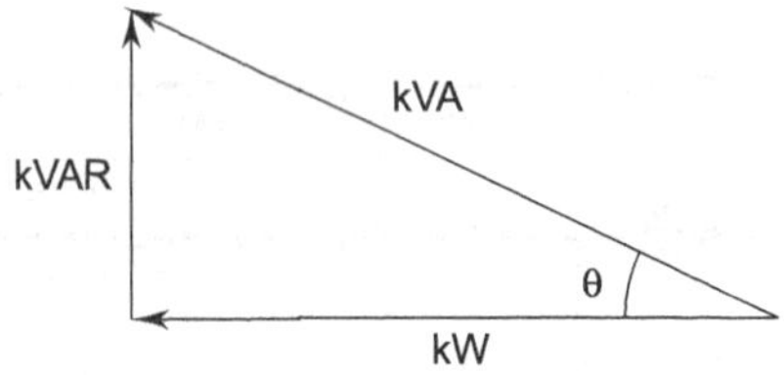

Fig. 4.20 *The power triangle*

$$\text{P.F.} = \frac{\text{kW}}{\text{kVA}} = \cos\theta \qquad \text{...(4.100)}$$

$$\text{kVA} = \sqrt{\text{kW}^2 + \text{kVAR}^2} \qquad \text{...(4.101)}$$

For a voltage waveform different waveforms of current and the corresponding power factor values are shown in Fig. 4.21 at different phase shift.

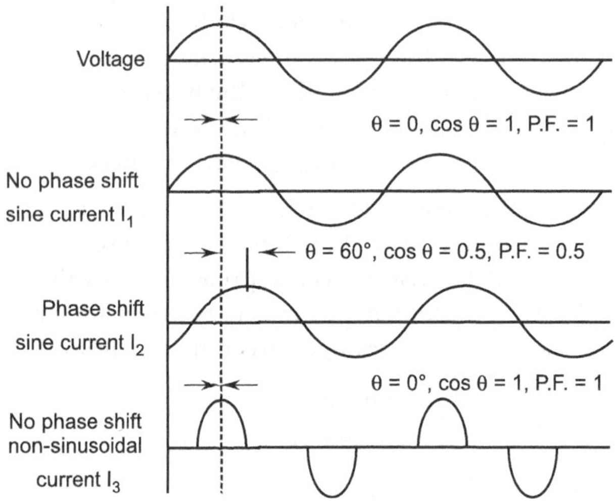

Fig. 4.21

4.9.1 Reasons for Power Factor Measurement

Power factor measurement is important for two main reasons. First, an overall power factor of less than 1 means that an electricity supplier has to provide more generating capacity than actually is required.

For example, consider an office building drawing 200 A at 400 V.

The supply transformer and backup UPS must be rated at

$$200 \text{ A} \times 400 \text{ V} = 80 \text{ kVA}.$$

But if the power factor of the loads is only 0.6, then only

$$80 \text{ kVA} \times 0.6 = 48 \text{ kVA}$$

of real power are being consumed. In other words, if the power factor were 1, the supply capacity (transformers, cables, switchgear, UPS) could be considerably smaller.

Secondly, the current waveform distortion that contributes to reduced power factor is a cause of voltage waveform distortion and overheating in the neutral cables of three-phase systems.

Primarily for this reason, international standards such as IEC 61000-3-2 have been established to control current waveform distortion by introducing limits for the amplitude of current harmonics.

4.9.2 Causes of Low Power Factor

Since power factor is defined as the ratio of kW to kVA, we see that low power factor results when kW is small in relation to kVA. Remembering our second analogy, this would occur when kVAR (Mr. Bond shoulder height) is large.

What causes a large kVAR in a system? The answer is inductive loads. Inductive loads (which are sources of reactive power) include:

- Transformers
- Induction motors
- Induction generators (wind mill generators)
- High intensity discharge (HID) lighting

These inductive loads constitute a major portion of the power consumed in industrial complexes. Reactive power (kVAR) required by inductive loads increases the amount of apparent power (kVA) in your distribution system. This increase in reactive and apparent power results in a larger angle (measured between kW and kVA). Recall that, as 0 increases, cosine 0 (or power factor) decreases.

4.9.3 Problems of Low Power Factor

Some of the problems occur due to low power factor are:

- Reduces equipment operation by decreasing voltage.
- Decrease energy efficiency.
- Increases line losses.
- Forced for costly upgrades.

4.9.4 Power Factor Improvement

We have seen that sources of reactive power (inductive loads) decrease power factor, such as:

- Transformers
- Induction motors
- Induction generators (wind mill generators)
- High intensity discharge (HID) lighting

Similarly, consumers of reactive power increase power factor, such as:

- Capacitors
- Synchronous generators
- Synchronous motors

Thus, it comes as no surprise that one way to increase power factor is to add capacitors to the system. This, and other ways of increasing power factor are listed as follows:

1. Installing capacitors (kVAR Generators)

   Installing capacitors decreases the magnitude of reactive power (kVAR or foam), thus increasing your power factor. Here is how it works. Reactive power (kVARS), caused by inductive loads, always acts at a 90° angle to working power (kW).

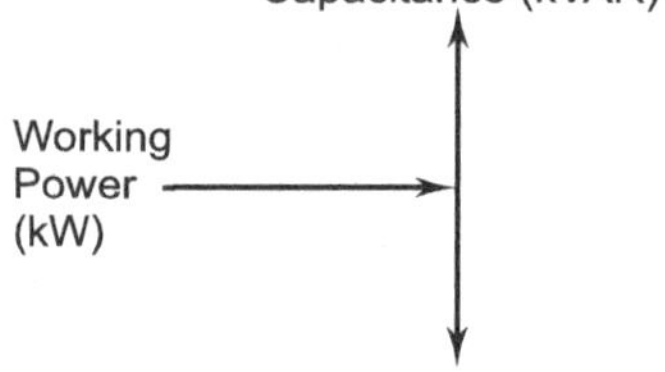

Fig. 4.22

   Inductance and capacitance react 180° to each other. Capacitors store kVARS and release energy opposing the reactive energy caused by the inductor.

   The presence of both a capacitor and inductor in the same circuit results in the continuous alternating transfer of energy between the two.

   Thus, when the circuit is balanced, all the energy released by the inductor is absorbed by the capacitor.

2. Minimizing operation of idling or lightly loaded motors.

   We already talked about the fact that low power factor is caused by the presence of induction motors. But, more specifically, low power factor is caused by running induction motors lightly loaded.

3. Avoiding operation of equipment above its rated voltage.

4. Replacing standard motors as they burn out with energy-efficient motors.

   Even with energy-efficient motors, power factor is significantly affected by variations in load. A motor must be operated near its rated load in order to realize the benefits of a high power factor design.

4.9.5 Power Factor Improvement by Connecting a Capacitor in Parallel to the Load

Initially let us assume that load connected in an AC circuit be of R-L type (Fig. 4.23).

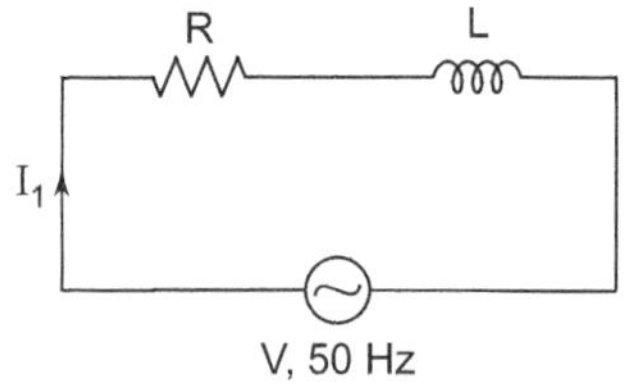

Fig. 4.23

Let total current in the circuit be denoted by I_1 and suppose that power factor of the circuit in present condition by $\cos \phi_1$, *i.e.*

Condition - I

$$\text{Total current} = I_1$$
$$\text{Power factor} = \cos \phi_1$$
$$\text{Applied voltage} = V$$

Phasor diagram will be as shown in Fig. 4.24.

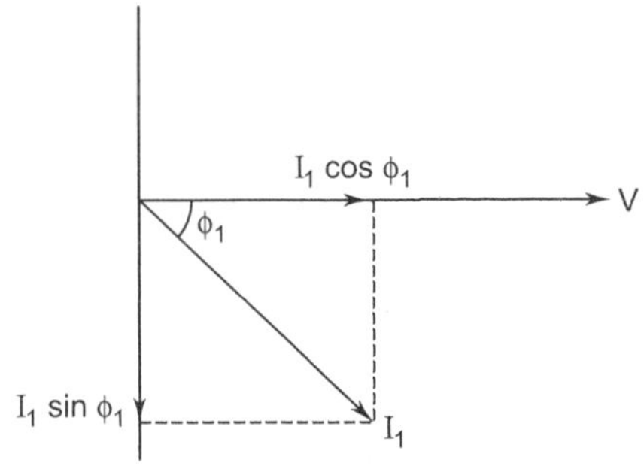

Fig. 4.24

Condition - II

When a capacitor is connected in parallel to the given load, the new power factor will be, say, $\cos \phi_2$.

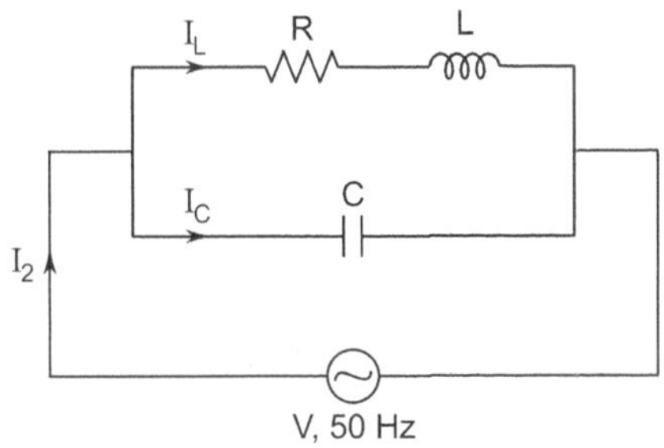

Fig. 4.25

In such a condition

Total current $= I_2$

Current through the load $= I_L = I_1$ (as voltage remains some across the load)

Current through the capacitor $= I_C$

New improved power factor $= \cos \phi_2$

When we connect a capacitor in parallel,

- Voltage remains same
- Power taken by the load remains same
- Current through the load remains same
- Current taken from the supply will reduce *i.e.*, (I_2 becomes smaller than I_1)
- As we connect the capacitor in parallel, the phase angle between the supply voltage (V) and the total circuit current (I_2) will reduce (as shown in phasor diagram of Fig. 4.26) and hence power factor will improve.

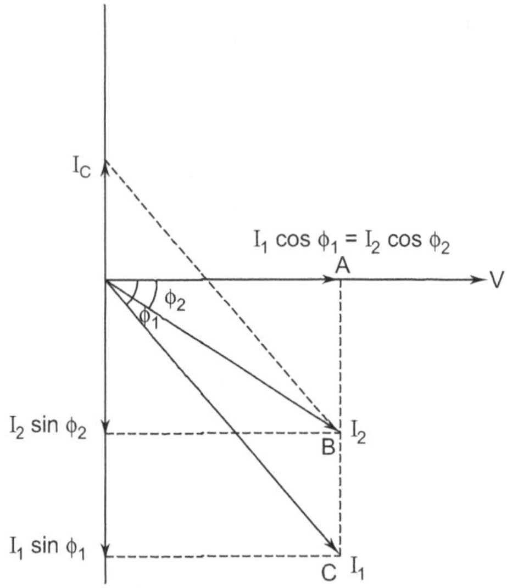

Fig. 4.26

It is clear from the phasor diagram that active component of both total current of the circuit remains same *i.e.*,

$$\boxed{I_1 \cos \phi_1 = I_2 \cos \phi_2} \quad ...(4.102)$$

or $$I_2 = \frac{I_1 \cos \phi_1}{\cos \phi_2} \quad ...(4.103)$$

The current through the capacitor (I_C) can be determined as follow:

$$AB = I_2 \sin \phi_2 \quad ...(4.104)$$

$$BC = I_C \quad ...(4.105)$$

$$AC = I_1 \sin \phi_1 \quad ...(4.106)$$

Also, $$AC = AB + BC \quad ...(4.107)$$

$$BC = AC - AB \quad ...(4.108)$$

From equations (4.104), (4.105) and (4.106)

$$\boxed{I_C = I_1 \sin \phi_1 - I_2 \sin \phi_2} \quad ...(4.109)$$

Equation (4.109) gives expression for the current following through the capacitor.

The rating of capacitor is given by

$$V = I_C X_C \quad ...(4.110)$$

or $$X_C = \frac{V}{I_C} = \frac{1}{\omega C} \quad ...(4.111)$$

or $$\boxed{C = \frac{I_C}{V\omega}} \quad ...(4.112)$$

SOLVED PROBLEMS

Problem 4.1. *A non-inductive resistance of 10 Ω is connected in series with an inductive coil across 200 V, 50 Hz AC supply. The current drawn by the series combination is 10 A. The resistance of the coil is 2 Ω. Determine*

(i) inductance of the coil

(ii) power factor

(iii) voltage across the coil.

Solution. Total resistance of the circuit (Fig. 4.27)

$$R = \text{Non-inductive resistance} + \text{resistance of coil}$$
$$= 10 + 2 = 12\,\Omega$$

Voltage drop across the resistance of the whole circuit,

$$V_R = IR = 10 \times 12 = 120 \text{ V}$$

Let the voltage drop across the inductance of the coil be V_L volts.

Now, $$\sqrt{V_R^2 + V_L^2} = \text{Supply voltage, } V_S$$

or $$V_L = \sqrt{V_S^2 - V_R^2}$$
$$= \sqrt{200^2 - 120^2} = 160 \text{ V}$$

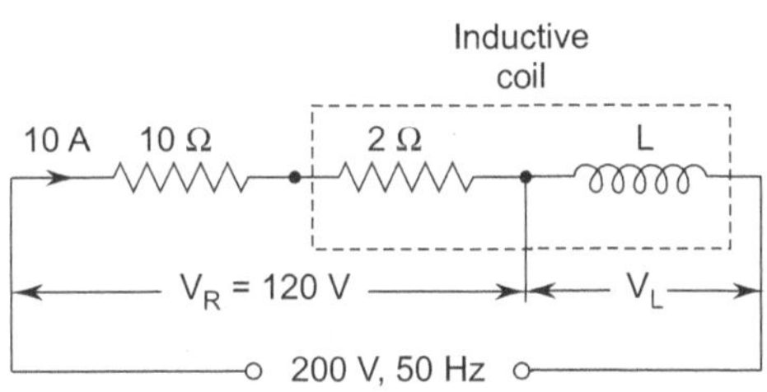

Fig. 4.27

Inductive reactance of the coil, $X_L = \dfrac{V_L}{I} = \dfrac{160}{10} = 16\,\Omega$

(*i*) Inductance of coil, $L = \dfrac{X_L}{2\pi f} = \dfrac{16}{2\pi \times 50} = \mathbf{0.051\ H\ or\ 51\ mH.}$ **Ans.**

(*ii*) Power factor of the coil $= \cos\left(\tan^{-1}\dfrac{16}{2}\right) = \mathbf{0.124\ (lagging)}$ **Ans.**

Power factor of the circuit $= \dfrac{R}{Z} = \dfrac{12}{\sqrt{12^2 + 16^2}} = \mathbf{0.6\ (lagging)}$ **Ans.**

(*iii*) Voltage across the coil, $V_C = I\sqrt{2^2 + 16^2} = 10 \times \sqrt{260} = \mathbf{161.245\ V.}$ **Ans.**

Problem 4.2 *A voltage wave e (t) = 141.4 sin 120t produces a current i (t) = 14.14 sin 120t + 7.07 cos (120t + 30°) in a circuit. Determine (i) the resultant time expression of current, (ii) the power factor and the power delivered by the source.*

Solution. Current $i(t) = 14.14 \sin 120t + 7.07 \cos(120t + 30°)$

$= 14.14 \sin 120t + 7.07 \sin(120t + 30° + 90°)$

Maximum value of current, $I_{r\,max} = \sqrt{(\text{X-component})^2 + (\text{Y-component})^2}$

$$= \sqrt{(14.14 \cos 0° + 7.07 \cos 120°)^2 + (14.14 \sin 0° + 7.07 \sin 120°)^2}$$

$$= \sqrt{(14.14 - 3.535)^2 + (6.1228)^2} = 12.25 \text{ A}$$

Phase angle of resultant current is given as

$$\theta = \tan^{-1} \frac{\text{Y-component}}{\text{X-component}} = \tan^{-1} \frac{6.1228}{10.605} = 30°$$

(*i*) So the resultant time expression of current is

$i(t) = \mathbf{12.25 \sin (120t + 30°)}$ **Ans.**

Phase angle between V and I is $\phi = 30°$ (leading)

(*ii*) Power factor = $\cos 30° = \mathbf{0.866}$ **(leading)** **Ans.**

Power delivered by source, $P = V_{rms} I_{rms} \cos\phi = \frac{V_{max}}{\sqrt{2}} \cdot \frac{I_{max}}{\sqrt{2}} \cos\phi$

$$= \frac{V_{max} I_{max}}{2} \cos\phi$$

$$= \frac{141.4 \times 12.25}{2} \cos 30° = \mathbf{750\ W}$$ **Ans.**

Problem 4.3. *A metal-filament lamp, rated at 750 W, 100 V is to be connected in series with a capacitance across a 230 V, 50 Hz supply. Calculate the value of capacitance required. Draw phasor diagram.*

Solution. Let a pure capacitance of C farads be connected in series with the lamp, as shown in Fig. 4.28 (*a*).

Since in this case voltage across the capacitor lags behind the current by 90° or $\frac{\pi}{2}$ radians while that across the lamp will be in phase with it, from phasor diagram shown in Fig. 4.28 (*b*), we have

$$V_C = \sqrt{(230)^2 - (100)^2} = 207 \text{ V}$$

$$I = \frac{P}{V} = \frac{750}{100} = 7.5 \text{ A}$$

$$X_C = \frac{V_C}{I} = \frac{207}{7.5} = 27.6\ \Omega$$

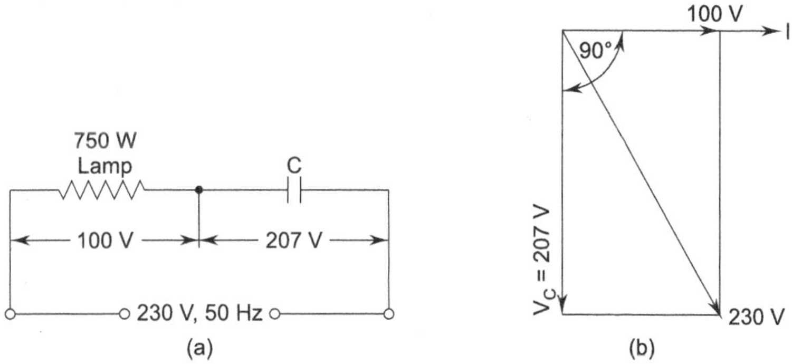

Fig. 4.28

and $$C = \frac{1}{2\pi f \cdot X_C} = \frac{1}{2\pi \times 50 \times 27.6} = \mathbf{115.33\ \mu F}$$ **Ans.**

Problem 4.4. *A 120 V, 60 W lamp is to be operated on 220 V, 50 Hz supply mains. In order that lamp should operate on correct voltage, calculate value of*

(a) non-inductive resistance, *(b) pure inductance.*

Solution. A lamp is consider as pure resistance

Rated power of lamp, P = 60 W

Rated voltage of lamp, V = 120 V

current drawn by the lamp when operated on rated voltage

$$I = \frac{P}{V} = \frac{60}{120} = 0.5 \text{ A}$$

Resistance of the lamp, $R_{Lamp} = \frac{V}{I} = \frac{120}{0.5} = 240\ \Omega$

(*a*) When a non-inductive resistance (means pure resistance) R, is put in series with the lamp [Fig. 4.29 (*a*)]

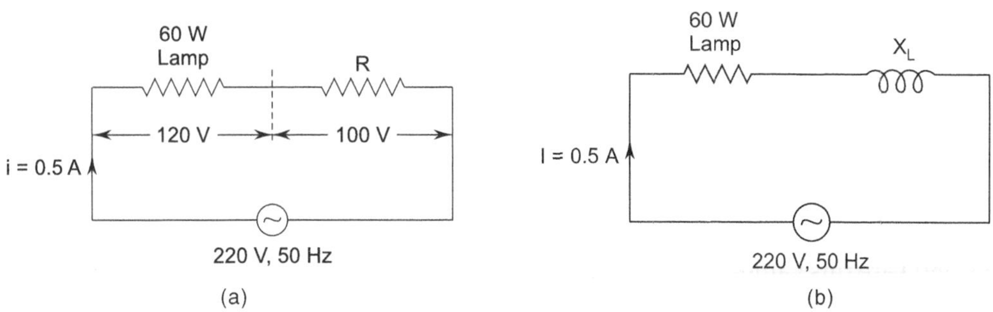

Fig. 4.29

$$V = I\,(R_{LAMP} + R)$$

or $$220 = 0.5\,(240 + R)$$

or $$R = \mathbf{200\ \Omega}$$ **Ans.**

(*b*) When a pure inductance, X_L is connected in series with the lamp [Fig. 4.29(*b*)]

$$I = \frac{V}{Z}$$

or $$0.5 = \frac{220}{\sqrt{R^2 + X_L^2}} = \frac{200}{\sqrt{(200)^2 + X_L^2}}$$

or $$X_L = 368.78$$

or $$2\pi f L = 368.78$$

or $$L = \frac{368.78}{2\pi \times 50} = \mathbf{1.173\ Henry}$$ **Ans.**

Problem 4.5. *A series RLC circuit has 100 Ω resistor, 0.318 H inductor and C, $v = 230 \times \sqrt{2}\sin \omega t$ V, $i = 2.3\sqrt{2}\sin \omega t$ A.*

Find (i) C (ii) V_L (iii) power. Take ω = 314.15 r/second.

Solution. Resistance, $R = 100\ \Omega$

Inductive reactance, $X_L = \omega L = 314.5 \times 0.318 = 100\ \Omega$ [∵ $\omega = 314.5$]

Impedance of the circuit (Fig. 4.30)

$$Z = \frac{V_m}{I_m} = \frac{230\sqrt{2}}{2.3\sqrt{2}} = 100\ \Omega$$

(*i*) Let capacitive reactance be X_C

Now, $$Z = \sqrt{R^2 + (X_L \sim X_C)^2}$$

or $$100 = \sqrt{(100)^2 + (100 \sim X_C)^2}$$

or $$X_C = 100\ \Omega$$

or $$\frac{1}{\omega C} = 100$$

or $$C = \mathbf{31.8\ \mu F}$$ **Ans.**

(*ii*) Voltage across inductor

$$V_L = I_{rms} X_L = \frac{I_m}{\sqrt{2}} . X_L$$

$$= \frac{2.3\sqrt{2}}{\sqrt{2}} \times 100 = \mathbf{230\ V}$$ **Ans.**

(*iii*) Power

$$\text{Power} = VI \cos \phi$$

where V and I are rms values

The phase difference between voltage and current is 0°.

Hence, $$\cos \phi = \cos 0° = 1$$

$$V_{rms} = \frac{V_m}{\sqrt{2}} = \frac{230\sqrt{2}}{\sqrt{2}} = 230$$

$$I_{rms} = \frac{I_m}{\sqrt{2}} = \frac{2.3\sqrt{2}}{2} = 2.3$$

$$P = VI\cos\phi$$

$$= 230 \times 2.3 \times 1 = \mathbf{529\ W}$$ **Ans.**

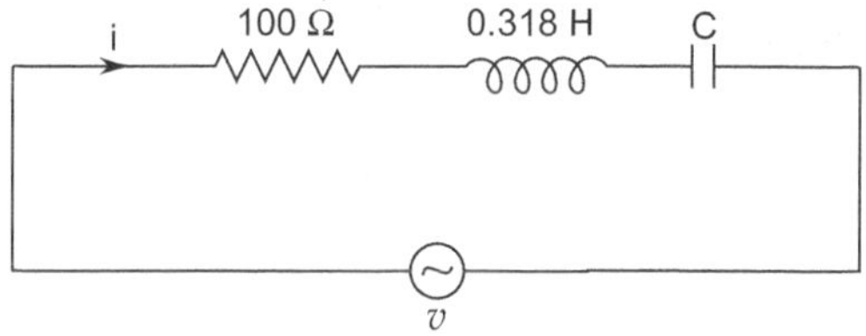

Fig. 4.30

Problem 4.6. *A 100 V, 80 W lamp is to be operated on 230 V, 50 Hz AC. Calculate the inductance to be connected in series with the lamp. Lamp can be taken as a pure resistance.*

Solution. Current drawn by the lamp when operated on rated voltage,

$$I = \frac{100}{80} = 1.25\ A.$$

Let pure inductance of L Henry be connected in series with the lamp, the voltage across inductance,

$$V_L = \sqrt{(230)^2 - (100)^2} = 207.123\ V$$

and $$L = \frac{V_L}{2\pi f I} = \frac{207.123}{2\pi \times 50 \times 1.25} = \mathbf{0.527\ H.}$$ **Ans.**

Problem 4.7. *A series AC circuit has a resistance of 15 Ω and inductive reactance of 10 Ω. Calculate the value of a capacitor which is connected across this series combination so that system has unity power factor. The frequency of AC supply is 50 Hz.*

Solution. Series circuit resistance, $R = 15\ \Omega$

Series circuit reactance, $X_L = 10\ \Omega$

Series circuit conductance, $$G = \frac{R}{R^2 + X_L^2} = \frac{15}{(15)^2 + (10)^2} = \frac{15}{325}\ S$$

Series circuit susceptance, $$B = \frac{X}{R^2 + X_L^2} = \frac{10}{15^2 + 10^2} = \frac{10}{325}\ S\ \text{(inductive)}$$

The power factor of the whole circuit will be unity only if capacitive susceptance of the capacitor to be connected across the series combination is equal to the equivalent inductive susceptance of the series combination *i.e.*, equal to $\frac{10}{325}$ S.

Reactance of the capacitance connected across the series combination,

$$X_C = \frac{1}{B_C} = \frac{1}{10/325} = 32.5\ \Omega$$

Capacitance of the required capacitor,

$$C = \frac{1}{2\pi f X_C} = \frac{1}{2\pi \times 50 \times 32.5} = \mathbf{98\ \mu F} \qquad \textbf{Ans.}$$

Problem 4.8. *A resistance and an inductance are connected in series across a voltage $v = 283 \sin 314t$. The current expression is found to be $4 \sin\left(314t - \frac{\pi}{4}\right)$. Find the values of resistance, inductance and power factor.*

Solution. Circuit voltage, $v = 283 \sin 314t$ V

Circuit current, $i = 4 \sin\left(314t - \frac{\pi}{4}\right)$

From the equations of circuit voltage and circuit current it is seen that current, I, Lags behind the applied voltage V, by $\frac{\pi}{4}$ radian or 45°.

So phase angle, $\phi = 45°$ (lagging)

Power factor of the circuit, $PF = \cos \phi = \cos 45° = \mathbf{0.707}$ **(lagging)** **Ans.**

Circuit impedance, $Z = \frac{V_{max}}{I_{max}} = \frac{283}{4} = 70.75\ \Omega$

Circuit resistance, $R = Z \cos \phi = 70.75 \times 0.707 = \mathbf{50\ \Omega}$ **Ans.**

Circuit reactance, $X_L = Z \sin \phi = 70.75 \times \sin 45°$

or $X_L = 70.75 \times 0.707 = 50\ \Omega$

Frequency, $f = \frac{314}{2\pi} = 50$ Hz

Inductance, $L = \frac{X_L}{2\pi f} = \frac{50}{2\pi \times 50} = \mathbf{0.1592\ H.}$ **Ans.**

Problem 4.9. *The voltage and current through a circuit element are*

$$v = 50 \sin (314t + 55°)\ V$$
$$i = 10 \sin (314t + 325°)\ A$$

Find the value of power drawn by the element.

Solution. Phase angle between voltage and current = 325° – 55° = 270° *i.e.*, current leads the voltage by 270° or voltage leads current by 90° *i.e.*, current lags behind the voltage by 90°.

The phasor diagram* is shown in Fig. 4.31.

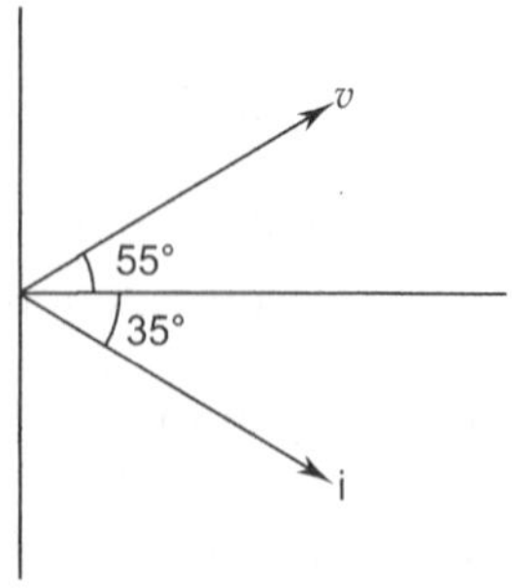

Fig. 4.31

*If both the alternating quantities (*i.e.*, current as well as voltage) have some value of phase angle, then take *x*-axis line as reference vector.

Power drawn by the element, $$P = VI\cos\phi = \frac{V_{max}}{\sqrt{2}} \times \frac{I_{max}}{\sqrt{2}} \cos\phi$$

$$= \frac{50}{\sqrt{2}} \times \frac{10}{\sqrt{2}} \times \cos 90° = \mathbf{0.} \qquad \textbf{Ans.}$$

Problem 4.10. *A 100 ohm resistance is carrying a sinusoidal current given by 3 cos ωt. Determine:*

(i) instantaneous power taken by resistance,

(ii) average power.

Solution. (*i*) Instantaneous power taken by the resistance,

$$p = vi = iR \times i = 100 \times 3\cos\omega t \times 3\cos\omega t$$

$$= 900\cos^2\omega t = \mathbf{450\,(1 + \cos 2\omega t)\ W} \qquad \textbf{Ans.}$$

(*ii*) Average power $= \mathbf{450\ W}$ **Ans.**

Problem 4.11. *The voltage and current through a circuit element are*

$$v = 100\sin(314t + 45°)\ V$$

$$i = 10\sin(314t + 315°)\ A.$$

(i) Identify the circuit element

(ii) Find its value

(iii) Obtain expression for power.

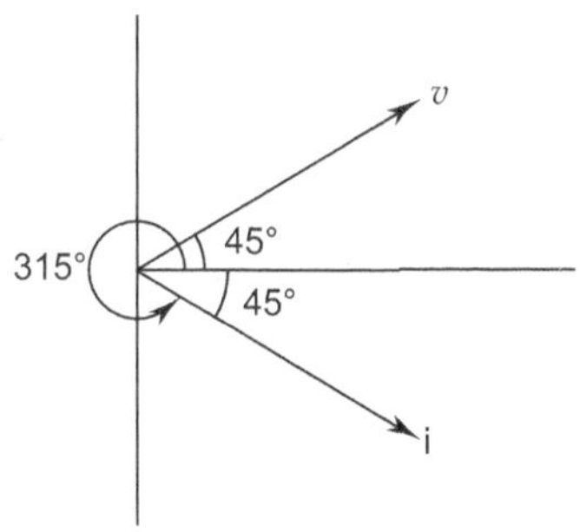

Fig. 4.32

Solution. (*i*) From the phasor diagram (Fig. 4.32) it is clear that current I lags voltage by an angle 90°. Hence the circuit element is an inductor.

(*ii*) Inductive reactance of the inductor,

$$X_L = \frac{V_{rms}}{I_{rms}} = \frac{100/\sqrt{2}}{10/\sqrt{2}} = \mathbf{10\ \Omega} \qquad \textbf{Ans.}$$

and $$L = \frac{X_L}{\omega} = \frac{10}{314} = \mathbf{0.0318\ H\ or\ 31.8\ mH.} \qquad \textbf{Ans.}$$

(*iii*) Expression for power is

$$P = v.\,i$$

$$= 100\sin(314\,t + 45°) \times 10\sin(314\,t - 45°)$$

$$= 1000 \sin(314\,t + 45°) \sin(314\,t - 45°)$$

$$= 1000\ \frac{\sin(628\,t)}{2}$$

$$P = \mathbf{500\ sin\ (628\ t)}$$ **Ans.**

Problem 4.12. *Determine the following in the circuit shown in Fig. 4.33.*

(i) The current phasors I, I_1 and I_2.

(ii) Active power dissipated in the three resistive branches.

(iii) Power factor of the circuit.

Solution. Impedance of branch I of parallel circuit,

$$Z_1 = R_1 + jX$$
$$= (3 + j4)\ \Omega$$
$$= 5 \angle 53.13°\ \Omega$$

Impedance of branch II of parallel circuit, $Z_2 = (5 - j12)\ \Omega$

$$= 13 \angle -67.38°\ \Omega$$

Impedance of series branch, $Z_s = (8 + j6)\ \Omega$

$$= 10 \angle 36.87°\ \Omega$$

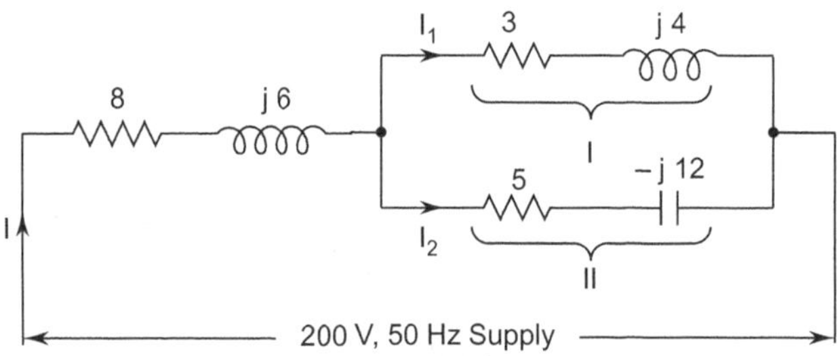

Fig. 4.33

Equivalent impedance of parallel circuit, $\boldsymbol{Z_p} = \dfrac{Z_1 \times Z_2}{Z_1 + Z_2} = \dfrac{5 \angle 53.13° \times 13 \angle -67.38°}{(3 + j4) + (5 - j12)}$

$$= \frac{65 \angle -14.25°}{(8 - j8)} = \frac{65 \angle -14.25°}{11.314 \angle -45°}$$

or $Z_p = 5.745 \angle 30.75°\ \Omega$ (in polar form)

or $Z_p = (4.937 + j2.937)\ \Omega$ (in rectangular form)

Total impedance of the circuit, $Z = Z_s + Z_p = (8 + j6) + (4.937 + j2.937)\ \Omega$

$$= (12.937 + j8.937)\ \Omega = 15.7237 \angle 34.64°\ \Omega$$

Let the supply voltage be the reference phasor *i.e.*, $V = 200 \angle 0°$ V

(*i*) Circuit current, $I = \dfrac{V}{Z} = \dfrac{200 \angle 0°}{15.7237 \angle 34.64°}$

$$= \mathbf{12.72 \angle -34.64°\ A}$$ **Ans.**

Power factor of the circuit, $\cos\phi = \cos(-34.64°) = \mathbf{0.8228\ (lagging)}$ **Ans.**

Current in branch I, $$I_1 = I\frac{Z_2}{Z_1 + Z_2} = \frac{12.72\angle-34.64° \times 13\angle-67.38°}{(3+j4)+(5-j12)}$$

$$= \frac{165.36\angle-102.02°}{11.314\angle-45°}$$

$$= \mathbf{14.616\angle-57.02°\ A}$$ **Ans.**

Current in branch II, $$I_2 = I\frac{Z_1}{Z_1 + Z_2} = \frac{12.72\angle-34.64 \times 5\angle 53.13°}{11.314\angle-45°}$$

$$= \mathbf{5.62\angle 63.49°\ A}$$ **Ans.**

(*ii*) Active power dissipated in:

Series branch $= I^2R_s = 12.72^2 \times 8 = \mathbf{1{,}294.4\ W}$

Parallel branch 1 $= I_1^2 R_1 = 14.616^2 \times 3 = \mathbf{641\ W}$

Parallel branch 2 $= I_2^2 R_2 = 5.62^2 \times 5 = \mathbf{158\ W}$

Total power loss, $= 1{,}294.4 + 641 + 158 = \mathbf{2{,}093.4\ W}$ **Ans.**

(*iii*) Power factor of the circuit, $\cos\phi = \frac{P}{VI} = \frac{2{,}093.4}{200 \times 12.72} = \mathbf{0.8228\ (lagging)}$ **Ans.**

Problem 4.13. *Fig. 4.34 shows a series-parallel circuit. Find:*

(*i*) *Admittance of each parallel branch.*

(*ii*) *Total circuit impedance.*

(*iii*) *Supply current and power factor.*

(*iv*) *Total power supplied by the source.*

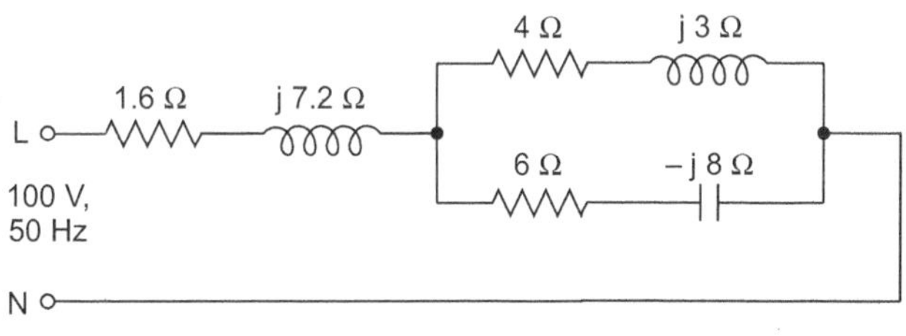

Fig. 4.34

Solution. (*i*) Admittance of parallel branches,

or $$Y_1 = \frac{1}{Z_1} = \frac{1}{4+j3} = \frac{4-j3}{4^2+3^2}$$

or $Y_1 = (0.16 - j0.12)$ S

$Y_1 = \mathbf{0.2\angle-36.87°\ S}$ **Ans.**

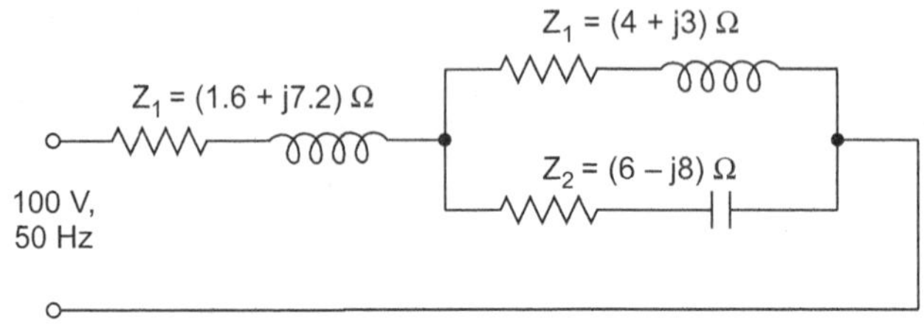

Fig. 4.35

$$Y_2 = \frac{1}{Z_2} = \frac{1}{6-j8} = \frac{6+j8}{6^2+8^2} = \textbf{(0.06 + } \boldsymbol{j}\textbf{0.08) S or 0.1 } \angle\textbf{53.13° S} \quad \textbf{Ans.}$$

Admittance of parallel circuit,

$$Y_p = Y_1 + Y_2 = (0.16 - j0.12) + (0.06 + j0.08)$$

or

$$Y_p = 0.22 - j0.04$$

Impedance of the parallel circuit,

$$Z_p = \frac{1}{Y_p} = \frac{1}{0.22 - j0.04} = \frac{0.22 + j0.04}{(0.22)^2 + (0.04)^2}$$

or

$$Z_p = (4.4 + j0.8)\ \Omega$$

(*ii*) Total circuit impedance,

$$Z = Z_s + Z_p = (1.6 + j7.2) + (4.4 + j0.8)$$
$$= (6 + j8)\ \Omega$$
$$= \mathbf{10 \angle 53.13°\ \Omega} \quad \textbf{Ans.}$$

(*iii*) Supply current,

$$I = \frac{V}{Z} = \frac{100}{10} = \mathbf{10\ A} \quad \textbf{Ans.}$$

Power factor, $\cos\phi = \cos 53.13° = \mathbf{0.6\ (lagging)}$ **Ans.**

(*iv*) Total power supplied by the source, $P = VI\cos\phi = 100 \times 10 \times 0.6 = \mathbf{600\ W}$ **Ans.**

Problem 4.14. *The parallel circuit shown in Fig. 4.36 is connected across a single phase 100 V, 50 Hz AC supply. Calculate*

(*i*) *the branch currents*
(*ii*) *the total current*
(*iii*) *the supply power factor*
(*iv*) *the active and reactive power supplied by the supply.*

Solution. Admittance of branch 1,

$$Y_1 = \frac{1}{Z_1} = \frac{1}{(8+j6)} = (0.08 - j0.06)\ S$$

Admittance of branch 2,

$$Y_2 = \frac{1}{Z_2} = \frac{1}{(6-j8)} = (0.06 + j0.08)\ S$$

Total admittance of the circuit,

$$Y = Y_1 + Y_2 = (0.08 - j0.06) + (0.06 + j0.08)$$
$$= (0.14 + j0.02) \text{ S}$$

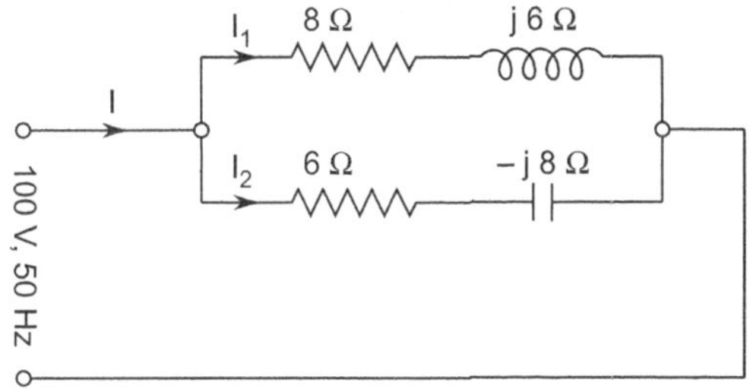

Fig. 4.36

(*i*) Branch currents, $I_1 = V \times Y_1 = (100 + j0)(0.08 - j0.06)$

or $\mathbf{I_1 = (8 - j6) \text{ A or } 10 \angle -36.87° \text{ A}}$ **Ans.**

$I_2 = V \times Y_2 = (100 + j0)(0.06 + j0.08)$

or $\mathbf{I_2 = (6 + j8) \text{ A or } 10 \angle 53.13° \text{ A}}$ **Ans.**

(*ii*) Total current, $I = V \times Y = (100 + j0)(0.14 + j0.02)$

or $\mathbf{I = (14 + j2) \text{ A or } 14.42 \angle 8.13° \text{ A}}$ **Ans.**

(*iii*) Supply power factor $= \cos \phi = \cos 8.13° = \mathbf{0.99 \text{ (leading)}}$ **Ans.**

(*iv*) Active power supplied, $P = VI \cos \phi = 100 \times 14.142 \times 0.99 = \mathbf{1400 \text{ W}}$ **Ans.**

Reactive power supplied, $Q = VI \sin \phi = 100 \times 14.142 \sin 8.13°$

or $Q =$ **200 VAR (capacitive)** **Ans.**

Problem 4.15. *An inductive coil of (6 + j8) Ω impedance is connected to 100 V, 50 Hz AC supply. It is desired to improve power factor of supply current to 0.8 lagging by connecting a capacitor*

(*i*) *in series of the coil*

(*ii*) *in parallel of the coil. Determine the value of the capacitance in each case.*

Solution. (*i*) Let a capacitor of capacitance C be connected in series with the coil so as to raise the power factor of supply current to 0.8 lagging, as shown in Fig. 4.37.

Resistance of the circuit, $R = 6 \ \Omega$

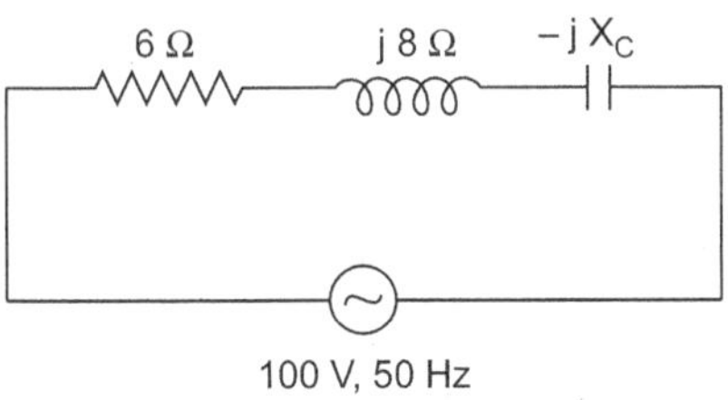

Fig. 4.37

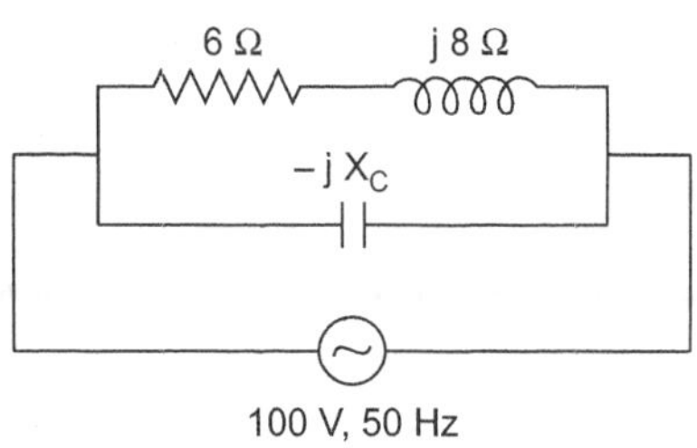

Fig. 4.38

Inductive resistance of the series circuit,

$$X = 8 - \frac{1}{2\pi f C}$$

Improved power factor of the circuit,

$$\cos\phi = 0.8 \text{ lagging}$$

$$\tan\phi = \tan\cos^{-1} 0.8 = 0.75$$

Since $\frac{X}{R} = \tan\phi$

So, $$\frac{8 - \frac{1}{2\pi f C}}{6} = 0.75$$

or $$\frac{1}{2\pi f C} = 8 - 6 \times 0.75 = 3.5\ \Omega$$

or $$C = \frac{1}{2\pi f \times 3.5} = \frac{1}{2\pi \times 50 \times 3.5} = \mathbf{909.46\ \mu F}$$ **Ans.**

(*ii*) Let a capacitor of capacitance C be connected across the coil so as to raise the power factor of supply current to 0.8 lagging as shown in Fig. 4.38.

Conductance of coil, $$G = \frac{R}{R^2 + X^2} = \frac{6}{6^2 + 8^2} = 0.06\ S$$

Susceptance of coil, $$B_L = \frac{X}{R^2 + X^2} = \frac{8}{6^2 + 8^2} = 0.08\ S \text{ (inductive)}$$

Capacitive susceptance of capacitor C,

$$B_C = 2\pi f C$$

Since $\frac{B}{G} = \tan\phi$

or $$\frac{B_L - B_C}{G} = 0.75$$

or $$B_L - B_C = 0.75 \times G = 0.75 \times 0.06 = 0.045\ S$$

or $$B_C = B_L - 0.045 = 0.08 - 0.045 = 0.035\ S$$

or $$2\pi f C = 0.035$$

or $$C = \frac{0.035}{2\pi \times 50} = \mathbf{111.4\ \mu F}$$ **Ans.**

Problem 4.16. *In the circuit shown in Fig. 4.39, the reactance of the capacitor C_1 is 4 Ω, the reactance of capacitor C_2 is 8 Ω and the reactance of L is 8 Ω. A sinusoidal voltage of 120 V is applied to the circuit. Find*

(*i*) *current in each branch*

(*ii*) *power loss in the circuit.*

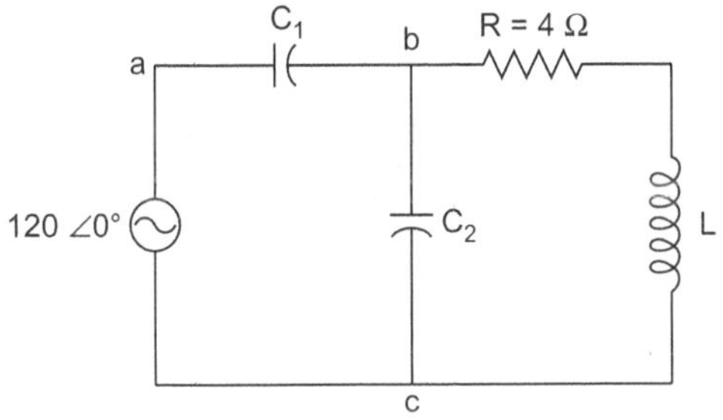

Fig. 4.39

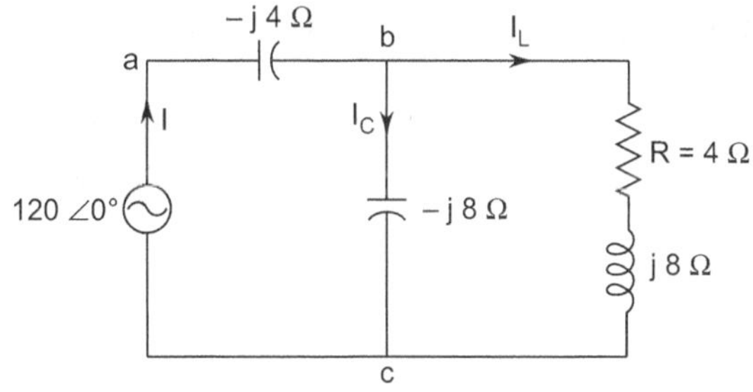

Fig. 4.40

Solution. The circuit given in Fig. 4.39 is redrawn with given values marked on it and is shown in Fig. 4.40.

Impedance of parallel branch, $Z_{bc} = \dfrac{1}{\dfrac{1}{-j8} + \dfrac{1}{(4+j8)}} = \dfrac{-j8 \times (4+j8)}{-j8+4+j8}$

$$= \frac{64 - j32}{4} = (16 - j8)$$

$$= 17.89 \angle -26.565° \ \Omega$$

Total impedance of circuit, $Z = Z_{ab} + Z_{bc} = (-j4) + (16 - j8)$

$$= (16 - j12) = 20 \angle -36.87° \ \Omega$$

Circuit current, $I = \dfrac{V}{Z} = \dfrac{120 \angle 0°}{\angle -36.87°} = 6 \angle 36.87° \text{ A}$

$$= (4.8 + j3.6) \text{ A}$$

Current in branch *ab*, $I_{ab} = I = \mathbf{(4.8 + j3.6)\ A}$ **Ans.**

Voltage, $V_{ab} = I_{ab}\, Z_{ab} = 6 \angle 36.87° \times 4 \angle -90° \text{ V}$

or $V_{ab} = 24 \angle -53.13° \text{ V} = \mathbf{(14.4 - j19.2)\ V}$ **Ans.**

Voltage, $V_{bc} = I \times Z_{bc} = 6 \angle 36.87° \times 17.89 \angle -26.565° \text{ V}$

$= 107.33 \angle 10.305° \text{ V} = \mathbf{(105.6 + j19.2)\ V}$ **Ans.**

Current in 8 Ω capacitor $I_C = \dfrac{V_{bc}}{-j8} = \dfrac{107.33 \angle 10.305°}{8 \angle -90°} = 13.416 \angle 100.305°$

or $I_C = \mathbf{(-2.4 + j\,13.2)\ A}$ **Ans.**

Current in 4 Ω resistor (or current in 8 Ω inductor) will be

$$I_L = \frac{V_{bc}}{(4+j8)} = \frac{107.33 \angle 10.305°}{8.944 \angle 63.435°} = 12 \angle -53.13° \text{ A}$$

or $I_L = \mathbf{(7.2 - j9.6)\ A}$ **Ans.**

Power loss in the circuit, $P = V \times I \times \cos\phi = 120 \times 6 \times \cos 36.87° = \mathbf{576\ W}$ **Ans.**

Problem 4.17. *Two impedances $Z_1 = (10 + j5)$ and $Z_2 = (8 + j6)$ are connected in parallel across a voltage of $V = 200 + j0$. Calculate the circuit current, power factor and reactive power.*

Solution. Equivalent impedance of parallel combination,

$$Z = \frac{Z_1 Z_2}{Z_1 + Z_2} = \frac{(10 + j5)(8 + j6)}{(10 + j5) + (8 + j6)} = \frac{(10 + j5)(8 + j6)}{(18 + j11)}$$

$$= \frac{11.18 \angle 26.56° \times 10 \angle 36.87°}{21.095 \angle 31.43°} = 5.3 \angle 32° \ \Omega$$

Circuit current, $I = \frac{V}{Z} = \frac{200 \angle 0°}{5.3 \angle 32°} = 37.736 \angle -32° \text{ A}$

$I =$ **37.736 A** **Ans.**

Phase angle, $\phi = -32°$

Power factor $= \cos \phi = \cos(-32°) =$ **0.848 (lagging)** **Ans.**

Reactive power, $Q = VI \sin \phi = 200 \times 37.736 \times \sin(-32°)$

$=$ **– 4,000 Watts or 4 kW (lagging)** **Ans.**

Problem 4.18. *Obtain power factor of a two branch parallel circuit where the first branch has $Z_1 = (2 + j4)$ and second $Z_2 = (6 + j0)$. To what value must the 6 Ω resistor be changed to result in overall power factor 0.9 lagging?*

Solution. Consider Fig. 4.41 (*i*) Admittance of branch I, $Y_1 = \frac{1}{Z_1} = \frac{1}{2 + j4} = \frac{2 - j4}{2^2 + 4^2} = (0.1 - j0.2) \text{ S}$

Admittance of branch II, $Y_2 = \frac{1}{6 + j0} = (0.167 + j0) \text{ S}$

Equivalent admittance of the circuit, $Y = Y_1 + Y_2 = (0.1 - j0.2) + (0.167 + j0) = (0.267 - j0.2) \text{ S}$

The magnitude of equivalent admittance, $Y = \sqrt{(0.267)^2 + (0.2)^2} = 0.333 \text{ S}$

Power factor of the circuit, $\cos \phi = \frac{0.267}{0.333} =$ **0.8 (lagging)** **Ans.**

(*ii*) Now consider Fig. 4.42. Let resistance of the resistor of Z_2 be changed to R ohms so as to make power factor of whole circuit 0.9 (lagging)

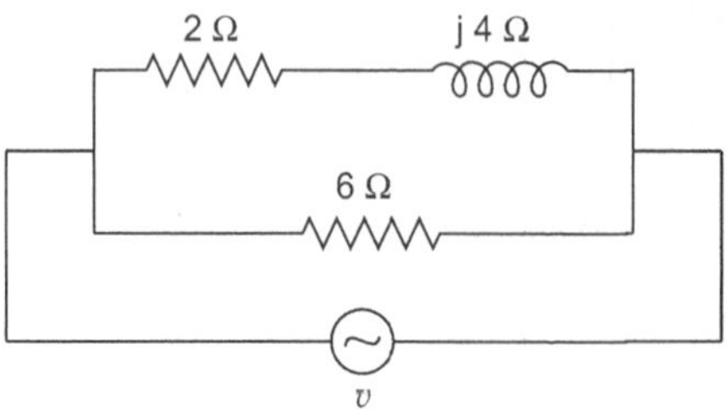

Fig. 4.41

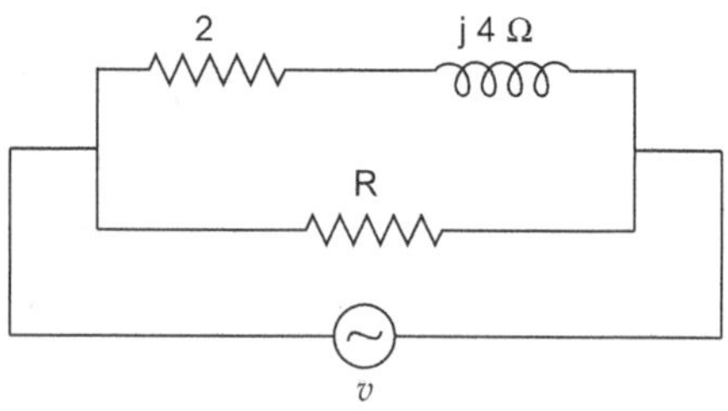

Fig. 4.42

Now total conductance of the circuit, $G = \left(\frac{1}{R} + 0.1\right) \text{S}$

Total susceptance of the circuit, $B' = B = -0.2$ S

Phase angle of the circuit, $\phi' = \cos^{-1} 0.9 = -25.84°$, (lagging)

Phase angle of the circuit is given as $\phi' = \tan^{-1} \dfrac{B'}{G'}$

or $$\tan^{-1} \frac{0.2}{\frac{1}{R}+0.1} = -25.84$$

or $$R = \mathbf{3.2\ \Omega.}$$ **Ans.**

Problem 4.19. *Consider an electric circuit shown in Fig. 4.43. Determine (i) the current and power consumed in each branch, (ii) the supply current and power factor.*

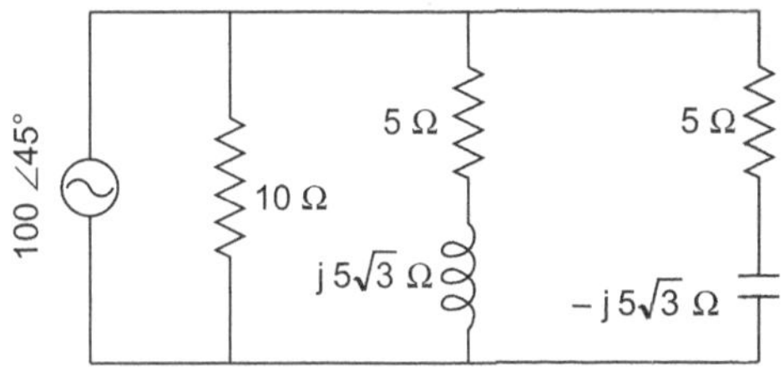

Fig. 4.43

Solution. (*i*) Current in 10 Ω branch,

$$I_1 = \frac{V}{R} = \frac{100\angle 45°}{10\angle 0°} = \mathbf{10\angle 45°\ A}$$ **Ans.**

Current in inductive branch, $$I_2 = \frac{V}{Z_2} = \frac{100\angle 45°}{5+j5\sqrt{3}}$$

or $$I_2 = \frac{100\angle 45°}{10\angle 60°} = \mathbf{10\angle -15°\ A}$$ **Ans.**

Current in capacitance branch, $$I_3 = \frac{V_3}{Z_3} = \frac{100\angle 45°}{5-j5\sqrt{3}}$$

or $$I_3 = \frac{100\angle 45°}{10\angle -60°} = \mathbf{10\angle 105°\ A}$$ **Ans.**

Power consumed in resistive branch, $P_1 = I_1^2 R = 10^2 \times 10$

$= \mathbf{1{,}000\ W}$ **Ans.**

Power consumed in inductive branch, $P_2 = I_2^2 R_2 = 10^2 \times 5$

$= \mathbf{500\ W}$ **Ans.**

Power consumed in capacitive branch, $P_3 = I_3^2 R_2 = 10^2 \times 5$

(*ii*) Supply current, $I = I_1 + I_2 + I_3 = 10\angle 45° + 10\angle -15° + 10\angle 105°$

$= (7.071 + j7.071) + (9.659 - j2.588) + (-2.588 + j9.659)$

$$= 14.142 + j14.142 = \mathbf{20 \angle 45° \ A}$$ **Ans.**

Power factor, $\cos \phi = \cos(45° - 45°) = \mathbf{1.0}$ **Ans.**

Example 4.20. *For an AC circuit the voltage and current are given by*

$$v = 200 \sin 377t \text{ V and}$$

$$i = 8 \sin (377t - 30°) \text{ A}.$$

Determine:

(a) The power factor *(b) True power* *(c) Apparent power* *(d) Reactive power.*

Solution.

(*a*) Given $v = 200 \sin 377t$

$$i = 8 \sin (377t - 30°)$$

It is clear from voltage and current equations that the current is lagging the voltage by an angle 30°. Hence

$$\text{Power factor} = \cos(\phi) = \cos(30°) = \mathbf{0.866 \ (Lagging)}$$ **Ans.**

(*b*) True power or active power

$$P_{true} = V_{rms} I_{rms} \cos \phi$$

$$V_{rms} = \frac{V_m}{\sqrt{2}} = \frac{200}{\sqrt{2}}$$

$$I_{rms} = \frac{I_m}{\sqrt{2}} = \frac{8}{\sqrt{2}}$$

Therefore the true power (or real power) will be

$$P_{true} = \frac{200}{\sqrt{2}} \times \frac{8}{\sqrt{2}} \times 0.866$$

or $P_{true} = \mathbf{692.8 \ W}$ **Ans.**

(*c*) Apparent power

$$P_{app} = V_{rms} I_{rms}$$

Therefore the apparent power will be

$$P_{app} = \frac{200}{\sqrt{2}} \times \frac{8}{\sqrt{2}}$$

or $P_{true} = \mathbf{800 \ VA}$ **Ans.**

(*d*) Reactive power

$$P_{rective} = V_{rms} I_{rms} \sin \phi$$

$$P_{true} = \frac{200}{\sqrt{2}} \times \frac{8}{\sqrt{2}} \times \sin 30°$$

or $P_{true} = \mathbf{400 \ VAR}$ **Ans.**

Problem 4.21. *For the circuit shown in Fig. 4.44, R = 2 Ω, L= 0.02 H and V = 200 V.*

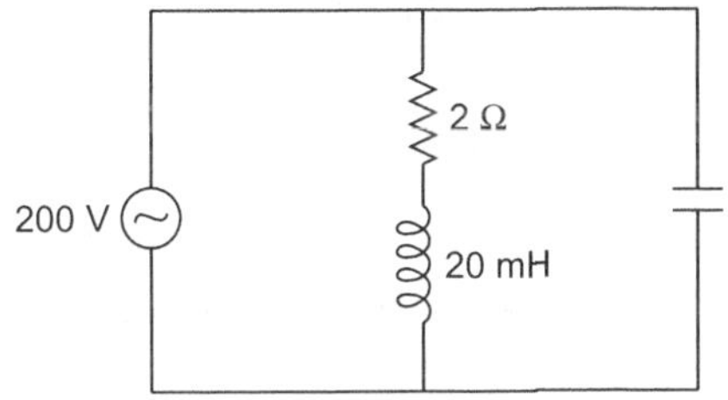

Fig. 4.44

What capacitance must the condenser have in order that the minimum current may occur at

(a) 25 Hz?

(b) 100 Hz?

Find also the current and the voltage across the condenser in each case.

Solution. Minimum current occurs at resonant frequency, hence

(*a*) When the frequency is 25 Hz

$$f_r = \frac{1}{2\pi}\sqrt{\frac{1}{LC} - \left(\frac{R}{L}\right)^2}$$

or $$25 = \frac{1}{2\pi}\sqrt{\frac{1}{0.02 \times C} - \left(\frac{2}{0.02}\right)^2}$$

or $$C = \mathbf{1442\ \mu F}$$ **Ans.**

(*b*) When the frequency is 100 Hz

$$f_r = \frac{1}{2\pi}\sqrt{\frac{1}{LC} - \left(\frac{R}{L}\right)^2}$$

or $$100 = \frac{1}{2\pi}\sqrt{\frac{1}{0.02 \times C} - \left(\frac{2}{0.02}\right)^2}$$

or $$C = \mathbf{123.5\ \mu F}$$ **Ans.**

Problem 4.22. *A resistance of 10 Ω is connected in series with an inductance of 0.05 H and capacitance of 300 μF to a 100 V AC supply. Calculate the value of current and power factor when the frequency is*

(a) 25 Hz *(b) 100 Hz.*

Solution.

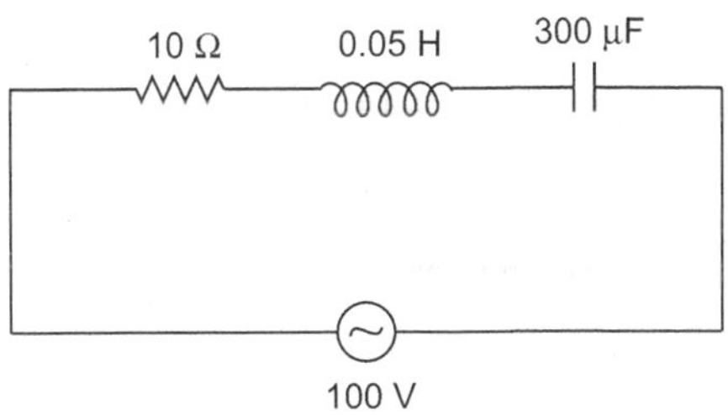

Fig. 4.45

(*a*) When frequency is 25 Hz

$$X_L = \omega L = 2\pi f L = 2\pi \times 25 \times 0.05 = 7.854\ \Omega$$

$$X_C = \frac{1}{\omega C} = \frac{1}{2\pi f C} = \frac{10^6}{2\pi \times 25 \times 300} = 21.22\Omega$$

Impedance, $$Z = R + j(X_L - X_C) = 10 + j(7.854 - 21.22)$$

or $$Z = 10 - j13.366 = 16.693 \angle -53.2°\ \Omega$$

Current, $$I = \frac{V}{Z} = \frac{100\angle 0°}{16.693\angle -53.2°} = \mathbf{5.99 \angle 53.2°\ A}$$ **Ans.**

Power factor, $\cos\phi = \cos 53.2° =$ **0.973 (Leading)** **Ans.**

(*b*) When frequency is 100 Hz

$$X_L = \omega L = 2\pi f L = 2\pi \times 100 \times 0.05 = 31.42\ \Omega$$

$$X_C = \frac{1}{\omega C} = \frac{1}{2\pi f C} = \frac{10^6}{2\pi \times 100 \times 300} = 5.31\Omega$$

Impedance, $$Z = R + j(X_L - X_C) = 10 + j(31.42 - 5.31)$$

or $$Z = 10 + j26.11 = 27.96 \angle 69.04°\ \Omega$$

Current, $$I = \frac{V}{Z} = \frac{100\angle 0°}{27.96\angle 69.04°} = \mathbf{3.58 \angle -69.04°\ A}$$ **Ans.**

Power factor, $\cos\phi = \cos 69.04° =$ **0.357 (Lagging)** **Ans.**

Problem 4.23. *Determine the power factor and the power consumed by the circuit shown in Fig. 4.46.*

Solution.

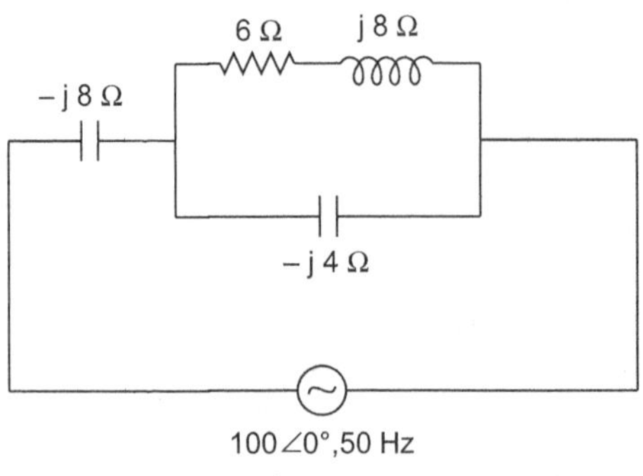

Fig. 4.46

(*a*) Total impedance of the circuit will be

$$Z = \frac{(6 + j8)(-j4)}{6 + j8 - j4} + (-j8)$$

or $$Z = \frac{40\angle -36.869°}{7.21\angle 33.69°} + (8\angle -90°)$$

or $$Z = 1.847 - j13.232 = 13.36 \angle -82.054°\ \Omega$$

Total current will be, $$I = \frac{V}{Z} = \frac{100\angle 0°}{13.36\angle -82.054°}$$

or $$I = 7.485 \angle 82.054° \text{ A}$$

Power factor, $\cos\phi = \cos(82.054) = \mathbf{0.138\ (Leading)}$ **Ans.**

(*b*) Power consumed

$$P = V_{rms} I_{rms} \cos\phi$$

$$P_{true} = 100 \times 7.485 \times 0.138$$

or $$P = \mathbf{103.293\ Watt}$$ **Ans.**

Problem 4.24. *A coil having a resistance of 6 Ω and an inductance of 0.0255 H is connected across a 230 V, 50 Hz AC supply. Calculate:*

(*i*) *Current*

(*ii*) *Power factor*

(*iii*) *Active power*

(*iv*) *Reactive power*

(*v*) *Apparent power*

(*vi*) *It is desired to improve the power factor to 0.8. What value of capacitance is to be connected in series with R?*

(*vii*) *What is the reduction in reactive power after connecting the capacitor?*

Solution.

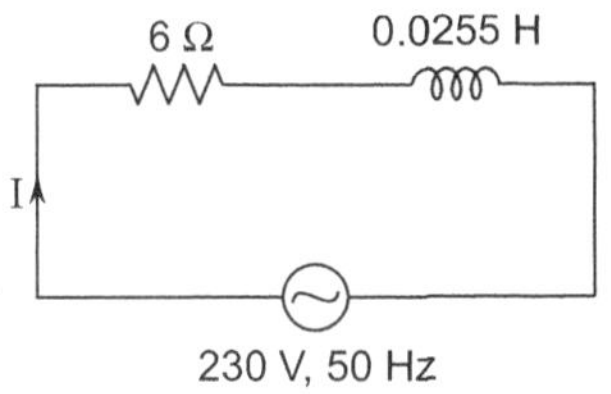

Fig. 4.47

Inductive reactance, $X_L = \omega L = 2\pi f L$

$$= 2\pi \times 50 \times 0.0255 = 8.011\ \Omega$$

Impedance, $Z = R + jX_L$

$$= (6 + j\,8.011)\ \Omega$$

or $$Z = 10 \angle 53.19°\ \Omega$$

(*i*) Current $$I = \frac{V}{Z} = \frac{230 \angle 0°}{10 \angle 53.19°}$$

$$I = \mathbf{23 \angle -53.19°\ A}$$ **Ans.**

(*ii*) Power factor $$\cos\phi = \frac{R}{Z} = \frac{6}{10}$$

or $$\cos\phi = \mathbf{0.6\ (lagging)}$$ **Ans.**

(*iii*) Active power

$$P_{act} = VI\cos\phi$$

$$= 230 \times 23 \times 0.6$$

or $$P_{act} = 3174 \text{ W} = \mathbf{3.174\ kW}$$ **Ans.**

(*iv*) Reactive power

$$P_{rec} = VI \sin \phi$$
$$= 230 \times 23 \times \sin 53.13$$

or $$P_{rec} = 4232 \text{ VAR} = \mathbf{4.23\ kVAR}$$ **Ans.**

(*v*) Apparent power, $$P_{app} = VI$$
$$= 230 \times 23 = 5290 \text{ VA}$$

or $$P_{app} = \mathbf{5.29\ kVA}$$ **Ans.**

(*vi*) To improve the power factor to 0.8, the value of capacitance will be

$$Z = \sqrt{R_2 + (X_L - X_C)^2}$$

$$\cos \phi = \frac{R}{Z}$$

or $$0.8 = \frac{6}{\sqrt{6^2 + (8 - X_C)^2}}$$

or $$36 + (8 - X_C)^2 = \frac{(6)^2}{(0.8)^2}$$

or $$X_C = 3.5\ \Omega$$

or $$X_C = \frac{1}{\omega C} = \frac{1}{2\pi f C}$$

or $$C = \frac{1}{2\pi \times 50 \times 3.5}$$

$$C = \mathbf{909.46\ \mu F}$$ **Ans.**

(*vii*) After inserting the capacitor in series, the new current will be

$$Z = 6 + (8 - 3.5)j = 6 + 4.5\,j\ \Omega$$

or $$Z = 7.5 \angle 36.86°\ \Omega$$

$$I = \frac{V}{Z} = \frac{230 \angle 0°}{7.5 \angle 36.86°}$$

or $$I = 30.67 \angle -36.86° \text{ A}$$

Reactive power will be

$$P_{rec} = VI \sin \phi$$

or $$P_{rec} = 230 \times 30.67 \times \sin 36.86$$

$$P_{rec} = 4231.48 \text{ VAR}$$

Reduction in reactive power is = 4232 – 4231.48

$$= \mathbf{0.5149\ VAR}$$ **Ans.**

Problem 4.25. *Two impedances $Z_1 = (10 + j15)\ \Omega$ and $Z_2 = (6 - j8)\ \Omega$ are connected in parallel. The total current applied is 15 A. What is the power taken by each impedance?*

Solution.

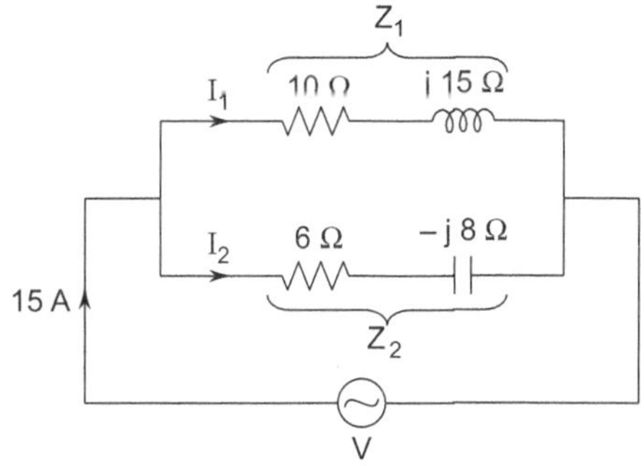

Fig. 4.48

$$Z_1 = (10 + j\,15)\ \Omega = 18.03 \angle 56.3^\circ\ \Omega$$
$$Z_2 = (6 - j\,8)\ \Omega = 10 \angle -53.13^\circ\ \Omega$$

Total impedance of the circuit will be

$$Z = Z_1 \,||\, Z_2 = \frac{Z_1 \times Z_2}{Z_1 \times Z_2}$$
$$= \frac{(18.03 \angle 56.3^\circ) \times (10 \angle -53.13^\circ)}{10 + j15 + 6 - j8}$$
$$= \frac{180.3 \angle 3.17^\circ}{16 + 7j} = \frac{180.3 \angle 3.17^\circ}{17.46 \angle 23.63^\circ}$$
$$Z = 10.33 \angle -20.46^\circ\ \Omega$$

Applied voltage will be

$$V = IZ = 15 \times 10.33 \angle -20.46^\circ$$
$$V = 155 \angle -20.46^\circ\ V$$

Current through impedance Z_1 will be

$$I_1 = \frac{V}{Z_1} = \frac{155 \angle -20.46}{18.03 \angle 56.3}$$

or $$I_1 = 8.6 \angle -76.76^\circ\ A$$

Current through impedance Z_2 will be

$$I_2 = \frac{V}{Z_2} = \frac{155 \angle -20.46^\circ}{10 \angle -53.13^\circ}$$

or $$I_2 = 15.5 \angle 32.67^\circ\ A$$

Power consumed by impedance Z_1 will be

$$P_1 = V I_1 \cos \phi_1$$
$$= 155 \times 8.6 \times \cos (56.3^\circ)$$
$$P_1 = \mathbf{739.6\ W}$$ **Ans.**

Power consumed by impedance Z_2 will be

$$P_2 = V I_2 \cos \phi_2$$
$$= 155 \times 15.5 \times \cos (53.13^\circ)$$
$$P_2 = \mathbf{1441.5\ W}$$ **Ans.**

Problem 4.26. *A non-inductive load takes 20 A at 200 V. Calculate the inductance of the inductor to be connected in series in order that the same current can be supplied from 230 V, 50 Hz mains. Also determine the phase angle between the 230 V supply and current. Neglect resistance of the reactor.*

Solution.

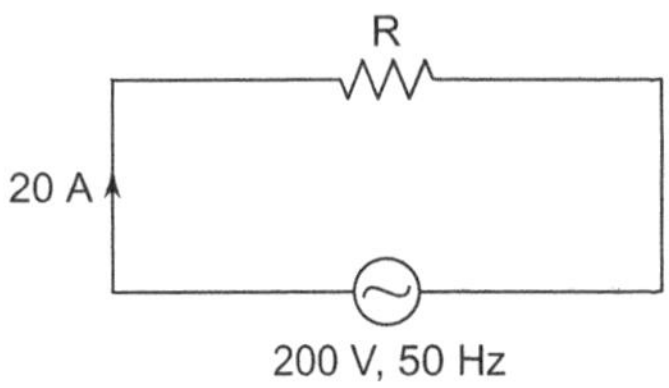

Fig. 4.49

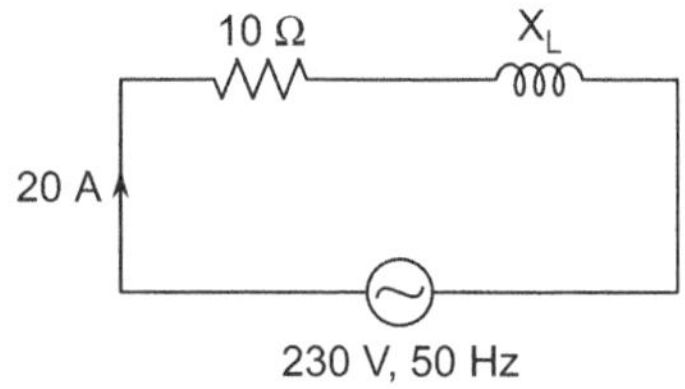

Fig. 4.50

From Fig. 4.49,

$$R = \frac{200}{20} = 10\ \Omega$$

In Fig. 4.50,

$$Z = \frac{V}{I} = \frac{230}{20}$$

$$= 11.5\ \Omega$$

$$\sqrt{R^2 + X_L{}^2} = 11.5$$

or

$$X_L = \sqrt{11.5^2 - 10^2}$$

$$X_L = \mathbf{5.679\ \Omega} \qquad \textbf{Ans.}$$

or

$$\omega L = 5.679$$

$$L = \frac{5.679}{2\pi \times 50} = \mathbf{18.08\ mH} \qquad \textbf{Ans.}$$

Impedance Z can be expressed in polar form as

$$Z = 10 + j\,5.679 = 11.5 \angle 29.59°\ \Omega$$

Hence, the phase angle between supply voltage and current will be

$$\phi = \mathbf{29.59°} \qquad \textbf{Ans.}$$

Problem 4.27. *Refer to the circuit shown in Fig. 4.51, find*

(a) rms value of current

(b) power discipated

(c) power factor

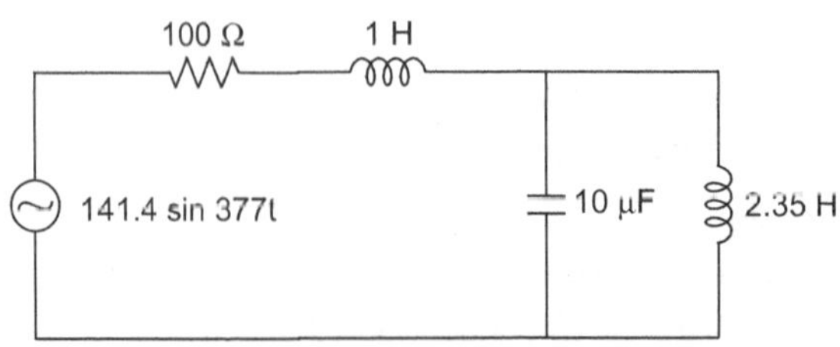

Fig. 4.51

Solution. In the given circuit

$$\omega = 377$$
$$R = 100\ \Omega$$
$$L_1 = 1\text{ H}$$
$$X_{L_1} = \omega L_1 = 377 \times 1 = 377\ \Omega$$
$$C = 10\ \mu F$$
$$X_C = \frac{1}{\omega C} = \frac{1}{377 \times 10 \times 10^{-6}} = 265.26\ \Omega$$
$$L_2 = 2.35\text{ H}$$
$$X_{L_2} = \omega L_2 = 377 \times 2.35 = 885.95\ \Omega$$
$$V_m = 141.4\text{ V}$$

Total impedance of the circuit will be

$$Z = (R + jX_{L1}) + [(jX_{L2}) \| (-jX_C)]$$
$$= (100 + j377) + \left[\frac{(j885.95) \times (-j\,265.26)}{j\,885.95 - j265.26}\right]$$
$$= (100 + j377) + \left[\frac{-j^2\ 235007.097}{j\,620.69}\right]$$
$$= 100 + j\,377 - j\,378.62$$
$$Z = 100 - j\,1.622$$

or $$Z = 100.013\ \angle -0.93°\ \Omega$$

(*a*) rms value of line current

$$I_m = \frac{V_m}{Z} = \frac{141.4\ \angle 0°}{100.013\ \angle -0.93°}$$
$$I_m = 14.122\ \angle\ 0.93°\text{ A}$$
$$I_{rms} = \frac{I_m}{\sqrt{2}} = \frac{14.122}{\sqrt{2}}$$
$$I_{rms} = \mathbf{9.985\ A}$$ **Ans.**

(*b*) Power discipated

$$P = VI\cos\phi = V_{rms} \cdot I_{rms}\cos\phi$$
$$= \frac{141.4}{\sqrt{2}} \cdot \frac{14.122}{\sqrt{2}} \times \cos(0.93)$$
$$P = \mathbf{998.29\ W}$$ **Ans.**

(*c*) Power factor

$$\text{P.F.} = \cos\phi = \cos 0.93°$$
$$\text{P.F.} = \mathbf{0.99\ (Lagging)}$$ **Ans.**

Problem 4.28. *A 10 Ω resistor, a 15.9 mH inductor and a 159 μF capacitor are connected in parallel to a 200 V, 50 Hz supply. Calculate the supply current and power factor.*

Solution. Fig. 4.52 shows the parallel circuit.

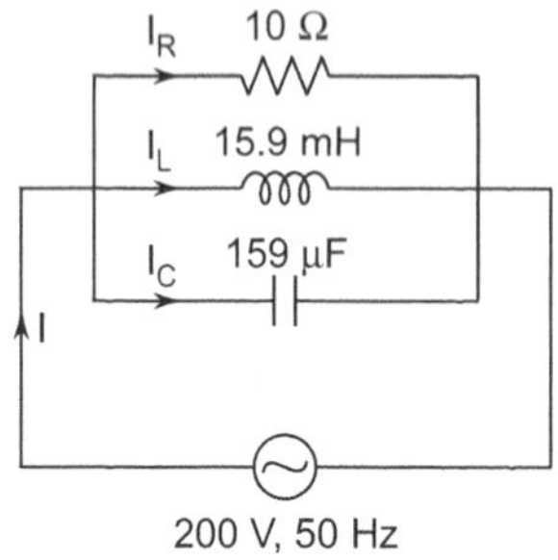

Fig. 4.52

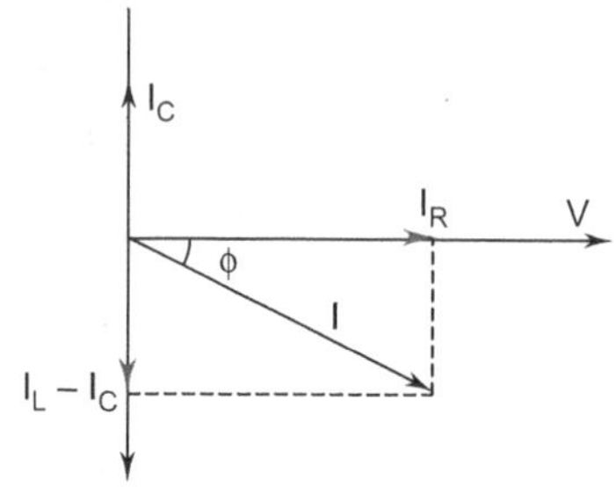

Fig. 4.53

$$R = 10\ \Omega,\ L = 15.9\ \text{mH},\ C = 159\ \mu\text{F}$$

$$X_L = 2\pi fL = 2\pi \times 50 \times (15.9 \times 10^{-3}) = 5\ \Omega$$

$$X_C = \frac{1}{2\pi f C} = \frac{1}{2\pi \times 50 \times 159 \times 10^{-6}} = 10\ \Omega$$

Since the elements R, L and C are in parallel, voltage is same across each element. The currents through R, L and C are

$$I_R = \frac{V}{R} = \frac{200}{10} = 20\ \text{A}$$

$$I_L = \frac{V}{X_L} = \frac{200}{5} = 40\ \text{A}$$

$$I_C = \frac{V}{X_C} = \frac{200}{20} = 10\ \text{A}$$

The phasor I_R is in phase with V.
The phasor I_L lags behind V by 90°.
The phasor I_C leads V by 90°.
The phasor diagram is shown in Fig. 4.53.

$$I_R = \frac{V}{R},\ I_L = \frac{V}{jX_L},\ I_C = \frac{V}{-jX_C}$$

$$I = I_R + I_L + I_C = \frac{V}{R} + \frac{V}{jX_L} + \frac{V}{-jX_C}$$

$$= \frac{200}{10} + \frac{200}{j5} - \frac{200}{j20} = 20 - j40 + j10 = 20 - j30$$

or $\quad I = \mathbf{36 \angle -56.3°\ A}$ **Ans.**

Power factor, $\quad \cos\phi = \cos(-56.3°)$

$= \mathbf{0.555\ (lagging)}$ **Ans.**

Problem 4.29. *A resistance of 10 Ω is connected in series with an inductance of 0.05 H and capacitance of 300 μF to a 100 V AC supply. Calculate the value of current and power factor when the frequency is*

(*a*) *25 Hz*

(*b*) *100 Hz.*

Solution.

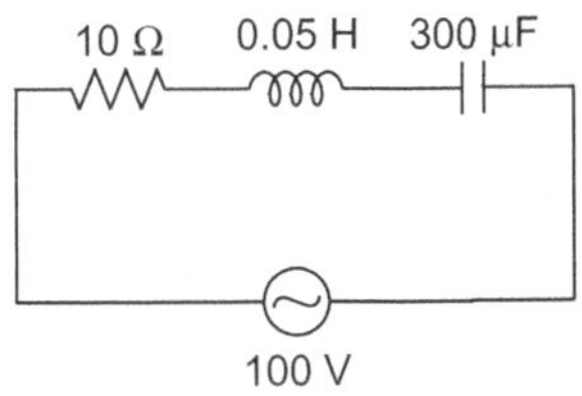

Fig. 4.54

(*a*) When the frequency is 25 Hz

$$X_L = \omega L = 2\pi f L = 2\pi \times 25 \times 0.05 = 7.854\ \Omega$$

$$X_C = \frac{1}{\omega C} = \frac{10^6}{2\pi \times 25 \times 300} = 21.22\ \Omega$$

Impedance, $Z = R + j(X_L - X_C) = 10 + j(7.854 - 21.22)$

or $Z = 10 - j13.366 = 16.693 \angle -53.3°\ \Omega$

Current, $I = \dfrac{V}{Z} = \dfrac{100 \angle 0°}{16.693 \angle 53.2°} = \mathbf{5.99 \angle 53.2°\ A}$ **Ans.**

Power factor, $\cos\phi = \cos 53.2° =$ **0.973 (leading)** **Ans.**

(*b*) When frequency is 100 Hz

$$X_L = \omega L = 2\pi f L = 2\pi \times 100 \times 0.05 = 31.42\ \Omega$$

$$X_C = \frac{1}{\omega C} = \frac{1}{2\pi f C} = \frac{10^6}{2\pi \times 100 \times 300} = 5.31\ \Omega$$

Impedance $Z = R + j(X_L - X_C) = 10 + j(31.42 - 5.31)$

or $Z = 10 + j\,26.11 = 27.96 \angle 69.04°\ \Omega$

Current, $I = \dfrac{V}{Z} = \dfrac{100 \angle 0°}{27.96 \angle 69.04°} = \mathbf{3.58 \angle -69.04°\ A}$ **Ans.**

Power factor, $\cos\phi = \cos 69.04° =$ **0.357 (lagging)** **Ans.**

Problem 4.30. *An inductive coil of (6 + j8) Ω impedance is connected to 100 V, 50 Hz AC supply. It is desired to improve the power factor of supply current to 0.8 lagging by connecting a capacitor in parallel of the coil. Determine the value of the capacitance.*

Solution.

Condition-I

When the coil is connected alone in the circuit,

$$Z_1 = 6 + j8 = 10 \angle 53.13°\ \Omega$$

$$I_1 = \frac{V}{Z_1} = \frac{100 \angle 0°}{10 \angle 53.13°} = 10 \angle -53.13°\ A$$

$$\cos\phi_1 = \cos 53.13 = 0.6 \text{ (lagging)}$$

$$\phi_1 = 53.13°$$

$$\sin\phi_1 = 0.799$$

Condition-II

When the capacitor is connected across the coil

$$\cos\phi_2 = 0.8 \text{ (lagging)}$$
$$\phi_2 = 36.87°$$
$$\sin\phi_2 = 0.6$$

The new total current in the circuit will be

$$I_1 \cos\phi_1 = I_2 \cos\phi_2$$

or
$$I_2 = \frac{10\times 0.6}{0.8} = 7.5 \text{ A}$$

Current through the capacitor will be

$$I_C = I_1 \sin\phi_1 - I_2 \sin\phi_2$$
$$= 10\times 0.799 - 7.5\times 0.6$$
$$I_C = 3.49 \text{ A}$$

Capacitor rating will be

$$C = \frac{I_C}{V\omega} = \frac{3.49}{100\times 2\pi\times 50}$$

$$C = \mathbf{111.1\ \mu F}$$ **Ans.**

Problem 4.31. *A single phase induction motor operates on 230 V, 50 Hz AC supply and takes 22 A at 0.76 power factor lagging.*

It is desired to raise the power factor to 0.92 (lagging) by connecting a capacitor in parallel. Determine:

(a) The new supply current
(b) Current through the capacitor
(c) Capacitance of the capacitor
(d) kVAR of the capacitor.

Solution.

Condition-I

$$I_1 = 22 \text{ A}$$
$$\cos\phi_1 = 0.76 \text{ (lagging)}$$
$$\phi_1 = 40.536°,\ \sin\phi_1 = 0.65$$

Condition-II

$$\cos\phi_2 = 0.92$$
$$\phi_2 = 23.074,\ \sin\phi_2 = 0.392$$

(a) The new supply current will be

$$I_1 \cos\phi_1 = I_2 \cos\phi_2$$

or
$$I_2 = \frac{I_1 \cos\phi_1}{\cos\phi_2} = \frac{22\times 0.76}{0.92}$$

$$I_2 = \mathbf{18.174\ A}$$ **Ans.**

(b) Current through the capacitor will be

$$I_C = I_1 \sin\phi_1 - I_2 \sin\phi_2$$
$$= 22\times 0.65 - 18.174\times 0.392$$
$$= 14.3 - 7.124$$
$$= \mathbf{7.076\ A}$$ **Ans.**

(c) Capacitor ratting

$$C = \frac{I_C}{V\omega} = \frac{7.076}{230 \times 2\pi \times 50}$$

$$C = \mathbf{97.928\ \mu F}$$ **Ans.**

(d) kVAR of the capacitor

$$= VI_C$$
$$= 230 \times 7.076$$
$$= \mathbf{1.627\ kVAR}$$ **Ans.**

EXERCISE

1. Two AC voltages are represented by

$$v_1(t) = 30 \sin (314\,t + 45^\circ)$$
$$v_2(t) = 60 \sin (314\,t + 60^\circ)$$

Calculate the resultant voltage $v(t)$ and express in the form $v(t) = V_m \sin (\omega t + \phi)$.

2. Draw a phasor diagram showing the following voltages:

$$v_1 = 100 \sin 500\,t$$
$$v_2 = 200 \sin (500\,t + 45^\circ)$$
$$v_3 = \cos 500\,t$$

Find r.m.s value of resultant voltage.

3. In an AC circuit the current is given by $i = 22 \sin (314 - \pi/6)$ A. If the voltage of the circuit is the reference quantity, determine
   (i) power factor
   (ii) r.m.s. value of current
   (iii) frequency of current.
4. A resistance and an inductance are connected in series across a voltage $v = 283 \sin 314\,t$. The current expression is found to be $i = 4 \sin (314\,t - \pi/4)$. Find the values of resistance, inductance and power factor.
5. A non-inductive resistance of 10 Ω is connected in series with an inductive coil across 200 V, 50 Hz AC supply. The current drawn by the series combination is 10 A. The resistance of the coil is 2 Ω. Determine
   (i) inductance of the coil
   (ii) power factor
   (iii) voltage across the coil.
6. A 120 V, 60 V lamp is to be operated on 220 V, 50 Hz supply mains. In order that lamp should operate on correct voltage, calculate the value of
   (i) non-inductive resistance
   (ii) pure inductance.
7. A 120 V, 100 W lamp is to be connected to a 220 V, 50 Hz AC supply. What value of pure inductance should be connected in series in order that lamp is run on the rated voltage?
8. An AC voltage $e(t) = 141.4 \sin 120t$ is applied to a series R-C circuit. The current through the circuit is obtained as

$$i(t) = 14.14 \sin 120t + 7.07 \cos (120t + 30^\circ).$$

Determine:

(*i*) values of resistance and capacitance

(*ii*) power factor

(*iii*) power developed by the source.

9. A metal-filament lamp, rated at 750 W, 100 V is to be connected in series with a capacitance across a 230 V, 50 Hz supply. Calculate the value of capacitance required. Draw phasor diagram.

10. A coil of resistance 8 Ω and inductance 0.12 Henry is connected in series with a condenser of capacitance 140 micro-farads across 230 volts, 50 Hz supply. Determine:

(*i*) impedance of the entire circuit,

(*ii*) current flowing through the condenser,

(*iii*) power factor of the circuit,

(*iv*) voltage across the condenser.

11. A coil of resistance 12 Ω and inductance 0.05 H, a non-inductive resistance of 20 Ω resistance and a loss-free 40 μF capacitor are connected across a 240 V, 50 Hz sinusoidal supply. Calculate:

(*i*) the current, and

(*ii*) the power factor of the circuit.

12. A series R-L-C circuit consisting of a resistance of 20 Ω, inductance 0.2 H and capacitive of 150 μF is connected across a 230 V, 50 Hz source. Calculate:

(*i*) the impedance

(*ii*) the current

(*iii*) the magnitude and nature of the power factor

(*iv*) the frequency of supply to be adjusted to make power factor unity.

13. A load having impedance of $(1 + j)$ Ω is connected to an AC voltage represented as

$$v = 20\sqrt{2}\ \cos(\omega t + 10°) \text{ volt.}$$

(*i*) find the current in load, expressed in the form of $i = I_m \sin(\omega t + \phi)$ A.

(*ii*) find real power consumed by the load.

14. Obtain the power factor of a two branch parallel circuit where the first branch has $Z_1 = (2 + j4)$ and second $Z_2 = (6 + j0)$. To what value must the 6 Ω resistor be changed to result in the overall power factor 0.9 lagging?

15. An inductive coil of $(6 + j8)$ Ω impedance is connected to 100 V, 50 Hz AC supply. It is desired to improve power factor of supply current to 0.8 lagging by connecting a capacitor

(*i*) in series of the coil

(*ii*) in parallel of the coil.

Determine the value of the capacitance in each case.

16. Three sinusoidal voltages acting in series are given by

$$v_1 = 10 \sin 440\, t$$

$$v_2 = 10\sqrt{2} \sin (440\, t - 45°)$$

$$v_3 = 20 \cos 440\, t.$$

Determine:

(*i*) an expression for the resultant voltage,

(*ii*) the frequency, and

(*iii*) rms value of resultant voltage.

17. The voltage and current through a circuit element are:

$$v = 100 \sin (314\, t + 45°)$$

$$i = 10 \sin (314\, t + 315°)$$

(*i*) identify the circuit element,
(*ii*) find its value,
(*iii*) find the power factor of the circuit,
(*iv*) nature of power factor, and
(*v*) expression for power.

18. A coil of resistance 12 Ω and inductance 0.05 H, a non-inductive resistor of 20 Ω and a loss free 40 μF capacitor are connected across a 240 V, 50 Hz sinusoidal supplies. Calculate:
(*i*) the current,
(*ii*) the power factor of the circuit, and
(*iii*) nature of power factor.

19. The instantaneous value of two alternating voltages are represented by

$$v_1 = 60 \sin \theta$$
$$v_2 = 40 \sin (\theta - 60°).$$

Derive expression for instantaneous value of
(*i*) the sum, and
(*ii*) the differences of these voltages.

20. A resistance and inductance are connected in series across a voltage $v = 283 \sin 314\,t$. The current expression is found to be $i = 4 \sin (314\,t - 45°)$. Find the value of resistance, inductance and power factor.

21. A two element series circuit is connected across an AC source $v = 300 \cos (314\,t + 20°)$ Volts. The current drawn is $i = 15 \cos (314\,t - 10°)$ Amperes. Determine the
(*i*) circuit impedance magnitude,
(*ii*) phase angle, and
(*iii*) what is the average power drawn?

22. An alternating voltage is given by $v = 141.4 \sin (314\,t)$. Find
(*i*) frequency,
(*ii*) rms value,
(*iii*) average value,
(*iv*) the instantaneous value of voltage when t is 3 m sec, and
(*v*) the time taken to voltage to reach 100 V for the first time after passing through zero value.

23. An alternating current of frequency 50 Hz has a maximum value of $200\sqrt{2}$ Ampere. Reckoning the time from the instant the current is zero and becoming positive, finds the time taken by the current to reach a value of 141.4 Ampere for a first and a second time.

24. Draw the phasor diagram showing the following voltages:

$$v_1 = 100 \sin 500\,t$$
$$v_2 = 200 \sin (500\,t + 60°)$$
$$v_3 = -\,50 \cos 500\,t$$
$$v_4 = 150 \sin (500\,t - 45°)$$

Find the resultant voltage and its rms value.

25. A current has the following steady state values in ampere for equal intervals of time changing instantaneously from one value to the next: 0, 10, 20, 30, 20, 10, 0, – 10, – 20, – 30, – 20, – 10, 0 etc. Calculate:
(*i*) average value,
(*ii*) rms value,

(*iii*) form factor, and

(*iv*) peak factor.

26. Two impedances $Z_1 = (10 + j5)$ ohms and $Z_2 = (8 + j6)$ ohms are connected in parallel across a 200 V, 50 Hz AC supply. Calculate:

(*i*) the circuit currents,

(*ii*) the power factor of the circuit,

(*iii*) nature of power factor,

(*iv*) real power, and

(*v*) reactive power.

27. Prove that the power drawn in an alternating circuit is given by VI cos ϕ where ϕ is the angle between V an I. Also differentiate between the terms Active or Real Power, Reactive Power and Apparent Power, mentioning therein the units and expression of each.

28. An alternating current of 1.5 A flows in a circuit when applied voltage is 300 V. The power consumed is 225 W. Find the resistance and reactance of the circuit.

29. A 100 V, 80 W lamp is to be operated on 230 V, 50 Hz AC supply. Calculate the inductance to be connected in series with the lamp for the above operation. Lamp can be taken as pure resistance.

30. In an AC circuit, the current is given by $i = 22 \sin (314\ t - \pi/6)$ amp. If the voltage of the circuit is the reference quantity, determine:

(*i*) power factor,

(*ii*) rms value of the current, and

(*iii*) frequency of the current.

31. A resistance and an inductance are connected in series across a voltage $v = 283 \sin (314\ t)$. The current expression is found to be by $i = 4 \sin (314\ t - \pi/4)$ amp. Find the value of resistance, inductance and power factor.

32. Two impedances $Z_1 = (10 + j15)\ \Omega$ and $Z_2 = (6 - j8)\ \Omega$ are connected in parallel. The total current supplied is 15 A. What is the power taken by each impedance?

33. A sinusoidal voltage is applied to three branches in parallel. The current in three branches are:

$$i_1 = 14.14 \sin (\omega t - 45°),$$
$$i_2 = 28.3 \sin (\omega t - 60°),$$
$$i_3 = 7.07 \sin (\omega t + 60°).$$

Find the effective value of current drawn from the supply.

34. An impedance of $(4 - j10)\ \Omega$ is connected in parallel with an Impedance of $(6 + j8)\ \Omega$. The circuit is fed from a 230 V, 50 Hz supply. Find the current through each branch, total circuit current, circuit impedance and power factor.

35. A circuit has a resistance of 60 Ω and a capacitive reactance of 8 Ω in series. A voltage $v = 141 \sin 314t$ is applied. Find:

(*i*) value of impedance,

(*ii*) expression for current,

(*iii*) power factor,

(*iv*) rms value of voltage and current, and

(*v*) reactive power.

36. A voltage $v = 141 \sin (314t + \pi/3)$ is applied to a capacitor of 100 μF. Find the rms current, power dissipated and power factor. Write expression for instantaneous current.

□ □ □

5

Resonance in AC Circuits

5.1 INTRODUCTION

"Resonance is the term employed for describing the steady state operation of circuit or system at that frequency for which the resultant response is in time phase with the source function even though the energy storing elements are present in the system".

The opposition to flow of AC current offered by inductor L and capacitor C not only depend on the value of L and C but also on frequency of supply voltage.

As discussed in previous chapter, inductive reactance is directly proportional to supply frequency and capacitive reactance is inversely proportional to supply frequency.

Now as we increase the frequency of supply voltage, the value of inductive reactance (X_L) keeps on increasing and that of capacitive reactance (X_C) keeps on decreasing causing the net impedance in the circuit to get changed.

An AC circuit containing R, L and C is said to be at resonance when the circuit impedance becomes purely resistive *i.e.*, when the circuit power factor becomes unity or when the total current drawn by the circuit is in phase with the applied voltage.

5.2 RESONANCE IN SERIES AC CIRCUIT

Consider a series RLC circuit, shown in Fig. 5.1.

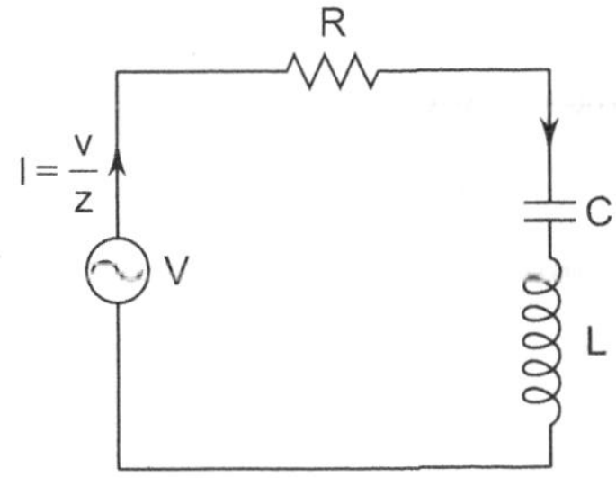

Fig. 5.1

As discussed in previous chapter, inductive reactance is given by

$$X_L = \omega L = 2\pi f L \quad \ldots(5.1)$$

And capacitive reactance is given by

$$X_C = \frac{1}{\omega C} = \frac{1}{2\pi f C} \quad \ldots(5.2)$$

The total impedance of the circuit will be,

$$Z = \sqrt{R^2 + (X_L \sim X_C)^2} \quad \ldots(5.3)$$

or

$$Z = \sqrt{R^2 + X^2} \quad \ldots(5.4)$$

where

$$X = X_L \sim X_C \quad \ldots(5.5)$$

= Net reactance of the circuit

Let such a circuit be connected across an AC sources of constant voltage V, but of frequency varying from zero to infinity.

There would be a certain frequency of applied voltage which would make X_L equal to X_C in magnitude. In that case the net reactance will be

$$X = X_L \sim X_C = 0 \quad \ldots(5.6)$$

or

$$X_L = X_C \quad \ldots(5.7)$$

And the impedance of the circuit will become

$$Z = \sqrt{R^2 + (X_L \sim X_C)^2} \quad \text{[Eqn. (5.3)]}$$

or

$$Z = \sqrt{R^2 + (0)^2} \quad \text{[From equation (5.7)]}$$

or

$$\boxed{Z = R} \quad \ldots(5.8)$$

Under this condition, the circuit is said to be in electrical resonance and the frequency at which this occur is termed as ***Resonant Frequency, f_r***.

5.3 RESONANT FREQUENCY

The frequency at which resonance occur is termed as resonant frequency, denoted by f_r.

At resonance

$$X_L = X_C \quad \text{[Eqn. (5.7)]}$$

as

$$X_L = \omega L = 2\pi f L \quad \text{[Eqn. (5.1)]}$$

and

$$X_C = \frac{1}{\omega C} = \frac{1}{2\pi f C} \quad \text{[Eqn. (5.2)]}$$

From equations (5.1) and (5.2), equation (5.7) will become

$$2\pi f_r L = \frac{1}{2\pi f_r C} \quad \ldots(5.9)$$

As these expressions for reactances are considered at resonance, hence the frequency notation is taken as f_r in place of f.

or $$f_r^2 = \frac{1}{(2\pi)^2 LC} \qquad ...(5.10)$$

or $$\boxed{f_r = \frac{1}{2\pi}\sqrt{\frac{1}{LC}} \text{ Hz}} \qquad ...(5.11)$$

5.4 SALIENT FEATURE ABOUT SERIES RESONANCE

Some of the important features of series resonance are:

1. At resonance, the series RLC circuit is also known as ***acceptor circuit*** because a circuit accepts currents at one particular frequency but rejects the current at all other frequencies.
2. Such circuits are used in radio receivers.
3. When at resonance the inductive reactance become equal to the capacitive reactance
   $$X_L = X_C \qquad [\text{Eqn. (5.7)}]$$
   Then overall impedance of the circuit will become
   $$Z = \sqrt{R^2 + (X_L \sim X_C)^2} \qquad [\text{Eqn. (5.3)}]$$
   or $$Z = R \qquad [\text{Eqn. (5.8)}]$$
   i.e., ***impedance*** of the circuit become ***minimum***.
4. The ***current*** at resonance become ***maximum***.
   $$I_m = \frac{V}{Z} = \frac{V}{R} \qquad ...(5.12)$$
5. As Z become equal to R, the current will be in the phase with applied voltage and hence power factor of the circuit become ***unity***.
6. Circuit becomes ***purely resistive***.
7. If R is not present in the circuit then the circuit, at resonance, will act like short-circuit.
8. The voltage drop across inductor is equal to the voltage drop across capacitor and is maximum.
9. Overall voltage drop across L and C is zero because they are 180° out of phase (voltages).
10. Since in this resonance, the voltage across L or C is maximum, it is also known as the ***voltage resonance***.
11. The frequency at which this occur is termed as resonant frequency, and is given by
    $$f_r = \frac{1}{2\pi}\sqrt{\frac{1}{LC}} \text{ Hz} \qquad [\text{Eqn. (5.11)}]$$
12. Phasor diagram of RLC series circuit at the time of resonance will be

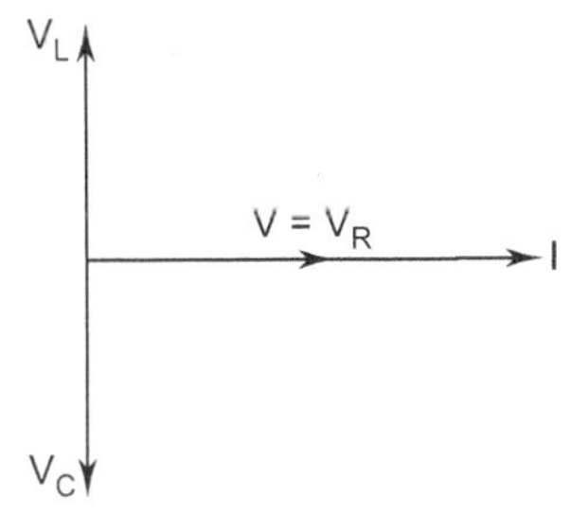

5.5 GRAPHICAL REPRESENTATION OF SERIES RESONANCE

Now we will discuss the change in characteristics of various parameters with change in frequency.

Keeping the applied alternating voltage at constant magnitude, but varying the applied frequency we have

1. **Resistance, (R)**

   It is independent of f; hence, it is represented by a **straight line** (Fig. 5.2).

2. **Inductive Reactance, (X_L)**

   $$X_L = \omega L = 2\pi f L \qquad \text{[Eqn. (5.1)]}$$

   X_L is directly proportional to f *i.e.*, X_L increase linearly with f. Hence its graph is a **straight line** passing through the origin (Fig. 5.3).

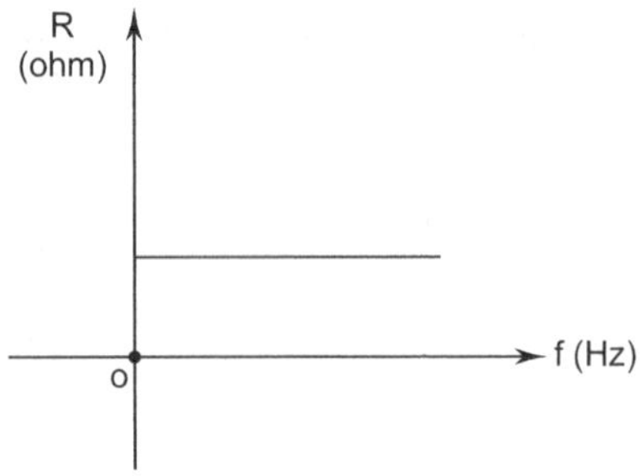

Fig. 5.2

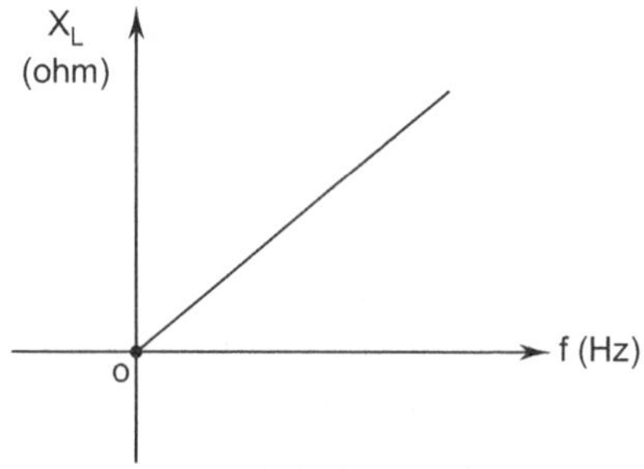

Fig. 5.3

3. **Capacitive Reactance, (X_C)**

   $$X_C = \frac{1}{\omega C} = \frac{1}{2\pi f C} \qquad \text{[Eqn. (5.2)]}$$

   X_C is inversely proportional to f. Its graph is a **rectangular hyperbola** which lie in the fourth quadrant because X_C is regarded negative (Fig. 5.4).

4. **Net Reactance, (X)**

   $$X = X_L \sim X_C \qquad \text{[Eqn. (5.5)]}$$

   Its graph is a **hyperbola** (not rectangular) and crosses the X-axis at a point which represents resonant frequency f_r (Fig. 5.5).

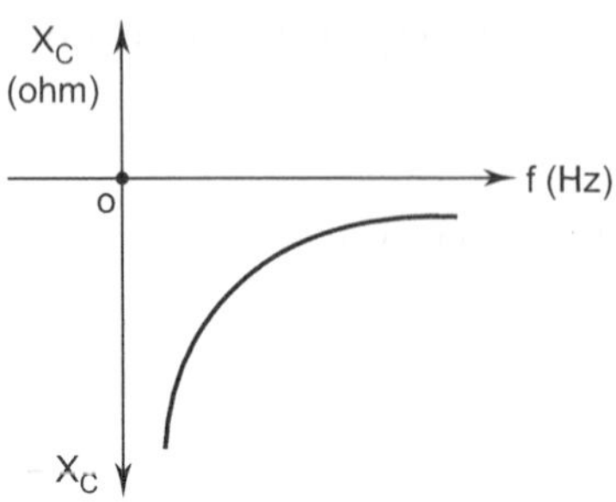

Fig. 5.4

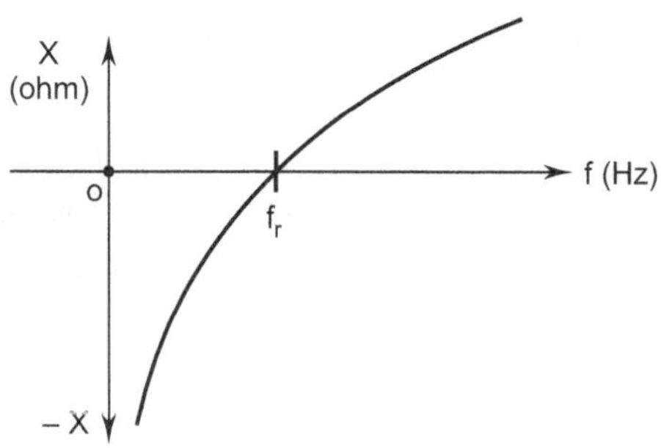

Fig. 5.5

5. **Circuit Impedance, (Z)**

$$Z = \sqrt{R^2 + (X_L \sim X_C)^2} \quad \text{[Eqn. (5.3)]}$$

or

$$Z = \sqrt{R^2 + X^2} \quad \text{[Eqn. (5.4)]}$$

At low frequencies Z is large because X_C is large and also at high frequencies Z is a large because X_L is large. At and near the resonant frequency, Z is minimum. It is **inverted bell** shape curve (Fig. 5.6).

6. **Current, (I)**

It is reciprocal of the circuit impedance. It is a **bell shaped** curve (Fig. 5.7).

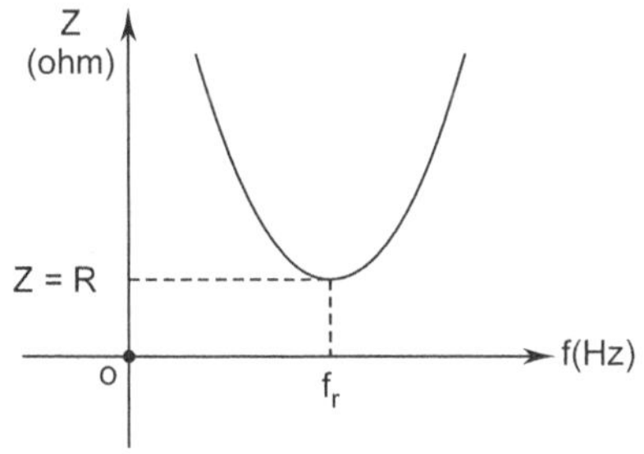

Fig. 5.6

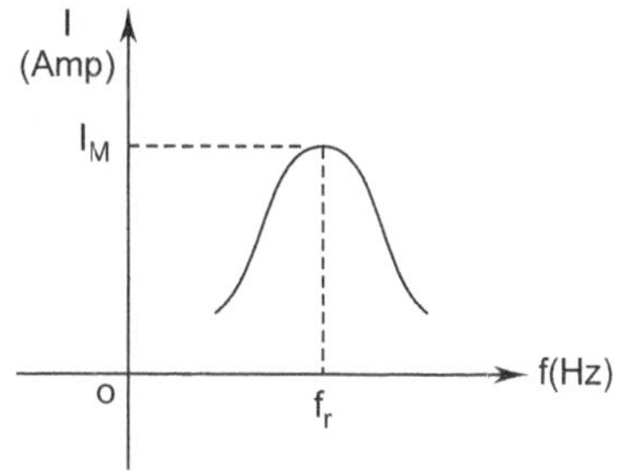

Fig. 5.7

The complete series resonance curve is shown in Fig. 5.8.

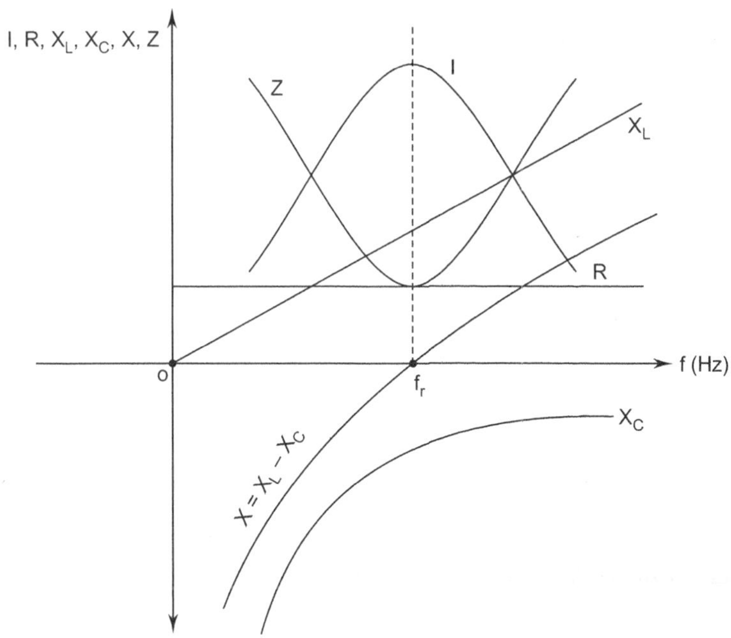

Fig. 5.8

5.6 BANDWIDTH

To describe the width of the resonance curve the term bandwidth is used. For a series RLC circuit bandwidth is defined as *"the range of frequency for which the power delivered to the resistance (R)* *is greater than or equal to half of the maximum power ($P_m/2$), where P_m is the maximum power delivered to R at resonance".*

From the shape of the resonance curve (Fig. 5.9), it is clear that there are two frequencies for which the power delivered to R is half the power at resonance. Therefore these frequencies are referred to as half power point frequencies.

The maximum power delivered to R at Resonance is given by

$$P_m = I_m^2 R \quad \text{...(5.13)}$$

At half power points, the power delivered to R is half the maximum power delivered to R at resonance. Let that power at half power points be denoted by P_1 and P_2 and are given by

$$P_1 = \frac{1}{2} I_m^2 R = \left(\frac{I_m}{\sqrt{2}}\right)^2 R = I_1^2 R \quad \text{...[5.14 (a)]}$$

Similarly

$$P_2 = \frac{1}{2} I_m^2 R = \left(\frac{I_m}{\sqrt{2}}\right)^2 R = I_2^2 R \quad \text{...[5.14 (b)]}$$

The magnitude of current at each half power ordinate is same and hence

$$I_1 = I_2 = \frac{I_m}{\sqrt{2}} = 0.707\, I_m \quad \text{...(5.15)}$$

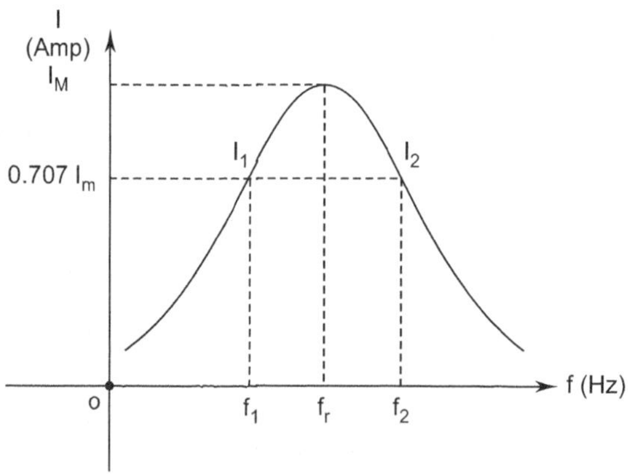

Fig. 5.9

Hence another definition for bandwidth can be given as,

"The bandwidth is that range of frequencies over which magnitude of the current is equal to or greater than 0.707(or 70.7%) of the current at resonance." Mathematically

Bandwidth

$$(\Delta f) = f_2 - f_1 = \frac{\omega_2 - \omega_1}{2\pi} \quad \text{...(5.16)}$$

Bandwidth can also be represented by the expression

$$\text{B.W.} = \frac{\text{Resonant frequency}}{\text{Quality factor}} = \frac{f_r}{Q} \quad \text{...(5.17)}$$

5.7 CUT-OFF FREQUENCIES

The cut-off frequencies or half power points can be defined as

"Those frequencies or points at which the net reactance (X_L - X_C) becomes equal to the resistance (R) of the circuit."

Or

"Those frequencies or points at which the current in the circuit becomes 0.707 (or 70.7%) of the maximum value of current at resonance."

Or

"Those frequencies or points at which the power dissipated in the circuit becomes half of the maximum value of power dissipated at resonance."

At resonance the impedance of circuit is given by

$$Z = \frac{V}{I_m} \quad ...(5.18)$$

Also at resonance, $Z_r = R$ [Eqn. (5.8)]

From equation (5.15) it is clear that the current at half power points is $\frac{1}{\sqrt{2}}$ the maximum value and hence the impedance at half power point will be

$$Z_{1,2} = \frac{V}{I_m/\sqrt{2}} \quad ...(5.19)$$

or

$$Z_{1,2} = \frac{V}{I_m}\sqrt{2} = \sqrt{2}\,R \quad ...(5.20)$$

Hence, at half power points, the impedance of the tuned circuit must be $\sqrt{2}$ times its impedance at resonance.

Since, $Z = \sqrt{R^2 + (X_L \sim X_C)^2}$ [Eqn. (5.3)]

or $Z = \sqrt{R^2 + X^2}$ [Eqn. (5.4)]

Therefore by comparing equations (5.4) and (5.20), we get

$$\sqrt{2}\,R = \sqrt{R^2 + X^2} \quad ...(5.21)$$

or

$$X = R \quad ...(5.22)$$

Calling the reactance at lower half power point as X_1 and as V_C leads V_L initially, we have

$$X_1 = X_C - X_L = R \quad ...(5.23)$$

or

$$R = \frac{1}{\omega_1 C} - \omega_1 L \quad ...(5.24)$$

The above equation may also be written as

$$\omega_1^2 + \frac{R}{L}\omega_1 - \frac{1}{LC} = 0 \qquad ...(5.25)$$

or
$$\omega_1 = -\frac{R}{2L} \pm \sqrt{\left(\frac{R}{2L}\right)^2 + \frac{1}{LC}} \qquad ...(5.26)$$

The negative frequency is meaning less, so is discarded. So the equation for lower half power frequency is

$$\omega_1 = -\frac{R}{2L} + \sqrt{\left(\frac{R}{2L}\right)^2 + \frac{1}{LC}} \qquad ...(5.27)$$

or
$$\boxed{f_1 = -\frac{R}{4\pi L} + \frac{1}{2\pi}\sqrt{\left(\frac{R}{2L}\right)^2 + \frac{1}{LC}}} \qquad ...(5.28)$$

Equation (5.28) represents lower cut-off frequency (or they are also known as lower half power frequencies or edge frequencies).

Similarly the upper half power frequency can be calculated. Calling the reactance at upper half power point as X_2 and as V_L leads V_C after crossing the point when resonant frequency occurs, we have

$$X_2 = X_L - X_C = R \qquad ...(5.29)$$

or
$$R = \omega_2 L - \frac{1}{\omega_2 C} \qquad ...(5.30)$$

or
$$\omega_2 - \frac{R}{L}\omega_2 - \frac{1}{LC} = 0 \qquad ...(5.31)$$

or
$$\omega_2 = \frac{R}{2L} + \sqrt{\left(\frac{R}{2L}\right)^2 + \frac{1}{LC}} \qquad ...(5.32)$$

or
$$\boxed{f_2 = \frac{R}{4\pi L} + \frac{1}{2\pi}\sqrt{\left(\frac{R}{2L}\right)^2 + \frac{1}{LC}}} \qquad ...(5.33)$$

Equation (5.33) represent upper cut-off frequency.

So the equation for bandwidth will become

$$\text{Bandwidth} = \Delta f = f_2 - f_1 = \frac{\omega_2 - \omega_1}{2\pi} \qquad \text{[Eqn. (5.16)]}$$

From equations (5.28) and (5.33) we have

$$= \left[\frac{R}{4\pi L} + \frac{1}{2\pi}\sqrt{\left(\frac{R}{2L}\right)^2 + \frac{1}{LC}}\right] - \left[-\frac{R}{4\pi L} + \frac{1}{2\pi}\sqrt{\left(\frac{R}{2L}\right)^2 + \frac{1}{LC}}\right] \qquad ...(5.34)$$

or
$$\boxed{\text{B.W.} = f_2 - f_1 = \frac{R}{2\pi L}} \qquad ...(5.35)$$

Lower half power frequency can also be expressed as

$$\boxed{f_1 = f_r - \frac{\text{B.W.}}{2}} \quad \text{...(5.36)}$$

or

$$\boxed{f_1 = f_r - \frac{R}{4\pi L}} \quad \text{...(5.37)}$$

And upper half power frequency can also be expressed as

$$\boxed{f_2 = f_r + \frac{\text{B.W.}}{2}} \quad \text{...(5.38)}$$

or

$$\boxed{f_2 = f_r + \frac{R}{4\pi L}} \quad \text{...(5.39)}$$

Also,

$$\boxed{f_r = \sqrt{f_1 f_2}} \quad \text{...(5.40)}$$

5.8 SELECTIVITY

For low resistance circuit, the resonance curve is sharply peaked and such a circuit is said to be sharply resonant or highly selective.

On the other hand, high resistance circuit have flat resonant curve and is said to have poor selectivity.

Selectivity of different resonant circuits is compared in terms of their bandwidth.

Consider an RLC series circuit having the following specifications:

$$V_{rms} = 5\,\text{mV}$$
$$C = 2\,\text{nF}$$
$$L = 5\,\mu\text{F}$$

For different values of R the current Vs frequency plot is shown in Fig. 5.10.

Note that as the resistance of the circuit decreases, the resonance curve peak becomes more sharp.

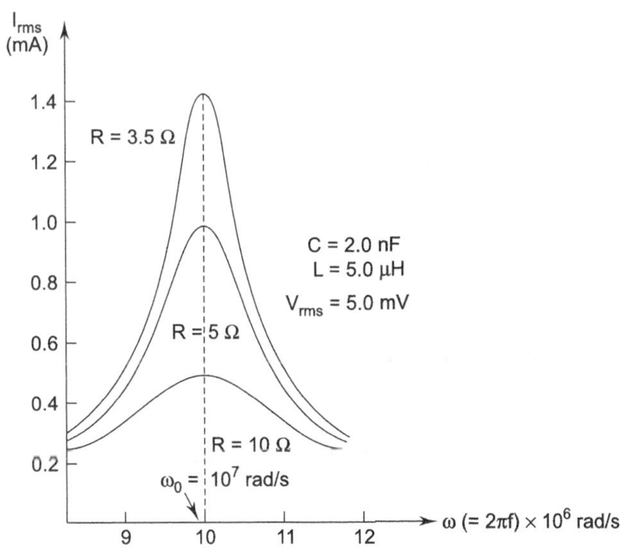

Fig. 5.10

5.9 QUALITY FACTOR

The quality factor is defined as ***"the voltage magnification produced across L or C in the circuit at resonance"***. At resonance, current has maximum value

$$I_m = \frac{V}{R} \qquad \text{[Eqn. (5.12)]}$$

First we will consider the voltage magnification across inductor, L.

At resonance the voltage across inductor will be

$$V_{Lr} = I_m X_{Lr} \qquad \text{...(5.41)}$$

where X_{Lr} is the inductive reactance at resonance

According to the definition, the quality factor will be

$$Q_o = \frac{\text{Voltage across inductor at resonance}}{\text{Voltage across the circuit}} \qquad \text{...[5.42(a)]}$$

From equations (5.12) and (5.41)

$$Q_o = \frac{V_{Lr}}{V} = \frac{I_m X_{Lr}}{I_m R} \qquad \text{...[5.42(b)]}$$

or

$$Q_o = \frac{\omega_r L}{R} \qquad \text{...(5.43)}$$

or

$$\boxed{Q_o = \frac{2\pi f_r L}{R}} \qquad \text{...(5.44)}$$

Now considering the voltage magnification across capacitor, C.

At resonance the voltage across capacitor will be

$$V_{Cr} = I_m X_{Cr} \qquad \text{...(5.45)}$$

where X_{Cr} is the capacitive reactance at resonance

Voltage magnification [From equations (5.12) and (5.45)] will be

$$Q_o = \frac{V_{Cr}}{V} = \frac{I_m X_{Cr}}{I_m R} \qquad \text{...(5.46)}$$

or

$$Q_o = \frac{1}{\omega_r CR} \qquad \text{...(5.47)}$$

or

$$\boxed{Q_o = \frac{1}{2\pi f_r CR}} \qquad \text{...(5.48)}$$

Also,

$$\boxed{Q_o = \frac{1}{R}\sqrt{\frac{L}{C}}} \qquad \text{...(5.49)}$$

Also,

$$Q_o = \frac{\text{Resonant frequency}}{\text{Bandwidth}} = \frac{f_r}{\text{B.W.}} \qquad \text{...(5.50)}$$

Also,

$$Q_o = \frac{f_r}{f_2 - f_1} \qquad \text{...(5.51)}$$

An example of the application of resonant circuits is the selection of AM radio stations by the radio receiver. The selectivity of the tuning must be high enough to discriminate strongly against stations above and below in carrier frequency, but not so high as to discriminate against the "sidebands" created by the imposition of the signal by amplitude modulation.

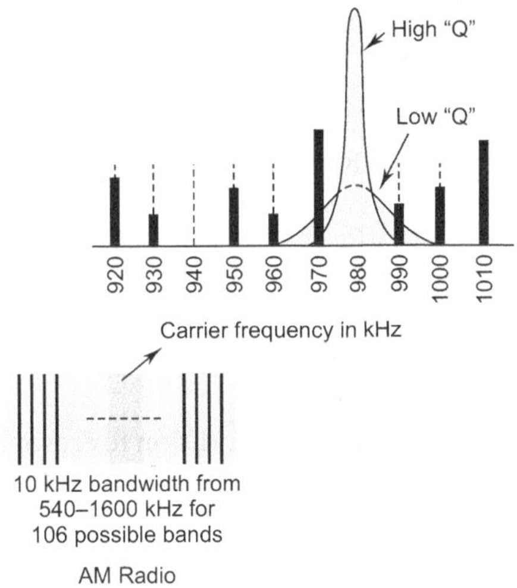

Fig. 5.11

5.10 RESONANCE IN PARALLEL CIRCUIT

Considering the practical case of a coil in parallel with a capacitor (Fig. 5.12)

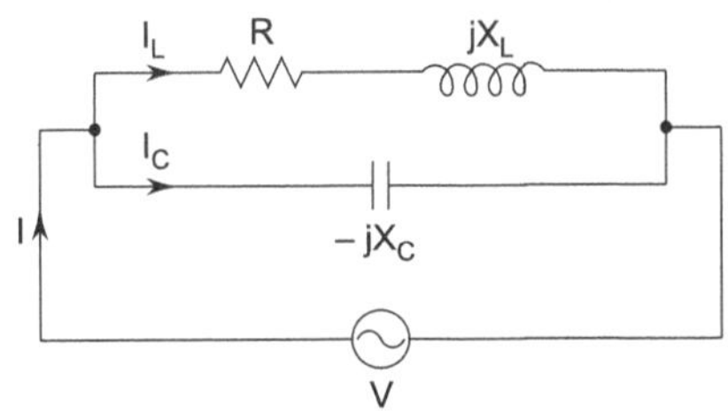

Fig. 5.12

Such a circuit is said to be in electrical resonance, when the reactive (or wattless) component of line current become zero.

The frequency at which it happens is known as resonant frequency.

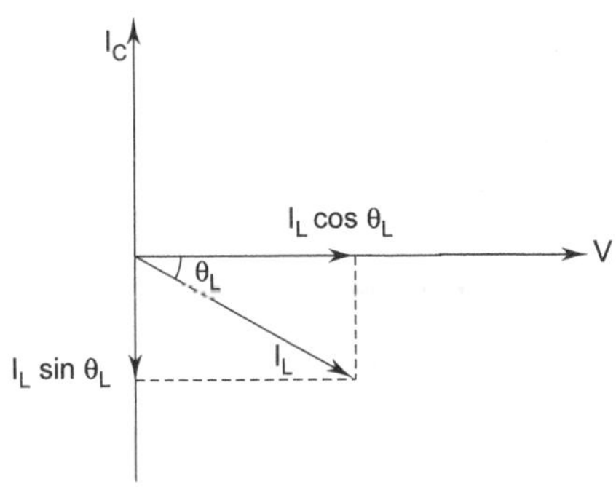

Fig. 5.13

From the phasor diagram (Fig. 5.13), net reactive or wattless component

$$= I_C - I_L \sin\theta_L \quad ...(5.52)$$

As at resonance its value is zero, hence

$$I_C - I_L \sin\theta_L = 0 \quad ...(5.53)$$

or

$$I_C = I_L \sin\theta_L \quad ...(5.54)$$

Now,

$$I_L = \frac{V}{Z_L} = \frac{V}{\sqrt{R^2 + X_L^2}} \quad ...(5.55)$$

$$I_C = \frac{V}{X_C} \quad ...(5.56)$$

$$\sin\theta_L = \frac{X_L}{Z_L} = \frac{X_L}{\sqrt{R_L + X_L^2}} \quad ...(5.57)$$

Hence from equations (5.55), (5.56) and (5.57) condition for resonance become

$$\frac{V}{X_C} = \frac{V}{Z_L} \times \frac{X_L}{Z_L} \quad ...(5.58)$$

or

$$X_L \times X_C = Z_L^2 \quad ...(5.59)$$

as

$$X_L = 2\pi f_r L \quad ...(5.60)$$

and

$$X_C = \frac{1}{2\pi f_r C} \quad ...(5.61)$$

Therefore, equation (5.59) will become

$$2\pi f_r L \times \frac{1}{2\pi f_r C} = Z_L^2 \quad ...(5.62)$$

or

$$Z_L^2 = \frac{L}{C} \quad ...(5.63)$$

as

$$Z_L = \sqrt{R^2 + X_L^2} \quad ...(5.64)$$

Therefore, equation (5.63) will become

$$R^2 + X_L^2 = \frac{L}{C} \quad ...(5.65)$$

or

$$R^2 + (2\pi f_r L)^2 = \frac{L}{C} \quad ...(5.66)$$

or

$$(2\pi f_r L)^2 = \frac{L}{C} - R^2 \quad ...(5.67)$$

or

$$(2\pi f_r)^2 = \frac{L}{C} - \left(\frac{R}{L}\right)^2 \quad ...(5.68)$$

or

$$\boxed{f_r = \frac{1}{2\pi}\sqrt{\frac{1}{LC} - \left(\frac{R}{L}\right)^2}} \quad ...(5.69)$$

5.10.1 Current at Parallel Resonance

Since wattless component of the current is zero, the circuit current is equal to active component only

$$I = I_L \cos\theta_L = \frac{V}{Z_L} \cdot \frac{R}{Z_L} \qquad ...(5.70)$$

or
$$I = \frac{VR}{Z_L^2} \qquad ...(5.71)$$

as
$$Z_L^2 = \frac{L}{C} \qquad \text{[From equation (5.63)]} \qquad ...(5.72)$$

Therefore,
$$I = \frac{VR}{L/C} \qquad ...(5.73)$$

or
$$\boxed{I = \frac{V}{L/CR}} \qquad ...(5.74)$$

5.10.2 Impedance at Parallel Resonance

The denominator L/CR is known as the ***Equivalent or Dynamic Impedance*** of the parallel circuit

$$\boxed{Z_r = \frac{L}{CR}} \qquad ...(5.75)$$

It should be noted that the impedance at resonance is *resistive* only. Since current is minimum at resonance, L/CR must, therefore, represent the maximum impedance of the circuit. In fact, parallel resonance is a condition of maximum impedance or minimum admittance.

Note. *Another method to find the resonance in any parallel circuit is that*

(*i*) *First find the total impedance of the circuit.*

(*ii*) *Then put imaginary part of the total impedance equal to zero.*

(*iii*) *Put the frequency as resonant frequency (f_r) and solve the equation to find the resonant frequency.*

5.11 SALIENT FEATURE OF PARALLEL RESONANCE

Some of the salient features of parallel resonance are:

1. Reactive or wattless component of current is zero.
2. The admittance is equal to conductance.
3. Dynamic impedance is $\frac{L}{CR}$. It is purely resistive and maximum in magnitude.
4. Line current is minimum and is equal to $\frac{V}{L/CR}$ and is in phase with applied voltage.
5. Power factor of the circuit is unity.
6. Parallel resonance is also known as ***anti-resonance***.

7. It is also known as ***current resonance*** because current between the branches (circulating current) is many times the line current.
8. The circuit at resonance is also called as ***rejecter circuit*** because at resonant frequency, the line current is minimum or it almost rejects it.
9. The circuit at resonance is also called as ***tank circuit*** as an inductor is connected in parallel to the capacitor.
10. The frequency at which the net susceptance curve crosses the frequency axis is called the resonant frequency.

5.12 GRAPHICAL REPRESENTATION OR PARALLEL RESONANCE

Now we will discuss the change in characteristics of various elements and parameters with change in frequency.

1. **Inductive Susceptance (B_L)**

   The effects of variation of frequency on the susceptance of the inductive branch is

$$B_L = -\frac{1}{X_L} = -\frac{1}{2\pi f L} \qquad \text{... (5.76)}$$

   It is inversely proportional to the frequency of the applied voltage. Hence it is represented by rectangular hyperbola (Fig. 5.14) drawn in the fourth quadrant (because it is assumed negative).

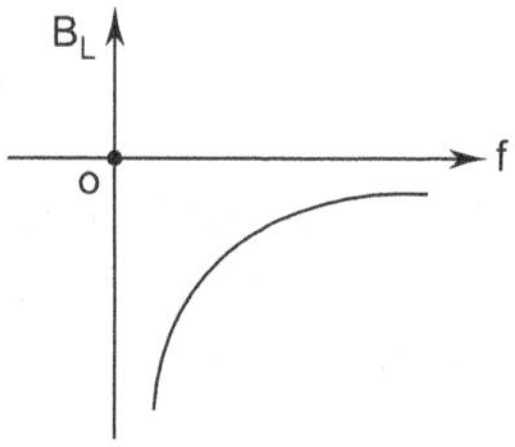

Fig. 5.14

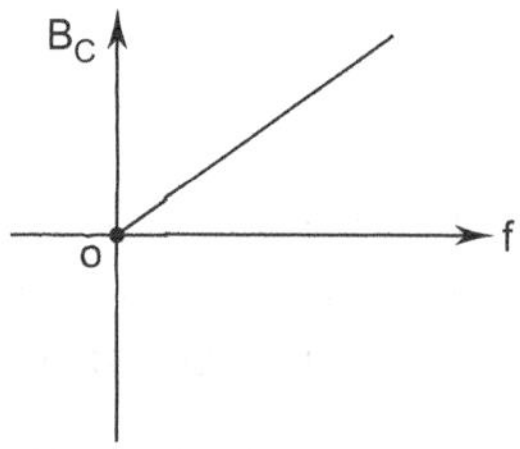

Fig. 5.15

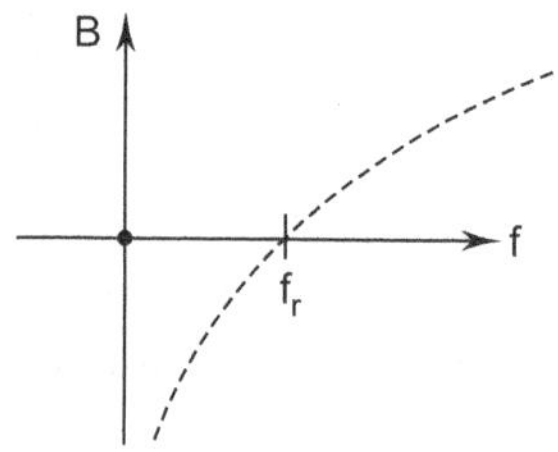

Fig. 5.16

2. **Capacitive Susceptance (B_C)**

$$B_C = \frac{1}{X_C} = \omega C = 2\pi f C \qquad \text{...(5.77)}$$

   It increases with increase in the frequency of applied voltage. Hence it is represented by a straight line drawn in the first quadrant (Fig. 5.15).

3. **Net Susceptance (B)**

   It is the difference of two susceptances and is represented by dotted hyperbola (Fig. 5.16).

$$B = B_L - B_C$$

4. **Current (I)**

   Current at resonance is minimum (Fig. 5.17), hence such a circuit (when used in radio work) is sometimes known as rejecter circuit because it rejects (or takes minimum current of) that frequency to which it resonates

$$I = \frac{V}{L/CR} \qquad \text{[Eqn. (5.74)]}$$

This resonance is often referred to as current resonance also because the current circulating between the two branches is many times greater than the line current taken from the supply. The graph of the current is inverted bell shape type.

5. **Circuit Impedance (Z)**

Since current is minimum at resonance, L/CR must, therefore, represent the maximum impedance of the circuit [Fig. 5.17 (*a*)].

$$Z_r = \frac{L}{CR} \qquad \text{[Eqn. (5.75)]}$$

In fact, parallel resonance is a condition of maximum impedance or minimum admittance.

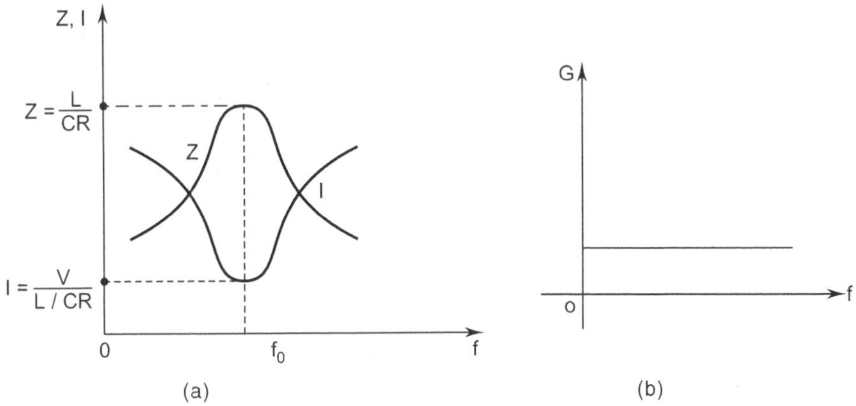

Fig. 5.17

6. **Conductance (G)**

The conductance, G is independance of frequency, hence it is a straight line as shown in Fig. 5.17 (*b*).

The complete resonance curve is shown in Fig. 5.18.

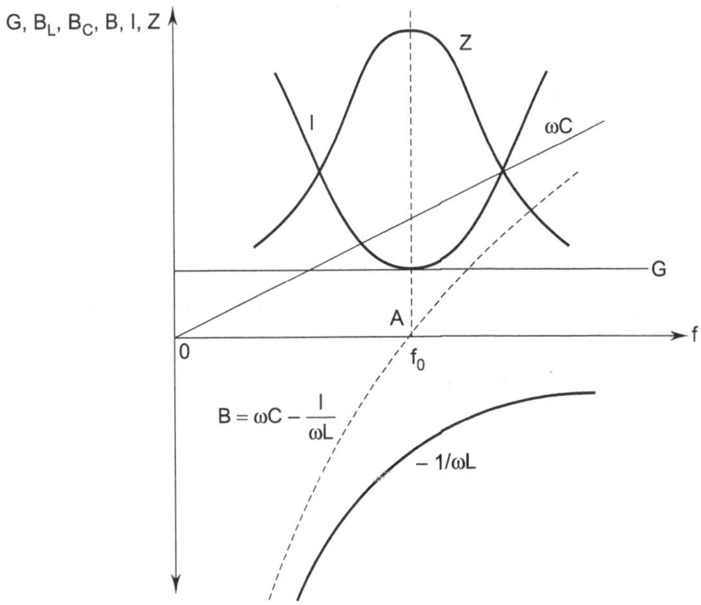

Fig. 5.18

5.13 BANDWIDTH OF A PARALLEL RESONANT CIRCUIT

The bandwidth of a parallel circuit is defined in the same way as that for a series circuit.

$$\text{Bandwidth} = \frac{R}{2\pi L} \quad \text{...(5.78)}$$

This circuit also has upper and lower half-power frequencies where power dissipated is half of that at resonant frequency.

5.14 QUALITY-FACTOR* OF A PARALLEL CIRCUIT

It is already discussed in article 5.12 that parallel resonance is often referred to as current resonance because the current circulating between two branches is many times greater than the line current taken from supply.

That is why, the quality factor in case of parallel resonance is defined as, "the current magnification produced in between the two parallel branches".

Q-factor can also be defined as "the ratio of the current circulating between its two branches to the line current drawn from the supply."

The circulating current between capacitor and coil branches is I_C.

$$I_C = \frac{V}{X_C} = \frac{V}{1/\omega C} = \omega CV \quad \text{...(5.79)}$$

And line current is

$$I = \frac{V}{L/CR} = \frac{VCR}{L} \quad \text{...(5.80)}$$

$$\therefore \qquad \text{Q-factor} = \frac{I_C}{I} \quad \text{...[5.81 }(a)\text{]}$$

From equations (5.79) and (5.80)

$$Q_o = \frac{\omega CV}{VCR/L} \quad \text{...[5.81}(b)\text{]}$$

or

$$\boxed{Q_o = \frac{\omega_r L}{R}} \quad \text{...[5.81}(c)\text{]}$$

$$Q = \frac{2\pi f_0 L}{R} \quad \text{...(5.82)}$$

Now resonant frequency when R is negligible is [from equation (5.69)]

$$f_r = \frac{1}{2\pi\sqrt{LC}} \quad \text{...(5.83)}$$

And the Q-factor is given by

$$Q = \frac{1}{R}\sqrt{\frac{L}{C}} \quad \text{...(5.84)}$$

Note that, the expressions for Q-factor for parallel resonances are same as that of series resonance.

---

* For separate Q-factor of both parallel branches refer to Solved Problem 5.8.

5.15 COMPARISON BETWEEN SERIES AND PARALLEL RESONANCE

Series Resonance	Parallel Resonance
1. Reactive component of line current becomes zero.	1. Reactive component of line current becomes zero.
2. At resonance the series RLC circuit is known as acceptor circuit and it exibits a property of current magnification.	2. At resonance the parallel RLC circuit is known a rejector circuit and it exhibits a property of voltage magnification.
3. The line current is maximum and is equal to $I = \frac{V}{R}$	3. The line current is minimum and is equal to $I = \frac{V}{L/CR}$
4. The impedance of the circuit is minimum and is equal to $Z = R$	4. The impedance of the circuit becomes maximum and is equal to $Z = \frac{L}{CR}$
5. Resonant frequency is $f_r = \frac{1}{2\pi}\sqrt{\frac{1}{LC}}$	5. Resonant frequency is $f_r = \frac{1}{2\pi}\sqrt{\frac{1}{LC} - \left(\frac{R}{L}\right)^2}$
6. At the time of resonance the circuit becomes purely resistive.	6. At the time of resonance the parallel RLC circuit also becomes purely resistive.

5.16 APPLICATIONS OF RESONANCE EFFECT

Some of the applications of resonance are:

- Most common application is tuning. For example, when we tune a radio to a particular station, the LC circuits are set at resonance for that particular carrier frequency.
- A series resonant circuit provides voltage magnification.
- A parallel resonant circuit provides current magnification.
- A parallel resonant circuit can be used as load impedance in output circuits of RF amplifiers. Due to high impedance, the gain of amplifier is maximum at resonant frequency.
- A parallel resonant circuit can be used in induction heating.

SOLVED PROBLEMS

Problem 5.1. *A coil of resistance 40 Ω and inductance 0.75 H are in a series circuit. The resonant frequency is 55 Hz. If the supply is 250 V, 50 Hz. Find:*

(i) Line current.
(ii) Power factor.
(iii) Nature of power factor.
(iv) Power consumed.
(v) Voltage across the coil.

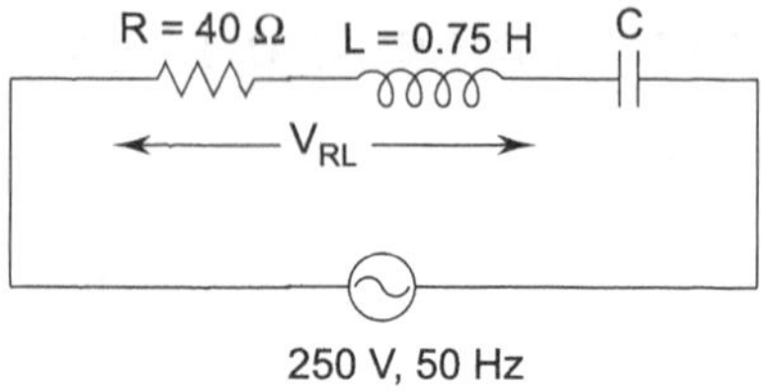

Fig. 5.19

Solution. At resonance, frequency is given by

$$f_r = \frac{1}{2\pi}\sqrt{\frac{1}{LC}}$$

or $$C = \frac{1}{(2\pi f_r)^2 L} = \frac{1}{(2\pi \times 55)^2 \times 0.75}$$

or $$C = 11.165\ \mu F.$$

At 250 V, 50 Hz

Reactance of the circuit will be

$$X = \omega L - \frac{1}{\omega C} = 2\pi f L - \frac{1}{2\pi f C}$$

$$= 2\pi \times 50 \times 0.75 - \frac{1}{2\pi \times 50 \times 11.165 \times 10^{-6}}$$

or $$X = -49.48\ \Omega$$

Impedance of the circuit will be

$$Z = \sqrt{R^2 + X^2} = \sqrt{(40)^2 + (49.48)^2}$$

$$Z = 63.63\ \Omega$$

(*i*) Line current, $I = \frac{V}{Z} = \frac{250}{63.63}$ = **3.93 A** **Ans.**

(*ii*) Power factor, $\cos\phi = \frac{R}{Z} = \frac{40}{63.63}$ = **0.629.** **Ans.**

(*iii*) Nature of power factor is **Leading,** because the reactance of the circuit, X is negative and hence circuit is capacitive.

(*iv*) Power consumed, $P = I^2 R = (3.93)^2 \times 40$

P = **618 Watt.** **Ans.**

(*v*) Voltage across coil, $V_{RL} = I \times Z_{RL} = I \times \sqrt{R^2 + X_L^2}$

$$= 3.93 \times \sqrt{(40)^2 + (2\pi \times 50 \times 0.75)^2}$$

V_{RL} = **939.2 V** **Ans.**

Problem 5.2. *If the bandwidth of a resonant circuit is 10 kHz and the lower half power frequency is 120 kHz, find out the value of the upper half power frequency and the quality factor of the circuit.*

Solution. Bandwidth of the resonant circuit,

$$\Delta f = 10 \text{ kHz}$$

Lower half power frequency, $f_1 = 120$ kHz

Since lower half-power frequency, $f_1 = f_r - \dfrac{\Delta f}{2}$

Resonant frequency, $f_r = f_1 + \dfrac{\Delta f}{2} = 120 + \dfrac{10}{2} = 125$ kHz

Upper half-power frequency, $f_2 = f_r + \dfrac{\Delta f}{2} = 125 + \dfrac{10}{2} = \mathbf{130\ kHz}$ **Ans.**

Quality factor of the resonant circuit, $Q_r = \dfrac{f_r}{\Delta f} = \dfrac{125}{10} = \mathbf{12.5}$ **Ans.**

Problem 5.3. *A 20 Ω resistor is connected in series with an inductor and a capacitor, across a variable frequency 25 V supply. When the frequency is 400 Hz, the current is at its maximum value of 0.5 A and the potential difference across the capacitor is 150 V. Calculate the resistance and inductance of the inductor.*

Solution. Supply voltage, $V_{rms} = 25$ V

Frequency, $f = 400$ Hz

Maximum current in a series RLC circuit will flow under resonant condition and current under resonant condition is given as

$$I_m = \frac{V}{R}$$

So circuit resistance, $R = \dfrac{V}{I_m} = \dfrac{25}{0.5} = \mathbf{50\ \Omega}$ **Ans.**

Voltage across the capacitor, $V_C = \dfrac{I_m}{2\pi f C}$

So circuit capacitance, $C = \dfrac{I_m}{2\pi f V_C} = \dfrac{0.5}{2\pi \times 400 \times 150} = 1.3263\ \mu F$

Since under resonant condition, $X_L = X_C$

or $2\pi f L = \dfrac{1}{2\pi f C}$

∴ Inductance of inductor, $L = \dfrac{1}{4\pi^2 f^2 C} = \dfrac{1}{4\pi^2 \times (400)^2 \times 1.3263 \times 10^{-6}} = \mathbf{0.119\ H}$ **Ans.**

Problem 5.4. *Voltage across resistance, inductance and capacitance connected in series are 3 V, 4 V and 5 V respectively. If supply voltage has 50 Hz frequency what is the magnitude of the supply voltage ? Find the resonant frequency of this series RLC circuit.*

Solution.

Voltage across resistance, $V_R = IR = 3\text{ V}$ ...(*i*)

Voltage across inductor, $V_L = IX_L = I\omega L = I\,2\pi f L = 4\text{ V}$ ...(*ii*)

Voltage across capacitor, $V_C = IX_C = \dfrac{I}{\omega C} = \dfrac{I}{2\pi f C} = 5\text{ V}$ ...(*iii*)

Supply voltage, $V = \sqrt{V_R^2 + (V_C - V_L)^2}$

$= \sqrt{3^2 + (5-4)^2} = \sqrt{10} = \mathbf{3.162\text{ V}}$ **Ans.**

Dividing equation (*ii*) by equation (*iii*), we get

$$2\pi f L \cdot I \times \frac{2\pi f C}{I} = \frac{4}{5}$$

or $4\pi^2 f^2 LC = 0.8$

or $LC = \dfrac{0.8}{4\pi^2 \times 50^2} = 8.1051 \times 10^{-6}$

Resonant frequency, $f_r = \dfrac{1}{2\pi}\sqrt{\dfrac{1}{LC}} = \dfrac{1}{2\pi}\sqrt{\dfrac{1}{8.1051\times 10^{-6}}}$

$f_r = \mathbf{55.9\text{ Hz.}}$ **Ans.**

Problem 5.5. *For the circuit shown below, determine*

(*i*) *Resonant frequency.*
(*ii*) *Total impedance of the circuit at resonance.*
(*iii*) *Bandwidth.*
(*iv*) *Quality factor.*

Solution. $R = 25\ \Omega$, $L = 0.5\text{ H}$ and $C = 5 \times 10^{-6}\text{ F}$

(*i*) Resonant frequency, $f_r = \dfrac{1}{2\pi}\sqrt{\dfrac{1}{LC} - \dfrac{R^2}{L^2}}$

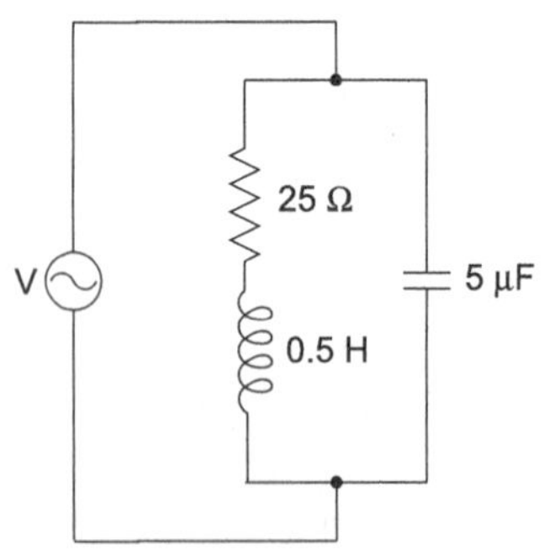

Fig. 5.20

or $$f_r = \frac{1}{2\pi}\sqrt{\frac{1}{0.5\times5\times10^{-6}} - \frac{25^2}{(0.5)^2}} = \mathbf{100.34\ Hz} \quad \textbf{Ans.}$$

(*ii*) Impedance of the circuit at resonance,

$$Z = \frac{L}{CR} = \frac{0.5}{5\times10^{-6}\times25} = \mathbf{4{,}000\ \Omega} \quad \textbf{Ans.}$$

(*iii*) Bandwidth $$= \frac{R}{2\pi L} = \frac{25}{2\pi\times0.5} = \mathbf{7.958\ Hz} \quad \textbf{Ans.}$$

(*iv*) Quality factor, $$Q = \frac{1}{R}\sqrt{\frac{L}{C}} = \frac{1}{25}\sqrt{\frac{0.5}{5\times10^{-6}}} = \mathbf{12.65} \quad \textbf{Ans.}$$

Problem 5.6. *A series RLC circuit has R = 10 Ω, L = 0.1 H and C = 8 μF. Determine:*

(*i*) *Resonant frequency.*

(*ii*) *Q-factor of the circuit at resonance.*

(*iii*) *The half power frequencies.*

Solution.

(*i*) Resonant frequency, $$f_r = \frac{1}{2\pi}\sqrt{\frac{1}{LC} - \frac{R^2}{L^2}}$$

$$= \frac{1}{2\pi}\sqrt{\frac{1}{0.1\times8\times10^{-6}} - \frac{10^2}{(0.1)^2}} = \mathbf{33.37\ Hz} \quad \textbf{Ans.}$$

(*ii*) Q-factor of circuit at resonance $$= \frac{1}{R}\sqrt{\frac{L}{C}} = \frac{1}{10}\sqrt{\frac{0.1}{8\times10^{-6}}} = \mathbf{11.18} \quad \textbf{Ans.}$$

(*iii*) Half-power frequencies, $$f_1 = f_r - \frac{R}{4\pi L} = 33.37 - \frac{10}{4\pi\times0.1} = \mathbf{25.41\ Hz} \quad \textbf{Ans.}$$

$$f_2 = f_r + \frac{R}{4\pi L} = 33.37 + \frac{10}{4\pi\times0.1} = \mathbf{41.33\ Hz.} \quad \textbf{Ans.}$$

Problem 5.7. *A 10 mH coil is connected in series with a loss free capacitor to a variable frequency source of 20 V. The current in the circuit has maximum value of 0.2 A at a frequency of 100 kHz. Calculate:*

(*i*) *The value of capacitance.*

(*ii*) *The Q-factor of the coil.*

(*iii*) *The half power frequencies.*

Solution.

(*i*) Under resonance condition current will be maximum

i.e., $$I_m = \frac{V}{R} \text{ or } R = \frac{V}{I_m}$$

$$R = \frac{20}{0.2} = 100\ \Omega$$

Maximum current occurs at resonant frequency, hence

$$f_r = 100 \times 10^3 \text{ Hz}$$

$$f_r = \frac{1}{2\pi}\sqrt{\frac{1}{LC}}$$

or $$10^5 = \frac{1}{2\pi}\sqrt{\frac{1}{10 \times 10^{-3}\, C}}$$

or $$C = \mathbf{253.3\ pF}$$ **Ans.**

(*ii*) Q-factor, $$Q_o = \frac{\omega_r L}{R} = \frac{2\pi f_r L}{R}$$

or $$Q_o = \frac{2\pi \times 10^5 \times 10 \times 10^{-3}}{100}$$

or Q-factor = **62.83.** **Ans.**

(*iii*) Half power frequencies

Lower half power frequency, $$f_1 = f_r - \frac{R}{4\pi L}$$

or $$f_1 = 10^5 - \frac{100}{4\pi \times 0.01}$$

or $$f_1 = \mathbf{99.204\ kHz}$$ **Ans.**

Upper half power frequency, $$f_2 = f_r + \frac{R}{4\pi L}$$

or $$f_2 = 10^5 + \frac{100}{4\pi \times 0.01}$$

or $$f_2 = \mathbf{100.796\ kHz.}$$ **Ans.**

Problem 5.8. *Calculate the value of C which results in resonance for the circuit shown in Fig. 5.15 when frequency is 1,000 Hz and find Q-factor for each branch.*

Solution. Admittance of inductive branch, $\mathbf{Y_{RL}} = \dfrac{1}{4 + j8} = (0.05 - j0.1)$ Siemen ℧

Admittance of capacitive branch, $\mathbf{Y_{RC}} = \dfrac{1}{5 - jX_C} = \left(\dfrac{5}{25 + X_C^2} + j\dfrac{X_C}{25 + X_C^2}\right)$ Siemen ℧

For resonance, immaginary parts of admittances must be equal in magnitude *i.e.*,

$$\frac{X_C}{25 + X_C^2} = 0.1$$

or $$0.1\,X_C^2 - X_C + 2.5 = 0$$

or $$X_C = 5\,\Omega$$

and $$C = \frac{1}{2\pi f X_C} = \frac{1}{2\pi \times 1{,}000 \times 5} = \mathbf{31.83\ \mu F}$$ **Ans.**

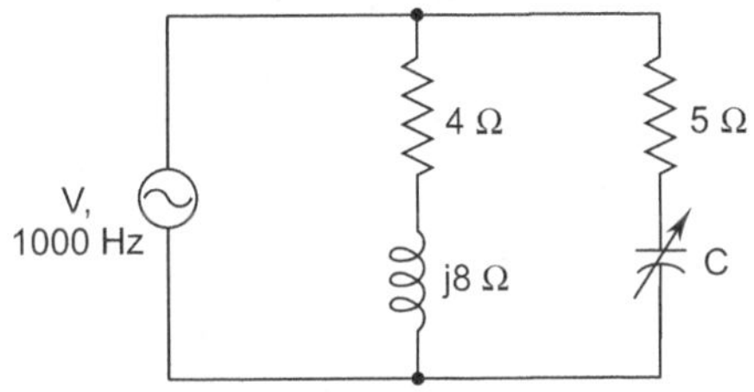

Fig. 5.21

Q-factor for R-L branch $= \dfrac{\omega L}{R} = \dfrac{8}{4} = \mathbf{2}$ **Ans.**

Q-factor for R-C branch $= \dfrac{1}{\omega RC} = \dfrac{1}{\omega CR} = \dfrac{5}{5} = \mathbf{1.0}$ $\left[\because \dfrac{1}{\omega C} = X_C = 5\,\Omega\right]$ **Ans.**

Problem 5.9. *A 46 mH inductive coil has a resistance of 10 Ω.*

(i) How much current will it draw if connected across a 100 V, 60 Hz source?

(ii) Determine the value of the capacitance that must be connected across the coil to make the power factor of the overall circuit unity.

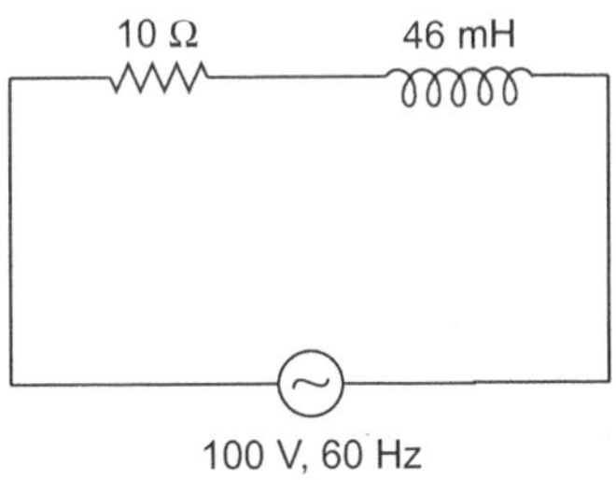

Fig. 5.22

Solution.

(i) $$X_L = \omega L = 2\pi f L = 2\pi \times 60 \times 0.046$$
$$= 17.34\ \Omega$$

Impedance, $$Z = R + jX_L = 10 + j17.34 = 20.01\angle 60°$$

Current, $$I = \frac{V}{Z} = \frac{100\angle 0°}{20.01\angle 60°} = 4.99\angle -60°$$

or Current, $$I = \mathbf{5\ A}$$ **Ans.**

(*ii*) The value of capacitance for the overall power factor of the circuit to be unity will be calculated as follow:

Before connecting capacitor, the power factor of the circuit is

$$\cos \phi_1 = \cos 60^\circ = 0.5$$

Now after connecting the capacitor the power factor of the circuit becomes unity. *i.e.*

$$\cos \phi_2 = 1$$

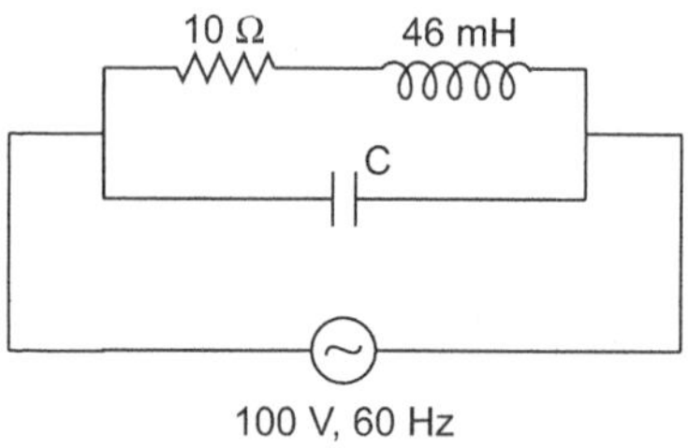

Fig. 5.23

Connecting a capacitor in parallel is one of the methods to improve the power factor of the circuit.

But in both the cases the power should remain the same, *i.e.*

$$P_1 = P_2$$

or $$V_1 I_1 \cos \phi_1 = V_2 I_2 \cos \phi_2$$

As in both the cases $V_1 = V_2 = V = 100$ V

Hence, $$I_1 \cos \phi_1 = I_2 \cos \phi_2$$

or $$5 \times 0.5 = I_2 \times 1$$

or $$I_2 = 2.5 \text{ A}$$

Power factor unity means the circuit is at resonance. Therefore the dynamic impedance of the circuit is

$$Z = \frac{L}{CR} = \frac{V}{I} = \frac{100}{2.5} = 40\ \Omega$$

or $$C = \frac{L}{ZR} = \frac{0.046}{40 \times 10} = \mathbf{115\ \mu F}$$ **Ans.**

Alternate method : In this case the condition for resonance is obtained by changing the value of capacitance and not by changing the frequency. Hence the frequency remains same and it will become resonant frequency. Therefore

$$f_r = \frac{1}{2\pi}\sqrt{\frac{1}{LC} - \left(\frac{R}{L}\right)^2}$$

or $$60 = \frac{1}{2\pi}\sqrt{\frac{1}{0.046 \times C} - \left(\frac{10}{0.046}\right)^2}$$

or $$C = \mathbf{115\ \mu F}$$ **Ans.**

Problem 5.10. *For the circuit shown in Fig. 5.24, R = 2 Ω, L = 0.02 H and V = 200 V.*

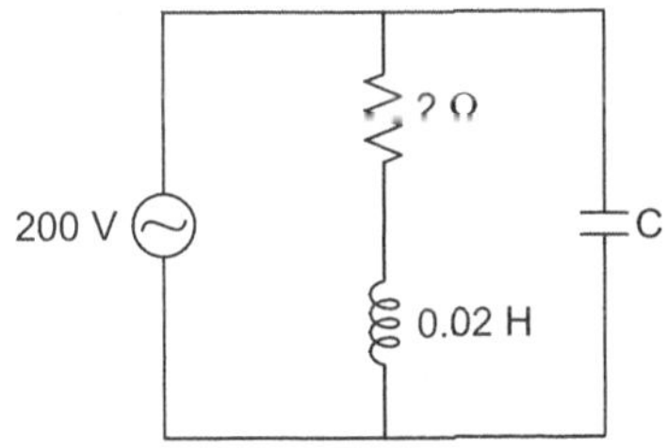

Fig. 5.24

What capacitance must the condenser have in order that the minimum current may occur at

(a) *25 Hz?*

(b) *100 Hz?*

Solution.

(*a*) When the frequency is 25 Hz

$$f_r = \frac{1}{2\pi}\sqrt{\frac{1}{LC} - \left(\frac{R}{L}\right)^2}$$

or $$25 = \frac{1}{2\pi}\sqrt{\frac{1}{0.02 \times C} - \left(\frac{2}{0.02}\right)^2}$$

or $$C = \mathbf{1442\ \mu F}$$ **Ans.**

(*b*) When the frequency is 100 Hz

$$f_r = \frac{1}{2\pi}\sqrt{\frac{1}{LC} - \left(\frac{R}{L}\right)^2}$$

or $$100 = \frac{1}{2\pi}\sqrt{\frac{1}{0.02 \times C} - \left(\frac{2}{0.02}\right)^2}$$

or $$C = \mathbf{123.5\ \mu F.}$$ **Ans.**

EXERCISE

1. A series AC circuit has a resistance of 15 Ω and inductive reactance of 10 Ω. Calculate the value of a capacitor which is connected across this series combination so that system has unity power factor. The frequency of AC supply is 50 Hz.
2. Derive an expression for the resonance frequency of the parallel resonance circuit, one branch consisting of inductor L, resistance R and other branch contain capacitor C. Also draw the Phasor diagram for the same.
3. Find the expression for the resonance frequency for the parallel circuit shown in Fig. 5.25.

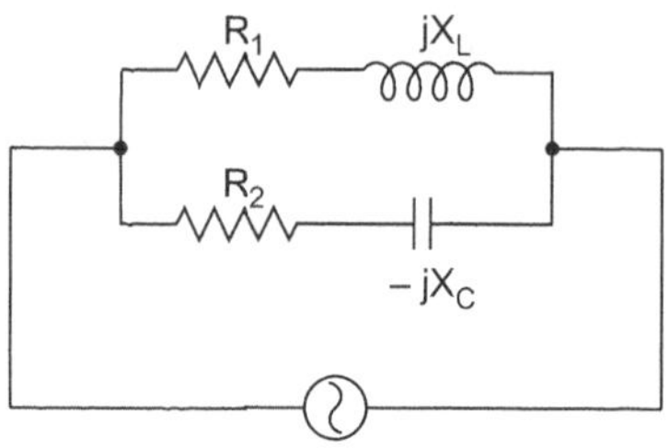

Fig. 5.25

4. Find the expression for the resonance frequency for the parallel circuit shown in Fig. 5.26.

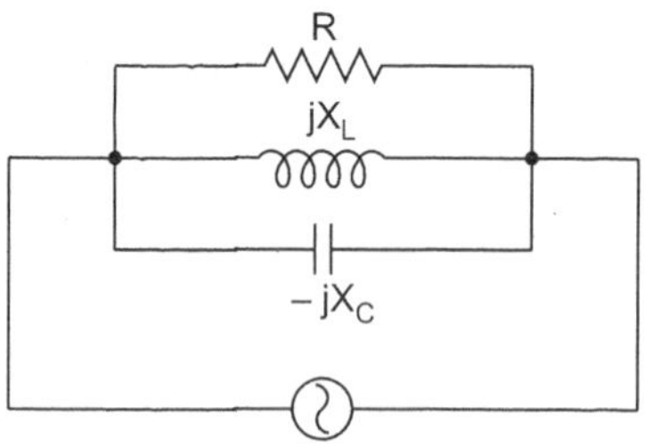

Fig. 5.26

5. A series RLC circuit has R = 10 Ω, L = 0.1 H and C = 8 μF. Determine:
   (*i*) resonant frequency,
   (*ii*) Q-factor of the circuit at resonance, and
   (*iii*) the half-power frequencies.
6. A 20 Ω resistor is connected in series with an inductor and a capacitor, across a variable frequency 25 V supply. When the frequency is 400 Hz, the current is as its maximum value of 0.5 A and the potential difference across the capacitor is 150 V. Calculate the resistance and inductance of the inductor.
7. Derive the condition for resonance in series RLC circuit.
8. Define quality factor. What is the significance of quality factor in both the cases *i.e.*, series AC circuit and parallel AC circuits?
9. What is selectivity? On which factor does it depends.
10. Give the salient features of series resonance.
11. Give the salient feature of parallel resonance.
12. What are the different applications of resonance?

□ □ □

6

Magnetic Circuit

6.1 INTRODUCTION

Before going on to understand a magnetic circuit we must first understand what is meant by a magnetic field. When a magnet is kept on a table and a small iron object is placed near it, it simply goes and sticks to the magnet. The force with which the piece is pulled decreases as its distance from the magnet increases. This phenomenon can be explained if you consider that the space around the magnet is permeated with a type of force which influences the iron piece near it. This force field is known as the magnetic field which can be visualized to be in the form of magnetic lines of flux. These are just imaginary lines and do not exist actually but can only be experienced. Fig. 6.1 shows these imaginary lines of magnetic flux surrounding a magnet.

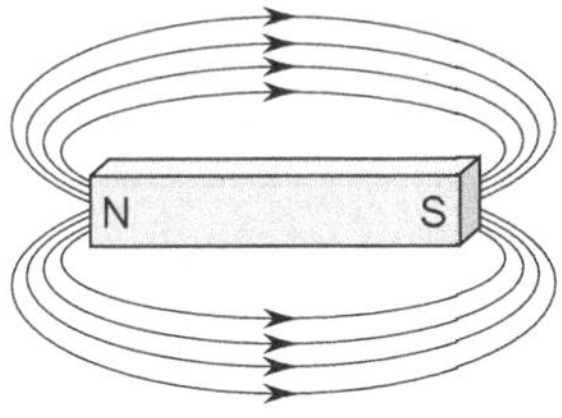

Fig. 6.1

There are several characteristics of lines of force which are as follows:

- *Direction:* Any line of flux always points towards the north seeking pole of a magnet provided it is not under the influence of another magnetic medium except the magnetic field of the earth.
- *Closed Line:* All lines of flux form a closed loop which emerge from the North Pole into the South Pole of the magnet.
- *Independent Region:* Lines of flux never intersect each other.
- *Repulsion:* Lines of flux which are in the same direction always repel each other.

6.2 THE MAGNETIC CIRCUIT

The above discussion about magnetic field and lines of flux now leads us to the definition of a magnetic circuit. A magnetic circuit can be described as a complete closed path followed by any group of lines of magnetic flux. A magnetic circuit is quite similar to an electrical circuit in terms of various parameters which are associated with the magnetic circuit.

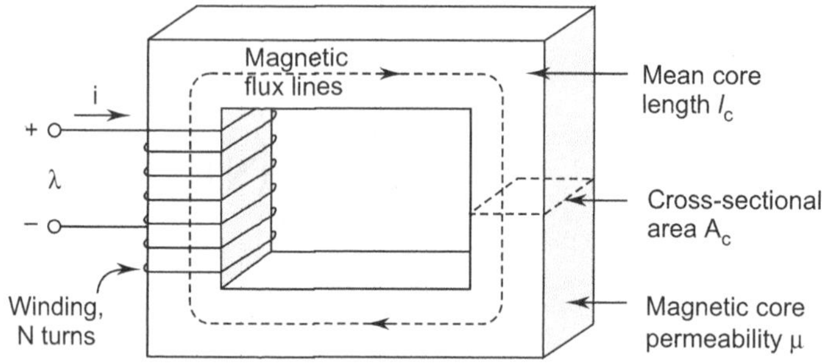

Fig. 6.2

Just as electromotive force or e.m.f. makes electric current flow, similarly magneto-motive force or m.m.f makes the magnetic flux possible in a magnetic circuit. The units of m.m.f are ampere-turns, where turns signify the number of electric coil turns which is generating that m.m.f. Since number of turns is a dimensionless quantity, effectively the units of m.m.f are same as that of current, namely amperes.

The m.m.f that exists per unit length of a magnetic circuit, provided the circuit is homogeneous and of a uniform cross-sectional area, is termed as Magnetic Field Strength. In the earlier days it was also known as magnetizing force but this term is obsolete now. The units of magnetic field strength are amperes/metre.

Of course these are just a few basic concepts associated with a magnetic circuit. We will learn more about this topic in different articles as well.

A sectional view of the magnetic circuit of Fig. 6.2 is given in Fig. 6.3.

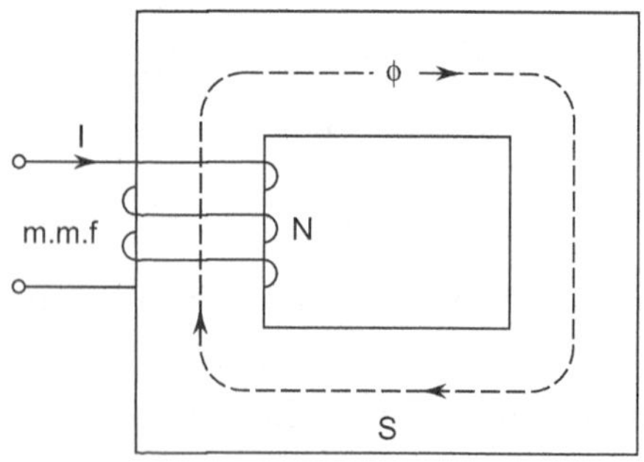

Fig. 6.3

The equivalent magnetic circuit of Fig. 6.3 is shown in Fig. 6.4 and its analogous electrical circuit is shown in Fig. 6.5.

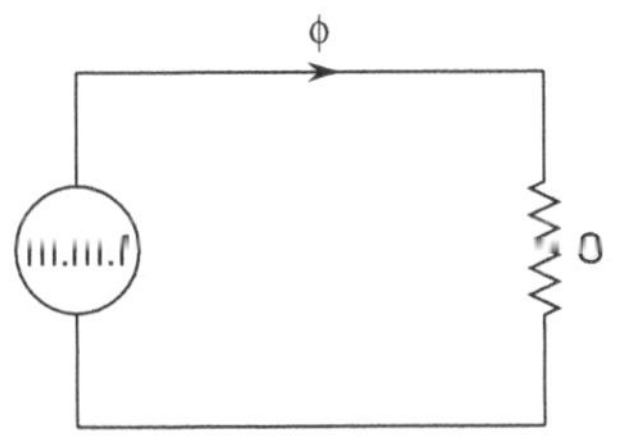

Fig. 6.4

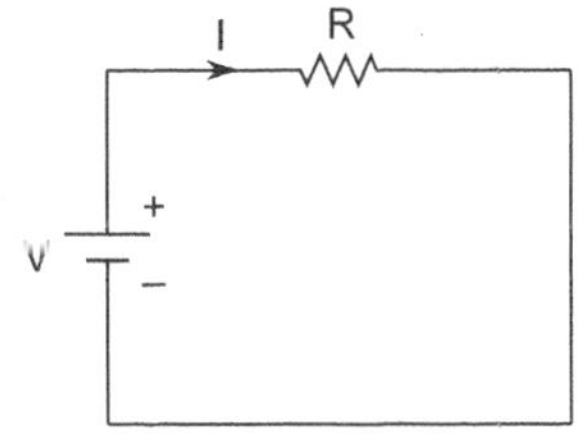

Fig. 6.5

6.3 ELEMENTARY CONCEPTS AND DEFINITIONS

Some of the basic terms used in explaining a magnetic circuit in a more *elucid* way will be discussed next.

6.3.1 Permeability

Permeability is defined as ***"the degree of magnetization of a material in response to a magnetic field"***. Permeability is a constant of proportionality that exists between magnetic induction and magnetic field intensity and is expressed as

$$\mu = \mu_o \, \mu_r \qquad \ldots(6.1)$$

where μ_o = Permeability of free space $(4\pi \times 10^{-7})$

μ_r = Relative permeability.

6.3.2 Magnetic Flux Density

Magnetic flux density is defined as ***"the amount of magnetic flux through a unit area taken perpendicular to the direction of the magnetic flux"***. It is also called as magnetic induction.

It is a vector quantity measuring the strength and direction of the magnetic field around a magnet or an electric current. Magnetic flux density is equal to magnetic field strength times the magnetic permeability in the region in which the field exists.

Electric charges moving through a magnetic field are subject to a force described by the equation

$$F = q\vec{v} \times \vec{B} \qquad \ldots(6.2)$$

where q is the amount of electric charge,

v is the velocity of the charge,

B is the magnetic flux density at the position of the charge, and

$\times$ is the vector product.

Magnetic flux density can also be understood as, ***"the density of magnetic lines of force, or magnetic flux lines, passing through a specific area"***. It is measured in units of weber/metre^2 (or Tesla).

$$B = \frac{\phi}{A} \quad (Wb/m^2) \qquad \ldots(6.3)$$

6.3.3 Magnetic Field Intensity

"Magnetic field strength (H) is the amount of magnetizing force". It is proportional to the length of a conductor and the amount of electrical current passing through the conductor. It is measured in units of Ampere-Turns/metre.

$$H = \frac{NI}{l} \qquad \ldots(6.4)$$

Magnetic field strength is a vector quantity whose magnitude is the strength of a magnetic field at a point in the direction of the magnetic field at that point. Flux density (B), the amount of magnetism induced in a body, is a function of the magnetizing force (H).

$$H = \frac{B}{\mu} \quad (AT/m) \qquad \ldots(6.5)$$

6.3.4 Magneto Motive Force

"Magneto motive force is any physical cause that produces magnetic flux". It is analogous to electromotive force or voltage in electricity.

In this context, the word 'force' is used in a general sense of 'work potential', and is analogous to, but distinct from mechanical force measured in Newtons.

The standard definition of magneto motive force involves current passing through an electrical conductor, which accounts for the magnetic fields of electromagnets. Permanent magnets also exhibit magneto motive force, but for different reasons.

The unit of magneto motive force is Ampere-Turn (AT), represented by a steady, direct electric current of one ampere flowing in a single-turn loop of electrically conducting material in a vacuum.

The gilbert (Gi), is the CGS unit of magneto motive force. The gilbert is defined differently, and is a slightly smaller unit than the ampere-turn. To convert from ampere-turns to gilberts, multiply by 1.25664. The unit is named after William Gilbert, an English physician and natural philosopher.

The magneto motive force in an inductor is given by

$$\text{m.m.f} = N \times I \quad AT \qquad \ldots(6.6)$$

where N is number of turns of the coil
I is current in the coil.

6.3.5 Reluctance

Magnetic reluctance or "Magnetic Resistance" is analogous to resistance in an electrical circuit (although it does not dissipate magnetic energy).

Magnetic reluctance is the property of magnetic circuit which offers opposition to the flow of magnetic flux.

In likeness to the way an electric field causes an electric current to follow the path of least resistance, a magnetic field causes magnetic flux to follow the path of least magnetic reluctance.

The reluctance (S) of a magnetic circuit is equal to ratio of 'magneto motive force' (m.m.f) in a passive magnetic circuit and the magnetic flux in this circuit.

$$S = \frac{\text{m.m.f}}{\phi} \quad (AT/Wb) \qquad \ldots(6.7)$$

where S is reluctance in ampere-turns per weber
m.m.f is magneto motive force in ampere-turns
ϕ is magnetic flux in webers.

Reluctance of a uniform magnetic circuit can be calculated as

$$S = \frac{1}{\mu}\frac{l}{A} = \frac{1}{\mu_o \mu_r}\frac{l}{A} \text{ AT/Wb} \quad \ldots(6.8)$$

where l is length of the circuit in metres.

μ_o is permeability of free space in H/m.

μ_r is relative magnetic permeability of the material, (dimensionless).

A is cross-sectional area of the circuit in square metres.

6.3.6 Permeance

"Permeance is a measure of ability of a magnetic circuit to conduct magnetic flux". It is reciprocal of reluctance. Its SI derived unit is the Henry (same as the unit of inductance, although two concepts are distinct).

$$P = \frac{1}{S} = \frac{\mu A}{l} \text{ H} \quad \ldots(6.9)$$

6.3.7 Fringing

Let us consider a magnetic circuit, having a small air gap. Useful flux present in the air-gap tends to buldge out (Fig. 6.6), thereby increasing the effective area of the gap and reducing flux density in the gap. This effect is called as fringing.

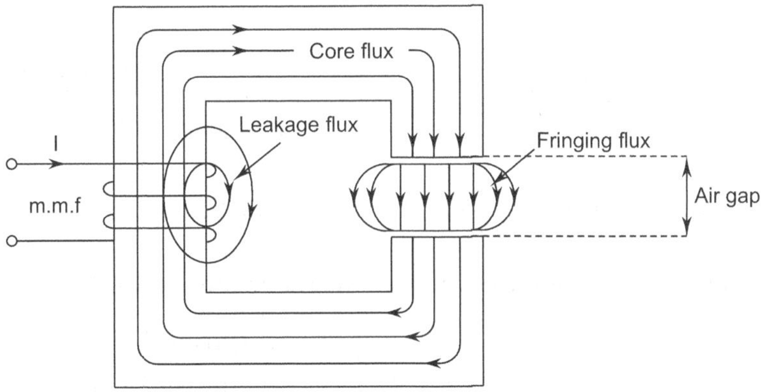

Fig. 6.6

6.3.8 Leakage Flux and Leakage Factor

It is not possible to limit flux path through the core entirely. Some flux always leaks to surrounding environment. Leakage flux is the flux that doesn't follow the desired path in a magnetic circuit. The leakage flux is shown in Fig. 6.6.

$$\text{Leakage flux} = \text{Total flux} - \text{Useful flux} \quad \ldots(6.10)$$

Leakage flux is represented by leakage factor 'λ' which is given by

$$\lambda = \frac{\text{Total flux}}{\text{Useful flux}} \quad \ldots(6.11)$$

For convenience in calculations, iron is assumed to carry whole of the flux throughout its entire length.

Hence, Leakage factor, $\lambda = \dfrac{\text{Flux in iron path}}{\text{Flux in air-gap}}$ ...(6.12)

6.4 ANALOGIES BETWEEN ELECTRIC AND MAGNETIC CIRCUITS

Even though there are a lot of differences between electric and magnetic circuits, they have some similarities also. Those similarities and the differences are listed next.

Table 6.1 Comparison between Magnetic and Electric Circuit

S.No.	Magnetic Circuit	Electric Circuit
1.	The closed path followed by magnetic flux is called magnetic circuit.	The closed path followed by electric current is called electric circuit.
2.	Excitation force is m.m.f.	Excitation force is e.m.f.
3.	Response is flux. Flux $(\phi) = \dfrac{\text{m.m.f (AT)}}{\text{Reluctance } (s)}$	Response is current. Current $(\text{I}) = \dfrac{\text{e.m.f } (v)}{\text{Reluctance (R)}}$
4.	m.m.f (AT) = ϕS	e.m.f (V) = IR
5.	Reluctance, $\text{S} = \dfrac{l}{\mu \text{A}}$	Resistance, $\text{R} = \rho \dfrac{l}{\text{A}}$
6.	Permeance $= \dfrac{1}{\text{Reluctance}}$	Conductance $= \dfrac{1}{\text{Resistance}}$
7.	Flux density, $\text{B} = \dfrac{\phi}{\text{A}}$ Wb/m^2	Current density, $\text{J} = \dfrac{\text{I}}{\text{A}}$ A/m^2
8.	Magnetic field intensity, $\text{H} = \dfrac{\text{m.m.f}}{l}$ AT/m	Electric field intensity, $\text{E} = \dfrac{\text{e.m.f}}{l}$ V/m

Table 6.2 Differences between Magnetic and Electric Circuit

S.No.	Magnetic Circuit	Electric Circuit
1.	Strictly speaking, magnetic flux does not flow.	Electric current actually flows in electric circuits.
2.	Energy is required only for creating the flux intensity but not for maintaining it.	Electric circuit requires continuous expenditure of energy when current flows.
3.	There is no magnetic insulator to oppose the flux.	There are a number of electric insulators, which opposes the flow of current.
4.	Flux can pass through air.	Current cannot flow through air.
5.	Residual flux persists after removal of m.m.f source.	Current does not remains in the circuit after removal of e.m.f source.
6.	Magnetic reluctance is not constant for given material. This is due to the fact that permeability or reluctance depends on the total flux or flux density.	Resistance of electric circuit is practically constant for constant temperature. It does not depend on current strength.

6.5 HYSTERESIS AND EDDY CURRENT LOSSES

Time-varying fluxes produce power losses in ferro-magnetic materials, known as *core losses*. These are also called *iron losses*. These losses occur in DC machines, AC machines, transformers, reactors, etc. The iron losses are classified into two types:

- Eddy current loss
- Hysteresis loss

6.5.1 Eddy Current Loss

Eddy currents are circulating currents which are induced in the core material. These currents are created by the difference in potential existing through out the body of the material because of the changing flux. Fig. 6.7 shows the nature of these circulating eddy currents.

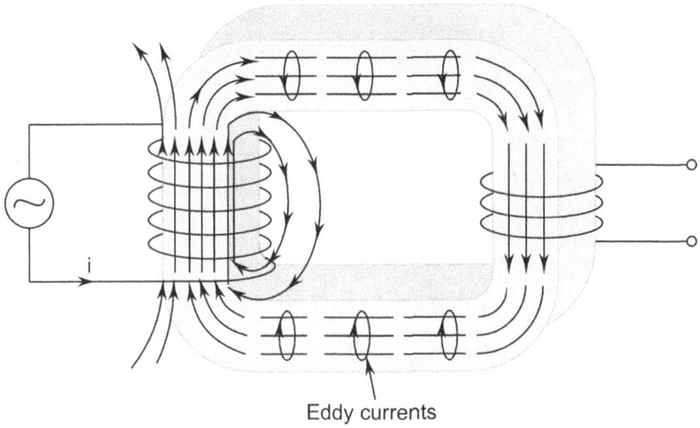

Fig. 6.7

If the magnetic circuit is composed of solid iron, the resulting power loss is appreciable because the circulating current encounters little resistance. To increase the resistance encountered by these eddy currents, the core is sliced to very thin laminations, insulated by varnish or oxide and then stacked together. A laminated core is shown in Fig. 6.8.

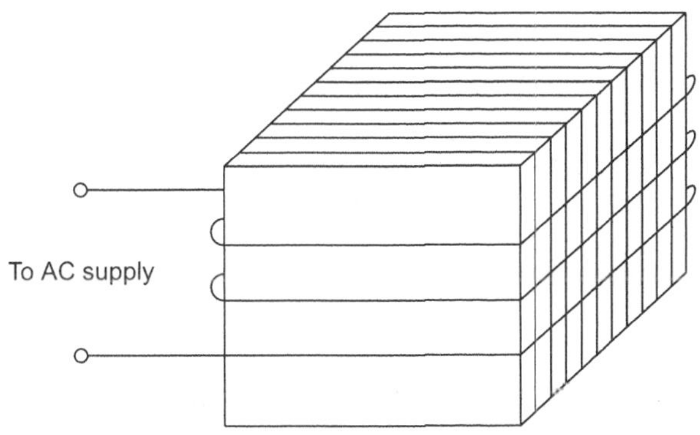

Fig. 6.8

The eddy current loss is given by the expression

$$P_e = K_e f^2 B_m^2 \tau^2 v \qquad \text{...(6.13)}$$

where, τ is the thickness of the laminations, and v is the volume of the core. If the core material volume and thickness are constant, the eddy current loss can be written as:

$$\boxed{P_e = K_e f^2 B_m^2} \qquad \text{...(6.14)}$$

where K_e is Eddy current Constant.

6.5.2 Hysteresis Loss

Consider that a magnetic core (Fig. 6.9) is supplied with an alternating current (voltage) to its windings. The wave shape of the applied voltage is shown in Fig. 6.10. As the current (or voltage) traces through one cycle starting from the origin, the B-H curve goes through OZCZ'R' (Fig. 6.11).

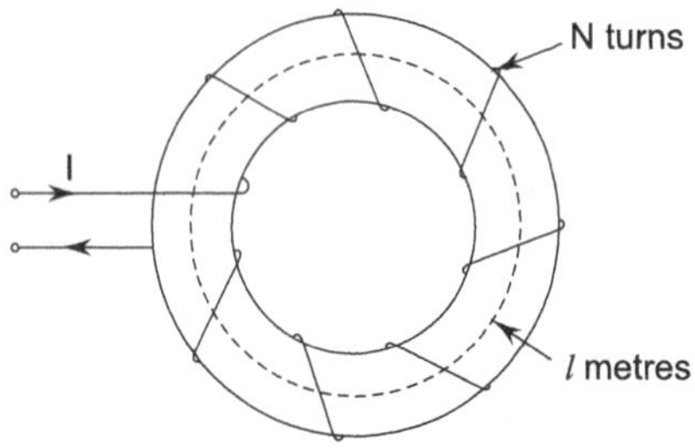

Fig. 6.9

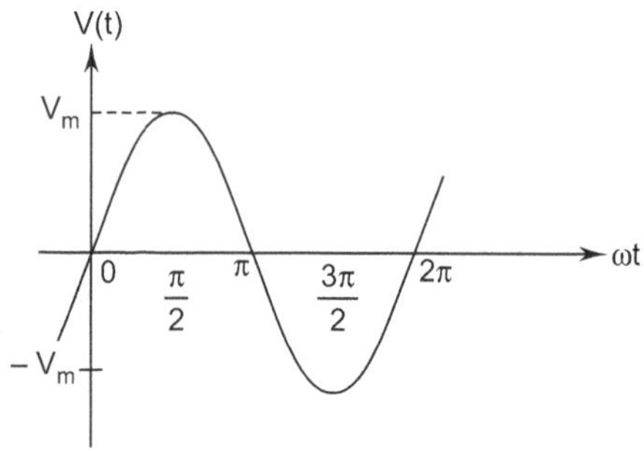

Fig. 6.10

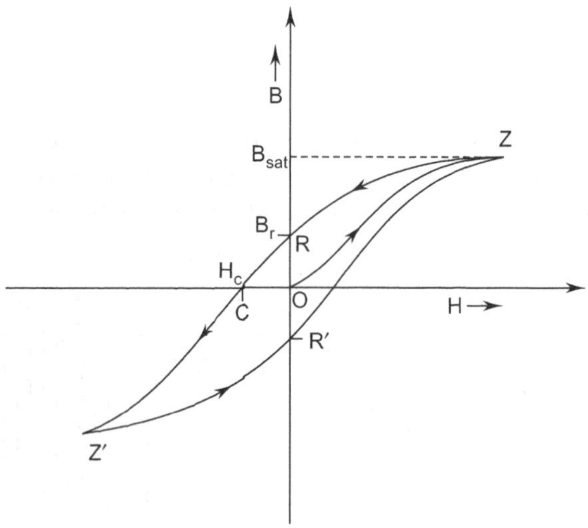

Fig. 6.11

After several cycles, the B-H characteristic becomes the loop as shown in Fig. 6.12. This is called the hysteresis loop. During half of the cycle, the electrical source gives energy to the magnetic circuit (core) and during the other half, the core returns part of this energy to the source. The shaded area in the loop represents the energy lost in the core during one cycle.

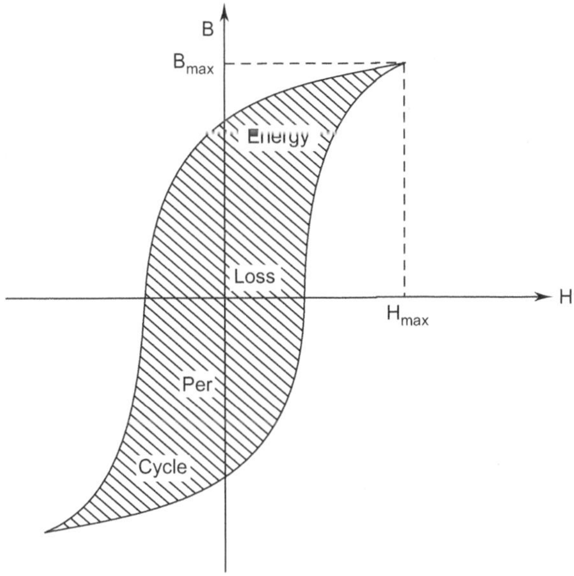

Fig. 6.12

The hysteresis loss expression is given by

$$\boxed{P_h = K_h f B_m^{\eta}} \quad \ldots(6.15)$$

where K_h is hysteresis loss constant
f is the frequency of the supply
B_m is the maximum flux density
η is called the Steinmetz exponent and has the range 1.5 – 2.5 depending on the material of the core.

Problem 6.1. *The eddy current loss in a certain electrical apparatus operating at its rated voltage and rated frequency of 240 V and 25 Hz is 642 W. Determine the eddy current loss if the apparatus is connected to a 60 Hz supply, whose voltage is such that the flux density becomes 62% of the rated voltage.*

Solution. Case I when $f_1 = 25$ Hz

$$P_{e1} = 642 \text{ W}$$

$$P_{e1} = K_e f_1^2 B_{m1}^2 \quad \ldots(i)$$

Case I when $f_2 = 60$ Hz

$$P_{e2} = ?$$

$$B_{m2} = 62\% \text{ of } B_{m1}$$

$$P_{e2} = K_e f_2^2 B_{m2}^2 \quad \ldots(ii)$$

Dividing equation (*ii*) by equation (*i*), we get

$$\frac{P_{e2}}{P_{e1}} = \frac{K_e f_2^2 B_{m2}^2}{K_e f_1^2 B_{m1}^2} \quad \ldots(iii)$$

or

$$P_{e2} = P_{e1}\left[\frac{f_2}{f_1}\right]^2 \left[\frac{B_{m2}}{B_{m1}}\right]^2$$

or $$P_{e2} = 642\left[\frac{60}{25}\right]^2\left[\frac{0.62}{1}\right]^2$$

or $$P_{e2} = \mathbf{1.42\ kW}$$ **Ans.**

Problem 6.2. *The hysteresis loss in a certain electrical apparatus operating at its rated voltage and rated frequency of 240 V and 25 Hz is 846 W. Determine the hysteresis loss if the apparatus is connected to a 60 Hz supply, whose voltage is such that the flux density becomes 62% of the rated voltage. Assume the Steinmetz exponent to be 1.4.*

Solution. Case I when $f_1 = 25$ Hz

$P_{h1} = 846$ W

$$P_{h1} = K_h f_1 B_{m1}^n \quad \ldots(i)$$

Case I when $f_2 = 60$ Hz

$P_{h2} = ?$

$B_{m2} = 62\%$ of B_{m1}

$n = 1.4$

$$P_{k2} = K_h f_2 B_{m2}^n \quad \ldots(ii)$$

Dividing equation (*ii*) by equation (*i*), we get

$$\frac{P_{h2}}{P_{h1}} = \frac{K_h f_2 B_{m2}^n}{K_h f_1 B_{m1}^n} \quad \ldots(iii)$$

$$P_{h2} = P_{h1}\left[\frac{f_2}{f_1}\right]\left[\frac{B_{m2}}{B_{m1}}\right]^n$$

or $$P_{e2} = 846\left[\frac{60}{25}\right]^2\left[\frac{0.62}{1}\right]^{1.4}$$

or $$P_{e2} = \mathbf{1.04\ kW}$$ **Ans.**

6.6 MUTUAL COUPLING

When interchanging of energy takes place from one circuit to other, the circuits are said to be mutually coupled. The coupling between two circuits may be conductive, electromagnetic or electrostatic. The conductive and electrostatic couplings are shown in Fig. 6.13 and Fig. 6.14 respectively. In Fig. 6.13, R_{12} is the mutual resistance between circuits 1 and 2. In Fig. 6.14, the two circuits are coupled electro statically through the capacitor C.

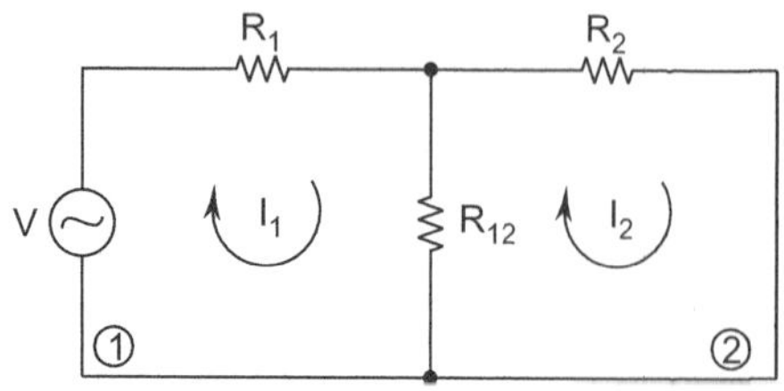

Fig. 6.13

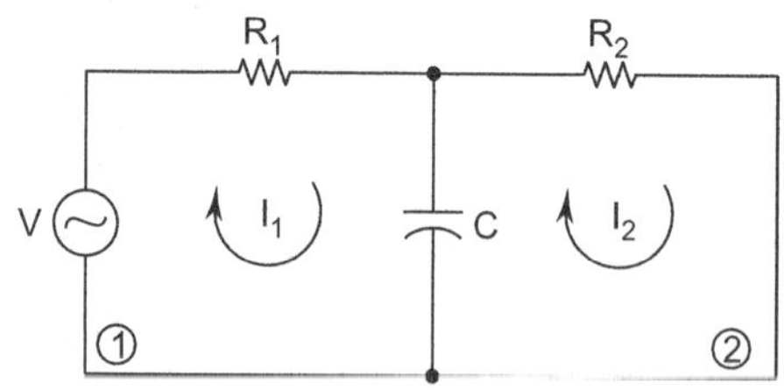

Fig. 6.14

6.7 MAGNETIC COUPLING

Fig. 6.15 shows magnetic coupling between two coils. In this case portions of the magnetic flux establish by one coil interlinks the other and the energy is transferred from one circuit to the other through the medium of magnetic flux that is common to both.

From Fig. 6.15, if ϕ_1 is the total flux produced by i_1 and ϕ_2 is the total flux produced by i_2, ϕ_1 and ϕ_2 can be written as follow

$$\phi_1 = \phi_{11} + \phi_{12} \quad \text{...(6.16)}$$

$$\phi_2 = \phi_{22} + \phi_{21} \quad \text{...(6.17)}$$

where, ϕ_{11} is the flux which is linked to coil 1.

ϕ_{22} is the flux which is linked to coil 2.

ϕ_{12} is the mutual flux that is linked to the turns of coil 2.

ϕ_{21} is the mutual flux that is linked to the turns of coil 1.

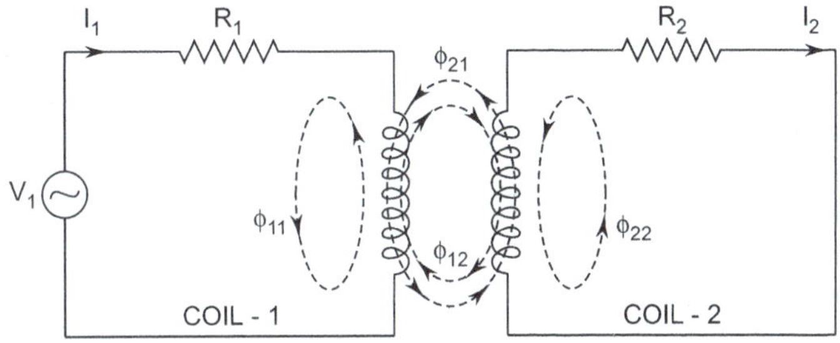

Fig. 6.15

6.7.1 Self-Inductance

Imagine a coil of N turns as shown in Fig. 6.16 connected to a battery. It is found that whenever an effort is made to increase current (and hence flux) through it, it is always opposed by the instantaneous production of counter e.m.f of self-induction. Energy required to overcome this opposition is supplied by the battery.

If, now an effort is made to decrease the current (and hence the flux), then again it is delayed due to the production of self-induced e.m.f. this time in the opposite direction. ***"This property of the coil due to which it opposes any increase or decrease of current or flux through it is known as self-Inductance".*** It is quantitatively measured in terms of coefficient of self-induction, L. This property is analogous to inertia in a material body.

We know by experience that initially it is difficult to set a heavy body into motion, but once in motion, it is equally difficult to stop it. Similarly, in a coil having large self-induction, it is initially difficult to establish a current through it, but once established, it is equally difficult to withdraw it. Hence, self-induction is sometimes analogously called electrical inertia or electromagnetic inertia.

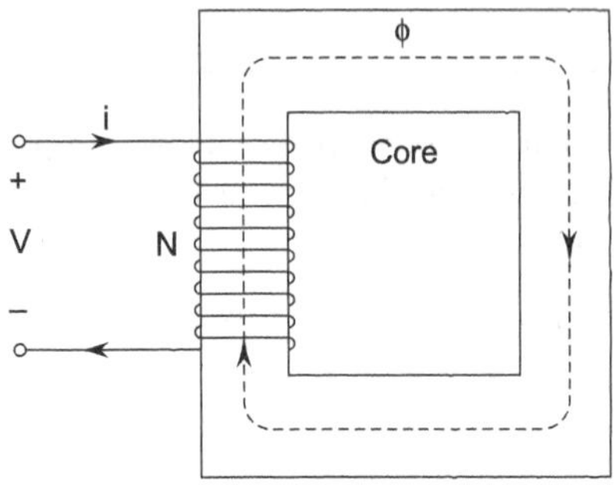

Fig. 6.16

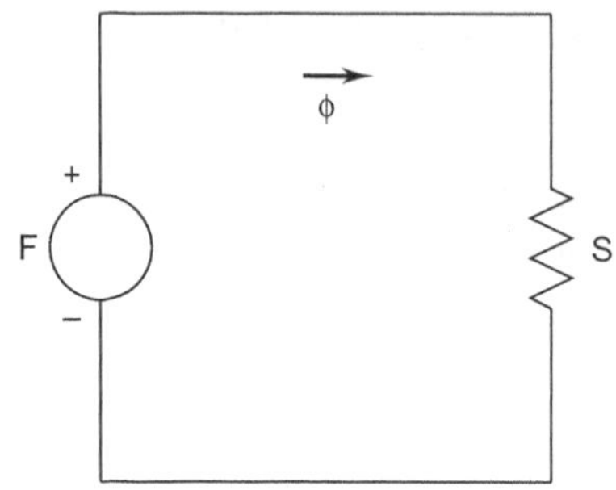

Fig. 6.17

Self-inductance (L) is defined as

$$L = \frac{\text{Flux linkage}}{\text{Current}} \quad \ldots(6.18)$$

or

$$L = \frac{\psi}{I} = \frac{N\phi}{I} \quad \ldots(6.19)$$

where, ψ is the flux linkage.

As

$$S = \frac{\text{m.m.f}}{\phi} = \frac{NI}{\phi} \quad \ldots(6.20)$$

or

$$\phi = \frac{NI}{S} \quad \ldots(6.21)$$

From equation (6.21) the expression for self-inductance L [Equation (6.19)] will become

$$\boxed{L = \frac{N^2}{S}} \quad \ldots(6.22)$$

Energy supplied to the inductor is given by,

$$W = \int_0^t p dt = \int_0^t v i dt \quad \ldots(6.23)$$

or

$$W = \int_0^t L \frac{di}{dt} i dt = \int_0^t L i di \quad \ldots(6.24)$$

or

$$\boxed{W = \frac{1}{2} L i^2} \quad \ldots(6.25)$$

6.7.2 Mutual Inductance

In starting of this Article (Fig. 6.15) we have discussed that any change of current in coil 1 is always accompanied by the production of mutually-induced e.m.f. in coil 2. Mutual inductance may, therefore, be defined as, ***"the ability of one coil (or circuit) to produce an e.m.f. in a nearby coil by induction when current in the first coil changes"***. This action being reciprocal, the second coil can also induce an e.m.f. in the first when current in the second coil changes. This ability of reciprocal induction is measured in terms of coefficient of mutual induction, M.

For explaining mutual inductance let us consider a magnetic circuit having two coils with N_1 and N_2 number of turns (Fig. 6.18).

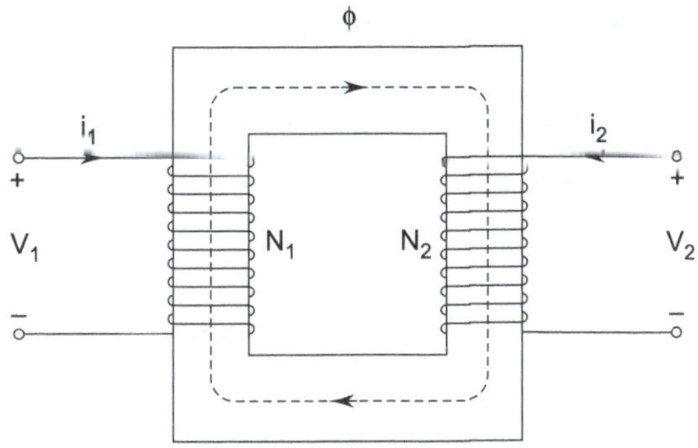

Fig. 6.18

The mutual inductance between the two windings, (if $i_1 \neq 0$ and $i_2 = 0$) M_{21}, is defined as

$$M_{21} = \left.\frac{\text{Flux linking circuit 2 while current in 1}}{\text{current in circuit 1}}\right|_{i_2=0} \quad \text{...[6.26(a)]}$$

$$M_{21} = \frac{\psi_{21}}{i_1} = \frac{N_2\phi}{i_1} \quad \text{...[6.26(b)]}$$

or
$$M_{21} = \frac{N_2\left[N_1 i_1/S\right]}{i_1} \quad \text{...[6.26(c)]}$$

or
$$\boxed{M_{21} = \frac{N_1N_2}{S}} \quad \text{...(6.27)}$$

Similarly, if $i_1 = 0$ and $i_2 \neq 0$, then

$$M_{12} = \left.\frac{\text{Flux linking circuit 1 while current in 2}}{\text{current in circuit 2}}\right|_{i_1=0} \quad \text{...(6.28)}$$

$$M_{12} = \frac{\psi_{12}}{i_2} = \frac{N_1\phi}{i_2} \quad \text{...(6.29)}$$

or
$$M_{12} = \frac{N_2\left[N_2 i_2/S\right]}{i_2} = \frac{N_1N_2}{S} \quad \text{...(6.30)}$$

6.7.3 Coefficient of Coupling

As discussed in starting of this article that a portion of magnetic flux establish by one coil interlinks the other and energy is transferred from one circuit to the other through medium of magnetic flux that is common to both. Let us consider that portion of magnetic flux which interlinks with other coil be denoted by K, where K being less than unity. This K is called as coefficient of coupling and is given by the expression.

$$\boxed{K = \frac{M}{\sqrt{L_1L_2}}} \quad \text{...(6.31)}$$

6.8 MAGNETIC CIRCUIT CALCULATIONS

For a solenoid, the magnetic field intensity is given by

$$H = \frac{NI}{l} \quad ...(6.32)$$

Also, $$B = \mu H = \mu \frac{NI}{l} \quad \text{[From equation (6.32)]} \quad ...(6.33)$$

Total flux produce is given as

$$\phi = BA \quad ...(6.34)$$

From equation (6.33), equation (6.34) will become

$$\phi = \mu \frac{NI}{l} A \quad ...(6.35)$$

$$= \frac{NI}{l/\mu A} = \frac{NI}{S} \quad ...(6.36)$$

$$\phi = \frac{\text{m.m.f}}{\text{Reluctance}} \quad ...(6.37)$$

$$\boxed{\phi = \frac{NI}{S}} \quad ...(6.38)$$

Now using all above relations defined and derived, we will proceed for magnetic circuit calculations.

6.8.1 Magnetic Circuit having Square Cross-section and no Air Gap

The magnetic circuit shown in Fig. 6.19 (*a*) has a square cross-section area and has no air gap. The calculations for such a circuit is as follows:

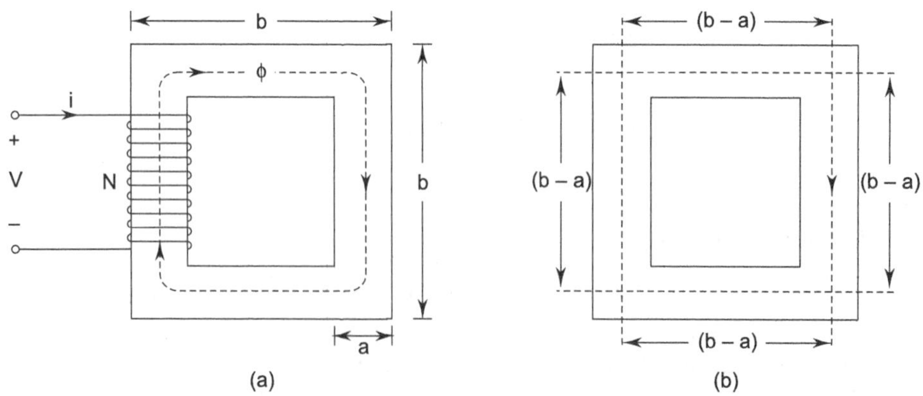

Fig. 6.19

Step 1. Calculation of mean length, l [from Fig. 6.19(*b*)]

$$l = 4 \times \text{Centre to centre distance} \quad ...(6.39)$$

or $$l = 4 \times (b - a) \quad ...(6.40)$$

Step 2. Calculation of cross-sectional area, A

$$A = a \times a \ m^2 \quad \ldots(6.41)$$

Step 3. Calculation of reluctance for material, S_m

$$S_m = \frac{l}{\mu A} \quad \ldots(6.42)$$

where $\mu = \mu_o \mu_r$ ...(6.43)

and μ_o = Permeability of free space ($4\pi \times 10^{-7}$)

μ_r = Relative permeability of material.

Step 4. Calculation of m.m.f

$$\text{m.m.f} = \phi S_m \quad \ldots(6.44)$$

Step 5. Calculation of excitation current, I

$$\text{m.m.f} = N \times I \quad \ldots(6.45)$$

or

$$I = \frac{\text{m.m.f}}{N} \quad \ldots(6.46)$$

6.8.2 Magnetic Circuit having Circular Cross-section and no Air Gap

The magnetic circuit shown in Fig. 6.20 (*a*) has a circular cross-section area and has no air gap. The calculations for this circuit is as follows:

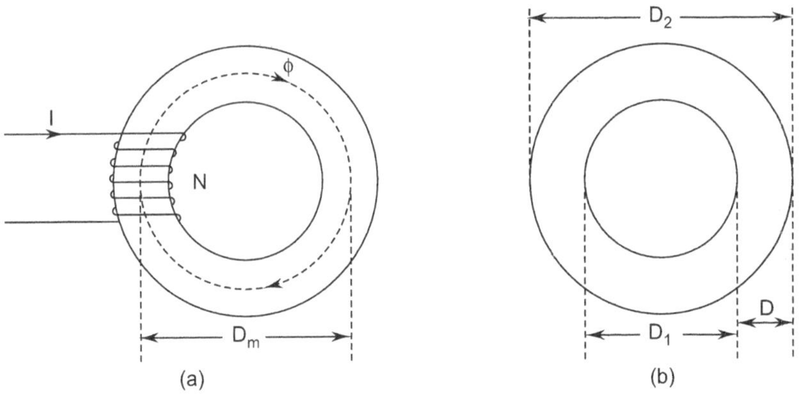

Fig. 6.20

Step 1. Calculation of mean length, l

$$l = 2\pi R_m \quad \ldots(6.47)$$

or

$$l = \pi D_m \quad \ldots(6.48)$$

where R_m = mean radius

D_m = mean diameter

$$D_m = D_1 + \left(\frac{D_2 - D_1}{2}\right) = \frac{D_1 + D_2}{2} \quad \ldots(6.49)$$

Step 2. Calculation of cross-sectional area, A

$$A = \pi r^2 = \frac{\pi D^2}{4} \ m^2 \quad \ldots(6.50)$$

where D is the diameter of the circular ring shown in Fig. 6.20(*b*)

Step 3. Calculation of reluctance for material, S_m

$$S_m = \frac{l}{\mu A} \quad \text{...(6.51)}$$

Step 4. Calculation of m.m.f

$$\text{m.m.f.} = \phi S_m \quad \text{...(6.52)}$$

Step 5. Calculation of excitation current, I

$$\text{m.m.f.} = N \times I \quad \text{...(6.53)}$$

or

$$I = \frac{\text{m.m.f}}{N} \quad \text{...(6.54)}$$

6.8.3 Magnetic Circuit having Rectangular Cross-section and an Air Gap

The magnetic circuit shown in Fig. 6.21(*a*) has a rectangular cross-section area and it has air-gap also. The calculations for this circuit are as follows:

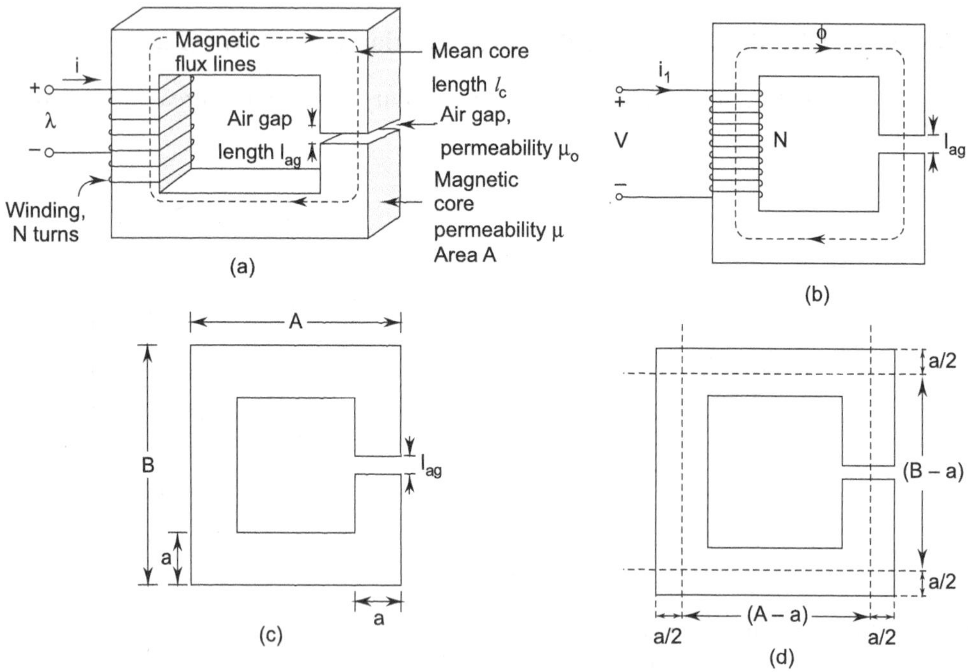

Fig. 6.21

Step 1. Calculation of mean length of material, l_m

$$l_m = (4 \times \text{centre to centre distance}) - l_{ag} \quad \text{...(6.55)}$$

or

$$l_m = 2\,(A - a) + 2\,(B - a) - l_{ag} \quad \text{...(6.56)}$$

where l_{ag} = length of air gap

Step 2. Calculation of cross-sectional area, A

$$A = a \times a \; m^2 \quad \text{...(6.57)}$$

Step 3. Calculation of reluctance for material, S_m

$$S_m = \frac{l_m}{\mu A} = \frac{l_m}{\mu_o \mu_r A} \quad \text{...(6.58)}$$

Step 4. Calculation of reluctance for air gap, S_{ag}

$$S_{ag} = \frac{l_{ag}}{\mu A} = \frac{l_{ag}}{\mu_o A} \quad \text{...(6.59)}$$

as μ_r = Relative permeability of air = 1.

Step 5. Calculation of total reluctance, S

$$S = S_m + S_{ag} \quad \text{...(6.60)}$$

Step 6. Calculation of m.m.f

$$\text{m.m.f.} = \phi S \quad \text{...(6.61)}$$

Step 7. Calculation of excitation current, I

$$\text{m.m.f.} = N \times I \quad \text{...(6.62)}$$

$$I = \frac{\text{m.m.f}}{N} \quad \text{...(6.63)}$$

6.8.4 Magnetic Circuit having Circular Cross-section and an Air Gap

The magnetic circuit having circular cross-section area with air gap is shown in Fig. 6.22 (*a*). The calculations for this circuit are as follows:

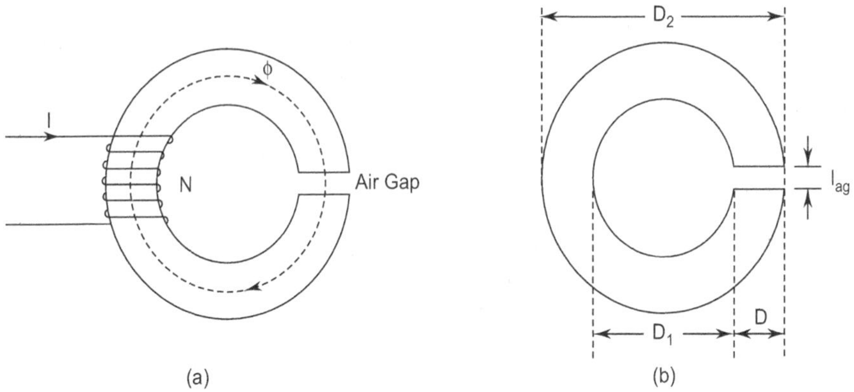

Fig. 6.22

Step 1. Calculation of mean length of material, l_m

$$l_m = 2\pi R_m - l_{ag} \quad \text{...(6.64)}$$

or

$$l_m = \pi D_m - l_{ag} \quad \text{...(6.65)}$$

where R_m = mean radius

D_m = mean diameter

l_{ag} = length of air gap

$$D_m = D_1 + \left(\frac{D_2 - D_1}{2}\right) = \frac{D_1 + D_2}{2} \quad \text{...(6.66)}$$

Step 2. Calculation of cross-sectional area, A

$$A = \pi r^2 = \frac{\pi D^2}{4} \, m^2 \quad \text{...(6.67)}$$

where, D is the diameter of the circular ring shown in Fig. 6.22(*b*)

Step 3. Calculation of reluctance for material, S_m

$$S_m = \frac{l_m}{\mu A} = \frac{l_m}{\mu_o \mu_r A} \qquad \ldots(6.68)$$

Step 4. Calculation of reluctance for air gap, S_{ag}

$$S_{ag} = \frac{l_{ag}}{\mu A} = \frac{l_{ag}}{\mu_o A} \qquad \ldots(6.69)$$

as μ_r = relative permeability of air = 1.

Step 5. Calculation of total reluctance, S

$$S = S_m + S_{ag} \qquad \ldots(6.70)$$

Step 6. Calculation of m.m.f

$$\text{m.m.f} = \phi S \qquad \ldots(6.71)$$

Step 7. Calculation of excitation current, I

$$\text{m.m.f} = N \times I \qquad \ldots(6.72)$$

$$I = \frac{\text{m.m.f}}{N} \qquad \ldots(6.73)$$

6.8.5 Parallel Magnetic Circuit having Rectangular Cross-section and Air Gap

A parallel magnetic circuit having rectangular cross-section area and with an air gap is shown in Fig. 6.23 (*a*). Its electrical equivalent circuit is also shown in Fig. 6.23 (*b*). The calculations for this circuit are as follows:

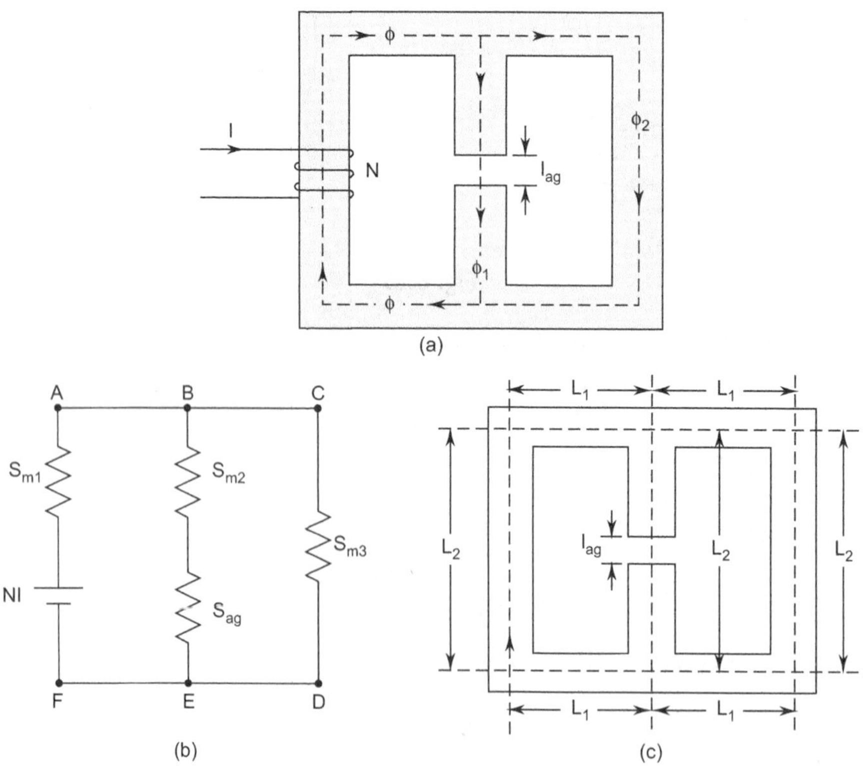

Fig. 6.23

Step 1. Calculation of mean length of material,

From Figs. 6.23(*b*) and 6.23(*c*) mean lengths of the three parallel branches are

$$\left.\begin{aligned} l_{m1} &= L_1 + L_2 + L_1 \quad &(BA + AF + FE) \\ l_{m2} &= L_2 - l_{ag} \quad &(BE - l_{ag}) \\ l_{m3} &= L_1 + L_2 + L_1 \quad &(BC + CD + DE) \end{aligned}\right\} \quad ...(6.74)$$

where l_{ag} = Length of air gap

Step 2. Calculation of cross-sectional area, A

$$A = a \times a \quad m^2 \quad ...(6.75)$$

Step 3. Calculation of reluctance for material, S_m

$$\left.\begin{aligned} S_{m1} &= \frac{l_{m1}}{\mu A} \\ S_{m2} &= \frac{l_{m2}}{\mu A} \\ S_{m3} &= \frac{l_{m3}}{\mu A} \end{aligned}\right\} \quad ...(6.76)$$

Step 4. Calculation of reluctance for air gap, S_{ag}

$$S_{ag} = \frac{l_{ag}}{\mu A} = \frac{l_{ag}}{\mu_o A} \quad ...(6.77)$$

As μ_r = Relative permeability of air = 1.

Step 5. Calculation of total reluctance, S

$$S = S_{m1} + \left[S_{m3} \,||\, (S_{m2} + S_{ag})\right] \quad ...(6.78)$$

Step 6. Calculation of m.m.f

$$\text{m.m.f} = \phi S \quad ...(6.79)$$

Step 7. Calculation of excitation current, I

$$\text{m.m.f} = N \times I \quad ...(6.80)$$

$$I = \frac{\text{m.m.f}}{N} \quad ...(6.81)$$

6.8.6 Magnetic Circuit having Rectangular Cross-section and two Air Gap

A magnetic circuit having two air gaps and rectangular cross-section area is shown in Fig. 6.24(*a*). Electrical equivalent circuit of such a magnetic circuit is shown in Fig. 6.24 (*c*). The calculations for such a circuit are as follows:

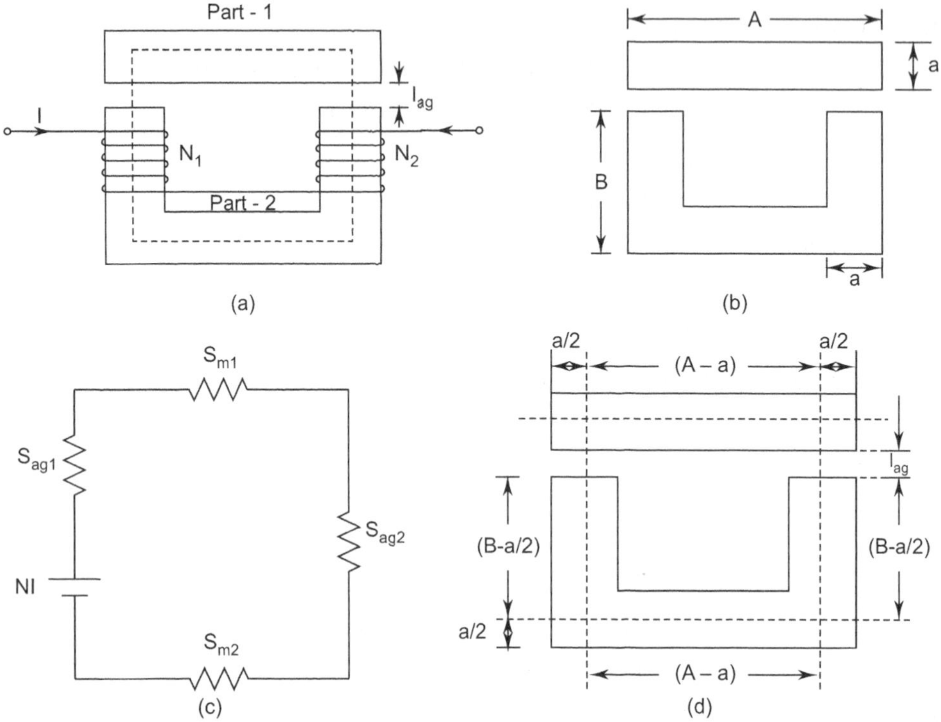

Fig. 6.24

Step 1. Calculation of mean length of material,

From Fig. 6.23(*d*) mean lengths of two parts and two air gaps will be mean lengths of part 1

$$L_{m1} = A - a \quad \text{...(6.82)}$$

Mean lengths of part 2

$$L_{m2} = \left(B - \frac{a}{2}\right) + (A - a) + \left(B - \frac{a}{2}\right) \quad \text{...(6.83)}$$

Mean lengths of 2 air gaps

$$L_{m3} = 2l_{ag} \quad \text{...(6.84)}$$

where l_{ag} = Length of air gap

Step 2. Calculation of cross-sectional area, A

$$A = a \times a \quad m^2 \quad \text{...(6.85)}$$

Step 3. Calculation of reluctance for material, S_m

$$S_{m1} = \frac{L_{m1}}{\mu A} \quad \text{...(6.86)}$$

$$S_{m2} = \frac{L_{m2}}{\mu A} \quad \text{...(6.87)}$$

Note. *The reluctances of part 1 and part 2 will be calculated separately if they have different relative permeabilities.*

If they have same relative permeabilities, then mean length will be

$$L_m = L_{m1} + L_{m2} - L_{m3} \quad \text{...(6.88)}$$

or

$$L_m = (A - a) + \left(B - \frac{a}{2}\right) + (A - a) + \left(B - \frac{a}{2}\right) - 2l_{ag} \quad \text{...(6.89)}$$

And reluctance for the material will be

$$S_m = \frac{L_m}{\mu A} \quad \ldots(6.90)$$

Step 4. Calculation of reluctance for air gaps, S_{ag}

$$S_{ag} = \frac{L_{m3}}{\mu A} = \frac{L_{m3}}{\mu_o A} \quad \ldots(6.91)$$

As μ_r = Relative permeability of air = 1.

Step 5. Calculation of total reluctance, S

$$S = S_{m1} + S_{m2} + S_{ag} \quad \ldots(6.92)$$

Step 6. Calculation of m.m.f.

$$\text{m.m.f.} = \phi S \quad \ldots(6.93)$$

Step 7. Calculation of excitation current, I

$$\text{m.m.f.} = N \times I \quad \ldots(6.94)$$

where, $N = N_1 + N_2$

$$I = \frac{\text{m.m.f.}}{N} \quad \ldots(6.95)$$

6.8.7 Composite Magnetic Circuit Having Circular Cross-section and Air-gap between two Different Materials

A magnetic circuit made up of composite material and having an air gap is shown in Fig. 6.25 (*a*) Let us consider, for example, that the ring is made up of cast iron and steel.

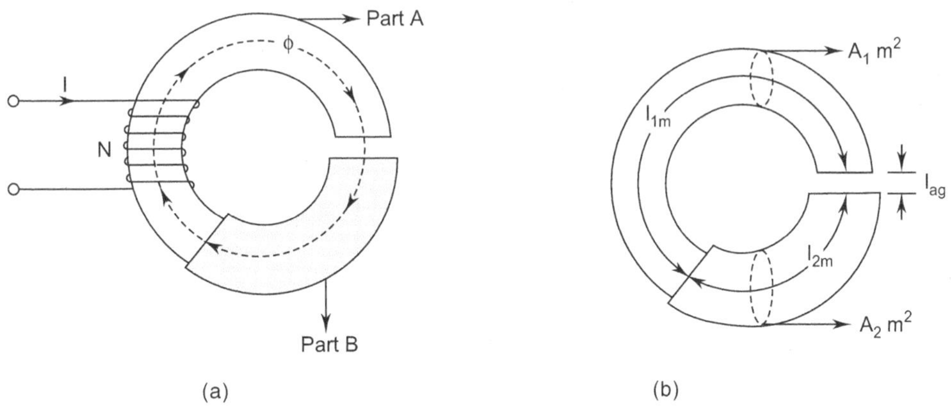

Fig. 6.25

For Part-A

Mean length of the material = $l_1\ m$

Relative permeability of the material = μ_{r1}

Cross-sectional area = $A_1\ m^2$

For Part-B

Mean length of material = $l_2\ m$

Relative permeability of material = μ_{r2}

Cross-sectional area = $A_2 \, m^2$

The calculations of such a circuit are as follows:

Step 1. Calculation of reluctance

For Part -A, reluctance of the material

$$S_{m1} = \frac{l_1}{\mu . A_1} = \frac{l_1}{\mu_o . \mu_{r1} . A_1} \qquad ...(6.96)$$

For Part-B, reluctance of material

$$S_{m2} = \frac{l_2}{\mu . A_2} = \frac{l_2}{\mu_o . \mu_{r2} . A_2} \qquad ...(6.97)$$

For Air-gap, reluctance will be

$$S_{ag} = \frac{l_{as}}{\mu .(A_1 \text{ or } A_2)^*} \qquad ...(6.98)$$

Step 2. Calculation of total reluctance, S

$$S = S_{m1} + S_{m2} + S_{ag} \qquad ...(6.99)$$

Step 3. Calculation of m.m.f.

$$\text{m.m.f} = \phi \, S \qquad ...(6.100)$$

Step 4. Calculation of excitation current, I

$$I = \frac{\text{m.m.f}}{N} \qquad ...(6.101)$$

Note. *If air gap is given between two similar material parts, then consider the area of that material while calculating the reluctance of air gap.*

SOLVED PROBLEMS

Problem 6.3. *For AC excited magnetic circuit of Fig. 6.26, calculate excitation current and induced e.m.f. of the coil having 500 turns to produce a core flux of 0.6 sin (314 t) mWb. Take relative permeability as μ_r = 3775 and area of cross-section as 9 cm^2.*

Solution.

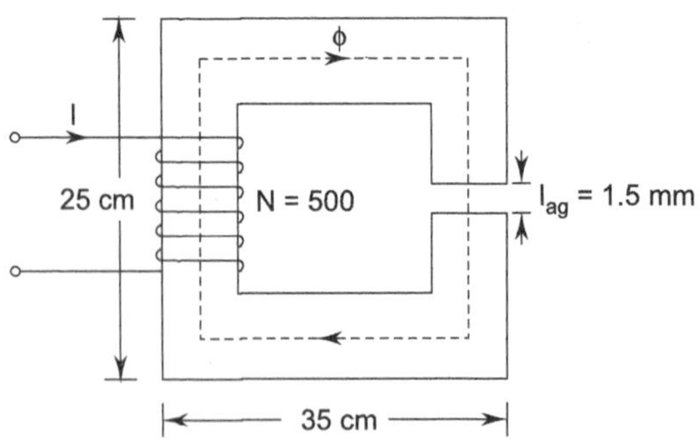

Fig. 6.26

* It is obvious that reluctance of air gap will be more than the reluctance of material. Mainly reluctance of air gap depends upon length of air gap and area of the material.

In this case, length is constant. Hence reluctance of air gap will be more when area of the material considered will be less. Hence while calculating reluctance for air gap, take that material area which is smaller.

Area of cross-section, $A = a \times a = 3 \times 3 = 9 \text{ cm}^2$

$a = 3 \text{ cm} = 3 \times 10^{-2} \text{ m}$

$\phi_m = 0.6 \text{ m Wb}$

$l_{ag} = 1.5 \text{ mm} = 1.5 \times 10^{-3} \text{ m}$

$\mu_r = 3775$

Mean length of material, $l_m = 2(35-3) + 2(25-3) - 0.15$

$= 107.85 \text{ cm} = 107.85 \times 10^{-2} \text{ m}$

Reluctance for material, $$S_m = \frac{l_m}{\mu A} = \frac{107.85 \times 10^{-2}}{3775 \times 4\pi \times 10^{-7} \times 9 \times 10^{-4}}$$

$$= 2.53 \times 10^5 \text{ AT/Wb}$$

Reluctance for air gap, $$S_{ag} = \frac{l_{ag}}{\mu A} = \frac{l_{ag}}{\mu_o A} = \frac{1.5 \times 10^{-3}}{4\pi \times 10^{-7} \times 9 \times 10^{-4}}$$

$$= 13.26 \times 10^5 \text{ AT/Wb}$$

Total reluctance, $S = S_m + S_{ag} = (2.53 + 13.26) \times 10^5$

$= 15.79 \times 10^5 \text{ AT/Wb}$

$\text{m.m.f.} = \phi S = 0.6 \times 10^{-3} \times 15.79 \times 10^5$

or $\text{m.m.f.} = 947.4 \text{ AT}$

Excitation current, $I = \dfrac{\text{m.m.f.}}{N} = \dfrac{947.4}{500} = \mathbf{1.895\ A}$ **Ans.**

r.m.s value of induced e.m.f in the coil

$E_{rms} = 4.44 f N \phi_m$

$E_{rms} = 4.44 \times \dfrac{314}{2\pi} \times 500 \times 0.6 \times 10^{-3} = \mathbf{66.6\ V}$ **Ans.**

Problem 6.4. *For rectangular magnetic core shown in Fig. 6.27, an air gap of 2 mm is cut across one of its limb. Calculate the excitation current of the coil, having 1000 turns, to produce a core flux of 4 mWb. The relative permeability is $\mu_r = 2000$ and area of cross-section is 16 cm^2.*

Solution.

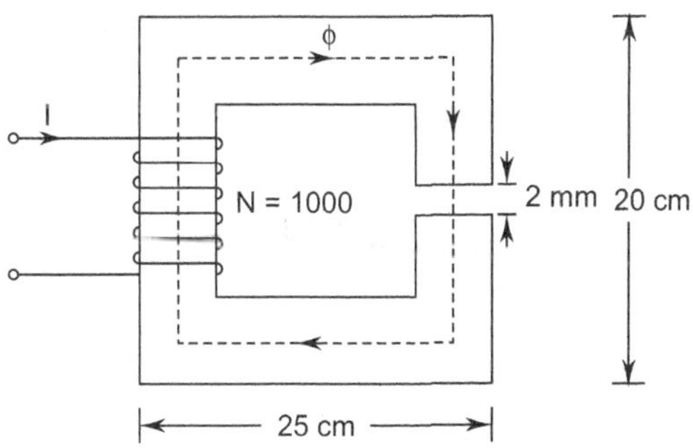

Fig. 6.27

Area of cross-section, $A = a \times a = 16 \text{ cm}^2$

$a = 4 \text{ cm} = 4 \times 10^{-2} \text{ m}$

$\phi_m = 4 \text{ mWb}$

$l_{ag} = 2 \text{ mm} = 2 \times 10^{-3} \text{ m}$

$\mu_r = 2000$

Mean length of material, $l_m = 2(20-4) + 2(25-4) - 0.2$

$= 73.8 \text{ cm} = 73.8 \times 10^{-2} \text{ m}$

Reluctance for material, $$S_m = \frac{l_m}{\mu A} = \frac{73.8 \times 10^{-2}}{2000 \times 4\pi \times 10^{-7} \times 16 \times 10^{-4}}$$

$= 1.835 \times 10^5 \text{ AT/Wb}$

Reluctance for air gap, $$S_{ag} = \frac{l_{ag}}{\mu_o A} = \frac{2 \times 10^{-3}}{4\pi \times 10^{-7} \times 16 \times 10^{-4}}$$

$= 9.947 \times 10^5 \text{ AT/Wb}$

Total reluctance, $S = S_m + S_{ag} = (1.835 + 9.947) \times 10^5$

$= 11.782 \times 10^5 \text{ AT/Wb}$

$\text{m.m.f.} = \phi S = 4 \times 10^{-3} \times 11.782 \times 10^5$

or $\text{m.m.f.} = 4712.8 \text{ AT}$

Excitation current, $$I = \frac{\text{m.m.f}}{N} = \frac{4712.8}{1000} = \mathbf{4.713\ A}$$ **Ans.**

Problem 6.5. *A wrought iron bar 30 cm long and 2 cm in diameter is bent into a circular shape as given in Fig. 6.28, an air gap of 1 mm is cut across one of its limb. Area of cross-section of core is $\pi \times 10^{-4}$ m^2. Calculate excitation current of the coil, having 500 turns, to produce a core flux of 0.5 mWb. The relative permeability, μ_r = 4000.*

Solution.

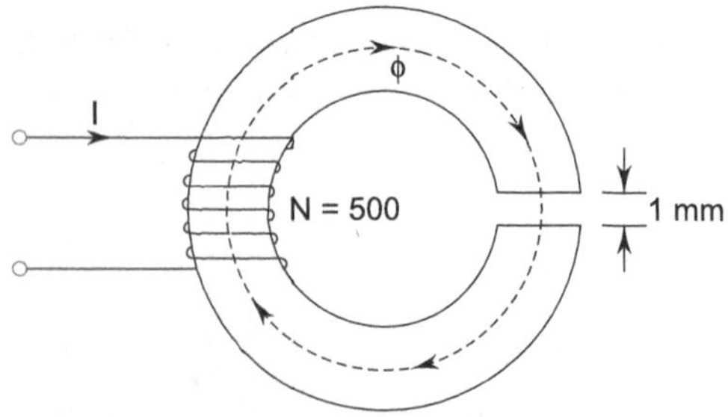

Fig. 6.28

Area of cross-section, $A = \pi \times 10^{-4} \text{ m}^2$

$\phi_m = 0.5 \text{ mWb}$

$l_{ag} = 1 \text{ mm} = 1 \times 10^{-3} \text{ m}$

$$\mu_r = 4000$$

Mean length of material, $l_m = 30 \text{ cm} = 0.3 \text{ m}$

Reluctance for material, $$S_m = \frac{l_m}{\mu A} = \frac{30 \times 10^{-2}}{4000 \times 4\pi \times 10^{-7} \times \pi \times 10^{-4}}$$

$$= 1.899 \times 10^5 \text{ AT/Wb}$$

Reluctance for air gap, $$S_{ag} = \frac{l_{ag}}{\mu_o A} = \frac{1 \times 10^{-3}}{4\pi \times 10^{-7} \times \pi \times 10^{-4}}$$

$$= 25.33 \times 10^5 \text{ AT/Wb}$$

Total reluctance, $$S = S_m + S_{ag} + (1.899 + 25.330) \times 10^5$$

$$= 27.229 \times 10^5 \text{ AT/Wb}$$

$$\text{m.m.f.} = \phi S = 0.5 \times 10^{-3} \times 27.229 \times 10^5$$

or $$\text{m.m.f.} = 1361.45 \text{ AT}$$

Excitation current, $$I = \frac{\text{m.m.f.}}{N} = \frac{1361.45}{500} = \mathbf{2.723\ A}$$ **Ans.**

Problem 6.6. *A magnetic circuit frame build is of iron of square cross-section of 3 cm (Fig. 6.29). Each air gap is 2 mm wide and each coil is carrying a current of 1 A and is wound with 1000 turns each. The relative permeability of part 1 and part 2 may be taken as μ_{r1} = 1000 and μ_{r2} = 1200 respectively. Calculates:*

(i) Reluctance for part 1
(ii) Reluctance for part 2
(iii) Reluctance for the air gaps
(iv) Core flux
(v) Flux density.

Solution.

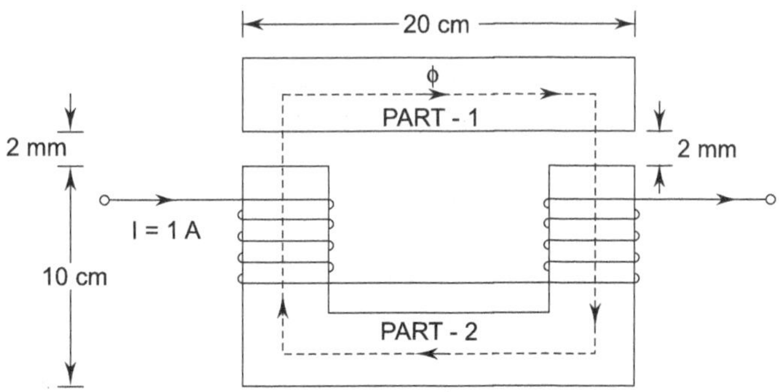

Fig. 6.29

$$\mu_{r1} = 1000$$
$$\mu_{r2} = 1200$$
$$a = 3 \text{ cm} = 3 \times 10^{-2} \text{ m}$$

Mean length of material,
Mean lengths of part 1, $l_{m1} = 20 - 3 = 17 \text{ cm} = 17 \times 10^{-2} \text{ m}$

Mean lengths of part 2, $l_{m2} = \left(10 - \frac{3}{2}\right) + (20 - 3) + \left(10 - \frac{3}{2}\right)$

$= 34 \text{ cm} = 34 \times 10^{-2} \text{ m}$

Mean lengths of 2 air gaps, $l_{m3} = 2 \times 2 = 4 \text{ mm} = 4 \times 10^{-3} \text{ m}$

Cross-sectional area, A $\quad A = 3 \times 3 = 9 \text{ cm}^2 = 9 \times 10^{-4} \text{ m}^2$

Reluctance for material, S_m

(*i*) Reluctance for part 1

$$S_{m1} = \frac{l_{m1}}{\mu A} = \frac{17 \times 10^{-2}}{1000 \times 4\pi \times 10^{-7} \times 9 \times 10^{-4}}$$

$$= \mathbf{1.503 \times 10^5 \ AT/Wb} \qquad \textbf{Ans.}$$

(*ii*) Reluctance for part 2

$$S_{m2} = \frac{l_{m2}}{\mu A} = \frac{34 \times 10^{-2}}{1200 \times 4\pi \times 10^{-7} \times 9 \times 10^{-4}}$$

$$= \mathbf{2.505 \times 10^5 \ AT/Wb} \qquad \textbf{Ans.}$$

(*iii*) Reluctance for air gaps, S_{ag}

$$S_{ag} = 2x\, S_{m3} = \frac{2x\, l_{m3}}{\mu_o A} = \frac{4 \times 10^{-3}}{4\pi \times 10^{-7} \times 9 \times 10^{-4}}$$

$$= \mathbf{35.367 \times 10^5 \ AT/Wb} \qquad \textbf{Ans.}$$

(*iv*) Total reluctance, S $\quad S = S_{m1} + S_{m2} + S_{ag}$

$$S = (1.503 + 2.505 + 35.367) \times 10^5$$

$$= 39.376 \times 10^5 \text{ AT/Wb}$$

$$\text{m.m.f.} = NI = 2 \times 1000 \times 1 = 2000 \text{ AT}$$

Total flux, $\quad \phi = \frac{\text{m.m.f}}{\text{Reluctance}} = \frac{2000}{39.376 \times 10^5}$

$$= \mathbf{5.08 \times 10^{-4} \ Wb} \qquad \textbf{Ans.}$$

(*v*) Flux density, $\quad B = \frac{\phi}{A} = \frac{5.08 \times 10^{-4}}{9 \times 10^{-4}}$

$$= \mathbf{0.564 \ Wb/m^2} \qquad \textbf{Ans.}$$

Problem 6.7. *An electromagnet has an gap of 4 mm and flux density in the gap is 1.3 Wb/m*^{2}*. Determine the ampere-turns for the gap.*

Solution. Magnetising force $\quad H = \frac{B}{\mu} = \frac{B}{\mu_o \cdot \mu_r} \qquad \ldots(i)$

Also, $\quad H = \frac{NI}{l_{ag}} \qquad \ldots(ii)$

or $$NI = H \cdot l_{ag}$$

or $$NI = \frac{B}{\mu_o \cdot \mu_r} \cdot l_{ag}$$ [From equation (*i*)]

or $$NI = \frac{1.3}{4\pi \times 10^{-7} \times 1} \times 4 \times 10^{-3}$$

or $$NI = \mathbf{4140\ AT}$$ **Ans.**

Problem 6.8. *An iron ring of mean length 50 cm and relative permeability 300 has an air gap of 1 mm. If the ring is provided with a winding of 200 turns and a current of 1 A is allowed to flow through it, find the flux density across the air gap.*

Solution. Total ampere-turns = A.T. required for air gap + A.T. requird for iron ring

$$= \frac{B}{\mu_o} \cdot l_{ag} + \frac{B}{\mu_o \cdot \mu_r} \cdot l_m = B\left(\frac{l_{ag}}{\mu_o} + \frac{l_m}{\mu_o \cdot \mu_r}\right)$$

$$= B\left(\frac{10^{-3}}{4\pi \times 10^{-7}} + \frac{50 \times 10^{-2}}{4\pi \times 10^{-7} \times 300}\right) = B\ (2122.066)$$

Total ampere-turns provided

$$= N \times I = 200 \times 1$$

$$= 200\ AT$$

Hence, $$200 = B\ (2122.066)$$

or $$B = \frac{200}{2122.066} = 0.0942$$

or $$B = \mathbf{94.2\ mWb/m^2}$$ **Ans.**

Problem 6.9. *A coil of 1000 turns is wound on a laminated core of steel having a cross-section of 5 cm^2. The core has an air gap of 2 mm cut at right angle. What value of current is required to have an air gap flux density of 0.5 T ? Permeability of steel may be taken as infinity. Determine the coil inductance.*

Solution.

$$N = 1000$$

$$A = 5\ cm^2 = 5 \times 10^{-4}\ m^2$$

$$l_{ag} = 2\ mm$$

$$B_m = 0.5\ T$$

$$\mu_r\ (\text{steel}) = \infty$$

Reluctance for material, $$S_m = \frac{1}{\mu_o \cdot \mu_r} \cdot \frac{l_m}{A} = \frac{1}{4\pi \times 10^{-7} \times \infty} \cdot \frac{l_m}{5 \times 10^{-4}} = 0$$

Reluctance for air gap, S_{ag}

$$S_{ag} = \frac{1}{\mu_o} \cdot \frac{l_{ag}}{A} = \frac{1}{4\pi \times 10^{-7}} \cdot \frac{2 \times 10^{-3}}{5 \times 10^{-4}}$$

$$= 31.831 \times 10^5$$

Total reluctance $S = S_m + S_{ag} = 0 + 31.831$

$$S = 31.831 \times 10^5$$

$$\text{m.m.f} = \phi S = B \times A \times S$$

$$= 0.5 \times 5 \times 10^{-4} \times 31.831 \times 10^5 = 795.775$$

$$I = \frac{\text{m.m.f}}{N} = \frac{795.775}{1000}$$

$I = \mathbf{0.796\ A}$ **Ans.**

Coil inductance $L = \dfrac{N\phi}{i} = \dfrac{NB \cdot A}{i}$

or $L = \dfrac{1000 \times 0.5 \times 5 \times 10^{-4}}{0.796}$

or $L = \mathbf{0.314\ H.}$ **Ans.**

Problem 6.10. *Two coils having 100 and 500 turns respectively are wound on a core with $\mu = 4000\ \mu_o$. Effective core length is 60 cm and core area is 9 cm^2. Find mutual inductance between two coils.*

Solution. $N_1 = 100$

$$N_2 = 50$$

$$A = 9\ \text{cm}^2 = 9 \times 10^{-4}\ \text{m}^2$$

$$l_m = 60\ \text{cm} = 60 \times 10^{-2}\ \text{m}$$

$$\mu_o = 4\pi \times 10^{-7}$$

$$\mu = 4000\ \mu_o$$

i.e., $\mu_r = 4000$

$$S_m = \frac{l_m}{\mu A} = \frac{60 \times 10^{-2}}{4\pi \times 10^{-7} \times 4000 \times 9 \times 10^{-4}}$$

$$S_m = 132.629 \times 10^3$$

Mutual Inductance between two coils

$$M = \frac{N_1 \cdot N_2}{S_m} = \frac{100 \times 50}{132.629 \times 10^3}$$

$M = \mathbf{37.7\ mH.}$ **Ans.**

Problem 6.11. *The core of the magnetic circuit has mean length of 40 cm and uniform cross-section area of 4 cm^2. The relative permeability of core material is 1000. An air gap of 1 mm is cut in core and 1000 turns are wound on core. Determine the inductance of coil, if fringing is negligible.*

Solution.

Area of cross-section $A = a \times a = 4\ \text{cm}^2 = 4 \times 10^{-4}\ \text{m}^2$

$$l_{ag} = 1\ \text{mm} = 10^{-3}\ \text{m}$$

$$\mu_r = 1000$$

$$N = 1000$$

Mean length of material, $l_m = 40 \text{ cm} = 0.4 \text{ m}$

Reluctance for material, $$S_m = \frac{l_m - l_{ag}}{\mu A} = \frac{0.4 - 10^{-3}}{1000 \times 4\pi \times 10^{-7} \times 4 \times 10^{-4}}$$

$$= 793.78 \times 10^3 \text{ AT/Wb}$$

Reluctance for air gap, $$S_{ag} = \frac{l_{ag}}{\mu_o A} = \frac{10^{-3}}{4\pi \times 10^{-7} \times 4 \times 10^{-4}}$$

$$= 1989.436 \times 10^5 \text{ AT/Wb}$$

Total reluctance, $$S = S_m + S_{ag} = (793.78 + 1989.436) \times 10^3$$

$$= 2783.216 \times 10^3 \text{ AT/Wb}$$

Inductance of the coil, $$L = \frac{N^2}{S} = \frac{(1000)^2}{2783.216 \times 10^3} = \mathbf{359\ mH}$$ **Ans.**

Problem 6.12. *The total core loss (hysteresis plus eddy current) for a specimen of magnetic sheet steel is found to be 1800 W at 60 Hz. If the flux density is kept constant and frequency of supply increased 50%, the total core loss is found to be 3000 W. Compute separate hysteresis and eddy current losses at both frequencies.*

Solution. Hysteresis loss are proportional directly to frequency

i.e., $$P_h \propto f$$

or $$P_h = K_1 f \quad \ldots(1)$$

Eddy current loss are proportional to square of the frequency

i.e., $$P_e \propto f^2$$

or $$P_e = K_2 f^2 \quad \ldots(2)$$

Total core losses are given by

$$P_c = P_h + P_e \quad \ldots(3)$$

or from equations (1) and (2), we get

$$P_c = K_1 f + K_2 f^2 \quad \ldots(4)$$

At 60 Hz frequency, from equation (4)

$$1800 = K_1 60 + K_2 (60)^2$$

or $$K_1 + 60 K_2 = 30 \quad \ldots(5)$$

At 90 Hz frequency (*i.e.,* 50% increased frequency)

$$3000 = 90 K_1 + (90)^2 K_2$$

or $$3K_1 + 270 K_2 = 100 \quad \ldots(6)$$

Solving equations (5) and (6) simultaneously, we get

$$K_1 = 23.33$$

$$K_2 = \frac{1}{9}$$

(*a*) When frequency is 60 Hz.

Hysteresis loss $= K_1 \times 60 = 23.33 \times 60$

or P_{h_1} = **1400 W** **Ans.**

Eddy current $= K_2 (60)^2 = \frac{1}{9} \times 3600$

or P_{e_1} = **400 W** **Ans.**

(*b*) When frequency is 90 Hz

Hysteresis loss $= K_1 \times 90 = 23.33 \times 90$

or P_{h_2} = **2100 W** **Ans.**

Eddy current Loss $= K_2 (90)^2 = \frac{1}{9} \times 8100$

or P_{e_2} = **900 W** **Ans.**

EXERCISE

1. The hysteresis and eddy current losses in a machine operating at 50 Hz are 250 W and 100 W, respectively. At what frequency will iron losses be halved? Assume that voltages are adjusted accordingly so the flux densities remain unchanged.
2. The core loss in an iron-core reactor is 600 W of which 400 W is hysteresis loss. The applied voltage is 220 V and frequency is 60 Hz. What is the core loss if the frequency is 10,000 Hz? What is the core loss if the frequency is 60 Hz and the voltage is 440 V? Assume the Steinmetz exponent to be 2.
3. A mild steel ring (Fig. 6.30) having a cross-sectional area of 500 mm^2 and a mean circumference of 400 mm have a coil of 200 turns wound uniformly around it. Calculate:
   (*i*) reluctance of the ring.
   (*ii*) current required to produce a flux of 800 μWb in the cring.
   Assume that μ_r = 370 for flux density of 1.6 T.

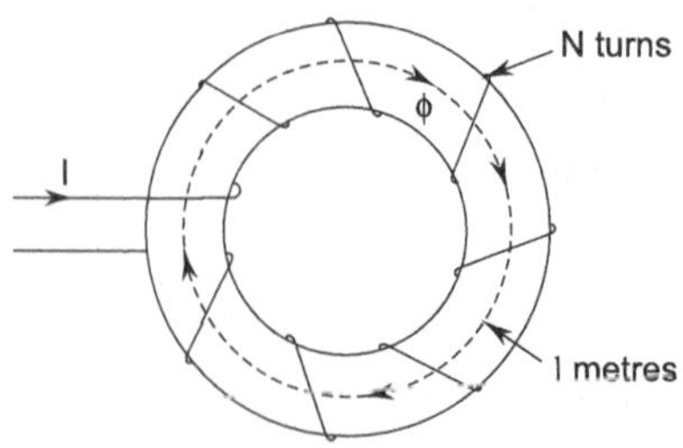

Fig. 6.30

4. A magnetic core has a circular cross-sectional area of 12.9×10^{-4} m^2, a mean circumference of 0.254 m and an air gap length of 3.175×10^{-3} m. A 350 turn coil is wound around the core and a current of 5 A is applied to the coil. Calculate:
   (*i*) reluctance of the magnetic circuit, (*ii*) flux density in the gap, and
   (*iii*) inductance of the coil.
   Assume permeability of iron to be infinite; and neglect fringing.
5. A ring (Fig. 6.31) of mean diameter 21 cm and cross-sectional area of 10 cm^2 is made up of semicircular sections of cast steel and cast iron. If each joint has reluctance equal to an air-gap of 0.2 mm, find the amp-turn required to produce a flux of 5×10^{-4} Web in the magnetic circuit. Consider relative permeability of steel and iron to be 825 and 165, respectively; neglect leakage and fringing.

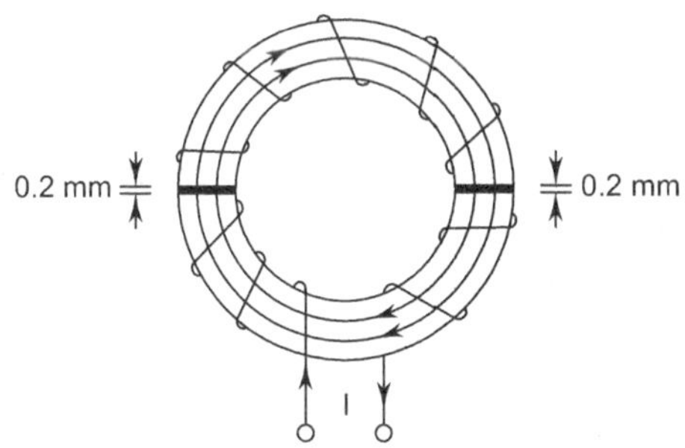

Fig. 6.31

6. Consider the magnetic circuit shown in Fig. 6.32. The core has reluctance of 16,580 A/Web. If an air-gap of 0.5 mm is cut in the leg, find the self and mutual inductances of the two coils. Given $N_1 = 500$, $N_2 = 1000$; area of the core = 12 cm^2.

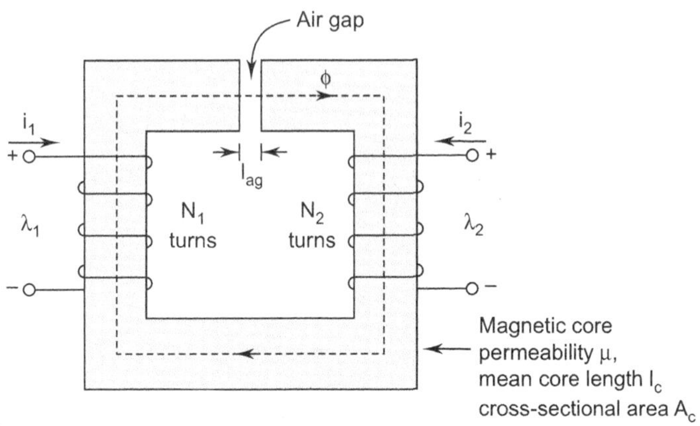

Fig. 6.32

7. The magnetic circuit frame shown in Fig. 6.33 is built of iron having cross-section area of 9 cm^2. Each air gap is 1.5 mm wide and each of the coils is wound with 800 turns. The relative permeability of part A and B may be taken as 1050 and 1250 respectively. Calculate:
   (*i*) Reluctance of part A.
   (*ii*) Reluctance of two air gaps.
   (*iii*) Total reluctance of complete magnetic path.

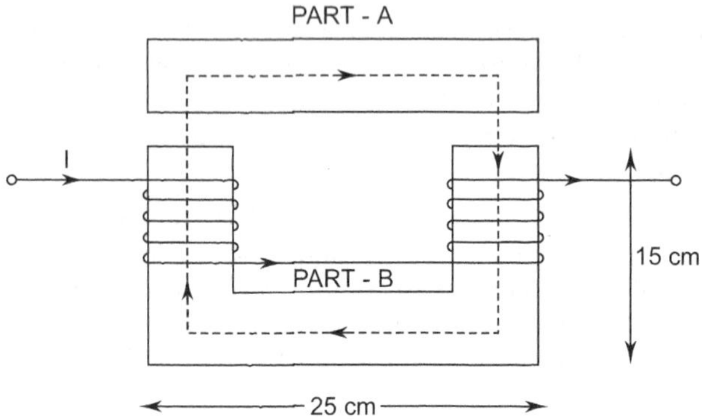

Fig. 6.33

8. What do you understand by self-inductance? Derive its expression.
9. What do you understand by mutual inductance? Derive its expression.
10. What is coefficient of coupling?
11. Define hysteresis and eddy current losses.

□ □ □

7

Electrical Measuring Instruments

7.1 INTRODUCTION

Electrical measuring instruments are used to measure magnitude of various electrical quantities such as current, voltage, resistance, power, energy etc. A fixed quantity is taken as 'unit' in order to measure the magnitude of electrical quantities. The process of comparing any quantity with a fixed unit for recording its magnitude is known as 'measurement' and the overall system involved for this purpose is called as the 'Measuring System'. In this chapter we will discuss the various electrical measuring instruments and their working and principles associated with them.

7.2 CLASSIFICATION OF ELECTRICAL MEASURING INSTRUMENTS

All electrical measuring instruments can be classified as follow:

(A) Absolute Instruments
(B) Secondary Instruments
- → Integrating instruments
- → Recording instruments
- → Indicating instruments
    - → Moving coil
    - → Moving iron type
    - → Hot wire
    - → Electrostatic
    - → Induction type

These instruments will be discussed in detail in later sections.

7.3 OPERATING PRINCIPLES OF ELECTRICAL MEASURING INSTRUMENTS

Most of the electrical quantities for *e.g.*, current, voltage, power etc., cannot be observed physically. These quantities are converted into a mechanical force and then the force is measured.

This mechanical force will move the pointer over the calibrated scale, indicating the value of electrical quantity to be measured.

This conversion from electrical quantity under measurement into mechanical force is done by using the following effects:

1. Magnetic effect
2. Thermal (or heating) effect
3. Electrodynamics effect
4. Electromagnetic effect
5. Electrostatic effect
6. Chemical effect.

7.4 ESSENTIAL TORQUES OF AN INDICATING INSTRUMENT

The analog instruments like ammeter, voltmeter and wattmeter consist essentially of a moving system with a pointer attached to it.

For satisfactory operation, the moving system is pivoted by jewelled bearings and is subjected to the following three distinct torques:

1. Deflecting Torque
2. Controlling Torque
3. Damping Torque.

7.4.1 Deflecting Torque

The torque which is responsible for the movement of the pointer from its zero position to the required deflection is called deflecting torque (T_d). The deflecting torque can be produced by magnetic, thermal or induction effects of current. In general, it is based upon operating principle of the instrument. A deflecting torque is required to overcome the inertia, damping effect and controlling effect of moving system.

7.4.2 Controlling Torque

With the help of deflecting torque, the pointer deflection will take place on calibrated scale but to stop the pointer at definite position, controlling torque (T_c) comes into action. As deflection of the pointer increases, the controlling torque also increases and stops the pointer at measured value.

Controlling torque is also known as restoring torque *i.e.*, it brings back the pointer to its zero position when deflecting torque is withdrawn. The pointer attains a steady position when controlling torque becomes numerically equal to deflecting torque *i.e.*, $T_d = T_c$.

Controlling torque or restoring torque can be achieved in measuring instruments by following two methods:

(*i*) Spring Control (*ii*) Gravity Control.

(*i*) **Spring Control :** In this method, two spiral hair springs are used (Fig. 7.1). They are made up of phosphor-bronze and spiraled in opposite direction.

One end of both the springs is attached to the body of the instrument and other end is attached to moving system.

Use of two springs also avoids error due to temperature variations. When the instrument is not in use, the two springs are in their natural position without any tension or compression. When the instrument is connected in the circuit for measurement, deflecting torque acts and the pointer moves on calibrated scale. One of the springs is unwound while other gets twisted. The resultant movement in springs provides the controlling torque.

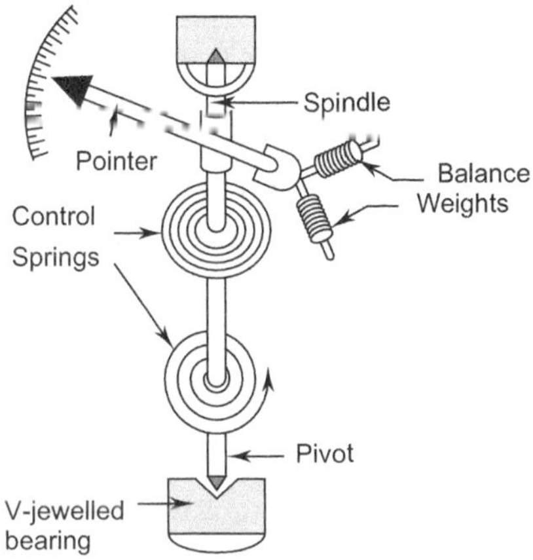

Fig. 7.1

More the deflection more is the twist and hence greater will be controlling torque. Thus the controlling torque is directly proportional to deflection of the moving system, *i.e.*,

$$T_c \propto \theta \quad \text{...(7.1)}$$

And it is also clear that the pointer comes to rest when controlling torque becomes numerically equal to deflecting torque *i.e.*,

$$T_d = T_c \quad \text{... (7.2)}$$

But deflection or deflection torque depends upon current, I, hence

$$T_d \propto I \quad \text{...(7.3)}$$

So from equations (7.1) and (7.3)

$$I \propto \theta \quad \text{...(7.4)}$$

(*ii*) **Gravity Control:** In gravity control method two adjustable weights $\mathbf{W}_c$ and $\mathbf{W}_b$ are attach to the moving system. The controlling torque can be varied by adjusting the position of control weight ($\mathbf{W}_c$) upon the arm (Fig. 7.2).

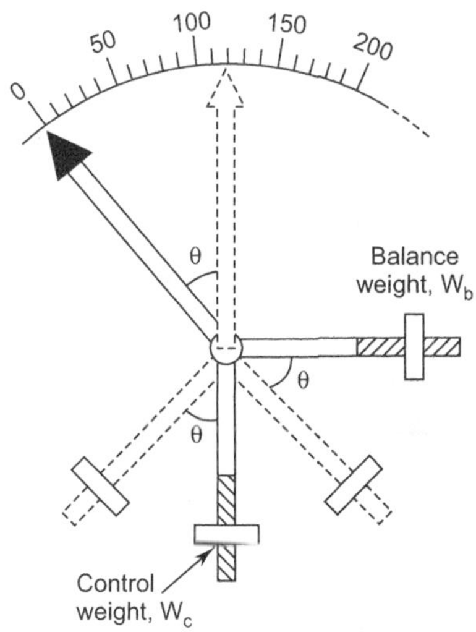

Fig. 7.2

Under the action of deflecting torque, the pointer is deflected through an angle θ from its zero position to the dotted position as shown in Fig. 7.2. Due to gravitational force acting on the weight, it would try to come back to its original position (vertical) and will produce controlling torque.

Let θ = angle through which pointer (and hence weights) is deflected

l = distance of controlling weight from axis of rotation

W_c = control weight

During deflection the controlling weight, W_c, can be resolved into two component, *i.e.*, $W_c \cos\theta$ and $W_c \sin\theta$. The component $W_c \sin\theta$ provides controlling torque.

$$\therefore \quad T_c = W_c \sin\theta \times l \quad \text{...(7.5)}$$

As W_c and l have fixed values hence

$$T_c \propto \sin\theta \quad \text{...(7.6)}$$

The pointer comes to rest when controlling torque becomes numerically equal to deflecting torque *i.e.*,

$$T_d = T_c \quad \text{[Eqn. (7.2)]}$$

But deflection or deflecting torque depends upon the current, I, hence

$$T_d \propto I \quad \text{[Eqn. (7.3)]}$$

So from equations (7.2), (7.3) and (7.6)

$$I \propto \sin\theta \quad \text{...(7.7)}$$

7.4.3 Damping Torque

When moving system of the instrument is subjected to deflecting and controlling torques, due to its inertia, the pointer oscillates around its final position before it settle down there. This type of oscillation makes it difficult to obtain accurate readings quickly.

With the help of damping torque, oscillations can be damped out and the pointer can be brought to final position quickly. The damping should oppose the movement of pointer but should be, at same time, depends on it so that it ceases to operate once the pointer comes to rest and hence does not affects its final deflection.

For developing damping torque, commonly used methods are:

(*i*) Air friction damping (*ii*) Eddy current damping

(*iii*) Hydraulic damping.

(*i*) **Air Friction Damping:** In this method, light aluminum piston, which is attached to moving system, moves in an air chamber closed at one end as shown in Fig. 7.3.

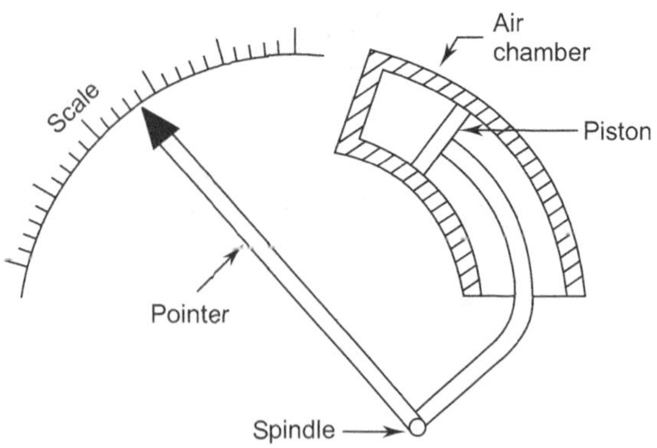

Fig. 7.3

There is very little clearance between the piston and the air chamber. When the spindle moves due to a deflecting torque, the piston also move along with it, say out of cylinder, air pressure inside the cylinder fall thus the outward movement of the piston is opposed by the outside air at atmospheric pressure. Similarly, when piston is moved into the cylinder, air trapped inside the chamber is compressed and opposes the movement. Thus, suction and compression action on the air produces necessary damping torque inside and thereby the oscillation of pointer is arrested.

(*ii*) **Eddy Current Damping:** In eddy current damping, a thin disc of conducting but non-magnetic material like aluminum or copper is attached to the instrument spindle as shown in Fig. 7.4.

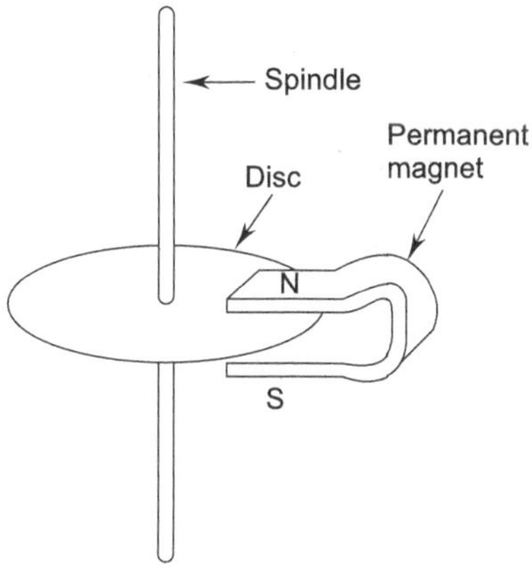

Fig. 7.4

The disc is allowed to move along with spindle inside the magnetic field provided by a permanent magnet. The movement of disc causes an eddy e.m.f to get induced in it causing an eddy current to flow. The flux set up eddy current interacting with magnetic field exerts a force on disc. The force by Lenz's law, opposes the movement of disc. Thus necessary damping torque is provided.

(*iii*) **Hydraulic Damping:** In hydraulic damping or fluid friction damping the vanes are connected to the moving system and are kept in oil filled pot (Fig. 7.5) which has high viscosity. When pointer tries to oscillates, the vanes and fluid viscosity opposes the motion of pointer. Hence the required damping torque is produce.

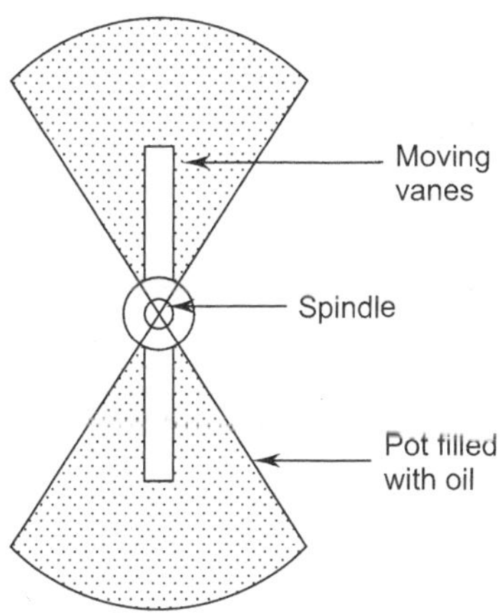

Fig. 7.5

In this method no extra care is required as in case of air friction damping.

But this method is not suitable for portable instruments because there can be leakage of oil. It has to be always used in vertical position only.

7.5 MOVING COIL INSTRUMENTS

There are two type of moving coil instruments, *viz.*

(*i*) Permanent Magnet type, and

(*ii*) Dynamometer type.

7.5.1 Permanent Magnet Type Moving Coil Instruments

Fig. 7.6 shows the main parts of permanent magnet type moving coil instrument. It consist of a coil of very thin insulated copper wire wound on an aluminum former and placed in field of a strong horseshoe-shaped permanent magnet.

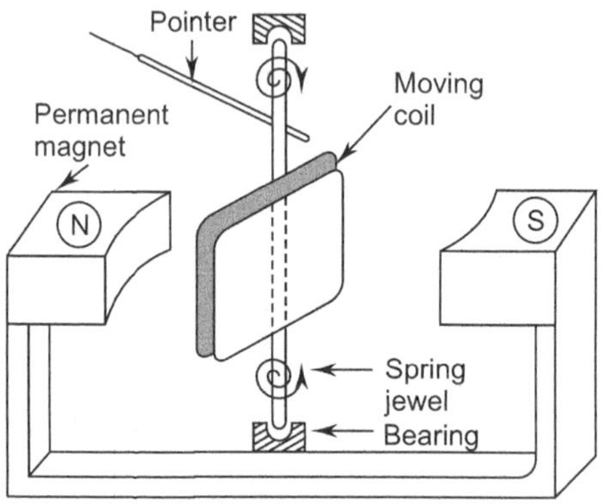

Fig. 7.6

The aluminum former carrying the coil is fitted to shaft. The shaft ends are housed in jewel bearings to facilitate frictionless movement of former together with the coil. The current to the coil fed through two spiral spring, one on either side of the coil. The spring in addition to serving as lead wires, also provide the controlling torque. A pointer is attached to shaft and entire moving system is carefully balanced by suitable counter weights.

When a current pass through the coil, a deflecting torque is produced causing the moving system, including the pointer, to swing through angle proportional to the current. The pointer is usually provided with a knife edge and is arranged to move over graduated dial. A mirror is also provided on dial besides the graduations to facilitate in observing the readings without parallax error. The deflecting torque T_d is given by

$$T_d \propto I \qquad \ldots(7.8)$$

The controlling torque T_c offered by the two spiral springs

$$T_c \propto \theta \qquad \ldots(7.9)$$

The pointer reaches a steady position when

$$T_d = T_c \qquad \ldots(7.10)$$

or

$$I \propto \theta \qquad \ldots(7.11)$$

Therefore, scale graduation in moving coil instruments will be uniform throughout the range. The instruments cannot be used in AC circuits.

7.5.2 Dynamometer Type Moving Coil Instruments

In this type, the operating magnetic field is produced by air-cored electromagnets excited by two fixed coils as shown in Fig. 7.7. In all other aspects, they are similar to the permanent magnet moving coil instrument.

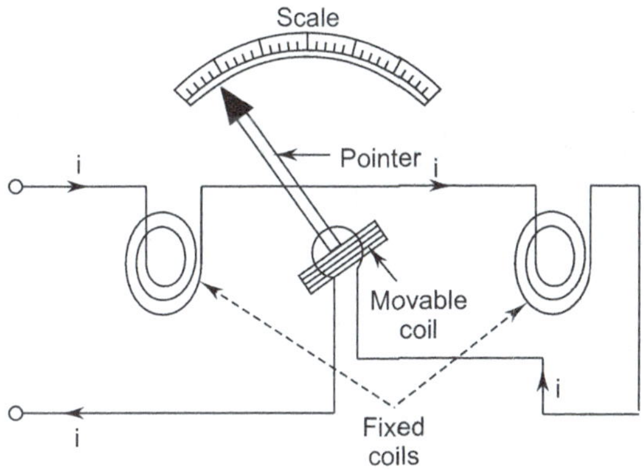

Fig. 7.7

As force on the moving coil is proportional to flux produced by fixed coils (and hence current through it, I_f) and current in the moving coil, I_m, hence the deflecting torque will be

$$T_d \propto I_f I_m \qquad \text{...(7.12)}$$

Therefore, it is seen from above equation that the meter deflection is not affected when currents through the coils are reversed. Thus, this type of instrument can be used in both AC and DC circuits.

7.6 MOVING IRON INSTRUMENTS

There are two type of moving iron instruments, *viz.*

(*i*) Attraction type and
(*ii*) Repulsion type.

7.6.1 Attraction Type Moving Iron Instruments

In this type of instrument, a stationary field coil carries the current proportional to the current to be measured in case of ammeter or voltage to be measured in case of voltmeter. The current carrying coil produces a magnetic field.

A small iron piece placed nearby and fitted to the spindle of instrument gets attracted by the magnetic instrument field and turns towards the coil (Fig. 7.8). This causes the spindle and pointer attached to it get the required deflection. The pointer moves over a graduated scale.

As the force of attraction due to an electromagnet varies in proportion to square of the flux density, the torque in this type of instrument is proportional to square of the current through the field coil.

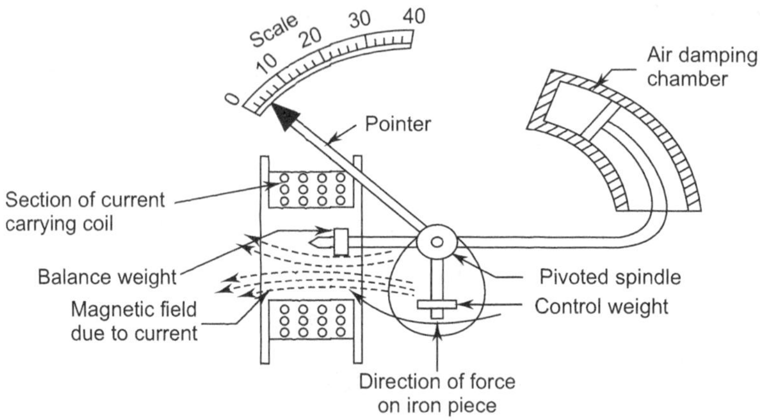

Fig. 7.8

$$T_d \propto I^2 \qquad ...(7.13)$$

The controlling torque T_c offered by two spiral springs

$$T_c \propto \theta \qquad \text{[Eqn. (7.9)]}$$

The pointer reaches a steady position when

$$T_d = T_c \qquad \text{[Eqn. (7.10)]}$$

or

$$\theta \propto I^2 \qquad \text{[Eqn. (7.11)]}$$

Therefore, scale graduation in moving iron instruments will be non-uniform throughout the range. Hence, moving iron instruments can be used in both DC and AC currents.

7.6.2 Repulsion Type Moving Iron Instruments

The principal of this instrument is that two iron pieces kept within close proximity in a magnetic field get magnetized to the same polarity and exert a force of repulsion within themselves.

As seen in Fig. 7.9, there are two vanes of iron inside the field coil. One of the vanes is rigidly mounted to the frame of instrument and other vane is attached to the spindle.

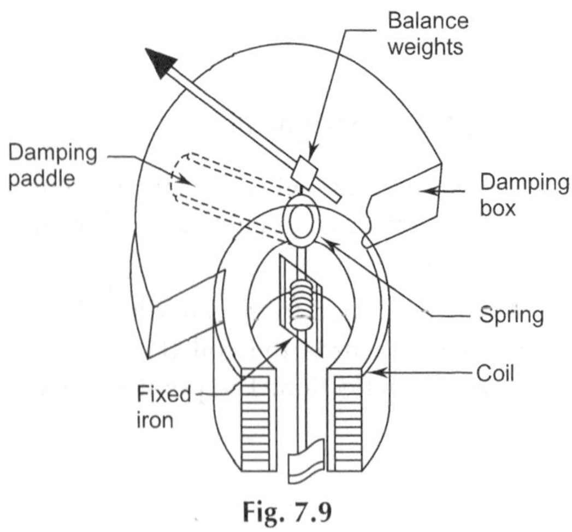

Fig. 7.9

When current flows through the coil, both vanes are magnetized with same polarity and there results a force of repulsion between them.

As a result of this force of repulsion, the movable vane rotates causing a corresponding deflection in the pointer. It is evident that, in this case too, the deflecting torque is proportional to square of the current.

$$T_d \propto I^2 \qquad ...(7.14)$$

Hence, the instruments can be used in both DC and AC circuits.

7.7 RANGE EXTENSION OF VOLTMETER AND AMMETER

7.7.1 Extension of Range of Ammeter

An ammeter, being an instrument to be connected in series with the circuit, must have a low resistance so that it will not cause a voltage change in it. The control springs which takes current to the moving coil are capable of carrying current to a few milli-amperes, and in some cases to micro-ampere. Therefore, if the moving coil instrument has to be used as an ammeter capable of measuring a large current, it should be used in conjunction with suitable shunts as shown in Fig.7.10.

A shunt is, essentially, a low resistor usually made of manganin strips so that its resistance practically remains constant irrespective of temperature changes and is connected in parallel with the moving coil ammeter.

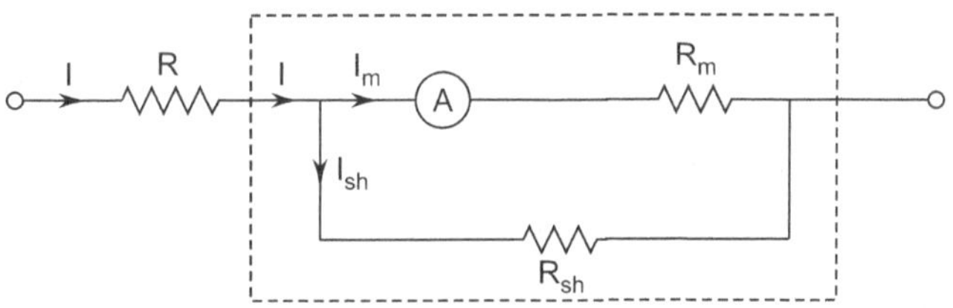

Fig. 7.10

In Fig. 7.10,

I – Current which is to be measured by meter.

R – Resistance or load through which current is to be measured

I_m – Current through meter

R_m – Resistance of meter

I_{sh} – Current through shunt resistance

R_{sh} – External resistance connected in shunt (or in parallel to meter)

Then current to be measured

$$I = I_m + I_{sh} \qquad ...(7.15)$$

As the meter and the shunt resistance are in parallel hence voltage across both will be same. Therefore

$$I_m R_m = I_{sh} R_{sh} \qquad ...(7.16)$$

or

$$R_{sh} = \frac{I_m}{I_{sh}} R_m \qquad ...(7.17)$$

or
$$R_{sh} = \frac{I_m}{I - I_m} R_m \qquad \ldots(7.18)$$

Equation (7.18) gives the value of shunt resistance to be used in parallel with meter for measuring current of I ampere.

7.7.2 Use as Voltmeter

A voltmeter should have a large resistance so that its power consumption, when connected across the supply lines, should be as low as possible.

Therefore, a moving coil instrument can be used as a voltmeter by connecting high resistance in series with the instrument as shown in Fig. 7.11. The value of external resistance, R_s must be adjusted so that total resistance limits the current within the maximum meter current I_m. In Fig. 7.11,

- **I** – Total current
- **R** – Resistance or load across which voltage is to be measured
- **I_m** – Current through meter
- **R_m** – Resistance of meter
- **I_L** – Current through load resistance
- **R_s** – External resistance connected in series with meter.

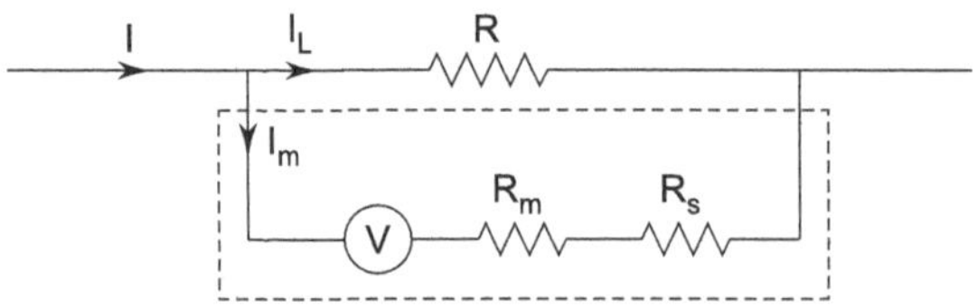

Fig. 7.11

Total resistance of branch in which meter is connected is

$$R_T = (R_m + R_s) = \frac{V}{I_m} \qquad \ldots[7.19(a)]$$

or
$$R_s = \frac{V}{I_m} - R_m \qquad \ldots[7.19(b)]$$

Equation [7.19 (*b*)] gives the value of external resistance to be used in series with meter for measuring voltage across the load resistance.

7.8 DYNAMOMETER TYPE WATTMETER

Wattmeter measure the active-power, P, in a circuit. In DC circuits, P= VI and may be measured either by wattmeter or by ammeter and voltmeter and then their product. In AC circuits, the average power at any intake is given as

$$P = vi \qquad \ldots(7.20)$$

and the active power is given by

$$p = vi \cos \phi \quad \ldots(7.21)$$

which can be measured by a wattmeter only. The dynamometer type wattmeter measures active-power both in AC and DC circuits.

The dynamometer type wattmeter is similar in construction to dynamometer type MC instrument discussed in section 7.4. The two fixed coils carry the current in the circuit and are known as current coils. The moving coil carries a current proportional to the voltage in circuit and the coil is called a pressure coil (or voltage coil). Fig. 7.12 shows the coil connection and Fig. 7.13 shows the schematic connection.

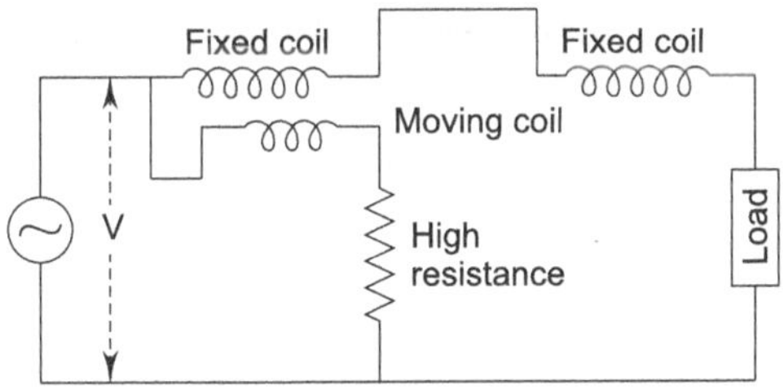

Fig. 7.12

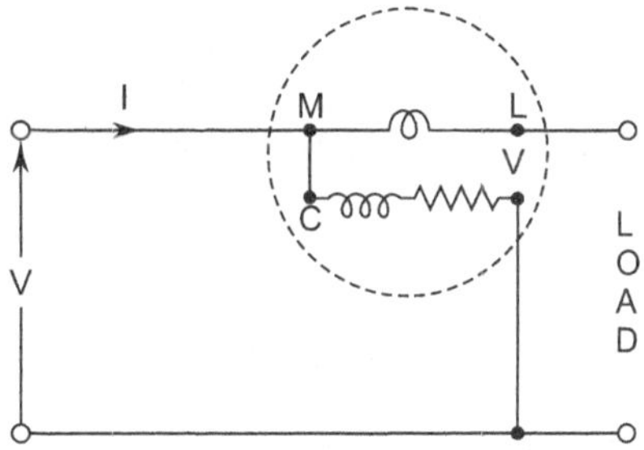

Fig. 7.13

7.9 INDUCTION TYPE ENERGY METER

Energy meter is an integrating instrument and takes into account both the quantities, *i.e.*, the power and time, the product of which gives the energy. An energy meter keeps a record of total energy consumed in a circuit during a particular period but it doesn't give any idea about the variation in the rate of energy consumption during that period. Energy meter is used for commercial measurement of energy. It directly indicates the energy consumed in kWh.

$$1 \text{ unit} = 1 \text{ kWh}$$

$$1 \text{ unit} = 1000 \text{ kWh}$$

The induction type energy meter works on the principle of electromagnetic induction. When an alternating current flows through two suitably located coils, a rotating magnetic field is produced. Eddy currents are generated in an Aluminium disc which cuts this flux. By the interaction of rotating magnetic field and eddy currents, a torque is produced which causes the disc to rotate. The number of rotations of the disc is proportional to the energy consumed by the load in a certain time interval (usually 1 hr).

A single-phase induction type energy meter consists of the followings systems:

(*i*) Driving system

(*ii*) Moving system

(*iii*) Braking/Damping system

(*iv*) Registering system.

7.9.1 Driving System

The driving system of energy meter consists of two electromagnets, a shunt magnet and a series magnet as shown in Fig. 7.14. The shunt magnet is excited by the pressure coil which carries current coil excited by the load current. Short circuited copper bands are provided on the lower part of the central limb of the shunt magnet. These loops are called as power factor compensators.

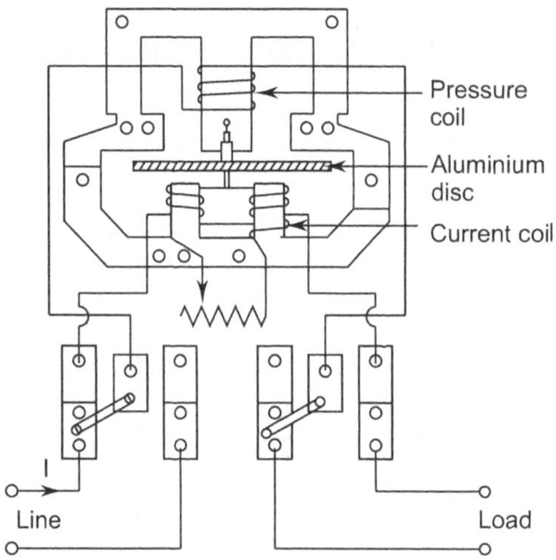

Fig. 7.14

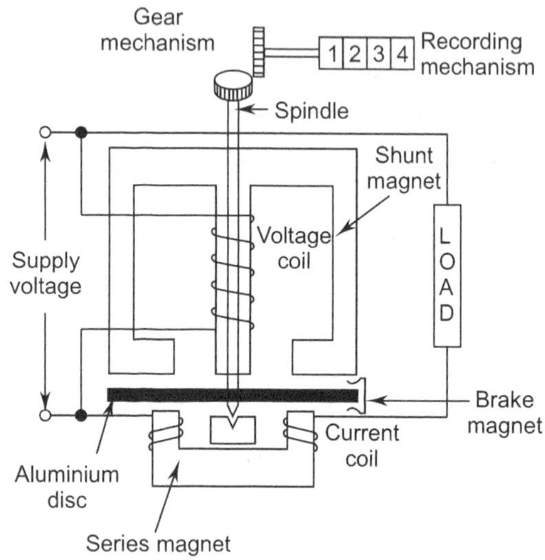

Fig. 7.15

7.9.2 Moving System

The moving system consists of thin aluminium disc mounted on the spindle. The disc is placed in the air-gap between the shunt and series magnets.

The current proportional to supply voltage in the shunt magnet produces a flux ϕ_{sh} which inturn produces an eddy current i_{sh} in the aluminium disc. The brake magnet provides eddy current damping to the moving system. The amount of braking or damping torque can be adjusted by adjusting the position of the magnet.

7.9.3 Braking/Damping System

The braking system consists of a permanent magnet called as brake magnet. It is placed near the edge of the aluminum disc (Fig. 7.15). The brake magnet provides eddy current damping to the moving system. The amount of braking or damping torque can be adjusted by adjusting the position of the magnet.

7.9.4 Registering System

The spindle is connected through gears to a counting mechanism (Fig. 7.15). This mechanism keeps on register the number of revolutions of the disc calibrated in kWh energy consumption.

Problem 7.1. *A moving coil instrument requires 20 mA and 100 mV for full scale deflection. What is the value of the shunt required to use the above meter to read up to 10 A?*

Solution.

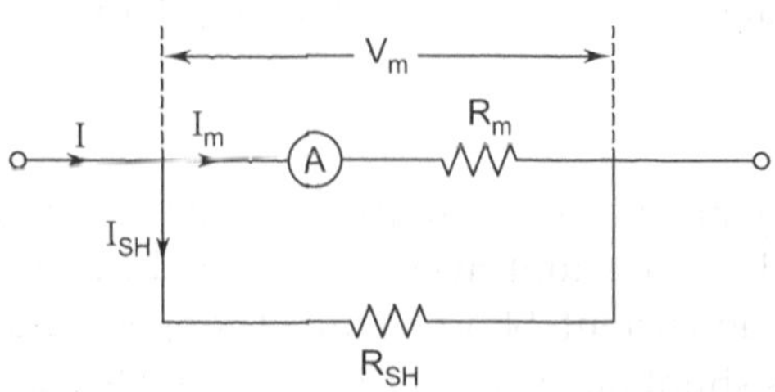

Fig. 7.16

$$I_m = 20\text{ mA}$$
$$V_m = 100\text{ mV}$$
$$I = 10\text{ A}$$
$$R_{SH} = ?$$

$$R_{SH} = \frac{V_m}{I_{SH}} = \frac{V_m}{I - I_m}$$

$$R_{SH} = \frac{10 \times 10^{-3}}{10 - 20 \times 10^{-3}} = 0.01002\ \Omega$$

$R_{SH} = \mathbf{0.01002\ \Omega.}$ **Ans.**

Problem 7.2. *A moving coil instrument has 50 Ω coil resistance and gives maximum deflection at 5 mA. Find the external resistance to be connected in series with the instrument so as to read up to 300 V.*

Solution.

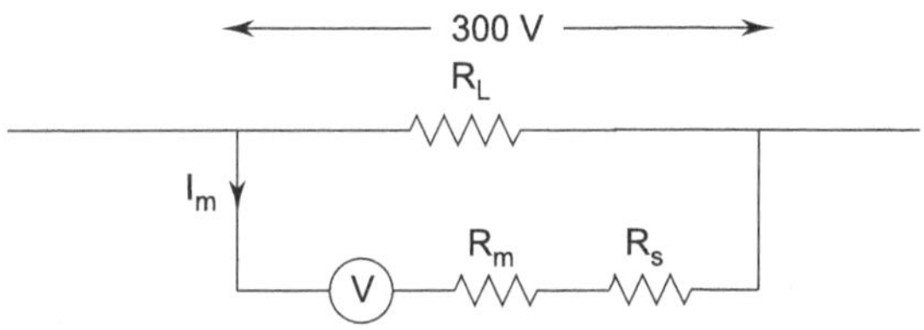

Fig. 7.17

$$I_m = 5\text{ mA} = 5 \times 10^{-3}\text{ A}$$
$$V = 300\text{ V}$$
$$R_m = 50\ \Omega$$
$$R_s = ?$$

The voltage across the meter and external resistance should be 300 V.

Hence, $$(R_m + R_s) = \frac{V}{I_m} = \frac{300}{5 \times 10^{-3}}$$

$$R_m + R_s = 60000\ \Omega$$

or $$R_s = 60000 - R_m = 60000 - 50$$

$R_s = \mathbf{59950\ \Omega.}$ **Ans.**

Problem 7.3. *The torque of an ammeter varies as the square of the current through it. If current of 10 A produces a deflection of 90°, what deflection will occur for a current of 5 A where the instrument is (a) spring controlled (b) gravity controlled ?*

Solution. The deflecting torque is proportional to the square of current

∴ $$T_d \propto I^2$$

(*a*) In spring control instruments, the controlling torque

$$T_c \propto \theta$$

∴ Deflection, $$\theta \propto I^2 \text{ as } T_d = T_c$$

or $$\frac{\theta_2}{\theta_1} = \left(\frac{I_2}{I_1}\right)^2$$

So, for a current of 5 A, the deflection will be

$$\theta_2 = \theta_1 \times \left(\frac{I_2}{I_1}\right)^2$$

Here $\theta_1 = 90°, I_1 = 10\text{ A, and } I_2 = 5\text{ A}$

$$\therefore \quad \theta_2 = 90° \times \left(\frac{5}{10}\right)^2 = 90° \times \frac{1}{4} = 22.5°$$

$$\theta_2 = \mathbf{22.5°}$$ **Ans.**

(*b*) In gravity controlled instruments, controlling torque $T_c \propto \sin\theta$ and $T_d \propto I^2$

$$\therefore \quad \sin\theta \propto I^2$$

or $$\frac{\sin\theta_2}{\sin\theta_1} = \left(\frac{I_2}{I_1}\right)^2$$

Then, for a current of 5 A, the deflection will be

$$\theta_2 = \sin^{-1}\left[\left(\frac{I_2}{I_1}\right)^2 \sin\theta_1\right] = \sin^{-1}\left[\left(\frac{5}{10}\right)^2 \times \sin 90°\right]$$

$$= \sin^{-1}\left[\frac{1}{4} \times 1\right] = \sin^{-1}(0.25) \qquad [\because \sin 90° = 1]$$

$$\theta_2 = \mathbf{14.5°}.$$ **Ans.**

Problem 7.4. *A current* $i = 14.1 \sin\left(314t + \frac{\pi}{3}\right)$ *is measured with the help of moving iron ammeter. What will be the reading of this ammeter?*

Solution. $I_m = 14.1\text{ A}$

However moving iron ammeter reads r.m.s. value.

Thus reading of ammeter, $I_m = \frac{14.1}{\sqrt{2}} = \mathbf{10\ A}$ **Ans.**

Problem 7.5. *A voltage* $v = 150 \sin(314t)$ *is measured with the help of moving iron instruments. What will be the reading of this voltmeter ?*

Solution. $V_m = 150\text{ V}$

Voltmeter will read r.m.s value

$\therefore$ Reading of voltmeter $V_{rms} = \frac{150}{\sqrt{2}} = \mathbf{106.06\ V}$ **Ans.**

Problem 7.6. *The resistance of a moving coil voltmeter is 12,000 Ω. The moving coil has 100 turns and is 4 cm long and 3 cm wide. The flux density in the air gap is 6 × 10^{-2} wb/m^2. Find the deflection produced by 300 V, if the spring control gives a deflection of one degree for a torque of 25 × 10^{-7} Nm.*

Solution. Given : Applied voltage, $V = 300$ V

Resistance of the coil, $R = 12{,}000\ \Omega$

∴ Current flowing through the coil, $I = \dfrac{V}{R} = \dfrac{300}{12{,}000} = 0.025$ A

Number of turns, N = 100 ; Length of former, l = 4 cm and = 0.04 m

Width of former, r = 3 cm = 0.03 m; Field strength, B = 0.06 Wb/m^2

∴ Deflecting torque, T_d is given by

$$T_d = N\,B\,I\,l\,r = 100 \times 0.06 \times 0.025 \times 0.04 \times 0.03$$

[Note that here $2r$ = 0.03 m]

or
$$T_d = 0.00018 \text{ Nm.}$$

Given that 25×10^{-7} Nm produces deflection = 1°

∴ Deflection produced by
$$300\text{ V} = \frac{1}{25 \times 10^{-7}} \times 0.00018$$

or
$$\theta = \frac{0.00018}{25 \times 10^{-7}} = \mathbf{72°}$$
Ans.

Problem 7.7. *Full-scale deflection of moving iron meter is obtained when carrying a current of 100 mA. Show how it can be used to measure voltage up to 150 V and current up to 20 A. Given that resistance of instrument is 20 ohm.*

Solution. $I_m = 100$ mA, $R_m = 20\ \Omega$

(*a*) For measurement of 150 V the series resistance called multiplier should be connected in series with meter.

Value of series resistance

$$R_{se} = \frac{V}{I_m} - R_m = \frac{150}{100 \times 10^{-3}} - 20 = \mathbf{1480\ \Omega}$$
Ans.

(*b*) For measurement of 20 amp. current, shunt resistance should be connected :

Value of shunt

$$R_{SH} = \frac{I_m \cdot R_m}{I - I_m} = \frac{100 \times 10^{-3} \times 20}{20 - 100 \times 10^{-3}} = \mathbf{0.1005\ \Omega.}$$
Ans.

Problem 7.8. *A moving coil instrument gives a full-scale deflection of 20 mA when a potential difference of 50 mV is applied. Calculate the series resistance to measure 500 V on full scale.*

Solution. Instrument resistance, $R_m = \dfrac{\text{PD across terminals}}{\text{Instrument current}}$

$$= \frac{50 \times 10^{-3}}{20 \times 10^{-3}} = 2.5\ \Omega$$

Series resistance required for full-scale deflection for V = 500 V

$$R = \frac{V}{I_m} - R_m = \frac{500}{20 \times 10^{-3}} - 2.5 = \mathbf{2497.5\ \Omega}. \qquad \textbf{Ans.}$$

Problem 7.9. *A moving coil instrument gives full-scale deflection with 10 mA and has resistance of 50 Ω. Calculate the resistance to be connected in series/parallel with instrument in order that it may be used as*

(i) 5 A ammeter. *(ii) 200 V voltmeter.*

Solution. $R_m = 50\ \Omega, I_m = 10 \times 10^{-3}$ amp.

(*i*) Resistance of shunt required to measure 5 amp. current,

$$R_{SH} = \frac{I_m \cdot R_m}{I - I_m} = \frac{10 \times 10^{-3} \times 50}{5 - 10 \times 10^{-3}} = \mathbf{0.1002\ \Omega} \qquad \textbf{Ans.}$$

(*ii*) Resistance of multiplier to measure 200 volt.

$$R_{se} = \frac{V}{I_m} - R_m = \frac{200}{10 \times 10^{-3}} - 50 = \mathbf{19950\ \Omega}. \qquad \textbf{Ans.}$$

Problem 7.10. *A PMMC instrument gives full-scale reading of 25 mA when a potential difference across its terminals is 15 mV. Show how it can be used as*

(i) an ammeter for the range of (0 – 150) A

(ii) a voltmeter for the range of (0 – 150) V.

Solution. Instrument resistance, $R_m = \frac{\text{PD across terminals}}{\text{Instrument current}}$

$$= \frac{15 \times 10^{-3}}{25 \times 10^{-3}} = 0.6\ \Omega$$

(*i*) To measure 150 A current, *i.e.*, I = 100 A, multiplying power of shunt is given by

$$m = \frac{I}{I_m} = \frac{100}{25 \times 10^{-3}} = 4000$$

Shunt resistance $R_{SH} = \frac{R_m}{m-1} = \frac{0.06}{4000-1} = \mathbf{\frac{1}{6666.5}}\ \Omega.$ **Ans.**

(*ii*) To measure 150 V, series resistance is given by

$$R_{se} = \frac{V}{I_m} - R_m = \frac{150}{25 \times 10^{-3}} - 0.6 = \mathbf{5999.4\ \Omega}. \qquad \textbf{Ans.}$$

Problem 7.11. *A moving coil instrument gives a full-scale deflection of 10 mA when a P.D. 10 mV is applied across the terminal. Show how you will use the instrument to measure :*

(i) Current up to 100 A *(ii) Voltage up to 500 V.*

Solution. Instrument resistance, $R_m = \frac{\text{P.D. across terminals}}{\text{Instrument current}} = \frac{100 \times 10^{-3}}{10 \times 10^{-3}} = 10\ \Omega$

(*i*) Multiplying power, $m = \frac{I}{I_m} = \frac{100}{10 \times 10^{-3}} = 10{,}000$

Shunt resistance for 100 A, $R_{SH} = \frac{R_m}{m-1} = \frac{10}{10,000-1} = \mathbf{1 \times 10^{-3}\ \Omega}$ **Ans.**

(*ii*) Series resistance, $R_{se} = \frac{V}{I_m} - R_m = \frac{500}{10 \times 10^{-3}} - 10 = \mathbf{49,990\ \Omega}$. **Ans.**

Problem 7.12. *A moving coil instrument gives full-scale deflection with 10 mA and has resistance of 50.0 Ω. Calculate the resistance required to be put in series/parallel with the instrument in order that it may be used as*

(i) 10 A ammeter *(ii) 100 V voltmeter.*

Solution. $R_m = 50\ \Omega,\ I_m = 10\ \text{mA} = 10 \times 10^{-3}\ \text{A}$

(*i*) For ammeter : Shunt resistance R_s is connected in parallel.

$$R_s = \frac{I_m \cdot R_m}{I - I_m} = \frac{10 \times 10^{-3} \times 50}{10 - 10 \times 10^{-3}} = \mathbf{0.05\ \Omega}$$ **Ans.**

(*ii*) For voltmeter : Resistance R is connected in series.

$$R = \frac{V}{I_m} - R_m = \frac{100}{10 \times 10^{-3}} - 50 = \mathbf{9950\ \Omega}.$$ **Ans.**

Problem 7.13. *A moving coil instrument gives a full-scale deflection of 10 mA and P.D. across its terminals is 100 mV. Calculate (a) shunt resistance for full-scale deflection corresponding to 200 A, (b) series resistance for full-scale reading corresponding to 1,000 V.*

Solution. Instrument resistance, $R_m = \frac{\text{P.D. across terminals}}{\text{Instrument current}} = \frac{100 \times 10^{-3}}{10 \times 10^{-3}} = 10\ \Omega$

(*a*) Shunt resistance required for full-scale deflection for I = 200 A

$$R_{SH} = \frac{R_m}{\frac{I}{I_m} - 1} = \frac{10}{\frac{200}{10 \times 10^{-3}} - 1} = \mathbf{5.0025 \times 10^{-4}\ \Omega}$$ **Ans.**

(*b*) Series resistance required for full-scale deflection for V = 1,000 V

$$R_{se} = \frac{V}{I_m} - R_m = \frac{1,000}{10 \times 10^{-3}} - 10 = \mathbf{99,990\ \Omega}.$$ **Ans.**

Problem 7.14. *Two voltmeters have the same range 0–400 V. The internal impedances are 30000 Ω and 20000 Ω. If they are connected in series and 600 V be applied across them what will be their readings?*

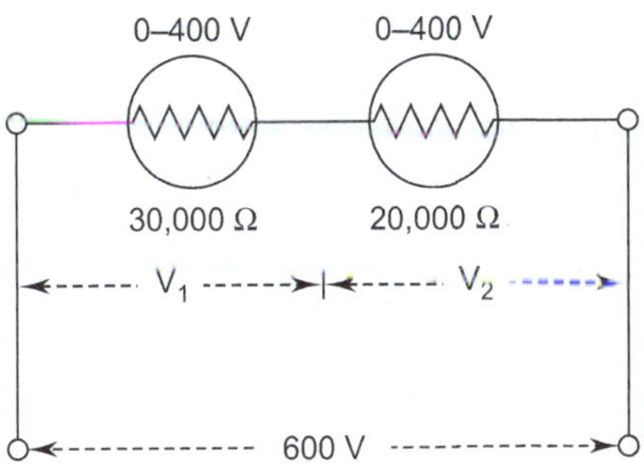

Fig. 7.18

Solution. Let the readings of two voltmeters be V_1 and V_2 respectively.

$$\therefore \quad V_1 + V_2 = V$$

From voltage divider rule

$$V_1 = \frac{R_1}{R_1 + R_2} \cdot V$$

$$= \frac{30{,}000}{30{,}000 + 20{,}000} \times 600 = \mathbf{360\ V} \qquad \textbf{Ans.}$$

$$V_2 = V - V_1 = \mathbf{840\ V.} \qquad \textbf{Ans.}$$

Problem 7.15. *Two ammeters, one with full-scale current of 1 mA and internal resistance of 100 ohm, and the other a full-scale current of 10 mA and internal resistance of 25 ohm, are in parallel. What is the total current these two meters can carry without any meter readings getting out of scale?*

Solution. Let V_1 and V_2 be the voltage across 100 Ω ammeter and 25 Ω ammeter respectively.

$$\therefore \quad V_1 = I_1 R_1 = 1 \times 10^{-3} \times 100 = 0.1\ V$$

and $$V_2 = I_2 R_2 = 10 \times 10^{-3} \times 25 = 0.25\ V.$$

But in parallel combination, voltage across each meter remains same. Therefore, for safety reason the voltage across 100 Ω meter cannot increase beyond 0.1 volt. Hence 0.1 volt will be a common voltage across each meter.

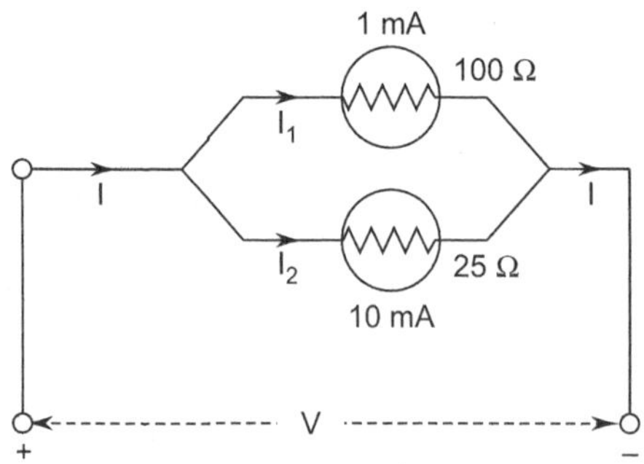

Fig. 7.19

Resistance of parallel meters, $$R_{eq} = \frac{100 \times 25}{100 + 25} = 20\ \Omega$$

$$I = \frac{0.1}{20} = \mathbf{0.005\ A\ or\ 5\ mA.} \qquad \textbf{Ans.}$$

Problem 7.16. *A moving coil instrument has a resistance of 2 ohm and it reads up to 250 V when a resistance of 500 ohm is connected in series with it. Find the current range of the instrument when it is used as an ammeter with the coil connected across a shunt resistance of 2 milli-ohm.*

Solution. Given $$R_m = 2\ \Omega,\ R_{se} = 5000\ \Omega,\ V = 250\ V$$

$\therefore$ Full-scale deflection current, $$I_m = \frac{V}{R_{se} + R_m} = \frac{250}{5000 + 2} = 0.0499\ A$$

When a 2×10^{-3} Ω shunt is connected across the meter, the circuit becomes as shown in Fig. 7.20.

We know that $$I_m \cdot R_m = I_{SH} \cdot R_{SH}$$

or
$$I_{SH} = I_m \times \frac{R_m}{R_{SH}}$$

$$= 0.0499 \times \frac{2}{2 \times 10^{-3}} = 49.98 \text{ A}$$

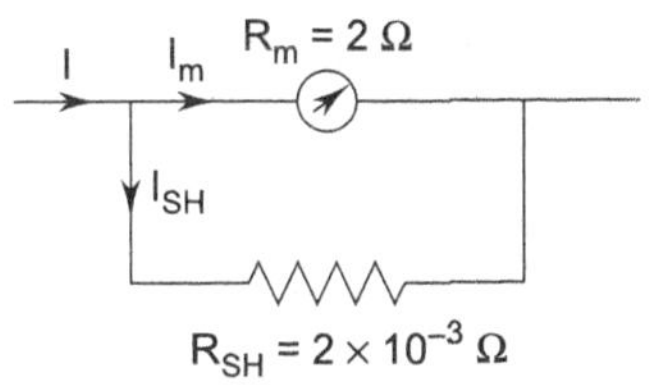

Fig. 7.20

Hence current range of instrument = Full-scale deflection current

$$= I_m \times I_{SH} = 0.0499 + 49.98 = \mathbf{50\ A}. \qquad \textbf{Ans.}$$

Problem 7.17. *A moving coil instrument has a resistance of 10 Ω and gives a full-scale deflection when carrying 50 mA. Show how the instrument can be used to measure voltages up to 750 V and current up to 100 A.*

Solution.

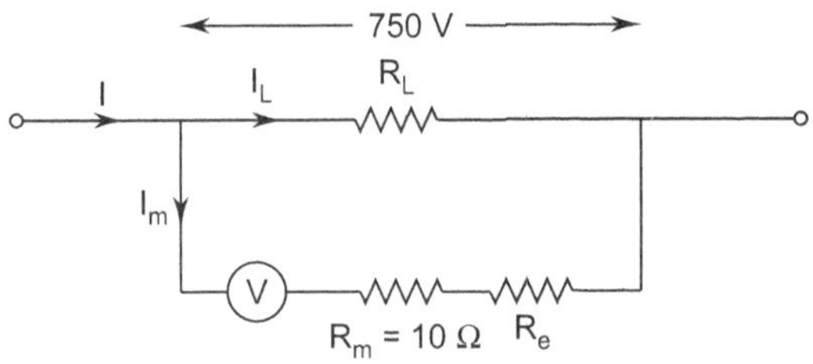

Fig. 7.21

(*i*) The instrument can be used to measured voltage up to 750 V when a series resistance is connected in series with the instrument. The value of that series resistance can be calculated with the help of the circuit diagram shown in Fig. 7.21.

$$750 = I_m (10 + R_e)$$

or
$$750 = 50 \times 10^{-3} (10 + R_e)$$

or
$$R_e = \mathbf{14990\ \Omega} \qquad \textbf{Ans.}$$

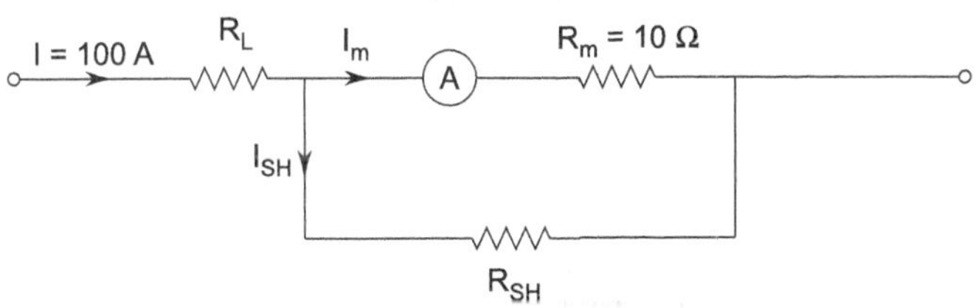

Fig. 7.22

(*ii*) The instrument can be used to measure current up to 100 Ampere when a shunt (Parallel) resistance is connected with the instrument. The value of that shunt resistance can be calculated with the help of the circuit diagram shown in Fig. 7.22.

As both the branches are in parallel hence the voltage drop across both will be same. So

$$I_{SH}R_{SH} = I_m R_m$$

or $$(100 - 0.05)\, R_{SH} = 0.05 \times 10$$

or $$R_{SH} = \mathbf{0.005\ \Omega}$$ **Ans.**

Problem 7.18. *A moving coil instrument gives full-scale deflection with 20 mA. The resistance of coil is 4 Ω. What will be the value of series resistance needed for the instrument to read upto 30 V?*

Solution.

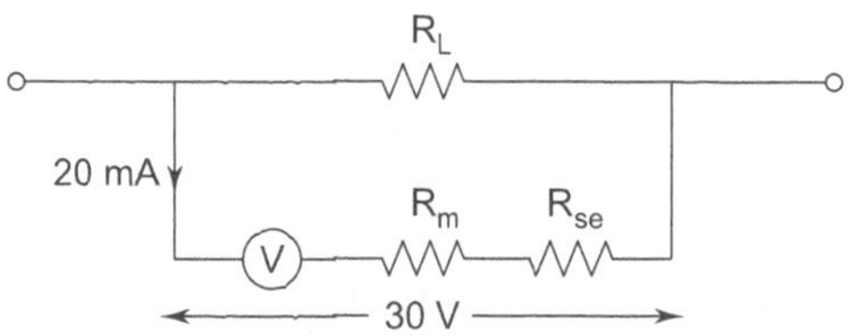

Fig. 7.23

Given $$R_m = 4\ \Omega$$

$$R_{se} = ?$$

As the current through both the resistances (*i.e.*, R_m and R_{se}) will be same, hence

$$V = I\,(R_m + R_{se})$$

or $$30 = 20 \times 10^{-3}\,(4 + R_{se})$$

or $$R_{se} = 1500 - 4 = \mathbf{1496\ \Omega}$$ **Ans.**

Problem 7.19. *A moving coil milli-ammeter having a resistance of 8 Ω gives full-scale deflection when a current of 5 mA is passed through it. Explain how this instrument can be used for measurement of:*

(a) Current upto 2 A

(b) Voltage upto 8 V.

Solution.

(a) Use of instrument as a Ammeter

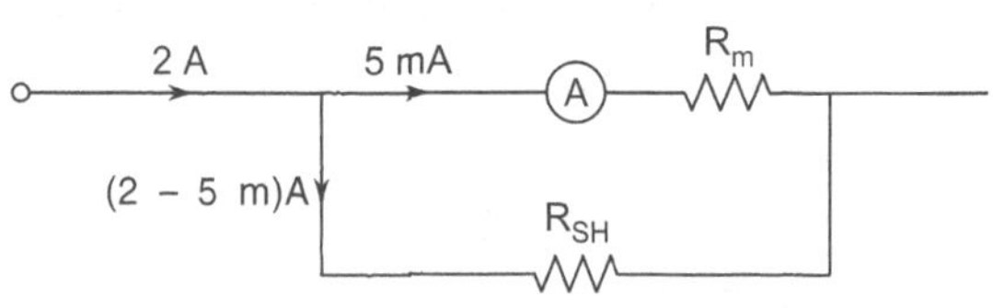

Fig. 7.24

$$R_m = 8\ \Omega$$

$$R_{SH} = ?$$

To measure the current upto 2A with the instrument, a shunt resistance is required. Voltage drop across both the resistance should be same *i.e.*

$$5 \times 10^{-3} \times R_m = (2 - 5 \times 10^{-3}) \times R_{SH}$$

or $$R_{SH} = \frac{5 \times 10^{-3} \times 8}{2 - 5 \times 10^{-3}} = \frac{0.04}{1.995}$$

$$R_{SH} = \mathbf{0.02005\ \Omega}$$ **Ans.**

(*b*) *Use of instrument as voltmeter*

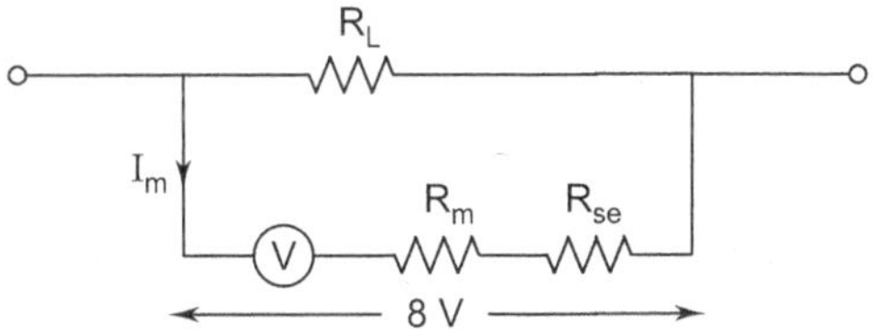

Fig. 7.25

A voltmeter is connected in parallel. To increase the range of the voltmeter a resistance is required in series with the voltmeter. The current through both the resistance will be same *i.e.,*

$$I_m R_m + I_m R_{se} = 8$$

$$5 \times 10^{-3} [8 + R_{se}] = 8$$

or $$R_{se} = 1600 - 8 = \mathbf{1592\ \Omega}$$ **Ans.**

Problem 7.20. *A moving coil instrument gives full-scale deflection with 20 mA. The resistance of the coil is 4 Ω. It is desired to convert this instrument into an ammeter to read upto 2 A. Find the resistance of the shunt to be connected.*

Solution.

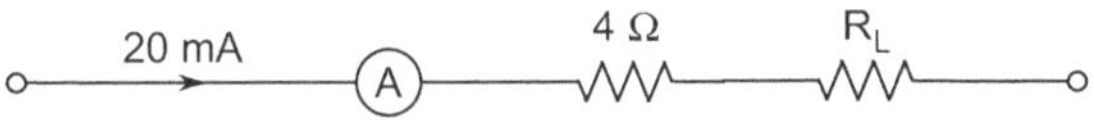

Fig. 7.26

As an Ammeter:

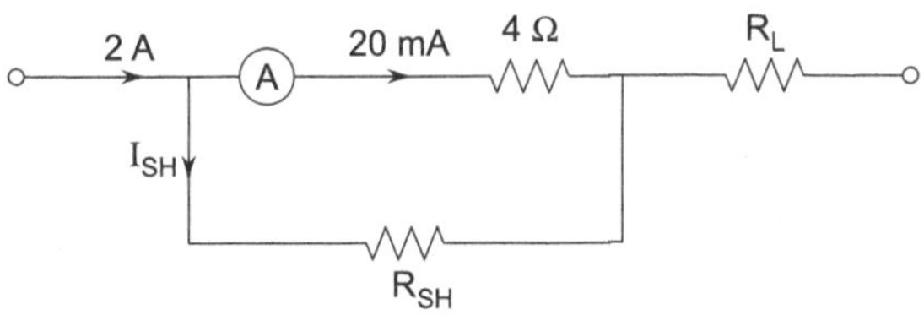

Fig. 7.27

$$I_{SH} = 2 - 20 \times 10^{-3}$$

$$20 \times 10^{-3} \times 4 = I_{SH} . R_{SH}$$

or $$R_{SH} = \frac{20 \times 10^{-3} \times 4}{2 - 20 \times 10^{-3}} = \frac{0.08}{1.98}$$

$$= 0.0404$$

$$R_{SH} = \mathbf{40.4\ m\ \Omega}$$ **Ans.**

As an Voltmeter:

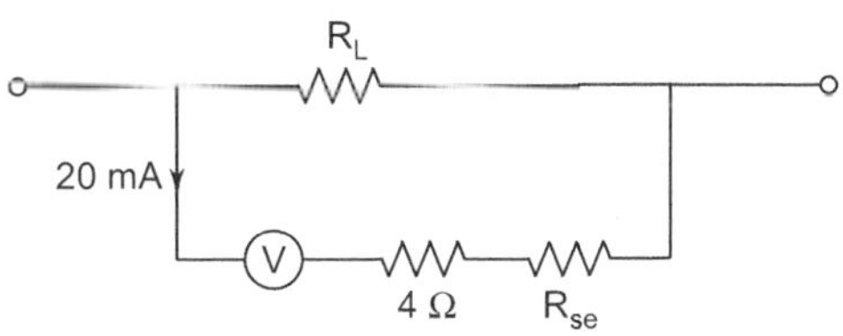

Fig. 7.28

to read upto a voltage of 30 Volt

$$30 = 20 \times 10^{-3} (4 + R_{se})$$

or $$R_{se} = 1496\ \Omega$$

Problem 7.21. *A single phase energy meter has a constant of 1200 revolutions/kWh. When a load of 200 Ω is connected, the disc rotates at 4.2 revolutions/minute. If the load is ON for 10 hours, how many units are recorded as an error? Also find the percentage error.*

Solution.

Given,

$$\text{Meter constant} = 1200 \text{ rev/kWh}$$

$$\text{Load} = 200 \text{ W}$$

$$\text{Disc rotations} = 4.2 \text{ revolution/min}$$

$$\text{Time} = 10 \text{ hours}$$

Energy consumed in one hour

$$= \frac{200}{1000} = 0.2 \text{ kWh}$$

Number of revolution for true energy in one hour

$$= \text{meter constant} \times 0.2$$

$$= 1200 \times 0.2$$

$$= 240$$

In actual the number of revolutions made by the meter in one hour

$$= 4.2 \times 60$$

$$= 252$$

(*a*) No. of units recorded as an error

$$= 252 - 240 = \mathbf{12 \text{ units}}$$ **Ans.**

In one hour 12 units are recorded as an error

In 10 hour $12 \times 10 = 120$ units are as an error

(*b*) Percentage error

$$\% \text{ error} = \frac{\text{Actual revolution} - \text{Rev. for true energy}}{\text{Revolutes for true energy}}$$

$$= \frac{252 - 240}{240} \times 100$$

$$\% \text{ error} = \mathbf{5\%}$$ **Ans.**

EXERCISE

1. Explain the principle and construction of Attraction Type Moving Iron Instruments. Write their merits and demerits also.
2. Describe the following in case of Measuring Instruments:
   (*i*) Deflecting Torque
   (*ii*) Controlling Torque
   (*iii*) Damping Torque.

3. A Moving Coil Instrument gives full-scale deflection with 10 mA and has a resistance of 50 Ω. Calculate the resistance to be connected in series/parallel with the instrument in order that it may be used as
   (*i*) 5 ampere ammeters
   (*ii*) 200 volt voltmeter.
4. Explain the principle and construction of Attraction Type Moving Coil Instruments.
5. Explain working of single phase induction type of energy meter with neat diagram.
6. Describe the working principle of Permanent Magnet Moving Coil (PMMC) Instruments. Why is the scale linear?
7. How range is extended in moving coil ammeter and voltmeter?
8. A moving coil instrument gives full-scale deflection with 15 mA. The resistance of coil is 5 Ω. It is desired to convert this instrument into an ammeter to read up to 2 A. How to achieve it? Further how to make this instrument to read up to 30 volts.
9. A 50 A, 230 V meter on full load test makes 61 revolutions in 37 seconds. If the normal disc speed is 500 revolutions per kWh, find the percentage error.
10. Explain the working principle of Induction type wattmeter with the help of diagram.
11. Compare merit and demerits of moving iron instruments and dynamometer type instrument. Which one is superior and why?
12. A moving coil instrument having a resistance of 50 Ω has a full-scale deflection of 1 mA. Calculate:
    (*i*) Shunt resistance to convert the instrument into an ammeter of 2 A range.
    (*ii*) Net resistance of the meter.
13. A single-phase energy meter has a constant speed of 1300 revolutions/kWh. The disc revolves at 3.5 rpm when a load of 150 W is connected to it. If the load is on for 11 hours, how many units are recorded as error? What is the percentage error?

□ □ □

8

TRANSFORMER (SINGLE PHASE)

8.1 INTRODUCTION

"A Transformer is a static electromagnetic energy conversion device, which transfers electrical energy from one electrical circuit to another electrical circuit, without any change in supply frequency".

The medium of transfer of energy is magnetic field. The voltage can be step up or step down but the current will be step down or step up respectively.

As transformer is a static device, *i.e.*, it has no moving parts; it has, therefore, highest possible efficiency among all electrical machines. In general some of the important applications of transformer are:

- For increasing or decreasing voltage and/or current level.
- For isolating one circuit from another circuit.
- For isolating DC while permitting the flow of AC between the two circuits.
- For matching the impedance of source and its load for maximum power transfer in electronics and control circuits.

An elementary transformer consists of a soft iron or silicon steel core and two winding placed on it (Fig. 8.1). The windings are insulated from core as well as from each other.

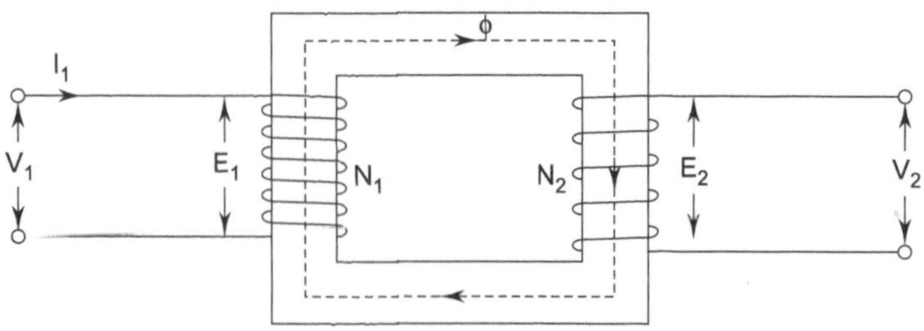

Fig. 8.1

8.2 TYPES OF TRANSFORMERS

Transformers are classified on the following three basis:

(*a*) Position of windings with respect to core

 (*i*) Core-type transformer (*ii*) Shell-type transformer

(*b*) Transformation ratio

 (*i*) Step-up transformer (*ii*) Step-down transformer

(*c*) Type of service

 (*i*) Distribution transformer (*ii*) Power transformer.

8.3 CONSTRUCTION

According to the construction, transformers are of two types namely core type and shell type. The difference between these two types of transformers is by the manner in which windings are wound around the magnetic core.

Beside other parts, the two essential parts of a transformer are:

- Magnetic core
- Primary and secondary windings

8.3.1 Magnetic Core

The magnetic core is stack of thin silicon steel laminations (Fig. 8.2). In order to reduce eddy current losses in the core these laminations are insulated from each other by thin layer of varnish.

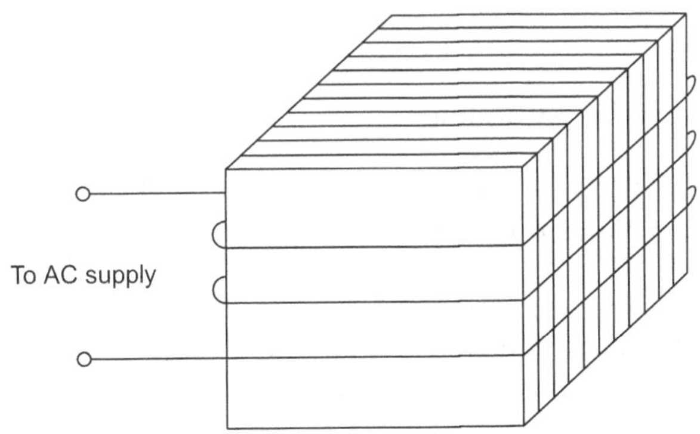

Fig. 8.2

For reducing the core losses nearly all transformer have their magnetic core made from cold-rolled grain-oriented silicon steel (C.R.G.O.). The vertical portions of the core are usually called as ***Limbs or Legs*** and top and bottom portions are called ***Yoke***. In case of core type we have two legs and two yokes.

In case of shell type we have three legs and two yokes.

In case of core type transformer, the core is rectangular in shape and having uniform cross-section. It is made up of '*L*' shape laminated strips shown in Fig. 8.3. The laminations have butt joints. The sequence of laminations while combining for making core is given in Fig. 8.3 (*a*) and (*b*).

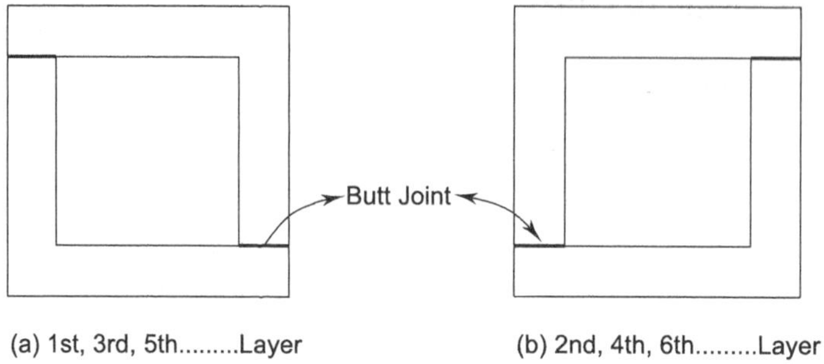

Fig. 8.3 *Frame of L-shaped Lamination*

In case of shell type transformer, the core is divided into two parts each having rectangular shape and having uniform cross-section. It is made up of **'E'** and **'I'** shape laminated strips shown in Fig. 8.4. The laminations have butt joints. The sequence of laminations while combining for making core is give in Fig. 8.4(*a*) and (*b*).

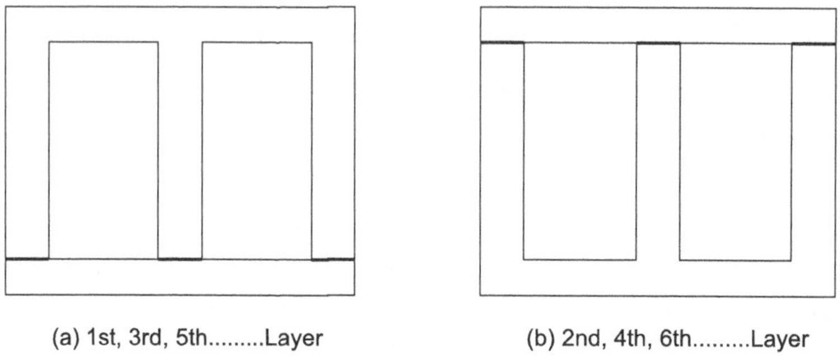

Fig. 8.4 *Frame of E-shaped and I-shaped Lamination*

8.3.2 Primary and Secondary Windings

In core type transformer the winding surround a considerable part of steel core (Fig. 8.5).

For a given output and voltage ratings, core type transformer requires less iron but more conductor material.

In case of shell type transformer, sandwich type windings are used. The high voltage and low voltage windings are split into a number of sections where each high voltage section lies between two low voltage sections. Such type of sandwich windings is shown in Fig. 8.6.

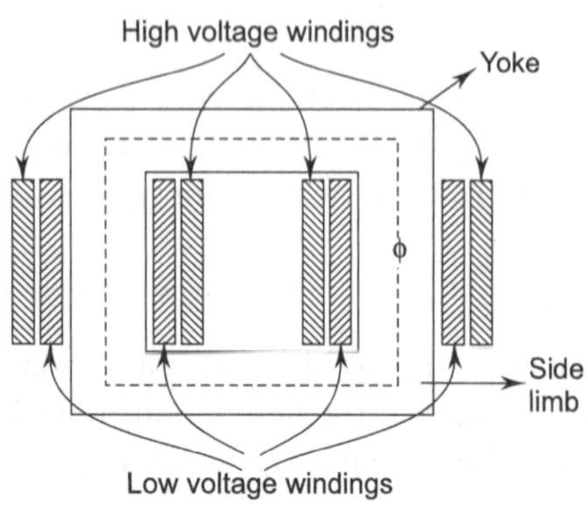

Fig. 8.5

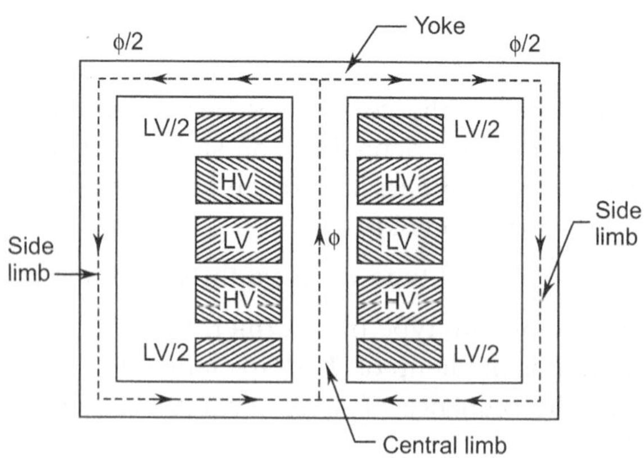

Fig. 8.6

8.4 WORKING PRINCIPLE

A transfer is based on the principle of "Electromagnetic Induction". When an alternating voltage (V_1) is applied across the primary side of the transformer, a current (I_1) flows in the primary winding. This primary current (I_1) produces the magnetic flux (ϕ) in the transformer core, as the current in the primary coil is alternating the magnetic flux set up in the transformer core will also be alternating in nature, that is, its magnitude and direction will change with time.

As per Faraday's law of electromagnetic induction, a statically self induced e.m.f is set up in the primary coil given by

$$e_1 = -N_1 \frac{d\phi}{dt} \qquad \text{...(8.1)}$$

where, N_1 is the number of turns in the primary coil and ϕ is the alternating flux linking the primary coil. The induced e.m.f in primary coil is nearly equal and opposite to the applied voltage V_1.

As leakage of flux has been assumed to be negligible, the same magnetic flux, which is alternating in nature, links with each and every turn of the secondary coil. As such, a statically mutual-induced e.m.f is setup in the secondary coil, the magnitude of which, as per Faraday's laws of electromagnetic induction, is given by

$$e_2 = -N_2 \frac{d\phi}{dt} \qquad \text{...(8.2)}$$

where, N_2 is the number of turns in secondary coil and ϕ is common alternating flux linking the secondary coil.

If the load circuit is complete, mutually induced e.m.f in the secondary coil will circulate the current in the load circuit.

8.5 E.M.F. EQUATION

When an alternating voltage is applied across the primary of transformer, it takes a magnetizing current and as a result of this current, a flux is established in the transformer core. The flux ϕ is uniformly distributed over transformer core section and is linked with all turns of the primary and secondary windings.

The main flux ϕ established in the core is alternating in nature. Hence, an e.m.f. is induced in the primary winding of transformer, which is given by

$$e_1 = -N_1 \frac{d\phi}{dt} \qquad \text{[Eqn. (8.1)]}$$

where, N_1 represents the number of turns in primary.

The main flux ϕ can be expressed as

$$\phi = \phi_m \cos \omega t \qquad \text{...(8.3)}$$

where, ϕ_m is the maximum value of main flux. Substituting equation (8.3) in equation (8.1) we get

$$e_1 = -N_1 \frac{d}{dt}(\phi_m \cos \omega t) \quad \ldots(8.4)$$

or $$e_1 = -N_1 \omega\phi_m \sin \omega t \quad \ldots(8.5)$$

The induced e.m.f. e_1 is maximum when $\sin \omega t = 1$ and that maximum value of induced e.m.f. is given by

$$E_{1,\,max} = N_1 \omega\phi_m \quad \ldots(8.6)$$

Hence, r.m.s value of induced e.m.f in the primary is given by

$$E_{1,\,rms} = \frac{E_{1,\,max}}{\sqrt{2}} = \frac{N_1\omega\phi_m}{\sqrt{2}} = \frac{N_1 2\pi f \phi_m}{\sqrt{2}} \quad \ldots(8.7)$$

or $$\boxed{E_{1,\,rms} = 4.44 f\phi_m N_1} \quad \text{Volts} \quad \ldots(8.8)$$

where, f is frequency of supply in Hz.

Similarly, e.m.f induced in secondary winding with N_2 turns is given by

$$\boxed{E_{2,\,rms} = 4.44 f\phi_m N_2} \quad \text{Volts} \quad \ldots(8.9)$$

On comparing equations (8.3) and (8.5) it is clear that induced e.m.f in primary winding E_1 (as well as in secondary winding E_2) lags main flux ϕ by 90°.

On dividing equation (8.9) by equation (8.8), we get

$$\frac{E_2}{E_1} = \frac{N_2}{N_1} = K \quad \ldots(8.10)$$

Also, $$\boxed{\frac{V_2}{V_1} = \frac{E_2}{E_1} = \frac{N_2}{N_1} = \frac{I_1}{I_2} = K = \frac{1}{a}} \quad \ldots(8.11)$$

where a = Turn ratio

K = Transformation ratio.

8.6 PHASOR DIAGRAM

The relations between voltages, currents, voltage drops, induced e.m.f's etc., related to both primary and secondary windings can be expressed with the help of circuit equations. These equations when represented with phasors on a single reference point are termed as phasor diagram of a transformer. Some of the important features to be remember while drawing the phasor diagram are:

- As flux ϕ is constant and links with both windings, it is consider as reference vector.
- The induced e.m.f always opposes the cause due to which it comes into existence. Hence induced e.m.f in primary winding, E_1 opposes the input voltage V_1 and hence $\vec{V_1} = -\vec{E_1}$.
- The induced e.m.f in primary winding, E_1 and the induced e.m.f in secondary winding, E_2 both have same direction.
- While deriving the expression for induced voltage, it is clear from equations (8.3) and (8.5)

  $$\phi = \phi_m \cos \omega t \quad \text{[Eqn. (8.3)]}$$
  $$e_1 = N_1 \omega \phi_m \sin \omega t \quad \text{[Eqn. (8.5)]}$$

  that induced e.m.f. E_1 and E_2 lags flux ϕ by an angle $\pi/2$.

8.6.1 Case-I

Let us consider a transformer at ***no-load*** in which

- **No** winding resistance.
- **No** leakage reactance.
- **No** core loss.
- Secondary side **opens.**

The circuit diagram is shown in Fig. 8.7. The equations for primary and secondary sides can be written as follow:

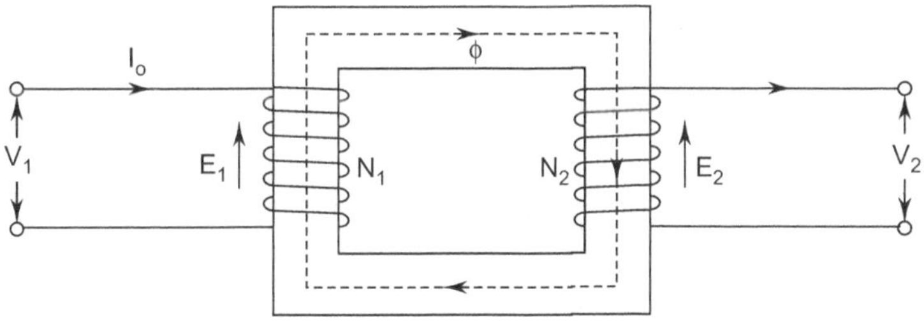

Fig. 8.7

- The primary current (denoted by I_o) is no load current, as secondary side is open circuited. Hence

$$\vec{I_1} = \vec{I_o} \quad \ldots(8.12)$$

- The secondary current (denoted by I_2) is zero, as secondary side is open circuited

$$I_2 = 0 \quad \ldots(8.13)$$

- From equations (8.3) and (8.5)

$$\phi = \phi_m \cos \omega t \quad \text{[Eqn. (8.3)]}$$

$$e_1 = N_1 \, \omega \, \phi_m \sin \omega t \quad \text{[Eqn. (8.5)]}$$

  it is clear that induced e.m.f lags the flux ϕ by an angle $\pi/2$.

- Induced e.m.f. in primary winding opposes the input voltage V_1 and hence

$$\vec{V_1} = -\vec{E_1} \quad \ldots(8.14)$$

- Induced e.m.f. in secondary winding, E_2, acts as a source for the secondary side circuit and secondary terminal voltage, V_1 (which is basically voltage drop across the load), will be equal to the secondary induced e.m.f as secondary is open circuited.

$$\vec{E_2} = \vec{V_2} \quad \ldots(8.15)$$

- No load current I_o will be in phase with the main flux ϕ, as there are no core losses and this primary no load current is totally responsible for the generation of flux, hence the phasor diagram will be as shown in Fig. 8.8.

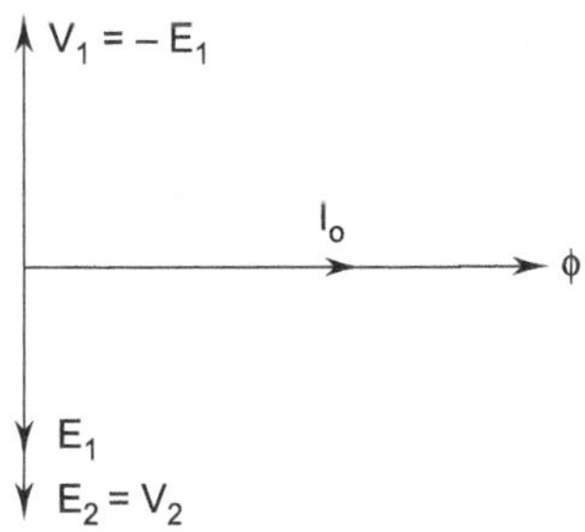

Fig. 8.8

8.6.2 Case-II

Now consider a transformer at **no-load** in which

- Winding resistances of both windings **included**.
- **No** leakage reactance.
- **No** core loss.
- Secondary side **opens**.

The circuit diagram is shown in Fig. 8.9. The equations for primary and secondary sides can be written as follow:

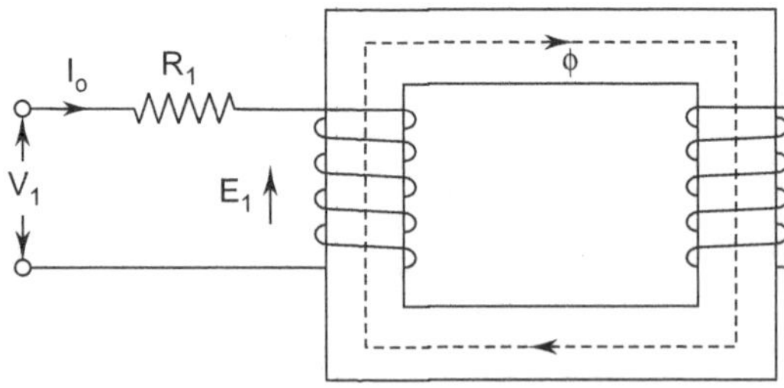

Fig. 8.9

- The primary current (denoted by I_o) is no load current, as secondary side is open circuited. Hence

$$\vec{I_1} = \vec{I_o} \qquad \text{[Eqn. (8.12)]}$$

- The secondary current (denoted by I_2) is zero, as secondary side is open circuited

$$I_2 = 0 \qquad \text{[Eqn. (8.13)]}$$

- Induced e.m.f lags the flux by an angle $\pi/2$ (*i.e.*, E_1, E_2 lags ϕ by 90°)
- Induced e.m.f in primary winding opposes input voltage V_1 and as resistance of the primary side is included hence primary side equation will become

$$V_1 = I_o R_1 + (-E_1) \qquad \ldots(8.16)$$

- Induced e.m.f in secondary winding, E_2, will be equal to the secondary terminal voltage V_2, as secondary is open circuited.

$$\vec{E_2} = \vec{V_2} \qquad \text{[Eqn. (8.15)]}$$

- No load current I_o is in phase with flux, ϕ as there are no core losses.

  Hence the phasor diagram will be as shown in Fig. 8.10.

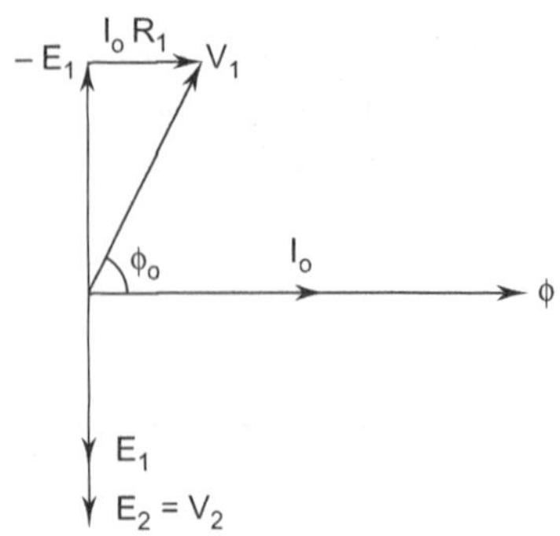

Fig. 8.10

- The phasor (I_oR_1) is a voltage drop across resistance; hence the current and the voltage related to this phasor will be in phase. That is the reason the phasor (I_oR_1) is drawn always parallel to the current phasor (I_o) and in the same direction as that of phasor (I_o).

8.6.3 Case-III

Next consider a transformer at **no-load** in which

- Winding resistances of both windings **included**.
- Core loss **included**.
- **No** leakage reactance.
- Secondary side **opens.**

The circuit diagram is shown in Fig. 8.11. The equations for primary and secondary sides can be written as follow:

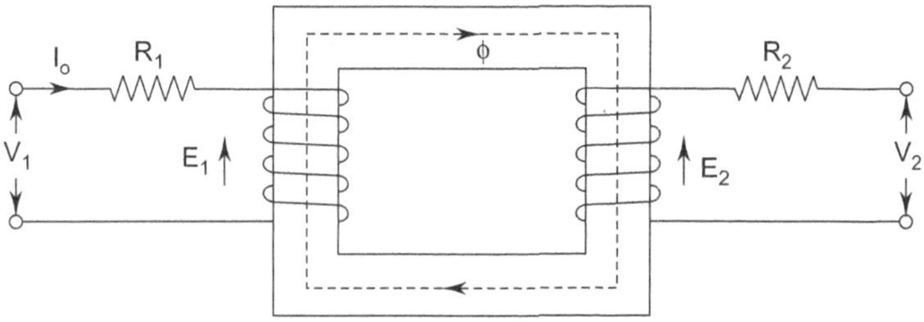

Fig. 8.11

- The secondary current is zero, as secondary side is open circuited

$$I_2 = 0 \qquad \text{[Eqn. (8.13)]}$$

- E_1, E_2 lags ϕ by 90°
- Induced e.m.f in primary winding opposes the input voltage V_1 and as the resistance of the primary side is included hence primary side equation will become

$$V_1 = I_oR_1 + (-E_1) \qquad \text{[Eqn. (8.16)]}$$

- As secondary is open circuited.

$$\overrightarrow{E_2} = \overrightarrow{V_2} \qquad \text{[Eqn. (8.15)]}$$

- No-load current I_o now will be not in phase with flux, ϕ as the core losses will now be considered.

  The primary no-load current (I_o), along with the production of main flux (ϕ), is also responsible for the core losses. So the no-load primary input current is not wholly reactive (*i.e.*, not at 90° behind V_1) but lags it by an angle ϕ_o (< 90°).

  Hence the primary no load current (I_o) has two components:

  (*i*) One is in phase with V_1. (Strictly speaking, this component is in phase with E_1. As, when the resistance and rectance of primary side are not consider and also the losses are not taken into account, then $V_1 = -E_1$). This is known as active or working or iron loss component I_ω because it mainly supplies the iron loss (and plus a small quantity of primary copper loss). The angle ϕ_o is with vertical component of I_o, hence

$$I_\omega = I_o \cos \phi_o \qquad \ldots(8.17)$$

  (*ii*) The other component is in quadrature with V_1 (or E_1) (in phase with the main flux ϕ). It is known as magnetizing component I_μ because its function is to sustain the alternating flux in the core. It is watt less.

$$I_\mu = I_o \sin \phi_o \quad \text{...(8.18)}$$

Hence no-load primary current will now be

$$\vec{I_o} = \vec{I_\mu} + \vec{I_\omega} \quad \text{...(8.19)}$$

or

$$I_o = \sqrt{I_\mu^2 + I_\omega^2} \quad \text{...(8.20)}$$

Without core loss and without winding resistances the phase diagram will be as shown in Fig. 8.12 (*a*) and with winding resistances and core loss the phasor diagram will be as shown in Fig. 8.12 (*b*).

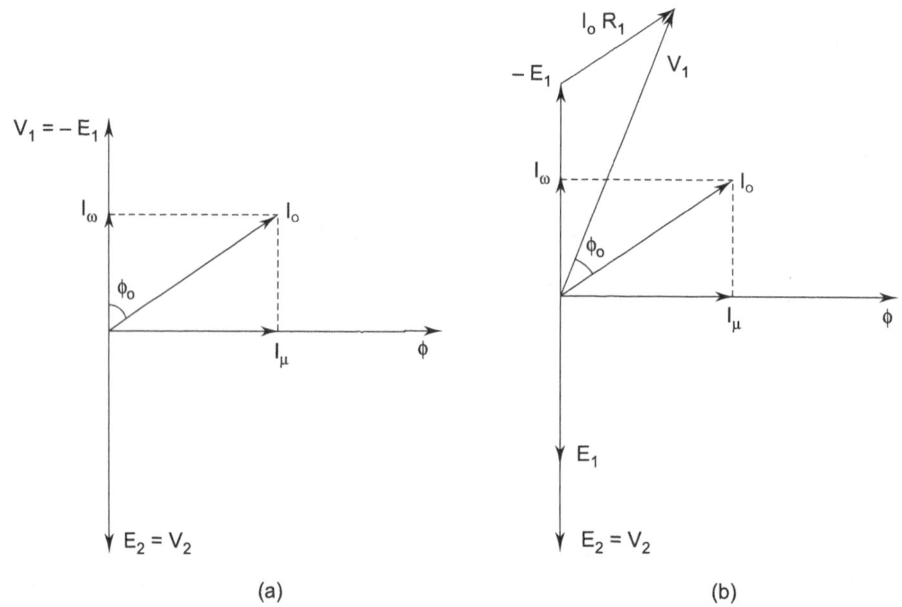

Fig. 8.12

The following points should be noted carefully:

- The no-load primary current I_o is very small as compared to the full-load primary current, I_1. It is about 2 to 5% of the full-load current.
- As I_o is very small, the no-load primary copper loss is negligibly small which means that no-load primary input is practically responsible for the iron loss and flux generation in the transformer.
- Again, the phasor (I_oR_1) is a voltage drop across resistance; hence the current and the voltage related to this phasor will be in phase. That is the reason the phasor (I_oR_1) is drawn always parallel to the current phasor (I_o) and in the same direction as that of phasor (I_o).

8.6.4 Case-IV

Next consider a transformer at **no-load** in which

- Winding resistances of both windings **included**.
- Core loss **included**.
- Leakage reactance **included**.
- Secondary side **opens.**

The circuit diagram is shown in Fig. 8.13. The equations for primary and secondary sides can be written as follow:

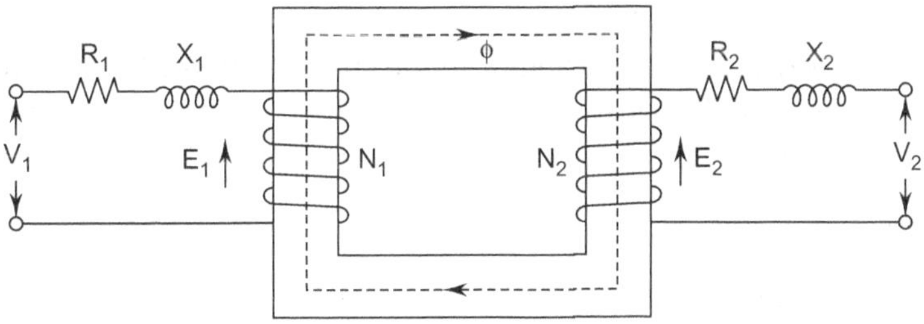

Fig. 8.13

- The secondary current is zero, as secondary side is open circuited

$$I_2 = 0 \qquad \text{[Eqn. (8.13)]}$$

- E_1, E_2 lags ϕ by 90°
- As secondary is open circuited,

$$\overrightarrow{E_2} = \overrightarrow{V_2} \qquad \text{[Eqn. (8.15)]}$$

- As the resistance and inductive reactance of the primary side are included, hence primary side equation will become

$$V_1 = R_1 I_o + jX_1 I_o + (-E_1) \qquad \text{[Eqn. (8.16)]}$$

- No-load current I_o

$$\overrightarrow{I_o} = \overrightarrow{I_\mu} + \overrightarrow{I_\omega} \qquad \text{[Eqn. (8.19)]}$$

Hence, the phasor diagram will be as shown in Fig. 8.14.

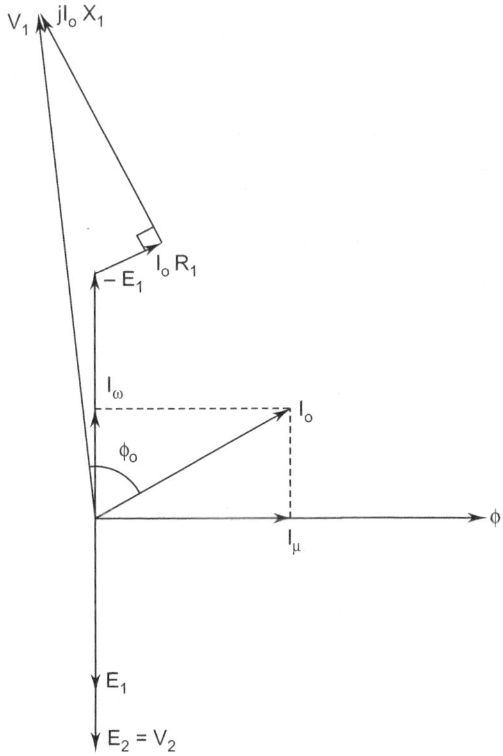

Fig. 8.14

The following points should be noted carefully:

- The phasor (I_oR_1) is drawn always parallel to the current phasor (I_o) and in the same direction as that of phasor (I_o).
- The phasor (I_oX_1) is drawn perpendicular to the current phasor (I_o) and in the anticlockwise direction at an angle of 90° because (I_oX_1) is a voltage drop across a pure inductor. Hence the phasor (I_oX_1) will always be drawn perpendicular to the primary current phasor I_o.

8.6.5 Transformer on Load

Let us consider a transformer on load in which

- **No** winding resistance.
- **No** leakage reactance.
- **No** core loss.
- Secondary side is **loaded**.

The circuit diagram is shown in Fig. 8.15.

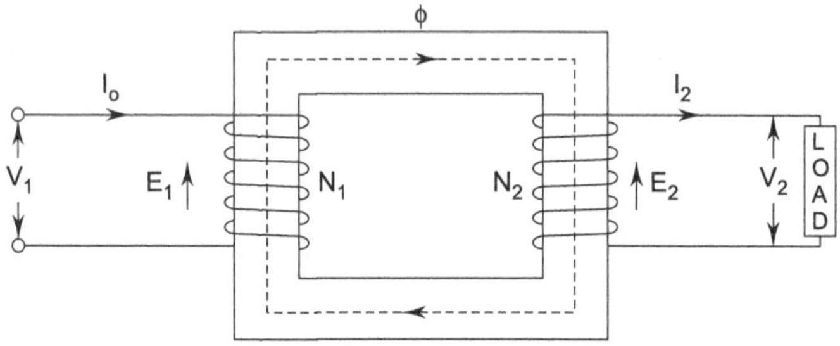

Fig. 8.15

- When the secondary side was open circuited, $I_2 = 0$.
- Now, the secondary side is loaded. So the secondary current I_2 is set up.
- The magnitude and phase of I_2 with respect to V_2 is determined by the ***characteristics of the load.***

    If the Load is ***resistive,*** current I_2 is ***in phase*** with V_2.

    If the Load is ***inductive,*** current I_2 ***lags*** behind V_2.

    If the load is ***capacitive,*** current I_2 ***leads*** the voltage V_2.

- Now, due to this secondary current I_2, an m.m.f (N_2I_2) will be produced in the secondary winding.
- Due to this m.m.f (N_2I_2), a flux, ϕ_2 is induced in the core shown in Fig. 8.16.

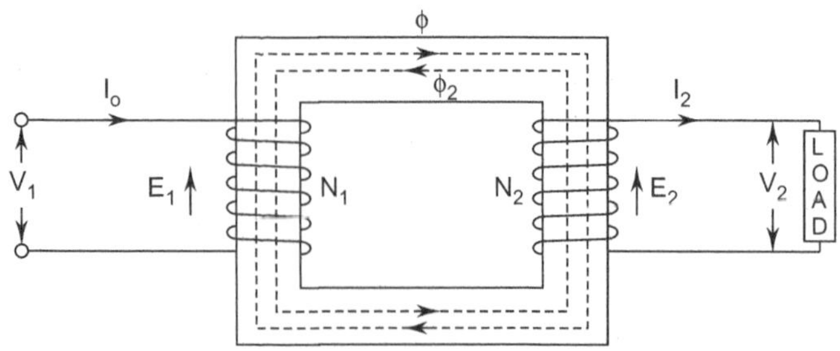

Fig. 8.16

- Now, according to Lenz's law the direction of this flux ϕ_2 is such that it opposes the main flux ϕ.
- This secondary flux ϕ_2 weakens the main flux ϕ.
- If ϕ reduces then E_1 and V_1 will reduce.
- Since the applied voltage V_1 is constant, so main flux ϕ has to be remain constant.
- So the m.m.f (N_2I_2) (and hence ϕ_2) have to be compensate from somewhere else.
- For compensating m.m.f (N_2I_2), we require extra m.m.f from primary side and hence an extra current from the primary side is to be drawn.
- This current is denoted as I_2'. This current I_2' will produce a flux ϕ_2' (Fig. 8.17), which is equal and opposite to the flux ϕ_2 and there for cancel out that flux ϕ_2.

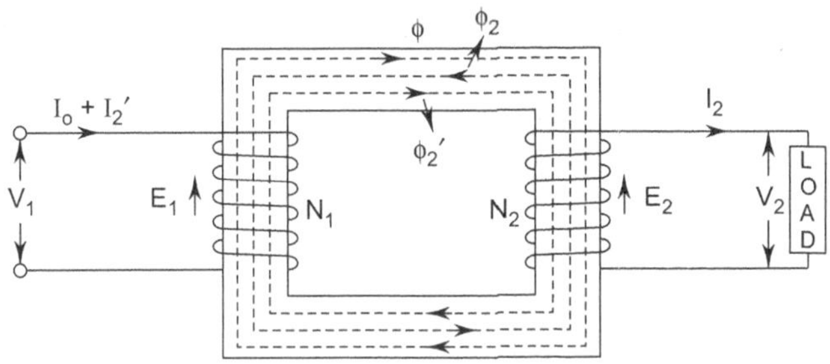

Fig. 8.17

- Hence the overall flux in the core will be the main flux ϕ (Fig. 8.18).

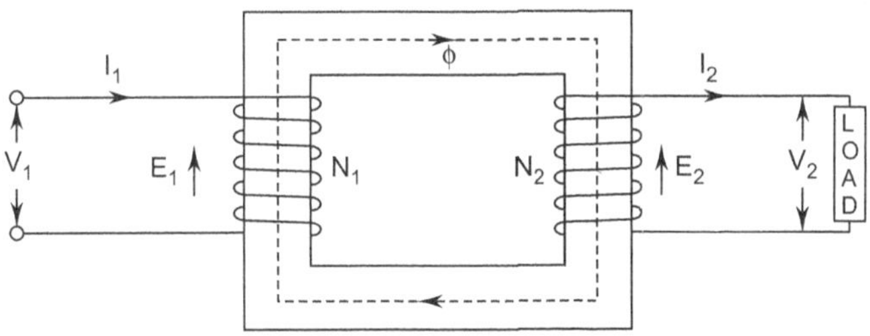

Fig. 8.18

- It means that when secondary side is loaded, then the primary current should have two components:
  * I_o, the no-load or exciting current and
  * I_2', ***load component*** of input current which compensates the effects of load current I_2 on primary side. ***(As Load component I_2' will produce a flux ϕ_2', which is equal and opposite to the flux ϕ_2 and there for cancel out that flux ϕ_2, hence the magnitude of I_2' is equal and opposite to that of I_2.)***

i.e.,
$$\vec{I}_1 = \vec{I}_o + \vec{I}_2' \qquad \text{...(8.21)}$$

8.6.6 Case-V

Let us consider a transformer **on load** in which

- **No** winding resistance.
- **No** leakage reactance.

- **No** core loss.
- Secondary side is **loaded with**
  - (*a*) **Resistive load**
  - (*b*) **Inductive load**
  - (*c*) **Capacitive load**

The circuit diagram is shown in Fig. 8.19.

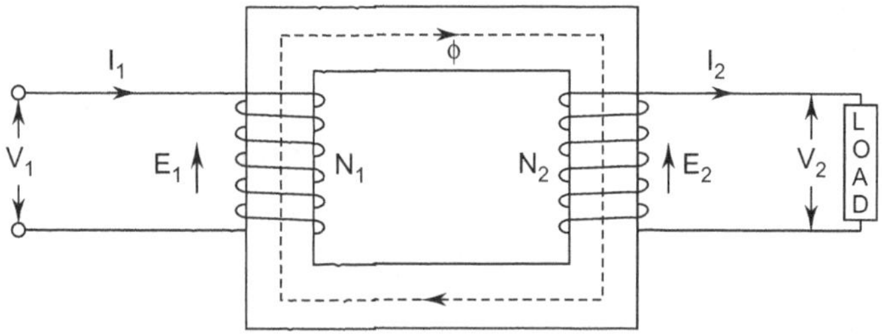

Fig. 8.19

The equations for primary and secondary sides can be written as follow:

- I_2' is equal and opposite to that of I_2.
- E_1, E_2 lags ϕ by 90°.
- As secondary side is loaded, but resistance and inductive reactance of the secondary side are excluded, hence

$$\vec{E_2} = \vec{V_2} \quad \text{[Eqn. (8.15)]}$$

- Primary side equation will be

$$V_1 = (-E_1) \quad \text{[Eqn. (8.16)]}$$

- The primary on load current will be

$$I_1 = I_o + I_2' \quad \text{[Eqn. (8.21)]}$$

Hence the phasor diagram for different load will be as shown in Fig. 8.20.

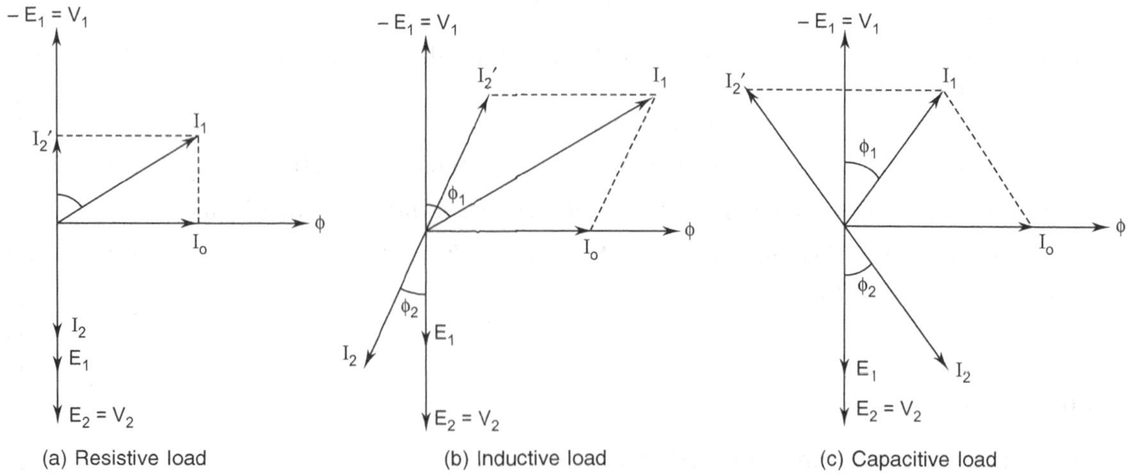

Fig. 8.20

8.6.7 Case-VI

Let us consider a transformer in which

- Winding resistances of both windings **included**.
- Core loss **included**.
- Leakage reactance **included**.
- Secondary side is **loaded with**
  - (*a*) **Resistive load**
  - (*b*) **Inductive load**
  - (*c*) **Capacitive load**

The circuit diagram is shown in Fig. 8.21.

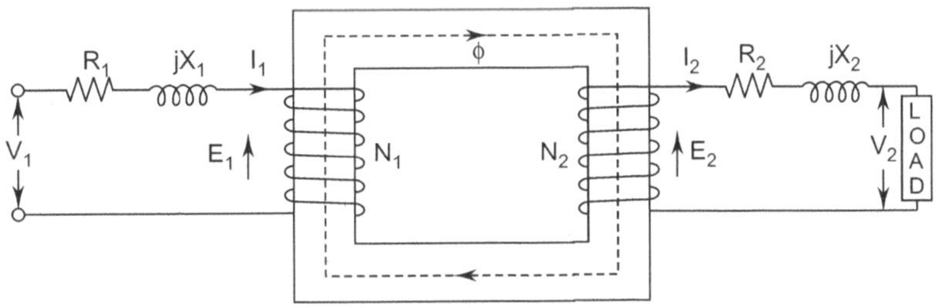

Fig. 8.21

The equations for primary and secondary sides can be written as follow:

- I_2' is equal and opposite to that of I_2.
- E_1, E_2 lags ϕ by 90°.
- As secondary side is loaded, and resistance and inductive reactance of the secondary side are also included, hence

$$E_2 = I_2R_2 + jI_2X_2 + V_2 \qquad \text{[Eqn. (8.15)]}$$

- The primary on load current will be

$$I_1 = I_o + I_2' \qquad \text{[Eqn. (8.21)]}$$

- As the resistance and inductive reactance of the primary side are included, hence primary side equation will become

$$V_1 = R_1I_o + jX_1I_o + (-E_1) \qquad \text{[Eqn. (8.16)]}$$

- No-load current I_o

$$\vec{I_o} = \vec{I_\mu} + \vec{I_\omega} \qquad \text{[Eqn. (8.19)]}$$

For drawing the final phasor diagram, one can follow the following steps.

Step 1. Draw the ϕ, E_1, E_2 and $(-E_1)$ first (Fig. 8.22).

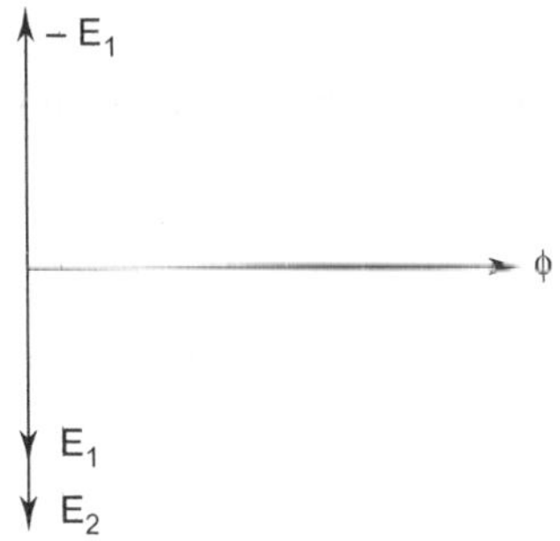

Fig. 8.22

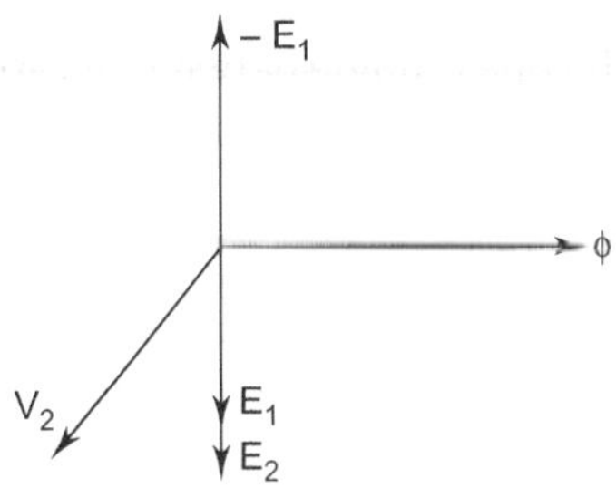

Fig. 8.23

Step 2. Draw V_2 behind E_2 and consider it as reference phasor for secondary side (Fig. 8.23).

Step 3. According to the type of Load, draw the I_2 phasor (Fig. 8.24).

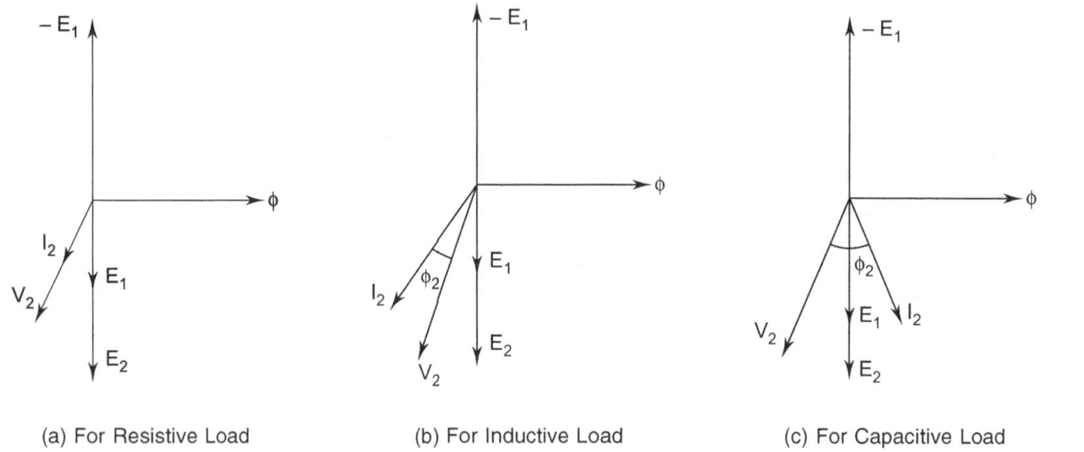

Fig. 8.24

Step 4. Next draw phasors for secondary side equation* (Fig. 8.25), *i.e.*,

$$E_2 = I_2R_2 + jI_2X_2 + V_2$$

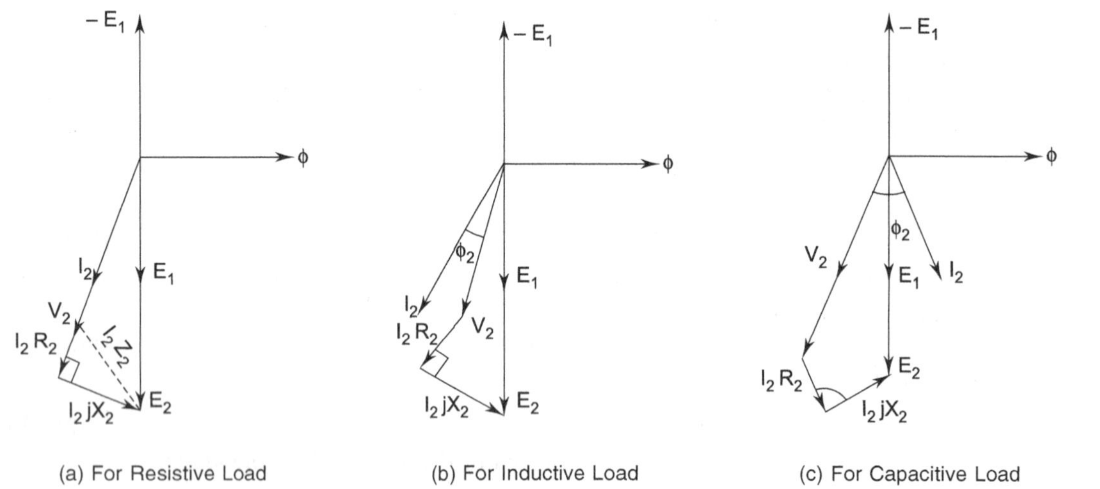

Fig. 8.25

Show the phase angle between V_2 and I_2. With this secondary side phasor diagram is complete.

Step 5. Draw the no load current phasor I_o, which is almost constant (Fig. 8.26).

---

*Note that the phasor I_2R_2 is drawn always parallel and in the same direction to that I_2 and the phasor jI_2X_2 is always drawn perpendicular to the I_2 phasor.

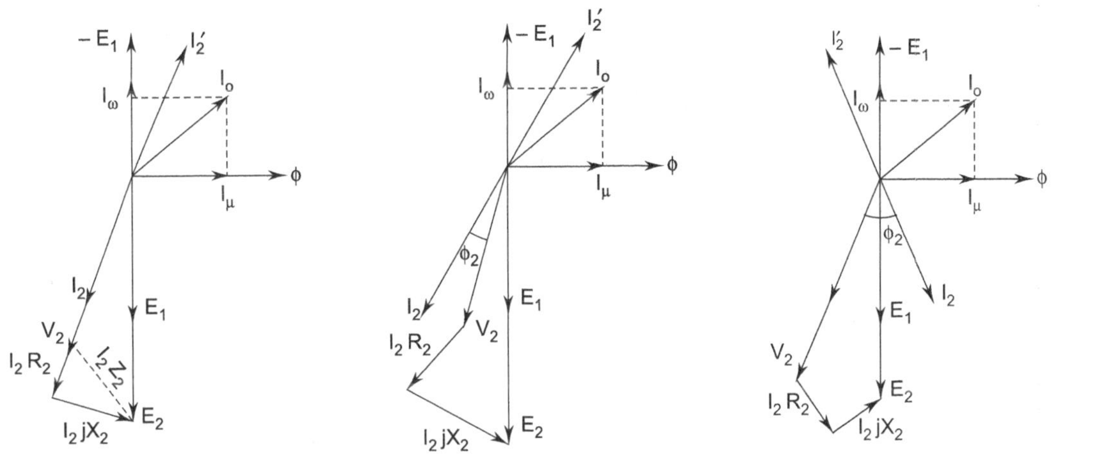

(a) For Resistive Load (b) For Inductive Load (c) For Capacitive Load

Fig. 8.26

Step 6. Draw the I_2' equal and opposite to I_2 (Fig. 8.26).

Step 7. Take the resultant of I_o and I_2' and draw the I_1 phasor (Fig. 8.27).

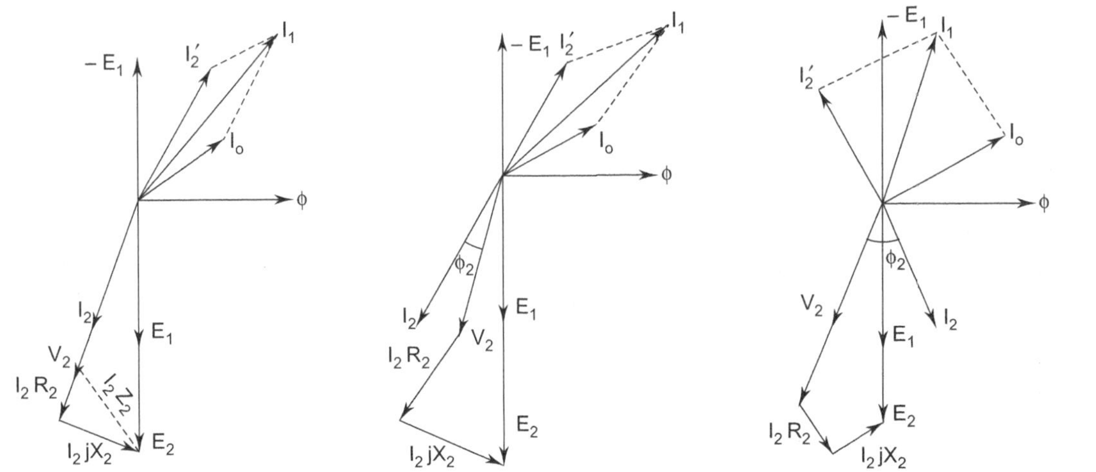

(a) For Resistive Load (b) For Inductive Load (c) For Capacitive Load

Fig. 8.27

Step 8. According to I_1 phasor, draw the primary side equation *i.e.,*

$$V_1 = R_1I_1 + jX_1I_1 + (-E_1)$$

Hence the phasor diagram for different load will be

(*a*) For resistive load (Fig. 8.28) secondary current I_2 is ***in phase*** with secondary terminal voltage V_2.

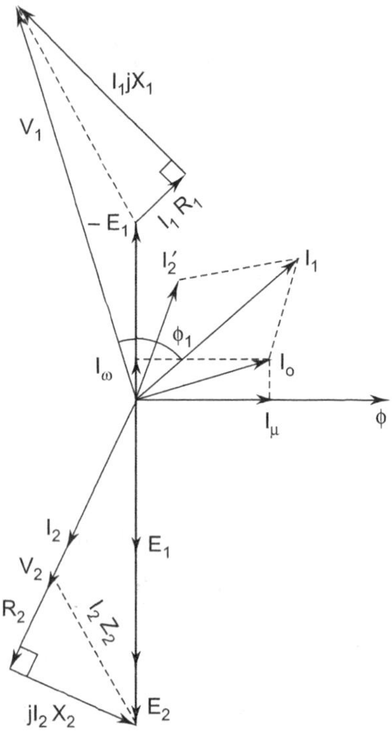

Fig. 8.28

(*b*) For inductive load (Fig. 8.29) secondary current I_2 ***lags*** behind the secondary terminal voltage V_2 by an angle ϕ_2.

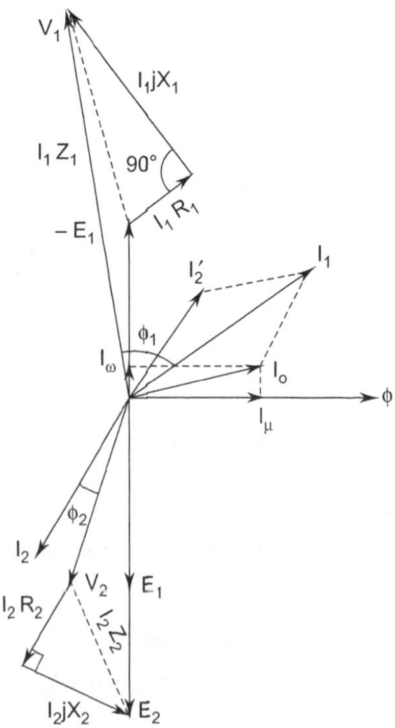

Fig. 8.29

(*c*) For capacitive load (Fig. 8.30) secondary current I_2 is ***leads*** the secondary terminal voltage V_2 by an angle ϕ_2.

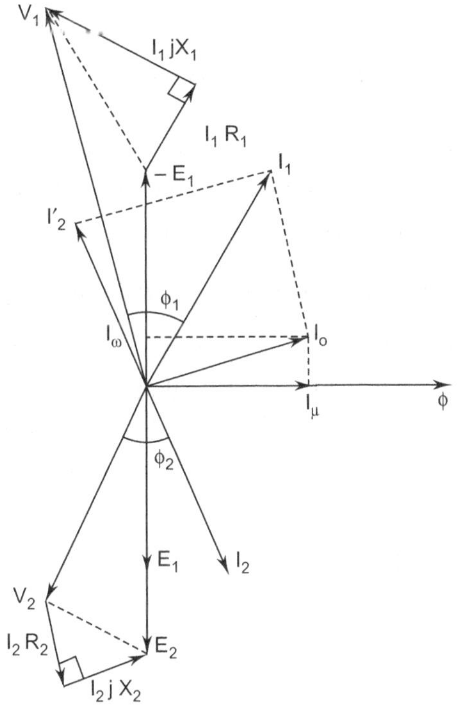

Fig. 8.30

8.7 EQUIVALENT CIRCUIT

The transformer can be resolved into an equivalent circuit in which the resistances and leakage reactances of both side of the transformer are imagined to be external to the winding. And rest whatever is left will have only function then is to transform the voltage.

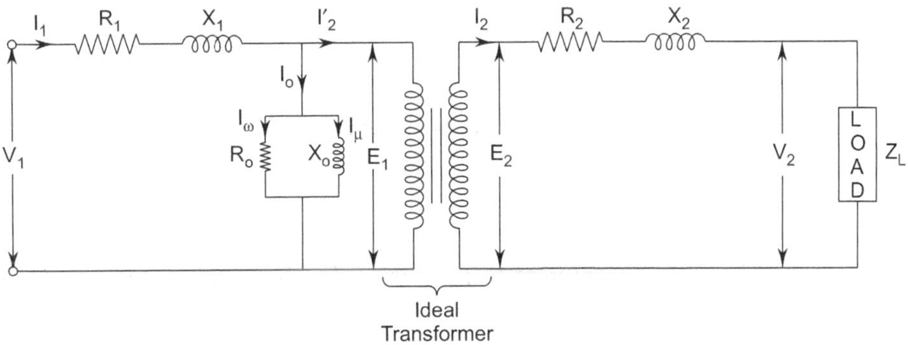

Fig. 8.31

The equivalent circuit shown in Fig. 8.31 is drawn on the basis of the following equations:

$$E_2 = I_2R_2 + jI_2X_2 + V_2 \qquad \text{[Eqn. (8.15)]}$$

$$V_1 = R_1I_o + jX_1I_o + (-E_1) \qquad \text{[Eqn. (8.16)]}$$

$$\vec{I_o} = \vec{I_\mu} + \vec{I_\omega} \quad \text{[Eqn. (8.19)]}$$

$$\vec{I_1} = \vec{I_o} + \vec{I_2'} \quad \text{[Eqn. (8.21)]}$$

For the above equivalent circuit, the calculations will become complicated for finding various parameters of the transformer.

To make transformer calculations simpler, it is preferable to transfer all the parameters (say resistance, reactance, voltage, current, etc.) from one side to other.

In that case, we would have to work in one winding only, which is more convenient.

8.7.1 Transfer of Secondary Side Parameters to Primary Side

Let us transfer secondary side parameters to primary side.

During this transformation one should remember that the effect of the parameters of secondary side should remain same, when they are transferred to primary side.

As the no-load current I_o is 3 to 5% of full load primary current I_1, hence it can be neglected as compared to I_2'. In that case

$$I_1 \approx I_2' \quad \text{...(8.22)}$$

In general

$$\frac{N_2}{N_1} = \frac{I_1}{I_2} \quad \text{...(8.23)}$$

From equation (8.22)

$$N_1 I_2' = N_2 I_2 \quad \text{...(8.24)}$$

or

$$I_2' = \frac{N_2}{N_1} . I_2 \quad \text{...(8.25)}$$

or

$$\boxed{I_2' = K . I_2} \quad \text{...(8.26)}$$

Let the resistance of secondary side, which is shifted to primary side, is referred as R_2'

The copper loss in the secondary is

$$P_{2,Cu} = I_2^2 R_2 \quad \text{...(8.27)}$$

The current in the primary side is I_1. Hence if R_2' is the equivalent resistance in primary which would have caused the same loss as R_2 in secondary then the copper loss in the primary winding is

$$P_{2,Cu}' = I_1^2 R_2' \quad \text{...(8.28)}$$

For shifting secondary side parameters to primary side, these two copper losses should be equal

$$I_1^2 R_2' = I_2^2 R_2 \quad \text{...(8.29)}$$

or

$$R_2' = \frac{I_2^2}{I_1^2} . R_2 = \left(\frac{I_2}{I_1}\right)^2 R_2 \quad \text{...(8.30)}$$

or $$\boxed{R_2' = \frac{R_2}{K^2}} \qquad \text{...(8.31)}$$

Similarly, the secondary leakage reactance with reference to primary is denoted by X_2' and is given by

$$\boxed{X_2' = \frac{X_2}{K^2}} \qquad \text{...(8.32)}$$

The secondary induced e.m.f. with reference to primary, is denoted by E_2' is

$$\frac{E_2'}{E_2} = \frac{N_1}{N_2} \qquad \text{...(8.33)}$$

or $$E_2' = \frac{N_1}{N_2} E_2 \qquad \text{...(8.34)}$$

or $$\boxed{E_2' = \frac{E_2}{K}} \qquad \text{...(8.35)}$$

Similarly secondary terminal voltage V_2' will be

$$\boxed{V_2' = \frac{V_2}{K}} \qquad \text{...(8.36)}$$

The load impedance with reference to primary side is denoted by Z_L' and is given by

$$\boxed{Z_L' = \frac{Z_L}{K^2}} \qquad \text{...(8.37)}$$

Further simplification of the equivalent circuit can be made by transferring the exciting circuit to the left as shown in Fig. 8.32.

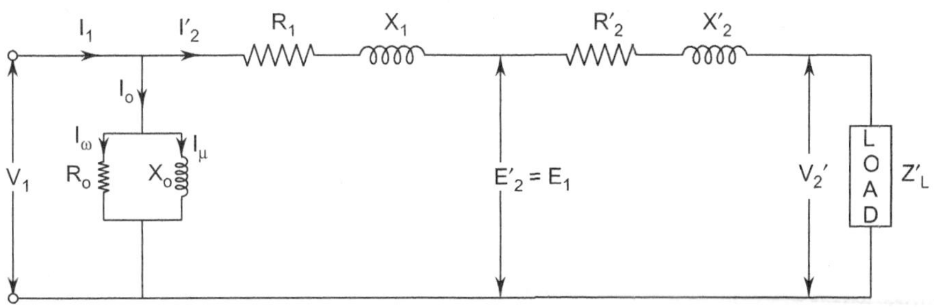

Fig. 8.32

And then combining the series resistances and reactances separately (Fig. 8.33).

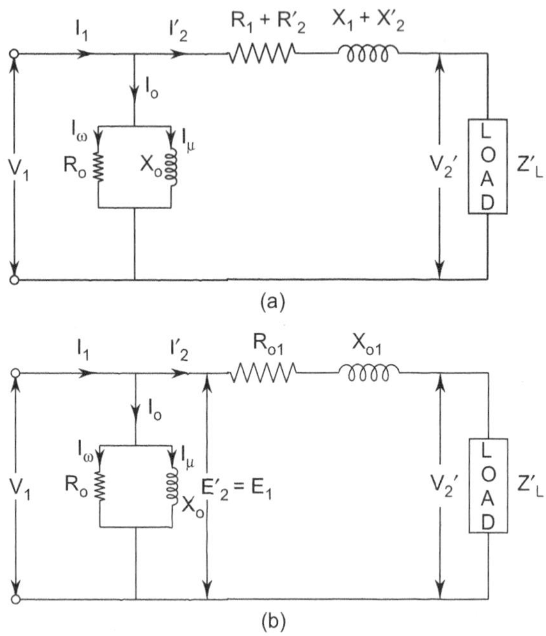

Fig. 8.33

where

$$\boxed{R_{o1} = R_1 + R_2' = R_1 + \frac{R_2}{K^2}} \quad ...(8.38)$$

= Equivalent resistance of the transformer with reference to primary side.

$$\boxed{X_{o1} = X_1 + X_2' = X_1 + \frac{X_2}{K^2}} \quad ...(8.39)$$

= Equivalent reactance of the transformer with reference to primary side.

$$\boxed{Z_{o1} = R_{o1} + jX_{o1}} \quad ...(8.40)$$

= Equivalent impedance of the transformer with reference to primary side.

This is all about change in parameters of secondary side, when they are shifted to primary side.

8.7.2 Transfer of Primary Side Parameters to Secondary Side

The primary side parameters can be transferred to secondary side in the same manner, as the secondary side parameters are shifted to primary side.

One should remember that while shifting the parameters from one side to other, the effect of the parameters must remain same.

The equivalent circuit with reference to secondary side is shown in Fig. 8.34.

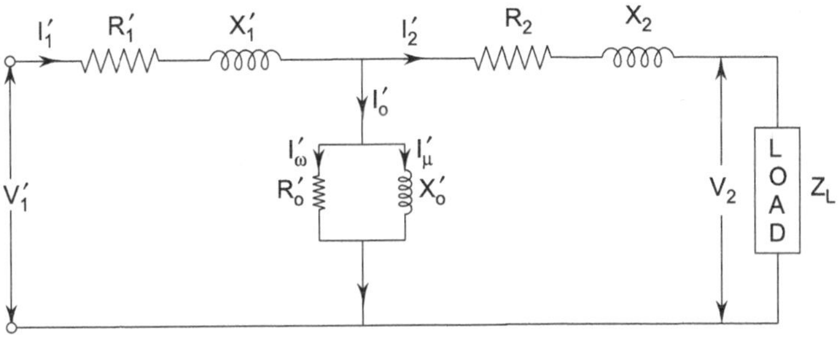

Fig. 8.34

After approximation, *i.e.*, after shifting the tank circuit in the begining, the equivalent circuit of a transformer with reference to secondary side will be similar to the circuit shown in Fig. 8.35.

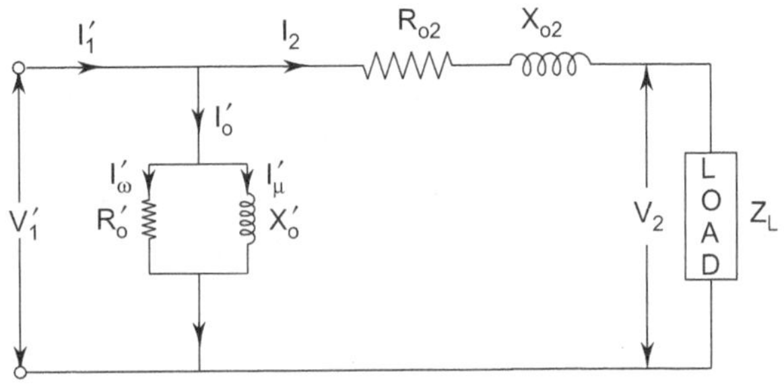

Fig. 8.35

The expression for change in parameters of primary side are as follows:

Primary side resistance with reference to secondary side will be

$$R'_1 = K^2 R_1 \quad \ldots(8.41)$$

Similarly the primary side reactance with reference to secondary side will be

$$X'_1 = K^2 X_1 \quad \ldots(8.42)$$

Equivalent resistance referred to secondary side will be

$$R_{o2} = R_2 + R'_1 = R_2 + K^2 R_1 \quad \ldots(8.43)$$

Equivalent reactance referred to secondary side

$$X_{o2} = X_2 + X'_1 = X_2 + K^2 X_1 \quad \ldots(8.44)$$

The tank circuit components with reference to secondary side are

$$R'_o = K^2 R_o \quad \ldots(8.45)$$

$$X'_o = K^2 X_o \quad \ldots(8.46)$$

$$I'_\mu = \frac{I_\mu}{K} \quad \ldots(8.47)$$

$$I'_\omega = \frac{I_\omega}{K} \quad \ldots(8.48)$$

and

$$I'_o = \frac{I_o}{K} \quad \ldots(8.49)$$

The primary current with reference to secondary side will be

$$I'_1 = \frac{I_1}{K} \qquad \text{...(8.50)}$$

The primary voltage with reference to secondary side will be

$$V'_1 = KV_1 \qquad \text{...(8.51)}$$

8.8 LOSSES

There are mainly two types of losses occur in a single phase transformer:

- Core Loss
- Copper Losses.

8.8.1 Core Loss

- As the nomenclature indicates, these losses occur in core which is, mostly made up of iron. That is why, core losses are also known as iron loss.
- Iron losses are constant losses because these are practically independent of load. They mainly depend upon the supply frequency and supply voltage.
- Core loss is further classified into two types:

(*a*) Eddy Current Loss

(*b*) Hysteresis Loss.

(*a*) **Eddy Current Loss**

Eddy currents are circulating currents which are induced in the core material. These currents are created by the difference in potential existing throughout the body of the material because of the changing flux. Fig. 8.36 shows the nature of these circulating eddy currents.

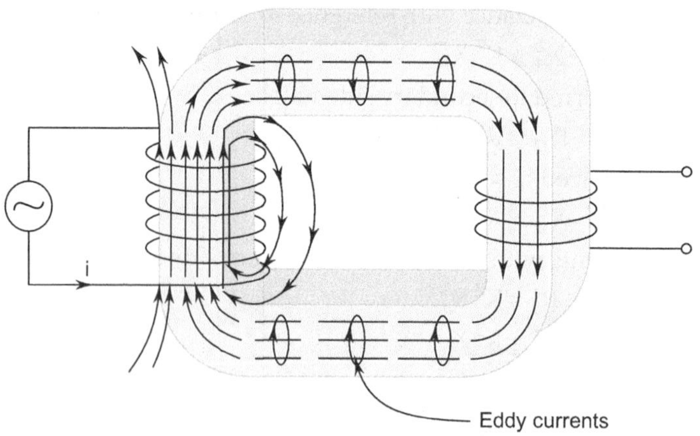

Fig. 8.36

If the magnetic circuit is composed of solid iron, resulting power loss is appreciable because the circulating current encounters little resistance. To increase the resistance encountered by these eddy currents, core is sliced to very thin laminations, insulated by varnish or oxide and then stacked together.

The eddy current loss is given by expression

$$P_e = K'_e f^2 B_m^2 \tau^2 v \qquad \ldots(8.52)$$

where, τ is thickness of laminations,

v is volume of core.

K_e is eddy current constant.

If the core material and thickness are constants, the eddy current loss can be written as

$$\boxed{P_e = K_e f^2 B_m^2} \qquad \ldots(8.53)$$

(*b*) **Hysteresis Loss**

The flux in a magnetic core of transformer is alternating in nature. So it magnetizes and demagnetizes the core in each complete cycle. Due to this, there is a loss in energy in magnetic domains, which is called hysteresis loss.

This can be reduced by using CRGO core. (Cold Rolled Grain Oriented Silicon Steel).

The hysteresis loss is given by the expression

$$\boxed{P_h = K_h f B_m^n} \qquad \ldots(8.54)$$

where, B_m is maximum flux density,

f is frequency of supply,

K_h is hysteresis loss constant,

n is called the Steinmetz exponent and has range 1.5–2.5 depending on the material of core.

8.8.2 Copper Loss

The primary and secondary windings are mostly made up of copper and heat loss occur in these winding are termed as Copper losses. Copper losses are also known as ohmic/windage/I^2R losses.

These losses are given by

$$\boxed{P_{Cu} = I_1^2 R_1 + I_2^2 R_2} \qquad \ldots(8.55)$$

or if the secondary parameters are shifted to primary side, then

$$\boxed{P_{Cu} = I_1^2 R_{o1}} \qquad \ldots(8.56)$$

And if the primary parameters are shifted to secondary side, then

$$\boxed{P_{Cu} = I_2^2 R_{o2}} \qquad \ldots(8.57)$$

Note that the ***transformer is always rated in kVA*** because it is seen that iron (or core) losses depends upon supply voltage while the copper losses depend on the current. The losses do not dependent on phase angle between voltage and current.

Hence the rating of transformer is expressed as a kVA.

8.9 EFFICIENCY

Efficiency of a transformer is defined as the ratio of output of transformer to its input.

$$\eta = \frac{\text{Power output}}{\text{Power input}} \quad ...(8.58)$$

or

$$\eta = \frac{V_2 I_2 \cos\phi_2}{V_1 I_1 \cos\phi_1} \quad ...(8.59)$$

As we know that

$$\text{Power Input} = \text{Power Output} + \text{Losses} \quad ...(8.60)$$

Therefore,

$$V_1 I_1 \cos\phi_1 = V_2 I_2 \cos\phi_2 + P_i + P_{Cu} \quad ...(8.61)$$

From equation (8.61), equation (8.59) will become

$$\eta = \frac{V_2 I_2 \cos\phi_2}{V_2 I_2 \cos\phi_2 + P_i + P_{Cu}} \quad ...(8.62)$$

Equation (8.62) represents general expression for finding the efficiency of a transformer.

The main effect of any type of change in load, mainly affects the load current (or secondary current I_2). Actually this secondary current I_2 is directly proportional to the fraction of load.

If the load becomes half means load current becomes half. If the load becomes x fraction of rated load, then it means the new current becomes x times of rated current.

This dependancy of secondary current (I_2) on fraction of load (x) also affect copper losses. Copper losses are referred to the secondary side are given by

$$P_{Cu} = I_2^2 R_{o2} \quad \text{[Eqn. (8.57)]}$$

That is, copper losses are directly proportional to the square of the secondary current I_2. Hence the copper losses will vary as the x^2 fraction of rated load.

$$P_{Cu} \propto I_2^2 \quad ...(8.63)$$

So the efficiency at x fraction of load is given by

$$\boxed{\eta = \frac{x V_2 I_2 \cos\phi_2}{x V_2 I_2 \cos\phi_2 + P_i + x^2 P_{Cu}}} \quad ...(8.64)$$

Equation (8.64) represents the general expression for finding the efficiency of a transformer at any fraction (x) of full load.

8.9.1 Condition for Maximum Efficiency

The transformer efficiency will be maximum when,

$$\frac{d\eta}{dx} = 0 \quad ...[8.65(a)]$$

or $$\frac{d}{dx}\left(\frac{xV_2I_2\cos\phi_2}{xV_2I_2\cos\phi_2+P_i+x^2P_{Cu}}\right)=0 \qquad \ldots[8.65(b)]$$

or $$\frac{x(V_2I_2\cos\phi_2)(V_2I_2\cos\phi_2+2xP_{Cu})-(V_2I_2\cos\phi_2)(xV_2I_2\cos\phi_2+P_i+x^2P_{Cu})}{(xV_2I_2\cos\phi_2+P_i+x^2P_{Cu})^2}=0 \qquad \ldots[8.65(c)]$$

or $$\boxed{P_i = x^2P_{Cu}} \qquad \ldots(8.66)$$

i.e., maximum efficiency will occur when variable losses (x^2 P_{Cu}) will becomes equal to constant losses (P_i). On full-load, x = 1, hence from equation (8.66) it is clear that maximum efficiency at full-load will occur, when full-load variable losses (P_{Cu}) will become equal to constant losses (P_i).

The kVA for maximum efficiency are given by

$$(kVA)_{max.\eta} = (kVA)_{rated} \times x \qquad \ldots(8.67)$$

or $$(kVA)_{max.\eta} = (kVA)_{rated}\sqrt{\frac{P_i}{P'_{Cu\ rated}}} \quad \text{[From equation (8.66)]} \qquad \ldots(8.68)$$

8.9.2 All-Day Efficiency

The all-day efficiency of a single phase transformer is given by

$$\boxed{\eta_{all\ day} = \left.\frac{\text{Output in kWh}}{\text{Input in kWh}}\right|_{\text{For 24 hours}}} \qquad \ldots(8.69)$$

To find this all day efficiency (or as it is also known as Energy Efficiency) we have to know the load cycle on transformer *i.e.*, how much and how long the transformer is loaded during 24 hours.

8.10 REGULATION

The voltage regulation of a transformer is defined as the ratio of difference of secondary terminal voltage from no-load to full load ($E_2 - V_2$), to the no load secondary terminal voltage (E_2).

$$\%\ \text{Voltage Regulation} = \frac{E_2 - V_2}{E_2} \times 100 \qquad \ldots(8.70)$$

Consider that the primary side parameters are shifted to secondary side (Fig. 8.37). In that case the equivalent resistances and reactances with reference to the secondary side will be

$$R_{o2} = R_2 + R'_1 = R_2 + K^2R_1 \qquad \text{[Eqn. (8.43)]}$$

and $$X_{o2} = X_2 + X'_1 = X_2 + K^2X_1 \qquad \text{[Eqn. (8.44)]}$$

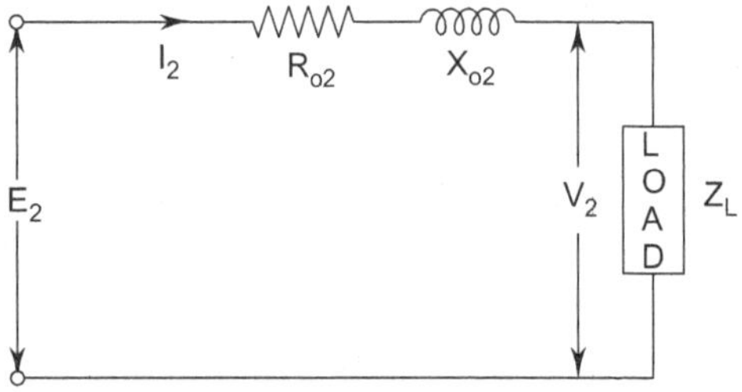

Fig. 8.37

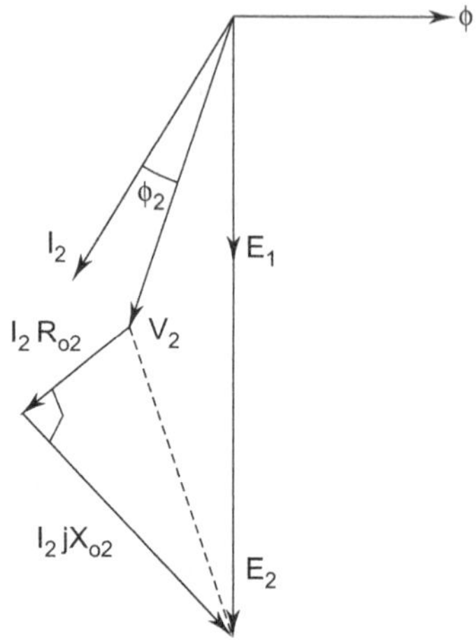

Fig. 8.38

Now the secondary side voltage equation can be written as

$$E_2 = I_2R_{o2} + I_2jX_{o2} + V_2 \quad \text{...(8.71)}$$

The above equation is represented in phasor diagram shown in Fig. 8.38. Equation (8.71) can also be written as

$$E_2 - V_2 = I_2 (R_{o2} + jX_{o2}) \quad \text{...(8.72)}$$

or

$$E_2 - V_2 = I_2 (R_{o2} \cos \phi_2 + jX_{o2} \sin \phi_2) \quad \text{...(8.73)}$$

or

$$E_2 - V_2 = I_2R_{o2} \cos \phi_2 + I_2X_{o2} \sin \phi_2 \quad \text{...(8.74)}$$

where

$$\phi_1 = \phi_2 = \phi \quad \text{(approx.)} \quad \text{...(8.75)}$$

Now, the voltage regulation expression in equation (8.70) will become

$$\boxed{\% \text{ V.R.} = \frac{I_2R_{o2} \cos \phi \pm I_2X_{o2} \sin \phi}{E_2} \times 100} \quad \text{...(8.76)}$$

Here Positive (+) sign is for inductive load (lagging) and
Negative (–) sign is for capacitive load (leading).

The above expression for voltage regulation can also be expressed as

$$\text{V.R.} = \frac{I_2R_{o2}}{E_2} \cos \phi \pm \frac{I_2X_{o2}}{E_2} \sin \phi \quad \text{...(8.77)}$$

or

$$\text{V.R.} = \epsilon_R \cos \phi \pm \epsilon_X \sin \phi \quad \text{...(8.78)}$$

where

$$\epsilon_R = \frac{I_2R_{o2}}{E_2} = \text{Per unit resistance drop} \quad \text{...(8.79)}$$

and

$$\epsilon_X = \frac{I_2X_{o2}}{E_2} = \text{Per unit reactance drop} \quad \text{...(8.80)}$$

When primary side parameters are provided, and we have to find the voltage regulation, in that case, the voltage regulation expression will become

$$\% \text{ V.R.} = \frac{I_1R_{o1} \cos \phi \pm I_1X_{o1} \sin \phi}{V_1} \times 100 \quad \text{...(8.81)}$$

or $$\% \text{ V.R.} = \left[\frac{I_1 R_{o2}}{V_2}\cos\phi \pm \frac{I_1 X_{o2}}{V_2}\sin\phi\right] \times 100 \qquad \text{...(8.82)}$$

or $$\% \text{ V.R.} = \left(E_R \cos\phi \pm E_X \sin\phi\right) \times 100 \qquad \text{...(8.83)}$$

where $$E_R = \frac{I_1 R_{o1}}{V_1} \times 100 = \text{Percentage resistance drop} \qquad \text{...(8.84)}$$

and $$E_X = \frac{I_1 X_{o1}}{V_1} \times 100 = \text{Percentage reactance drop} \qquad \text{...(8.85)}$$

[**Note.** *If not multiplied by 100, then E_R and E_X will denote per unit values.*]

8.11 TRANSFORMER TESTS

To determine performance and parameters of the equivalent circuit of the transformer such as core loss, copper loss, equivalent resistances, reactances etc., different tests are performed. Some of these tests are:

1. Open Circuit Test
2. Short Circuit Test
3. Polarity Test
4. Sumpner's Back-to-Back Test.

8.11.1 Open Circuit Test

Open circuit test is also known as no-load test. In this test::

- One winding of the transformer - usually high voltage winding, is left open. (Note that it is not necessary that high voltage winding is always left open).
- Other winding is connected to supply of normal voltage and frequency.
- A wattmeter (W), a voltmeter (V) and an ammeter (A) are connected on low voltage winding side (as shown in Fig. 8.39).

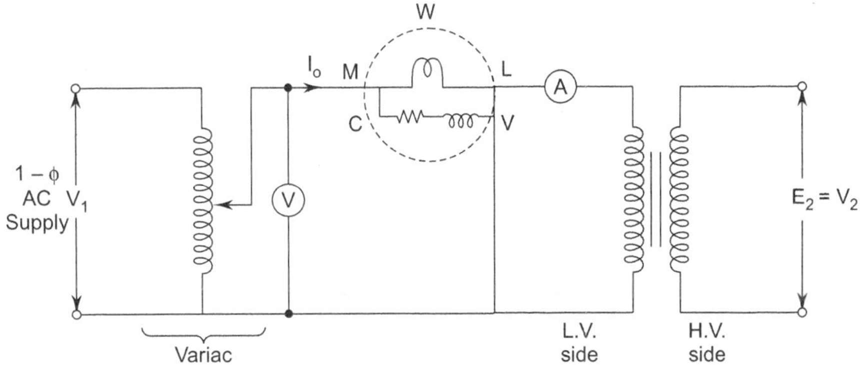

Fig. 8.39

- With normal voltage applied to primary, normal flux will setup in the core, hence normal iron losses will occur which are recorded by wattmeter 'W' *i.e.*,

  Iron Loss (P_i), $W_i = V_1 I_o \cos \phi_o$ ...(8.86)

where V_1 is primary terminal voltage measured by voltmeter 'V'.

I_o is primary no load current measured by ammeter 'A'.

- From equation (8.86) no load power factor can be calculated as

$$\cos\phi_o = \frac{W_i}{V_1 I_o} \quad \ldots(8.87)$$

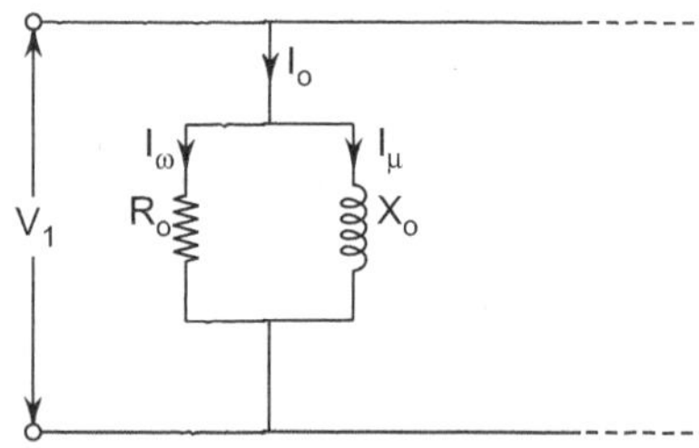

Fig. 8.40

- The tank circuit components (Fig. 8.40) can be calculated as

$$I_\omega = I_o \cos\phi_o \quad \ldots(8.88)$$

$$I_\mu = I_o \sin\phi_o \quad \ldots(8.89)$$

$$R_o = \frac{V_1}{I_\omega} \quad \ldots(8.90)$$

$$X_o = \frac{V_1}{I_\mu} \quad \ldots(8.91)$$

8.11.2 Short Circuit Test

While performing short circuit test on a single phase transformer,

- One winding of the transformer, mainly low voltage side, is short circuited (Fig. 8.41).

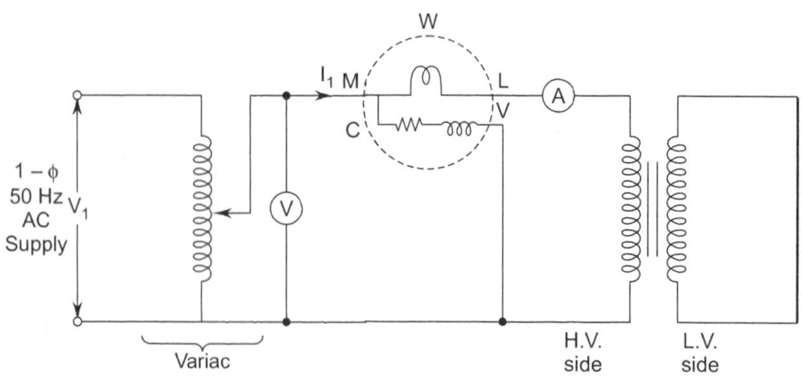

Fig. 8.41

- As the secondary side (LV) is short circuited, it may draw a very large current at rated voltage and transformer may burnt out. So this test is not performed at rated voltage.
- For getting full load or rated ohmic loss (or copper loss) this test is performed at rated current.

- Rated current may be drawn at only 2–10% of the rated voltage. So the reading of the voltmeter is 2 to 10% of rated voltage.
- The equivalent circuit of transformer is shown in Fig. 8.42.

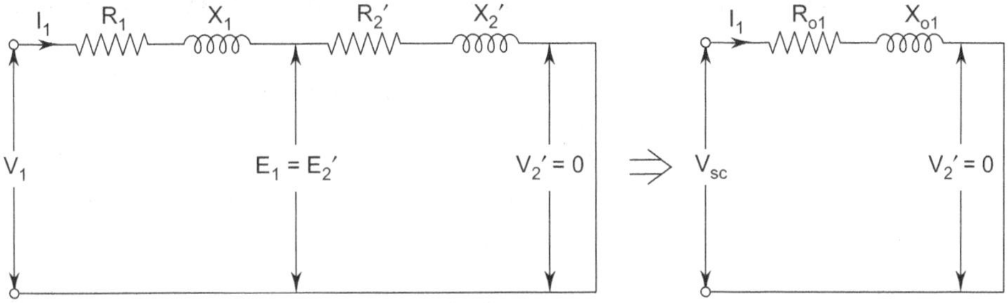

Fig. 8.42

Under short circuit conditions the wattmeter will show full-load copper loss given by

Copper Loss (P_c), $$W_c = I_1^2 R_{01} \quad \ldots(8.92)$$

where, I_1 is the reading of ammeter. Knowing the values of W_c and I_1 we can calculate the primary equivalent resistance as

$$R_{01} = \frac{W_c}{I_1^2} \quad \ldots(8.93)$$

From the voltmeter and ammeter readings the equivalent impedance will be

$$Z_{01} = \frac{V_{sc}}{I_1} \quad \ldots(8.94)$$

The equivalent reactance can be calculated as

$$Z_{01} = \sqrt{R_{01}^2 + X_{01}^2} \quad \ldots(8.95)$$

or $$X_{01} = \sqrt{Z_{01}^2 - R_{01}^2} \quad \ldots(8.96)$$

Important points regarding short circuit test

In case of short circuit test:

- In general, the test should be performed on high voltage side and low voltage side should be short-circuited.
- Low voltage side (LV) will have low voltage ratting but the current is high on LV side.
- High voltage side (HV) will have high voltage ratting but the current is low on HV side.
- As the short circuit test is performed at rated current, hence, when LV side is short circuited the test will be performed on HV side (which has low rated current as compare to LV side rated current) and that will provide safe operating conditions for the instrument.
- The short circuit test can also be performed on LV side and HV side can be short circuited.
- But now, if the test is performed on rated current of LV side, which is greated then the HV side rated current, there is a danger of burn out of HV side windings.

- Hence, it is always preferred and convenient that the short circuit test should always be performed on rated current of HV side (which is of low value as compared to LV side rated current).
- Now, when the short circuit test is performed on LV side and HV side is short circuited, even then the test should be performed on rated current of HV side.
- The short circuit test gives us copper losses. If the applied current is full-load rated current, then the copper losses so obtained will also be full-load copper loss. But when, the test is performed on LV side, and the current is applied according of rated current of HV side, then the copper losses so obtained are not equal to full-load or rated copper loss.
- In such a case, the rated copper losses, can be obtained as

$$P_{Cu(rated)} = \left(\frac{\text{LV side rated current}}{\text{HV side rated current}}\right)^2 \times P_{Cu,(obtained)}$$

8.11.3 Polarity Test

- Similar polarity ends of the two windings of a transformer are those ends that acquires simultaneously positive or negative polarity of e.m.f's induced in them.
- These are indicated by the dot convention.
- Usually the ends of the LV winding are labelled with a small letter of the alphabet and are suffixed 1 and 2 (for example a_1 and a_2).
- While the HV winding ends are labelled by the corresponding capital letters and are suffixed 1 and 2 (for example A_1 and A_2).

Method 1:

In determining the relative polarity of the two windings of a transformer, the two windings are connected in series across a voltmeter, while one of the windings is excited from a suitable voltage source as shown in Fig. 8.43.

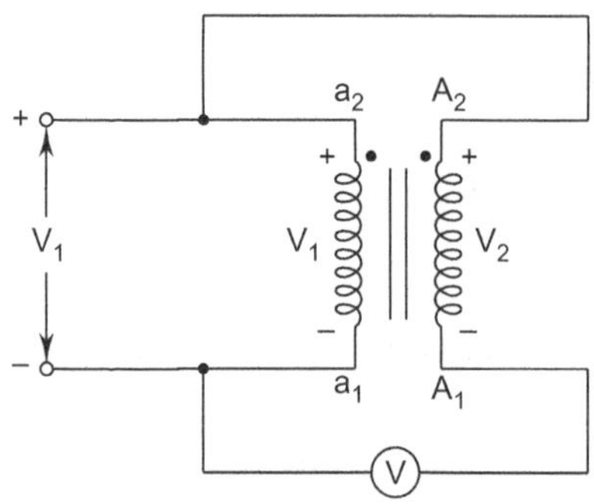

Fig. 8.43

If the polarities of the windings are as marked on the diagram, the voltmeter should read

$$V - V_1 \sim V_2 \qquad \text{...(8.97)}$$

If the voltmeter reads

$$V = V_1 + V_2 \qquad \text{...(8.98)}$$

then the polarity markings of one of the windings must be interchanged.

Method 2:

This method is used for field testing of a transformer.

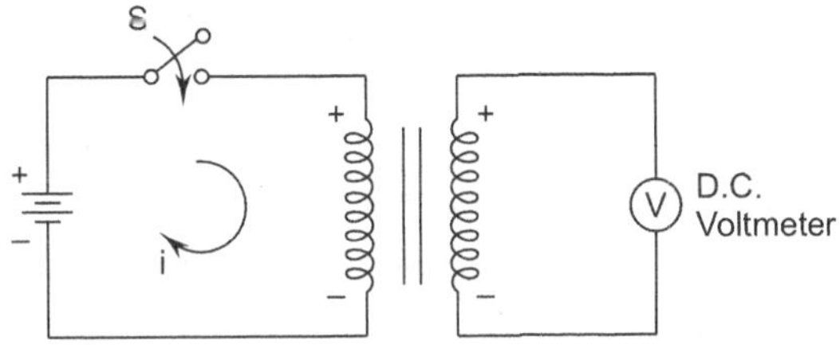

Fig. 8.44

This test can be done with simple setup as shown in Fig. 8.44.

As the switch on the primary side is closed, the primary current increases and so do the flux linkages of both the windings, inducing e.m.f. in them.

The positive polarity of this induced e.m.f in the primary winding will be at that end to which the battery is connected (according to Lenz's law).

If the voltmeter gives a positive deflection (as the voltmeter is of permanent magnet type) then the polarities assumed as shown in the above figure are correct.

If the voltmeter is showing negative deflection then one of the winding polarities must be reversed.

8.11.4 Sumpner's or Back-to-Back Test

For performing Sumpner's or Back-to-Back test, two **similar** transformers are required.

- This test provides data for finding the **regulation, efficiency** and **heating under load conditions.**

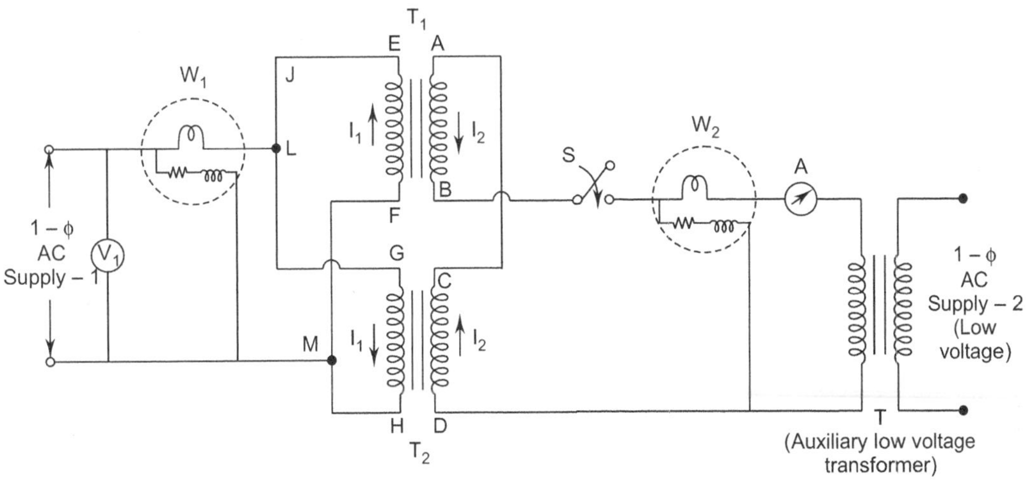

Fig. 8.45

- One transformer is loaded on the other and both are connected to the supply.
- The power taken from the supply is that necessary for supplying the losses of both the transformer, and the negligibly small loss in the control circuit.

- As shown in the circuit diagram of Fig. 8.45, primaries of the two transformers are connected in parallel across the same AC supply.
- With switch 'S' open, the wattmeter W, reads the core loss for the two transformers.
- The secondaries of the two transformers are so connected that their potentials are in opposition to each other.
- 'T' is an **auxiliary low voltage transformer,** which can be adjusted to give a variable voltage and hence current in the secondary loop circuit.
- By proper adjustment of T, full load secondary current I_2 can be made to flow.
- I_2 flows from D to C and then from A to B.
- Flow of I_1 is confined to the loop FEJLGHMF, and it does not pass through W_1.
- Hence W_1 continues to read core loss.
- W_2 measures full load Cu Loss. (or at any other load current I_2)
- Clearly, the power taken in is twice the losses of a single transformer.

8.12 AUTO-TRANSFORMER

An auto-transformer is a transformer with one winding only.

A part of this single winding is common to both primary and secondary windings (Fig. 8.46).

Obviously, in this transformer the primary and secondary windings are not electrically 'isolated' from each other as in case of a 2-winding transformer.

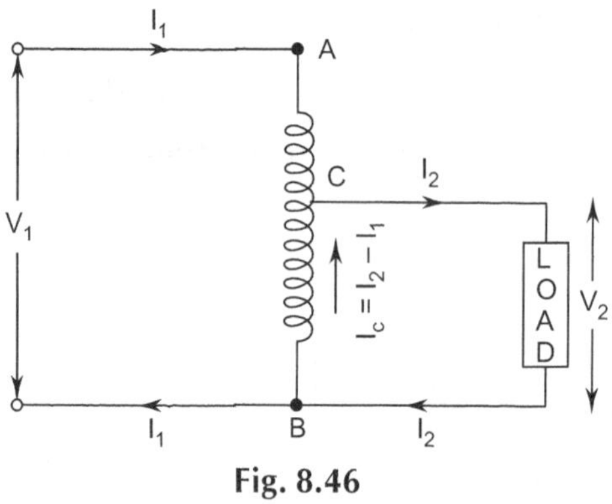

Fig. 8.46

AB = Primary windings, N_1 turns

BC = Secondary winding, N_2 turns

$$\frac{V_2}{V_1} = \frac{N_2}{N_1} = \frac{I_1}{I_2} = K \qquad \ldots(8.99)$$

The current in section CB is vector difference of I_2 and I_1.

As the two currents are practically in phase opposition, the resultant current is

$$I_c = I_2 - I_1$$

where I_2 is greater than I_1.

As compared to an ordinary 2-winding transformer of same output, an auto-transformer has higher efficiency but smaller size. Moreover, its voltage regulation is also superior.

8.12.1 Saving of Copper

Volume and hence weight of copper, is proportional to length and area of the cross-section of the conductors. Now,

Length $\propto$ Number of turns

Cross-section $\propto$ Current

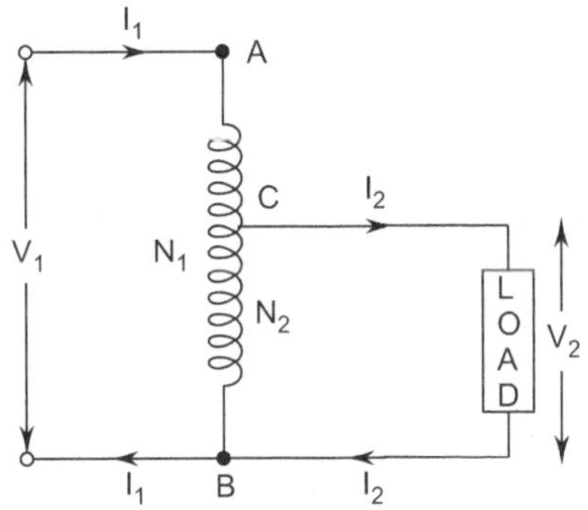

Fig. 8.47

With reference to Fig. 8.47 :

Weight of copper in section AC is $\propto (N_1 - N_2)\, I_1$

Weight of copper in section BC is $\propto N_2\, (I_2 - I_1)$

$\therefore$ Total weight of Cu in auto-transformer

$$\propto (N_1 - N_2)\, I_1 + N_2\, (I_2 - I_1) \qquad \ldots(8.100)$$

If the two winding transformer has to perform the same duty, then

Weight of copper on its primary $\propto N_1 I_1$

Weight of copper on its secondary $\propto N_2 I_2$.

$\therefore$ Total weight of Cu in 2-winding transformer

$$\propto N_1 I_1 + N_2 I_2 \qquad \ldots(8.101)$$

Taking ratio of equations (8.100) and (8.101), we get

$$\frac{\text{Weight of Cu in auto-transformer}}{\text{Weight of Cu in 2-winding transformer}} = \frac{(N_1 - N_2) I_1 + N_2 (I_2 - I_1)}{N_1 I_1 + N_2 I_2} \qquad \ldots(8.102)$$

or

$$= \frac{N_1 I_1 - N_2 I_1 + N_2 I_2 - N_2 I_1}{N_1 I_1 + N_2 I_2} \qquad \ldots(8.103)$$

or

$$= \frac{(N_1 I_1 + N_2 I_2) - 2N_2 I_1}{(N_1 I_1 + N_2 I_2)} \qquad \ldots(8.104)$$

$$= 1 - \frac{2N_2 I_1}{(N_1 I_1 + N_2 I_2)} \qquad \ldots(8.105)$$

Multiplying and dividing by $N_1 I_1$, we get

$$= 1 - \frac{2\frac{N_2}{N_1}}{1 + \frac{N_2}{N_1} \cdot \frac{I_2}{I_1}} \qquad \ldots(8.106)$$

As from equation (8.99), we have

$$\frac{N_2}{N_1} = \frac{I_1}{I_2} = K \qquad \ldots(8.107)$$

Inserting equation (8.107) in equation (8.106), we get

$$\frac{\text{Weight of copper in auto-transformer}}{\text{Weight of copper in 2-winding transformer}} = 1 - K \qquad \ldots(8.108)$$

Hence, weight of copper in auto-transformer = (1 – K) weight of copper in ordinary transformer

$$W_a = (1-K)\, W_o \qquad \ldots(8.109)$$

Saving $= W_o - W_a$

or $= W_o - (1-K)\, W_o$ ...(8.110)

or Saving $= KW_o$

Saving $= K \times$ (Weight of Cu in ordinary X'mer) ...(8.111)

8.12.2 Conductive Power and Inductive Power

Let the terminal A, of circuit diagram shown in Fig. 8.48, is positive with respect to terminal C, at the instant shown.

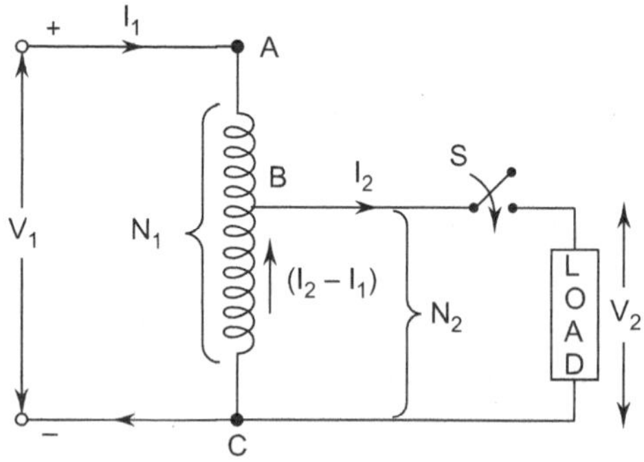

Fig. 8.48

- At no-load, the exciting current flow from A to C and it establishes a working m.m.f directed vertically downward in the core.
- When the switch S is closed, the current in winding B.C. must flow from C to B, in order to create an m.m.f. opposing the working m.m.f, as per Lenz's law.
- In winding AC, I_1 flows from A to C.
- In winding BC, I_2 flows from C to B.
- Therefore, the current in winding BC is I_1 from B to C and I_2 from C to B.

- Here the current I_2 is greater than I_1 because

$$V_2 < V_1 \quad \ldots(8.112)$$

and their m m f's are opposing each other at every instant.

- Therefore,

$$I_{CB} = I_2 - I_1 \quad \ldots(8.113)$$

$$\text{m.m.f. of winding } AB = I_1 (N_1 - N_2) \quad \ldots(8.114)$$

$$= (I_2 - I_1) N_2$$

$$= I_{CB} \times N_2 \quad \ldots(8.115)$$

$$= \text{m.m.f of winding CB} \quad \ldots(8.116)$$

It is, therefore seen that the transformer action takes place between winding AB and winding BC.

In otherwords, the volt-amperes across winding AB are transferred by transformer action to the load connected across winding BC

$$\therefore \text{ Transformed, } VA = V_{AB} \cdot I_{AB} \quad \ldots(8.117)$$

$$= (V_1 - V_2) I_1 \quad \ldots(8.118)$$

$$\text{Total input VA} = V_1 \cdot I_1 \quad \ldots(8.119)$$

$$= \text{Output VA} \quad \ldots(8.120)$$

Dividing (8.118) by equation (8.119), we get

$$\frac{\text{Tranformed VA}}{\text{Input VA}} = \frac{(V_1 - V_2) I_1}{V_1 \cdot I_1} \quad \ldots(8.121)$$

$$= (1 - K) \quad \ldots(8.122)$$

Transformed VA means that input VA, which is transferred to the load by 'Induction' action *i.e.*, Power transferred inductively = $(1 - K) \times$ Input ...(8.123)

The rest of the volt-amperes required for the output, are transferred 'conductively', directly from the input.

i.e., Power transferred conductively = Total Input VA – Inductively transferred VA ...(8.124)

$$= V_1 I_1 - (1 - K) V_1 I_1 \quad \ldots(8.125)$$

$$= V_1 I_1 - V_1 I_1 + K V_1 I_1 \quad \ldots(8.126)$$

$$= K V_1 I_1 \quad \ldots(8.127)$$

Power transferred conductively = $K \times$ Input ...(8.128)

8.12.3 Conversion of 2-winding Transformer into Auto-transformer

A 2-winding transformer can be converted into an auto-transformer either step down or step up. Let us take an example of 20 kVA, 2400/240 V transformer. If we employ 'additive' polarity between the high voltage and low voltage side, we get a step up auto-transformer (Fig. 8.49).

If, we use subtractive polarity, we get a step down auto-transformer (Fig. 8.50).

Additive Polarity :

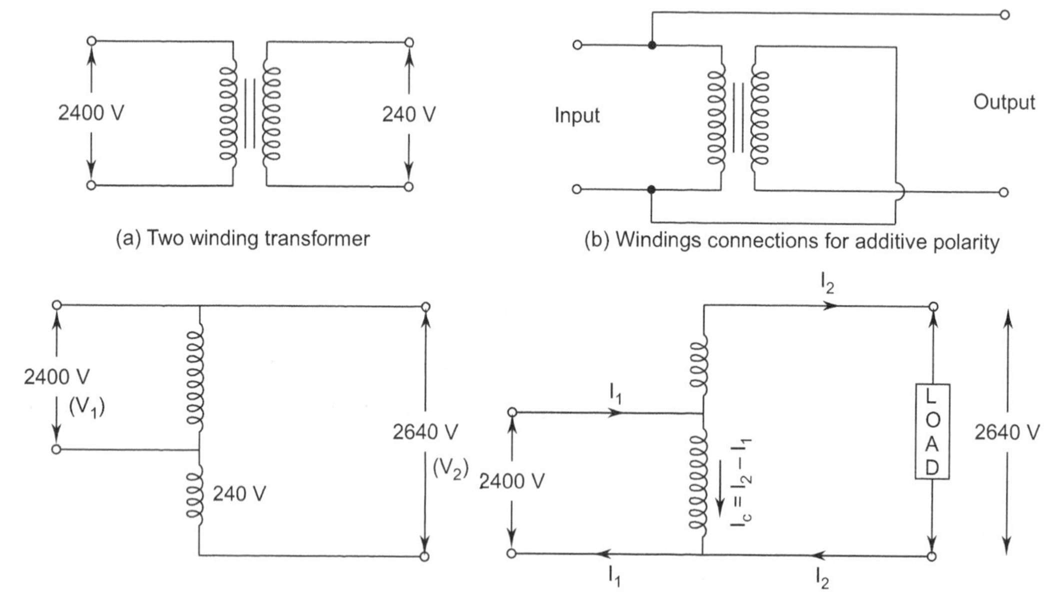

(a) Two winding transformer (b) Windings connections for additive polarity

(c) Step up auto-transformer (d) With current directions

Fig. 8.49

Subtractive Polarity :

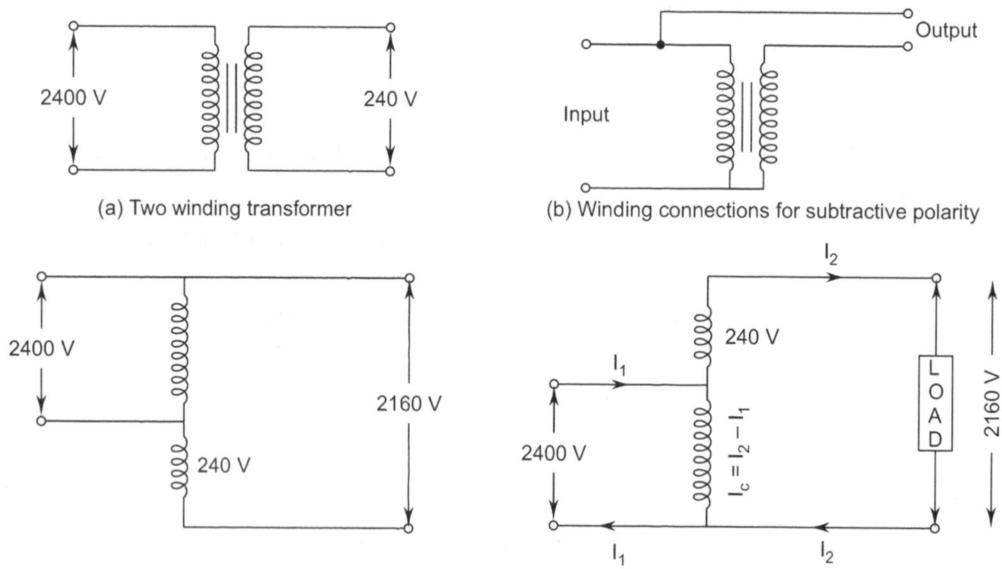

(a) Two winding transformer (b) Winding connections for subtractive polarity

(c) Step down transformer (d) With current directions

Fig. 8.50

8.12.4 Advantages

1. As we know that, saving of copper in auto-transformer is 'K' times the copper required in 2-winding transformer, hence less copper is required in case of an auto-transformer.
2. As less copper is required, hence less ohmic losses and core losses will occur. Correspondingly the efficiency of an auto-transformer is higher than efficiency of 2-winding transformer.

3. Leakage reactance is less in case of an auto-transformer, hence it has superior voltage regulation.
4. Due to single winding, less weight as well as less space will be there.

8.12.5 Disadvantages

1. If the ratio of transformation K differs far from unity, the economic advantages of auto-transformer over 2-winding transformer decreases.
2. The main disadvantage of an auto-transformer is due to the direct electrical connection between low-voltage and high-voltage sides.
3. The short circuit current in an auto-transformer is higher than that in a corresponding two-winding transformer.

8.12.6 Applications of Auto-transformer

Auto-transformers are mostly used when K is nearly equal to unity and where there is no objection to electrical connections between primary and secondary. Hence such transformers are used:

1. To give small boost to a distribution cable to correct the voltage drop.
2. As auto-starter transformers to give upto 50 to 60% of full voltage to an induction motor during starting.
3. As furnance transformers for getting a convenient supply to suit the furnance winding from a 230 V supply.
4. As interconnecting transformers in 132 kV/330 kV system.
5. In control equipment for 1-phase and 3-phase electrical locomotives.

SOLVED PROBLEMS

Problem 8.1. *The efficiency of a 400 kVA, single phase transformer is 98.77% at full-load 0.8 power factor and 99.13% at half-full load unity power factor. Find :*

(i) Iron losses at full and half full-loads *(ii) Cu losses at full and half full-loads.*

Solution. Let the iron loss and full-load copper losses be P_i and P_{Cu} respectively.

At full-load and 0.8 power factor

Output power, $P = 400 \times 0.8 = 320$ kW

Transformer efficiency, $\eta = 98.77\%$ or 0.9877

Transformer input power $= \dfrac{P}{\eta} = \dfrac{320}{0.9877} = 323.985$ kW

Transformer losses, $(P_i + P_{Cu})$ = Input – output = 323.985 – 320 = 3.985 kW ...(*i*)

At half-full-load and unity power factor

Output power, $P' = \dfrac{1}{2} \times 400 \times 1 = 200$ kW

Transformer efficiency, $\eta' = 99.13\%$ or 0.9913

Transformer input power $= \dfrac{P'}{\eta'} = \dfrac{200}{0.9913} = 201.755$ kW

Total losses $= P_i + \left(\frac{1}{2}\right)^2 P_{Cu} = \text{Input} - \text{output}$

$= 201.755 - 200 = 1.755$ kW

or $$P_i + \frac{1}{4} P_{Cu} = 1.755 \text{ kW} \qquad ...(ii)$$

Solving equations (*i*) and (*ii*) we have

(*i*) Iron loss at full-load = Iron loss at half-load

$= P_i =$ **1.0117 kW.** **Ans.**

(*ii*) Copper loss at full-load, P_{Cu} = **2.9733 kW.** **Ans.**

Copper loss at half-full-load

$$= \left(\frac{1}{2}\right)^2 \times 2.9733 = \mathbf{0.743325\ kW.}$$ **Ans.**

Problem 8.2. *In a 25 kVA, 2000 V/200 V transformer the iron and copper losses are 200 W and 400 W respectively. Calculate the efficiency at half load and 0.8 power factor lagging. Determine also the maximum efficiency and the corresponding load.*

Solution. Full-load copper loss, $P_{Cu} = 400 \text{ W} = 0.4 \text{ kW}$

Iron loss, $P_i = 200 \text{ W} = 0.2 \text{ kW}$

Output at half-full load and 0.8 p.f. $= \frac{1}{2} \times 25 \times 0.8 = 10$ kW

Copper losses at half-full load $= \left(\frac{1}{2}\right)^2 \times 0.4 \text{ kW} = 0.1 \text{ kW}$

So efficiency at half-full load and 0.8 p.f.,

$$\eta = \frac{P_{out}}{P_{in}} \times 100 = \frac{10}{10 + 0.1 + 0.2} \times 100 = \mathbf{97.087\%}$$ **Ans.**

Load corresponding to maximum efficiency

$$\text{kVA}_{max} = \text{Rated kVA} \times \sqrt{\frac{P_i}{P_{Cu}}} = 25 \times \sqrt{\frac{0.2}{0.4}} = \mathbf{17.68\ kVA.}$$ **Ans.**

Maximum efficiency assuming power factor unity

$$\eta_{max} = \frac{17.68 \times 1.0}{17.68 + 0.2 + 0.2} \times 100 = \mathbf{97.788\%}$$ **Ans.**

Problem 8.3. *A voltage $v = 200 \sin 314t$ is applied to the transformer winding in a no-load test. The resulting current is found to be*

$$i = 3 \sin (314t - 60°)$$

Determine the core loss and the parameters of no-load approximate equivalent circuit.

Solution. No-load current, $$I_o = \frac{3}{\sqrt{2}} = 2.12 \text{ A}$$

Core loss = Input at no-load $= V I_o \cos \phi_o$

$$= \frac{200}{\sqrt{2}} \times \frac{3}{\sqrt{2}} \cos 60° = \mathbf{150\ W}$$ **Ans.**

Energy component of no-load current, $I_\omega = I_o \cos \phi_o = \frac{3}{\sqrt{2}} \cos 60° = 1.06$ A

Magnetising component of no-load current, $I_\mu = \sqrt{I_o^2 - I_\omega^2} = \sqrt{(2.12)^2 - (1.06)^2} = 1.836$ A

No-load resistance, $R_o = \frac{V}{I_\omega} = \frac{200/\sqrt{2}}{1.06} = \mathbf{133.42\ \Omega}$ **Ans.**

No-load reactance, $X_o = \frac{V}{I_\mu} = \frac{200/\sqrt{2}}{1.836} = \mathbf{77\ \Omega.}$ **Ans.**

Problem 8.4. *A single phase 440/110 V transformer takes a no-load current of 4 A and 0.2 power factor lagging. If the secondary supplies a current of 100 A at a power factor of 0.8 lagging determine (i) the current taken by the primary winding, (ii) the magnetising reactance and resistance representing core losses.*

Solution.

Transformation ratio, $K = \frac{V_2}{V_1} = \frac{110}{440} = 0.25$

Primary no-load current, $I_o = 4$ A

No-load power factor, $\cos \phi_o = 0.2$ (lagging)

or $\phi_o = \cos^{-1} 0.2 = 78.46°$

Secondary Current, $I_2 = 100$ A

Load power factor, $\cos \phi_2 = 0.8$ (lagging)

or $\phi_2 = \cos^{-1} 0.8 = 36.87°$

Secondary equivalent current, $I_2' = KI_2$

$= 0.25 \times 100 = 25$ A

The primary no-load current and secondary current can be expressed in polar and rectangular form as

$\bar{I}_o = 4 \angle 78.46°$ A

or $\bar{I}_o = (0.8 + j\,3.919)$ A

and $\bar{I}_o = 25 \angle 36.87°$

or $\bar{I}_2' = (20 + j15)$A

(*i*) The primary full load current is given by

$\bar{I}_1 = \bar{I}_o + \bar{I}_2'$

or $I_1 = 0.8 + j\,3.919 + 20 + j\,15$

$= (20.8 + 18.919\,j)$ A

or $I_1 = \mathbf{28.12 \angle 42.29°\ A}$ **Ans.**

(*ii*) Energy component of no-load current, $I_\omega = I_o \cos \phi_o = 4 \times 0.2 = 0.8$ A

Magnetising component of no-load current,

$I_\mu = I_o \sin \phi_o = 4 \times 0.98 = 3.92$ A

Resistance representing core losses, $R_o = \frac{V_1}{I_\omega} = \frac{440}{0.8} = \mathbf{550\ \Omega}$ **Ans.**

Magnetising reactance, $X_o = \frac{V_1}{I_\mu} = \frac{440}{3.92} = \mathbf{112.25\ \Omega}$ **Ans.**

Problem 8.5. *A 30 kVA, 2000/200 V, single phase, 50 Hz transformer has a primary resistance of 3.5 Ω and reactance of 4.5 Ω. The secondary resistance and reactance are 0.015 Ω and 0.02 Ω respectively. Find*

(i) Equivalent resistance referred to primary

(ii) Equivalent reactance referred to primary

(iii) Equivalent Impedance referred to primary

(iv) Total copper loss in the transformer.

Solution. Transformation ratio, $K = \frac{V_2}{V_1} = \frac{200}{2,000} = 0.1$

Full-load primary current, $I_1 = \frac{30 \times 1,000}{2,000} = 15\ A$

(*i*) Equivalent resistance referred to primary, $R_{o1} = R_1 + \frac{R_2}{K^2} = 3.5 + \frac{0.015}{(0.1)^2} = \mathbf{5\ \Omega}$ **Ans.**

(*ii*) Equivalent reactance referred to primary, $X_{o1} = X_1 + \frac{X_2}{K^2} = 4.5 + \frac{0.02}{(0.1)^2} = \mathbf{6.5\ \Omega}$ **Ans.**

(*iii*) Equivalent impedance referred to primary, $Z_{o1} = \sqrt{(R_{o1})^2 + (X_{o1})^2}$

$= \sqrt{(5)^2 + (6.5)^2} = \mathbf{8.2\ \Omega}$ **Ans.**

(*iv*) Total copper loss in transformer $= I_1^2 R_{o1} = (15)^2 \times 5 = \mathbf{1,125\ Watts.}$ **Ans.**

Problem 8.6. *A single phase transformer having ratings 100 kVA, 11 kV/220 V, 50 Hz gave the following test results:*

OC test :	*220 V*	*45 A*	*2 kW*
SC test :	*500 V*	*9.09 A*	*3 kW*

Determine equivalent circuit parameters referred to LV side.

Solution. Assuming low-voltage winding as primary

Energy component of no-load current, $I_\omega = \frac{P_o}{V} = \frac{2 \times 1,000}{220} = 9.091\ A$

Magnetising component of no-load current, $I_\mu = \sqrt{I_o^2 - I_\omega^2} = \sqrt{(45)^2 - (9.091)^2} = 44.07\ A$

Equivalent circuit parameter, $R_o = \frac{V_1}{I_\omega} = \frac{220}{9.091} = \mathbf{24.2\ \Omega}$ **Ans.**

Equivalent circuit parameter, $X_o = \frac{V_1}{I_\mu} = \frac{220}{44.07} = \mathbf{4.992\ \Omega}$ **Ans.**

Equivalent impedance referred to HV side, $Z_{o2} = \frac{V_{sc}}{I_{sc}} = \frac{500}{9.09} = 55\ \Omega$

Equivalent resistance referred to HV side, $R_{o2} = \frac{W_{sc}}{I_{sc}^2} = \frac{3 \times 1,000}{(9.09)^2} = 36.3\ \Omega$

Equivalent reactance referred to HV side, $X_{o2} = \sqrt{(Z_{o2})^2 - (R_{o2})^2} = \sqrt{(55)^2 - (36.3)^2} = 41.31\ \Omega$

Transformation ratio, $K = \frac{11,000}{220} = 50$

Equivalent resistance referred to LV side $R_{o1} = \frac{R_{o2}}{K^2} = \frac{36.3}{(50)^2} = \mathbf{0.0145\ \Omega}$ **Ans.**

Equivalent reactance referred to LV side $X_{o1} = \frac{X_{o2}}{K^2} = \frac{41.31}{(50)^2} = \mathbf{0.0165\ \Omega}$ **Ans.**

Problem 8.7. *A 250/500 V, single-phase transformer gave the following test results:*

Short-circuit test (LV short) :	*20 V*	*12 A*	*100 W*
Open-circuit test (HV open) :	*250 V*	*1 A*	*80 W*

Determine the efficiency of the transformer when the output is 12 A, 500 V at 0.85 power factor lagging.

Solution. Output current, $I = 12\ A$

Output voltage, $V = 500\ V$

Power factor, $\cos\phi = 0.85$ (lagging)

Output power, $P_{out} = VI\cos\phi = 500 \times 12 \times 0.85 = 5,100\ W$

Iron loss, $P_i = 80\ W$

Copper loss, $P_{Cu} = 100\ W$ (with 12 A in HV winding)

So copper loss, $P_{Cu} = 100\ W$

Transformer efficiency, $$\eta = \frac{\text{Output}}{\text{Output} + P_i + P_{Cu}} \times 100$$

$$= \frac{5,100}{5,100 + 80 + 100} \times 100 = \mathbf{96.59\%}.$$ **Ans.**

Problem 8.8. *In a 50 kVA, 1-phase 3300/230 V, transformer, the iron losses and full-load copper losses were found to be as 500 W and 650 W respectively. Calculate efficiency at half load and 0.6 power factor. Determine maximum efficiency and corresponding load.*

Solution. Iron loss, $P_i = 500\ W$

Copper loss at full load, $P_{Cu} = 650\ W$

Output at half-full load $= \frac{1}{2} \times \text{rated kVA} \times \text{p.f.}$

$$= \frac{1}{2} \times 50 \times 0.6 = 15\ kW = 15,000\ W$$

Input at half-full load $= P_{out} + P_i + \left(\frac{1}{2}\right)^2 P_{Cu}$

$$= 15{,}000 + 500 + \frac{650}{4}$$

$$= 15{,}622.5 \text{ W}$$

Efficiency at half-full load $= \dfrac{P_{out}}{P_{in}} \times 100 = \dfrac{15{,}000}{15{,}622.5} \times 100 = \mathbf{95.77\%}$ **Ans.**

Load corresponding to maximum efficiency $= \text{Rated kVA} \sqrt{\dfrac{P_i}{P_{Cu}}}$

$$kVA_{max} = 50 \times \sqrt{\frac{500}{650}} = 43.853 \text{ kVA}$$

Maximum efficiency, assuming unity p.f. $\eta_{max} = \dfrac{43{,}853 \times 1.0}{43{,}853 + 500 + 500} \times 100$

$= \mathbf{97.77\%}$ **Ans.**

Problem 8.9. *Find the all-day efficiency of 500 kVA distribution transformer whose copper loss and iron loss at full load are 4.5 kW and 3.5 kW respectively. During a day of 24 hours, it is loaded as under:*

No. of hours	*Loading in kW*	*Power factor*
06	*400*	*0.80*
10	*300*	*0.75*
04	*100*	*0.80*
04	*0*	—

Solution. First we convert the load in kW to load in kVA and find the corresponding copper losses (as copper losses varies according to load).

Load in kW	**Load in kVA = kW/p.f.**	**Copper Loss**
400	$\frac{400}{0.8} = 500$ kVA	4.5 kW
300	$\frac{300}{0.75} = 400$ kVA	$\left(\frac{400}{500}\right)^2 \times 4.5 = 2.88$ kW
100	$\frac{100}{0.8} = 125$ kVA	$\left(\frac{125}{500}\right)^2 \times 4.5 = 0.281$ kW
0	0	0

Total copper loss in 24 hours

$$= (6 \times 4.5) + (10 \times 2.88) + (4 \times 0.281)$$

$$P_{Cu} = 56.924 \text{ kWh.}$$

Total iron loss in 24 hours

$$P_i = 3.5 \times 24^* = 84 \text{ kWh.}$$

Transformer output in 24 hours (in kWh)

$$= (6 \times 400) + (10 \times 300) + (4 \times 100)$$

$$= 5800 \text{ kWh.}$$

---

* As iron loss do not depends upon load and they will take place in the transformer throughout the day.

Hence all efficiency will be

$$\eta_{\text{all day}} = \frac{5800}{5800 + 84 + 56.924} \times 100$$

or $$\eta_{\text{all day}} = \mathbf{97.627\%}$$ **Ans.**

Problem 8.10. *Two similar 250 kVA, single phase transformers gave the following results when tested by back-to-back method :*

Mains Wattmeter $W_1 = 5\ kW$

Secondary series circuit wattmeter $W_2 = 7.5\ kW$ *(at full-load current)*

Find out the individual transformer efficiencies at 75% full-load and 0.8 power factor leading.

Solution. Full load copper loss of each transformer $= \frac{7.5}{2} = 3.75$ kW.

Copper-loss at 75% of full load (for each transformer) $= \left(\frac{3}{4}\right)^2 \times 3.75$

$= 2.11$ kW

Iron loss of each transformer $= \frac{5}{2} = 2.5$ kW.

Output of each transformer = 250 kVA.

Output of each transformer at 75% of full load and at 0.8 power factor

$$= 250 \times \frac{75}{100} \times 0.8$$

$= 150$ kW.

Hence efficiency at 75% of full load, 0.8 p.f. will be

$$\eta = \frac{150}{150 + 2.5 + 2.11} \times 100$$

$$\eta = \mathbf{97\%.}$$ **Ans.**

Problem 8.11. *The ohmic values of the circuit parameters of a transformer, having a turn's ratio of 5, are*

$R_1 = 0.5\ \Omega,\ R_2 = 0.021\ \Omega$

$X_1 = 3.2\ \Omega,\ X_2 = 0.12\ \Omega$

$R_o = 350\ \Omega,\ X_o = 98\ \Omega$

Draw the approximate equivalent circuit of the transformer referred to secondary. Show the numerical values of the circuit parameters.

Solution.

$$a = 5$$

$$K = \frac{1}{a} = \frac{1}{5} = 0.2$$

The equivalent circuit is shown in Fig. 8.51 and the equivalent circuit parameters are calculated as follow.

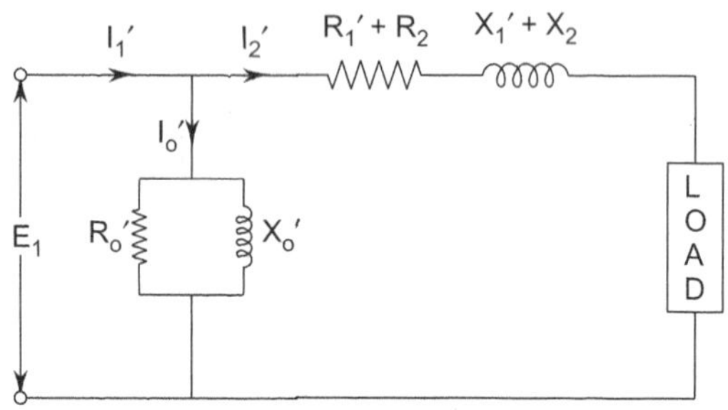

Fig. 8.51

Equivalent resistance referred to secondary

$$R_{o2} = R_2 + R_1' = R_2 + \frac{R_1}{a^2} = R_2 + K^2 R_1$$

or $$R_{o2} = 0.021 + (0.2)^2 \times 0.5 = \mathbf{0.041\ \Omega}$$ **Ans.**

Equivalent reactance referred to secondary

$$X_{o2} = X_2 + X_1' \approx X_2 + \frac{X_1}{a^2} = X_2 + K^2 X_1$$

or $$X_{o2} = 0.12 + (0.2)^2 \times 3.2 = \mathbf{0.248\ \Omega}$$ **Ans.**

Equivalent impedance referred to secondary

$$Z_{o2} = R_{o2} + jX_{o2} = \mathbf{(0.041 + \mathit{j}0.248)\ \Omega}$$ **Ans.**

Equivalent tank circuit resistance referred to secondary

$$R_o' = K^2 R_o = (0.2)^2 \times 350 = \mathbf{14\ \Omega}$$ **Ans.**

Equivalent tank circuit resistance referred to secondary

$$X_o' = K^2 X_o = (0.2)^2 \times 98 = \mathbf{3.92\ \Omega}$$ **Ans.**

Problem 8.12. *The following test results were obtained on a 20 kVA, 2200/220 V transformer.*

OC test (LV):	*220 V*	*1.1 A*	*125 W*
SC test (HV):	*52.7 V*	*8.4 A*	*287 W*

Assuming power factor to be 0.8 (lagging), determine efficiency at half and full load.

Solution.

From open circuit test

$$P_i = 125 \text{ W}$$

The rated current or full load current of HV side is

$$I_1 = \frac{20{,}000}{2{,}200} = 9.09 \text{ A}$$

The short circuit test is performed at 8.4 Ampere. Hence the full load copper loss will be

$$P_{Cu} = \left(\frac{9.09}{8.4}\right)^2 287 = 336 \text{ W}$$

(*i*) Efficiency at half load will be

$$\eta = \frac{x \text{ kVA} \cos\phi}{x \text{ kVA} \cos\phi + P_i + x^2 P_{Cu}}$$

or
$$\eta_{1/2\,\text{load}} = \frac{\frac{1}{2} \times 20 \times 10^3 \times 0.8}{\frac{1}{2} \times 20 \times 10^3 \times 0.8 + 125 + \left(\frac{1}{2}\right)^2 336}$$

or $\eta_{1/2\,\text{load}} =$ **97.454%** **Ans.**

(*ii*) Efficiency at full load will be

$$\eta_{\text{full load}} = \frac{20 \times 10^3 \times 0.8}{20 \times 10^3 \times 0.8 + 125 + 336}$$

or $\eta_{1/2\,\text{load}} =$ **97.199%** **Ans.**

Problem 8.13. *A 100 kVA single phase transformer operating at 0.9 power factor has 90% maximum efficiency. What will be its iron losses?*

Solution.

Given Power factor = 0.9

Max. efficiency = 90% = 0.9

Input = 100 kVA

At maximum efficiency iron losses are equal to copper losses.

$$\eta_{max} = \frac{\text{kVA} \cos\phi}{\text{kVA} \cos\phi + P_i + P_i}$$

or
$$0.9 = \frac{100 \times 10^3 \times 0.9}{100 \times 10^3 \times 0.9 + 2P_i}$$

$$1.8\ P_i = 90{,}000 - 81000$$

$$1.8\ P_i = 9000$$

or $P_i =$ **5000 W** **Ans.**

Problem 8.14. *A 20 kVA, 2000/200 V, single phase, 50 Hz transformer has a primary resistance of 1.5 Ω and a reactance of 2 Ω. The secondary resistance and reactance are 0.015 Ω and 0.02 Ω respectively. The no load current of transformer is 1 A at 0.2 power factor.*

Determine:

(*i*) *Equivalent resistance referred to primary*

(*ii*) *Equivalent reactance referred to primary*

(*iii*) *Equivalent impedance referred to primary*

(*iv*) *Supply current*

(*v*) *Total copper loss*

(*vi*) *Draw approximate equivalent circuit.*

Solution.

Given 20 kVA, 2000 V/200 V

$$f = 50 \text{ Hz} \qquad K = \frac{200}{2000} = 0.1$$

$$R_1 = 1.5\,\Omega \qquad X_1 = 2\,\Omega$$

$$R_2 = 0.015\,\Omega \qquad X_2 = 0.02\,\Omega$$

$$I_o = 1\text{ A}, \qquad \cos\phi_o = 0.2$$

(*i*) Equivalent resistance, referred to primary

$$R_{o1} = R_1 + R_2' = R_1 + \frac{R_2}{K^2}$$

$$= 1.5 + \frac{0.015}{(0.1)^2} = 1.5 + 1.5$$

or $$R_{o1} = \mathbf{3\,\Omega}$$ **Ans.**

(*ii*) Equivalent reactance referred to primary

$$X_{o1} = X_1 + X_2' = X_1 + \frac{X_2}{K^2}$$

$$= 2 + \frac{0.02}{(0.1)^2} = 2 + 2$$

or $$X_{o1} = \mathbf{4\,\Omega}$$ **Ans.**

(*iii*) Equivalent impedance refer to primary

$$Z_{o1} = R_{o1} + jX_{o1} = 3 + 4j$$

or $$Z_{o1} = \mathbf{5\angle 53.13^\circ\,\Omega}$$ **Ans.**

(*iv*) Supply current

$$\cos\phi_o = 0.2$$

$$\phi_o = \cos^{-1} 0.2 = 78.46^\circ$$

The no load current I_o in polar and rectangular form can be expressed as:

$$I_o = 1\angle 78.46 = (0.2 + j\,0.97)\text{ A}$$

$$I_2 = \frac{\text{kVA}}{V_2} = \frac{20{,}000}{200}$$

or $$I_2 = 100\text{ A}$$

$$I_2' = K\,I_2 = 0.1 \times 100$$

or $$I_2' = 10\text{ A}$$

Taking equivalent impedance angle as primary phase angle, the load component of input current, I_2' can be expressed in polar and rectangular form as:

$$I_2' = 10\angle 53.13^\circ = (6 + j\,7.99)\text{ A}$$

Supply current,

$$I_1 = I_o + I_2'$$
$$= 0.2 + j\,0.97 + 6 + j\,7.99$$
$$= 6.2 + j\,8.96$$

or $$I_1 = \mathbf{10.89 \angle 55.31° \ A}$$ **Ans.**

(*v*) Primary current, $I_1 = \dfrac{20,000}{2000} = 10\,A$

Total copper loss,

$$P_{Cu} = I_1^2 R_{o1} = (10)^2 \times 3$$
$$= \mathbf{300\ W}$$ **Ans.**

(*vi*) Approximate equivalent circuit is shown in Fig. 8.52.

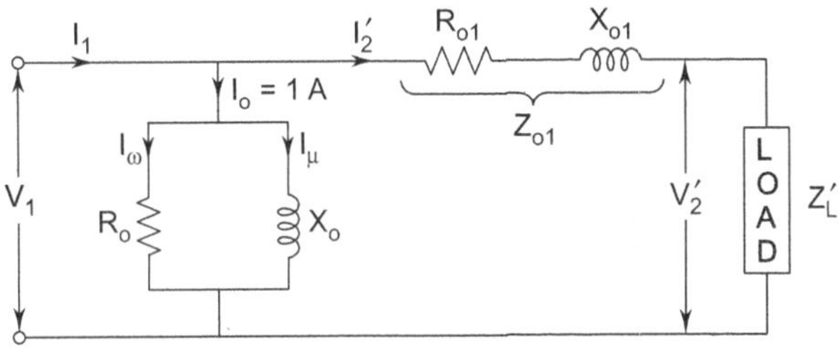

Fig. 8.52

Problem 8.15 *A transformer has a primary winding of 600 turns and a secondary winding of 150 turns. When the load current on the secondary is 60 A at 0.8 power factor lagging, the primary current is 20 A at 0.707 power factor lagging.*

Determine the no-load current of the transformer and its phase with respect to voltage.

Solution.

Given $N_1 = 600$ $\quad N_2 = 150$

$I_1 = 20$ A $\quad \cos\phi_1 = 0.707$

$I_2 = 60$ A $\quad \cos\phi_2 = 0.8$

$$K_1 = \frac{N_2}{N_1} = \frac{150}{600} = 0.25$$

$$\cos\phi_1 = 0.707 \quad \Rightarrow \quad \phi_1 = \cos^{-1} 0.707 = 45°$$
$$\cos\phi_2 = 0.8 \quad \Rightarrow \quad \phi_2 = \cos^{-1} 0.8 = 36.87°$$

$$\bar{I}_1 = 20 \angle 45° = (14.14 + j\,14.14)\ A$$

$$\bar{I}_2 = 60 \angle 36.87° = (48 + j\,36)\ A$$
$$I_2' = K.\,I_2 = 0.25 \times (48 + j\,36)$$
$$= (12 + j\,9)\ A$$

The primary full load current is given by

$$\bar{I}_1 = \bar{I}_o + \bar{I}_2'$$

or $$\bar{I}_o = \bar{I}_1 - I_2'$$

$$= (14.14 + j\,14.14) - (12 + j9)$$

or $$\bar{I}_o = (2.14 + 5.14j)\text{ A}$$

or $$\bar{I}_o = \mathbf{5.57 \angle 67.4° \text{ A}}$$ **Ans.**

Hence no-load current of the transformer is 5.57 A and it lags behind the voltage by an angle of 67.4°.

Problem 8.16. *In a 50 kVA, 1100/220 V single phase transformer, the iron and copper losses at full load are 350 W and 425 W respectively. Calculate the efficiency at*

(i) full load unity power factor

(ii) half load unity power factor.

Solution.

$$P_i = 350 \text{ W}$$

$$P_{Cu} = 425 \text{ W}$$

(*i*) Efficiency at full load unity power factor

$$\eta = \frac{x \text{ kVA} \cos\phi}{x \text{ kVA} \cos\phi + P_i + x^2 P_{Cu}} \times 100$$

$$\eta_{FL} = \frac{1 \times 50 \times 10^3 \times 1}{1 \times 50 \times 10^3 \times 1 + 350 + (1)^2 . 425} \times 100$$

$$= \frac{50{,}000}{50{,}000 + 350 + 425} \times 100$$

$$\eta_{FL} = \mathbf{98.474\%}$$ **Ans.**

(*ii*) Efficiency at half load unity power factor

$$\eta_{HL} = \frac{\tfrac{1}{2} \times 50 \times 10^3 \times 1}{\tfrac{1}{2} \times 50 \times 10^3 \times 1 + 350 + \left(\frac{1}{2}\right)^2 425} \times 100$$

$$= \frac{25000}{25000 + 350 + 106.25} \times 100$$

$$\eta_{\frac{1}{2}\text{ Load}} = \mathbf{98.208\%}$$ **Ans.**

Problem 8.17. *A 100 kVA, 1100/220 V, 60 Hz transformer has a high voltage winding resistance of 0.1 Ω and a leakage reactance of 0.3 Ω. The low voltage winding resistance is 0.004 and the leakage reactance is 0.012 Ω. The source is applied to high voltage side. Find:*

(a) Equivalent winding resistance referred to high voltage side.

(b) Equivalent winding reactance referred to high voltage side.

(c) Equivalent winding resistance referred to low voltage side.

(d) Equivalent winding reactance referred to low voltage side.

(e) Equivalent resistance drop in volts and in percent of rated winding voltages expressed in terms of the primary quantities.

(*f*) *Equivalent reactance drop in volts and in percent of rated winding voltages expressed in terms of the primary quantities.*

(*g*) *Equivalent leakage impedances of the transformer referred to primary side.*

(*h*) *Equivalent leakage impedances of the transformer referred to secondary side.*

Solution.

Given

$$R_1 = 0.1\,\Omega \quad R_2 = 0.004\,\Omega$$
$$X_1 = 0.3\,\Omega \quad X_2 = 0.012\,\Omega$$
$$K = \frac{220}{1100} = 0.2$$

(*a*) Equivalent winding resistance refer to high voltage side

$$R_{o1} = R_1 + R_2' = R_1 + \frac{R_2}{K^2}$$
$$= 0.1 + \frac{0.004}{(0.2)^2} = 0.1 + 0.1$$
$$R_{o1} = \mathbf{0.2\,\Omega}$$ **Ans.**

(*b*) Equivalent winding reactance refer to high voltage side

$$X_{o1} = X_1 + X_2' = X_1 + \frac{X_2}{K^2}$$
$$= 0.3 + \frac{0.012}{(0.2)^2} = 0.3 + 0.3$$
$$X_{o1} = \mathbf{0.6\,\Omega}$$ **Ans.**

(*c*) Equivalent winding resistance refer to low voltage side

$$R_{o2} = R_2 + R_1' = R_2 + K^2 . R_1$$
$$= 0.004 + (0.2)^2 \times 0.1 = 0.004 + 0.004$$
$$R_{o2} = \mathbf{0.008\,\Omega}$$ **Ans.**

(*d*) Equivalent winding reactance refer to low voltage side

$$X_{o2} = X_2 + X_1' = X_2 + K^2 . X_1$$
$$= 0.012 + (0.2)^2 \times 0.3 = 0.012 + 0.012$$
$$X_{o2} = \mathbf{0.024\,\Omega}$$ **Ans.**

(*e*) % Resistance drop expressed in terms of primary quantities

$$\epsilon_{r_1} = \frac{I_1 R_{o1}}{V_1} \times 100$$

$$I_1 = \frac{100 \times 1000}{1100} = \frac{1000}{11} = 90.91\text{ A}$$

$$R_{o1} = 0.2\,\Omega$$
$$V_1 = 1100\text{ V}$$

$$\epsilon_{r_1} = \frac{I_1 R_{o1}}{V_1} \times 100 = \frac{90.91 \times 0.2}{1100} \times 100$$

or $\in_{r_1} = \mathbf{1.65\%}$ **Ans.**

(*f*) % Reactance drop expressed in terms of primary quantities

$$I_1 = 90.91 \text{ A}$$
$$X_{o1} = 0.6\,\Omega$$
$$V_1 = 1100 \text{ V}$$

$$\in_{x_1} = \frac{I_1 X_{o1}}{V_1} \times 100 = \frac{90.91 \times 0.6}{1100} \times 100$$

$$\in_{x_1} = \mathbf{4.96\%}$$ **Ans.**

(*g*) Equivalent leakage impedance refer to primary side

$$Z_{o1} = R_{o1} + j\,X_{o1}$$
$$= 0.2 + j\,0.6$$
$$Z_{o1} = \mathbf{0.632 \angle 71.57° \,\Omega}$$ **Ans.**

(*h*) Equivalent leakage impedance refer to secondary side

$$Z_{o2} = R_{o2} + j\,X_{o2}$$
$$= 0.008 + j\,0.024$$
$$Z_{o2} = \mathbf{0.0253 \angle 71.57° \,\Omega}$$ **Ans.**

EXERCISE

1. Why the efficiency of transformer is high as compared to other electrical machines?
2. What do you understand by the term ideal transformer?
3. When the load current in a transformer increases, how does the input current adjust to meet the new conditions? Discuss.
4. A 5 kVA, 500/250 V, 50 Hz single-phase transformer gave the following test results:

O.C. Test:	500 V	1 A	50 Watt	(L.V. side open)
S.C. Test:	12.5 V	5 A	15 Watt	(L.V. side short)

   Find:

   (*i*) equivalent impedance refers to secondary.

   (*ii*) efficiency while supplying half load at 0.8 power factor leading.

   (*iii*) value of maximum efficiency.

   (*iv*) voltage regulation at 0.8 power factor leading.
5. Prove that for a single-phase transformer
6. $(\text{kVA})_{\text{at max.}} = (\text{kVA})_{\text{rated}} \sqrt{\dfrac{P_{\text{core}}}{P_{\text{ohmic (rated)}}}}$
7. What does voltage regulation signifies. If the per unit resistance drop is ε_r and per unit Reactance drop is ε_x, prove that voltage regulation in a transformer is given by

$$\text{V.R.} = \varepsilon_r \cos\phi_2 \pm \varepsilon_x \sin\phi_2,$$

   where $\cos\phi_2$ is load power factor.

8. If in a transformer core the hysteresis and eddy current losses are 80 W and 50 W at normal voltage and frequency, calculate the losses when voltage and frequency are increased by 20%.

9. A 1-phase, 250/500 V transformer gave the following results:

Open circuit test:	250 V	1 A	80 W	(L.V. side)
Short circuit test:	20 V	12 A	100 W	(H.V. side)

Calculate:

(i) the equivalent circuit parameters.
(ii) show them on an equivalent circuit.
(iii) efficiency at ½ load 0.8 power factor leading.
(iv) the load at which maximum efficiency occurs.
(v) voltage regulation at 0.9 power factor leading.

10. Explain why the hysteresis loss and eddy current loss occur in a transformer. Explain how hysteresis loss and eddy current losses can be reduced in a transformer.

11. A single phase transformer has 350 primary and 1050 secondary turns. Primary is connected to 400 V, 50 Hz AC supply. If the net cross-sectional area of the core is 50 cm^2. Calculate:

(i) maximum value of flux density in the core.
(ii) induced e.m.f in the secondary winding.

12. A 400/200 V transformer takes 1 A at a power factor of 0.4 on no-load. If the secondary supplies a load current of 50 A at 0.8 lagging power factor, calculate the primary current.

13. A 15 kVA, 2200/110 V transformer has R_1 = 4.75 Ω, R_2 = 0.0045 Ω, X_1 = 2.6 Ω, X_2 = 2.6 Ω calculates:

(i) equivalent resistance referred to primary
(ii) equivalent resistance referred to secondary
(iii) equivalent reactance referred to primary
(iv) equivalent reactance referred to secondary
(v) copper losses on full load
(vi) efficiency while supplying full load at 0.9 power factor lagging
(vii) efficiency while supplying half load at 0.8 power factor leading
(viii) value of maximum efficiency
(ix) voltage regulation at 0.8 lagging and leading power factor.

Assume iron loss equal to 60 W.

14. A transformer has reactance drop 5% and a resistance drop of 2.5%. Find the lagging power factor at which voltage regulation is zero and maximum and value of this regulation.

15. Parameters of a 2300/230 V, 50 Hz transformer is given below:

R_1 = 0.286 Ω,	X_1 = 0.73 Ω,	R_o = 250 Ω
R_2 = 0.319 Ω,	X_2 = 0.73 Ω,	X_o = 1250 Ω

Secondary load impedance, $Z_L = 0.387 + 0.29j$. Find

(i) primary current
(ii) no load current
(iii) efficiency
(iv) voltage regulation

Normal voltage is applied across the primary.

□ □ □

9

ELECTROMECHANICAL ENERGY CONVERSION

9.1 INTRODUCTION

In the previous chapters, we saw how electrical power is transmitted by means of single-phase and three-phase circuits and how transformers are used to raise and lower the system voltages. Next, we should consider to what purpose the electrical power is used at the consumer's end.

Much electrical power is used to provide heat during winters, either for heating a room or for heating water in a geyser. During summers, we use electrical power to provide cooling, as in air-conditioners or refrigerators. Some heat conversion is used to provide light as in the common light bulb. But much electrical power is used to drive machines in industry and homes. Useful gadgets like fans, tape recorders, mixers, etc., all run with the help of a ***motor*** fitted inside them. Basically, these motors take energy from electrical power mains and convert it into mechanical energy which is used to turn the blades of the fan or juicers.

Reverse conversion takes place in a machine called ***generator***. You have a generator (also, called *dynamo*) in your car. While the car is running, the dynamo converts mechanical energy (drawn from the engine) into electrical energy which keeps on charging the battery.

Motors and generators are electromechanical energy conversion (EMEC) devices or machines.

Not all machines deal with large amount of energy. Some devices operate at very low power levels and are termed *transducers* (such as a microphone or a loudspeaker), which deal with 'signals' to operate electronic circuits.

9.2 COUPLED MAGNETIC FIELD

The process of electromechanical energy conversion is essentially *reversible* (except for a small amount which is lost as heat energy). In both the motor and the generator, it is the *magnetic field* that ***couples*** the electrical system and the mechanical system, as illustrated in Fig. 9.1. The magnetic field provides the mutual link between the two systems. The energy to be transferred from one system to the other is temporarily stored in the magnetic field and then released in the other system.

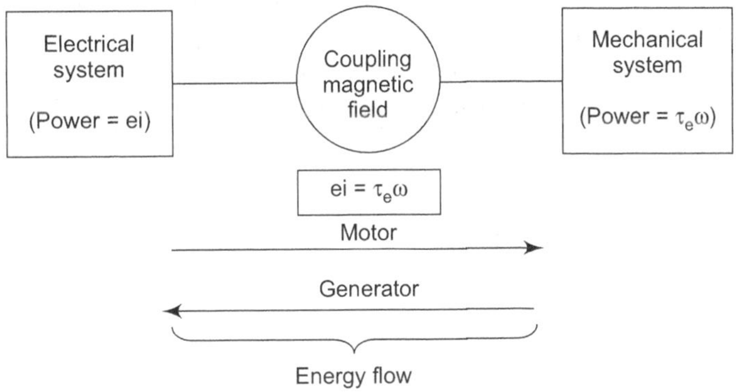

Fig. 9.1

The interchange of energy between the electrical system and mechanical system takes place through the conductors placed in a magnetic field. The primary quantities in the electrical system are *induced e.m.f* (e) and *current* (i). The analogous quantities in the mechanical system are *electromagnetic torque* (τ_e) and *angular speed* (ω), respectively.

When the electrical system is characterized by direct current, the machines are called *DC motors* and *DC generators*. Similarly, if the electrical system is characterized by alternating current, the machines are called *AC motors* and *AC generators*. Basically, the AC machines are not different from the DC machines. They differ merely in the constructional details. Only for our convenience, we discuss different type of machines in different Chapters.

Essentially, an electrical rotary machine has two parts—the fixed part, called the ***stator***, and the moving part, called the ***rotor***. Following are the two phenomena associated with the operation of a machine:

- Whenever the conductors of a coil cut across the magnetic flux (or are cut by it), an e.m.f. e is induced in it. This e.m.f. can supply a current i to an electrical load. Thus, an electrical power ei is generated. This is the *generator action.*
- Whenever a current-carrying conductors of a coil are placed (perpendicularly) in a magnetic flux, a force is experienced by each conductor giving rise to an electromagnetic torque τ_e. This torque can rotate a mechanical load at an angular speed ω. Thus, a mechanical power $\tau_e\omega$ is generated. This is the *motor action.*

Both the generator action and motor action go hand in hand in an electromechanical energy conversion machine. Only the direction of power flow decides whether the machine is a generator or a motor, as illustrated in Fig. 9.1. In a generator, the magnetic field converts the mechanical power $\tau_e\omega$ into electrical power ei. Reverse is the conversion in a motor.

Conservation of energy dictates that we must have

$$ei = \tau_e \omega \qquad \text{...(9.1)}$$

9.3 POWER CONSIDERATIONS FOR A GENERATOR

Consider a generator (Fig. 9.2) supplying power to an electrical load whose shaft is rotated by a prime mover (such as a water-turbine or a steam-turbine for large size generators (large size generators are usually called ***alternators***), and a petrol-engine or a diesel-engine for small size generators). The prime mover applies a mechanical torque τ_m on the rotor of the generator. The armature, placed in a magnetic field, is made to rotate at a speed ω. An e.m.f. e is induced in the coil. If the coil terminals AB are connected to an external electrical load R_L, a current i flows.

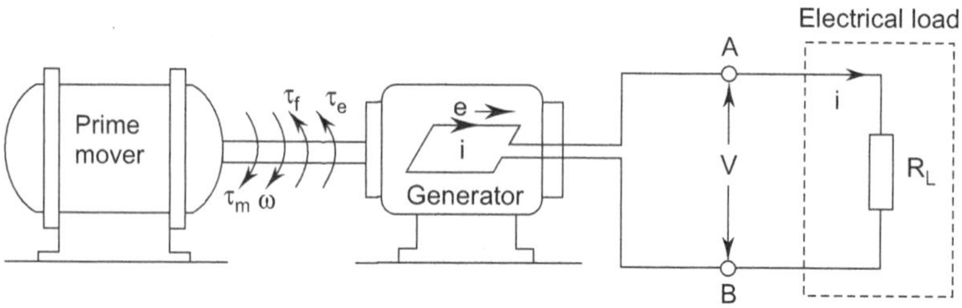

Fig. 9.2

The current-carrying coil in the armature kept in magnetic field experiences a torque, called electromagnetic torque τ_e. This is the *reaction torque* and it opposes the applied torque τ_m. In addition, there is also a frictional torque τ_f that opposes the applied torque τ_m.

Thus, the applied torque τ_m overcomes both the electromagnetic torque τ_e and the frictional torque τ_f, as shown in Fig. 9.2. That is,

$$\tau_m = \tau_e + \tau_f \qquad \text{...(9.2)}$$

Multiplying both sides by ω, we get

$$\tau_m \omega = \tau_e \omega + \tau_f \omega \qquad \text{...(9.3)}$$

This equation simply states that out of the total mechanical power $\tau_m \omega$ supplied by the prime mover, only the power $\tau_e \omega$ is converted by the generator into electrical power ei. The remaining power $\tau_f \omega$ is wasted as frictional loss.

Now, let us see what happens at the output side. If r is the resistance of the coil, the current i in the circuit is given as

$$i = \frac{e}{r + R_L} \qquad \text{...(9.4)}$$

or

$$e = ir + iR_L \qquad \text{...(9.5)}$$

Multiplying both sides by i, we get

$$ei = i^2 r + i^2 R_L \qquad \text{...(9.6)}$$

or

$$ei = i^2 r + iV \qquad \text{...(9.7)}$$

where, $V = iR_L$ is the terminal voltage. This equation simply states that out of the electrical power ei generated by the generator, only the power Vi is delivered to the electrical load R_L and the remaining power i^2r is wasted as copper-loss in the coil. Thus, the power flow diagram for a generator can be represented as shown in Fig. 9.3.

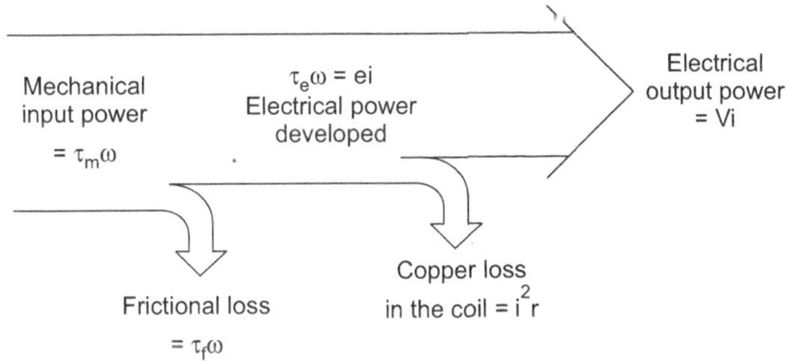

Fig. 9.3

9.4 POWER CONSIDERATIONS FOR A MOTOR

The same rotary machine can work as a motor. An electrical source of voltage V is connected to the terminals AB of the armature-coil as shown in Fig. 9.4 in which a motor is supplying power to a mechanical load. Current i flows through the conductors of the coil. Since these current-carrying conductors of the coil are placed in a magnetic field, an electromagnetic torque τ_e is developed. This causes the coil (and hence the rotor) to rotate at a speed ω. As a result, the mechanical load (say, a water pump) connected to the shaft of the motor also rotates at speed ω.

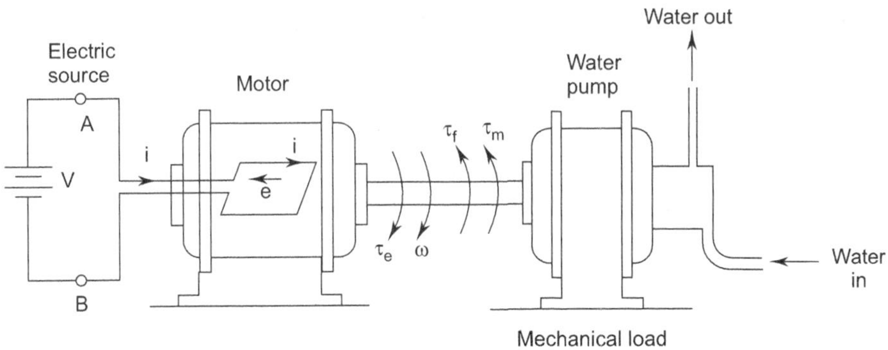

Fig. 9.4

Since the conductors of the coil are cut through the magnetic field, an e.m.f. e is induced. This is the *reaction e.m.f.* and it opposes the applied voltage V. The difference of the applied voltage V and the induced e.m.f. e is responsible for causing the current i to flow in the coil. That is,

$$i = \frac{V - e}{r} \qquad \ldots(9.8)$$

or

$$V = e + ir \qquad \ldots(9.9)$$

where, r is the resistance of the coil. Note that in a motor the direction of the induced e.m.f. e is opposite to that of the current i; whereas in a generator, the e.m.f. e and the current i are in the same direction. Multiplying both sides of eqn. (9.9) by i, we get

$$Vi = ei + i^2r \quad ...(9.10)$$

This equation simply states that out of the total power V*i* supplied by the electrical source, only the power *ei* is made available for conversion to mechanical power, the remaining power i^2r is wasted as copper-loss in the coil.

Now, next let us see what happens at the output side. The motor converts electrical power *ei* into mechanical power $\tau_e\omega$ [eqn.(9.1)]. Out of this mechanical power $\tau_e\omega$ developed by the motor, a portion $\tau_f\omega$ is lost in overcoming the friction; the remaining part $\tau_m\omega$ is supplied to the mechanical load. That is,

$$\tau_e\omega = \tau_f\omega + \tau_m\omega \quad ...(9.11)$$

Thus, the power flow diagram for an electrical motor can be represented as shown in Fig. 9.5.

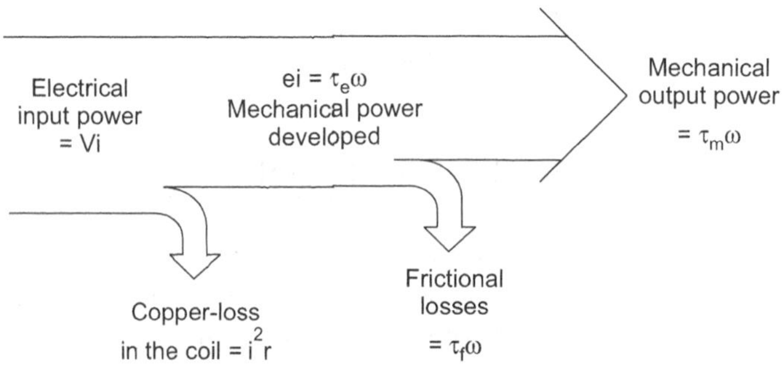

Fig. 9.5

9.5 HOW A GENERATOR DIFFERS FROM A MOTOR

With reference to Fig. 9.2 and Fig. 9.4, we can bring out the essential differences between a generator and a motor as follows:

- The directions of induced e.m.f *e* and current *i* are same in a generator but opposite in a motor (In fact, we can conclude that a circuit having *e* and *i* in the same direction acts as a source of electrical energy).
- The directions of electromagnetic torque τ_e and rotation ω of shaft are opposite in a generator but same in a motor.
- The directions of electromagnetic torque τ_e and friction torque τ_f are same in a generator but opposite in a motor.

9.6 HOW MECHANICAL FORCE IS DEVELOPED IN A MACHINE

A mechanical energy system must have a mechanical force associated with a displacement of its point of action. There are two ways in which a mechanical force can be developed in an electromagnetic system:

(*i*) by alignment, and
(*ii*) by interaction.

The force of alignment can be illustrated by the arrangement shown in Fig. 9.6. In the first case, the two poles are situated opposite to one another. The flux passes from one pole to the other. The two poles are attracted towards one another. The force of alignment acts in a direction that will increase the magnetic energy stored in the arrangement. In this case, the force will try to bring the poles together thereby decreasing the reluctance of the air-gap in the magnetic circuit. Consequently, this will increase the stored magnetic energy.

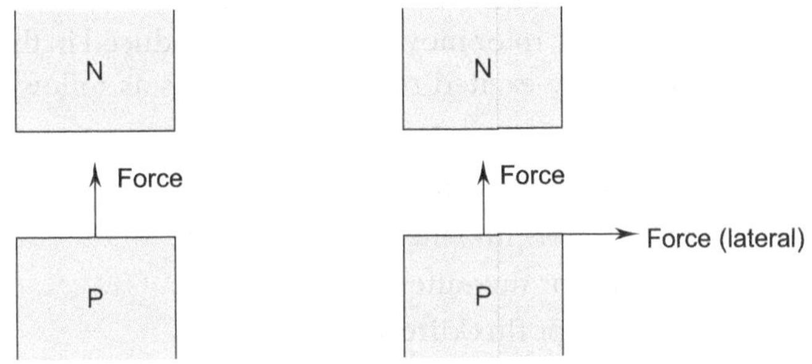

(a) Force of attraction (b) Lateral force of alignment

Fig. 9.6

In Fig. 9.6(*b*), the two poles are not situated opposite to one another. In addition to the force of attraction, there is a lateral force that tries to move the poles laterally. If the poles move laterally, the cross-sectional area of the air-gap will increase. This again will reduce the reluctance and consequently increase the stored magnetic energy.

The *force of alignment* can give rise to linear motion, as in an electromagnetic relay shown in Fig. 9.7(*a*). When the coil is energized, the armature is pulled in the direction indicated. Such relays are used to operate switches in many gadgets. The force of alignment can also be used to produce a rotary motion, as in the *reluctance motor* shown in Fig. 9.7(*b*). The lateral force of alignment on the ends of the rotating piece, termed the *rotor*, gives rise to rotating torque.

Fig. 9.7(*c*) shows a simple machine which uses the *force of interaction* to give rise to rotary motion. By passing current through the coil, it experiences a force on each conductors of the coil resulting in a torque about the axis of rotation.

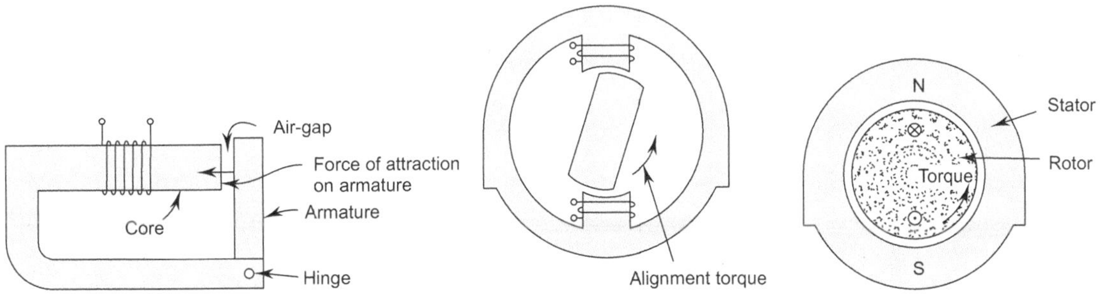

(a) Electromagnetic relay. (b) Reluctance motor. (c) Rotary machine.

Fig. 9.7

9.7 TYPE OF ROTARY MACHINES

The force of interaction finds applications in different types of ***rotary machines***. Essentially, an electrical rotary machine has two parts–the fixed part, called the ***stator***, and the moving part, called the ***rotor***. The machine has a *magnetic system* (usually excited by a current in winding; a ***winding*** is a series connection of coils) on one part and an *armature* winding on the other. When these windings are excited by an AC or DC currents, their fluxes interact with each other giving rise to rotating torque. And, when the rotor moves, an e.m.f. is induced in the armature. There exist three important categories of doubly-excited rotating machines as follows:

1. *Synchronous machines:* Stator flux-alternating current
   Rotor flux-direct current
2. *Asynchronous machines:* Stator flux-alternating current
   Rotor flux-alternating current
3. *Commutator machines:* Stator flux-direct current
   Rotor flux-direct current.

10

DC GENERATOR

10.1 INTRODUCTION

Converters that are used to transform mechanical energy into electrical energy are called Generator. In case the electrical power obtained is in DC form, then it is known as DC generator. The process of transformation is known as electromechanical energy conversion.

10.2 CONSTRUCTION OF A DC MACHINE

A DC machine mainly consists of two parts:

- Stator
- Rotor

These two main parts are further classified as shown in the flow chart.

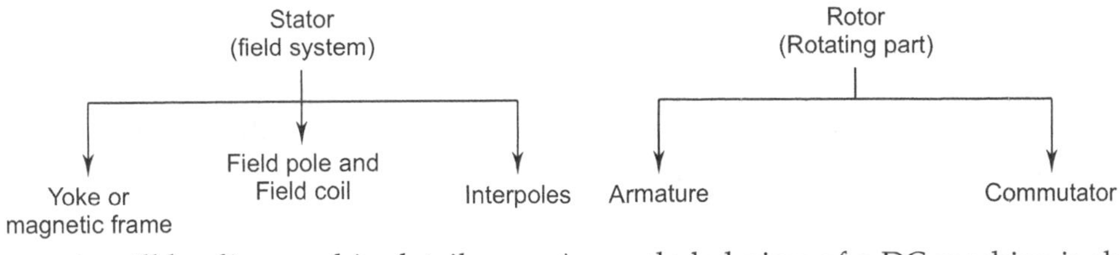

These parts will be discussed in detail now. An exploded view of a DC machine is shown in Fig. 10.1 and Fig. 10.2 shows a sketch of a typical DC machine.

Fig. 10.1

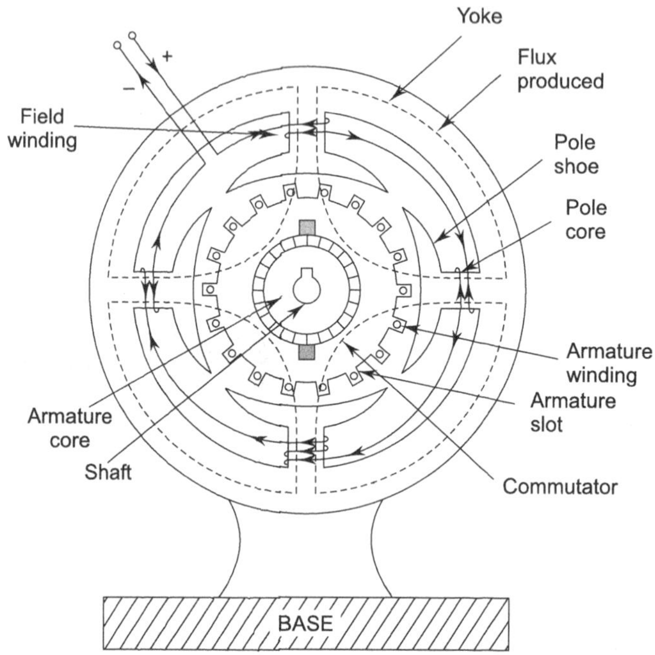

Fig. 10.2 *Construction of a DC Machine*

10.2.1 Armature and Commutator

The armature is the part of a DC machine that rotates and provides energy at the end of the shaft. It is basically an electromagnet, since it is a coil of wire that has to be specially designed to fit around core material on the shaft. The core of the armature is made of laminated steel and provides slots for the coils of wire to be pressed onto. Fig. 10.3(*a*) shows a sketch of a typical DC machine armature. Fig. 10.3(*b*) shows the laminated steel core of the armature without any coils of wire on it. This gives a better look at the core.

The armature core is made of laminated steel to prevent the circulation of eddy currents. If the core were solid, magnetic currents would be produced that would circulate in the core material near the surface and cause the core metal to heat up. These magnetic currents are called *eddy currents.* When laminated steel sections are pressed together to make the core, the eddy currents cannot flow from one laminated segment to another, so they are effectively cancelled out. The laminated core also prevents other magnetic losses called *flux losses.* These losses tend to make the magnetic field weaker so that more core material is required to obtain the same magnetic field strengths. The flux losses and eddy current losses are grouped together by designers and called *core losses.* The laminated core is designed to allow the armature's magnetic field to be as strong as possible since the laminations prevent core losses.

Notice that one end of the core has commutator segments. There is one commutator segment for each end of each coil. This means that an armature with four coils will have eight commutator segments. The commutator segments are used as a contact point between the stationary brushes and the rotating armature. When each coil of wire is pressed onto the armature, the end of the coil is soldered to a specific commutator segment. This makes an electrical terminal point for the current that will flow from the brushes onto the commutator segment and finally through the coil of wire. Fig. 10.3(*c*) shows the coil of wire before it is mounted in the armature slot, and Fig. 10.3(*d*) shows the coil mounted in the armature slot and soldered to the commutator segment.

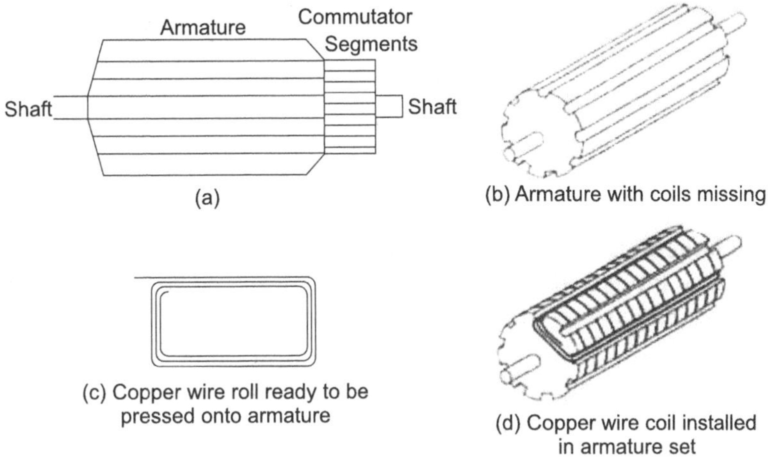

(a)

(b) Armature with coils missing

(c) Copper wire roll ready to be pressed onto armature

(d) Copper wire coil installed in armature set

Fig. 10.3

The shaft is designed so that the laminated armature segments can be pressed onto it easily. It is also machined to provide a surface for a main bearing to be pressed on at each end. The bearing will ride in the end plates and support the armature when it begins to rotate. One end of the shaft is also longer other, since it will provide the mounting shaft for motor's load to be attached. Some shafts have a key way or flat spot machined into them so that the load that is mounted on it can be secured. One must be careful when handling a motor so that one do not damage the shaft, since it must be smooth to accept the coupling mechanism. It is also possible to bend the shaft or cause damage to the bearings so that the motor will vibrate when it is operating at high speed. The commutator is made of copper. A thin section of insulation is placed between each commutator segment. This effectively isolates each commutator segment from all others.

10.2.2 Yoke and Poles

The armature is placed inside the Yoke (Magnetic frame) of motor where the field coils are mounted. When the field coils and armature coils become magnetized, the armature will begin to rotate. The field winding is made by coiling up a long piece of wire. The wire is mounted on laminated pole pieces called field poles. Similar to an armature, these poles are made of laminated steel or cast iron to prevent eddy current and other flux losses. Fig. 10.4(*a*) shows the location of the pole pieces inside a DC machine frame. Fig. 10.4(*b*) shows an individual pole piece. You can see that it is made of laminated sections. The field coils are wound around the pole pieces as shown in Fig. 10.4(*c*).

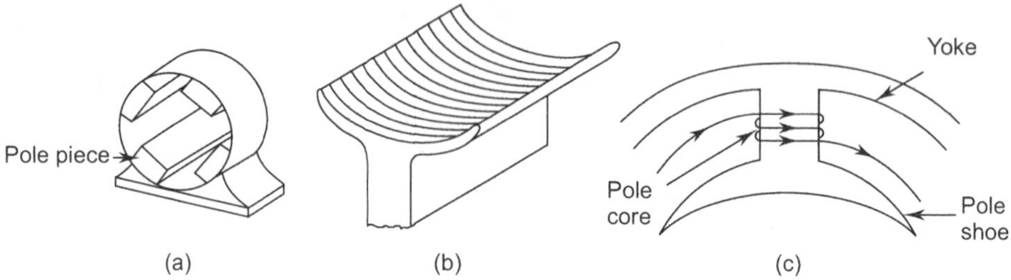

Fig. 10.4

The amount of wire that is used to make the field winding will depend on the type of machine that is being manufactured. A series machine uses heavy-gauge wire for its field winding so that it can handle very large field currents. Since the wire is of large gauge, the number of turns of wire in the coil will be limited. If the field winding is designed for a shunt machine, it will be made of small-gauge wire and many turns can be used.

After the coils are wound, they are coated for protection against moisture and other environmental elements. After they have been pressed onto the field poles, they must be secured with shims or bolts so that they are held rigidly in place. Remember: When current is passed through the coil, it will become strongly magnetized and attract or repel the armature magnetic poles. If the field poles are not rigidly secured, they will be pulled loose when they are attracted to the armature's magnetic field and then pressed back into place when they become repelled. This action will cause the field to vibrate and damage the outer protective insulation and cause a short-circuit or a ground condition between the winding and the frame of the machine.

The ends of the frame are machined so that the end plates will mount firmly into place. An access hole is also provided in the side of the frame or in the end plates so that the field wires can be brought to the outside of the machine, where DC voltage can be connected.

The bottom of the frame has the mounting bracket attached. The bracket has a set of holes or slots provided so that the machine can be bolted down and securely mounted on the machine it is driving. The mounting holes will be designed to specifications by frame size.

The dimensions for the frame sizes are provided in tables printed by motor manufacturers. Since these holes and slots are designed to a standard, you can predrill the mounting holes in the machinery before the machine is put in place. The slots are used to provide minor adjustments to the mounting alignment when the machine is used in belt-driven or chain-driven applications. It is also important to have a small amount of mounting adjustment when the machine is used in direct-drive applications. It is very important that the machine be mounted so that the armature shaft can turn freely and not bind with the load.

10.3 PRINCIPLE AND WORKING OF DC GENERATOR

An electrical generator is a machine which converts mechanical energy or mechanical power into electrical energy or electrical power.

The energy conversion is based on the principle of the production of dynamically induced e.m.f. *"Whenever a conductor is placed in magnetic field and is rotating in the field, it cuts magnetic flux, and according to Faraday's Law of Electromagnetic Induction, dynamically induced e.m.f is produced in it. This e.m.f. causes a current to flow if the conductor circuit is closed."*

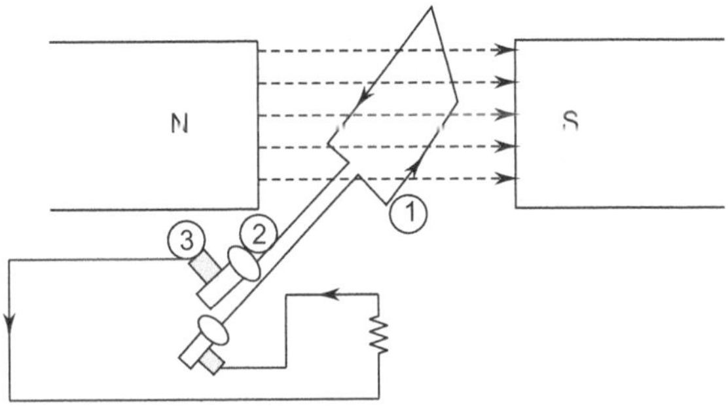

1. Conductor rotating in the magnetic field.
2. Slip rings.
3. Carbon brushes to collect current.

Fig. 10.5

As shown in Fig. 10.5, when the conductor placed in magnetic field is rotates, it cut line of *flux*, due to which an e.m.f. is induced in it. The circuit is close and hence a current start flowing in the closed circuit.

But this current is of alternating nature.

It is obviously, different from a direct current which continuously flows in one and the same direction.

For making the flow of current unidirectional (or direct) in the external circuit, the slip-rings are replaced by split-rings.

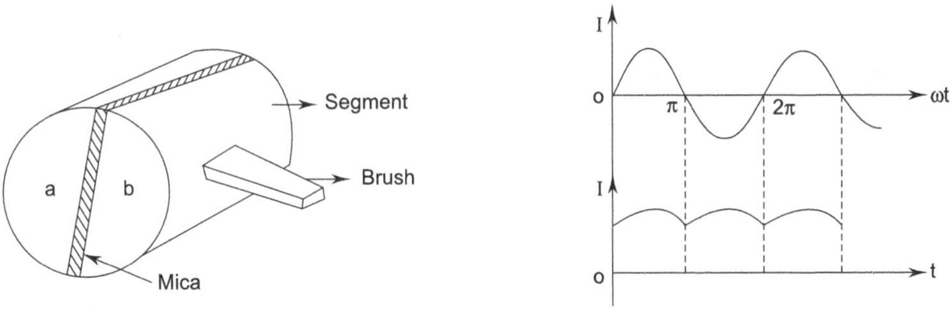

(a) (b)

Fig. 10.6

Now after using split rings, the current becomes unidirectional as shown in Fig. 10.7. Even now, we haven't got pure DC. That pure DC can be obtained by using rectifiers.

10.4 TYPES OF DC GENERATOR

DC generators are mainly of two types :

1. Separately excited DC Generator
2. Self excited DC Generator
   - Series wound
   - Shunt wound
   - Compound wound
     - Short shunt compound wound
     - Long shunt compound wound

Electrical equivalent circuit diagrams of these types of DC generators are shown in Figs. 10.7 to 10.11. The notation and symbols used while describing the electrical circuits are as follow:

- In all these circuit diagrams, the armature is represented by a circle and its terminals are marked as A and AA.
- Carbon brushes are shown by small blocks on two sides of the armature.
- The generated e.m.f.* developed in the armature winding is represented by symbol E_g.
- The field is represented by a coil and its terminals are marked as Z and ZZ.
- Different currents and voltages shown in the circuit diagrams are:

  I_f = Field current
  I_a = Armature current
  I_L = Load current
  E_g = Generated voltage
  V = Terminal voltage

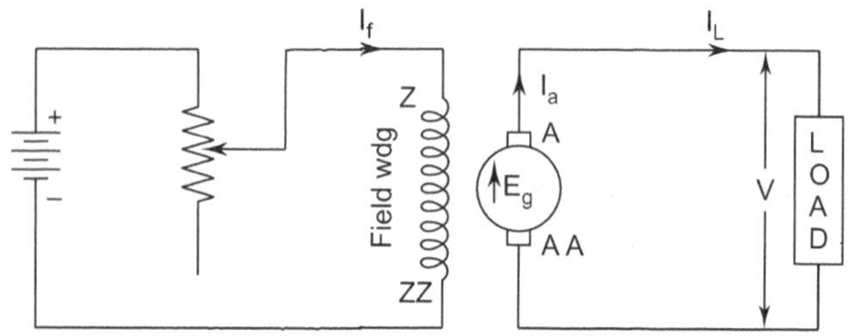

Fig. 10.7 *Separately excited DC Generator*

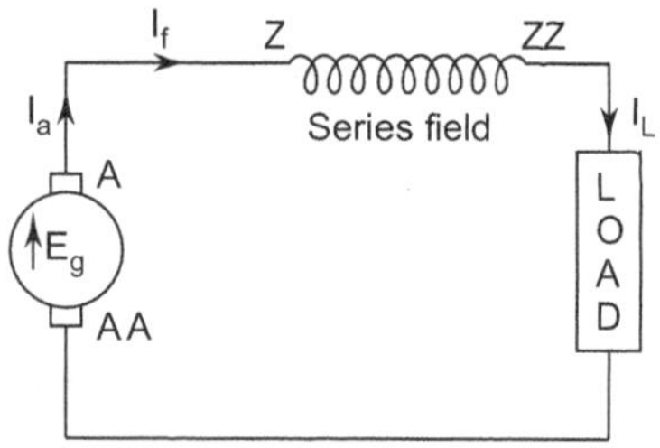

Fig. 10.8 *Series DC Generator*

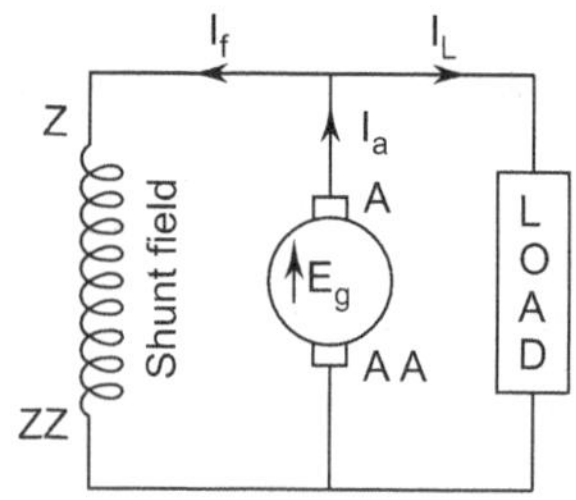

Fig. 10.9 *Shunt DC Generator*

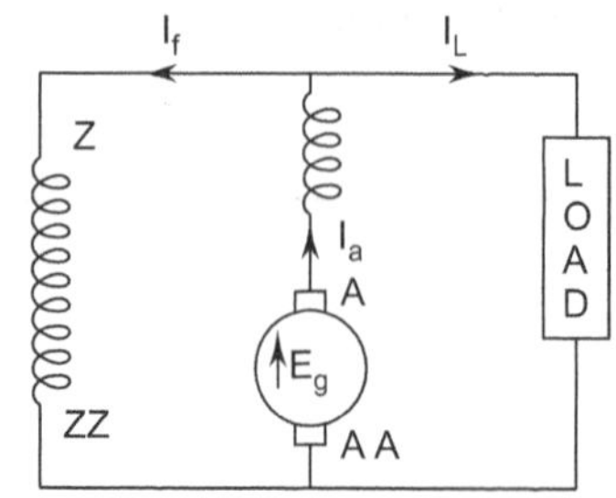

Fig. 10.10 *Long Shunt Compound wound DC Generator*

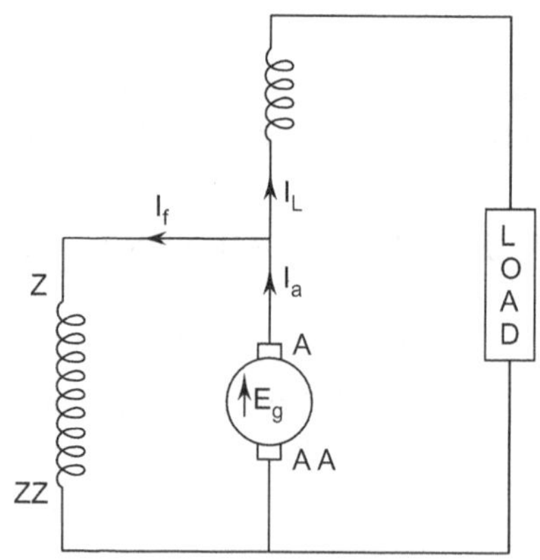

Fig. 10.11 *Short Shunt Compound wound DC Generator*

---

* Generated e.m.f. concept will be discussed in later section.

10.5 TYPES OF ARMATURE WINDINGS

Armature winding are of two types:

1. Lap Winding
2. Wave Winding

For Lap winding A = P (Number of Poles in the machine)

For Wave winding A = 2 (Number of Parallel paths)

[To remember L A P → A = P]

Note. That LAP winding is used for low voltage and high current machines whereas wave winding is used for high voltage and low current machines.

10.6 ARMATURE REACTION

By armature reaction, we mean the effect of magnetic field, set up by armature current, on the distribution of flux under main poles of a generator.

When there is no current in the armature conductors, and only field is excited, flux is distributed symmetrically as shown in Fig. 10.12.

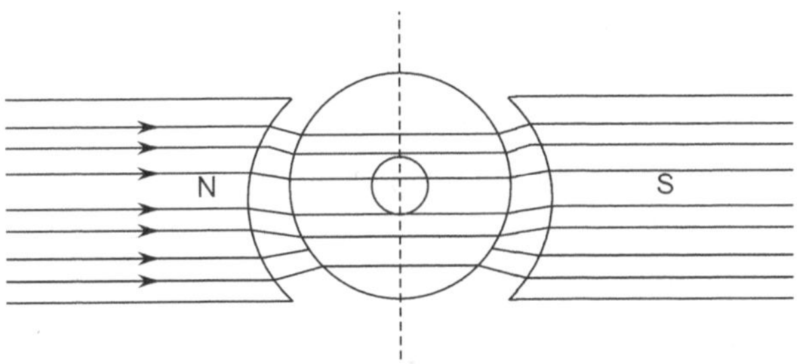

Fig. 10.12

When field is not excited and armature is carrying current in that case also the flux produced (Fig. 10.13) is symmetrical but in a direction of 90° to that of the previous flux produced.

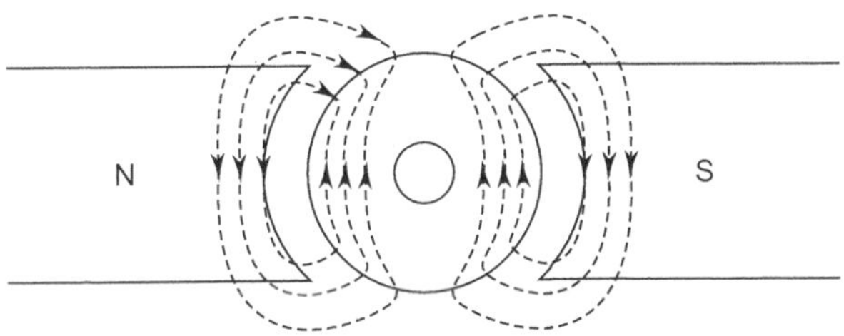

Fig. 10.13

Now consider the case of actual load condition, when field and armature both are excited, in that case the overall flux will be as shown in Fig. 10.14.

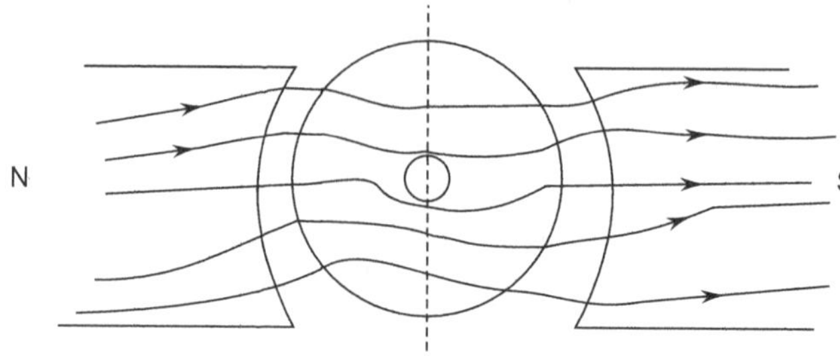

Fig. 10.14

Hence the armature magnetic field has two effects

1. It demagnetizes or weakens the main flux and
2. It cross-magnetizes or distorts it.

The first effect leads to reduced generated voltage and the second to the sparking at the brushes.

10.7 E.M.F. EQUATION OF A GENERATOR

Let

ϕ = Flux per pole in weber

Z = Total number of armature conductors

= Number of slots × number of conductors per slot

P = Number of generator pole

A = Number of parallel path in armature

N = Armature rotation in revolution per minute (*i.e.*, in rpm)

E = e.m.f. induced in any parallel path in armature

E_g = Generated e.m.f in any one of the parallel path

According to the Law of electromagnetic induction, the generated or induced e.m.f is proportional to rate of change of flux with respect to time and if only one conductor is taken into consideration then the induced e.m.f is given by

$$E_g = -\frac{d\phi}{dt} \qquad ...(10.1)$$

If there are 'Z' number of conductors then

$$E_g = -Z\frac{d\phi}{dt} \qquad ...(10.2)$$

Initially let us consider only one conductor, which is rotating inside a magnetic field having 'P' number of poles. If the conductor makes one complete rotation then to total flux cut by conductor

$$\text{Total flux} = \phi P \qquad ...(10.3)$$

If 'N' is the speed in rpm at which the conductor is rotating in that magnetic field then time taken by the conductor for one complete rotation.

Time for one complete rotation $= \frac{1}{N}$ minutes $= \frac{60}{N}$ seconds ...(10.4)

$\therefore$ Average e.m.f induce in one conductor $= \dfrac{\text{Total flux cut in one revolution}}{\text{Time taken for one complete revolution}}$ ...(10.5)

or
$$E_g = \frac{\phi P}{60/N} = \frac{\phi PN}{60} \quad ...(10.6)$$

If the machine has total 'Z' number of conductors and has 'A' number of parallel path, number of conductors in each parallel path is

$$\text{Number of conductor/Parallel path} = \frac{Z}{A} \quad ...(10.7)$$

The conductors in each parallel path are connected in series hence the total e.m.f is added.

Total e.m.f induced in all conductors will be

$$E_g = \frac{\phi PN}{60} \times \frac{Z}{A} \quad ...(10.8)$$

or
$$\boxed{E_g = \frac{\phi ZNP}{60A}} \quad ...(10.9)$$

10.8 DC GENERATOR CHARACTERISTICS

There are two types of characteristics associated with a DC generator.

1. No-Load Characteristics
2. Load Characteristics
   (*i*) Internal or Total Characteristics
   (*ii*) External Characteristics.

10.9 NO-LOAD CHARACTERISTICS

The No-Load Characteristics are also known as

Open Circuit Characteristics or

Magnetization Curve or

Magnetization Characteristics.

It gives the relation between the induced e.m.f. E_g (on Y-axis in volts) in the generator and the field current I_f (on X-axis in ampere) when the generator is driven at constant speed (generally at the rated speed) with the armature circuit kept open.

10.9.1 No-Load Characteristics of a DC Shunt Generator

- In order to build up voltage in a shunt generator, the generator's armature is first rotated at its rated speed (Fig. 10.15).

- At rated speed a small voltage is developed across the armature terminals AB, at noted condition.
- This small voltage is mainly due to the residual magnetism, which was initially present inside the field.
- This developed voltage is applied to the field (Z-ZZ) which is connected in parallel to the armature (A-AA).
- Due to this voltage a small current flows in the field-coil, as a result, the net flux increases which is equal to the sum of residual flux + flux due to current.
- Increase in flux increases the voltage across the terminals and again the current in field circuit and so on.
- This is a cumulative process as a result of which the terminal voltage increases till magnetic saturation occurs.
- That is why this generator was given the name self excited generator. The V-I characteristics are shown in Fig. 10.16.

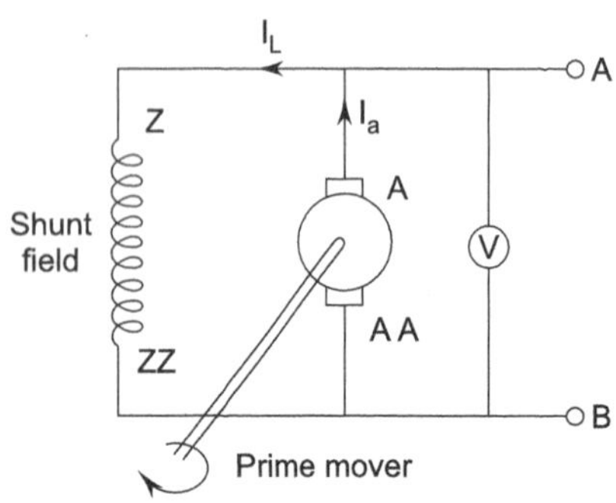

Fig. 10.15

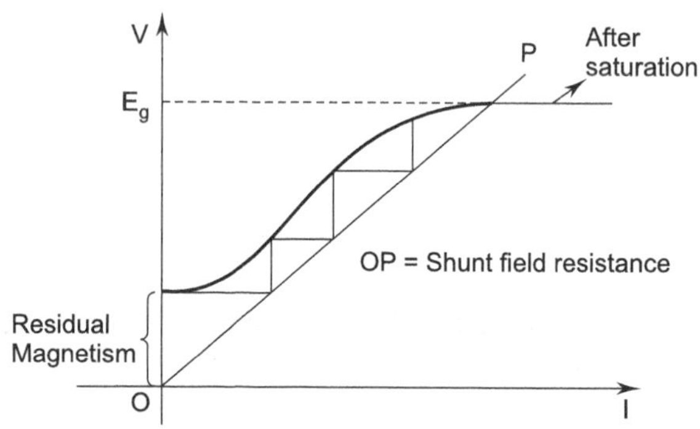

Fig. 10.16

The above description of no-load characteristics of a DC shunt generator also explains how the voltage is build up in a DC generator.

Reasons of failure of Voltage Buildup:

- There is no residual magnetism.
- Residual magnetism may have been reversed in polarity or field connections reversed.
- Field circuit resistance is above the critical field resistance.
- Driven speed is below the critical speed.
- Field circuit gets open circuited or faulty connections.

10.9.2 No-Load Characteristics of a Separately Excited DC Generator

The open circuit characteristic of a separately excited DC generator is obtained by varying the field current (*i.e.*, by varying the variable resistance in the field winding circuit). Keeping the load circuit open, the respective induced e.m.f across the armature is plotted for different value of field current (Fig. 10.17).

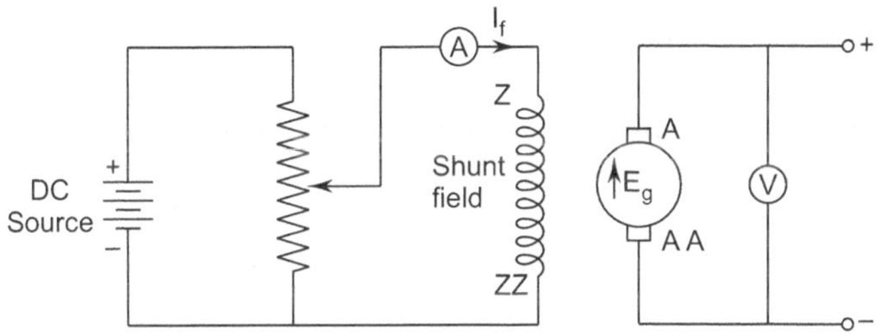

Fig. 10.17

The graph drawn (Fig. 10.18) between the field current and the no-load voltage induced across the armature terminals *i.e.*, I_f *vs* V, with generator driven at constant speed (generally the rated speed).

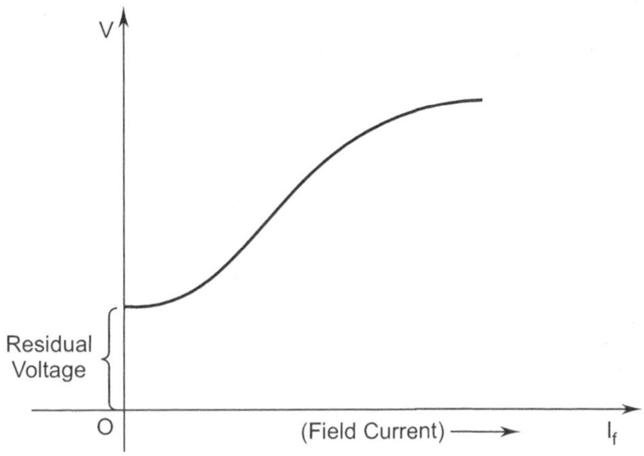

Fig. 10.18

With zero field current a small voltage is induced due to residual magnetism, hence the graph does not starts from origin. As the field current increases by means of variable resistance of field circuit the flux increases and hence the voltage also increases. This process continues till magnetic saturation occurs. The no-load characteristics of a DC generator is similar to the B-H curve of any magnetic material, hence the very name suggested magnetization curve.

As no load is applied to the generator, so armature terminals are open circuited, hence also called open circuit characteristics.

Note. *Whether it is shunt generator or series generator, to draw the no-load characteristics, the field must be disconnected from the armature and should be excited separately. Hence all kind of DC generators will have the same no-load characteristics.*

10.9.3 No-Load Characteristics of a DC Series Generator

In case of series generator, the open circuit characteristics cannot be obtained. However, if required, it is plotted by exciting the field separately. The so obtained characteristics will be similar to that of separately excited DC generator.

Note that a series generator excites by itself only if the load circuit is closed.

10.10 LOAD CHARACTERISTICS

The load characteristics of DC generator are further classified into two types:

- Internal characteristics
- External characteristics

The internal load characteristics, also known as Total Characteristics, give the relation between the on-load induced e.m.f E_g and the armature current I_a at a given speed of the generator and at a constant field current.

The external load characteristics, (also known as Performance Characteristics or Voltage regulating curve) gives the relation between the terminal voltage V and the load curve I_L for a given speed of the generator at a given field current.

10.10.1 Load Characteristics of a Separately Excited DC Generator

The following procedure is adopted to get the load characteristics.

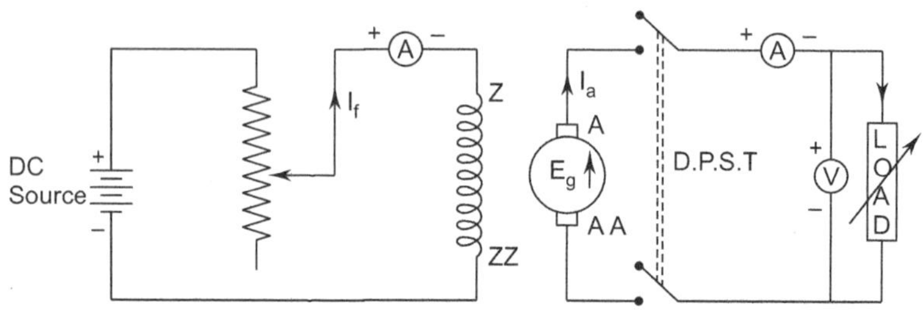

Fig. 10.19

- Keeping the load circuit open (load side D.P.S.T off), the generator is driven at rated speed (Fig. 10.19). (D.P.S.T. = Dual Port Single Terminal Switch).
- The field current, I_f is now adjusted with the help of variable resistance of the field circuit to get the rated terminal voltage.
- Then the load circuit is closed and load current I_L is increased gradually by increasing the electrical load in steps.
- The load current should be increases till the rated current is obtained.
- The corresponding terminal voltage V is noted and plotted against the armature current I_a. We get the external load characteristics as shown in Fig. 10.20.

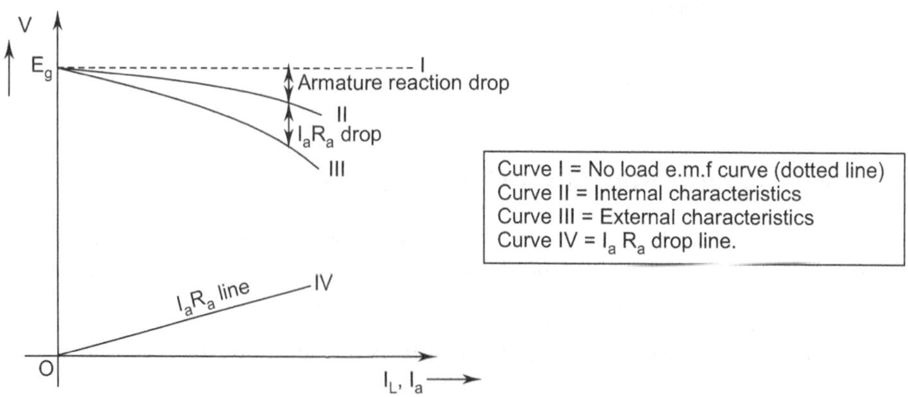

Fig. 10.20

With the increase in load current, the voltage drops in the load terminal is due to

- Armature resistance drop ($I_a R_a$)
- Armature reaction drop (Reduction in flux due to demagnetizing effect of armature reaction).

If 'E_g' is the actual voltage generated for a fixed field current and speed then,
For external characteristics

$$V = E_g - I_a R_a - \text{Armature reaction drop} \quad \ldots(10.10)$$

And for internal characteristics

$$V = E_g - \text{Armature reaction drop} \quad \ldots(10.11)$$

10.10.2 Load Characteristics of a DC Shunt Generator

- Initially the generator is run without load on its output terminal, with D.P.S.T. switch open.
- Now load is applied and is gradually increased (Fig. 10.21).

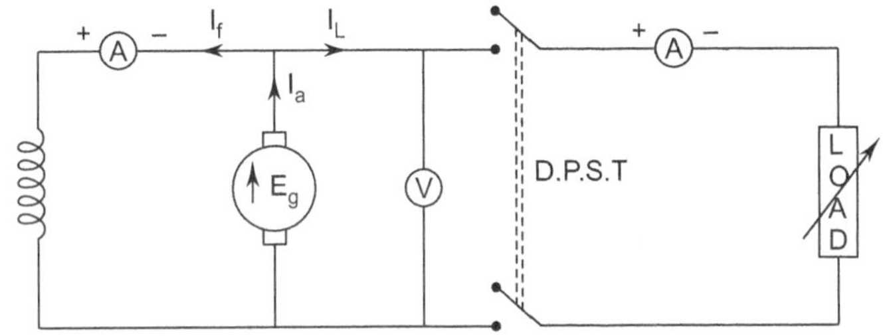

Fig. 10.21

- As the load is increased, the field current I_f will continuously reduced and the load current is increased due to fall in terminal voltage.
- The fall in the terminal voltage is because of the three reasons
  (*i*) Armature resistance drop ($I_a R_a$ drop).
  (*ii*) Armature reaction drop which reduces the air gap flux and consequently the generated voltage.
  (*iii*) Reduction in the field current and hence flux, due to fall in the terminal voltage.
- Hence due to the above reasons, shunt generator have more dropping characteristics than separately excited generators.

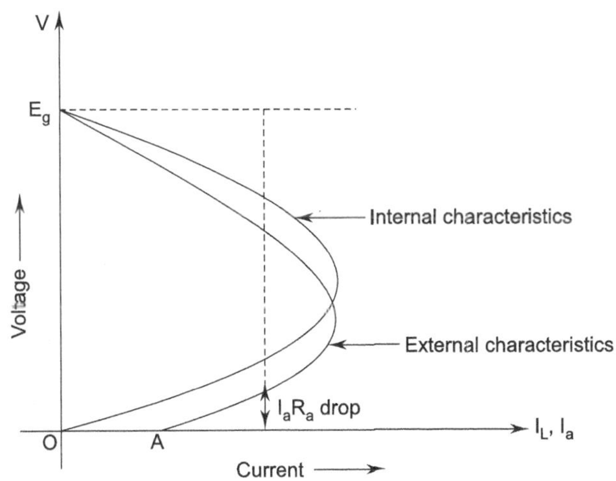

Fig. 10.22

- From the graph we can conclude that if any short-circuit occurs at the shunt generator terminals, then the load current automatically becomes zero or a small value, hence need no protective device for short-circuit. The shunt generators are therefore self protective against accidental short-circuit.

10.10.3 Load Characteristics of a DC Series Generator

- Initially the generator is run with a small load on its output terminal, with D.P.S.T. switch closed.
- Now load is gradually increased (Fig. 10.23).

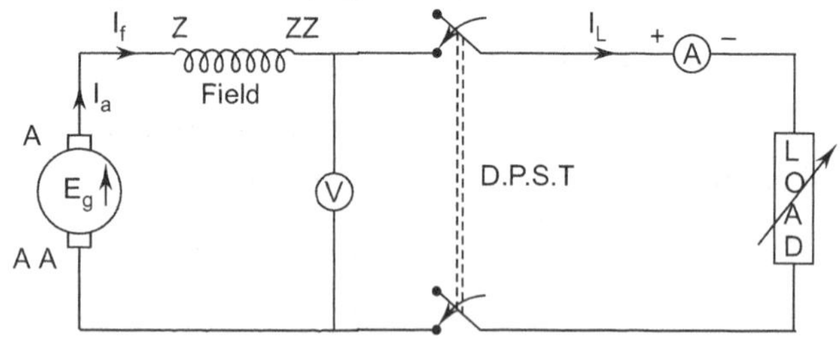

Fig. 10.23

- As in a series generator, the field circuit is connected in series with the armature, hence increase in load current means increase in the field current.
- This increase in the field current in turn increases the flux; as a result, the voltage is increased irrespective of any drop.
- Even though there is voltage drop but increase in the flux dominates the voltage drop, as a result the voltage increases with the increase in current.
- Hence series generator has rising voltage characteristics.

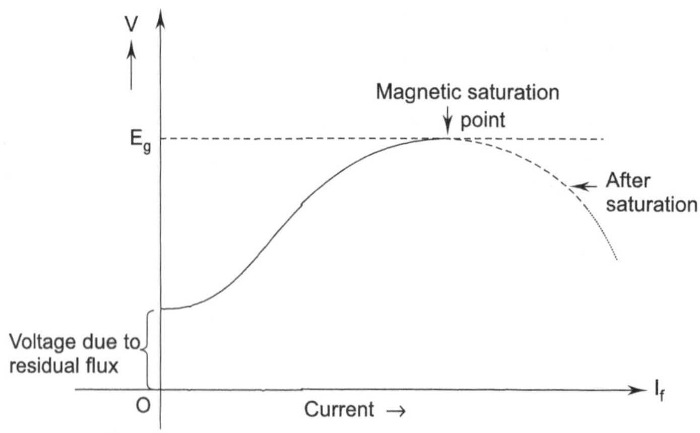

Fig. 10.24

- At zero load current, the voltage is not zero rather it has a small value due to residual magnetism.
- With increase in load current, the terminal or load voltage increases till magnetic saturation occurs.
- If the load current is increased even after magnetic saturation then the voltage decreases due to various drops as mentioned earlier. ($I_a R_a$, Armature Reaction etc.,)

10.10.4 Load Characteristic of DC Compound Generator

In shunt generator the terminal voltage decreases with increase in load current but in case of series generator the terminal voltage increases with increase in load current. Compound generator has both series and shunt field.

A. Load Characteristics of Cumulative Compound (or Long Shunt) Generator

In case of cumulative compound generator, shunt flux and series flux support each other.

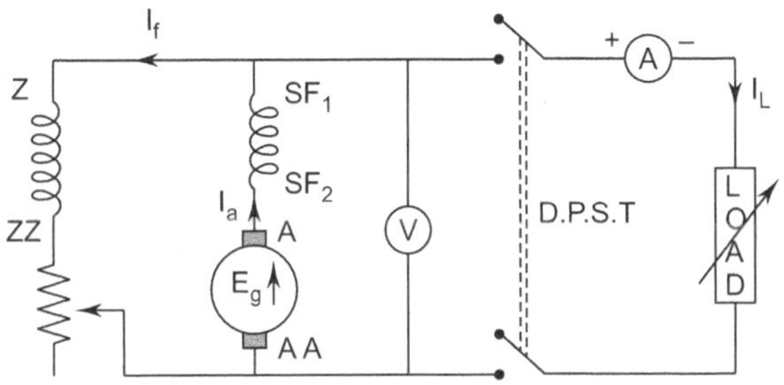

Fig. 10.25

As per the characteristics, a cumulative compound generator is of three types:

(*i*) Over compound

(*ii*) Level or Flat compound

(*iii*) Under compound.

(*i*) **Over Compound:**

If increase in series flux dominates the decrease in the shunt voltage drop and $I_a R_a$ drop then net voltage increases with increase in load current (Fig. 10.26).

Such cumulative generators are known as over compound generator. In this case large numbers of series turns are used.

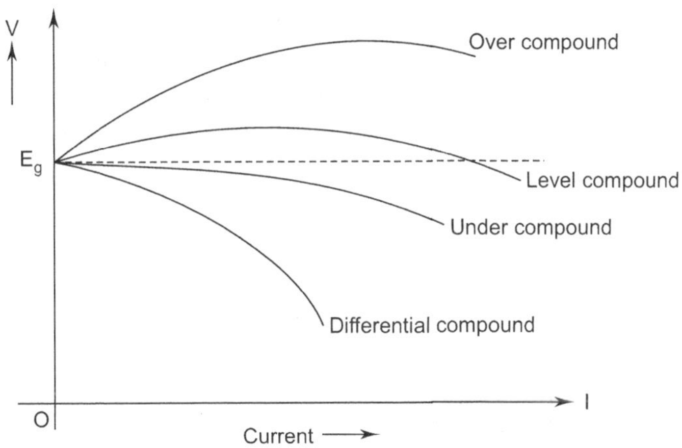

Fig. 10.26

(*ii*) **Level or Flat Compound:**

If increase in series flux and hence the corresponding voltage, just meet the voltage drop, then the terminal voltage remains the same from no-load to full-load and such compound generate is known as level or flat compound generator (Fig. 10.26).

(*iii*) **Under Compound:**

If the drop in shunt flux or drop in armature resistance dominates the increase in series flux, then the terminal voltage slightly decreases with increase in load current. Such compound generator is known as under compound generator (Fig. 10.26).

B. Load Characteristics of Differential Compound (or Short Shunt) Generator

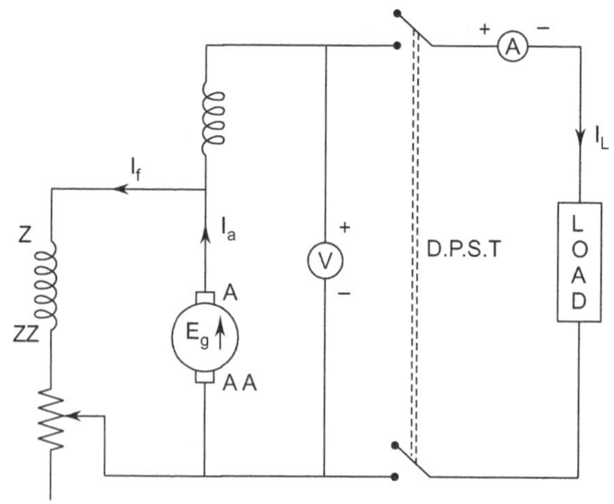

Fig. 10.27

In case of differential compound DC generator the shunt flux and the series flux opposes each other, as a result net flux and hence the terminal voltage fall rapidly (Fig. 10.26).

10.11 CRITICAL RESISTANCE AND CRITICAL SPEED

Run the DC machine as a shunt generator. Due to residual magnetism in the poles, some e.m.f and hence current, would be generated. This current while passing through the field coils will strengthen the magnetism of the poles (provided field coils are properly connected as regards polarity). This will increase the pole flux which will further increase the generated e.m.f. Increased e.m.f. means more current which further increases the flux and so on. This mutual reinforcement of e.m.f. and flux proceeds on till equilibrium is reached at some point like P (Fig. 10.28).

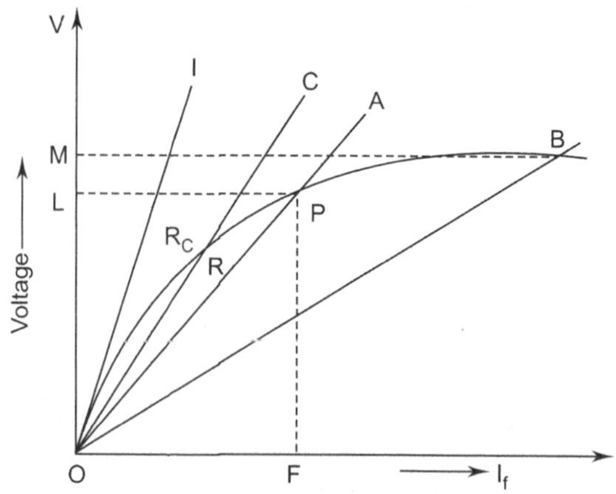

Fig. 10.28

The point lies on the resistance line OA of the field winding. Let R be the resistance of the field winding. Line OA is drawn such that its slope equals the field winding resistance *i.e.*, every point on this curve is such that volt/ampere = R.

The voltage OL corresponding to point P represents the maximum voltage to which the machine will build up with R as field resistance. OB represents smaller resistance and the corresponding voltage OM is slightly greater than OL. If field resistance is increased, then slope of the resistance line increased, and hence the maximum voltage to which the generator will build up at a given speed, decreases. If R is increased so much that the resistance line does not cut the O.C.C. at all (like OT), F then obviously the machine will fail to excite *i.e.*, there Fig. 10.28 will be no 'build up' of the voltage.

If the resistance line just lies along the slope, then with that value of field resistance, the machine will just excite. The value of the resistance represented by the tangent to the curve is known as ***critical resistance*** R_C for a given speed.

Critical speed of a shunt generator is that speed for which the given shunt field resistance represents critical resistance. In Fig. 10.29, curve 2 corresponds to critical speed because R_{sh} line is tangential to it.

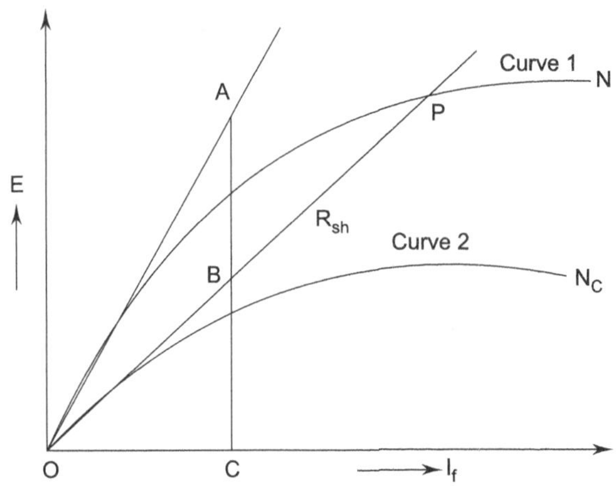

Fig. 10.29

Obviously $$\frac{BC}{AC} = \frac{N_C}{\text{Full speed}} = \frac{N_C}{N} \qquad ...(10.12)$$

Therefore, $$N_C = \frac{BC}{AC} \times \text{Full speed}\,(N) \qquad ...(10.13)$$

10.12 LOSSES

Different types of losses occur in a DC generator are listed below.

A. Copper Losses

The copper losses can be divided into three parts:

(*i*) **Armature Copper Loss**

$$P_a = I_a^2 R_a \qquad ...(10.14)$$

(*ii*) **Field Copper Loss**

$$P_{F,\,shunt} = I_{sh}^2 R_{sh} \quad \text{(in shunt winding)} \qquad ...(10.15)$$

$$P_{F,\,series} = I_{se}^2 R_{se} \quad \text{(in series winding)} \qquad ...(10.16)$$

(*iii*) **Brush Contact Loss**

This loss is due to the resistance of brush contact and is included in armature copper loss.

B. Iron Losses

These losses occur in armature and field cores. These are basically divided into two parts

(*i*) **Hysteresis Loss**

$$P_h = K_h f B_m^n \qquad ...(10.17)$$

(*ii*) **Eddy Current Loss**

$$P_e = K_e f^2 B_m^2 \qquad ...(10.18)$$

C. Mechanical Losses

These losses are due to air friction and windage loss of rotating armature.

D. Stray Losses

Iron loss and Mechanical loss are collectively called as stray losses or constant losses.

10.13 EFFICIENCY

Efficiency of a DC generator is given by

$$\eta = \frac{\text{Output electrical power}}{\text{Input mechanical power}} \qquad ...(10.19)$$

or

$$\% \eta = \frac{\text{VI(Watt)}}{\text{BHP} \times 735.5} \times 100 \qquad ...(10.20)$$

Maximum efficiency will occur when

$$\text{Variable losses} = \text{Constant losses} \qquad ...(10.21)$$

or

$$\text{Cu loss} = \text{Stray loss} \qquad ...(10.22)$$

10.14 VOLTAGE REGULATION

The Voltage Regulation of a DC generator is the change in the voltage when the load is reduce from rated value to zero, expressed as percentage of the rated load voltage.

$$\% \text{Regulation} = \frac{V_{NL} - V_{FL}}{V_{rated}} \times 100 \qquad ...(10.23)$$

10.15 APPLICATIONS OF DC GENERATORS

A. Separately Excited Generator

This is rarely used because it needs a separate source for excitation. It is used where self excited generators are relatively unsatisfactory.

B. Series Generators

- Used on a constant current source due to its rising V-I characteristics.
- For regenerative braking in locomotives
- Series arc lighting
- Series boosters (for *e.g.*, in railway services).

C. Shunt Generators

- As constant voltage source
- For ordinary lighting and power supply
- For charging batteries.

D. Cumulatively Compound Generators

- For motor driving
- For lamp loads
- For heavy power service such as electric railway.

E. Differentially Compound Generators

- As a constant current generator
- For arc welding.

SOLVED PROBLEMS

Problem 10.1. *A 4-pole DC shunt generator with lap-connected armature has field and armature resistances of 80 Ω and 0.1 Ω respectively. It supplies power to 50 lamps rated for 100 volts, 60 watts each. Calculate the total armature current and the generated e.m.f. by allowing a contact drop of 1 volt per brush.*

Solution. Total lamp load, P_L = Number of lamps × wattage of each lamp = 50 × 60 = 3,000 W

Terminal voltage, $V_T = 100$ V

Load current, $$I_L = \frac{P_L}{V_T} = \frac{3,000}{100} = 30 \text{ A}$$

Shunt field current, $$I_f = \frac{V_T}{R_f} = \frac{100}{80} = 1.25 \text{ A}$$

Total armature current, $I_a = I_L + I_f = 30 + 1.25 = \mathbf{31.25\ A}$ **Ans.**

Generated e.m.f., $E_g = V + I_a R_a + \text{brush drop}$

$= 100 + 31.25 \times 0.1 + 2 \times 1 = \mathbf{105.125\ V}$. **Ans.**

Problem 10.2. *A 4-pole DC shunt generator with wave wound armature has 40 slots each having 12 conductors. Armature resistance is 1 ohm and shunt field resistance is 200 ohm. The flux per pole is 25 mWb. If a load of 50 ohm is connected across the armature terminals, calculate the voltage across the load when the generator is driven at 1,000 rpm. What will be the load voltage if the generator is lap wound?*

Solution. *(a) With wave connected armature*

Generated e.m.f, $$E_g = \frac{\phi ZN}{60} \times \frac{P}{A} = \frac{25 \times 10^{-3} \times 40 \times 12 \times 1,000}{60} \times \frac{4}{2} = 400 \text{ V}$$

Let the terminal voltage be V_T volts

Shunt field current, $$I_f = \frac{V_T}{R_f} = \frac{V_T}{200} \text{ A}$$

Load current, $$I_L = \frac{V_T}{R_L} = \frac{V_T}{50} \text{ A}$$

Armature current, $$I_a = I_L + I_f = \frac{V_T}{50} + \frac{V_T}{200}$$

or $$I_a = \frac{V_T}{40} \text{ A}$$

Since generated e.m.f,

$$E_g = V_T + I_a R_a \text{ (neglecting brush drop)}$$

$$400 = V_T + \frac{V_T}{40} \times 1$$

or Terminal voltage, V_T = **390.24 V** **Ans.**

(*b*) *With lap connected armature*

Generated e.m.f, $$E_g = \frac{25 \times 10^{-3} \times 40 \times 12 \times 1000 \times 4}{60 \times 4}$$

or $$E_g = 200 \text{ V}$$

Terminal voltage, $V_T = \dfrac{200}{41/40}$ = **195.12 V** **Ans.**

Problem 10.3. *In case of a DC generator, what will be change in e.m.f. induced if flux is reduced by 20% and the speed is increased by 20%?*

Solution. Since induced e.m.f, $E \propto \phi N$

$\therefore$ $$E_1 = \phi_1 N_1 \quad ...(i)$$

and $$E_2 = \phi_2 N_2 \quad ...(ii)$$

Dividing equation (*ii*) by equation (*i*), we get

$$E_2 = E_1 \times \frac{\phi_2}{\phi_1} \times \frac{N_2}{N_1} = E_1 \times 0.8 \times 1.2 = 0.96 \, E_1.$$

So change in e.m.f. $= \dfrac{E_1 - E_2}{E_1} \times 100 = \dfrac{E_1 - 0.96E_1}{E_1} \times 100$ = **4% decrease**. **Ans.**

Problem 10.4. *A 20 kW, 200 V shunt generator has an armature resistance of 0.05 Ω and a shunt field resistance of 200 Ω. Calculate the power developed in the armature when it delivers rated output.*

Solution. Load current, $$I_L = \frac{P}{V_T} = \frac{20{,}000}{200} = 100 \text{ A}$$

Shunt field resistance, $R_f = 200\ \Omega$

Shunt field current, $$I_f = \frac{V_T}{R_f} = \frac{200}{200} = 1 \text{ A}$$

Armature current, $I_a = I_L + I_f = 100 + 1 = 101$ A

Generated e.m.f, $E_g = V_T + I_a R_a$

$= 200 + 101 \times 0.05$

$= 205.05$ V

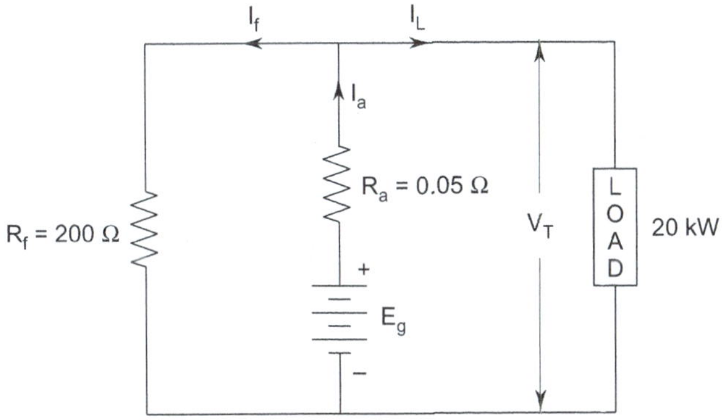

Fig. 10.30

Power developed, $P_g = E_g \times I_a = 205.05 \times 101 =$ **20,710 W or 20.71 kW** **Ans.**

Problem 10.5. *A 4-pole DC generator with wave connected armature has 41 slots, and 12 conductors/slot. Armature resistance is $R_a = 0.5\ \Omega$ and field resistance is $R_f = 200\ \Omega$. Flux per pole is 125 mWb. Speed is 1000 rpm. Calculate voltage across 10 Ω load resistance across the armature terminal.*

Solution. Generated e.m.f, $E_g = \dfrac{\phi ZN}{60} \times \dfrac{P}{A} = \dfrac{125 \times 10^{-3} \times 41 \times 12 \times 1{,}000}{60} \times \dfrac{4}{2} = 2{,}050$ V

Let the voltage across 10 Ω load resistance be V_T

Load current, $I_L = \dfrac{V_T}{10}$ A

Shunt field current, $I_{sh} = \dfrac{V_T}{R_{sh}} = \dfrac{V_T}{200}$ A

Armature current, $I_a = I_L + I_f = \dfrac{V_T}{10} + \dfrac{V_T}{200} = \dfrac{21\ V_T}{200}$ A

Since generated e.m.f, $E_g = V_T + I_a R_a V_T$ (neglecting brush contact voltage drop)

$$2{,}050 = V_T + \frac{21\ V_T}{200} \times 0.5 = \frac{421}{400}\ V_T$$

or $V_T = \dfrac{2{,}050 \times 400}{421} =$ **1,947.7 V** **Ans.**

Problem 10.6. *A DC generator has an armature e.m.f. of 100 V when the useful flux per pole is 20 mWb, and the speed is 800 rpm. Calculate the generated e.m.f.*

(i) with the same flux and a speed of 1,000 rpm,

(ii) with a flux per pole of 24 mWb and a speed of 900 rpm.

Solution. Generated e.m.f, $E_1 = 100$ V

Flux per pole, $\phi_1 = 20$ mWb $= 20 \times 10^{-3}$ Wb

Speed, $N_1 = 800$ rpm

Since induced e.m.f. is proportional to the product of flux per pole ϕ and armature speed N for a given machine *i.e.,*

$$E \propto \phi N$$

or

$$\frac{E_2}{E_1} = \frac{N_2}{N_1} \times \frac{\phi_2}{\phi_1}$$

(*i*) $$E_2 = \frac{N_2}{N_1} \times \frac{\phi_2}{\phi_1} \times E_1 = \frac{1,000}{800} \times 1 \times 100 = \mathbf{125\ V}$$ **Ans.**

($\because$ $N_2 = 1,000$ rpm and $\phi_2 = \phi_1$)

(*ii*) $$E_3 = \frac{N_3}{N_1} \times \frac{\phi_3}{\phi_1} \times E_1 = \frac{900}{800} \times \frac{24 \times 10^{-3}}{20 \times 10^{-3}} \times 100 = \mathbf{135\ V}$$ **Ans.**

($\because$ $N_3 = 900$ rpm and $\phi_3 = 24 \times 10^{-3}$ Wb)

Problem 10.7. *An 8-pole DC generator has 500 armature conductors and a useful flux of 0.05 Wb. What will be the e.m.f. generated, if it is lap-connected and runs at 1200 rpm? What must be the speed at which it is to be driven to produce the same e.m.f, if it is wave-wound?*

Solution. e.m.f. generated when the generator is lap-connected,

$$E_g = \frac{\phi ZN}{60} \times \frac{P}{A} = \frac{0.05 \times 500 \times 1,200}{60} \times \frac{8}{8} = \mathbf{500\ V}$$ **Ans.**

($\because$ In lap-connected armature, number of parallel paths, A = P = 8)

Speed of the generator, when wave-connected, to generate an e.m.f. of 500 V will be

$$N' = \frac{E_g \times 60}{\phi \times Z} \times \frac{A}{P} = \frac{500 \times 60}{0.05 \times 500} \times \frac{2}{8} = \mathbf{300\ rpm}$$ **Ans.**

($\because$ In wave-connected armature, A = 2)

Problem 10.8. *Calculate the voltage induced in the armature winding of 4-pole, lap wound DC machine having 728 active conductors and running at 1800 rpm. The flux per pole is 30 mWb. If the armature is designed to carry a maximum line current of 100 A, what is the maximum electromagnetic power developed by the armature?*

Solution.

$$P = 4$$
$$Z = 728$$
$$A = \text{Lap wound} = P = 4$$
$$N = 1800 \text{ rpm}$$
$$\phi = 30 \text{ mWb}$$

(*a*) $$E = \frac{\phi ZNP}{60\ A} = \frac{30 \times 10^{-3} \times 728 \times 1800 \times 4}{60 \times 4}$$

or $$E = \mathbf{655.2\ V}$$ **Ans.**

(*b*) When line current is 100 A, the electromagnetic power developed is

$$P_e = I_a E = 100 \times 655.2 = \mathbf{65.52\ MW}$$ **Ans.**

Problem 10.9. *A 4 pole lap wound DC generator generates 200 V at 1000 rpm. If this generator is now wave wound and runs at 500 rpm, what will be the generated voltage?*

Solution.

Given
$$P = 4$$
$$E_g = 200 \text{ V}$$
$$N_1 = 1000 \text{ rpm}$$
$$N_2 = 500 \text{ rpm}$$

For Lap wound:
$$E_{g_1} = \frac{\phi Z N_1 P}{60\ A}$$

or
$$200 = \frac{\phi Z \times 1000 \times 4}{60 \times 4}$$

(as for Lap wound, A = P)

or
$$\phi = \frac{12}{Z} \text{ Wb}$$

For wave wound:
$$E_{g_2} = \frac{\phi Z N_2 P}{60\ A}$$

or
$$E_{g_2} = \frac{{}^{12}\!/\!{}_{Z} \times Z \times 500 \times 4}{60 \times 2}$$

(as for wave wound A = 2)

$\therefore$
$$E_{g_2} = \mathbf{200\ V}$$ **Ans.**

As the number of parallel paths (A) decreases from 4 to 2, the speed also becomes half (*i.e.* from 1000 rpm to 500 rpm). Hence the generated e.m.f. will remain same.

Problem 10.10. *A 6 pole wave wound armature has 300 conductors and run at 1000 rpm. The e.m.f. generated on open circuit is 400 V. Find the usefull flux per pole.*

Solution.

$P = 6$	$N = 1000$ rpm
$A = 2$	$E_g = 400$ V
$Z = 300$	$\phi = ?$

$$E_g = \frac{\phi ZNP}{60 A} \text{ or } \phi = \frac{60\ A \times E_g}{ZNP} = \frac{60 \times 2 \times 400}{30 \times 1000 \times 6}$$

$$\phi = \mathbf{266.67\ mWb}$$ **Ans.**

Problem 10.11. *A 4-pole, lap-wound armature has 144 slots with two coil sides per slot, each coil having two turns. If the flux per pole is 20 mWb and the armature rotates at 720 rpm. What is the induced voltage?*

Solution.

$P = 4$ slots = 144

$A = P = 4$ coil sides = 2 per slot

$\phi = 20$ mWb each coil = 2 turns

$N = 720$ rpm

$E = ?$

$$Z = 2 \times 2 \times 144 = 576$$

$$E = \frac{\phi ZNP}{60\ A}$$

$$= \frac{20 \times 10^{-3} \times 576 \times 720 \times 4}{60 \times 4}$$

$E =$ **138.24 V** **Ans.**

EXERCISE

1. Derive the e.m.f. equation of a DC generator. What will be the change in e.m.f. induced if flux is reduced by 20% and the speed is increased by 20%?
2. Why does the terminal voltage of a DC shunt generator fall when it is loaded? Explain with the help of load characteristics.
3. What are the different losses in DC machine and on what factors they depend? Mention the methods to reduce them.
4. Explain the generator action of a DC machine.
5. Describe the open circuit characteristics of DC generators. How do you obtain the internal characteristics using external characteristics of a DC shunt generator?
6. Why the shunt generator having a more drooping characteristic then separately excited generator?
7. Draw the speed torque characteristics for DC series and compound generator.
8. What are the different types of DC generators and motors? Also write the application of each.
9. A DC shunt generator delivers an output of 100 kW at 500 V when running at 800 rpm. The armature and field resistance are 0.1 Ω and 100 Ω respectively. Calculate the speed of the same machine when running as a shunt motor and taking 80 kW input at 500 V allow 1 volt per brush contact drop.
10. Explain the terms critical resistance and critical speed of a DC shunt generator.
11. Derive the e.m.f. equation of a DC generator.

□ □ □

11

DC Motor

11.1 INTRODUCTION

Motors come in many sizes and types, but their basic function is the same. Motors of all types serve to convert electrical energy into mechanical energy. They can be found in VCR's, elevators, CD players, toys, robots, watches, automobiles, subway trains, fans, space ships, air conditioners, refrigerators, and many other places.

The **DC Motor** is the most commonly used actuator for producing movement and whose speed of rotation can easily be controlled, making them ideal for use in applications where speed control, servo type control, and/or positioning is required.

"The electric motor is the most convenient of all sources of motive power. It is clean and silent, starts instantly, and can be built large enough to drive the world's fastest trains or small enough to work a watch."

11.2 BASIC PRINCIPLE

An electric motor is a machine which converts electric energy into mechanical energy.

Its action is based on the principle that "when a current carrying conductor is placed in a magnetic field, it experiences a mechanical force whose direction is given by Fleming's Left Hand Rule and whose magnitude is given by F = BIL Newton."

There are different kinds of DC motors, but they all work on the same principles.

It will be easier to understand the operation of the DC motor from a basic diagram that shows the magnetic interaction between the rotating armature and the stationary field's coils. Fig. 11.1 to Fig. 11.3 shows three diagrams that explain the DC motor's operation in terms of the magnetic interaction.

In Fig. 11.1 you can see that a bar magnet has been mounted on a shaft so that it can spin.

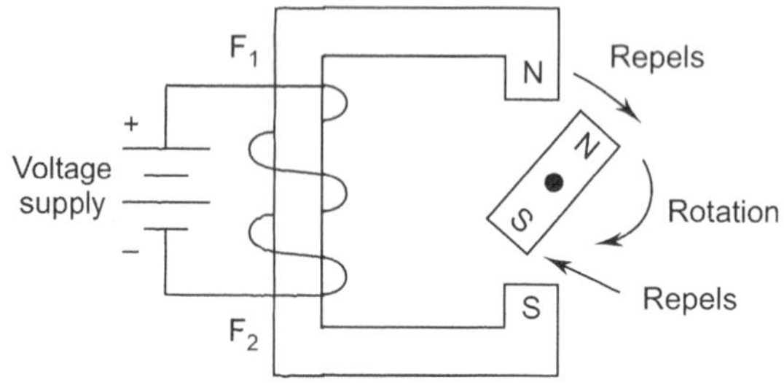

Fig. 11.1

The field winding is one long coil of wire that has been separated into two sections. The top section is connected to the positive pole of the battery and the bottom section is connected to the negative pole of the battery. It is important to understand that the battery represents a source of voltage for this winding. In the actual industrial-type motor this voltage will come from the DC voltage source for the motor. The current flow in this direction makes the top coil the north pole of the magnet and the bottom coil the south pole of the magnet.

The bar magnet represents the *armature* and the coil of wire represents *the field.* The arrow shows the direction of the armature's rotation. Notice that the arrow shows the armature starting to rotate in the clockwise direction. The north pole of the field coil is repelling the north pole of the armature, and the south pole of the field coil is repelling the south pole of the armature.

As the armature begins to move, the north pole of the armature comes closer to the south pole of the field, and the south pole of the armature is coming closer to the north pole of the field. As the two unlike poles near each other, they begin to attract. This attraction becomes stronger until the armature's north pole moves directly in line with the field's south pole, and its south pole moves directly in line with the field's north pole (Fig. 11.2).

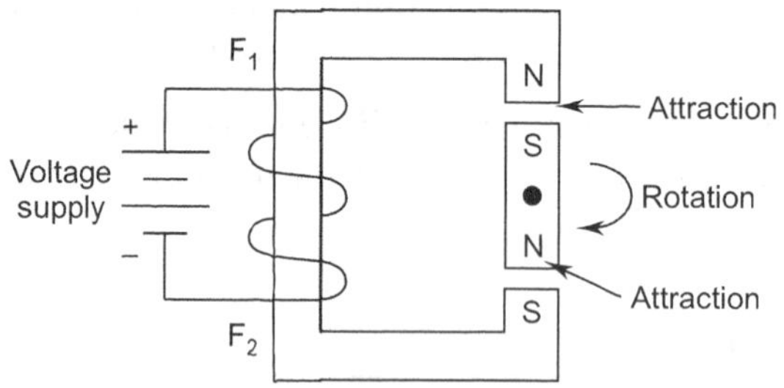

Fig. 11.2

When the opposite poles are at their strongest attraction, the armature will be "locked up" and will resist further attempts to continue spinning. For the armature to continue its rotation, the armature's polarity must be switched. Since the armature in this diagram is a permanent magnet, you can see that it would lock up during the first rotation and not work. If the armature is an electromagnet, its polarity can be changed by changing the direction of current flow through it. For this reason the armature must be changed to a coil (electromagnet) and a set of commutator segments must be added to provide a means of making contact between the rotating member and the stationary member. One commutator segment is provided for each terminal of the magnetic coil. Since this armature has only one coil, it will have only two terminals, so the commutator has two segments.

Since the armature is now a coil of wire, it will need DC current flowing through it to become magnetized. This presents another problem; since the armature will be rotating, the DC voltage wires cannot be connected directly to the armature coil. A stationary set of carbon brushes is used to make contact to the rotating armature. The brushes ride on the commutator segments to make contact so that current will flow through the armature coil.

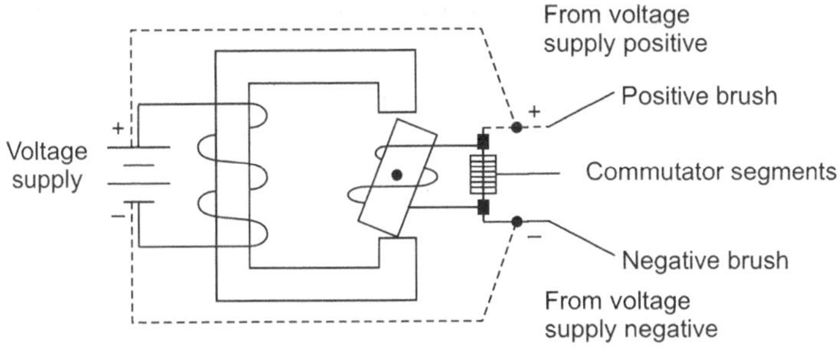

Fig. 11.3

In Fig. 11.3 you can see that the DC voltage is applied to the field and to the brushes. Since negative DC voltage is connected to one of the brushes, the commutator segment the negative brush rides on will also be negative. The armature's magnetic field causes the armature to begin to rotate. This time when the armature gets to the point where it becomes locked up with the magnetic field, the negative brush begins to touch the end of the armature coil that was previously positive and the positive brush begins to touch the end of the armature coil that was negative. This action switches the direction of current flow through the armature, which also switches the polarity of the armature coil's magnetic field at just the right time so that the repelling and attracting continues. The armature continues to switch its magnetic polarity twice during each rotation, which causes it to continually be attracted and repelled with the field poles.

This is a simple two-pole motor that is used primarily for instructional purposes. Since the motor has only two poles, the motor will operate rather roughly and not provide too much torque. Additional field poles and armature poles must be added to the motor for it to become useful for industry.

Constructionaly, there is no basic difference between a DC generator and a DC motor.

11.3 TYPES OF DC MOTORS

As the DC Machine can be operated both as motor as well as generator, hence the motor and generator both have same types. The basic types of DC motors used in industry today are:

- Separately excited
- Self-excited
    - → Series
    - → Shunt
    - → Compound
        - → Differentially Compound
        - → Cumulatively Compound

These different types of motors along with their equivalent electrical circuit diagrams and direction of current flow is discussed next.

- In all these circuit diagrams, the armature is represented by a circle and its terminals are marked as A and AA.
- Carbon brushes are shown by small blocks on two side of the armature.
- The back e.m.f.* developed in the armature winding is represented by symbol E_b, whose direction is shown upwards or in opposite direction to the applied voltage.
- The field is represented by a coil and its terminals are marked as Z and ZZ.
- Different currents and voltages shown in the circuit diagram are:

I_f = Field current
I_a = Armature current
I_L = Line current
V = Applied input voltage
E_b = Back e.m.f. developed in the armature winding.

11.3.1 Separately Excited DC Motor

Separately excited DC motors have field coil similar to those of a shunt wound machine, but the armature and field windings are excited by different supply sources (Fig. 11.4).

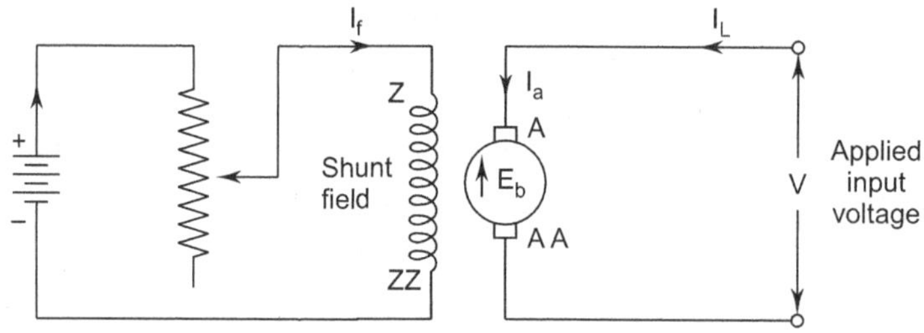

Fig. 11.4

11.3.2 DC Series Motor

The series motor provides high starting torque and is able to move very large shaft loads when it is first energized. Fig. 11.5 shows the wiring diagram of a series motor. From the diagram you can see that the field winding in this motor is wired in series with the armature winding. This is the attribute that gives the series motor its name.

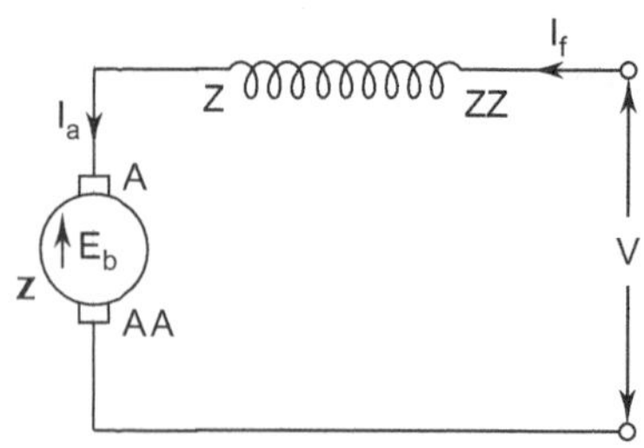

Fig. 11.5

Since the series field winding is connected in series with the armature, it will carry the same amount of current that passes through the armature.

*Concept of back e.m.f. will be discussed in later section.

Operation of the series motor is easy to understand. In Fig. 11.5 you can see that the field winding is connected in series with the armature winding. This means that power will be applied to one end of the series field winding and to one end of the armature winding (connected at the brush).

When voltage is applied, current begins to flow from negative power supply terminals through the series winding and armature winding. The armature is not rotating when voltage is first applied, and the only resistance in this circuit will be provided by the large conductors used in the armature and field windings. Since these conductors are so large, they will have a small amount of resistance. This causes the motor to draw a large amount of current from the power supply. When the large current begins to flow through the field and armature windings, it causes a strong magnetic field to be built. Since the current is so large, it will cause the coils to reach saturation, which will produce the strongest magnetic field possible.

11.3.3 DC Shunt Motor

DC shunt motor is different from series motor in that the field winding is connected in parallel with the armature instead of in series. One should remember from basic electrical theory that a parallel circuit is often referred to as a shunt. Since the field winding is placed in parallel with the armature, it is called a shunt winding and the motor is called a shunt motor. Fig. 11.6 shows a diagram of a shunt motor. Notice that the field terminals are marked Z and ZZ, and the armature terminals are marked A and AA.

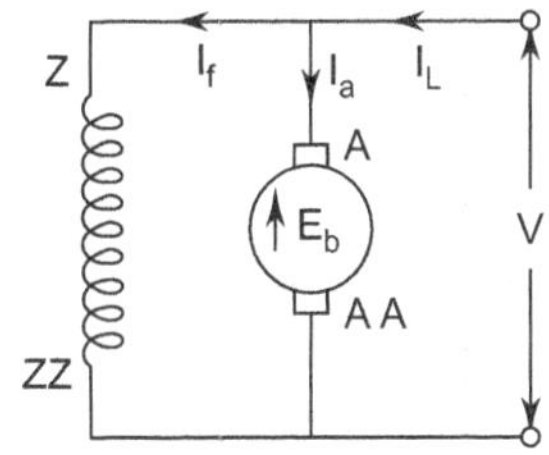

Fig. 11.6

11.3.4 DC Compound Motor

The DC compound motor is a combination of the series motor and the shunt motor. It has a series field winding that is connected in series with the armature and a shunt field that is in parallel with the armature. The combination of series and shunt winding allows the motor to have the torque characteristics of the series motor and the regulated speed characteristics of the shunt motor. Fig. 11.7 shows a diagram of the long shunt compound wound motor and Fig. 11.8 shows a diagram of the short shunt compound wound motor.

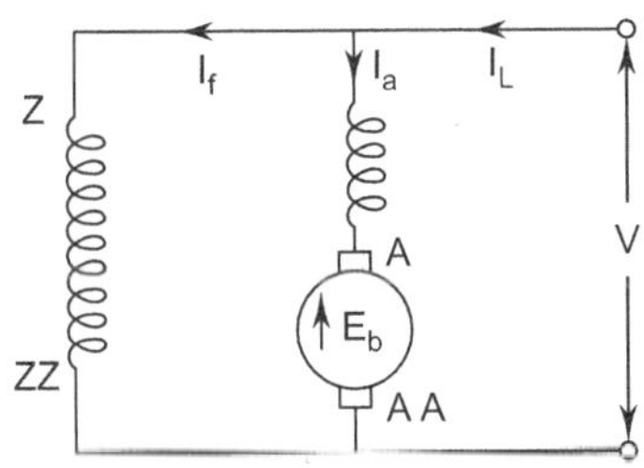

Fig. 11.7

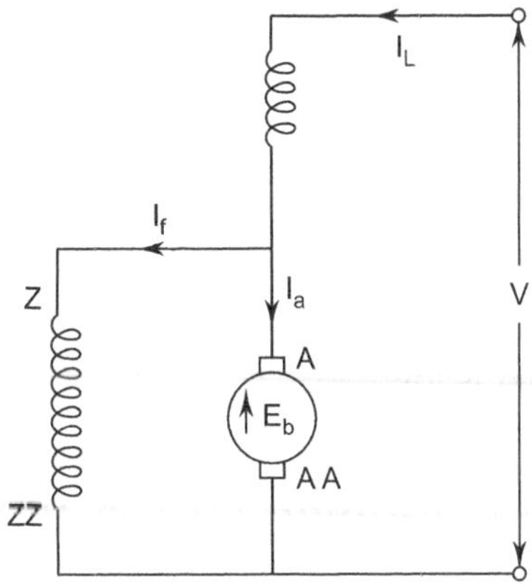

Fig. 11.8

(*i*) **Cumulative Compound Motor**

Cumulative compound motor is so called because the shunt field is connected so that its coils are aiding the magnetic fields of the series field and armature. The shunt winding can be wired as a *long shunt* or as a *short shunt.*

(*ii*) **Differential Compound Motor**

Differential compound motors use the same motor and windings as the cumulative compound motor, but they are connected in a slightly different manner to provide slightly different operating speed and torque characteristics. In differential compound motor the shunt field is connected so that its polarity is reversed to the polarity of the armature. Since the shunt field is still connected in parallel with only the armature, it is considered a short shunt. The shunt winding can be wired as a *long shunt* or as a *short shunt.*

11.4 CONCEPT OF BACK E.M.F.

When the motor armature, carrying current, rotates in the magnetic field, the armature conductors also rotate and hence cut the flux. In accordance with the laws of electromagnetic induction, e.m.f. is induced whose direction, as found by Fleming's right hand rule, is in opposition to the applied voltage (Fig. 11.9).

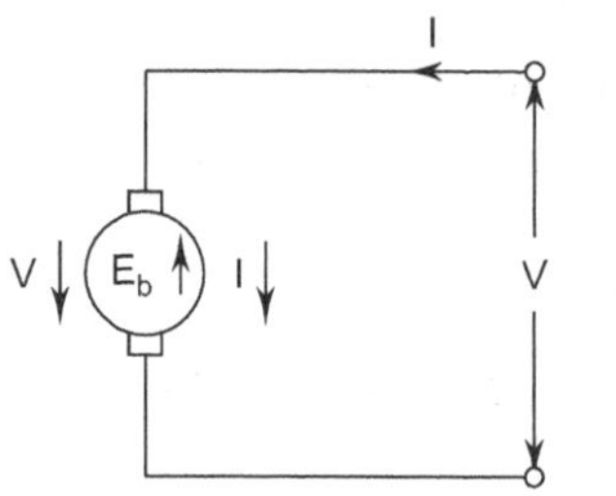

Fig. 11.9

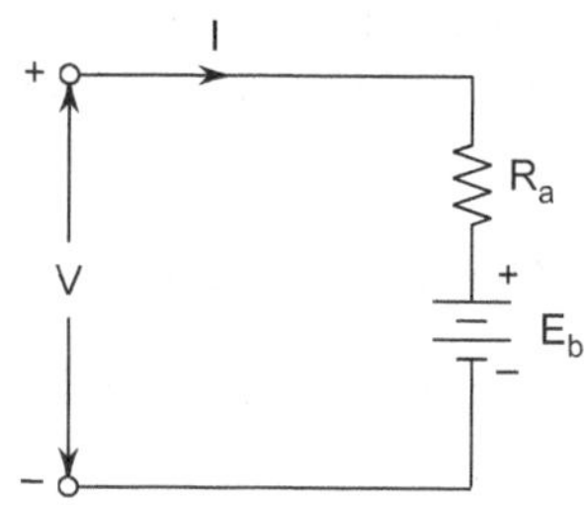

Fig. 11.10

Because of its opposing direction, it is referred to as counter e.m.f. or Back e.m.f., E_b.

The rotating armature generating the back e.m.f, E_b is like a battery of e.m.f, E_b put in opposition to the supply mains of V volts (Fig. 11.10).

Obviously, V has to drive against the opposition of E_b.

$$V = E_b + I_a R_a \qquad \text{...(11.1)}$$

where R_a is the armature winding resistance.

Among other factor, Back e.m.f. also depends upon speed.

If speed is high,	E_b is large,	I_a will be small
If speed is low,	E_b is small,	I_a will be large

11.5 MECHANICAL POWER DEVELOPED

Let

V = Supply voltage

E_b = Back e.m.f.

I_a = Armature current

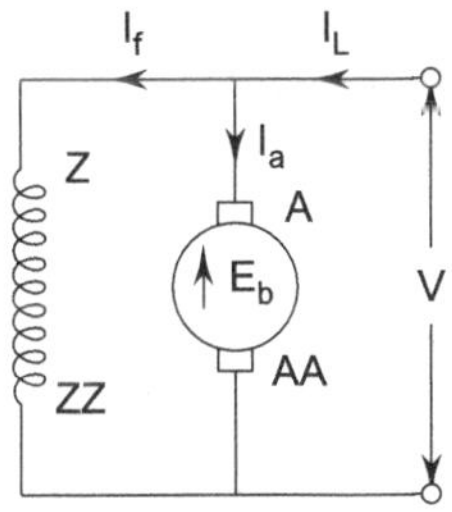

Fig. 11.11

From equation (11.1)

$$V = E_b + I_a R_a \qquad \text{[Eqn. (11.1)]}$$

Multiplying equation (11.1) with I_a on both sides, we get

$$E_b I_a = VI_a - I_a^2 R_a \qquad \ldots(11.2)$$

Comparing the equation (11.2) with the general input-output relation equation, we can conclude that

$E_b I_a$ is the electrical equivalent of mechanical power developed in the armature and hence will be denoted by Gross Mechanical Power (P_m).

VI_a is electrical input to armature.

$I_a^2 R_a$ is copper loss in the armature.

Hence the equation (11.2) can be written as

$$P_m = VI_a - I_a^2 R_a \qquad \ldots(11.3)$$

11.5.1 Condition for Maximum Power

The gross mechanical power will be maximum when

$$\frac{dP_m}{dI_a} = 0 \qquad \ldots(11.4)$$

or $$\frac{dP_m}{dI_a} = V - 2 I_a R_a = 0 \qquad \ldots(11.5)$$

or $$V = 2 I_a R_a \qquad \ldots(11.6)$$

or $$I_a R_a = \frac{V}{2} \qquad \ldots(11.7)$$

From equation (11.1)

$$E_b = V - I_a R_a \qquad \ldots(11.8)$$

Putting equation (11.7) in equation (11.8), we get

$$E_b = V - \frac{V}{2} \qquad \ldots(11.9)$$

or $$\boxed{E_b = \frac{V}{2}} \qquad ...(11.10)$$

Power will be maximum when back e.m.f. (E_b) developed is equal to half of the supply voltage (V).

11.6 EXPRESSION FOR BACK E.M.F.

The e.m.f. equation of a DC motor can be derived in the same way as that of the DC generator.

Let

ϕ = Flux per pole in Weber

Z = Total number of armature conductors

= Number of slots × Number of conductors per slot

P = Number of generator pole

A = Number of parallel path in armature

N = Armature rotation in revolution per minute (*i.e.*, in rpm)

E = e.m.f.. induced in any parallel path in armature

E_g = Generated e.m.f. in any one of the parallel path

According to the law of electromagnetic induction, the generated or induced e.m.f. is proportional to rate of change of flux with respect to time and if only one conductor is taken in to consideration then the induced e.m.f. is given by

$$E_b = -\frac{d\phi}{dt} \qquad ...(11.11)$$

If there are 'Z' number of conductors, then

$$E_b = -Z\frac{d\phi}{dt} \qquad ...(11.12)$$

Initially let us consider only one conductor, which is rotating inside a magnetic field having 'P' number of poles. If the conductor makes one complete rotation then to total flux cut by conductor

$$\text{Total flux} = \phi P \qquad ...(11.13)$$

If 'N' is the speed in rpm at which the conductor is rotating in that magnetic field, then time taken by the conductor for one complete rotation

$$\text{Time for one complete rotation} = \frac{1}{N}\text{ minutes} = \frac{60}{N}\text{ seconds} \qquad ...(11.14)$$

$$\therefore \text{ Average e.m.f. induce in one conductor} = \frac{\text{Total flux cut in one revolution}}{\text{Time taken for one complete revolution}} \qquad ...(11.15)$$

or $$E_b = \frac{\phi P}{60/N} = \frac{\phi PN}{60} \qquad ...(11.16)$$

If the machine has total 'Z' number of conductors and has 'A' number of parallel path, number of conductors in each parallel path is

$$\text{Number of conductor/Parallel path} = \frac{Z}{A} \quad \ldots(11.17)$$

The conductors in each parallel path are connected in series hence the total e.m.f. is added.

Total e.m.f. induced in all conductors will be

$$E_b = \frac{\phi PN}{60} \times \frac{Z}{A} \quad \ldots(11.18)$$

or

$$\boxed{E_b = \frac{\phi ZNP}{60\ A}} \quad \ldots(11.19)$$

11.7 TORQUE DEVELOPED

Let

T_a = Torque develop by the armature

ω = Speed in radian per second (rps) = $\frac{2\pi N}{60}$

N = Speed in radian per minute (rpm)

Mechanical power developed by armature is

$$P_m = \text{Torque} \times \text{Speed} \quad \ldots(11.20)$$

$$P_m = T_a \times 2\pi N \quad \ldots(11.21)$$

Also the electrical equivalent of this power is

$$P_m = E_b I_a \quad \ldots(11.22)$$

Equating equations (11.21) and (11.22), we get

$$T_a \times 2\pi N = E_b I_a \quad \ldots(11.23)$$

The back e.m.f. expression [from equation (11.19)] when speed is considered in rps will be

$$E_b = \frac{\phi ZNP}{A} \quad \ldots(11.24)$$

From equation (11.24), equation (11.23) will become

$$T_a \times 2\pi N = \frac{\phi ZNP}{A} I_a \quad \ldots(11.25)$$

or

$$T_a = \frac{1}{2\pi} Z\left(\frac{P}{A}\right) . \phi I_a \text{ Nm} \quad \ldots(11.26)$$

It is clear from above expression that

$$T_a \propto \phi I_a \quad \ldots(11.27)$$

As $\left(\frac{ZP}{2\pi A}\right)$ is a constant term.

11.8 SPEED OF A DC MOTOR

The expression for speed of a DC motor can be derived from the basic e.m.f. equation and expression.

From equation (11.8)

$$E_b = V - I_a R_a \qquad \text{[Eqn. (11.8)]}$$

Putting the general expression of e.m.f. in the above equation, we get

$$\frac{\phi ZNP}{60\ A} = V - I_a R_a \qquad \text{...(11.28)}$$

or

$$N = \frac{V - I_a R_a}{\phi} \cdot \left(\frac{60\ A}{ZP}\right) \qquad \text{...(11.29)}$$

or

$$N = K \frac{V - I_a R_a}{\phi} \qquad \text{...(11.30)}$$

where

$$K = \frac{60\ A}{ZP} \qquad \text{...(11.31)}$$

Hence the expression for speed of a DC motor will become

$$\boxed{N = K \frac{E_b}{\phi}} \qquad \text{...(11.32)}$$

11.9 DC MOTOR CHARACTERISTICS

DC motor characteristics are classified into three main type:

1. Torque Vs Armature Current Characteristics (T_a Vs I_a)
2. Speed Vs Armature Current Characteristics (N Vs I_a)
3. Speed Vs Torque Characteristics (N Vs T_a)

While discussing motor characteristics, the following two relations should always be kept in mind

$$T_a \propto \phi I_a \qquad \text{[Eqn. (11.27)]}$$

and

$$N \propto \frac{E_b}{\phi} \qquad \text{...(11.33)}$$

11.9.1 Characteristics of DC Series Motor

In case of series motor,

$$I_L = I_a = I_f = I \text{ (say)} \qquad \text{...(11.34)}$$

Flux is directly proportional to current

$$\phi \propto I \qquad \text{...(11.35)}$$

From equation (11.27) the torque can be expressed as

$$T \propto \phi I_a \qquad \text{[Eqn. (11.27)]}$$

As $\phi \propto I$, equation (11.25) will become

$$T \propto I^2 \quad \text{...(11.36)}$$

But after saturation ϕ becomes almost constant and the torque equation then reduces to

$$T \propto I \quad \text{...(11.37)}$$

From equations (11.36) and (11.37) we can easily draw the torque Vs armature current characteristics (T_a Vs I_a), which is shown in Fig. 11.12.

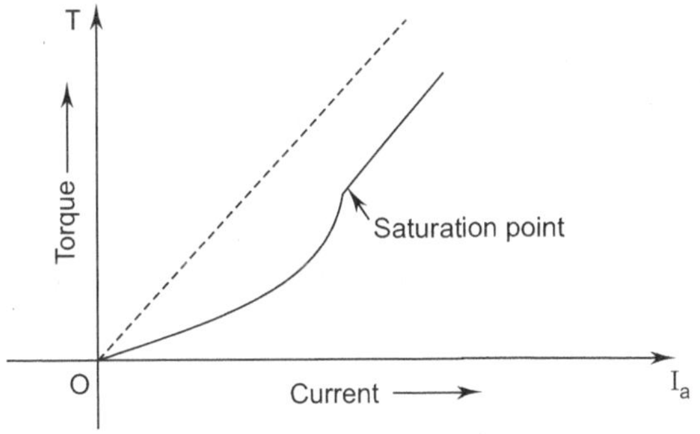

Fig. 11.12

$E_b = V - I_aR_a$ is almost constant because of negligible drop in armature resistance. Hence the speed will be

$$N \propto \frac{1}{\phi} \propto \frac{1}{I} \quad \text{...(11.38)}$$

From equation (11.36) we can easily draw the speed Vs armature current characteristics (N Vs I_a) (Fig. 11.13).

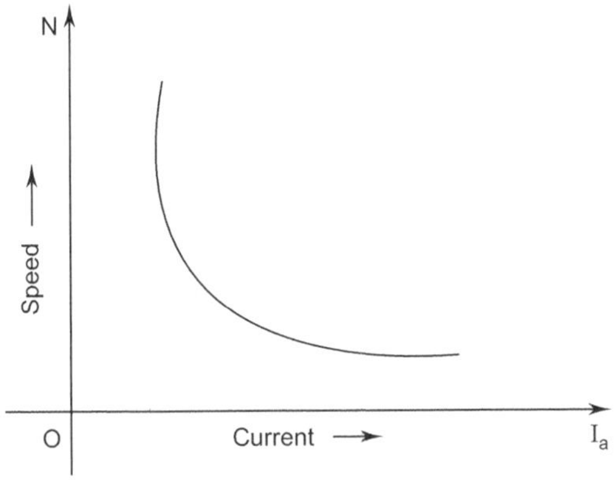

Fig. 11.13

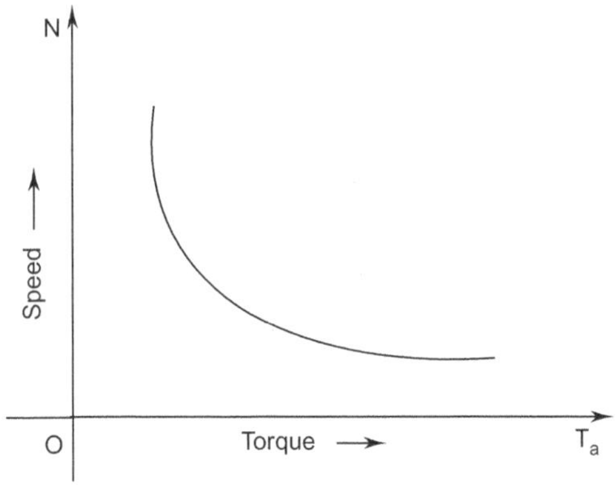

Fig. 11.14

From equations (11.37) and (11.38)

Hence, $$N \propto \frac{1}{T} \quad \text{...(11.39)}$$

From equation (11.39) we can easily draw the speed Vs torque characteristics (N Vs T_a) (Fig. 11.14).

11.9.2 Characteristics of DC Shunt Motor

In case of DC shunt motor

$$N \propto \frac{E_b}{\phi} \propto \frac{V - I_a R_a}{\phi} \propto \frac{V - IR_a}{\phi} \quad \text{...(11.40)}$$

where $I = I_L \approx I_a$

Load current = Armature current

In case of a shunt motor, as supply voltage is kept constant hence ϕ is almost constant.

Therefore, $N \propto V - IR_a$ ...(11.41)

i.e., with increase in the load current, $V - IR_a$ will decrease and hence speed of the shunt motor will decrease. So N Vs I_a characteristics will be droping in nature as shown in Fig. 11.15.

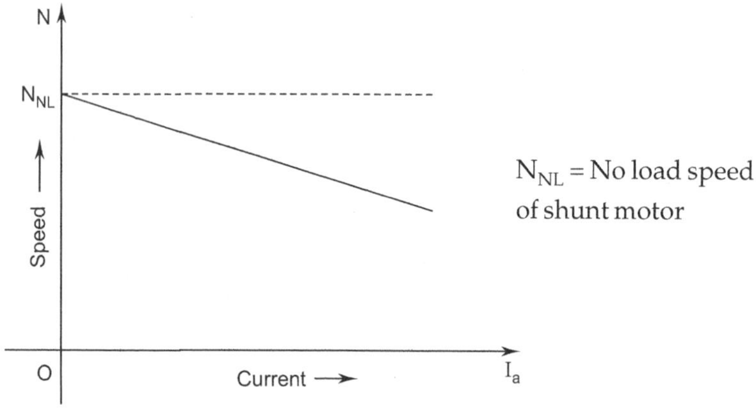

Fig. 11.15

As $T \propto \phi I_a \propto \phi I$ ...(11.42)

In case of shunt motor, supply voltage is kept constant, hence ϕ is almost constant.

Therefore, $T \propto I$ ...(11.43)

Hence, torque Vs armature current characteristics will be a straight line, passing through origin as shown in Fig. 11.16.

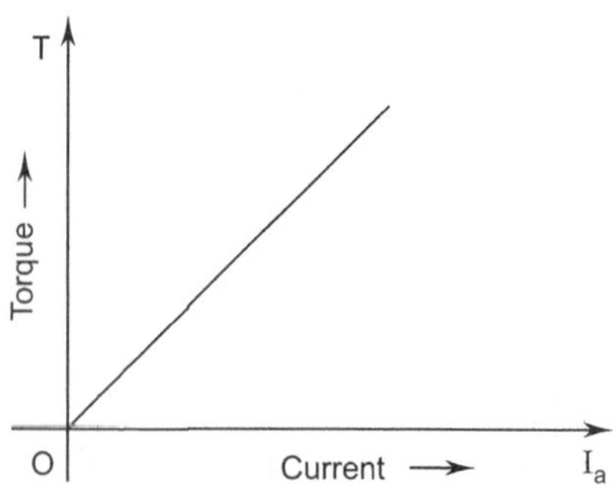

Fig. 11.16

Speed-torque characteristics of any motor is almost same as that of its speed-current characteristics.

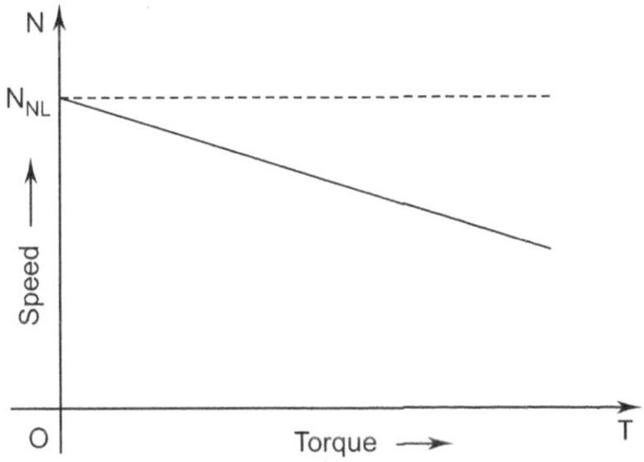

Fig. 11.17

11.9.3 Characteristics of DC Compound Motors

In case of cumulatively compound motor the total flux increases with increase in current. Therefore

$$N \propto \frac{E_b}{\phi} \text{ as } E_b \text{ is almost constant} \qquad \ldots(11.44)$$

$$N \propto \frac{1}{\phi} \qquad \ldots(11.45)$$

At the same time, for a differentially compound motor the total flux decreases with increase in load current. So as the flux decreases speed increases.

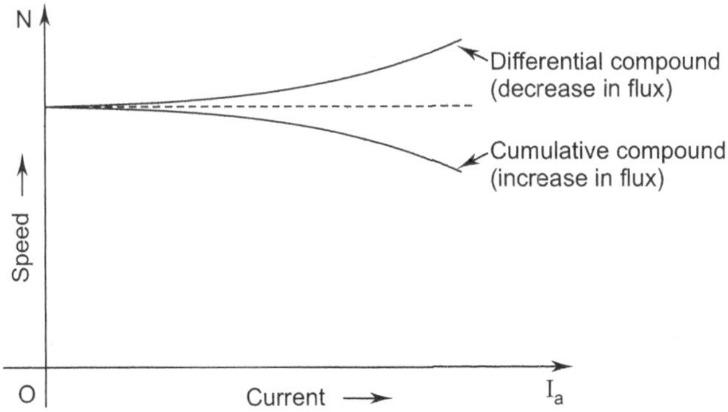

Fig. 11.18

In case of cumulative compound motor the flux due to series field winding increases and causes the torque greater than it would have been with shunt field-winding (Fig. 11.19).

In case of differential compound motor the resultant flux decreases with the increase in load, as a result the torque decreases as compared to the torque with shunt-field winding (Fig. 11.19).

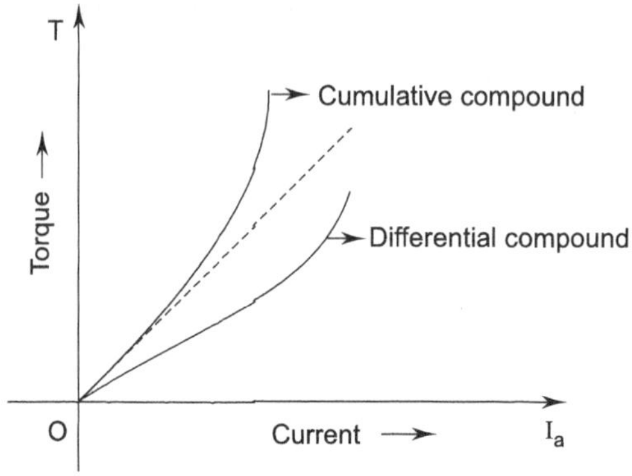

Fig. 11.19

As stated earlier, speed-torque characteristics of any motor is almost same as that of its speed-current characteristics (Fig. 11.20)

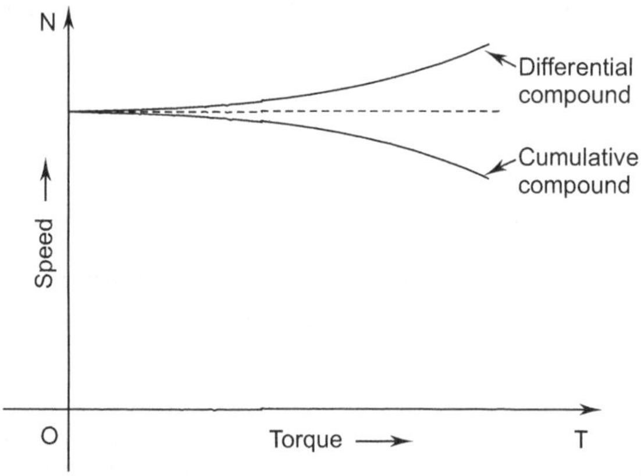

Fig. 11.20

11.10 SPEED CONTROL OF DC SHUNT MOTOR

Speed of a shunt motor is given by

$$N \propto \frac{E_b}{\phi} \propto \frac{V - I_a R_a}{\phi} \propto \frac{V - I_a R_a}{I_f} \left(\text{as } \phi \propto I_f\right) \quad \ldots(11.46)$$

For constant supply voltage and armature current the speed of the shunt motor can be varied by

(*i*) Changing the armature resistance R_a by putting another variable resistance in series with the armature. This method is known as armature control method.

(*ii*) Or by changing the field current I_f. This method is known as field control method.

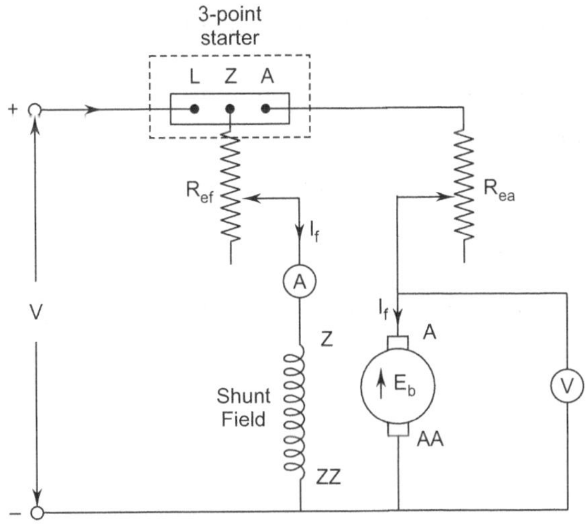

Fig. 11.21

where R_{ea} = Variable external resistance connected in series with the armature circuit.
R_{ef} = Variable external resistance connected in series with the field circuit.

In a 3-point starter, the three terminals are marked as L, Z and A.

Terminal L is connected to the positive terminal of the input supply.

Terminal Z is connected to the shunt field terminal 'Z'.

Terminal A is connected to the armature field terminal 'A'.

11.10.1 Field Control Method

In this method, a variable external resistance connected in series with the field circuit, R_{ea} is kept constant (not varied). Hence the armature voltage

$$E_a = V - I_a R_{ea} \text{ is constant} \qquad ...(11.47)$$

But the variable resistance **R_{ef} is varied**; (Fig. 11.21) as a result the shunt field current is varied.

$$N \propto \frac{1}{\phi} \propto \frac{1}{I_{sh}} \qquad ...(11.48)$$

As the field current will decrease, flux will also decrease which will increase the speed of the motor. Hence field control method is used for the control of speed above the base speed*.

11.10.2 Armature Control Method

As the supply voltage is normally constant, the voltage across the armature drops when any external resistance is connected in series with the armature circuit. As the armature voltages decreases the speed falls.

So by **varying the external resistance connected in series with armature, R_{ea},** the armature voltage can be varied (Fig. 11.21) and hence speed variation can be achieved.

This method gives speed control below the base speed only.

* Base speed is the speed of the motor when there is no external resistance present in the field circuit as well as armature circuit.

11.11 SPEED CONTROL OF DC SERIES MOTOR

Speed of a series motor can be varied mainly by two methods:

A. Flux control method.

B. Variable resistance in series with the motor.

11.11.1 Flux Control Method

Flux control method can be done by the following ways:

- Field diverters
- Armature diverters
- Parallel field coils
- Tapped field control

(*i*) **Field Diverters**

By changing the diverter resistance R_d, the current through the series field can be changed and hence field flux can be changed.

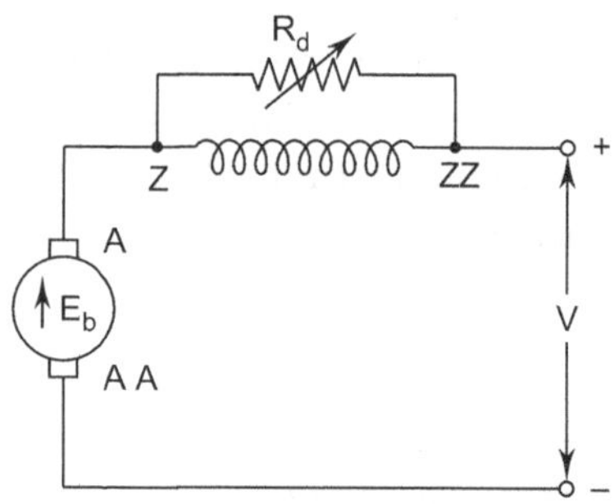

Fig. 11.22

As a result, the speed changes. This is a speed control method for obtaining speed above the base speed.

(*ii*) **Armature Diverters**

By changing the armature diverter resistance R_{ad}, the net supply current changes.

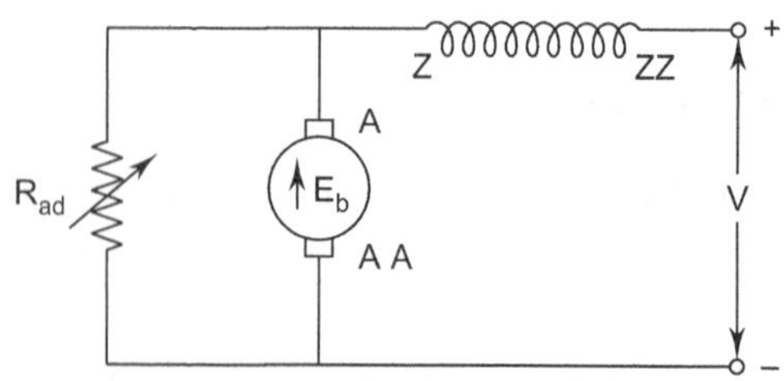

Fig. 11.23

The decrease in R_{ad} will result in increase in current, and hence flux increase. As flux increases speed drops.

This is a speed control below the base speed.

(*iii*) **Parallel Field Coils**

In this method, used for fan motors, several speeds can be obtained by regrouping the field coils.

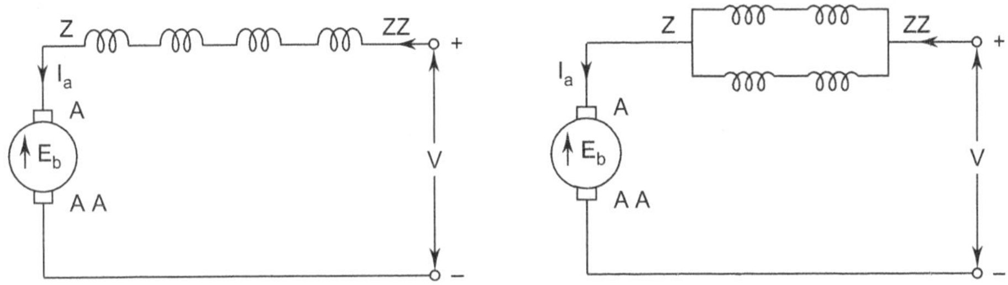

Fig. 11.24

(*iv*) **Tapped Field Control**

In this method the number of series field turns in the circuit can be changed. By means of tapping, the field flux can be changed and hence the speed.

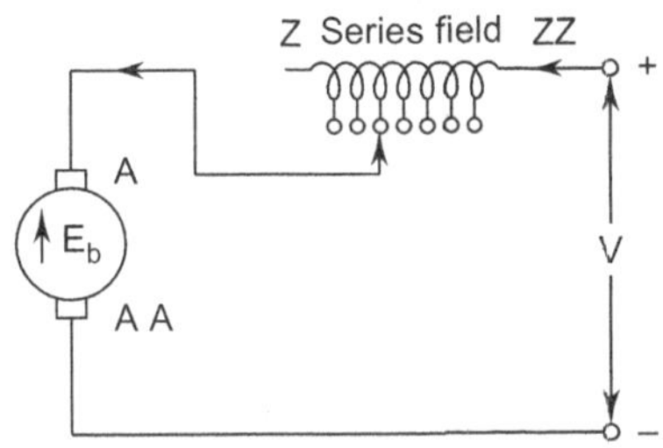

Fig. 11.25

11.11.2 Variable Resistance in Series with Motor

By increasing the resistance (R_n) in series with the armature, the voltage across the armature terminals decreases and hence the speed decreases. But this method leads to power loss in R_n.

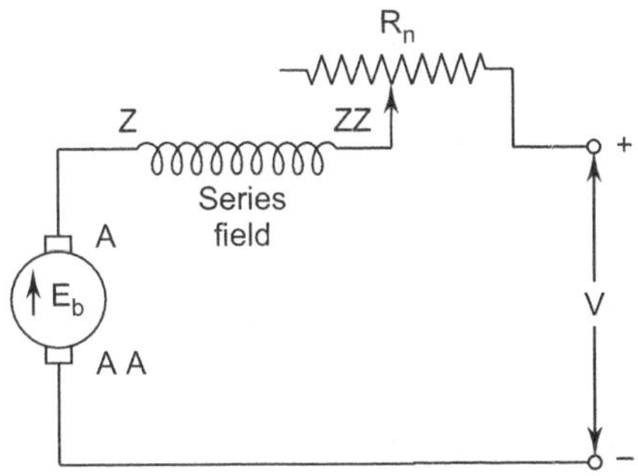

Fig. 11.26

11.12 APPLICATIONS

A. **Shunt Motor**

Shunt motor, being a constant speed motor, can be used as:

- For driving tiny shafts, lathes, fans, milling machines

- Printing presses
- Wood working machines
- Blowers
- Spinning and weaving machines
- Conveyors etc.

B. **Series Motor**

It is a variable speed motor, having very high starting torque and can be used for

- Locomotives
- Hoists
- Trolley cars
- Electromotive
- Elevators
- Air-compressors
- Vacuum cleaners
- Sewing machines etc.

C. **Compound Motor**

Cumulative compound wound DC motors are used in driving machines which are subjected to sudden applications of heavy load such as:

- Rolling mills
- Punching
- Shearing machine
- Lifts
- Mine-hoists.

SOLVED PROBLEMS

Problem 11.1. *The armature of a DC separately excited machine has a resistance of 0.1 Ω and is connected to a 230 V supply. Calculate the generated e.m.f. when it is running (i) as a generator giving 80 A, (ii) as a motor taking 60 A.*

Solution. As a generator : Generated e.m.f., $E_g = V + I_a R_a$

$= 230 + 80 \times 0.1 = \mathbf{238\ V}$ **Ans.**

As a motor : Back e.m.f., $E_b = V - I_a R_a$

$= 230 - 60 \times 0.1 = \mathbf{224\ V}$ **Ans.**

Problem 11.2. *A DC shunt motor runs at 600 rpm taking 60 A from a 230 V supply. Armature resistance is 0.2 ohm and field resistance is 115 ohms. Find the speed when the current through the armature is 30 A.*

Solution.

Shunt field current, $I_f = \dfrac{V}{R_f} = \dfrac{230}{115} = 2\text{ A}$

When line current, $I_{L_1} = 60\text{ A}$

Armature current, $I_{a_1} = I_{L_1} - I_f = 60 - 2 = 58$ A

Back e.m.f., $E_{b_1} = V - I_{a_1} R_a = 230 - 58 \times 0.2 = 218.4$ V

Speed, $N_1 =$ **600 rpm** **Ans.**

When armature current, $I_{a_2} = 30$ A

Back e.m.f., $E_{b_2} = V - I_{a_2} R_a = 230 - 30 \times 0.2$

$= 224$ V

As $E_b \propto N$

Speed, $$N_2 = \frac{E_{b_2}}{E_{b_1}} \times N_1$$

$$= \frac{218.4}{224} \times 600$$

$N_2 =$ **585 rpm** **Ans.**

Problem 11.3. *A DC shunt generator running at 1,200 rpm supplies a load of 60 kW at 250 V. Find the speed at which it runs as a shunt motor when taking 60 kW from 250 V supply. Take armature resistance as 0.1 Ω and field winding resistance as 50 Ω. Neglect brush drop.*

Solution. Shunt field current, $$I_f = \frac{\text{Terminal voltage, V}}{\text{Shunt field resistance, } R_f} = \frac{250}{50} = 5 \text{ A}$$

As generator : Load current, $$I_L = \frac{\text{Output in kW} \times 1{,}000}{V} = \frac{60 \times 1{,}000}{250} = 240 \text{ A}$$

Armature current, $I_{ag} = I_L + I_f = 240 + 5 = 245$ A

Generated e.m.f., $E_g = V + I_a R_a = 250 + 245 \times 0.1 = 274.5$ V

Speed, $N_g = 1{,}200$ rpm

As motor : Line current, $$I_L = \frac{60 \times 1{,}000}{250} = 240 \text{ A}$$

Armature current, $I_{am} = I_L - I_f = 240 - 5 = 235$ A

Back e.m.f. developed, $E_b = V - I_{am} R_a = 250 - 235 \times 0.1 = 226.5$ V

Speed of motor, $$N_m = \frac{E_b}{E_g} \times N_g = \frac{226.5}{274.5} \times 1{,}200 = \mathbf{990 \text{ rpm}}.$$ **Ans.**

Problem 11.4. *A 250 V DC shunt motor having an armature resistance of 0.25 Ω carries an armature current of 50 A and runs at 750 rpm. If the flux is reduced by 10%, find the speed. Assume that the torque remains the same.*

Solution. Supply voltage, $V = 250$ V

Armature current, $I_{a_1} = 50$ A

Back e.m.f., $E_{b_1} = V - I_{a_1} R_a = 250 - 50 \times 0.25 = 237.5$ V

Flux, $\phi_2 = 0.9\,\phi_1$ ($\because$ Flux has been reduced by 10%)

Since torque developed remains unchanged,

$$I_{a_2}\phi_2 = I_{a_1}\phi_1 \quad (\text{as } T \propto \phi I_a)$$

or Armature current, $I_{a_2} = I_a 1 \times \frac{\phi_1}{\phi_2} = 50 \times \frac{1}{0.9} = 55.5$ A

Back e.m.f., $E_2 = V - I_{a_2} R_a = 250 - 55.55 \times 0.25 = 236.11$ V

Speed,
$$N_2 = \frac{E_{b_2}}{E_{b_1}} \times \frac{\phi_1}{\phi_2} \times N_1$$
$$= \frac{236.11}{237.50} \times \frac{1}{0.9} \times 750$$
$$= \mathbf{828.5\ rpm} \qquad (\because E_b \propto \phi N)$$ **Ans.**

Problem 11.5. *A DC series motor whose combined resistance of armature and field circuit is 0.1 ohm is connected across 230 V supply mains. The armature takes 100 A and its speed is 1000 rpm. Find the speed when the armature takes 200 A and flux being increased by 20%.*

Solution. $E_{b_1} \propto \phi_1 N_1$

$\therefore$ $\left(V_t - I_{a_1} R_a\right) \propto \phi_1 N_1$

and, $\left(V_t - I_{a_2} R_a\right) \propto \phi_2 N_2$

or,
$$\frac{V_t - I_{a_1} R_a}{V_t - I_{a_2} R_a} = \frac{\phi_1 N_1}{\phi_2 N_2}$$

Here, $V_t = 230$ V, $I_{a_1} = 100$ A,

$I_{a_2} = 200$ A, $N_1 = 1000$,

$R_a = 0.1\ \Omega$ and $\phi_2 = 1.2\,\phi_1$

$$\therefore \quad \frac{230 - 100 \times 0.1}{230 - 200 \times 0.1} = \frac{\phi_1 \times 1000}{1.2\,\phi_1 \times N_2}$$

$$\therefore \quad N_2 = \frac{1000}{1.2} \times \frac{210}{220}$$

or $N_2 = \mathbf{795\ rpm.}$ **Ans.**

Problem 11.6. *A 230 V, DC shunt motor takes an armature current of 3.33 A at rated voltage and at a no-load speed of 1000 rpm. The resistance of the armature circuit and field circuit are respectively 0.3 Ω and 160 Ω. The line current at full load and rated voltage is 40 A. Calculate the full load speed if the armature reaction weakens the no-load flux by 4%.*

Solution. At no-load, the counter e.m.f. is,

$$E_{a_1} = V_t - I_{a_1} r_a$$
$$= 230 - 3.33 \times 0.3 = 229 \text{ V}.$$

Field current, $$I_f = \frac{230}{160} = 1.44 \text{ A}$$

At full load, $$I_{a_2} = I_L - I_f = 40 - 1.44 = 38.56 \text{ A}$$

∴ Counter e.m.f. at full load $$= E_{a_2} = 230 - 38.56 \times 0.3 = 218.43 \text{ V}$$

At full load the field flux, $$\phi_2 = 0.96\,\phi_1 \text{ (given)}$$

$$\frac{E_{a_1}}{E_{a_2}} = \frac{\phi_1 N_1}{\phi_2 N_2}$$

$$N_2 = \frac{E_{a_2}}{E_{a_1}} \times \frac{\phi_1}{\phi_2} \times N_1$$

$$= \frac{218.43}{229} \times \frac{\phi_1}{0.96\phi_1} \times 1000$$

or $$N_2 = \textbf{994 rpm.}$$ **Ans.**

Problem 11.7. *A 215 V DC machine has an armature resistance of 0.4 Ω. It is supplying 5 kW as a generator when run at 1000 rpm and is excited to give a terminal voltage of 215 V. At what speed would it run as a motor if it fed at the same terminal voltage draws same armature current but the flux per pole is increased by 10%.*

Solution. When the machine is running as a generator

Terminal voltage 215 V

$$R_a = 0.4\,\Omega, \quad N_1 = 1000 \text{ rpm}$$

Armature current $$I_{a_1} = \frac{5000}{215} = 23.26 \text{ A}$$

Induced e.m.f. $$E_1 = V + I_a R_a = 215 + 23.26 \times 0.4$$

or $$E_1 = 224.304 \text{ V}$$

Hence the equation of induced e.m.f. for generator is

$$E_1 = K\phi_1 N_1 \quad \text{...}(i)$$

When the machine is running as a motor

Terminal voltage 215 V (Same)

$$R_a = 0.4\,\Omega, \quad N_1 = ?$$

Armature current, $$I_{a_1} = 23.26 \text{ A (Same)}$$

Induced e.m.f., $$E_2 = V - I_a R_a = 215 - 23.26 \times 0.4$$

or $$E_1 = 205.696 \text{ V}$$

In case of motor the flux per pole is increased by 10% hence

$$\phi_2 = \frac{110}{100}\phi_1 = \frac{11}{10}\phi_1$$

Hence the equation of induced e.m.f. for motor is

$$E_2 = K\phi_2 N_2 \quad ...(ii)$$

Dividing equation (*ii*) by equation (*i*), we get

$$\frac{E_2}{E_1} = \frac{\phi_2 N_2}{\phi_1 N_1}$$

or
$$\frac{205.696}{224.304} = \frac{\frac{11}{10}\phi_1 N_2}{\phi_1 1000}$$

or
$$N_2 = \mathbf{917\ rpm}$$ **Ans.**

Problem 11.8. *A DC shunt generator delivers 50 kW at 250 V when running at 500 rpm. The armature and field resistances are 0.05 Ω and 125 Ω respectively. Calculate:*

(i) Speed

(ii) Developed torque

of the same machine when running as a shunt motor taking 50 kW at 250 V. Allow 1 volt per brush for contact drop.

Solution.

Condition-1

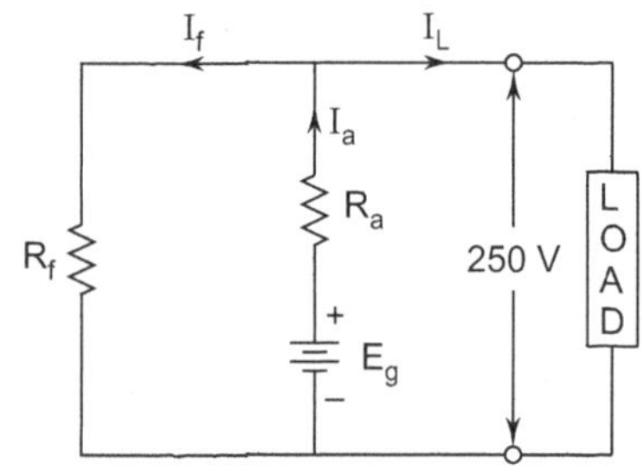

Fig. 11.27

Given, Electrical load = 50 kW

$$V_T = 250 \text{ V}$$
$$N_1 = 500 \text{ rpm}$$
$$R_a = 0.05\ \Omega$$
$$R_f = 125\ \Omega$$

The circuit diagram of the machine running as generator is shown in Fig. 11.27.

Field current, $$I_f = \frac{V_T}{R_f} = \frac{250}{125} = 2\text{ A}$$

Load current, $$I_L = \frac{50{,}000}{250} = 200\text{ A}$$

Armature current, $$I_a = I_L + I_f = 200 + 2 = 202\text{ A}$$

∴ e.m.f. developed by the generator will be

$\therefore$
$$E_g = V_T + I_a R_a$$
$$= 250 + 202 \times 0.05$$
$$E_g = 260.1 \text{ V}$$

Condition-2

When the same machine is running as a motor

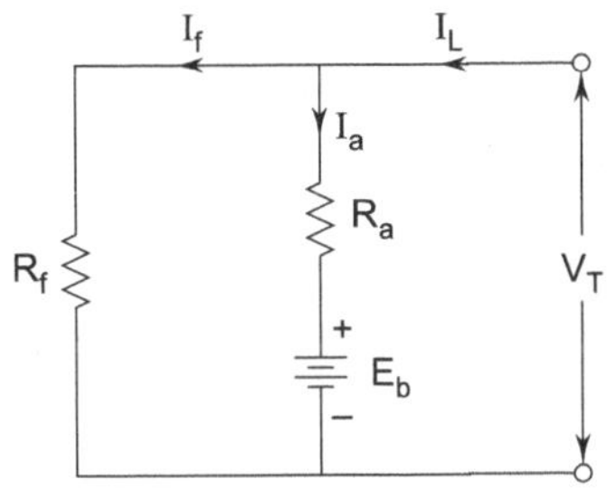

Fig. 11.28

Line/load current, $$I_L = \frac{50,000}{250} = 200 \text{ A}$$

Field current, $$I_f = \frac{V_T}{R_f} = \frac{250}{125} = 2 \text{ A}$$

Armature current, $$I_a = I_L - I_f$$
$$= 200 - 2 = 198 \text{ A}$$

$\therefore$ Back e.m.f. developed in the motor will be

$$E_b = V_T - I_a R_a - \text{Brush contact drop}$$
$$= 250 - 198 \times 0.05 - 2$$
$$E_b = 238.1 \text{ V}$$

(*i*) Speed of the machine, when running as motor will be

$$\frac{E_g}{E_b} = \frac{N_1}{N_2}$$

or $$\frac{260.1}{238.1} = \frac{500}{N_2}$$

or $$N_2 = \mathbf{458 \text{ rpm}}$$ **Ans.**

(*ii*) Developed torque

$$T_a = \frac{E_b I_a}{2\pi N/60}$$
$$= \frac{238.1 \times 198 \times 60}{2\pi \times 458}$$

or $$T_a = \mathbf{983 \text{ Nm}}$$ **Ans.**

Problem 11.9. *A 140 V DC shunt motor has an armature resistance of 0.02 Ω and a field resistance of 70 Ω. The full load line current is 40 A and the full load speed is 1800 rpm. If the brush contact drop is 3 V, find the speed of the motor at half load.*

Solution.

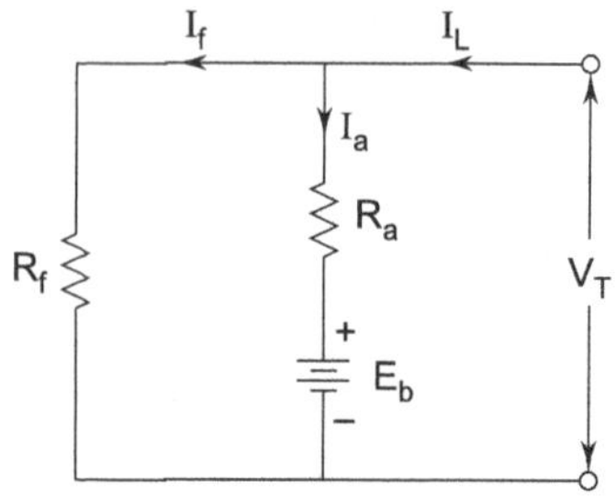

Fig. 11.29

Under full-load conditions:

$$R_a = 0.2\ \Omega$$
$$R_f = 70\ \Omega$$
$$V_T = 140\ V$$

Brush contact drop = 3 V

$$I_f = \frac{V_T}{R_f} = \frac{140}{70} = 2\ A$$
$$N_1 = 1800\ rpm$$
$$I_{L_1} = 40\ A$$
$$I_{a_1} = I_{L_1} - I_f = 40 - 2 = 38\ A$$
$$E_{b_1} = V_T - I_a R_a - \text{Brush contact drop}$$
$$E_{b_1} = 140 - 38 \times 0.2 - 3 = 129.4\ V$$

Under half-load conditions:

$$R_a = 0.2\ \Omega$$
$$R_f = 70\ \Omega$$
$$V_T = 140\ V$$

Brush contact drop = 3 V

$$I_f = \frac{V_T}{R_f} = \frac{140}{70} = 2\ A$$
$$I_{L_2} = 20\ A$$
$$I_{a_2} = I_{L_2} - I_f = 20 - 2 = 18\ A$$
$$E_{b_2} = V_T - I_{a_2} R_a - \text{Brush contact drop}$$
$$= 140 - 18 \times 0.2 - 3$$
$$= 133.4\ V$$

Now the speed of the motor at half-load will be

$$E_b \propto \phi N$$

As field current is constant and hence flux ϕ will remains constant therefore

$$E_b \propto N$$

or $$E_{b_1} \propto N_1$$

and $$E_{b_2} \propto N_2$$

$\therefore$ $$\frac{E_{b_2}}{E_{b_1}} = \frac{N_2}{N_1}$$

or $$N_2 = N_1 \frac{E_{b_2}}{E_{b_1}} = \frac{1800 \times 133.4}{129.4}$$

or $$N_2 = \mathbf{1856\ rpm}$$ **Ans.**

EXERCISE

1. A 20 HP, 230 V, 1200 rpm shunt motor has four poles, four parallel armature paths, and 868 armature conductors. The armature circuit resistance is 0.18 ohms. At rated speed and rated output the armature current is 70 amperes and the field current is 1.5 ampere. Calculate:
   (*i*) the electromagnetic torque
   (*ii*) the flux pre pole
   (*iii*) the rotational losses (or constant losses)
   (*iv*) the efficiency.
2. Derive the torque equation for a DC motor.
3. What are the different types of motors? Also write the application of each.
4. How can be the speed of a DC motor is varied? Explain the method by which you can vary the speed above and below the base speed.
5. A 20 HP, 230 V, 1150 rpm, 4 pole DC shunt motor has a total of 620 conductors arranged in a 2-parallel paths yielding an armature circuit resistance of 0.2 Ω, when it delivered rated power at rated speed it draws a line current of 74.8 amp and field current of 3.0 amp. Calculate:
   (*i*) torque developed
   (*ii*) rotational losses
   (*iii*) total losses.
6. A 200 V DC shunt motor has a R_a = 0.4 Ω and R_f = 200 Ω, when the armature is driving at 600 rpm at a constant load torque, armature takes 20 amp. It is desired to raise the speed from 600 rpm to 900 rpm by inserting a resistance in the shunt field circuit. Find the additional resistance in the field circuit.
7. Explain the speed torque characteristics of a DC shunt and series motors.
8. A 4-pole DC shunt motor working on 250 V takes a current of 2 A when running at no load at 1000 rpm. How much back e.m.f. is generated? What will be its e.m.f. and speed if the motor takes 51 A at a certain load? Armature and field resistances are 0.2 Ω and 250 Ω respectively.

□ □ □

12

Three-Phase Induction Motor

12.1 INTRODUCTION

A Three-Phase Induction Motor is a type of asynchronous AC motor where power is supplied to the rotating device by means of electromagnetic induction. An electric motor converts electrical power to mechanical power in its rotor (rotating part). There are several ways to supply power to the rotor. In a DC motor this power is supplied to the armature directly from a DC source, while in an AC motor this power is induced in the rotating device. An induction motor is sometimes called a *rotating transformer* because the stator (stationary part) is essentially the primary side of the transformer and the rotor (rotating part) is the secondary side. Induction motors are widely used, especially poly-phase induction motors, which are frequently used in industrial drives.

Induction motors are now the preferred choice for industrial motors due to their rugged construction, absence of brushes (which are required in most DC motors) and thanks to modern power electronics—the ability to control the speed of the motor.

Similar to a DC machine and synchronous machine, an induction machine can also be operated as both a motor and a generator. The mode of operation of a induction machine is determined by the speed of the rotating field in relation to the rotor.

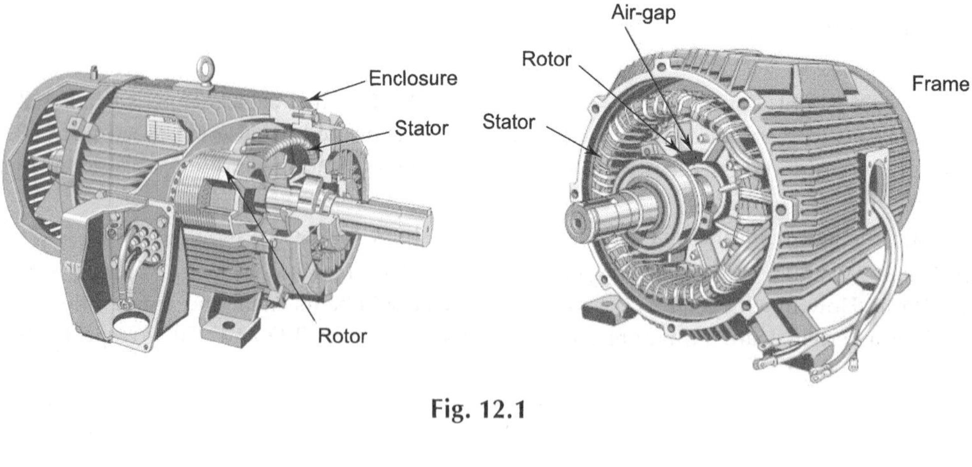

Fig. 12.1

12.2 HISTORY

The induction motor with a wrapped rotor was invented by Nikola Tesla in 1888 in the United States. In his scientific work, Tesla laid the foundations for understanding the way the motor operates. The induction motor with a cage was invented by Mikhail Dolivo-Dobrovolsky about a year later in Europe. Technological development in the field has improved to where a 100 HP (73.6 kW) motor from 1976 takes the same volume as a 7.5 HP (5.5 kW) motor did in 1897. Currently, the most common induction motor is the cage rotor motor.

12.3 TYPES

Three-phase induction motors are classified according to the construction of the rotor and are of two types:

(*i*) Squirrel Cage Induction Motor

(*ii*) Slip Ring Induction Motor.

The rotor of a squirrel cage induction motor resembles the cage of a squirrel from where it got its name. Among the two types it is the most widely used motor.

Slip ring induction motor employs wound rotor and are less extensively used then the squirrel cage type because of its initial high cost and maintenance cost.

12.4 CONSTRUCTION

An Induction motor consists of mainly two parts:

(*i*) Stator

(*ii*) Rotor.

(*i*) Stator. Stator core is made up of laminated steel stampings and has slots and teeth on its inner periphery to house stator windings. It consists of wound 'poles' that carry the supply current that induces a magnetic field in the conductor.

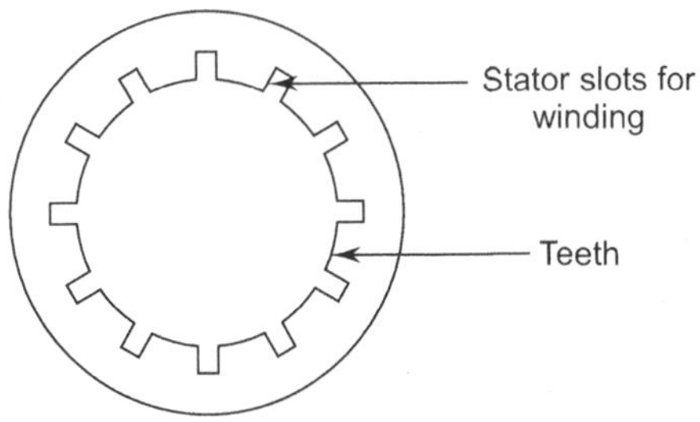

Fig. 12.2

Stator carries a three-phase winding having space displacement of 120° electrical. The three-phase winding is either star or delta connected and is fed from three-phase supply.

It is wound with definite number of poles, the exact number of poles being determined by the requirements of speed. The number of 'poles' can vary between motor types but the poles are always in pairs (*i.e.*, 2, 4, 6, etc.).

(*ii*) **Rotor.** As already discussed in the previous section that, three-phase induction motor are classified according to the construction of rotor. There are two types of rotor:

(*a*) Squirrel cage rotor [Fig. 12.3 (*a*)]

(*b*) Slip ring or wound rotor [Fig. 12.3 (*b*)]

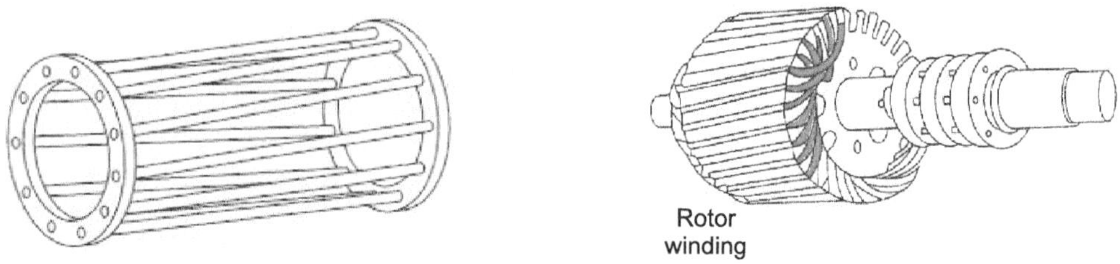

(a) Squirrel cage rotor (b) Slip ring rotor

Fig. 12.3

The most common rotor is a squirrel cage rotor. It is made up of bars of either solid copper (most common) or aluminium that span the length of the rotor, and are connected through a ring at each end. The rotor bars in squirrel cage induction motors are not straight, but have some skew to reduce noise and harmonics.

12.5 PRINCIPLE OF OPERATION

A three phase induction motor is based on the principle of mutual induction. The principle of operation can be understood by the following steps:

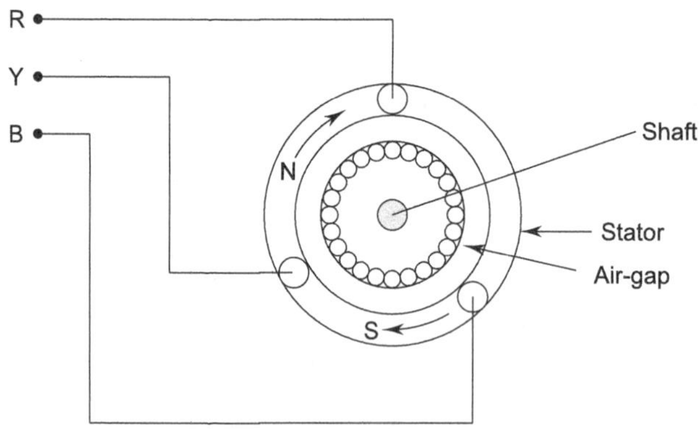

Fig. 12.4

- When the three-phase stator winding is energized from a three-phase supply, a rotating magnetic field is setup in stator.
- This rotating magnetic field rotates with synchronous speed $\left(N_s = \frac{120 f_s}{P}\right)$ with respect to stationary stator in the air-gap.
- This rotating magnetic field passes through the air-gap and cuts the stationary rotor conductors.

- Due to the relative speed between the rotating flux and the stationary rotor, e.m.f's are induced in the rotor conductors.
- If the rotor conductors are short-circuited, currents start flowing in the rotor conductors. According to Lenz's law the direction of the induced current is such that it opposes the 'cause'.
- The 'cause' is the relative speed between rotating field and stationary rotor.
- Hence, a rotor has a tendency to reduce relative speed.
- So rotor begins to move in the direction of rotating field and continues towards synchronous speed.
- The machine runs at a speed near but below synchronous speed.

If in case, the speed of rotor reaches to synchronous speed (*i.e.,* the speed of the field), relative speed becomes zero. Hence no e.m.f. no current and therefore no torque at synchronous speed. Hence rotor never reaches to synchronous speed.

12.6 COMPARISON BETWEEN SQUIRREL CAGE AND SLIP RING INDUCTION MOTOR

Squirrel Cage Induction Motor	Slip Ring Induction Motor
Advantages	*Advantages*
1. Rugged in construction	1. Much higher starting torque
2. No slip rings, brush gears etc.	2. Speed control
3. Minimum maintenance	3. Can be started directly online
4. Trouble free performance	4. Comparatively lesser starting current
5. Cheaper	
6. Comparatively higher efficiency	
7. Relatively better cooling conditions	
Disadvantages	*Disadvantages*
1. Low starting torque	1. Higher cost
2. Higher starting current	2. Comparatively lower efficiency
3. No speed control	3. Higher degree of maintenance
4. Needs a starter	4. Extra slip ring, brush gears required

12.7 SLIP

The rotor runs at a speed which is always less than the speed of the stator field, *i.e.,* synchronous speed. Let the synchronous speed is denoted by N_s and the rotor speed is denoted by N_r.

The difference between the synchronous speed and actual speed of rotor is known as ***'slip speed'***.

$$\text{slip speed} = N_s - N_r \qquad \ldots(12.1)$$

In actual, percentage change in slip speed is termed as 'slip'.

$$s = \frac{N_s - N_r}{N_s} \times 100 \quad \text{...(12.2)}$$

or $$N_r = N_s (1 - s) \quad \text{...(12.3)}$$

When the rotor is stationary, $N_r = 0$ and slip will be $s = 1$ or 100%

12.8 FREQUENCY OF ROTOR CURRENT

When rotor is stationary, frequency of the rotor current is same as supply frequency (*i.e.,* frequency of stator e.m.f.). But when rotor starts revolving, then frequency of the rotor current depends upon relative speed or on slip speed. Let at any slip speed,

The frequency of rotor current is denoted by f_r.

The stator frequency is denoted by f_s.

Then, field in the stator is rotating with synchronous speed which is given as

$$N_s = \frac{120 f_s}{P} \quad \text{...(12.4)}$$

And field in rotor is rotating with a speed which is given as

$$N_s - N_r = \frac{120 f_r}{P} \quad \text{...(12.5)}$$

Dividing equation (12.5) by equation (12.4), we get

$$\frac{N_s - N_r}{N_s} = \frac{f_r}{f_s} = s \quad \text{...(12.6)}$$

or $$f_r = sf_s \quad \text{...(12.7)}$$

12.9 INDUCED E.M.F. IN THE ROTOR CONDUCTORS

Similar to a transformer, the induced e.m.f. in the rotor and stator are given by

$$E_{stator} = 4.44 f_s N_1 \phi K_{w1} \quad \text{...(12.8)}$$

and $$E_{rotor} = 4.44 f_r N_2 \phi K_{w2} \quad \text{...(12.9)}$$

where N_1 and N_2 are stator turns per phase and rotor turns per phase respectively
f_s and f_r are stator frequency and rotor frequency respectively
K_{w1} and K_{w2} are stator winding factor and rotor winding factor respectively

Dividing equation (12.9) by equation (12.8), we get

$$E_r = sE_s \quad \text{...(12.10)}$$

12.10 STARTING TORQUE

The torque developed by the motor at instant of starting is called starting torque. Similar to the secondary winding of transformer, the rotor of an induction motor can be represented by a resistance and an inductive reactance but with a short-circuit (Fig. 12.5).

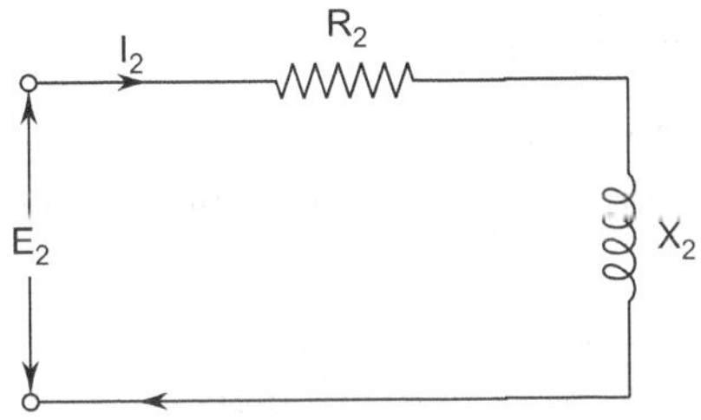

Fig. 12.5

Let E_2 = Rotor e.m.f. per phase at standstill

R_2 = Rotor resistance per phase

X_2 = Rotor reactance per phase at standstill

I_2 = Rotor current per phase at standstill

Now rotor impedance/phase at standstill will be

$$Z_2 = \sqrt{R_2^2 + X_2^2} \quad \text{...(12.11)}$$

The rotor current per phase will be

$$I_2 = \frac{E_2}{Z_2} = \frac{E_2}{\sqrt{R_2^2 + X_2^2}} \quad \text{...(12.12)}$$

Power factor of rotor circuit will be

$$\cos \phi_2 = \frac{R_2}{Z_2} = \frac{R_2}{\sqrt{R_2^2 + X_2^2}} \quad \text{...(12.13)}$$

The torque developed by rotor of an induction motor at time of starting is directly proportional to

(*i*) Rotor current, I_2

(*ii*) Stator flux per pole, Φ

(*iii*) Power factor of the rotor circuit, $\cos \phi_2$

i.e.,
$$T_{st} \propto \Phi I_2 \cos \phi_2 \quad \text{...(12.14)}$$

or
$$T_{st} = K \Phi I_2 \cos \phi_2 \quad \text{...(12.15)}$$

where K is proportionality constant, whose value will be determined later-on.

The flux per pole is proportional to the rotor induced e.m.f. at standstill *i.e.*,

$$\Phi \propto E_2 \quad \text{...(12.16)}$$

Hence equation (12.15) can also be written as

$$T_{st} = K E_2 I_2 \cos \phi_2 \quad \text{...(12.17)}$$

Putting equations (12.12) and (12.13) in equation (12.17), we get

$$T_{st} = K E_2 \frac{E_2}{\sqrt{R_2^2 + X_2^2}} \frac{R_2}{\sqrt{R_2^2 + X_2^2}} \quad \text{...(12.18)}$$

or
$$\boxed{T_{st} = K \frac{E_2^2 R_2}{R_2^2 + X_2^2}} \quad \text{...(12.19)}$$

12.10.1 Determination of Constant K

The constant K can be determined as follow:

Let per phase power input to rotor is denoted by P_g and is given by

$$P_g = E_2 I_2 \cos \phi_2 \qquad \ldots(12.20)$$

For three-phase, $$P_g = 3E_2 I_2 \cos \phi_2 \qquad \ldots(12.21)$$

This power input to the rotor is equivalent to mechanical output power

$$P_m = \text{Torque} \times \text{Speed} \qquad \ldots[12.22(a)]$$

or $$P_m = P_g = T \times 2\pi N_s \qquad \ldots[12.22(b)]$$

where N_s = Synchronous speed of rotating magnetic field in rps.

Comparing equation (12.21) and equation (12.22), we get

$$3E_2 I_2 \cos \phi_2 = T \times 2\pi N_s \qquad \ldots(12.23)$$

or $$T = \frac{3}{2\pi N_s} E_2 I_2 \cos\phi_2 \qquad \ldots(12.24)$$

Comparing equation (12.17) and equation (12.24), we get expression for constant K as

$$\boxed{K = \frac{3}{2\pi N_s}} \qquad \ldots(12.25)$$

Therefore, from equation (12.25), the starting torque expression will become

$$\boxed{T_{st} = \frac{3}{2\pi N_s} \frac{E_2^2 R_2}{R_2^2 + X_2^2}} \qquad \ldots(12.26)$$

12.10.2 Condition for Maximum Starting Torque

The starting torque will be maximum when

$$\frac{dT_{st}}{dR_2} = 0 \qquad \ldots(12.27)$$

or $$\frac{3E_2^2}{2\pi N_s} \left[\frac{\left[R_2^2 + X_2^2\right] - R_2 (2R_2)}{\left(R_2^2 + X_2^2\right)^2} \right] = 0 \qquad \ldots(12.28)$$

or $$R_2^2 + X_2^2 - 2R_2^2 = 0 \qquad \ldots(12.29)$$

or $$\boxed{R_2 = X_2} \qquad \ldots(12.30)$$

i.e., for maximum starting torque, the rotor resistance (R_2), must be equal to rotor reactance (X_2).

12.11 TORQUE UNDER RUNNING CONDITIONS

Under running conditions, when 'slip' comes into play, the rotor frequency, e.m.f. induced in the rotor circuit and rotor reactance will become

E_r = Rotor induced e.m.f. per phase under running conditions.

$E_r = sE_2$

f_r = Frequency of induced e.m.f. under running conditions

$f_r = sf_2$
X_r = Rotor reactance per phase under running conditions
$X_r = sX_2 = s2\pi fL$
R_r = Rotor resistance per phase under running conditions
$R_r = R_2$

Now rotor impedance per phase under running conditions will be

$$Z_r = \sqrt{R_r^2 + X_r^2} = \sqrt{R_2^2 + (sX_2)^2} \quad ...(12.31)$$

The rotor current per phase under running conditions will be

$$I_r = \frac{E_r}{Z_r} = \frac{sE_2}{\sqrt{R_r^2 + X_r^2}} = \frac{sE_2}{\sqrt{R_2^2 + (sX_2)^2}} \quad ...(12.32)$$

Power factor of rotor circuit under running conditions will be

$$\cos\phi_r = \frac{R_r}{Z_r} = \frac{R_2}{\sqrt{R_2^2 + (sX_2)^2}} \quad ...(12.33)$$

The torque developed by rotor of an induction motor under running conditions is directly proportional to

(*i*) Rotor current, I_r

(*ii*) Stator flux per pole, Φ

(*iii*) Power factor of the rotor circuit, $\cos\phi_r$

or
$$T_r = K\,\Phi I_r \cos\phi_r \quad ...(12.34)$$

The flux per pole is proportional to rotor induced e.m.f. at standstill *i.e.*,

$$\Phi \propto E_2 \quad \text{[Eqn. (12.16)]}$$

Hence equation (12.34) can also be written as

$$T_r = K\,E_2 I_r \cos\phi_r \quad ...(12.35)$$

Putting equations (12.25), (12.32) and (12.33) in equation (12.35), we get

$$T_r = \frac{3}{2\pi N_s} E_2 \frac{E_r}{\sqrt{R_r^2 + X_r^2}} \frac{R_r}{\sqrt{R_r^2 + X_r^2}} \quad ...(12.36)$$

or
$$\boxed{T_r = \frac{3}{2\pi N_s} \frac{sE_2^2 R_2}{R_2^2 + (sX_2)^2}} \quad ...(12.37)$$

Equation (12.37) represents expression for running torque at any slip's'.

12.11.1 Condition for Maximum Running Torque

The torque under running condition will be maximum when

$$\frac{dT_r}{ds} = 0 \quad ...(12.38)$$

or
$$\frac{3E_2^2R_2}{2\pi N_s}\left[\frac{\left[R_2^2 + (sX_2)^2\right] - s2\left(sX_2^2\right)}{\left(R_2^2 + (sX_2)^2\right)^2}\right] = 0 \quad ...(12.39)$$

or $$R_2^2 + s^2X_2^2 - 2s^2R_2^2 = 0 \qquad \text{...(12.40)}$$

or $$\boxed{R_2 = sX_2} \qquad \text{...(12.41)}$$

i.e., for maximum running torque, the rotor resistance (R_2), must be equal to rotor reactance (sX_2), under running conditions.

12.11.2 Expression for Maximum Running Torque

The torque under running condition will be maximum at that value of slip '*s*' which makes rotor reactance per phase equal to rotor resistance per phase. The slip corresponding to maximum running torque will be

$$s = \frac{R_2}{X_2} \qquad \text{...(12.42)}$$

Now putting $R_2 = sX_2$ in running torque expression of equation (12.37), we get

$$T_{r,\,\max} = \frac{3}{2\pi N_s} \frac{sE_2^2R_2}{R_2^2 + (sX_2)^2} \qquad \text{[Eqn. (12.37)]}$$

or $$T_{r,\,\max} = \frac{3}{2\pi N_s} \frac{sE_2^2(sX_2)}{(sX_2)^2 + (sX_2)^2} \qquad \text{...(12.43)}$$

or $$T_{r,\,\max} = \frac{3}{2\pi N_s} \frac{s^2E_2^2X_2}{2(sX_2)^2} \qquad \text{...(12.44)}$$

or $$\boxed{T_{r,\,\max} = \frac{3}{2\pi N_s} \frac{E_2^2}{2X_2}} \qquad \text{...(12.45)}$$

Equation (12.45) represents expression for maximum running torque. It is clear from the expression that maximum running torque is

- Independent of R_2
- Varies inversely as standstill reactance.
- Varies directly as square of the e.m.f. induced.

12.12 ROTOR POWER EXPRESSIONS AND RELATIONS

As we know that the mechanical power is given by

$$\text{Mechanical power} = \text{Torque} \times \text{Speed} \qquad \text{...(12.46)}$$

Now the power transferred from stator to rotor or input power to rotor or air-gap power is given by

$$P_g = T \times 2\pi N_s \qquad \text{...(12.47)}$$

where N_s = Synchronous speed of rotating magnetic field in rps.

Mechanical power developed by rotor will be

$$P_{mec} = T \times 2\pi N_r \qquad \text{...(12.48)}$$

where N_r = Speed of rotor, at full-load, in rps.

Note. *The rotor speed as well as synchronous speed is taken in* ***rps*** *and in most of the numerical problems these speeds are given in* ***rpm***.

Rotor copper loss will be

$$\text{Losses} = \text{Input} - \text{Output} \quad \text{...(12.49)}$$

$$P_{Cu} = P_g - P_{mec} \quad \text{...(12.50)}$$

or $$P_{Cu} = T \times 2\pi N_s - T \times 2\pi N_r \quad \text{...(12.51)}$$

or $$P_{Cu} = 2\pi T\,(N_s - N_r) \quad \text{...(12.52)}$$

As we know that the slip is given by

$$s = \frac{N_s - N_r}{N_s} \quad \text{...(12.53)}$$

or $$N_r = N_s\,(1 - s) \quad \text{...(12.54)}$$

or $$N_s - N_r = sN_s \quad \text{...(12.55)}$$

From equation (12.55) equation (12.52) will become

$$P_{Cu} = 2\pi Ts\,N_s \quad \text{...(12.56)}$$

or $$P_{Cu} = s\,P_g \quad \text{...(12.57)}$$

Putting equation (12.54) in equation (12.48) *i.e.,* in the expression of mechanical power P_{mec}, we get

$$P_{mec} = T \times 2\pi N_s\,(1 - s) \quad \text{...(12.58)}$$

or $$P_{mec} = (1 - s)\,P_g \quad \text{...(12.59)}$$

Now from equations (12.57), (12.59) it is clear that

$$\boxed{P_g : P_{mec} : P_{Cu} = 1 : (1 - s) : s} \quad \text{...(12.60)}$$

Alternative Method

Power transferred from stator to rotor or Input power to rotor is also given by

$$P_g = E_2 I_2 \cos\phi_2 \quad \text{...(12.61)}$$

where E_2, I_2 and $\cos\phi_2$ are the rotor induced e.m.f, rotor current and rotor power factor at running conditions which are given by

$$I_r = \frac{E_r}{Z_r} = \frac{sE_2}{\sqrt{R_r^2 + X_r^2}} = \frac{sE_2}{\sqrt{R_2^2 + (sX_2)^2}} \quad \text{...(12.62)}$$

$$\cos\phi_r = \frac{R_r}{Z_r} = \frac{R_2}{\sqrt{R_2^2 + (sX_2)^2}} \quad \text{...(12.63)}$$

$$Z_r = \sqrt{R_r^2 + X_r^2} = \sqrt{R_2^2 + (sX_2)^2} \quad \text{...(12.64)}$$

From equations (12.62), (12.63) and (12.64), equation (12.61) will become

$$P_g = \frac{I_2\sqrt{R_2^2 + (sX_2)^2}}{s} I_2 \frac{R_2}{\sqrt{R_2^2 + (sX_2)^2}} \quad \text{...(12.65)}$$

or $$P_g = \frac{I_2^2 R_2}{s} \quad \text{...(12.66)}$$

In general $$\text{Losses} = \text{Input} - \text{Output} \quad \text{...(12.67)}$$

or $$\text{Output} = \text{Input} - \text{Losses} \quad \text{...(12.68)}$$

Mechanical power developed by the rotor will be

$$P_{mec} = P_g - \text{Rotor copper loss} \quad \text{...(12.69)}$$

The rotor copper losses are given by

$$P_{Cu} = I_2^2 R_2 \quad ...(12.70)$$

Now from equations (12.66) and (12.70) equation (12.69) will become

$$P_{mec} = \frac{I_2^2 R_2}{s} - I_2^2 R_2 \quad ...(12.71)$$

or $$P_{mec} = I_2^2 R_2 \left(\frac{1-s}{s}\right) \quad ...(12.72)$$

or $$P_{mec} = (1-s) P_g \quad ...(12.73)$$

Now from equations (12.66), (12.70) and (12.73) it is clear that

$$\boxed{P_g : P_{mec} : P_{Cu} = 1 : (1-s) : s} \quad ...(12.74)$$

12.12.1 Power Flow Diagram

Fig. 12.6 shows the flow of power from stator terminals (electrical power input) to the power obtained at shaft of the motor (mechanical power output).

The useful power and losses at different stages are also mentioned.

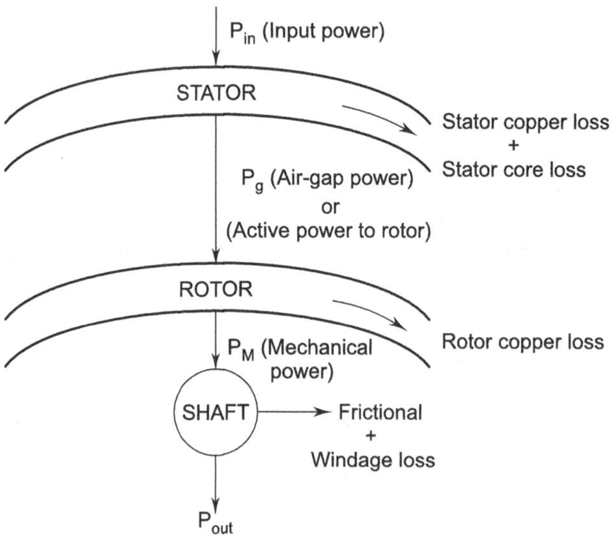

Fig. 12.6

12.13 TORQUE-SLIP CHARACTERISTICS

The expression for torque at any slip is given by

$$T_r = \frac{3}{2\pi N_s} \frac{s E_2^2 R_2}{R_2^2 + (sX_2)^2} \quad \text{[Eqn. (12.37)]}$$

or $$T_r = \frac{3}{2\pi N_s} \frac{E_2^2 \frac{R_2}{s}}{\left(\frac{R_2}{s}\right)^2 + X_2^2} \quad ...(12.75)$$

The torque-slip or torque-speed curve depends upon value of slip, and accordingly, the induction motor can have following three operating region.

12.13.1 Motoring Mode

Initially at time of starting, slip $s = 1$ and torque expression from equation (12.75) will become

$$T_{st} = \frac{3}{2\pi N_s} \frac{E_2^2 R_2}{R_2^2 + X_2^2} \qquad \text{[Eqn. (12.26)]}$$

This is starting torque which is shown in curve below (Fig. 12.7). The slip for motoring mode will lie between

$$1 \leq s \leq 0 \qquad \ldots(12.76)$$

When the slip is $s = 1$, at the time of starting, $X_2 >> \frac{R_2}{s}$, so that

$$T \propto \frac{1}{s} \qquad \ldots(12.77)$$

As slip decreases from '1' and moves towards '0', torque increases.

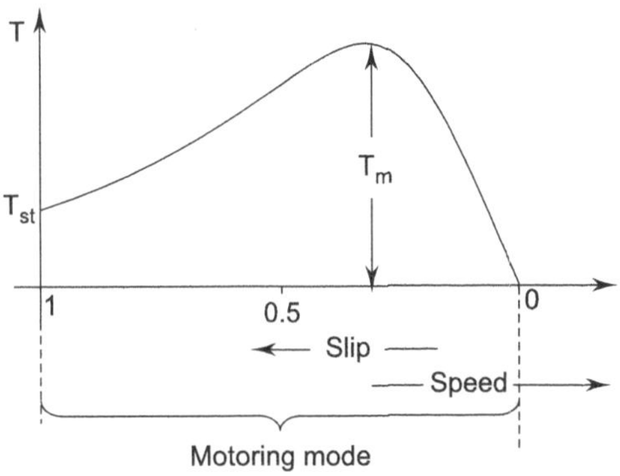

Fig. 12.7

The torque expression given in equation (12.75) will be maximum when

$$R_2 = sX_2 \qquad \text{[Eqn. (12.41)]}$$

And maximum running torque is given by

$$T_{r,\max} = \frac{3}{2\pi N_s} \frac{E_2^2}{2X_2} \qquad \text{[Eqn. (12.45)]}$$

The slip corresponding to maximum running torque will be

$$s = \frac{R_2}{X_2} \qquad \text{[Eqn. (12.42)]}$$

After attaining maximum torque, as slip decreases further, value $\frac{R_2}{s}$ increases and X_2 decreases. Now, in such a case, the torque expression will become

$$T = K E_2^2 \frac{s}{R_2} \qquad \ldots(12.78)$$

or $$T \propto s \qquad \ldots(12.79)$$

So as slip decreases, torque also decreases.

12.13.2 Generating Mode

After slip crosses '0' point, rotor speed (N_r) will become greater than synchronous speed (N_s) and hence slip becomes negative (Fig. 12.8).

The direction of rotation of rotor is in the direction of rotating magnetic field.

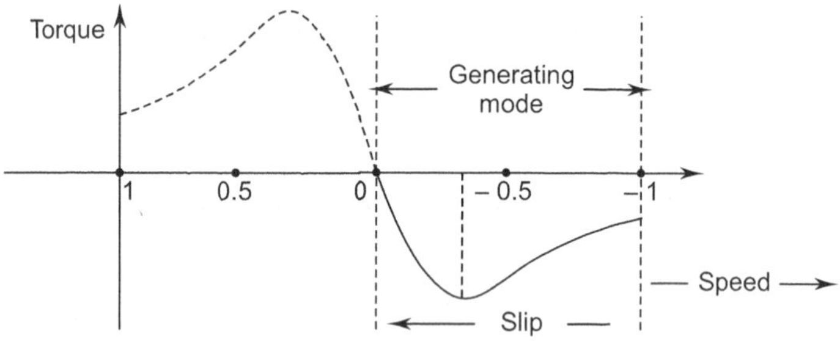

Fig. 12.8

12.13.3 Breaking Mode

- The motor is in running condition.
- The phase sequence is changed.
- The direction of rotating field suddenly becomes opposite to that of rotor rotation.
- Electromagnetic torque, 'T' now act in opposite direction to rotor rotation produces the breaking action.
- This is also called plugging (Fig. 12.9).

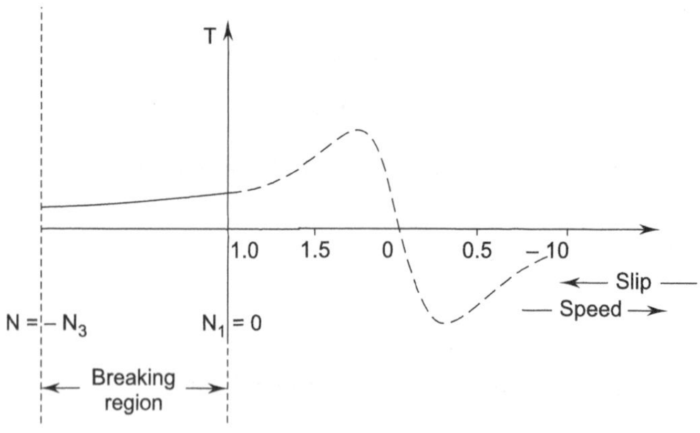

Fig. 12.9

The operating slip is greater than one as rotor speed (N_r) will be negative and hence

$$s = \frac{N_s - (-N_r)}{N_s} = \frac{N_s + N_r}{N_s} \qquad \ldots(12.80)$$

Motor can quickly stop by plugging, but supply must be disconnected before rotor starts rotating in opposite direction.

The complete torque-slip characteristics (or torque-speed characteristics) is shown in Fig. 12.10.

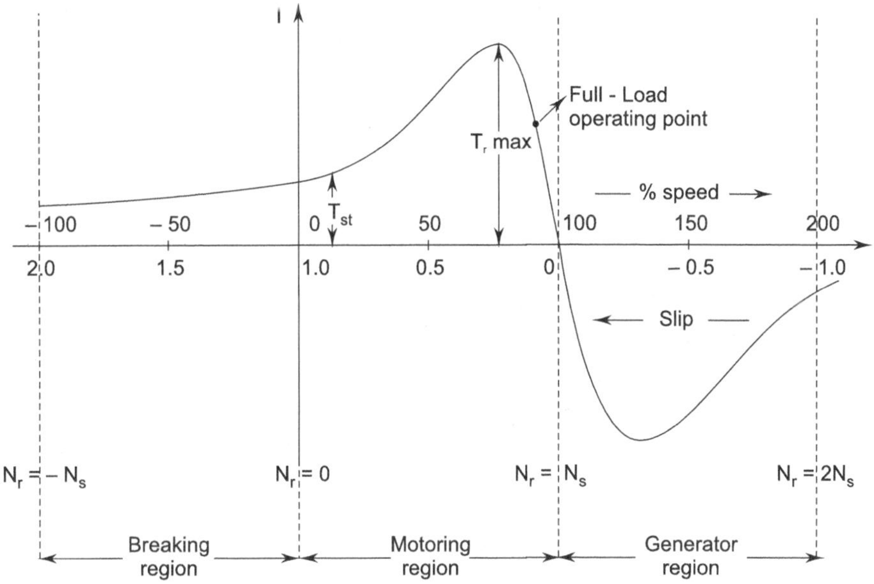

Fig. 12.10

12.14 EQUIVALENT CIRCUIT

As already discussed in introduction part that, an induction motor is also known as rotating transformer. The stator winding of induction motor is equivalent to primary winding of a transformer and the rotor winding of induction motor is equivalent to secondary winding of a transformer.

The differences between an induction motor and a transformer are that

- In case of induction motor one winding (rotor) is rotating.
- In induction motor, we have very high magnetizing current.

Similar to a transformer, let us consider the primary and secondary parameters of a three-phase induction as shown in equivalent circuit of Fig. 12.11.

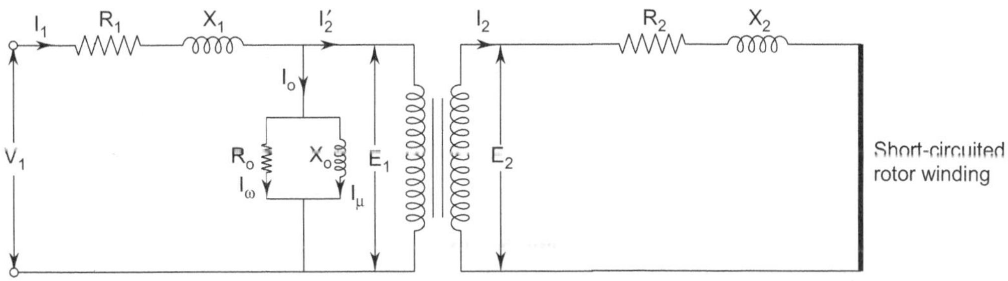

Fig. 12.11

The rotor equivalent circuit at running condition will be

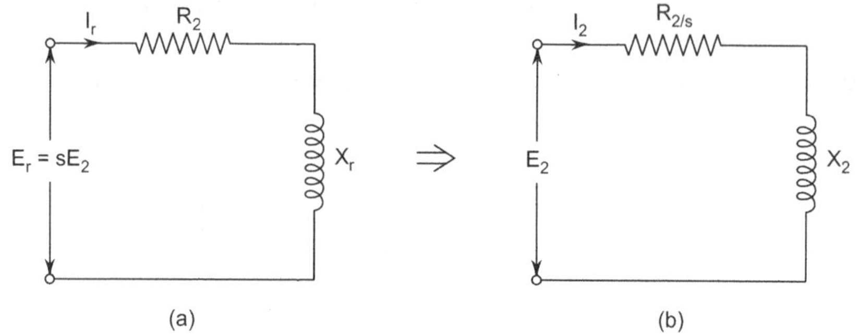

Fig. 12.12

The above rotor equivalent circuit can be modified as follow:

The rotor current (or secondary current) per phase under running conditions [Fig. 12.12(*a*)] will be

$$I_2 = I_r = \frac{E_r}{Z_r} = \frac{sE_2}{\sqrt{R_r^2 + X_r^2}} = \frac{sE_2}{\sqrt{R_2^2 + (sX_2)^2}} \qquad \text{[Eqn. (12.32)]}$$

or from Fig. 12.12(*b*)

$$I_2 = \frac{E_2}{\sqrt{\left(\frac{R_2}{s}\right) + (X_2)^2}} \qquad \text{...(12.81)}$$

The rotor resistance $\frac{R_2}{s}$ is a function of slip and can be split into two parts

$$\frac{R_2}{s} = \underbrace{R_2}_{\text{I}} + \underbrace{R_2\left(\frac{1}{s} - 1\right)}_{\text{II}} \qquad \text{...(12.82)}$$

- The first part R_2 is rotor resistance itself and is responsible for rotor copper loss.
- The second part $R_2\left(\frac{1}{s} - 1\right)$ is a **fictitious resistance**, which represents electrical equivalent of mechanical load on the motor.

Now the equivalent circuit of rotor will become as shown in Fig. 12.13.

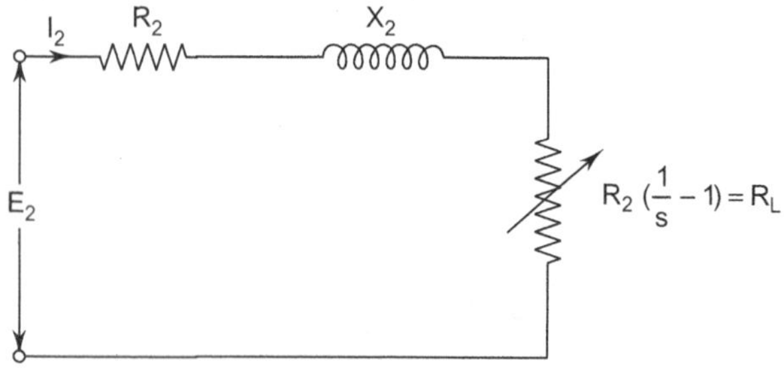

Fig. 12.13

The secondary side (rotor side) is short-circuited and hence secondary parameters are transferred to primary side. Including the above rotor equivalent circuit (Fig. 12.13) in the main equivalent circuit (Fig. 12.11) and shifting secondary parameters on primary side we have following equivalent circuit (Fig. 12.14).

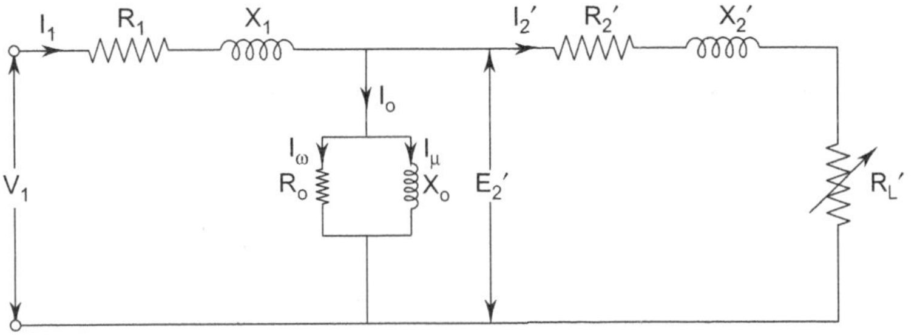

Fig. 12.14

The exciting circuit may be transferred to the left as in accuracy involved is negligible (Fig. 12.15).

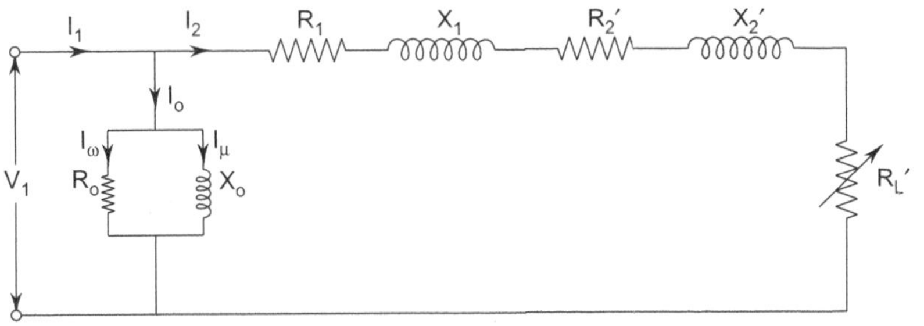

Fig. 12.15

and the approximate equivalent circuit of an induction motor is shown in Fig. 12.16.

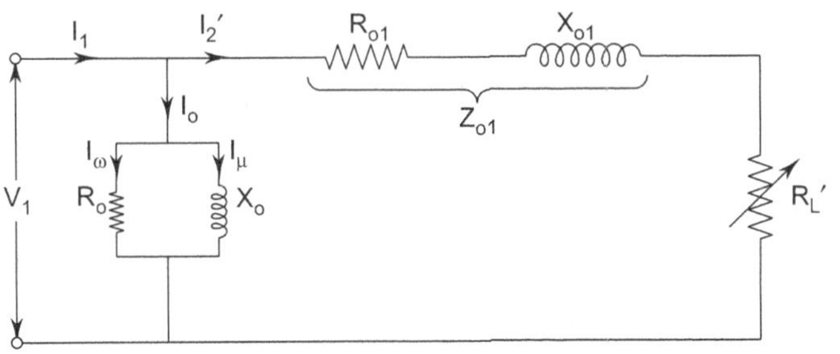

Fig. 12.16

where $$R_2' = \frac{R_2}{K^2} \qquad ...(12.83)$$

$$X_2' = \frac{X_2}{K^2} \qquad ...(12.84)$$

$$I_2' = KI_2 \qquad ...(12.85)$$

$$R_L' = \frac{R_L}{K^2} = \frac{R_2}{K^2}\left(\frac{1}{s}-1\right) = R_2'\left(\frac{1}{s}-1\right) \qquad ...(12.86)$$

$$R_{o1} = R_1 + R_2' \qquad ...(12.87)$$

$$X_{o1} = X_1 + X_2' \qquad ...(12.88)$$

$$Z_{o1} = \sqrt{R_{o1}^2 + X_{o1}^2} \qquad ...(12.89)$$

12.15 PHASOR DIAGRAM

To draw phasor diagram for induction motor, consider load as resistive. Note that V_2 will be considered if the fictitious resistance, $R_2\left(\frac{1}{s}-1\right)$, which represent mechanical load, is considered.

From approximate equivalent circuit of induction motor shown in Fig. 12.11, the equations for primary and secondary sides are

$$V_1 = I_1R_1 + jI_1X_1 + (-E_1) \quad \text{...(12.90)}$$

$$I_1 = I_o + I_2' \quad \text{...(12.91)}$$

$$I_o = I_\mu + I_\omega \quad \text{...(12.92)}$$

$$E_2 = I_2R_2 + jI_2X_2 + V_2 \quad \text{...(12.93)}$$

E_1, E_2 lags behind flux Φ by 90°

I_2' is equal and opposite to I_2

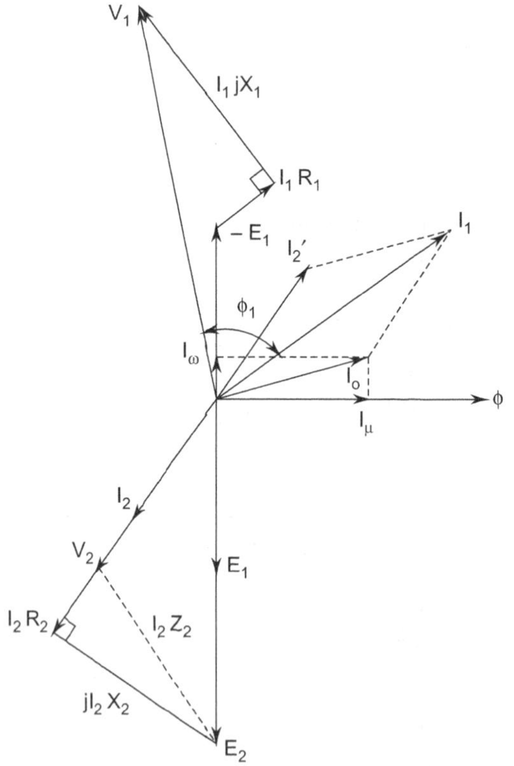

Fig. 12.17

12.16 STARTING METHODS OF THREE-PHASE INDUCTION MOTOR

In a three-phase induction motor, the induced e.m.f. in rotor circuit depends on slip of the induction motor and the magnitude of rotor current depends upon this induced e.m.f. When motor is started, the slip is equal to 1 as rotor speed is zero, so induced e.m.f in the rotor is large. As a result, a very high current flows through the rotor. This is similar to a transformer with the secondary coil short-circuited, which causes the primary coil to draw a high current from the mains. Similarly, when an induction motor starts, a very high current is drawn by the stator of the order of 5 to 9 times the full load current. This high current can damage motor windings and because it causes heavy line voltage drop, other appliances connected to the same line may be affected by voltage fluctuation. To avoid such effects, the starting current should be limited. A soft start starter is a device which limits the starting current by providing reduced voltage to the motor. Once the rotor speed increases, full rated voltage is given to it.

Following are methods for starting of induction motor:

1. Direct on line starting
2. Stator resistance starting
3. Auto transformer starting
4. Star delta starting

} For squirrel-cage induction motor

5. Rotor resistance starting } For slip-ring induction motor

In all these methods (method 2 to 4) the applied voltage to squirrel-cage induction motor is reduced during starting.

12.16.1 Direct On-Line Starting

$$\text{Rotor Input} = 2\pi N_s T \qquad \text{...(12.94)}$$

$$= KT \qquad \text{...(12.95)}$$

Rotor copper losses are given by

$$P_{Cu} = I_2^2 R_2 \qquad \text{...(12.96)}$$

or

$$P_{Cu} = s \times \text{Rotor input} \qquad \text{...(12.97)}$$

or

$$P_{Cu} = s \times KT \qquad \text{...(12.98)}$$

Comparing equations (12.96) and (12.98), we get

$$I_2^2 R_2 = sKT \qquad \text{...(12.99)}$$

As rotor resistance R_2 in case of squirrel-cage induction motor is constant hence

or

$$T \propto \frac{I_2^2}{s} \qquad \text{...(12.100)}$$

Now at starting moment, slip $s = 1$, and as $I_1 = I_2$

Hence,

$$T_{st} \propto I_2^2 \propto I_1^2 \propto I_{st}^2 \qquad \text{...(12.101)}$$

And at full-load let the

Full-load current = I_f

Full-load slip = s_f

Full-load torque = T_f

$$T_f \propto \frac{I_f^2}{s_f} \qquad \text{...(12.102)}$$

So from equations (12.101) and (12.102), we get

$$\frac{T_{st}}{T_f} = \left(\frac{I_{sc}}{I_f}\right)^2 s_f \qquad \text{...(12.103)}$$

Now taking an example when slip is 4% or

$$s = 0.04$$

$$I_{sc} = 7I_f$$

Then from equation (12.103) we have

$$\frac{T_{st}}{T_f} = (7)^2\, 0.04 = 1.96$$

or

$$T_{st} = 1.96\ T_f$$

Hence it is clear that when the starting current (or I_{sc}) is 7 times the full load current, the motor develops a starting torque which is only 1.96 times the full-load value. Therefore this method is useful for small motors only.

12.16.2 Stator Resistor Starting

This method is also known as Primary Resistors or Primary Rheostat or Primary Reactor starting.

By using external primary resistors, applied voltage can be reduced. Hence the initial current drawn by motor is reduced. However it should be noted that current varies direct as the voltage, whereas the torque varies square of the applied voltage. This is because when the applied voltage is reduced, the rotating flux Φ is reduced which in turn, decreases rotor e.m.f. and hence rotor current I_2.

Starting torque, which depends both on Φ and I_2 suffers on two counts when applied voltage is reduced.

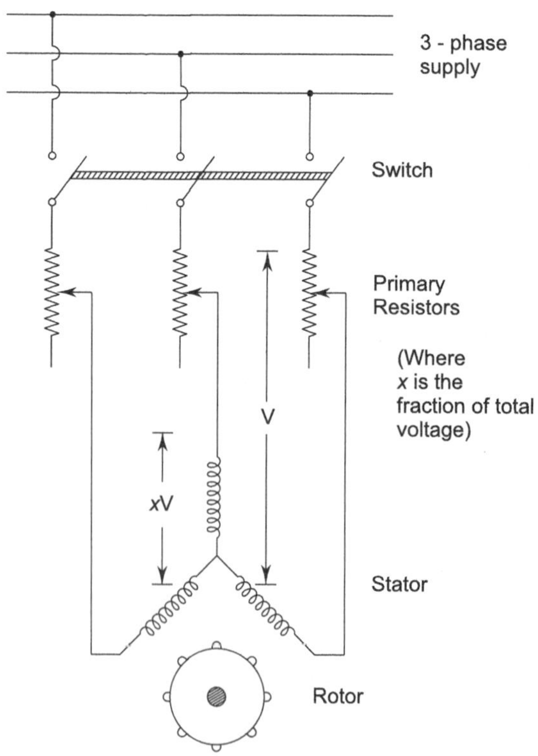

Fig. 12.18

Now, the voltage across winding of the stator and the external primary resistor connected in series with the winding be 'V' as shown in Fig. 12.18. There will be a voltage drop across resistor and hence a reduced voltage will appear across the winding which is say 'xV'. Hence the starting current will be

$$I_{st} = \frac{xV_1}{Z_{sc}} \quad \text{...(12.104)}$$

and the short circuit current will be

$$I_{sc} = \frac{V_1}{Z_{sc}} \quad \text{...(12.105)}$$

From equations (12.104) and (12.105) we have

$$I_{st} = xI_{sc} \quad \text{...(12.106)}$$

From equation (12.106) equation (12.103) will become

$$\frac{T_{st}}{T_f} = \left(\frac{I_{st}}{I_f}\right)^2 s_f = \left(\frac{xI_{sc}}{I_f}\right)^2 s_f \quad \text{...(12.107)}$$

or

$$\frac{T_{st}}{T_f} = x^2\left(\frac{I_{sc}}{I_f}\right)^2 s_f \quad \text{...(12.108)}$$

Hence it is seen from the above expression that the short-circuit current I_{sc} decreases by same fraction 'x' at which the applied voltage 'V' decreases. As the external resistor is cut-out of the circuit, the motor will speed up.

12.16.3 Auto Transformer Starting

This method can be used both for star and delta connected motors. With direct on line starting, the starting current is given by

$$I_{sc} = \frac{V_1}{Z_{sc}} \quad \text{...(12.109)}$$

or

$$I_{st} = xI_{sc} \quad \text{...(12.110)}$$

Now in a transformer, the ratio of current is inversely proportional to the voltage ratio *i.e.*,

$$\frac{I_1}{I_2} = \frac{V_2}{V_1} \quad \text{...(12.111)}$$

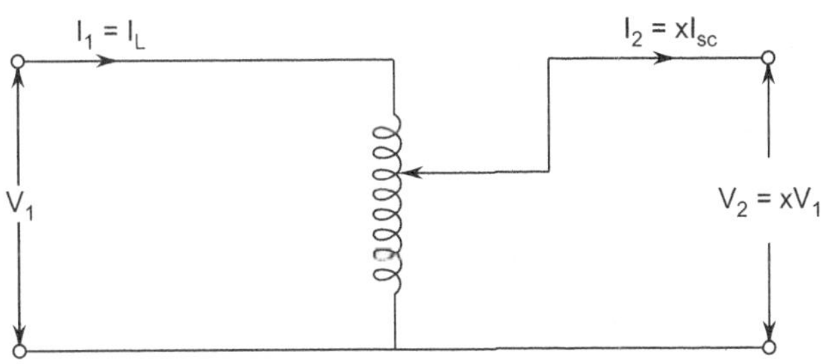

Fig. 12.19

or $$I_1V_1 = I_2V_2 \quad \text{...(12.112)}$$

or $$V_1I_L = (x\,I_{sc})(xV_1) \quad \text{...(12.113)}$$

or $$I_L = x^2\,I_{sc} \quad \text{...(12.114)}$$

So with the help of auto transformer starting, the line current is reduced by x^2 fraction.

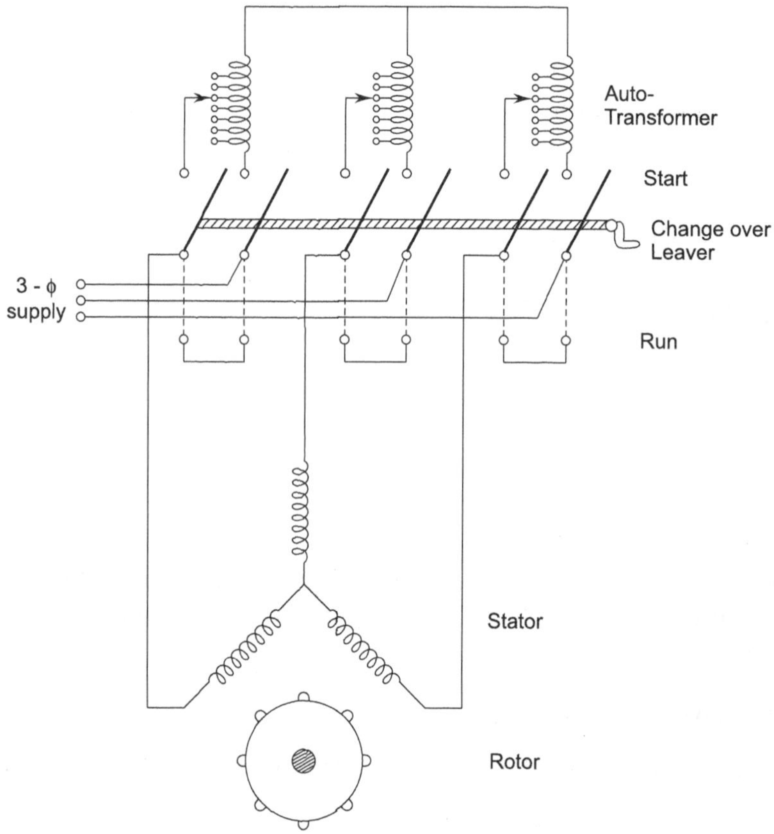

Fig. 12.20

Now, $$\frac{T_{st}}{T_{fl}} = \left(\frac{I_{st}}{I_{fl}}\right)^2 s_f = \left(\frac{xI_{sc}}{I_{fl}}\right)^2 s_{fl} \quad \text{...(12.115)}$$

or $$\frac{T_{st}}{T_{fl}} = x^2\left(\frac{I_{sc}}{I_f}\right)^2 s_{fl} \quad \text{...(12.116)}$$

From equation (12.114) putting $I_{sc} = I_L/x^2$, we get

or $$\frac{T_{st}}{T_{fl}} = x^2\left(\frac{I_L}{x^2 I_{fl}}\right)^2 s_f \quad \text{...(12.117)}$$

or $$\frac{T_{st}}{T_{fl}} = \frac{1}{x^2}\left(\frac{I_L}{I_{fl}}\right)^2 s_f \quad \text{...(12.118)}$$

So it is clear from the equation (12.118) that the starting torque are reduced to x^2 times to their corresponding values with direct on line (DOL) method.

12.16.4 Star-Delta Starting

A star-delta starter is used for a squirrel-cage motor designed to run normally on data connected stator winding.

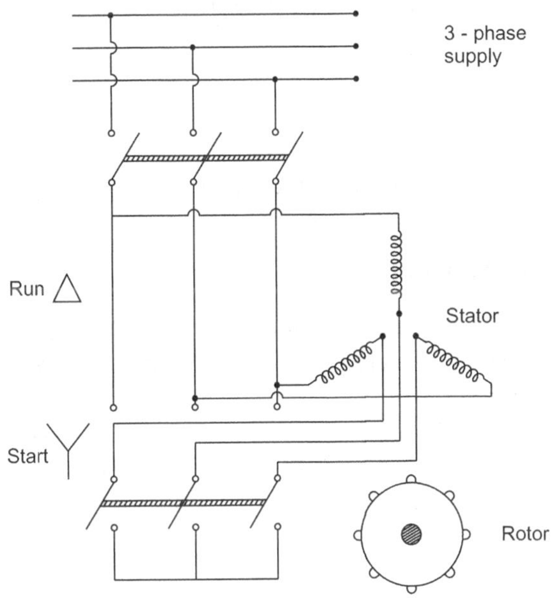

Fig. 12.21

It consists of a two-way switch which connects the motor in star for starting and then in delta for normal running.

In Star Connection

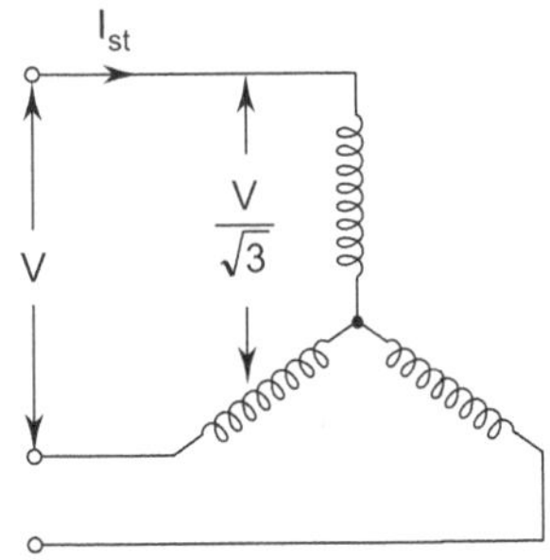

Fig. 12.22

$$\text{Line voltage} = \sqrt{3}\ \text{Phase voltage}$$

or $$V_L = \sqrt{3}\, V_{ph} \qquad \text{...(12.119)}$$

and $$I_{st} = I_{ph} = \frac{V_p}{Z} = \frac{V_L}{\sqrt{3}Z} \qquad \text{...(12.120)}$$

or $$I_{st} = \frac{V_L}{\sqrt{3}Z} = \frac{1}{\sqrt{3}} I_{sc} \qquad \text{...(12.121)}$$

or $$I_{st} = \frac{I_{sc}}{\sqrt{3}} \qquad \text{...(12.122)}$$

From the torque ratio expression

$$\frac{T_{st}}{T_{fl}} = \left(\frac{I_{st}}{I_{fl}}\right)^2 s_{fl} = \left(\frac{1}{\sqrt{3}}\frac{I_{sc}}{I_{fl}}\right)^2 s_{fl} \qquad \ldots(12.123)$$

or

$$\frac{T_{st}}{T_{fl}} = \frac{1}{3}\left(\frac{I_{sc}}{I_{fl}}\right)^2 s_{fl} \qquad \ldots(12.124)$$

i.e., the starting torque is reduced $\frac{1}{3}$rd and the starting current also decreases $\frac{1}{3}$rd.

12.16.5 Rotor Resistance Starting

Slip ring induction motors are started by rotor resistance starting method. In this method, a variable star-connected rheostat connected in the rotor circuit through slip rings and full voltage is applied to the stator winding.

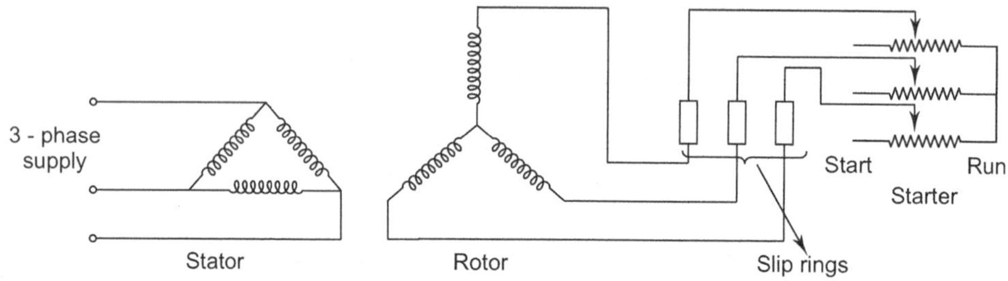

Fig. 12.23

In starting, full resistances are inserted in the rotor circuit and thus the supply current to the stator is reduced. The rotor begins to rotate and the external resistances are gradually cut-out of the circuit.

12.17 POWER FLOW DIAGRAM

A simplified power flow diagram of a three-phase induction motor is shown below.

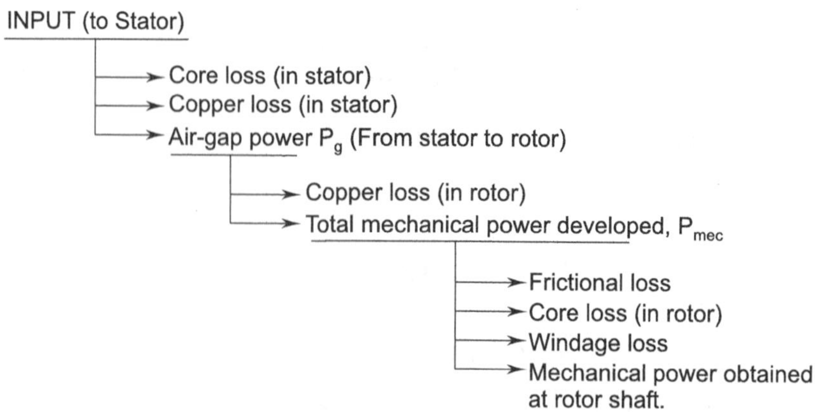

Note that the rotor core losses vary with rotor frequency because

Core loss = Eddy current loss + Hysteresis loss

where both eddy current loss and hysteresis loss depends upon frequency.

The frequency of the rotor depends upon the slip. This is the reason why the rotor core losses are included in total mechanical power developed P_g, rather than in air-gap power.

12.18 INDUCTION MOTOR'S TEST

The equivalent circuit parameters and hence the efficiently of a 3-phase induction motor can be determined by

(*i*) No-Load Test
(*ii*) Block Rotor Test

12.18.1 No-Load Test

- No-load test is also known as open circuit test.
- Rated voltage and frequency is applied across the stator terminals.
- The motor is run without any load.

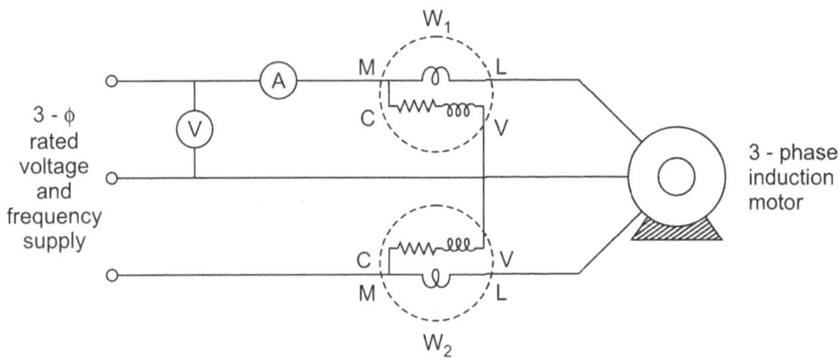

Fig. 12.24

- An ammeter, a voltmeter and two wattmeters are connected as shown in Fig. 12.24.
- The ammeter will give the reading of no-load input line current (I_o).
- The voltmeter measures the normal rated supply voltage (V_L).
- Total input power is given by the two wattmeters.
- In general, the power factor of the induction motor under no-load condition is less than 0.5.

$$\cos \phi < 0.5.$$

Hence, one wattmeter will show negative reading. So, in this case it is necessary to reverse the direction of current coil terminals to get the readings.

But total power will be the difference of the two wattmeter readings.

$$P_i = W_1 - W_2.$$

- Similar to a transformer no-load test, the exciting circuit parameters (R_o and X_o) can be determined as follow :

$$P_i = \sqrt{3}\, V_L\, I_o \cos \phi_o \qquad \text{...(12.125)}$$

From equation (12.125) no load power factor and hence the phase angle (ϕ_o) can be calculated as

$$\cos \phi_o = \frac{P_i}{\sqrt{3} V_L I_o} \quad \text{...(12.126)}$$

or

$$\phi_o = \cos^{-1} \frac{P_i}{\sqrt{3} V_L I_o} \quad \text{...(12.127)}$$

The no-load current, I_o, is a phasor sum of two components

$$I_\mu = I_o \sin \phi_o \quad \text{...(12.128)}$$

$$I_\omega = I_o \cos \phi_o \quad \text{...(12.129)}$$

From equations (12.128) and (12.129), R_o and X_o will be

$$R_o = \frac{V_L}{I_\omega} \quad \text{...(12.130)}$$

$$X_o = \frac{V_L}{I_\mu} \quad \text{...(12.131)}$$

- Stator copper loss $= 3\ I_o^2\ R_1$
- Stator core loss $= 3\ \dfrac{V_L^2}{R_o}$

  or = Input – Cu loss (stator) – Windage loss
- Fixed losses = Stator core loss + Friction loss + Windage loss.

12.18.2 Blocked Rotor Test

- Blocked rotor test is also known as short-circuit test or locked rotor test.
- In this test, the shaft of the motor is locked so that the rotor cannot move.

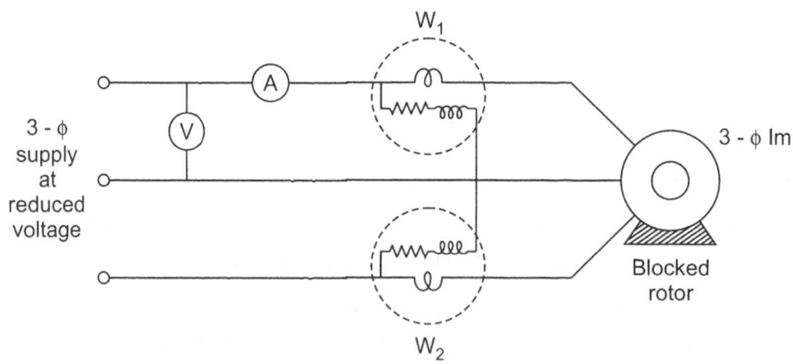

Fig. 12.25

- Similar to the short-circuit test on a transformer, a reduced voltage (up to 15 to 20% of normal value) is applied to the stator terminals and is so adjusted that full-load current flows in the stator.
- As the rotor is blocked (stationary)

$$s = 1. \quad \text{...(12.132)}$$

  Hence the equivalent circuit of the motor is exactly like a tranformer having a short-circuited secondary.
- The ammeter, voltmeter and wattmeters will read the current, voltage and power input.
- Short-circuit current $= I_1$

Stator applied voltage $= V_1$
Total power input $= P_{sc}$

$$P_{sc} = \sqrt{3}\ V_1 I_1 \cos\phi_1 \quad ...(12.133)$$

where $\cos\phi_1$ = power factor on short-circuit.

Equivalent resistance of motor referred to primary side (*i.e.*, stator)

$$R_{o1} = \frac{P_{sc}}{I_1^2} \quad ...(12.134)$$

Equivalent impedance (Z_{o1}) of the motor referred to stator

$$Z_{o1} = \frac{V_1}{I_1} \quad ...(12.135)$$

Equivalent rectance of the motor referred to stator

$$X_{o1} = \sqrt{Z_{o1}^2 - R_{o1}^2}\ . \quad ...(12.136)$$

12.19 Applications of Three-Phase Induction Motor

Some of the applications of both type of three-phase induction motors are listed below.

(A) Squirrel-cage Induction Motors

1. Used in applications where at the time of starting, less load torque is required such as
   - Fan
   - Pump load
2. Used in industrial applications where constant speed is required such as
   - Blowers
3. Used also in such an applications where required starting torque is high such as
   - Compressors
   - Conveyors
   - Crushers
4. Used for driving intermittent loads requiring rapid acceleration and high impact loads such as
   - Punch presses
   - Shears
5. Other applications are
   - Water pumps
   - Grinders
   - Lathe machines
   - Drilling machines
   - Printing machines

(B) Slip Ring Induction Motors

A slip ring or wound rotor induction motor is used for loads which require speed control or where severe starting conditions are there. They may be used for:

- Hoists
- Cranes
- Elevators
- Compressors etc.

SOLVED PROBLEMS

Problem 12.1. *A three-phase, 4-pole induction motor is supplied from 3-phase, 50 Hz AC supply. Calculate:*

(i) The synchronous speed.

(ii) The rotor speed when slip is 4%.

(iii) The rotor frequency when rotor runs at 600 rpm.

Solution.

(*i*) Synchronous speed, $N_s = \frac{120 \times f_s}{P} = \frac{120 \times 50}{4} = \mathbf{1{,}500\ rpm.}$ **Ans.**

(*ii*) Speed when slip is 4% or 0.04, $N = N_s(1 - s) = 1{,}500\,(1 - 0.04) = \mathbf{1{,}440\ rpm.}$ **Ans.**

(*iii*) Slip when the motor runs at 600 rpm $s' = \frac{1{,}500 - 600}{1{,}500} = 0.6$

Rotor frequency, $f_r = s'f = 0.6 \times 50 = \mathbf{3\ Hz.}$ **Ans.**

Problem 12.2. *A three-phase delta connected 440 volts, 50 Hz, 4-pole induction motor has a rotor stand still e.m.f. per phase of 130 volts. If the motor is running at 1,440 rpm, calculate for this speed : (i) the slip, (ii) the frequency of rotor induced e.m.f., (iii) the value of the rotor induced e.m.f. per phase, and (iv) stator to rotor turn ratio.*

Solution.

Synchronous speed of induction motor,

$$N_s = \frac{120 \times \text{supply frequency}}{\text{Number of poles}} = \frac{120 \times 50}{4} = 1{,}500 \text{ rpm}$$

Motor speed, $N_r = 1{,}440$ rpm

(*i*) Slip, $s = \frac{N_s - N_r}{N_s} = \frac{1{,}500 - 1{,}440}{1{,}500} = \mathbf{0.04\ or\ 4\%.}$ **Ans.**

(*ii*) Frequency of rotor induced e.m.f.

$= \text{Slip} \times \text{supply frequency} = sf = 0.04 \times 50 = \mathbf{2\ Hz.}$ **Ans.**

(*iii*) Rotor induced e.m.f. per phase at the speed of 1,440 rpm or at slip of 0.04

= Slip × induced e.m.f. per phase in rotor winding at standstill

$= s \times E_2 = 0.04 \times 130 = \mathbf{5.2\ V}$ **Ans.**

(*iv*) Stator-to-rotor turn-ratio $= \frac{\text{Induced e.m.f. per phase in stator}}{\text{Induced e.m.f. per phase in rotor at standstill}}$

$$= \frac{E_{P_1}}{E_{P_2}} = \frac{440}{130} = \mathbf{\frac{44}{13}}.$$ **Ans.**

Problem 12.3. *A three-phase, 440 V, 50 HP, 50 Hz induction motor runs at 1,450 rpm when it delivers rated output power. Determine :*

(i) Number of poles in the machine.

(ii) Speed of rotating air-gap field.

(iii) Rotor induced voltage if stator to rotor turns ratio is 1 : 0.80. Assume the winding factors are the same.

(iv) Frequency of rotor current.

Solution.

(*i*) The number of poles are always even in number and to give synchronous speed nearest (on higher side) to the actual speed of motor (*i.e.*, 1,450 rpm), then the motor must have 4 No. of poles. **Ans.**

(*ii*) Speed of rotating air gap field, $N_s = \dfrac{120 \times f_s}{P} = \dfrac{120 \times 50}{4} = \mathbf{1{,}500\ rpm}$ **Ans.**

(*iii*) Induced e.m.f in rotor at standstill $E_2 = K\,E_1 = K\,V_1 = 0.8 \times 440 = 352$ V

Slip, $s = \dfrac{N_s - N_r}{N_s} = \dfrac{1{,}500 - 1{,}450}{1{,}500} = 0.033333$

Rotor induced voltage $= sK\,V_1 = 0.03333 \times 352 = \mathbf{11.733\ V}$ **Ans.**

(*iv*) Frequency of rotor current, $f_r = sf = 0.03333 \times 50 = \mathbf{1.6666\ Hz.}$ **Ans.**

Problem 12.4. *A 3-phase, 50 Hz induction motor has 6 poles and operates with a slip of 5% at a certain load. Determine:*

(*i*) *The speed of the rotor with respect to the stator.*

(*ii*) *The frequency of rotor current.*

(*iii*) *The speed of the rotor magnetic field with respect to rotor.*

(*iv*) *The speed of the rotor magnetic field with respect to stator, and*

(*v*) *The speed of the rotor magnetic field with respect to the stator magnetic field.*

Solution.

Supply frequency, $f_s = 50$ Hz

Number of poles, $P = 6$

Slip, $s = 5\%$ or 0.05

Synchronous speed, $N_s = \dfrac{120 f_s}{P} = \dfrac{120 \times 50}{6} = 1{,}000$ rpm

(*i*) Speed of rotor w.r.t. the stator,

$$N_r = N_s(1 - s) = 1{,}000\,(1 - 0.05) = \mathbf{950\ rpm.}$$ **Ans.**

(*ii*) Frequency of rotor current, $f_r = sf = 0.05 \times 50 = \mathbf{2.5\ Hz}$ **Ans.**

(*iii*) Speed of rotor magnetic field with respect to the rotor

$$= \frac{120 f_r}{P} = \frac{120 \times 2.5}{6} = \mathbf{50\ rpm.}$$ **Ans.**

(*iv*) Speed of rotor magnetic field with respect to the stator $= N_s = \mathbf{1{,}000\ rpm.}$ **Ans.**

(*v*) Rotor field and stator field are revolving at the same speed of 1,000 rpm, therefore, speed of rotor field w.r.t. stator field is zero.

Problem 12.5. *A three-phase, 4-pole, 50 Hz, induction motor runs at 1,460 rpm. Determine its percentage slip.*

Solution.

Synchronous speed, $N_s = \dfrac{120 f_s}{P} = \dfrac{120 \times 50}{4} = 1{,}500$ rpm

Running speed of motor, $N_r = 1{,}460$ rpm

Slip, $s = \dfrac{N_s - N_r}{N_s} \times 100 = \dfrac{1{,}500 - 1{,}460}{1{,}500} \times 100 = \mathbf{2.667\%.}$ **Ans.**

Problem 12.6. *A three-phase slip ring 4-pole induction motor has rotor frequency of 2 Hz while connected to 400 V, 3-phase, 50 Hz supply. Determine slip and rotor speed.*

Solution.

Synchronous speed, $$N_s = \frac{120 f_s}{P} = \frac{120 \times 50}{4} = 1{,}500 \text{ rpm}$$

Slip, $$s = \frac{\text{Frequency of rotor e.m.f, } f_r}{\text{Supply frequency, } f_s} = \frac{2}{50} = \mathbf{0.04} \qquad \textbf{Ans.}$$

Speed of motor on load, $N_r = N_s (1 - s) = 1{,}500 (1 - 0.04) = \mathbf{1{,}440\ rpm.}$ **Ans.**

Problem 12.7. *A 60 Hz 3-phase induction motor has 2 poles and runs at 3510 rpm. Calculate:*

(*i*) *Synchronous speed, in rpm*

(*ii*) *Percentage slip.*

Solution.

(*i*) The synchronous speed will be

$$N_s = \frac{120 f}{P} = \frac{120 \times 60}{2} = \mathbf{3600\ rpm} \qquad \textbf{Ans.}$$

(*ii*) Percentage slip is given by

$$s = \frac{N_s - N_r}{N_s} \times 100$$

or $$s = \frac{3600 - 3510}{3600} \times 100 = \mathbf{2.5\%} \qquad \textbf{Ans.}$$

Problem 12.8. *An 8-pole, 50 Hz, 3-phase induction motor has rotor resistance and standstill reactance of 0.5 Ω and 5 Ω respectively. Calculate:*

(*i*) *The speed at which the torque is maximum*

(*ii*) *The ratio of maximum torque to starting torque.*

Neglect stator impedance.

Solution.

The synchronous speed is

$$N_s = \frac{120 f}{P} = \frac{120 \times 50}{8} = 750 \text{ rpm}$$

(*i*) Maximum torque occurs when

$$s = \frac{R_2}{X_2} = \frac{0.5}{5} = 0.1$$

The speed at which the torque is maximum

$$N_r = N_s(1 - s) = 750(1 - 0.1)$$

or $$N_r = \mathbf{675\ rpm} \qquad \textbf{Ans.}$$

(*ii*) Expression for maximum running torque is

$$T_{max} = K \frac{E_2^2}{2X_2}$$

Expression for starting torque is

$$T_{st} = K \frac{E_2^2 R_2}{R_2^2 + X_2^2}$$

Ratio of maximum torque to starting torque

$$\frac{T_{max}}{T_{st}} = \frac{R_2^2 + X_2^2}{2R_2 X_2} = \frac{0.5^2 + 5^2}{2 \times 0.5 \times 5}$$

$$\frac{T_{max}}{T_{st}} = \mathbf{5.05 \text{ p.u.}}$$ **Ans.**

Problem 12.9. *A 3-phase induction motor connected from a 3-phase, 50 Hz AC supply runs at 720 rpm and has 4% slip. What will be the number of poles of the motor?*

Solution.

Given $f = 50$ Hz, $N_r = 720$ rpm, $s = 4\% = 0.04$, $P = ?$

Slip expression is given by

$$s = \frac{N_s - N_r}{N_s}$$

or $$0.04 = \frac{N_s - 720}{N_s}$$

or $$N_s = 750 \text{ rpm}$$

Also, $$N_s = \frac{120 f}{P}$$

or $$P = \frac{120 f}{N_s} = \frac{120 \times 50}{750}$$

∴ $$P = \mathbf{8}$$ **Ans.**

Problem 12.10. *A three-phase, 50 Hz induction motor has a full-load speed of 960 rpm. Calculate:*

(a) Slip

(b) Frequency of rotor induced e.m.f.

(c) Number of poles

(d) Speed of rotor field w.r.t. rotor structure

(e) Speed of rotor field w.r.t. stator field.

Solution.

Given $f_s = 50$ Hz

$N_r = 960$ rpm

Motor synchronous speed will be near to (and greater than) the rotor speed, hence

$$N_s = \frac{120 f}{P}$$

$$= \frac{120 \times 50}{2} = 3000 \text{ rpm (Not possible)}$$

$$= \frac{120 \times 50}{4} = 1500 \text{ rpm (Not possible)}$$

$$= \frac{120 \times 50}{6} = 1000 \text{ rpm (Possible)}$$

$$= \frac{120 \times 50}{8} = 750 \text{ rpm (Not possible)}$$

so, the synchronous speed will be

$$N_s = 1000 \text{ rpm}$$

(*a*) Slip, $s = \frac{N_s - N_r}{N_s} = \frac{1000 - 960}{1000}$

$$= 0.04 = \mathbf{4\%}$$ **Ans.**

(*b*) Frequency of the rotor induced e.m.f.

$$f_r = s f_s = 0.04 \times 50$$

or $f_r = \mathbf{2\ Hz}$ **Ans.**

(*c*) Number of poles

As the synchronous speed near to and greater than the rotor speed comes out to be 1000 rpm at 6 poles hence

$$P = \mathbf{6}$$ **Ans.**

(*d*) Rotor speed = 960 rpm

Rotor field speed = 1000 rpm

Speed of rotor field with respect to rotor structure = 1000 – 960 = **40 rpm** **Ans.**

(*e*) Speed of rotor field = 1000 rpm

Speed of stator field = 1000 rpm

Speed of rotor field w.r.t. stator field

= 1000 – 1000 = **0 rpm** **Ans.**

Problem 12.11. *A 60 Hz induction motor has 2 poles and runs at 3510 rpm. Calculate:*

(a) The synchronous speed, and

(b) The percent slip.

Solution.

$$f = 60 \text{ Hz}$$
$$p = 2$$
$$N_r = 3510 \text{ rpm}$$

(*a*) Synchronous speed

$$N_s = \frac{120 f}{P} = \frac{120 \times 60}{2}$$

$$N_s = \mathbf{3600\ rpm}$$ **Ans.**

(*b*) Percentage slip

$$s = \frac{N_s - N_r}{N_s} \times 100$$

$$= \frac{3600 - 3510}{3600} \times 100$$

$$s = \mathbf{2.5\%}$$ **Ans.**

EXERCISE

1. A 6-pole wound rotor induction motor is excited by a three-phase, 60 Hz source. If the full-load speed is 1140 rpm, calculate the slip and the frequency of the rotor current under given conditions:
   (*i*) at standstill
   (*ii*) motor running at 500 rev/min in the same direction as the revolving field.
   (*iii*) motor running at 500 rev/min in the opposite direction to the revolving field.
   (*iv*) motor running at 2000 rev/min in the same direction as the revolving field.
2. A three-phase induction motor runs at 1140 rpm at full-load when supplied with power factor from a 50 Hz, 3-phase line. Calculate:
   (*i*) no. of poles
   (*ii*) slip at full-load
   (*iii*) frequency of rotor voltage
   (*iv*) speed of rotor field with respect to rotor.
   (*v*) speed of rotor field with respect to stator and stator field.
   (*vi*) speed of rotor at a slip of 10% and rotor frequency at this speed.
   (*vii*) if the applied voltage per phase is 230 V, find the rotor-induced e.m.f. at standstill and at 10% slip with stator to rotor turn ratio of 1 : 0.5.
3. A three-phase squirrel cage induction motor has maximum torque equal to the twice of full-load torque. R_2 = 0.2 ohms and X_2 = 2 ohms. Determine the ratio of starting torque to full-load torque, if it is started by
   (*i*) direct online starter
   (*ii*) star-delta starter
   (*iii*) auto-transformer starter with 70% tapping.
4. A 12-pole, three-phase alternator driven at a speed of 500 rpm supplies a power to an 8 pole, three-phase induction motor. If the slip of the motor is 0.03 p.u, calculate its speed.
5. A three-phase slip ring 4-pole induction motor, the rotor frequency is found to be 2 Hz while connected to a 400 V, 3-phase star connected supply. Determine the motor speed, if the stator to rotor turn ratio is 1 : 0.5. Find the rotor-induced e.m.f. at the above frequency 2 Hz. Assume rotor is also star connected.
6. A 4-pole, three-phase 50 Hz induction motor has a star connected rotor; the rotor has a resistance of 0.1 ohms per phase and standstill reactance of 2 ohms/phase. Induced e.m.f. between the slip rings is 100 V. If the full-load speed is 1460 rpm, calculate:
   (*i*) slip
   (*ii*) the slip at which maximum torque occurs
   (*iii*) e.m.f. induced in the rotor per phase
   (*iv*) rotor reactance per phase
   (*v*) rotor current
   (*vi*) rotor power factor.
7. A three-phase induction motor having a synchronous of 1200 rev/min draws 80 kW from a three-phase feeder. Copper losses and iron losses in the stator amount to 5 kW. If the motor runs at 1152 rev/min, calculate
   (*i*) active power transmitted to rotor
   (*ii*) rotor I^2R losses

(*iii*) mechanical power developed

(*iv*) mechanical power delivered to load, knowing that winding and friction losses are equal to 2 kW

(*v*) efficiency of the motor.

8. A 150 kW, 300 V, 50 Hz, 6 pole star connected induction motor has a star-connected slip ring rotor with a transformation ratio of 3.6 (stator/rotor). The rotor resistance is 0.1 ohms/phase and its per phase leakage inductance is 3.61 mH. The stator impedance may be neglected. Find

(*i*) starting current and torque on rated voltage with short-circuited slip rings.

(*ii*) necessary external resistance to reduce the rated-voltage starting current to 30 A and the corresponding starting torque.

9. A small squirrel cage induction motor has a starting current of six times the full-load current and a full load slip of 0.05. Find the p.u. of full load values, the line current and starting torque with the following methods of starting:

(*i*) direct switching.

(*ii*) stator resistance starting with motor current limited to 2 p.u.

(*iii*) auto-transformer starting with motor current limited to 2 p.u.

(*iv*) star-delta starting.

(*v*) what auto-transformer ratio would give 1 p.u. starting torque?

10. Explain the construction and working principle of 3-ϕ induction motor.

11. What is use of starter? Explain various methods of starting of three-phase induction motors.

12. Explain how slip frequency currents are set up in the rotor windings of a three phase induction motor. Define slip-speed of an induction motor.

13. Explain with the help of diagrams how a rotating magnetic field is produced in the air-gap of a three-phase induction motor.

14. Draw typical speed-torque characteristics of three-phase squirrel-cage induction motor. Mark the starting torque, maximum torque and normal operating region.

15. Three-phase induction motor cannot operate between slip =1 and slip at maximum torque. Explain it.

16. Develop the following relationship for three-phase induction motor

$$P_g : P_m : P_{Cu} = 1 : (1 - s) : s\,P_g$$

where P_g, P_{Cu} and P_m are respectively air-gap power, rotor copper loss and mechanical power developed.

17. Compare a three-ϕ induction motor with 1-ϕ induction motor on the basis of the following points:

(*i*) starting torque

(*ii*) slip torque characteristic

(*iii*) magnetic field

(*iv*) power factor and efficiency

(*v*) applications.

18. A 150 kW, 300 V, 50 Hz star connected induction motor has a star-connected rotor with a transformation ratio of 3.6 (stator/rotor). The rotor resistance is 0.1 ohms/phase and its per phase leakage reactance is 1.13 ohms and motor runs at 1140 rpm on full load when supplies with 50 Hz supply. Calculate:

(*i*) number of poles

(*ii*) speed of rotor field with respect to rotor and stator (*iii*) e.m.f. induced in the rotor at full load

(*iv*) starting current and torque on rated voltage with short-circuit rotor.

(*v*) if the maximum torque equal to the twice the full-load torque, determine ratio of starting torque to full load torque, if it is started with auto transformer with 70% tapping.

19. An 8-pole alternator runs at 750 rpm. It supplies power to a 6-pole, three-phase induction motor, which has a full load slip of 3%. Find the full load speed of the induction motor and the frequency of its rotor e.m.f. Explain the principle of electromechanical energy conversion.

20. Draw a typical torque slip characteristic of a three-phase induction motor at different resistances ($r_1 > r_2 > r_3$) and explain it in brief.

21. The induced e.m.f. between the slip-ring terminals of a three-phase induction motor, when the rotor is standstill is 100 V. The rotor winding are star connected and have resistance and standstill resistance of 0.05 and 0.1 per phase respectively. Calculate the rotor current and phase difference between rotor voltage and current at 4% slip and the slip at maximum torque.

□ □ □

13

SINGLE-PHASE INDUCTION MOTOR

13.1 INTRODUCTION

In households as well as in most of the commercial fields, single-phase AC is supplied. In all such installations, Single-Phase Induction Motors are one of the most popular tools and have a wide field of applications.

Fans, blowers, refrigerators, washing machines etc., are some of the most common applications of single-phase induction motors in daily use.

Single-phase induction motors are simple in construction, reliable, easy to repair and comparatively cheaper in cost as compare to that of 3-phase motors.

Fig. 13.1

13.2 TYPES OF SINGLE-PHASE INDUCTION MOTOR

Depending upon the starting arrangements provided with the motors, single-phase induction motors are of many types. Some common types are:

1. Split Phase Motor.
2. Shaded-Pole Motor.
3. Repulsion-Start Motor.

13.3 PRINCIPLE OF OPERATION

The principle of operation of a single-phase induction motor can be explained with the help of these two theories.

13.3.1 Cross-Field Theory

A single-phase induction motor physically looks similar to that of a 3-phase induction motors. Just like a 3-phase induction motors, the rotor of a single-phase induction motor is also provided with a squirrel-cage short-circuited winding.

Stator is provided with single-phase distributed winding.

- When this single-phase winding on stator is connected to a 1-ϕ AC supply, an alternating field is produced which is pulsating in nature and vary only with time.
- This alternating field produces an alternating flux (Fig. 13.2).
- This pulsating flux in turns produces an alternating e.m.f. in the rotor field.

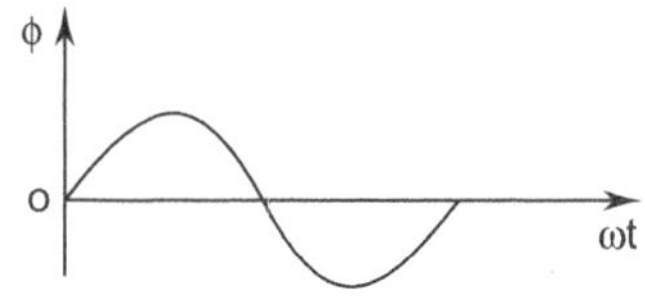

Fig. 13.2

- As the rotor winding is short-circuited, an alternating current is induced in it.
- According to Faraday's law, these current carrying conductor experiences a force, which tends to rotate the rotor in the direction of the field.
- As the flux and hence the current is alternating in nature, after every half-cycle, the polarity or direction of the induced e.m.f. is changed.
- And so is the direction of force or torque is changed after every half-cycle.
- Hence no torque is developed at starting and so the single-phase induction motor is not self starting.
- However, if by some external means, (for *e.g.*, a push by the hand) the rotor is rotated in either direction, then immediately a torque will develop and motor accelerates in that direction.

13.3.2 Double Revolving Field Theory

The principle of operation can also be understood with the help of double field revolving theory. The double field revolving theory is based on the Ferraris principle.

Ferraris Principle

According to Ferraris principle, *"the magnetic field produced by single-phase alternating current, can be replaced exactly by two rotating fields of half its amplitude travelling in opposite directions at synchronous speed".*

Lets us consider that the alternating flux, produced by the 1-ϕ alternating current, at any instant is given by

$$\Phi = \Phi_m \cos \omega t \qquad ...(13.1)$$

This alternating flux can be resolved into two components. First component of flux travelling in clockwise direction

$$\Phi_{cw} = \frac{\Phi_m}{2} \qquad ...(13.2)$$

Second component of flux travelling in counter-clockwise direction

$$\Phi_{acw} = \frac{\Phi_m}{2} \qquad ...(13.3)$$

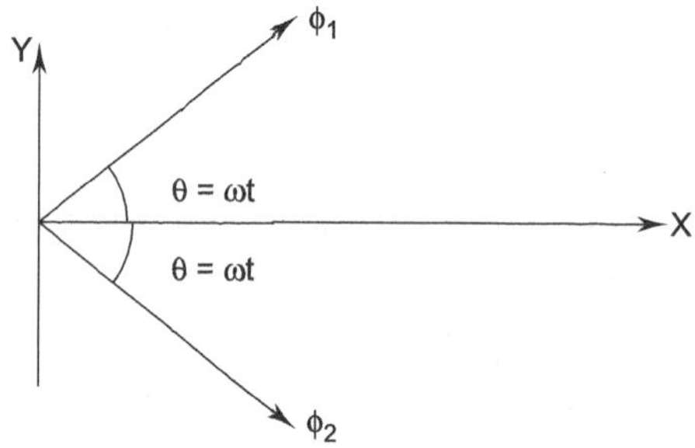

Fig. 13.3

Now assumed that the rotor is given a push in anyone direction say in clockwise direction, and is running with a speed of say $\mathbf{N}_r$ rpm.

The flux component revolving in the clockwise direction Φ_{cw} will have a synchronous speed

$$N_s = \frac{120f}{P} \qquad ...(13.4)$$

And the flux component revolving in the counterclockwise direction Φ_{ccw} will have a synchronous speed

$$N_s = \frac{-120f}{P} \qquad ...(13.5)$$

The per unit slip for the first component Φ_{cw} will be

$$s_{cw} = \frac{N_s - N_r}{N_s} = s \text{ (say)} \qquad ...(13.6)$$

$$s_{cw} = s \qquad ...(13.7)$$

Similarly, the per unit slip for the second component Φ_{ccw} will be

$$s_{ccw} = \frac{-N_s - N_r}{-N_s} = \frac{N_s + N_r}{N_s} \qquad ...(13.8)$$

From equations (13.5) and (13.6) equation (13.7) will become

$$s_{ccw} = 2 - s_{cw} \qquad ...(13.9)$$

or

$$\boxed{s_{ccw} = 2 - s} \qquad ...(13.10)$$

At the time of starting

$$s_{cw} = 1 \quad \text{...(13.11)}$$

And $$s_{ccw} = 2 - 1 = 1 \quad \text{...(13.12)}$$

Hence $$\boxed{s_{ccw} = 1} \quad \text{...(13.13)}$$

So at standstill the resultant torque will be

$$T_r = T_{cw} - T_{ccw} = 0 \quad \text{...(13.14)}$$

Hence a single-phase induction motor is self starting.

13.4 STARTING METHOD OF SINGLE-PHASE INDUCTION MOTOR

As it is already discussed that in a single-phase induction motor, only one winding on stator cannot produce any starting torque, hence some extra arrangement is required to produce any starting torque.

For this purpose:

- Single-phase winding on stator is split into two windings which are 90 electrical degree apart from each other (it means the two windings or coil have current 90° out of phase with each other).
- The two windings are connected in parallel.
- Impedance is connected in series with one of these windings.
- By proper selection of such impedance, the current may be made to differ by as much as 90° in the two coils.
- This arrangement can produce a rotating magnetic field similar to a 2-phase motor.

The method of achieving such an arrangement is known as split phase method. This phase splitting can be achieved by the following methods:

1. Resistance Start
2. Capacitor Start
3. Capacitor Start - Capacitor Run
4. Permanent Capacitor
5. Shaded Pole.

13.4.1 Resistance Start Single-Phase Induction Motor

In this starting method,

- Stator winding is split into two parts and are connected in parallel.
- One winding is considered as main winding and other one is considered as auxiliary winding.
- A large resistance and a centrifugal switch (C.F.S.) are connected in series with the auxiliary winding as shown in Fig. 13.4.

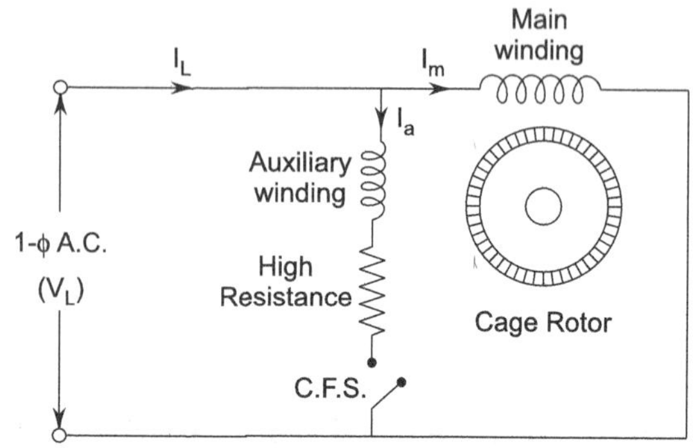

Fig. 13.4

- At the time of starting both windings draw current from the supply.
- The main winding has, as compared to auxiliary winding, less resistance and high reactance.
- Hence the current drawn by main winding, I_m, lags behind the applied voltage, V_L by a greater angle θ_m.
- The auxiliary winding has high resistance and low reactance and hence current drawn by the auxiliary winding, I_a, lags behind the applied voltage, V_L by a small angle say θ_a.

These currents and voltage are shown in phasor diagram in Fig. 13.5.

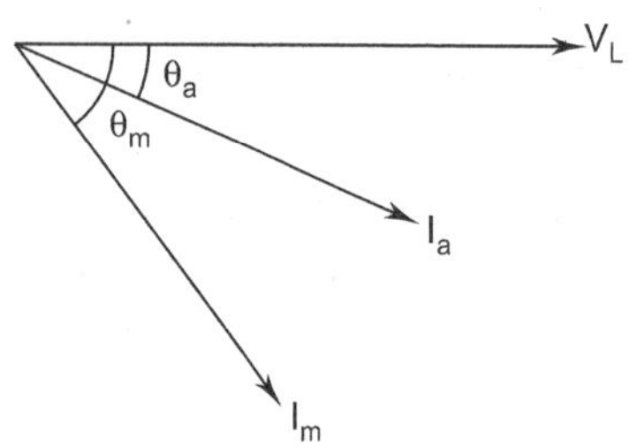

Fig. 13.5

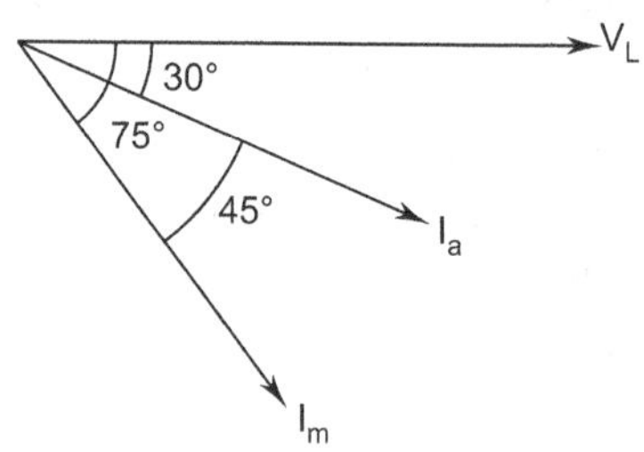

Fig. 13.6

Let us take an example (phasor diagram shown in Fig. 13.6).

- Let I_m lags behind the applied voltage V_L by 75° and
- Let I_a lags behind the applied voltage V_L by 30°.
- Hence the phase difference between the main winding current I_m and auxiliary winding current, I_a will be 75° – 35° = 45°.
- Since the two currents are due to same applied voltage but are different in magnitude and have a phase displacement, the arrangement is sufficient for producing rotating magnetic field and producing starting torque.

When motor picks up the 75 to 80% of synchronous speed, the auxiliary winding is cut-off from the circuit with the help of centrifugal switch.

If this auxiliary winding is not taken out of circuit, then it will result in overheating and burning of auxiliary winding. It also leads to noisy and inefficient performance.

13.4.2 Capacitor Start Single-Phase Inductance Motor

In capacitor start method,

- A capacitor is connected in series with the auxiliary winding as shown in Fig. 13.7.

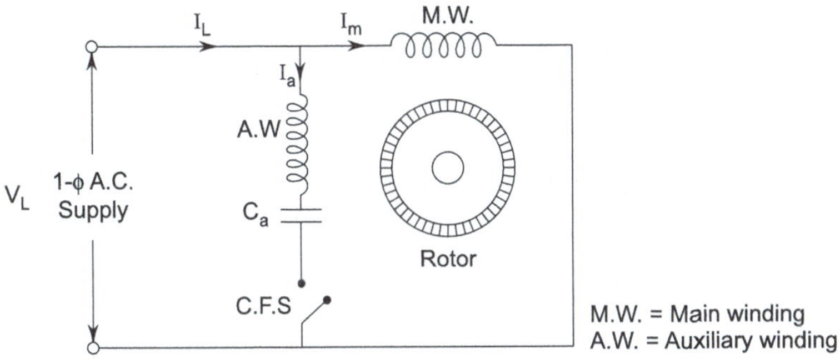

Fig. 13.7

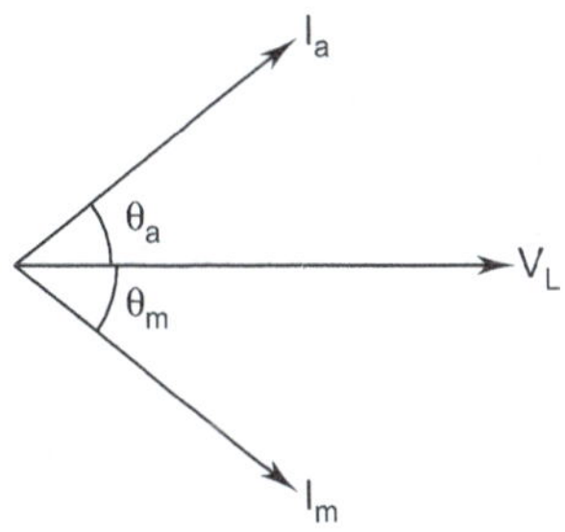

Fig. 13.8

- Due to this capacitor, the auxiliary winding current I_a, will leads the applied voltage V_2 by an angle say θ_a.
- The currents and voltage as reference vector are shown in the phasor diagram of Fig. 13.8.
- It is clear from the phasor diagram that the phase difference between the auxiliary winding current I_m is increased.
- When the motor attains 75 to 80% of synchronous speed, the auxiliary winding is disconnected from the circuit by centrifugal switch.
- The split phase capacitor start motors have good started torque but low power factor at rated load and rated speed.

13.4.3 Capacitor Run Single-Phase Induction Motor

In this method,

- A low capacitor is connected in series with the auxiliary winding and is not cut-out from the circuit even during running condition.
- Centrifugal switch is not required in this case.
- This motor is also known as permanent split capacitor single-phase induction motor.
- The split phase capacitor run motor have fairly good power factor at rated load but it does not have good starting torque.

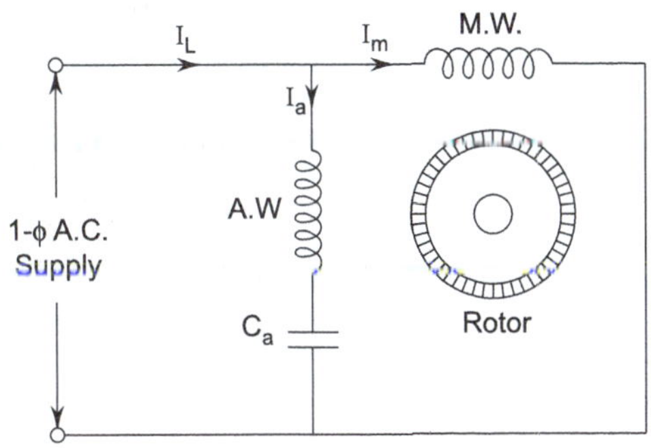

Fig. 13.9

13.4.4 Capacitor Start—Capacitor Run Single-Phase Induction Motor

As mentioned in the previous two types of single-phase induction motor (*i.e.*, in capacitor run single-phase induction motor and capacitor start-capacitor run single-phase induction motor) that

- Capacitor start single-phase induction motors have good started torque but low power factor at rated load and rated speed.
- The capacitor run single-phase induction motors have fairly good power factor at rated load but it does not have good starting torque.

In this motor

- Both these drawbacks are removed. In this method two capacitors are connected in parallel with each other connected in series with the auxiliary winding as shown in Fig. 13.10.

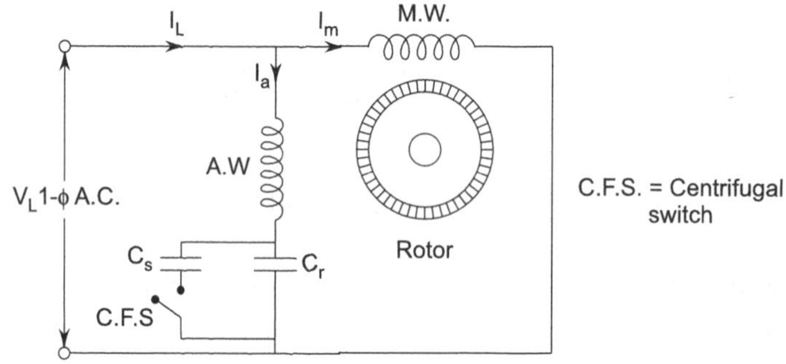

Fig. 13.10

- During starting both capacitors are in the circuit.
- The value of starting capacitor C_s is designed on the basis of starting torque requirement.
- The value of running capacitor C_r is designed on the basis of required running performance.
- As the motor pick up speed the starting capacitor C_s is cut-off from the circuit with the help of centrifugal switch.
- The running capacitor C_r and the auxiliary winding remains in the circuit during running conditions.

13.4.5 Shaded Pole Single-Phase Induction Motor

In this type of motor

- The stator is of salient pole type similar to the stator of a DC machine.
- The rotor is squirrel cage type.
- The poles are physically divided into two sections as shown in Fig. 13.11(*i*).
- One section is comparatively smaller than the other one as shown in Fig. 13.11(*ii*) below.

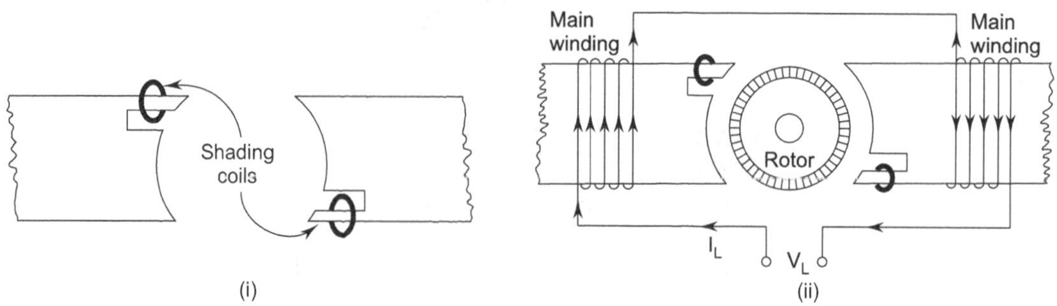

Fig. 13.11

- A heavy short-circuited copper coil is placed on the smaller section of the pole.
- When single-phase supply is given to the main winding, an alternating flux is produced in the pole.
- A portion of this main flux links with the shaded coil and induced a voltage in it.
- This large short-circuited current in turn produces a flux in the shaded portion.
- This shaded portion flux opposes the main flux or the unshaded portion flux.
- The flux in the shaded portion thus lags behind the flux in the unshaded portion. These two fluxes generate an unbalanced rotating magnetic field.
- And hence a starting torque is developed in the motor.
- These motors have poor starting torque, low power factor and poor efficiency.

13.5 DRAWBACKS

Some of the drawbacks of a single-phase induction motor are:

- Suitable for only fractional horse power range
- Low overload capacity
- Low efficiency
- Low power factor
- Low output
- Major drawback is that it does not have inherent self starting torque.

EXERCISE

1. Using Double-field revolving field theory, explain the principle of operation of a single-phase induction motor.
2. Describe shaded pole method of starting a single-phase induction motor.
3. Explain why single-phase induction motor has no starting torque. What modifications are being done to have starting torque in normal single-phase fan motors? Give brief explanations.
4. Briefly discuss different methods of starting single-phase induction motor.
5. Why there is problem of starting in single-phase induction motor? Describe one method of starting.
6. An 115 V, 50 Hz, 4-pole single-phase induction motor is rotating in clockwise direction at a speed of 1710 rpm. Determine it's per unit slip

   (*a*) in the direction of rotation

   (*b*) in the opposite direction.

   If the rotor resistance at standstill is 13.5 ohms, determine the effective rotor resistance in each branch.

□ □ □

14

THREE-PHASE SYNCHRONOUS MACHINE

14.1 INTRODUCTION

Three-phase synchronous machine can operate both as a motor as well as a generator. Similar to DC machines, three-phase synchronous machines are also doubly excited machine, *i.e.*, both stator and rotor windings are excited separately. A synchronous machine runs only at synchronous speed, which depends upon the frequency of supply and number of poles in the machine. Hence keeping the frequency constant (as numbers of poles of a machine are constant) the motor can be run at constant speed and a constant speed-torque characteristic can be obtained.

A synchronous motor is not inherently self-starting. It has to be run up to or nearly equal to synchronous speed by some auxiliary means. After reaching at or near to the synchronous speed the motor gets synchronize with the supply. It can be operated over a wide range of power factor, both lagging and leading.

Synchronous machine has same constructional features for both motor as well as for generator.

14.2 CONSTRUCTION

Basically a synchronous machine has two parts:

- Stator
- Rotor.

Stator

Stator, as the name indicates, is stationary member of the machine (Fig. 14.1). The armature winding is placed on stator. For low power rating, the armature winding can be placed on rotating part and for medium and high power ratings; the armature winding is placed on stator.

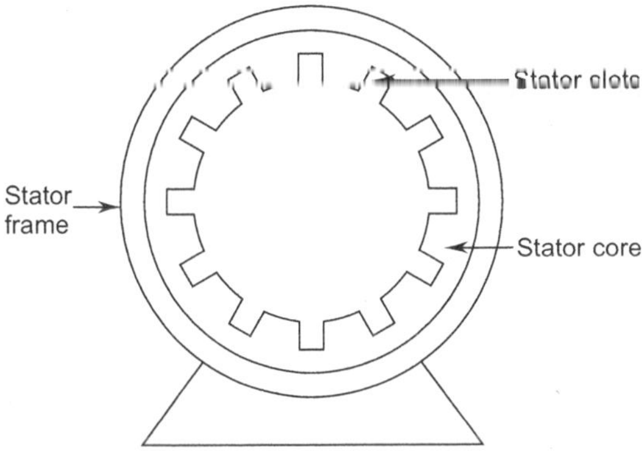

Fig. 14.1

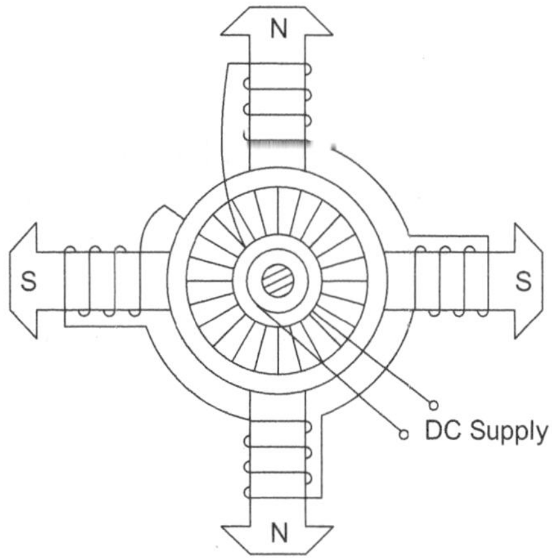

Fig. 14.2

The stator consist of a cast iron frame, which support the armature core (or stator core) having slots on its inner periphery for housing armature conductors.

Rotor

In synchronous machines two types of rotors are used:

(*i*) Salient-pole type

(*ii*) Smooth cylindrical type.

The field winding is placed on the rotor. This winding is excited with direct current (DC) supply.

(*i*) **Salient-Pole Type Rotor**

- Salient-pole type rotor is used in low and medium speed synchronous machines (Fig. 14.2).
- They have large diameter and short axial length.
- Poles are made up of thick steel laminations and are fixed to rotor by dove tail joints.
- The pole shoes cover $\frac{2}{3}$rd of pole pitch.

(*ii*) **Smooth Cylindrical Pole Type Rotor**

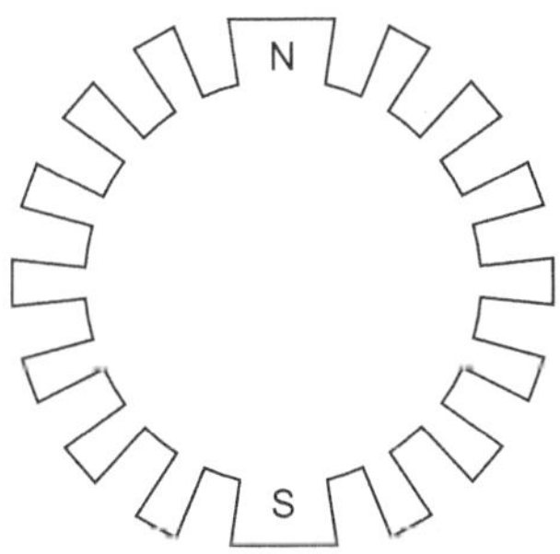

(i)

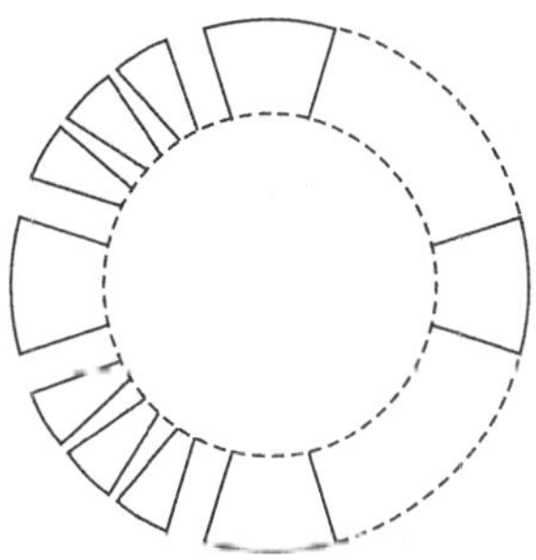

(ii)

Fig. 14.3

- For high speed synchronous machines cylindrical pole type rotor are used.
- It is made up of smooth solid forged steel.

- This forging has radial slots in the outer periphery.
- The slots over certain portion of the core are left unslotted to form pole faces as shown in Fig. 14.3.
- The rotor should have the same number of poles as that of stator.
- Opposite to that of salient-pole type rotor, cylindrical pole types rotor are of smaller diameter and have very long axial length.

14.3 SYNCHRONOUS GENERATOR

Synchronous machine are of two types: generator and motor. A synchronous generator is commonly known as alternator. First we will discuss synchronous generator. Nearly 90% of the power generation is done with the help of alternators.

14.4 BASIC PRINCIPLE OF ALTERNATOR

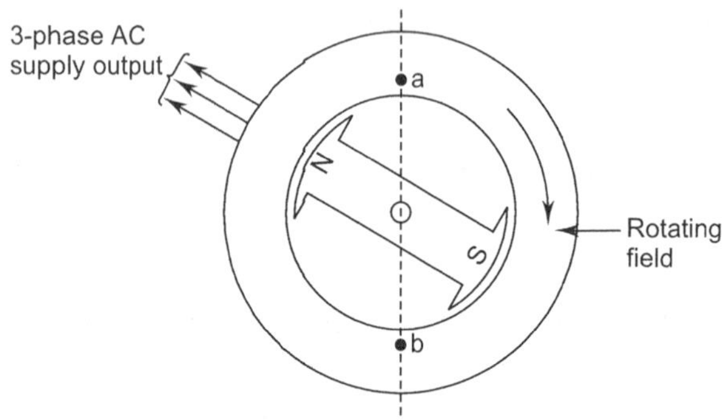

Fig. 14.4

- The rotor of the machine is provided DC supply and is rotated by means of some prime mover.
- The flux generated in the rotor conductor, sweeps past the air-gap and cuts the stator conductor.
- According to Faraday's law, an e.m.f. is induced in the stator (or armature) conductor. If the stator circuit is closed, the 3-phase alternating current will be start flowing in the circuit.
- As the rotor has same number of poles as that of stator, they are stationary with respect to each other.
- The frequency of the output will be same as that of the rotor speed.
- The *synchronous* speed is given by

$$N_s = \frac{120 f}{P} \quad \ldots(14.1)$$

where f is the frequency and P are the no. of poles.

14.5 FREQUENCY OF E.M.F. INDUCED IN ALTERNATOR

Let us assume that we have 6-poles synchronous machine.

Considered the armature conductor marked '**A**' in Fig. 14.5 situated at the center of the north pole.

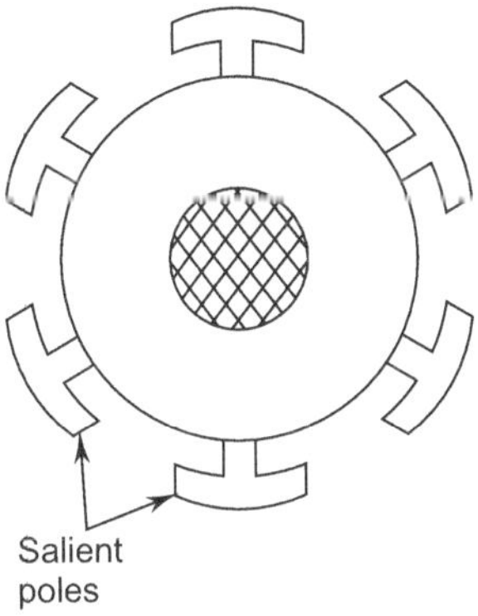

Fig. 14.5

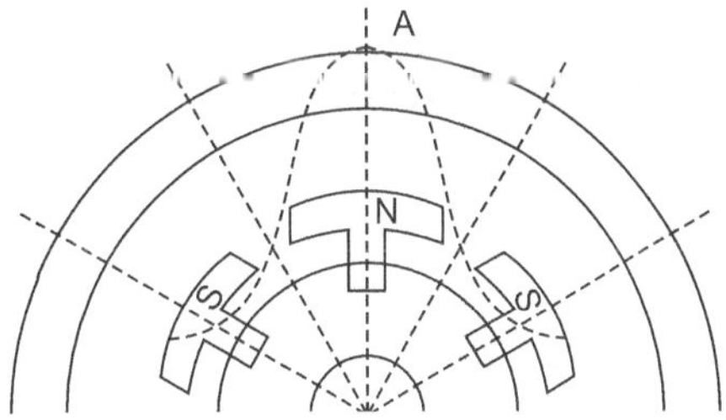

Fig. 14.6

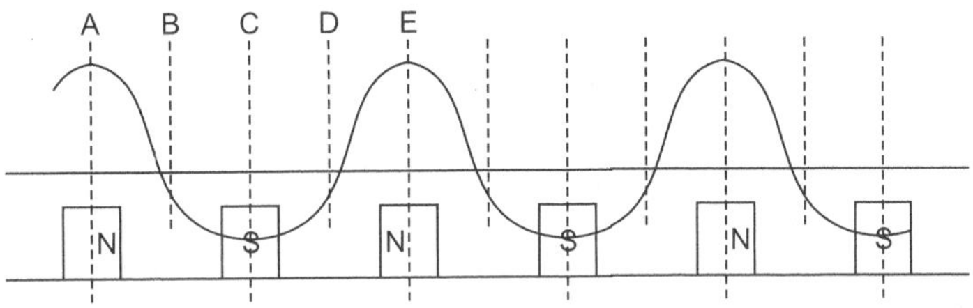

Fig. 14.7

Redrawing the Fig. 14.6 in horizontal position as shown in Fig. 14.7, for understanding the procedure better.

The conductor being situated at the place of maximum flux density will have maximum e.m.f. induced in it.

When the conductor is in inter polar gap (gap between two poles) say at point B, it has minimum e.m.f. induced in it because flux density is minimum there.

Again when the conductor is at center of south pole (say at point C) it has maximum e.m.f. induced in it, because flux density at C is maximum.

But the direction of e.m.f. when the conductor is over N-pole is opposite to that when it is over S-pole.

Now it is clear that one cycle of e.m.f. is induced in a conductor when one pair of pole passes over it.

In other words the e.m.f. in an armature conductor goes through one cycle in angular distance twice the pole pitch as shows in Fig. 14.7.

Let

Total no. of poles $= P$

Speed of rotor $= N_s$ rpm

Frequency of the generated e.m.f. $= f$ Hz

One cycle of e.m.f. is produced when a pair of poles passes past a conductor.

Number of cycle of e.m.f. produced in one revolution is equal to number of pair of poles or half the number of poles.

$$\text{Number of cycles/revolution} = \frac{P}{2} \quad \text{...(14.2)}$$

$$\text{Number of revolutions/sec.} = \frac{N_s}{60} \quad \text{...(14.3)}$$

$$\text{Frequency,} \quad f = \frac{P}{2} \times \frac{N_s}{60} \quad \text{...(14.4)}$$

or

$$\boxed{f = \frac{PN_s}{120}} \text{ Hz} \quad \text{...(14.5)}$$

14.6 E.M.F. EQUATION

Let

Flux per pole $= \Phi$

Total no. of poles $= P$

Speed of rotor $= N_s$

Frequency of e.m.f. induced $= f$

Number of conductor per phase $= Z$

Number of turns per phase $= T$

(One turn has two sides)

Form factor $= K_f$

(For sinusoidal wave, $K_f = 1.11$)

In general the e.m.f. induced is given by,

$$\text{e.m.f. induced} = \text{Total no. of conductor} \times \frac{\text{Rate of change of flux}}{\text{Per unit time}} \quad \text{...(14.6)}$$

$$e = N\frac{d\Phi}{dt} \quad \text{...(14.7)}$$

where N are the total number of conductors in general. For convenience, initially let us assume that a single conductor is rotated in the magnetic field. In such a case, e.m.f. induced per conductor will be

$$\text{E.m.f. induced per conductor,} \quad e = \frac{d\Phi}{dt} \quad \text{...(14.8)}$$

Flux per pole $= \Phi$

Total flux $= P\,\Phi$ ...(14.9)

$$\text{Time taken by rotor pole to take one revolution} = \frac{60}{N_s} \text{ sec} \quad \text{...(14.10)}$$

From equations (14.9) and (14.10) equation (14.8) will become

$$e = \frac{P\Phi}{60/N_s} \quad \text{...(14.11)}$$

or $$e = \frac{P\Phi N_s}{60} \qquad ...(14.12)$$

As the frequency of the induced e.m.f is given by [From equation (14.5)]

$$f = \frac{PN_s}{120} \text{ Hz} \qquad \text{[Eqn. (14.5)]}$$

or $$N_s = \frac{120 f}{P} \qquad ...(14.13)$$

Substituting equation (14.13) in equation (14.12), we get

$$e = \frac{\Phi P}{60} \times \frac{120 f}{P} \qquad ...(14.14)$$

or $$e = 2 f\Phi$$

If number of conductors per phase are Z, then average induced e.m.f. per phase will be

$$E_{av} = 2 f\Phi Z \qquad ...(14.15)$$

If T is the total number of turns, each turn has two sides or 2 conductors, then

$$Z = 2T \qquad ...(14.16)$$

Putting Z = 2T in equation (14.15) we get

$$E_{av} = 2 fP \times 2T \qquad ...(14.17)$$

or $$E_{av} = 4f\phi T \qquad ...(14.18)$$

As form factor for a sinusoidal wave is given by

$$\mathbf{K}_f = \frac{E_{rms}}{E_{av}} = 1.11 \qquad ...(14.19)$$

or $$\mathbf{E_{rms} = 1.11 \times E_{av}} \qquad ...(14.20)$$

Substituting equation (14.18) in equation (14.20), we get

$$E_{rms} = 4.44 f\Phi T \qquad ...(14.21)$$

For distributed winding, the r.m.s value of e.m.f per phase will be K_d times the above value, where K_d is the distribution factor. Therefore, equation (14.21) can be written as

$$E_{rms} = 4.44 f\Phi TK_d \qquad ...(14.22)$$

For short pitched winding, the r.m.s value of e.m.f per phase will be K_p times the above value, where K_p is the pitch or coil span factor. Therefore, equation (14.22) will become

$$E_{rms} = 4.44 f\Phi TK_dK_p \qquad ...(14.23)$$

or $$E_{rms} = 4.44 f\Phi TK_dK_p \qquad ...(14.24)$$

For a full pitched and concentric winding,

$$K_p = K_d = 1 \qquad ...(14.25)$$

As usual the alternator is normally star connected. In that case the line voltage per phase will be

$$E_L = \sqrt{3}\, E_{rms} \qquad ...(14.26)$$

From equation (14.26), equation (14.24) will become

$$E_L = \sqrt{3} \times 4.44\, K_dK_p f\Phi T \qquad ...(14.27)$$

14.7 ADVANTAGES OF HAVING STATIONARY ARMATURE AND ROTATING FIELD SYSTEM

Very high rated synchronous machines mainly have stationary armature and the field is placed on rotor *i.e.*, they have rotating field system.

Advantages of having stationary armature and rotating field system are:

- If the armature is rotating, we require 3 slip rings to collect current. Whereas now the field is rotating which is energized by DC supply and hence, we need two brushes only.
- The output terminals are fixed in case of stationary armature and hence output current can be taken out directly without having it to pass through brush contacts.
- In case of high voltages, it is difficult to insulate the slip rings from rotating shaft.
- It is easier to insulate stationary armature winding for high AC voltages.

14.8 SYNCHRONOUS MOTOR

Similar to a DC machine, a synchronous machine can also be operated both as a generator and as a motor. Synchronous motor only operates at synchronous speed. This synchronous speed depends on the frequency of the applied voltage and the number of poles. It can be operated over a wide range of power factor, both lagging and leading. Fig. 14.8 illustrates a 3-phase, 2-pole synchronous machine. The rotor has two poles and the stator has two poles per phase. In practice it is usual to have salient poles on the rotor, but the armature winding is housed in slots in the concave periphery of the stator. In the schematic diagram shown in Fig. 14.8, salient type stator poles have been shown for convenience only.

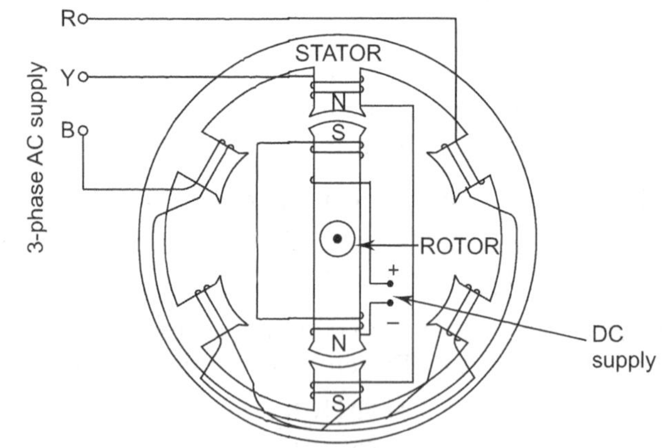

Fig. 14.8

A synchronous motor is inherently not self starting. It has to be run up to or nearly synchronous speed by some auxiliary means. After reaching at or near to the synchronous speed the motor get synchronize with the supply.

14.9 BASIC PRINCIPLE

To understand the basic principle of a synchronous motor, let us consider a 2-pole synchronous motor (Fig. 14.8).

- Three-phase supply is fed to the three-phase winding of the stator.
- A magnetic flux of constant magnitude but rotating at synchronous speed is produced in the stator.
- The stator poles (N_s and S_s) will rotate with synchronous speed $\left(N_s = \frac{120f}{P}\right)$ in clockwise direction.
- As the rotor is excited from DC supply, so the poles of the rotor retain the same polarity throughout but the polarity of the stator poles changes as it is connected to an AC supply mains.

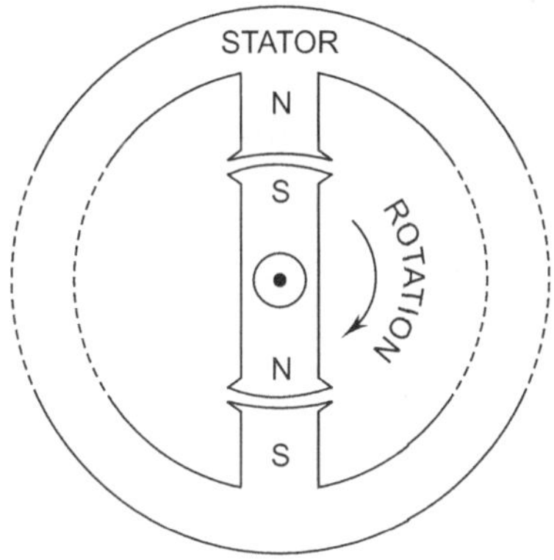

Fig. 14.9

- First of all let us consider the rotor as stationary and in the position shown in Fig. 14.9.
- At this instant the rotor S-pole is attracted to the stator N-pole and, therefore, the rotor tends to rotate in clockwise direction.
- After half a period (*i.e., after T/2 second whereas T = 1/f*) the polarity of stator poles is reversed but the polarity of the rotor poles remains the same (Fig. 14.10).

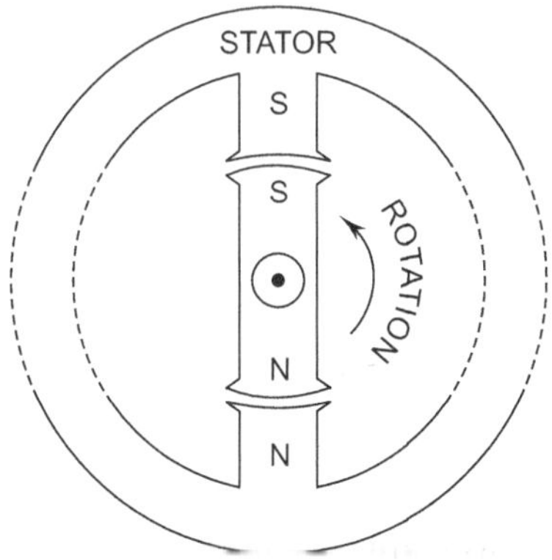

Fig. 14.10

- At this instant, the rotor S-pole is repelled by the stator S-pole, being similar in nature and, therefore, rotor tends to rotate in counterclockwise direction.
- Thus the torque acting on the rotor of a synchronous motor is not unidirectional but pulsating one and due to inertia of the rotor, it will not move in any direction.
- So the synchronous motor has got no self-starting torque.

Now let us consider the rotor rotating in clockwise direction, which is possible by external means, and in the position shown in Fig. 14.11. At this instant, as mentioned above, the rotor S-pole is attracted to stator N-pole and so the torque acts on the rotor in clockwise direction. After half a period the stator polarity is reversed *i.e.,* stator S-pole becomes N-pole and N-pole becomes S-pole but if the rotor is rotated at such a speed by some external means at the starting moment that rotor S-pole advances by a pole pitch so that it is again under the influence of stator N-pole, as shown in Fig. 14.11, the torque acting on the rotor will be again clockwise. Hence a continuous (unidirectional) torque will be obtained. Now if the external means is removed the rotor will continue to rotate in clockwise direction under the influence of clockwise continuous torque acting on the rotor.

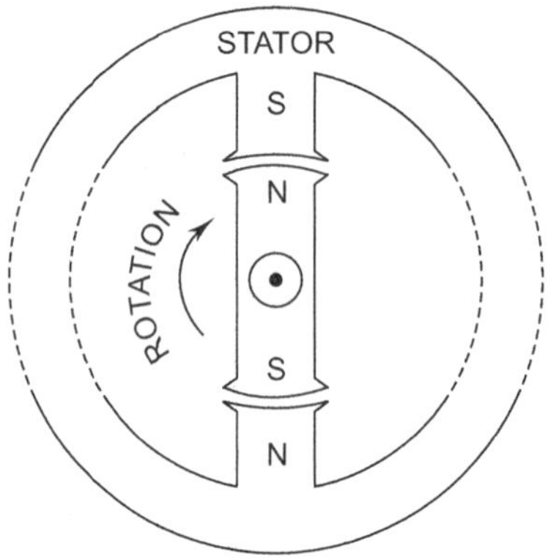

Fig. 14.11

Hence to obtain a continuous torque it is necessary that the rotor rotates at such a speed that it moves through the distance equal to pole pitch in half the period *i.e.,* in or second where f is the supply frequency. If there are 'P' poles then to complete one revolution it should take second or the rotor should rotate at a speed of x 60 revolutions per minute.

Hence to obtain a continuous torque the rotor should rotate at synchronous speed given by the expression

$$N_s = \frac{120f}{P}.$$

14.10 STARTING METHODS

As already discussed, a synchronous motor does not have a self-starting torque. It can develop synchronous torque only at synchronous speed. The methods generally used for developing the starting torque in synchronous motors are:

- Induction motor starting
- Starting by means of a wound rotor.

In ***induction motor starting method,*** the 3-phase supply at reduced voltage is applied to the stator of the synchronous motor which results in a magnetic field rotating at synchronous speed. This would cause an induced e.m.f and hence and induced current in the cage winding provided in the pole faces. The rotating field interacts with the induced current in the cage winding and thus develops sufficient torque to raise the speed of the rotor to about 95 percent of the synchronous speed. Till this stage, the rotor field winding is not excited. The rotor field winding is now connected to the proper DC supply to magnetize the rotor poles.

The ***wound rotor starting method*** of the synchronous motor is mainly based on the concept that the rotor of the synchronous motor is constructed similar to the wound rotor of an induction motor. When the 3-phase wound rotor of the synchronous motor is excited from DC source, it gives rise to alternate N and S poles on the rotor. As the direct current excitation is, fixed, the pole axes are also fixed relative to rotor winding. The machine is started by connecting the rotor winding to the variable resistors. When these are fully cut-out, the motor runs with a small slip. At this stage the connections are changed over which would connect the exciter in series with the rotor winding.

14.11 V-CURVES

The power drawn by 3-phase synchronous motor is given by

$$P = \sqrt{3}\, V_L I_L \cos\phi$$

where V_L is the line voltage,

I_L is the armature current, and

$\cos\phi$ is the power factor.

Since, input power P and supply voltage V_2 are constant, hence the armature current and power factor combination must remains constant.

The change in field excitation neither affects the speed of the motor nor the output of the motor but does affects the power factor and consequently armature current for constant supply voltage and constant input power.

When the synchronous motor is operated at constant supply voltage, it demands nearly constant flux. Both the DC source and AC source cooperate in establishing this resultant flux.

- When the DC excitation is weak, the flux produced by it is weak and needs strong thening, so the stator draws the current (which lays behind the supply voltage by almost 90°) so as to magnetize the air-gap to the extent needed to provide the necessary resultant flux.

  For constant applied voltage and constant input power, the active component of the current drawn form the supply mains should be constant. Hence the main variation will be in magnetizing components. This magnetizing component of current, being large part of total input current, results in low lagging power factor.
- With the increase in excitation to 100%, the magnetizing current drawn form the supply is reduced and hence the power factor improves. This results in reduction of armature current drawn from the supply.
- If the excitation is further increased such that power factor becomes unity, the stator draws only active component. The armature current draw will be minimum. The DC field circuit will be provide all the necessary current for producting flux.
- On increasing the DC excitation further, the field produced by it becomes stronger than it is required. So it needs to demagnetized. Hence the motor draws leading current to demagnetize the main field. This makes the power factor leading and lesser than unity and so the armature current increases.

Hence, Variation in excitation or in field current causes the variation in armature current and curves drawn between armature current and field current for different constant loads are known as V-curves. (These curves resembles the English letter V, hence so named). V-curves are shown in Fig. 14.12.

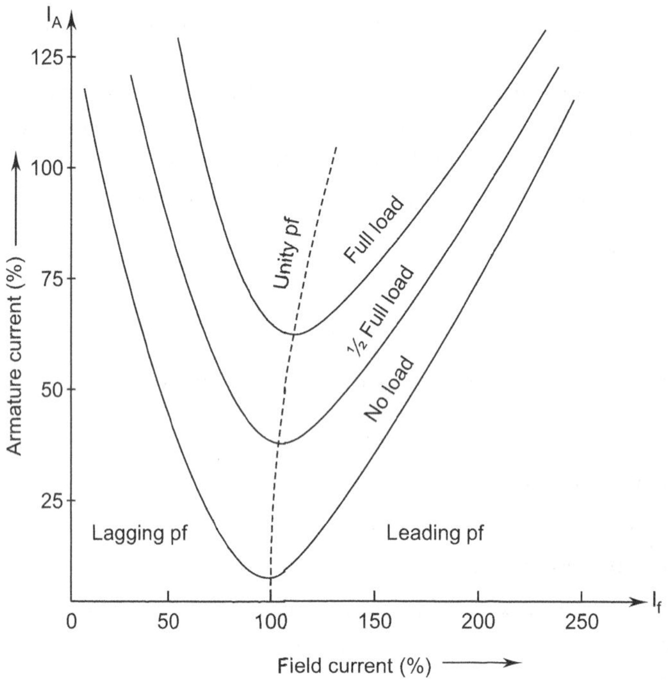

Fig. 14.12

Similarly, the variation of power factor with a variation in field current (or DC excitation) for a constant load gives 'Inverted V-curves', shown in Fig. 14.13.

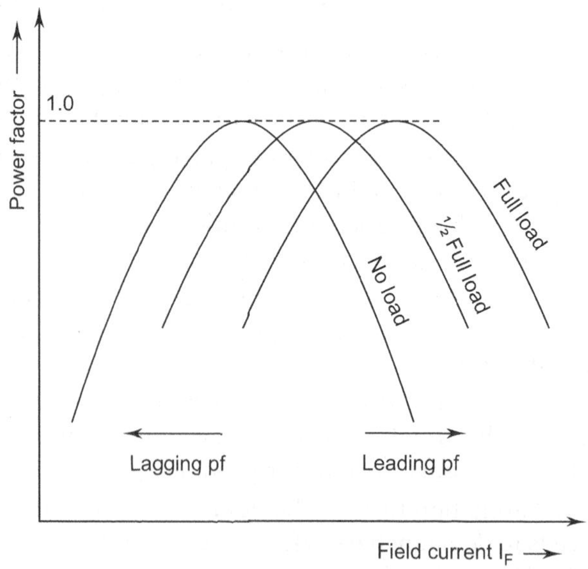

Fig. 14.13

The plot of variation of power factor as a function of field current is shown in Fig. 14.13 and is known as **inverted V-curves** of the synchronous motor.

14.12 POWER FACTOR IMPROVEMENT

Synchronous motor can be operated under a wide range of power factor both leading and lagging by controlling its excitation current.

On no-load synchronous motor draws very little current from the supply mains to meet the internal losses. With fixed excitation the input current increases with the increase in load. After the input current reaches maximum no further increase in load is possible. If the motor is further loaded, the motor will stop.

If field excitation is small, the stator inductance will be more so the motor draws lagging current and, therefore, power factor is low. If the excitation is increased the stator flux is neutralized by the rotor flux and, therefore, power factor becomes unity. In this position, the current and voltage are in phase and motor current will be minimum.

If excitation is further increased rotor flux increases. It neutralizes the stator flux and causes leading current. This property makes the motor extremely useful as a *power factor corrector*. By taking leading current it compensates the lagging current drawn by other equipment from the supply mains and thus brings the total current drawn from the supply mains more nearly in phase with the voltage.

14.13 MERITS

- These motors usually operate at higher efficiencies, especially in the low speed unity pf ranges.
- The power factor can be controlled very easily.
- The speed of a synchronous motor is constant and independent of load.
- An over-excited synchronous motor having a leading power factor can be operated in parallel with induction motors and other power apparatus operating at lagging power factors, thereby improving the power factor of the supply system.
- Electromagnetic power varies linearly with the voltage.
- These motors can be constructed with wider air-gaps than induction motors, which make them better mechanically.

14.14 DEMERITS

- The synchronous motor is inherently not self-starting motor and needs some arrangement for its starting and synchronizing.
- It requires DC excitation which must be supplied from external source.
- The cost per kW output is generally higher than that of an induction motor.
- It cannot be used for variable speed jobs as there is no possibility of speed adjustment.
- It cannot be started under load. Its starting torque is zero.
- It has a tendency to hunt.

- It may fall out of synchronism and stop when over-loaded.
- Collector rings and brushes are required.

14.15 APPLICATIONS

- The fields of applications of ***low speed*** synchronous motors (below 500 rpm) are drives such as electro-plating generators, reciprocating compressors when started unloaded, DC generators, centrifugal and screw type pumps, vacuum pumps, chippers, line shafts, rubber and band mills, ball and tube mills, metal rolling mills etc.
- Typical applications of ***high-speed*** synchronous motors (above 500 rpm) are drives such as fans, blowers, DC generators, line shafts, centrifugal pumps and compressors, reciprocating pumps and compressors, constant speed frequency changers, rubber and paper mills etc.
- Synchronous motors in ***very large sizes*** (as high as 10 MW size) operating at ultra-low speeds are employed to drive crushers, rotary kilns and variable speed ball mills.
- ***Power factor correction:*** They are used in power houses and substations in parallel to the bus-bars to improve the power factor. For this purpose they are run without mechanical load on them and over-excited.

  In factories having a large number of induction motors, or other power apparatus operating at lagging power, they are employed to improve the power factor.
- Voltage regulation.
- Constant speed, constant-load drives.
- Such motors are also used to regulate the voltage at the end of transmission lines.
- Because of higher efficiency possible with synchronous motors, they can be employed advantageously for the loads where constant speed is required.
- High power electronic converters generating very low frequencies made the operation of synchronous motors at ultra-low speeds.

EXERCISE

1. Differentiate between the constructional features of salient pole synchronous machine and round rotor synchronous machine. Why a salient pole synchronous generator is used in hydroelectric plants and round rotor synchronous generator in steam plants?
2. Why the synchronous motor is not self starting? Explain the various methods of starting these motors.
3. A 4-pole synchronous motor is connected to a 50 Hz, 40 V and 3-phase AC supply. What is the speed in rpm? Clearly explain. Will this speed depend upon voltage?
4. Discuss the advantages and disadvantages synchronous motors.
5. State the advantages of having rotating field system rather than a rotating armature system in synchronous machine.
6. What is the effect of change of excitation of a synchronous motor on its power factor for different loads?
7. Give the constructional details of salient pole synchronous machine.
8. Explain V-curves in detail for a synchronous machine. Give the application of synchronous machine.

9. Explain why the synchronous motor always runs at synchronous speed?
10. Derive the expression for the voltage induced in an alternator. A 4-pole alternator has armature with 25 slots and 8 conductors/slot and rotates at 1500 rpm and the flux/pole is 0.05 Wb. Calculate the e.m.f. generated, if winding factor is 0.96 and all conductors are in series.
11. Derive an expression for voltage induced in an alternator per phase consisting of a number of full pitch coils joined in series. Assume the air-gap flux to have sinusoidal distributions.
12. Explain how power factor of an inductive load can be improved employing synchronous motor?
13. Three-phase induction motor is self starting but 3-phase synchronous motor is not self starting. Explain why? Give application of 3-phase synchronous motor.

□ □ □

15

THREE-PHASE AC CIRCUITS

15.1 INTRODUCTION

In everyday's life, people normally encounter with single phase AC supply. Most of the appliances at home as well as in commercial field operate on single phase AC supply.

But this single phase AC supply arises from a 3-phase power system. Actually, the generation of power is done with 3-phase AC circuits (except with home or commercial low ratting generators).

15.2 REASON FOR USING THREE-PHASE SYSTEM

Some of the main reasons to use three-phase systems or network are:

- Three-phase power has constant magnitude whereas single phase power pulsate from zero to peak value at twice the supply frequency.
- A three-phase system can set up a rotating magnetic field in stationary winding. This cannot be done with a single phase current.
- For the same rating three-phase machines for example generators, motors, transformer etc., are smaller, simpler in construction and have better operating characteristics then single phase machines.
- To transmit the same amount of power over a fixed distance at a given voltage, the 3-phase system requires only three-fourth the weight of copper that is required by the single phase system.
- The voltage regulation of a three-phase transmission line is better than that of a single phase line.

15.3 GENERATION OF THREE-PHASE VOLTAGES

First we will discuss the generation of single-phase alternating voltage, which we have already studied in chapter 3.

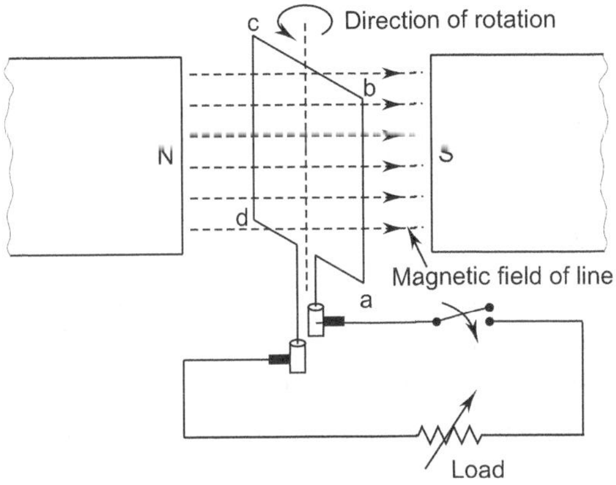

Fig. 15.1

Consider that a single conductor is placed in a magnetic field (Fig. 15.1).

According the Faraday's law of electromagnetic induction, *"when a conductor is rotated in a magnetic field, an e.m.f. is induced in the conductor. The magnitude of induced e.m.f. or voltage is directly proportional to the rate of change of flux linking the coil."*

For a single conductor the e.m.f. induced is shown in Fig. 15.2.

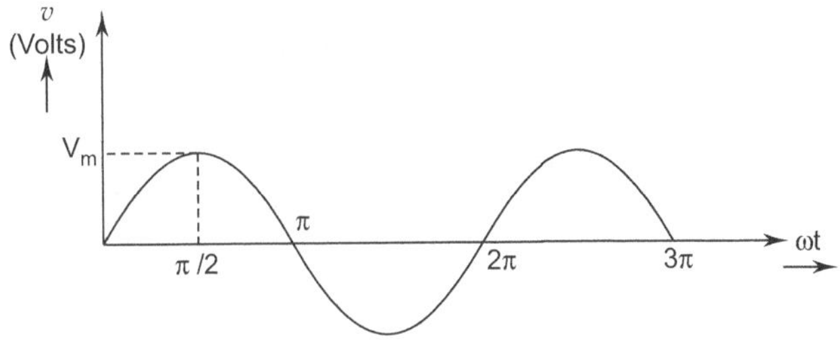

Fig. 15.2

Now consider a case where three coils, displaced 120 electrical degrees apart with each other, are placed in a magnetic field (Fig. 15.3).

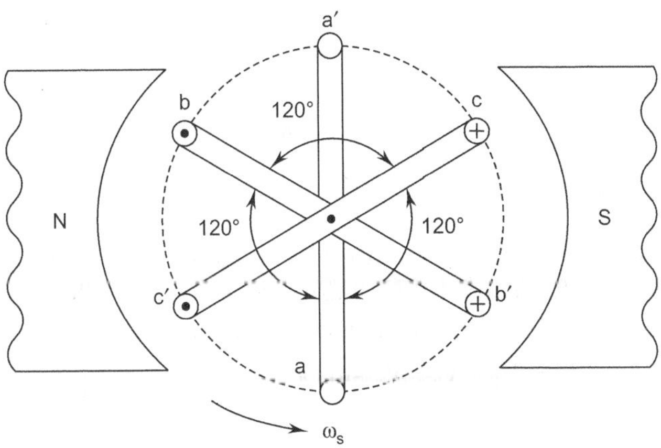

Fig. 15.3

First coil: *aa′*
Second coil: *bb′*
Third coil: *cc′*
These three coils are:

- Identical
- Has equal number of turns
- Wound with same type of wire having same cross-section area.

Now when these three coils are rotated in a magnetic field, e.m.fs induced in each coil has same magnitude, having same frequency and have similar wave forms.

But they are displaced from each other by an angle 120° as shown in Fig. 15.4.

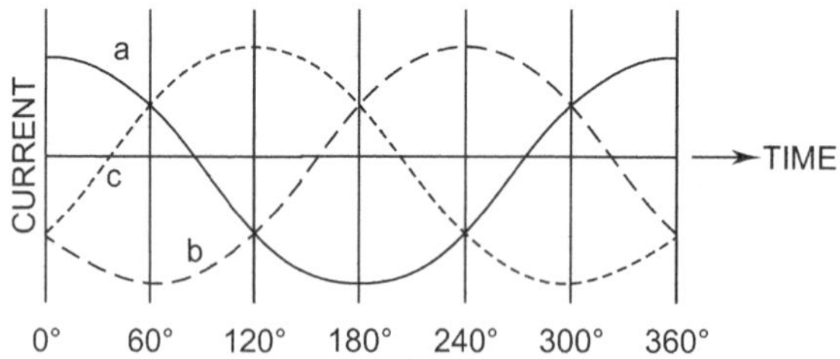

Fig. 15.4

The instantaneous value of these e.m.fs induced in the coils *aa′*, *bb′* and *cc′* are:

$$V_a = V_m \sin \omega t \quad ...(15.1)$$

$$V_b = V_m \sin (\omega t - 120°) \quad ...(15.2)$$

$$V_c = V_m \sin (\omega t - 240°) = V_m \sin (\omega t + 120°) \quad ...(15.3)$$

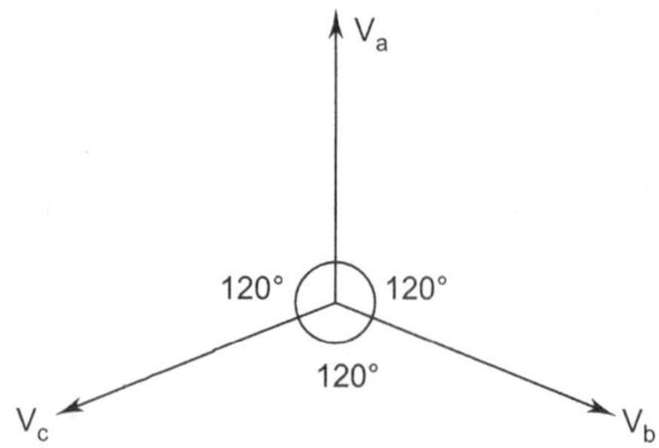

Fig. 15.5

15.3.1 Phase Sequence

Phase sequence is the order in which the induced voltage in each phase attains their maximum value. In technical terms, RYB is consider as phase sequence where 'R' stands for Red, 'Y' stands for Yellow and 'B' stands for Black or Blue. Hence, the above three e.m.fs can also be denoted as

$$V_a = V_R = V_1 = V_m \sin \omega t \quad ...(15.4)$$

$$V_b = V_Y = V_2 = V_m \sin (\omega t - 120°) \quad ...(15.5)$$

$$V_c = V_B = V_3 = V_m \sin (\omega t - 240°) = V_m \sin (\omega t + 120°) \quad ...(15.6)$$

The three voltages V_1, V_a and V_R are one and the same thing. These are simply notations which vary from text to text. We will now follow one notation *i.e.*, of V_R, V_Y and V_B.

15.3.2 The Sum of Three e.m.f. at Every Instant is Zero

Once again considering the above three e.m.fs equations with RYB notation:

$$V_R = V_m \sin \omega t \quad ...(15.7)$$

$$V_Y = V_m \sin (\omega t - 120°) \quad ...(15.8)$$

$$V_B = V_m \sin (\omega t - 240°) \quad ...(15.9)$$

Resultant of the three e.m.fs equation will be

$$V = V_R + V_Y + V_B \quad ...(15.10)$$

or

$$V = V_m [\sin (\omega t) + \sin (\omega t - 120^o) + \sin (\omega t - 240^o)] \quad ...(15.11)$$

or

$$V = V_m [\sin (\omega t) + \sin (\omega t - 120^o) + \sin (\omega t + 120^o)] \quad ...(15.12)$$

On solving the above trigonometric identity, we get resultant

$$V = 0 \quad ...(15.13)$$

15.3.3 Representation of e.m.f. in Polar and Rectangular Form

The three e.m.f., equations can be represented in polar and rectangular form as given below:

e.m.f.	Polar Form	Rectangular Form
V_R	$V_m \angle 0°$	$E_m (1 + j0)$
V_Y	$V_m \angle -120°$	$E_m (-0.5 - j0.866)$
V_B	$V_m \angle -240°$	$E_m (-0.5 + j0.866)$

15.4 INTERCONNECTION OF THREE-PHASES

Each coil of three-phase circuit has two terminals; therefore we have total number of 6 conductor's ends. The interconnections of these three coils are done in two ways:

(*i*) Star or Wye (Y) Connection

(*ii*) Delta or Mesh (Δ) Connection.

In case of three-phase generation, the alternator (or generator) windings as well as in case of three-phase power consumption in 3-phase load (for *e.g.*, 3-phase induction motor), the phases are connected in either star configuration or delta configuration.

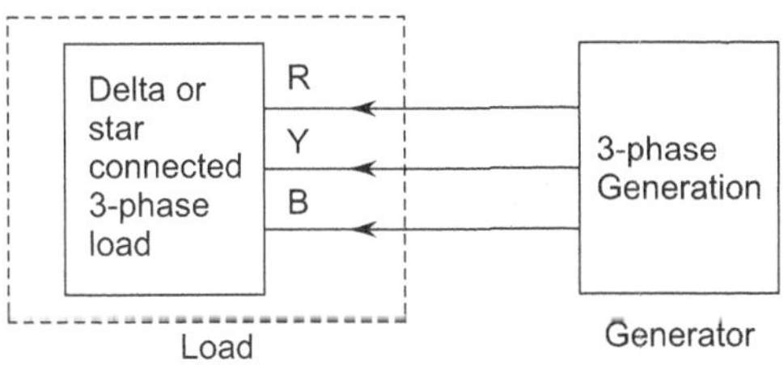

Fig. 15.6

In Fig. 15.6, note that the direction of currents are outgoing from the generating point and are incoming at power consumption point.

In all the upcoming sections, we will consider the load section (shown in dotted box) in which the current is always going into the load.

15.5 STAR OR WYE (Y) CONNECTION

Initially let us consider three isolated phases R, Y and B.

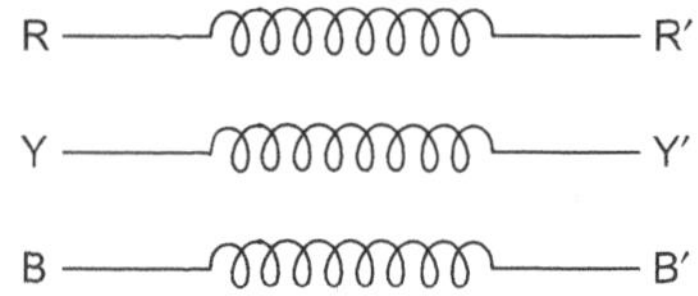

Fig. 15.7

These three-phases are connected together to form a neutral point, thus reducing the total number of terminals to three. Such a configuration of connection of three-phases is known as star or wye connection. Normally, it is denoted by capital 'Y'.

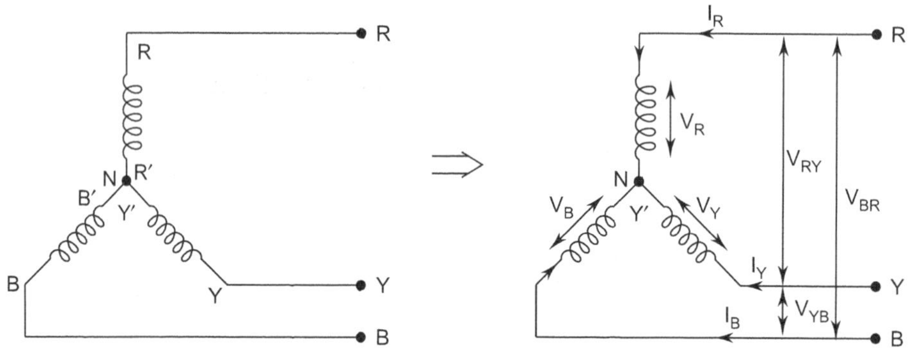

Fig. 15.8

The voltage induced in each winding is known as **phase voltage**, given by

$$V_R = V_Y = V_B = V_{ph} \qquad \ldots(15.14)$$

The current flowing in each winding is known as **phase current**, given by

$$I_R = I_Y = I_B = I_{ph} \qquad \ldots(15.15)$$

The voltage available between any pair of terminals (or lines) is known as **line voltage** and is given by

$$V_{RY} = V_{YB} = V_{BR} = V_L \qquad \ldots(15.16)$$

The current between any pair of terminals (or lines) is known as **line current** and is given by

$$\text{Line current} = I_L \qquad \ldots(15.17)$$

15.5.1 Line Voltage and Phase Voltage Relations

In case of three-phase star connection, the voltage between two lines (*i.e.*, line voltage) and voltage across one phase are different.

Next we will derive the relation between line voltage and phase voltage in case of star connection.

Let us consider a 3-phase star connected load as shown in Fig. 15.9. Different voltages and currents of the three-phase and three lines are shown in the figure. The parameters on the right side of the dotted line are all line parameters and the parameters on the left side of the dotted line are all phase parameters.

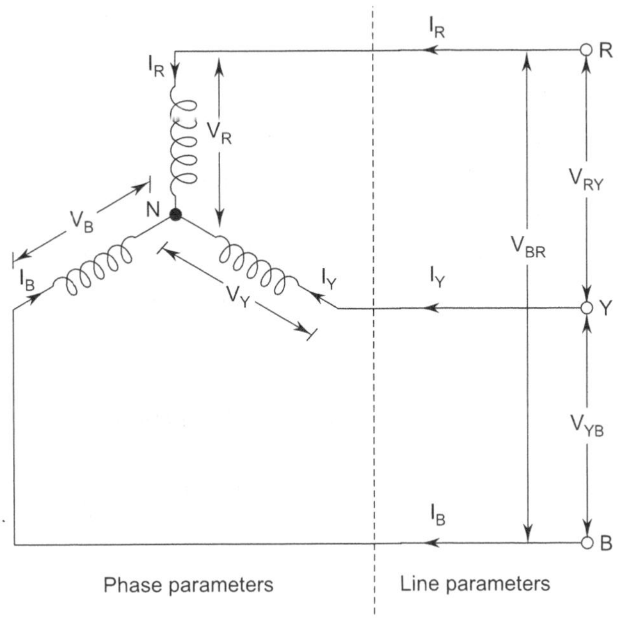

Fig. 15.9

The potential difference between line 1 and line 2 is

$$V_{RY} = V_R - V_Y \qquad \text{...(15.18)}$$

or

$$V_{RY} = V_R + (-V_Y) \qquad \text{...(15.19)}$$

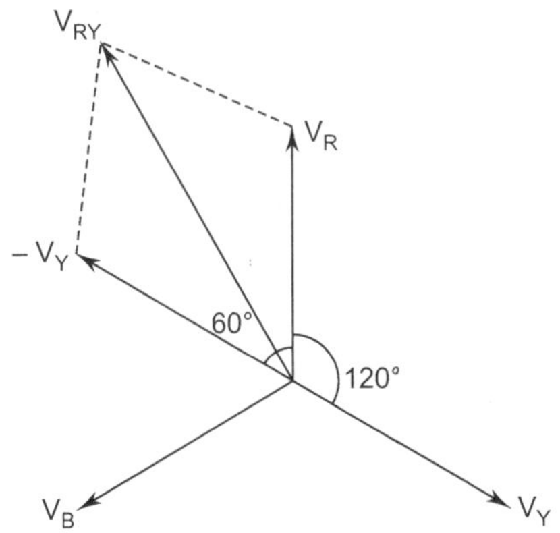

Fig. 15.10

The potential difference between line 1 and line 2, *i.e.*, V_{RY}, can be calculated by taking phasor sum of V_R and reversed of V_Y.

From the phasor diagram of phase voltages shown in Fig. 15.10, it is clear that the phasor V_{RY} will be given by the diagonal of the parallelogram.

Applying the law of parallelogram, in general, we have

$$V_{12} = \sqrt{|V_1|^2 + |-V_2|^2 + 2|V_1||-V_2|\cos\theta} \qquad \text{...(15.20)}$$

where θ is the angle between two vectors V_1 and V_2 which, in our case is

$$\theta = 60° \quad \text{...(15.21)}$$

$$V_1 = V_R \quad \text{...(15.22)}$$

$$V_2 = V_Y \quad \text{...(15.23)}$$

Hence, with the above notations equation (15.23) can be written as

$$V_{RY} = \sqrt{|V_R|^2 + |-V_Y|^2 + 2|V_R||-V_Y|\cos 60°} \quad \text{...(15.24)}$$

As the three-phase circuit is considered as balanced one, hence

$$V_R = V_Y = V_B = V_{ph} \quad \text{[Eqn. (15.14)]}$$

$$V_{RY} = V_{YB} = V_{BR} = V_L \quad \text{[Eqn. (15.16)]}$$

From equations (15.14) and (15.16) the equation (15.24) will become

$$V_L = \sqrt{|V_{ph}|^2 + |-V_{ph}|^2 + 2|V_{ph}||-V_{ph}|\frac{1}{2}} \quad \text{...(15.25)}$$

or

$$V_L = \sqrt{3V_{ph}^2} \quad \text{...(15.26)}$$

or

$$V_L = \sqrt{3}\, V_{ph} \quad \text{...(15.27)}$$

Hence in case of 3-phase balanced star network,

$$\boxed{\textit{Line voltage} = \sqrt{3}\ \textit{Phase voltage}} \quad \text{...(15.28)}$$

15.5.2 Line Current and Phase Current Relations

It is clear from Fig. 15.8 that the current is same in the line as well as in phase in case of 3-phase balanced star network, *i.e.*

$$I_L = I_{ph} \quad \text{...(15.29)}$$

or

$$\boxed{\textit{Line current} = \textit{Phase current}} \quad \text{...(15.30)}$$

15.5.3 Power Expressions

Assuming that in case of 3-phase balanced star network, the load is inductive and the current is lagging behind the voltage (phase-to-phase) by an angle ϕ, active power in each phase will be given by

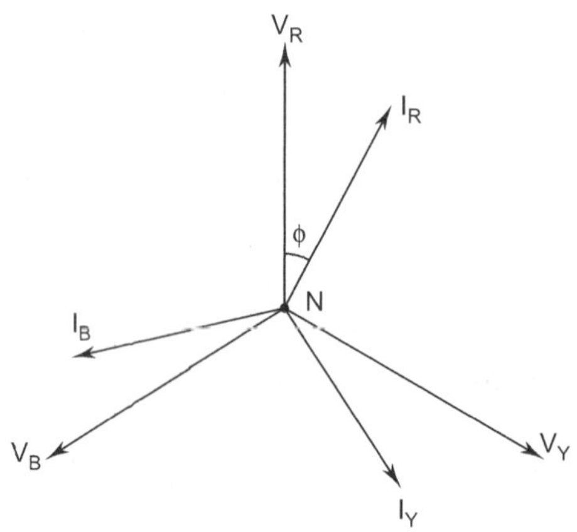

Fig. 15.11

$$P = V_{ph} I_{ph} \cos \phi \qquad ...(15.31)$$

And active power in 3-phase is given by

$$P = 3V_{ph} I_{ph} \cos \phi \qquad ...(15.32)$$

This is the expression for active power in 3-phase in term of phase values. From equations (15.27) and (15.29), equation (15.32) will become

$$P = 3 \frac{V_L}{\sqrt{3}} I_L \cos\phi \qquad ...(15.33)$$

or

$$\boxed{P = \sqrt{3}\, V_L I_L \cos\phi} \qquad ...(15.34)$$

[Note that ϕ is the angle between phase voltage and phase current and not between line voltage and line current.]

15.6 POWER MEASUREMENT IN STAR CONNECTED LOAD

In case of 3-phase circuits (whether star connected or delta connected), power can be measured by following three methods:

- One wattmeter method
- Two wattmeter method
- Three wattmeter method

Out of these three methods, one wattmeter method can be used only for a balanced star connected load. It cannot be used for an unbalanced load.

Whereas two wattmeter method and three wattmeter method can measure power in both balanced as well as unbalanced connected load.

15.6.1 One Wattmeter Method

In case of a 3-phase star connected balanced load power can be measured by connecting a single wattmeter.

The voltage or pressure coil of the wattmeter is connected between the line and neutral point.

The current coil is connected in series with the line as shown in Fig. 15.12.

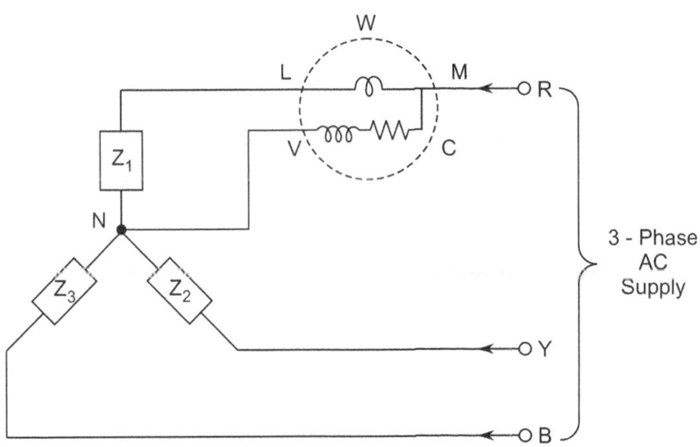

Fig. 15.12

The wattmeter connected will give the reading of power per phase. As the network is balanced one, hence power in 3-phase will be

$$\text{Power} = 3 \times \text{Wattmeter reading} \quad ...(15.35)$$

or

$$\text{Power} = 3 \times \text{Power per phase}$$

It is to be noted that this method is applicable only when the load is balanced one.

15.6.2 Three Wattmeters Method

In case of three-phases – four wire system, the current flows in the neutral wire also. In such a case two-wattmeter method cannot be used. In such a system three wattmeters are required to measure the power in three-phase unbalanced load.

The current coil of each wattmeter is connected in three lines and the voltage coils of all the three wattmeters are connected to the neutral wire or point as shown in Fig. 15.13.

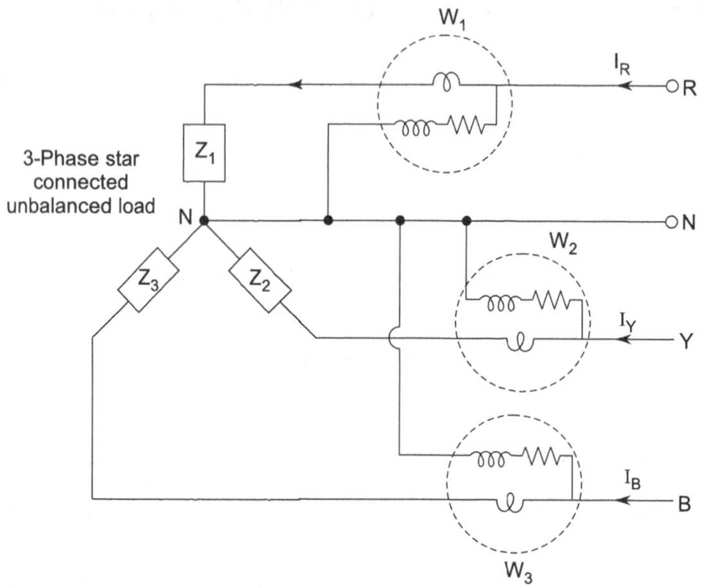

Fig. 15.13

Addition of the readings of the three wattmeters will give the total power consumed by the load *i.e.*,

$$P = W_1 + W_2 + W_3 \quad ...(15.36)$$

15.6.3 Two Wattmeters Method

With the help of two wattmeter method, power can be measured in a 3-phase star connected (balanced or unbalanced) load.

The current coils of the two-wattmeters are connected to any two lines of the network.

The voltage or pressure coils of the two wattmeters are connected to the third line as shown in Fig. 15.14.

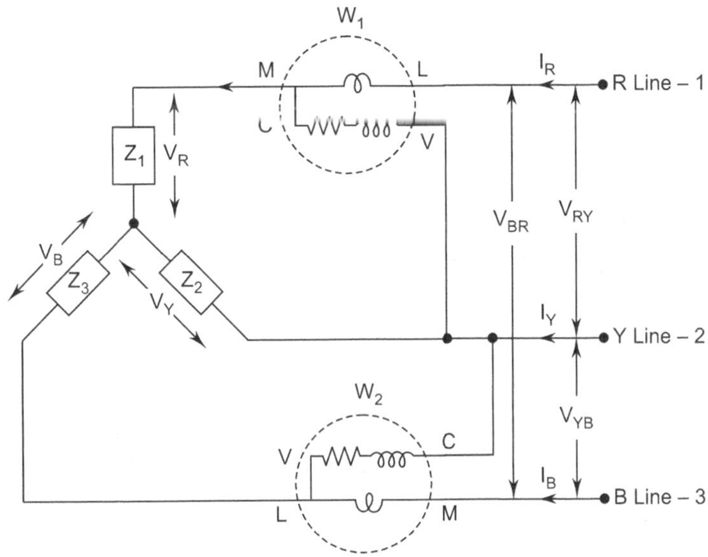

Fig. 15.14

While measuring the power, we will consider the instantaneous values. Power is mainly consumed by the load. That power is in terms of phase values. According to the circuit diagram shown in Fig. 15.14, let

Current through impedance, $Z_1 = I_R$

Current through impedance, $Z_2 = I_Y$

Current through impedance, $Z_3 = I_B$

Voltage across impedance, $Z_1 = V_R$

Voltage across impedance, $Z_2 = V_Y$

Voltage across impedance, $Z_3 = V_B$

Total instantaneous power in the load,

$$P = V_R I_R + V_Y I_Y + V_B I_B \quad \text{...(15.37)}$$

For Wattmeter W_1

Wattmeter W_1 is connected between line-1 (R) and line-2 (Y).

I_R = Current through wattmeter W_1

$V_{RY} = V_R - V_Y$ = Voltage across the voltage coil of wattmeter W_1

Instantaneous power measured by wattmeter W_1

$$P_1 = I_R \times V_{RY} = I_R (V_R - V_Y) \quad \text{...(15.38)}$$

For Wattmeter W_2

Wattmeter W_2 is connected between line-3 (B) and line-2 (Y).

I_B = Current through wattmeter W_2

$V^*_{BY} = V_B - V_Y$ = Voltage across the voltage coil of wattmeter W_2

Instantaneous power measured by wattmeter W_2

$$P_2 = I_B \times V_{BY} = I_B (V_B - V_Y) \quad \text{...(15.39)}$$

Total instantaneous power measured by the two-wattmeters will be

*Note that the voltage across the voltage coil of wattmeter W_2 is V_{BY} and not V_{YB} as the Wattmeter is connected between line-3 (B) and Line-2 (Y) not between Line-2 and Line-3.

$$P_1 + P_2 = I_R (V_R - V_Y) + I_B (V_B - V_Y) \quad ...(15.40)$$

or $$P_1 + P_2 = I_R V_R - I_R V_Y + I_B V_B - I_B V_Y \quad ...(15.41)$$

or $$P_1 + P_2 = I_R V_R + I_B V_B - V_Y (I_B + I_R) \quad ...(15.42)$$

According to Kirchhoff's current law

$$I_R + I_Y + I_B = 0 \quad ...(15.43)$$

or $$I_B + I_R = -I_Y \quad ...(15.44)$$

Putting equation (15.44) in equation (15.42), we get

$$P_1 + P_2 = I_R V_R + I_B V_B - V_Y (-I_Y) \quad ...(15.45)$$

$$P_1 + P_2 = I_R V_R + I_Y V_Y + I_B V_B \quad ...(15.46)$$

or $$P_1 + P_2 = W_1 + W_2 = P \quad ...(15.47)$$

From the above equation, (15.47) it is clear that in case of 3-phase star connected load; two-wattmeter method can measure the instantaneous power whether the load is balanced or unbalanced.

15.7 ANALYSIS OF TWO-WATTMETER METHOD

Assume that in case of 3-phase balanced star network, the load is inductive and the phase current (I_{ph}) is lagging behind the phase voltage (V_{ph}) by an angle ϕ. This is shown in the phasor diagram of Fig. 15.15.

The line current (I_L) is lagging behind the line voltage (V_L) by an angle θ_1, which is determined as follow:

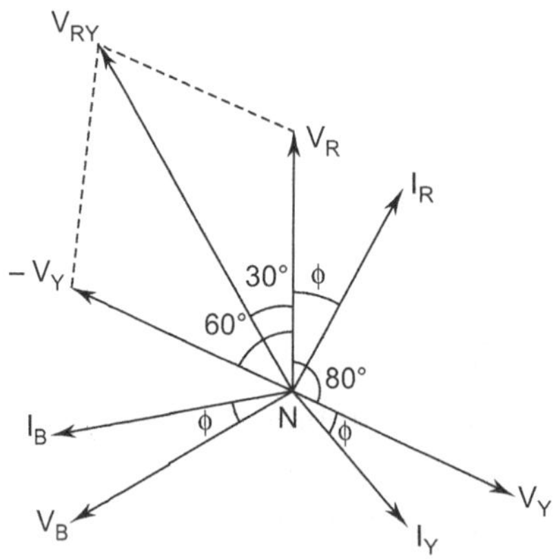

Fig. 15.15

From the above phasor diagram shown in Fig. 15.15,

Phase angle between V_R and $V_Y = 120^\circ$

Phase angle between V_R and $-V_Y = 60^\circ$

Phase angle between V_R and $V_{RY} = 30^\circ$

Phase angle between V_R and $I_R = \phi$

Hence, Phase angle between I_R and V_{RY}, $(\theta_1) = (30^\circ + \phi)$

Current through wattmeter $W_1 = I_R$

Voltage across the voltage coil of wattmeter $W_1 = V_{RY}$

Reading of wattmeter W_1 will be

$$W_1 = I_R V_{RY} \cos(30^\circ + \phi) \quad (15.48)$$

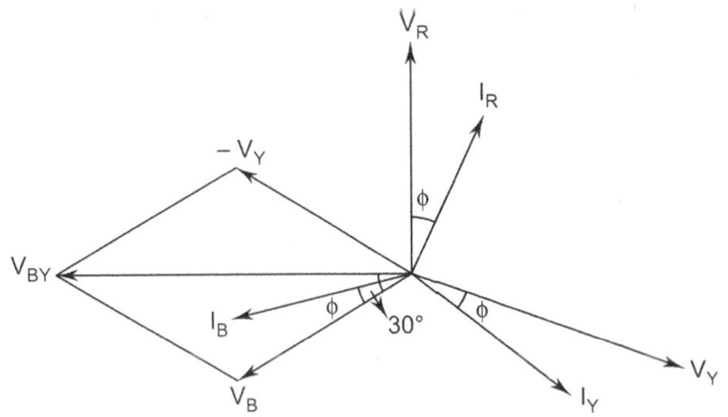

Fig. 15.16

The line current (I_L) is lagging behind the line voltage (V_L) by an angle θ_2, which is determined as follow. From the above phasor diagram shown in Fig. 15.16,

Phase angle between V_B and $V_Y = 120^\circ$

Phase angle between V_B and $-V_Y = 60^\circ$

Phase angle between V_B and $V_{BY} = 30^\circ$

Phase angle between V_B and $I_B = \phi$

Hence, Phase angle between I_B and V_{BY} (θ_2) = $(30^\circ - \phi)$

Current through Wattmeter $W_2 = I_B$

Voltage across the voltage coil of Wattmeter $W_2 = V_{BY}$

Reading of Wattmeter W_2 will be

$$W_2 = I_B V_{BY} \cos(30^\circ - \phi) \quad \text{...(15.49)}$$

Adding equations (15.48) and (15.49), we get

$$W_1 + W_2 = I_R V_{RY} \cos(30^\circ + \phi) + I_B V_{BY} \cos(30^\circ - \phi) \quad \text{...(15.50)}$$

As $$I_R = I_Y = I_B = I_{ph} = I_L \quad \text{...(15.51)}$$

and $$V_{RY} = V_{YB} = V_{BR} = V_L \quad \text{...(15.52)}$$

$\therefore$ $$W_1 + W_2 = I_L V_L \cos(30^\circ + \phi) + I_L V_L \cos(30^\circ - \phi) \quad \text{...(15.53)}$$

We will consider $V_{BY} = V_L$ (*Line voltage*) because here we are concern only with the magnitude and the sign of the voltage will be decided by the power factor, $\cos(30^\circ - \phi)$.

Hence equation (15.53) will become

or $$W_1 + W_2 = I_L V_L \left[\cos(30^\circ + \phi) + \cos(30^\circ - \phi)\right] \quad \text{...(15.54)}$$

or $$W_1 + W_2 = I_L V_L \left[\cos(30^\circ)\cos\phi - \sin(30^\circ)\sin\phi + \cos(30^\circ)\cos\phi + \sin(30^\circ)\sin\phi\right] \quad \text{...(15.55)}$$

or $$W_1 + W_2 = I_L V_L \left[\sqrt{3}\cos\phi\right] \quad \text{...(15.56)}$$

or $$W_1 + W_2 = \sqrt{3}\, I_L V_L \cos\phi \quad \text{...(15.57)}$$

From equations (15.48) and (15.49), it is clear from the power factors that smaller will be the angle bigger will be the wattmeter reading. Here W_2 will have smaller angle than W_1 hence wattmeter W_2 will read higher reading than wattmeter W_1. Hence that value will be taken first which is greater than other. So subtracting equations (15.49) and (15.48), we get

$$W_2 - W_1 = I_1 V_{12} \cos(30° - \phi) - I_3 V_{32} \cos(30° + \phi) \quad ...(15.58)$$

or $$W_2 - W_1 = I_L V_L \cos(30° - \phi) - I_L V_L \cos(30° + \phi) \quad ...(15.59)$$

or $$W_2 - W_1 = I_L V_L \left[\cos(30° - \phi) - \cos(30° + \phi)\right] \quad ...(15.60)$$

or $$W_2 - W_1 = I_L V_L \left[\cos(30°)\cos\phi + \sin(30°)\sin\phi - \cos(30°)\cos\phi + \sin(30°)\sin\phi\right] \quad ...(15.61)$$

or $$W_2 - W_1 = I_L V_L \left[2\sin 30° \sin\phi\right] \quad ...(15.62)$$

or $$W_2 - W_1 = V_L I_L \sin\phi \quad ...(15.63)$$

Dividing equation (15.63) by equation (15.57), we get

$$\frac{W_2 - W_1}{W_2 + W_1} = \frac{V_L I_L \sin\phi}{\sqrt{3} V_L I_L \cos\phi} \quad ...(15.64)$$

or $$\frac{W_2 - W_1}{W_2 + W_1} = \frac{1}{\sqrt{3}} \tan\phi \quad ...(15.65)$$

or $$\tan\phi = \frac{\sqrt{3}(W_2 - W_1)}{W_2 + W_1} \quad ...(15.66)$$

From equation (15.66) the phase angle between the line voltage and line current can be calculated and hence the power factor, cos ϕ can be obtained.

In general the above expression is given by

or $$\boxed{\tan\phi = \frac{\sqrt{3}(W_1 - W_2)}{W_1 + W_2}} \quad ...(15.67)$$

Note that, while using equation (15.67) in numerical problems, higher value should be taken as W_1 and the lower value should be taken as W_2.

15.8 VARIATION IN WATTMETER READINGS DUE TO DIFFERENT POWER FACTORS

The power factor of a circuit mainly depends upon nature of load. The load may be inductive, capacitive or resistive. So will be the nature of power factor.

As the power factor of the circuit changes, the reading of the wattmeters, used for power measurement, also changes.

Now we will discuss the change in wattmeter readings due to different power factor.

Case I. When power factor of the load is unity (cos ϕ = 1)

$$\cos\phi = 1 \quad ...(15.68)$$

or $$\phi = 0° \quad ...(15.69)$$

From equation (15.48)

$$W_1 = I_L V_L \cos(30° + \phi) = I_L V_L \cos(30° + 0) \quad ...(15.70)$$

or

$$W_1 = \frac{\sqrt{3}}{2} I_L V_L \quad ...(15.71)$$

And from equation (15.49)

$$W_2 = I_L V_L \cos(30° - \phi) = I_L V_L \cos(30° - 0) \quad ...(15.72)$$

or

$$W_2 = \frac{\sqrt{3}}{2} I_L V_L \quad ...(15.73)$$

It is clear from equations (15.71) and (15.73) that when the power factor of the load is ***unity***, then

- Both wattmeters read same
- Both wattmeters read positive
- Both wattmeters read half the total power

Case II. When power factor of the load is 0.5 (ϕ = 60°)

$$\cos\phi = 0.5 \quad ...(15.74)$$

or

$$\phi = 60° \quad ...(15.75)$$

From equation (15.48)

$$W_1 = I_L V_L \cos(30° + \phi) = I_L V_L \cos(30° + 60°) \quad ...(15.76)$$

or

$$W_1 = 0 \quad ...(15.77)$$

And from equation (15.49)

$$W_2 = I_L V_L \cos(30° - \phi) = I_L V_L \cos(30° - 60°) \quad ...(15.78)$$

or

$$W_2 = I_L V_L \cos(-30°) \quad ...(15.79)$$

or

$$W_2 = \frac{\sqrt{3}}{2} I_L V_L \quad ...(15.80)$$

It is clear from equations (15.77) and (15.80) that when the power factor of the load is 0.5, then

- Only one wattmeter will give reading.
- Other wattmeter read Zero.
- Total power of the load is given by the reading giving wattmeter

Case III. When power factor of the load is less than 0.5 (ϕ > 60°)

$$90° > \phi > 60°$$

or

$$0 < \text{P.F.} < 0.5$$

Taking an example of $\phi = 75°$

From equation (15.48)

$$W_1 = I_L V_L \cos(30° + \phi) = I_L V_L \cos(30° + 75°) \quad ...(15.81)$$

or

$$W_1 = I_L V_L \cos(105°) = I_L V_L \cos(90° + 15°) \quad ...(15.82)$$

or

$$W_1 = -I_L V_L \sin(15°) \quad ...(15.83)$$

And from equation (15.49)

$$W_2 = I_L V_L \cos(30° - \phi) = I_L V_L \cos(30° - 75°) \quad ...(15.84)$$

or

$$W_2 = I_L V_L \cos(-45°) \quad ...(15.85)$$

or

$$W_2 = I_L V_L \cos(45°) \quad ...(15.86)$$

It is clear from equations (15.83) and (15.86) that when the power factor of the load is less than 0.5, then

- One wattmeter will reads positive
- Other wattmeter read negative
- In such a case the pressure coil terminals or current coil terminals of the wattmeter (which is giving negative reading) must be reversed to obtain the reading
- So the total power will be

$$P = -W_1 + W_2$$

or $$P = W_2 - W_1 \quad ...(15.87)$$

Case IV. When power factor of the load is Zero (0) (ϕ = 90°)

$$\phi = 90^o$$

or $$\text{P.F.} = \cos\phi = 0$$

From equation (15.48)

$$W_1 = I_L V_L \cos(30° + \phi) = I_L V_L \cos(30° + 90°) \quad ...(15.88)$$

or $$W_1 = I_L V_L \cos(120°) = I_L V_L \cos(90° + 30°) \quad ...(15.89)$$

or $$W_1 = -I_L V_L \sin(30°) \quad ...(15.90)$$

or $$W_1 = -\frac{1}{2} I_L V_L \quad ...(15.91)$$

And from equation (15.49)

$$W_2 = I_L V_L \cos(30° - \phi) = I_L V_L \cos(30° - 90°) \quad ...(15.92)$$

or $$W_2 = I_L V_L \cos(-60°) \quad ...(15.93)$$

or $$W_2 = I_L V_L \cos(60°) \quad ...(15.94)$$

or $$W_2 = \frac{1}{2} I_L V_L \quad ...(15.95)$$

It is clear from equations (15.91) and (15.95) that when the power factor of the load is *zero (0)* then

- Both wattmeter will read equal
- One wattmeter will reads positive
- Other wattmeter read negative.
- In such a case the pressure coil terminals or current coil terminals of the wattmeter (which is giving negative reading) must be reversed to obtain the reading.
- So the total power will be

$$P = -W_1 + W_2 \quad ...(15.96)$$

As $$W_1 + W_2 = 0$$

Therefore, $$P = 0 \quad ...(15.97)$$

15.9 DELTA OR MESH (Δ) CONNECTION

Let us once again consider three isolated phases R, Y and B.

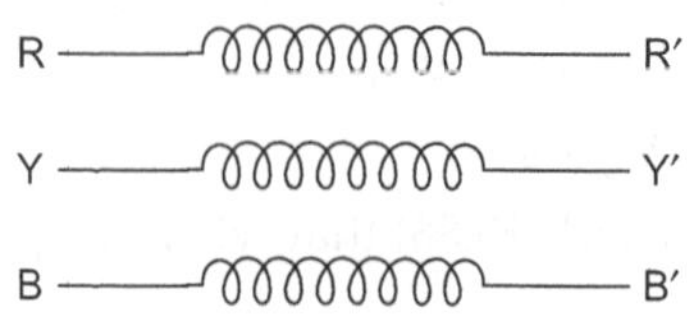

Fig. 15.17

In case of a three-phase network, when starting end of one coil is connected to the finishing end of other coil, delta configuration is obtained.

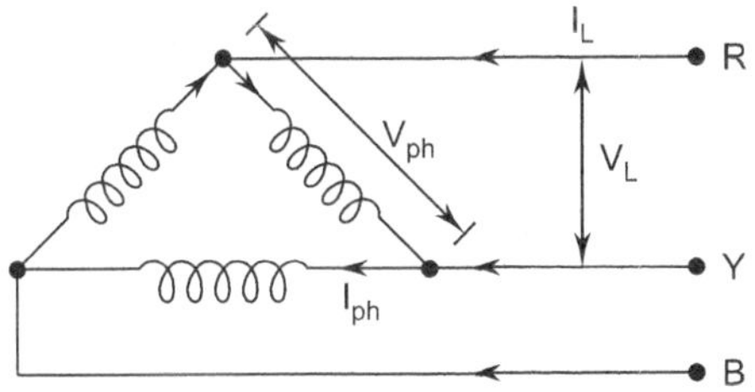

Fig. 15.18

The voltage induced in each winding is known as phase voltage

$$\text{Phase voltage} = V_{RY} = V_{YB} = V_{BR} = V_{ph} \quad \ldots(15.98)$$

The current flowing in each winding is known as phase current

$$\text{Phase current} = I_{RY} = I_{YB} = I_{BR} = I_{ph} \quad \ldots(15.99)$$

The voltage available between any pair of terminals (or Lines) is known as line voltage

$$\text{Line voltage} = V_{RY} = V_{YB} = V_{BR} = V_L \quad \ldots(15.100)$$

The current between any pair of terminals (or Lines) is known as line current

$$\text{Line current} = I_R = I_Y = I_B = I_L \quad \ldots(15.101)$$

15.9.1 Line Current and Phase Current Relation

Consider a delta connected load as shown in Fig. 15.19.

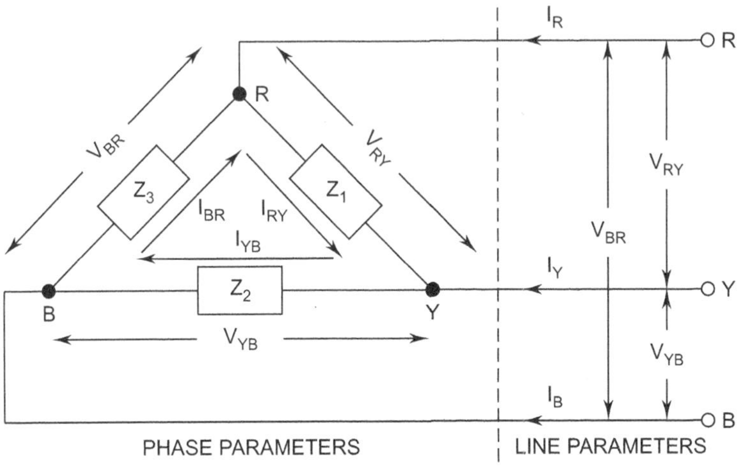

Fig. 15.19

Inside a delta (or a closed circuit) the currents are of circulating in nature. Here

I_{RY} is the current through the impedance Z_1 connected between phases R and Y. The direction of this current is from R to Y.

I_{YB} is the current through the impedance Z_2 connected between phases Y and B. The direction of this current is from Y to B.

I_{BR} is the current through the impedance Z_3 connected between phases B and R. The direction of this current is from B to R.

Applying Kirchhoff's current law at node R, we have

$$I_R + I_{BR} = I_{RY} \quad ...(15.102)$$

or $$I_R = I_{RY} - I_{BR} \quad ...(15.103)$$

or $$I_R = I_{RY} + (-I_{BR}) \quad ...(15.104)$$

Applying Kirchhoff's current law at node Y, we have

$$I_Y + I_{RY} = I_{YB} \quad ...(15.105)$$

or $$I_Y = I_{YB} - I_{RY} \quad ...(15.106)$$

or $$I_Y = I_{YB} + (-I_{RY}) \quad ...(15.107)$$

Applying Kirchhoff's current law at node B, we have

$$I_B + I_{YB} = I_{BR} \quad ...(15.108)$$

or $$I_B = I_{BR} - I_{YB} \quad ...(15.109)$$

or $$I_B = I_{BR} + (-I_{YB}) \quad ...(15.110)$$

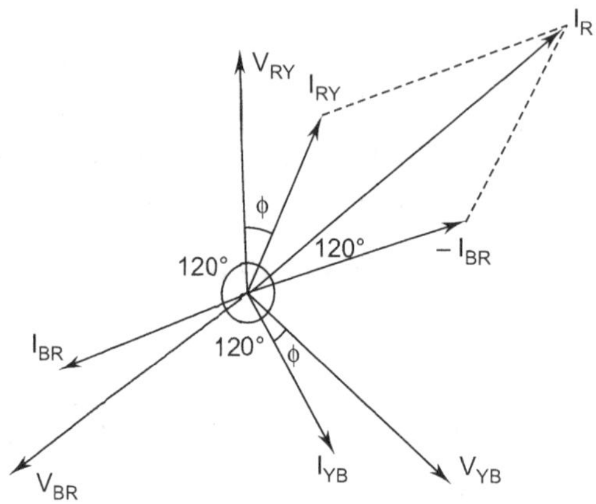

Fig. 15.20

By law of parallogram

$$|I_R| = \sqrt{|I_{RY}|^2 + |-I_{BR}|^2 + 2|I_{RY}||-I_{BR}|\cos 60^\circ} \quad ...(15.111)$$

as $$I_{RY} = I_{YB} = I_{BR} = I_{ph} \quad \text{[Eqn. (15.99)]}$$

and $$I_R = I_Y = I_B = I_L \quad \text{[Eqn. (15.101)]}$$

Hence equation (15.105) will become

$$|I_L| = \sqrt{|I_{ph}|^2 + |I_{ph}|^2 + 2|I_{ph}|^2 |I_{ph}|^2 \cos 60^\circ} \quad ...(15.112)$$

or $$|I_L| = \sqrt{2I_{ph}^2 + 2I_{ph}^2 \frac{1}{2}} \quad ...(15.113)$$

or $$I_L = \sqrt{3\,I_{ph}^2} \quad ...(15.114)$$

Hence $$I_L = \sqrt{3}\,I_{ph} \quad ...(15.115)$$

or $$\boxed{\textit{Line current} = \sqrt{3}\ \textit{Phase current}}$$ ...(15.116)

15.9.2 Line Voltage and Phase Voltage Relation

It is clear from Fig. 15.20 that the voltage across the three-phases *i.e.*, phase voltages are equal to the line voltages. Hence in case of Δ-connected load

$$\overline{\overrightarrow{V_{RY}}} = \overline{\overrightarrow{V_{YB}}} = \overline{\overrightarrow{V_{BR}}} = \overline{\overrightarrow{V_{ph}}} = V_L$$ ...(15.117)

or $$V_L = V_{ph}$$ ...(15.118)

or $$\boxed{\textit{Line voltage} = \textit{Phase voltage}}$$ ...(15.119)

15.9.3 Power Expressions

Assuming that in case of 3-phase balanced delta network, the load is inductive and the current is lagging behind the voltage (phase-to-phase) by an angle ϕ.

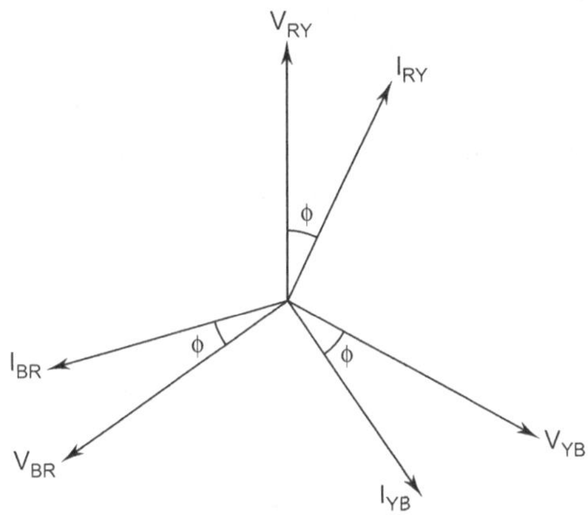

Fig. 15.21

Active power in each phase is given by

$$P = V_{ph}I_{ph} \cos \phi$$ ...(15.120)

And active power in 3-phase is given by

$$P = 3V_{ph}I_{ph} \cos \phi$$ ...(15.121)

This is the expression for active power in 3-phase in term of phase values. From equations (15.115) and (15.118), equation (15.121) will become

$$P = 3V_L \frac{I_L}{\sqrt{3}} \cos\phi$$ ...(15.122)

or $$P = \sqrt{3}\ V_L I_L \cos\phi$$ ...(15.123)

Note. ϕ is the angle between phase voltage and phase current and not between line voltage and line current.

15.10 POWER MEASUREMENT IN DELTA CONNECTED LOAD

Similar to the star connected 3-phase load, in case of delta connected 3-phase circuits also, power can be measured by following three methods:

- One wattmeter method
- Two wattmeter method
- Three wattmeter method

Out of these three methods, one wattmeter method can be used only for a balanced star connected load. It cannot be used for an unbalanced load.

Where as two wattmeter method and three wattmeter method can measure power in both balanced as well as unbalanced connected load.

15.10.1 One Wattmeter Method

There are two configurations for measuring power in a three-phase delta connected load.

In first configuration (Fig. 15.22), the current coil of wattmeter is connected in any one line and the pressure coil is connected alternatively to the other two lines. The two readings so obtained from the wattmeter are corresponds to those obtained by normal two-wattmeter method.

In second configuration, two similar resistances equivalent to the resistance of pressure coil are connected in each of the remaining two phases as shown in Fig. 15.23.

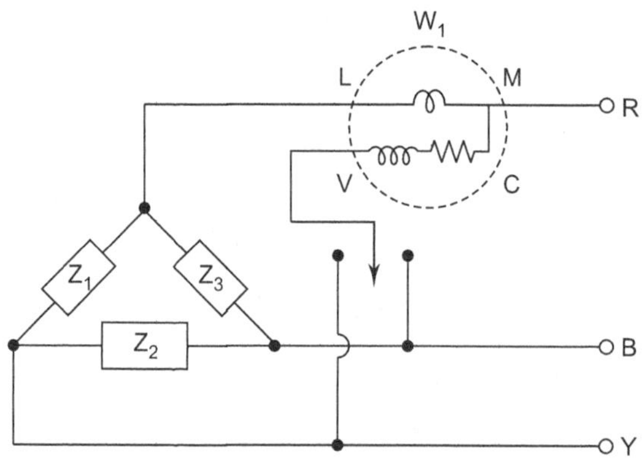

Fig. 15.22

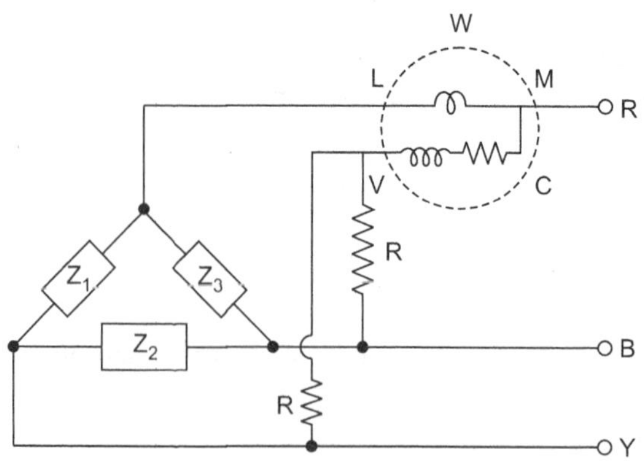

Fig. 15.23

The two resistances connected in each phase and the wattmeter resistance forms a balanced star connection.

In this way the power can be measured in a three-phase delta connected load with one wattmeter but it is not a convenient method for unbalanced loads.

15.10.2 Three Wattmeters Method

In case of three wattmeter method, the current coils of all the three wattmeter are inserted in series with the load and the voltage coils are connected at the junction of the two coils (Fig. 15.24).

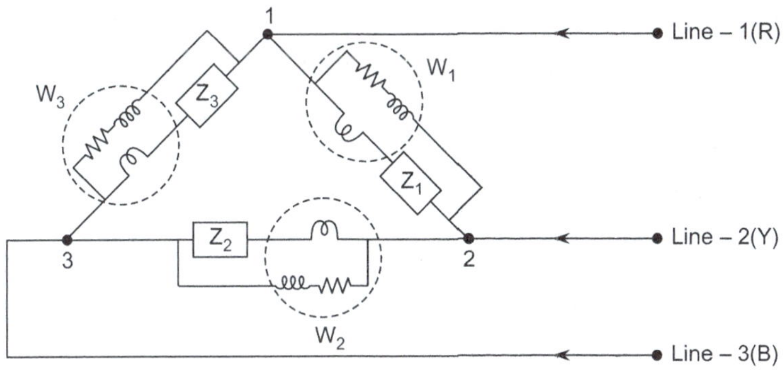

Fig. 15.24

The three wattmeter will then provide the total power of the load as

$$P = W_1 + W_2 + W_3 \qquad \text{...(15.124)}$$

The difficulty with this method is that under normal conditions it is not generally feasible to break into the phases of a delta connected load.

15.10.3 Two Wattmeters Method

The current coils of the two wattmeters are connected to any two lines of the network. The voltage or pressure coils of the two wattmeters are connected to the third line as shown in Fig. 15.25.

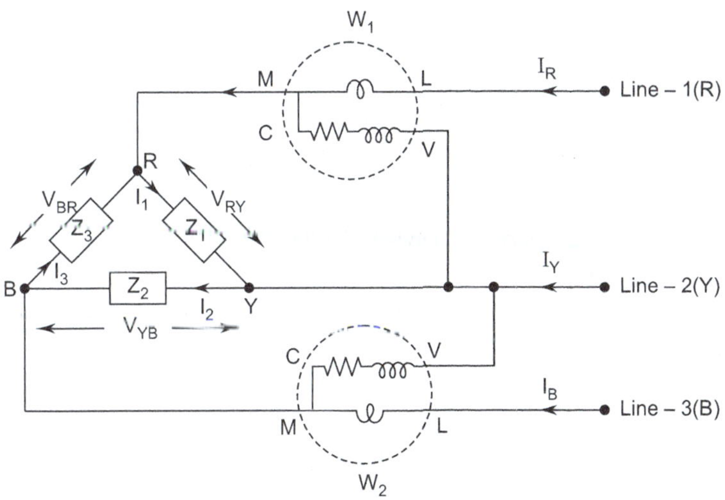

Fig. 15.25

While measuring the Power, we will consider the instantaneous values. Power is mainly consumed by the load. That power is in terms of phase values. According to the circuit diagram shown in Fig. 15.14, let

Current through load impedance, $Z_{12} = I_{RY}$
Current through load impedance, $Z_{23} = I_{YB}$
Current through load impedance, $Z_{23} = I_{BR}$
Voltage across load impedance, $Z_{12} = V_{RY}$
Voltage across load impedance, $Z_{23} = V_{YB}$
Voltage across load impedance, $Z_{23} = V_{BR}$
Total instantaneous power in the load

$$P = V_{RY}I_{RY} + V_{YB}I_{YB} + V_{BR}I_{BR} \quad \text{...(15.125)}$$

For Wattmeter W_1

Current through wattmeter, $W_1 = I_R$
Voltage across the voltage coil of wattmeter, $W_1 = V_{RY}$
Power measured by wattmeter W_1

$$P_1 = I_R \times V_{RY} \quad \text{...(15.126)}$$

Recalling equation (15.103) we have

$$I_R = I_{RY} - I_{BR} \quad \text{[Eqn. (15.103)]}$$

From equation (15.103), equation (15.126) will become

$$P_1 = V_{RY}(I_{RY} - I_{BR}) \quad \text{...(15.127)}$$

For Wattmeter W_2

Current through wattmeter, $W_2 = I_B$
Voltage across the voltage coil of wattmeter, $W_2 = V_{BY}$
Power measured by wattmeter W_2

$$P_2 = I_B \times V_{BY} \quad \text{...(15.128)}$$

Recalling equation (15.109) we have

$$I_B = I_{BR} - I_{YB} \quad \text{[Eqn. (15.109)]}$$

From equation (15.109), equation (15.128) will become

$$P_2 = V_{BY}(I_{BR} - I_{YB}) \quad \text{...(15.129)}$$

Total instantaneous power measured by the two-wattmeters can be obtained by adding equations (15.127) and (15.129)

$$P_1 + P_2 = V_{RY}(I_{RY} - I_{BR}) + V_{BY}(I_{BR} - I_{YB}) \quad \text{...(15.130)}$$

or

$$P_1 + P_2 = V_{RY}(I_{RY} - I_{BR}) - V_{YB}(I_{BR} - I_{YB}) \quad \text{...(15.131)}$$

(as $V_{BY} = -V_{YB}$)

or

$$P_1 + P_2 = V_{RY}I_{RY} - V_{RY}I_{BR} - V_{YB}I_{BR} + V_{YB}I_{YB} \quad \text{...(15.132)}$$

or

$$P_1 + P_2 = V_{RY}I_{RY} + V_{YB}I_{YB} - I_{BR}(V_{RY} + V_{YB}) \quad \text{...(15.133)}$$

As we know that, in a closed path, according to KVL, the sum of all the voltage drops and all the e.m.fs is zero, *i.e.*,

$$V_{RY} + V_{YB} + V_{BR} = 0 \quad \text{...(15.134)}$$

or

$$V_{RY} + V_{YB} = -V_{BR} \quad \text{...(15.135)}$$

From equation (15.135), equation (15.133) will become

$$P_1 + P_2 = V_{RY}I_{RY} + V_{YB}I_{YB} - I_{BR}(-V_{BR}) \quad ...(15.136)$$

or
$$P_1 + P_2 = V_{RY}I_{RY} + V_{YB}I_{YB} + I_{BR}V_{BR} \quad ...(15.137)$$

or
$$P_1 + P_2 = W_1 + W_2 = P \quad ...(15.138)$$

From equation, (15.138) it is clear that in case of 3-phase delta connected load; two-wattmeter method can measure the instantaneous power whether the load is balanced or unbalanced.

SOLVED PROBLEMS

Problem 15.1. *Each phase of a star-connected load consists of a resistance of 100 Ω in parallel with a capacitance 31.8 μF. Calculate the line current, power absorbed, total kVA and power factor of the load when connected to a 416 V, 3-φ, 4 wire 50 Hz supply.*

Solution. Phase voltage, $V_{ph} = \frac{V_L}{\sqrt{3}} = \frac{416}{\sqrt{3}} = 240.2 \text{ V}$

Taking phase voltage as reference phasor, $V_{ph} = 240.2 \angle 0° \text{ V}$

Admittance per phase, $Y_{ph} = \frac{1}{R} + j\omega C = \frac{1}{100} + j2\pi \times 50 \times 31.8 \times 10^{-6}$

$= (0.01 + j0.01) \text{ or } 0.01414 \angle 45° \text{ S}$

Line current, $I_L = I$

$= V_{ph} Y_{ph} = 240.2 \angle 0° \times 0.01414 \angle 45°$

$= 3.397 \angle 45° \text{ A}$

So line current, $I_L = \mathbf{3.397 \text{ A}}$ **Ans.**

Power factor, $\cos\phi = \cos 45° = \mathbf{0.707 \text{ (leading)}}$ **Ans.**

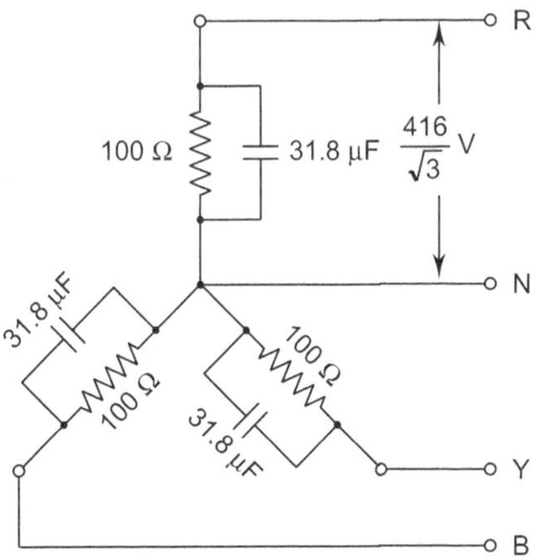

Fig. 15.26

Power absorbed, $P = \sqrt{3} V_L I_L \cos\phi = \sqrt{3} \times 416 \times 3.397 \times 0.707$

$= \mathbf{1,730 \text{ Watts or } 1.73 \text{ kW}}$ **Ans.**

Total kVA, $S = \frac{\sqrt{3} V_L I_L}{1,000} = \frac{\sqrt{3} \times 416 \times 3.397}{1,000} = \mathbf{2.45 \text{ kVA}}$ **Ans.**

Problem 15.2. *A balanced 3-phase star-connected load of 120 kW takes a leading current of 85 amp., when connected across a 3-phase 1,100 volts, 50 Hz supply. Obtain the values of the resistance, impedance and capacitance of the load per phase and also calculate the power factor of the load.*

Solution. Supply voltage, $V_L = 1{,}100$ V

Supply current, $I_L = 85$ A

Power supplied, $P = 120$ kW or 1,20,000 W

Power factor of the load, $$\cos\phi = \frac{P}{\sqrt{3}\,V_L I_L}$$

$$= \frac{1{,}20{,}000}{\sqrt{3}\times 1{,}100\times 85} = \mathbf{0.741\ (leading)} \quad \textbf{Ans.}$$

Impedance of the load per phase, $$Z = \frac{V_{ph}}{I_{ph}} = \frac{V_L/\sqrt{3}}{I_L}$$

$$= \frac{1{,}100/\sqrt{3}}{85} = \mathbf{7.472\ \Omega} \quad \textbf{Ans.}$$

Resistance of the load per phase, $R = Z\cos\phi$

$$= 7.472 \times 0.741 = \mathbf{5.536\ \Omega} \quad \textbf{Ans.}$$

Reactance of the load per phase, $X_C = \sqrt{Z^2 - R^2}$

$$= \sqrt{(7.472)^2 - (5.536)^2} = 5.02\ \Omega$$

Capacitance of the load per phase, $$C = \frac{1}{2\pi f X_C}$$

$$= \frac{1}{2\pi \times 50 \times 5.02} = \mathbf{634.3\ \mu F.} \quad \textbf{Ans.}$$

Problem 15.3. *A 3-ϕ, 3-wire Y-connected system has a voltage of 150 V between two lines. Each phase has impedance of $Z = 5\angle 30°\ \Omega$. Find :*

(i) Current in each phase

(ii) Total power

(iii) Draw phasor diagram.

Solution. Line voltage, $V_L = 150$ V

Phase voltage, $$V_{ph} = \frac{V_L}{\sqrt{3}} = \frac{150}{\sqrt{3}} = 86.6\ \text{V}$$

(*i*) Current in each phase, $$I_{ph} = \frac{V_{ph}}{Z_{ph}} = \frac{86.6}{5} = \mathbf{17.32\ A} \quad \textbf{Ans.}$$

Power factor of the circuit, $\cos\phi$ = cosine of phase angle

$= \cos 30° = 0.866$ (leading)

(*ii*) Total power,
$$P = 3\,V_{ph}\,I_{ph}\cos\phi$$
$$= 3 \times 86.6 \times 17.32 \times 0.866$$
$$= \mathbf{3{,}897\ W\ or\ 3.897\ kW}$$ **Ans.**

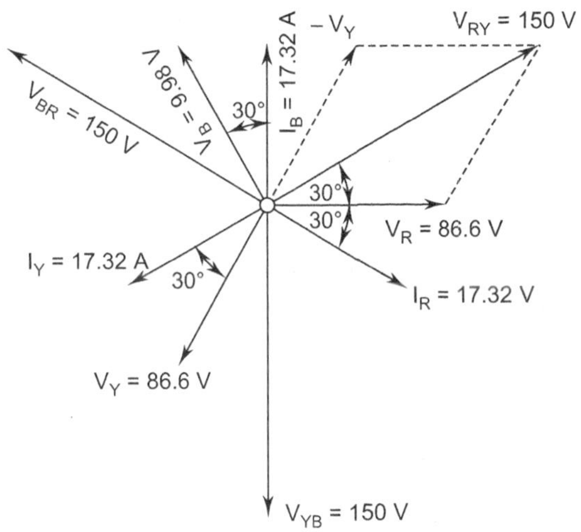

Fig. 15.27

(*iii*) Phasor diagram is shown in Fig. 15.27.

Problem 15.4. *A balanced 3-phase star-connected load of 18 kW taking a leading current of 60 amperes when connected across a 3-phase 440 V, 50 Hz supply. Find the values and nature of the load.*

Solution. Load supplied, $P = 18\text{ kW} = 18{,}000\text{ W}$

Line voltage, $V_L = 440\text{ V}$

Line current, $I_L = 60\text{ A}$

Phase voltage, $V_{ph} = \dfrac{V_L}{\sqrt{3}} = \dfrac{440}{\sqrt{3}} = 254\text{ V}$

Phase current, $I_{ph} = I_L = 60\text{ A}$

Load impedance, $Z = \dfrac{V_{ph}}{I_{ph}} = \dfrac{254}{60} = 4.234\ \Omega.$

Load power factor, $\cos\phi = \dfrac{P}{\sqrt{3}\,V_L\,I_L}$

$$= \frac{18{,}000}{\sqrt{3} \times 440 \times 60} = \mathbf{0.39365\ (leading)}$$ **Ans.**

(∵ Current is leading)

Resistance of load, $R = Z\cos\phi = 4.234 \times 0.39365 = \mathbf{1.6667\ \Omega}$ **Ans.**

Reactance of load, $X = Z\sin\phi = 4.234\sqrt{1-(0.39365)^2}$

$= 4.234 \times 0.9193 = \mathbf{3.89242\ \Omega}$ (capacitive) **Ans.**

(∵ Current is leading)

Problem 15.5. *If the phase voltage of a 3-phase star-connected system is 200 V, what will be the line voltages (a) when the phases are correctly connected, (b) when connections to one of phases are reversed ?*

Solution.

(*a*) When the phases are correctly connected *i.e.*, similar ends are connected together, as shown in Fig. 15.28 (*a*) all line voltages will be equal in magnitude and will be given as

Each line voltage, $V_L = \sqrt{3}\ V_{ph} = \sqrt{3} \times 200 = \mathbf{346.41\ V}$ **Ans.**

(*b*) When the connections to one of the phases are reversed, as shown in Fig. 15.28 (*a*), voltage in phase R is in reverse direction. The voltage phasor diagram for such a case is shown in Fig. 15.28 (*b*).

It should be noted that V_R has been drawn in reverse direction, so the angles between the three-phase voltages are of 60°, as shown in Fig. 15.28 (*b*).

So,
$$V_{RY} = V_R - V_Y$$
$$= 2\ V_{ph} \cos 60° = V_{ph} = \mathbf{200\ V}$$ **Ans.**
$$V_{YB} = V_Y - V_B$$
$$= 2V_{ph} \times \cos 30°$$

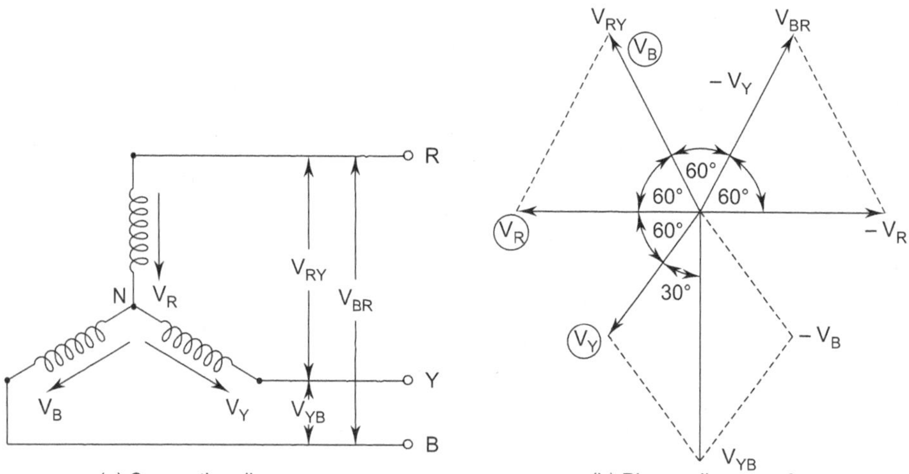

(a) Connection diagram

(b) Phasor diagram of line and phase voltage

Fig. 15.28

$$= 2 \times 200 \times \frac{\sqrt{3}}{2} = \mathbf{346.41\ V}$$ **Ans.**

and
$$V_{BR} = V_B - V_R = 2\ V_{ph} \cos 60° = \mathbf{200\ V}$$ **Ans.**

Problem 15.6. *A 3-phase, 400 V supply is connected to a 3-phase star-connected balanced load. The line current is 20 A and the power consumed by the load is 12 kW. Calculate the impedance of the load, phase current and power factor.*

Solution. Load supplied, $P = 12\ kW = 12{,}000\ W$

Line voltage, $V_L = 400\ V$

Line current, $I_L = 20\ A$

Phase voltage, $V_{ph} = \dfrac{V_L}{\sqrt{3}} = \dfrac{400}{\sqrt{3}} = 230.94 \text{ V}$

Phase current, $I_{ph} = I_L = \mathbf{20\ A}$ **Ans.**

Load impedance, $Z_L = \dfrac{V_{ph}}{I_{ph}} = \dfrac{230.94}{20} = \mathbf{11.547\ \Omega}$ **Ans.**

Load power factor, $\cos\phi = \dfrac{P}{\sqrt{3}\, V_L\, I_L} = \dfrac{12{,}000}{\sqrt{3} \times 400 \times 20} = \mathbf{0.866}$ **Ans.**

Problem 15.7. *Three similar coils each having a resistance of 8 ohm and an inductance of 0.0191 H in series in each phase is connected across a 400 V, three-phase, 50 Hz supply. Calculate the line current, power input, kVA and kVAR taken by the load.*

Solution. Resistance of coil, $R = 8\ \Omega$

Inductive reactance of the coil, $X_L = 2\pi f L = 2\pi \times 50 \times 0.0191 = 6\ \Omega$

Impedance of each coil, $Z = \sqrt{R^2 + X_L^2} = \sqrt{(8)^2 + (6)^2} = 10\ \Omega$

Power factor of coil, $\cos\phi = \dfrac{R}{Z} = \dfrac{8}{10} = 0.8$ (lagging) $(\because$ Circuit is inductive)

When coils are connected in star

Line current, $I_L = \text{Phase current}, I_{ph} = \dfrac{V_L/\sqrt{3}}{Z} = \dfrac{400}{\sqrt{3} \times 10} = 23.094 \text{ A}$

Power input, $P = \dfrac{\sqrt{3}\, V_L\, I_L \cos\phi}{1{,}000} = \dfrac{\sqrt{3} \times 400 \times 23.094 \times 0.8}{1{,}000} = \mathbf{12.8\ kW}$ **Ans.**

kVA taken by load $= \dfrac{\sqrt{3}\, V_L\, I_L}{1{,}000} = \dfrac{\sqrt{3} \times 400 \times 23.094}{1{,}000} = \mathbf{16\ kVA}$ **Ans.**

kVAR taken by load $= \dfrac{\sqrt{3}\, V_L\, I_L \sin\phi}{1{,}000}$

$= \dfrac{\sqrt{3} \times 400 \times 23.094 \times 0.6}{1{,}000} = \mathbf{9.6\ kVAR}$ **Ans.**

When coils are connected in delta

Line current, $I_L = \sqrt{3} \times \text{phase current} = \dfrac{\sqrt{3}\, V_L}{Z} = \dfrac{\sqrt{3} \times 400}{10} = \mathbf{69.282\ A}$ **Ans.**

Power input, $P = \dfrac{\sqrt{3}\, V_L\, I_L \cos\phi}{1{,}000} = \dfrac{\sqrt{3} \times 400 \times 69.282 \times 0.8}{1{,}000} = \mathbf{38.4\ kW}$ **Ans.**

kVA taken by load $= \dfrac{\sqrt{3}\, V_L\, I_L}{1{,}000} = \dfrac{\sqrt{3} \times 400 \times 69.282}{1{,}000} = \mathbf{48\ kVA}$ **Ans.**

kVAR taken by load $= \dfrac{\sqrt{3}\, V_L\, I_L \sin\phi}{1{,}000} = \dfrac{\sqrt{3} \times 400 \times 69.282 \times 0.6}{1{,}000} = \mathbf{28.8\ kVAR}$ **Ans.**

Problem 15.8. *A balanced star-connected load of (8 + j6) Ω per phase is connected to a balanced 3-phase, 400 V supply. Find the line current, power factor, power and total volt-amperes.*

Solution. Phase voltage, $V_{ph} = \dfrac{V_L}{\sqrt{3}} = \dfrac{400}{\sqrt{3}} = 231\ V$

Impedance per phase, $Z_{ph} = \sqrt{R_{ph}^2 + X_{ph}^2} = \sqrt{(8)^2 + (6)^2} = 10\ \Omega$

Phase current, $I_{ph} = \dfrac{V_{ph}}{Z_{ph}} = \dfrac{231}{10} = 23.1\ A$

Line current, $I_L = I_{ph} = \mathbf{23.1\ A}$ **Ans.**

Power factor, $\cos\phi = \dfrac{R_{ph}}{Z_{ph}} = \dfrac{8}{10} = \mathbf{0.8\ (lagging)}$ **Ans.**

Total power, $P = \sqrt{3}\, V_L\, I_L \cos\phi = \sqrt{3} \times 400 \times 23.1 \times 0.8 = \mathbf{12{,}800\ W}$ **Ans.**

Total volt-amperes, $S = \sqrt{3}\, V_L\, I_L = \sqrt{3} \times 400 \times 23.1 = \mathbf{16{,}000\ VA}$ **Ans.**

Also, total power, $P = 3\, I_{ph}^2\, R_{ph} = 3 \times (23.1)^2 \times 8 = \mathbf{12{,}800\ W.}$ **Ans.**

Problem 15.9. *A balanced Δ-connected load of (8 + j6) ohm per phase is connected to 3-phase 230 V supply. Find the line current.*

Solution. Impedance in each phase, $Z_{ph} = \sqrt{(8)^2 + (6)^2} = 10\ \Omega$

Phase current, $I_{ph} = \dfrac{V_{ph}}{Z_{ph}} = \dfrac{V_L}{Z_{ph}} = \dfrac{230}{10} = 23\ A$

Line current, $I_L = \sqrt{3}\, I_{ph} = \sqrt{3} \times 23 = \mathbf{39.84\ A}$ **Ans.**

Problem 15.10. *A star-connected three-phase load has a resistance of 8 Ω and an inductive reactance of 6 Ω in each phase. It is fed from a 400 V, 3-phase balanced supply. Determine:*

(i) Line current *(ii) Power factor*

(iii) Active power *(iv) Reactive power*

(v) Draw phasor diagram.

Solution. Phase voltage, $V_{ph} = \dfrac{V_L}{\sqrt{3}} = \dfrac{400}{\sqrt{3}} = 231\ V$

Impedance per phase, $Z_{ph} = \sqrt{R_{ph}^2 + X_{ph}^2} = \sqrt{(8)^2 + (6)^2} = 10\ \Omega$

(*i*) Line current, I_L = Phase current, I_{ph}

$$= \frac{V_{ph}}{Z_{ph}} = \frac{231}{10} = \mathbf{23.1\ A} \qquad \textbf{Ans.}$$

(*ii*) Power factor, $$\cos\phi = \frac{V_{ph}}{Z_{ph}} = \frac{8}{10}$$

$$= \mathbf{0.8\ (lagging)} \qquad \textbf{Ans.}$$

Phase angle, $\phi = \cos^{-1} 0.8 = 36.87°$ (lagging)

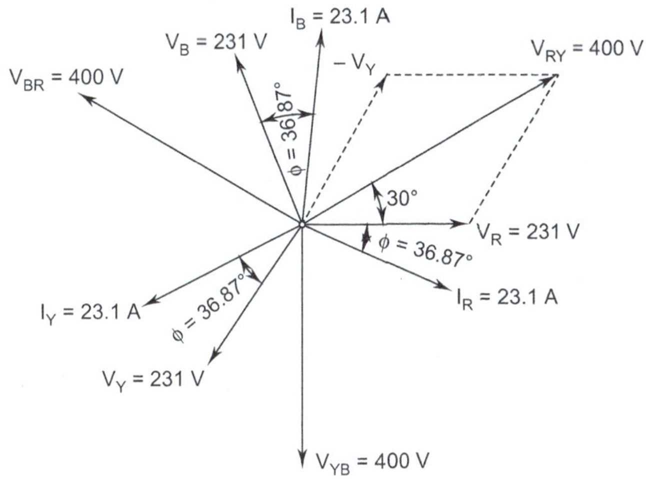

Fig. 15.29

(*iii*) Active power, $$P = \sqrt{3}\, V_L I_L \cos\phi$$

$$= \sqrt{3} \times 400 \times 23.1 \times 0.8 = \mathbf{12{,}800\ W} \qquad \textbf{Ans.}$$

(*iv*) Reactive power, $$Q = \sqrt{3}\, V_L I_L \sin\phi$$

$$= \sqrt{3} \times 400 \times 23.1 \times 0.6 = \mathbf{9{,}600\ VA} \qquad \textbf{Ans.}$$

Wattmeter readings, $W_1 = V_L I_L \cos(30° - \phi)$

$$= 400 \times 23.1 \cos(30° - 36.87°) = \mathbf{9{,}171.3\ W} \qquad \textbf{Ans.}$$

$W_2 = V_L I_L \cos(30° + \phi)$

$$= 400 \times 23.1 \cos(30° + 36.87°) = \mathbf{3{,}628.7\ W} \qquad \textbf{Ans.}$$

(*v*) Phasor diagram for currents and voltages is shown in Fig. 15.30.

Problem 15.11. *Three equal impedances, each consisting of R and L in series are connected in star and are supplied from a 400 volts, 50 Hz, 3-phase, 3-wire balanced supply system. The power input to the load is measured by 2-wattmeter method and the two-wattmeters read 3 kW and 1 kW. Determine the values of R and L connected in each phase.*

Solution. Reading of wattmeter, $W_1 = 3$ kW

Reading of wattmeter, $W_2 = 1$ kW

Total power, $P = W_1 + W_2 = 3 + 1 = 4$ kW

Power factor of the circuit, $\cos\phi = \cos\tan^{-1}\dfrac{W_1 - W_2}{W_1 - W_2}\sqrt{3} = \cos\tan^{-1}\dfrac{3-1}{3+1}\sqrt{3}$

$$= \cos\tan^{-1}\frac{\sqrt{3}}{2} = \cos 40.89° = 0.7559 \text{ (lagging)}$$

Line current, $I_L = \dfrac{P}{\sqrt{3}\,V\cos\phi} = \dfrac{4 \times 1{,}000}{\sqrt{3} \times 400 \times 0.7559} = 7.6376$ A

Impedance of the circuit per phase, $Z_{ph} = \dfrac{V_{ph}}{I_{ph}} = \dfrac{400/\sqrt{3}}{7.6376} = 30.237\ \Omega$

Resistance per phase, $R = Z_{ph}\cos\phi = 30.237 \times 0.7559 = \mathbf{22.856\ \Omega}$ **Ans.**

Reactance per phase, $X_L = \sqrt{Z_{ph}^2 - R^2} = \sqrt{(30.237)^2 - (22.856)^2} = 19.796\ \Omega$

Inductance per phase, $L = \dfrac{X_L}{2\pi f} = \dfrac{19.796}{2\pi \times 50} = \mathbf{0.063\ H\ or\ 63\ mH.}$ **Ans.**

Problem 15.12. *Three identical resistors of 20 Ω each are connected in star to a 415 V, 50 Hz, 3-phase supply. Calculate (i) the total power consumed, (ii) the total power consumed, if they are connected in delta, (iii) the power consumed, if one of the resistors is opened.*

Solution. Total power consumed from 415 V, 50 Hz 3-phase supply, when three 20 Ω resistors are connected in

(*i*) Star-connection, $P_S = \dfrac{3 \times \left(\dfrac{V_L}{\sqrt{3}}\right)^2}{R} = \dfrac{V_L^2}{R} = \dfrac{(415)^2}{20}$

$= \mathbf{8{,}611.25\ W\ or\ 8.6\ kW}$ **Ans.**

(*ii*) Delta-connection, $P_D = 3 \times \dfrac{V_L^2}{R} = 3 \times \dfrac{(415)^2}{20}$

$= \mathbf{25{,}833.75\ W\ or\ 25.83\ kW}$ **Ans.**

(*iii*) In the event of one of the three resistors getting open, power drawn in

(*a*) Star-connections, $P_S = \dfrac{2 \times \left(\dfrac{V_L}{2}\right)^2}{R} = \dfrac{V_L^2}{2R} = \dfrac{(415)^2}{2 \times 20}$

$= \mathbf{4{,}305.625\ W\ or\ 4.31\ kW}$ **Ans.**

(*b*) Delta-connections, $P_D = \dfrac{2 \times V_L^2}{R} = \dfrac{2 \times (415)^2}{20}$

$= \mathbf{17{,}222.5\ W\ or\ 17.22\ kW}$ **Ans.**

Problem 15.13. *In a 2 wattmeter method power measured was 30 kW at 0.7 pf lagging. Find the reading of each wattmeter.*

Solution. Total load supplied, P = 30 kW

Power factor, $\cos\phi = 0.7$ (lagging)

Phase angle, $\phi = \cos^{-1} 0.7 = 45.57°$ (lagging)

So, $$V_L I_L = \frac{\text{P in kW} \times 1{,}000}{\sqrt{3}\cos\phi} = \frac{30 \times 1{,}000}{\sqrt{3} \times 0.7} = 24{,}743.6 \text{ VA}$$

Reading of wattmeter, $W_1 = V_L I_L \cos(30° - \phi)$

$= 24{,}743.6 \cos(30° - 45.57°) = \mathbf{23{,}835\ W\ or\ 23.84\ kW}$ **Ans.**

Reading of wattmeter, $W_2 = V_L I_L \cos(30° + \phi)$

$= 24{,}743.6 \cos(30° + 45.57°) = \mathbf{6{,}165\ W\ or\ 6.17\ kW}$ **Ans.**

Problem 15.14. *A 3-phase voltage source has a phase voltage of 120 V and supplies star connected load having impedance 36 + j48 Ω per phase. Calculate:*

(a) The line voltage *(b) The line current*

(c) The power factor *(d) The total 3-phase power supplied to the load.*

Solution.

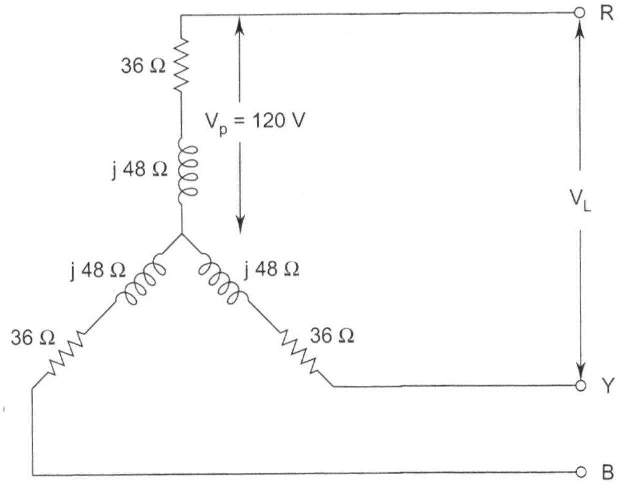

Fig. 15.30

(*a*) In case of star connection, line voltage,

$$V_L = \sqrt{3}\, V_{ph} = \sqrt{3} \times 120 = \mathbf{207.85\ V}$$ **Ans.**

(*b*) Line current, $$I_L = I_{ph} = \frac{V_{ph}}{Z_{ph}} = \frac{120}{36 + j48}$$

$$= \frac{120}{60\angle 53.13} = \mathbf{2\angle -53.13\ A}$$ **Ans.**

(*c*) Power factor, $\cos\phi = \cos 53.13 = \mathbf{0.6\ (Lagging)}$ **Ans.**

(*d*) Power, $P = 3\,V_{ph}I_{ph}\cos\phi = 3 \times 120 \times 2 \times 0.6$

or $P = \mathbf{432\ W}$ **Ans.**

Problem 15.15. *Three arms of a three-phase load each comprise an inductor of resistance 25 Ω and of inductance 0.15 H in the series with 120 μF capacitor. The supply is 415 V, 50 Hz. If load is delta connected, calculate:*

(*i*) *Line current* (*ii*) *Line voltage* (*iii*) *Power* (*iv*) *Power factor*

Solution.

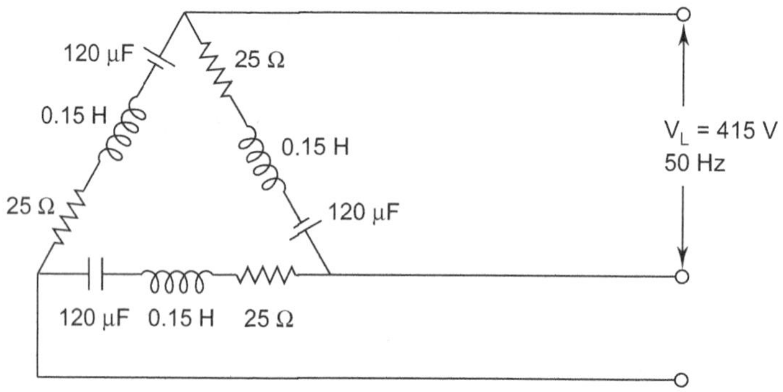

Fig. 15.31

$$X_L = \omega L = 2\pi f L = 2\pi \times 50 \times 0.15 = 47.124\ \Omega$$

$$X_C = \frac{1}{\omega C} = \frac{1}{2\pi f C} = \frac{10^6}{2\pi \times 50 \times 120} = 2.653\ \Omega$$

Impedance per phase, $Z_{ph} = R + j(X_L - X_C) = 25 + j(47.124 - 2.653)$

or $Z_{ph} = 25 + j44.47 = 51.02 \angle 60.66°$

In case of delta connected load,

Line voltage, $V_L = V_{ph} = 415\ V$

Phase current, $$I_{ph} = \frac{V_{ph}}{Z_{ph}} = \frac{415}{51.02\angle 60.66°} = 8.134\angle -60.66°$$

(*i*) Line current, $I_L = \sqrt{3}\,I_{ph} = \sqrt{3} \times 8.134 = \mathbf{14.08\ A}$ **Ans.**

(*ii*) Line voltage, $V_L = V_{ph} = \mathbf{415\ V}$ **Ans.**

(*iii*) Power, $P = 3\,V_{ph}I_{ph}\cos\phi = 3 \times 415 \times 8.134 \times \cos(60.66)$

or $P = \mathbf{4.96\ kW}$ **Ans.**

(*iv*) Power factor, $$\cos\phi = \frac{R_{ph}}{Z_{ph}} = \frac{25}{51.02} = \mathbf{0.49\ (Lagging)}$$ **Ans.**

Problem 15.16. *A balanced star connected inductive load is connected to a 400 V, 50 Hz AC supply. Two wattmeters used to measure supply power indicates 8000 W and 4000 W respectively. Determine:*

(*i*) *Line current*
(*ii*) *Impedance of each phase*
(*iii*) *Resistance of each phase*
(*iv*) *Inductance of each phase.*

Solution.

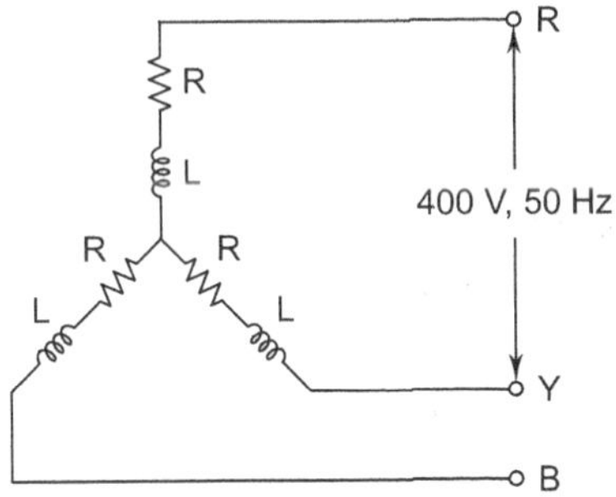

Fig. 15.32

Given
$$W_1 = 8000 \text{ W}$$
$$W_2 = 4000 \text{ W}$$
$$V_L = 400 \text{ V}$$
$$f = 50 \text{ Hz}$$

(*i*) Phase angle will be

$$\phi = \tan^{-1}\left[\sqrt{3}\left(\frac{W_1 - W_2}{W_1 + W_2}\right)\right]$$

or
$$\phi = \tan^{-1}\left[\sqrt{3}\left(\frac{8000 - 4000}{8000 + 4000}\right)\right]$$

or
$$\phi = 30°$$

As, the two wattmeter readings are expressed as

$$W_1 = V_L I_L \cos(30° - \phi)$$

or
$$8000 = 400 \times I_L \times \cos(30° - 30°)$$

or
$$I_L = \frac{8000}{400} = \mathbf{20\ A}$$ **Ans.**

(*ii*) In case of star connection

$$I_L = I_{ph} = 20 \text{ A}$$
$$V_L = \sqrt{3}\, V_{ph}$$

or
$$V_{ph} = \frac{V_L}{\sqrt{3}} = \frac{400}{\sqrt{3}} = 231 \text{ V}$$

Therefore, impedance of each phase will be

$$Z_{ph} = \frac{V_{ph}}{I_{ph}} = \frac{231}{20}$$

or $$Z_{ph} = \mathbf{11.55\ \Omega}$$ **Ans.**

(*iii*) Power factor of the circuit will be

$$\cos\phi = \cos 30°$$

or $$\text{p.f.} = 0.866 \text{ (lagging)}$$

Also, $$\cos\phi = \frac{R}{Z}$$

or $$R = Z\cos\phi = 11.55 \times 0.866$$

or $$R \simeq \mathbf{10\ \Omega}$$ **Ans.**

(*iv*) As impedance is given by

$$Z = \sqrt{R^2 + X_L{}^2}$$

or $$11.55 = \sqrt{(10)^2 + X_L{}^2}$$

or $$X_L = 5.779\ \Omega$$

or $$2\pi f L = 5.779$$

or $$L = \frac{5.779}{2\pi \times 50} = \mathbf{18.4\ H}$$ **Ans.**

Problem 15.17. *Three inductive coils, each with a resistance of 15 Ω and an inductance of 0.03 H are connected in star to 3-phase, 400 V, 50 Hz supply. Calculate:*

(*a*) *Phase current* (*b*) *Line current*

(*c*) *Total power absorbed.*

Solution.

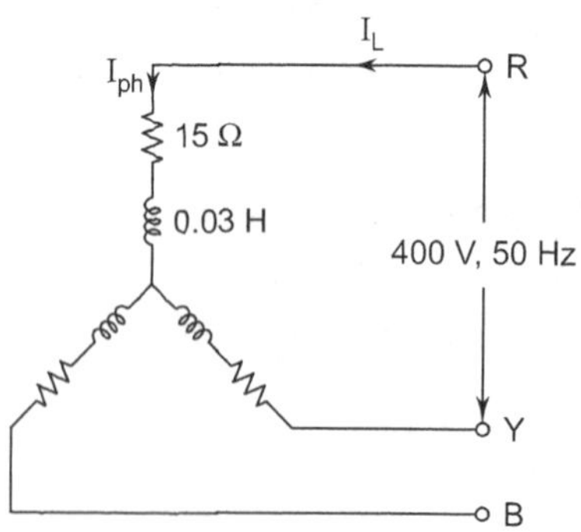

Fig. 15.33

$$X_L = \omega L = 2\pi f L$$

$$= 2\pi \times 50 \times 0.03 = 9.425\ \Omega$$

$$Z_{ph} = (15 + j\,9.425)\ \Omega$$

or $$Z_{ph} = 17.72 \angle 32.14°\ \Omega$$

In case of star connection

$$I_L = I_{ph}$$

$$V_L = \sqrt{3}\, V_{ph}$$

or $$V_{ph} = \frac{V_L}{\sqrt{3}} = \frac{400}{\sqrt{3}} = 230.94\text{ V}$$

(*a*) Phase current, $$I_{ph} = \frac{V_{ph}}{Z_{ph}} = \frac{230.94 \angle 0°}{17.72 \angle 32.14°}$$

or $$\mathbf{I_{ph} = 13.03 \angle -322.14° \text{ A}}$$ **Ans.**

(*b*) Line current, $$I_L = I_{ph} = \mathbf{13.03 \angle -32.14° \text{ A}}$$ **Ans.**

(*c*) Total power absorbed,

$$P = 3\, V_{ph} . I_{ph} . \cos\phi$$

$$= 3 \times 230.94 \times 13.03 \times \cos 32.14°$$

$$P = 2547.998 \text{ W}$$

or $$P = \mathbf{2.548 \text{ kW}}$$ **Ans.**

Problem 15.18. *A balanced delta connected load of impedance (16 + j 12) Ω per phase is connected to a 3-phase 400 V supply. Find the phase current, line current, power factor, power, reactive VA and total VA.*

Solution.

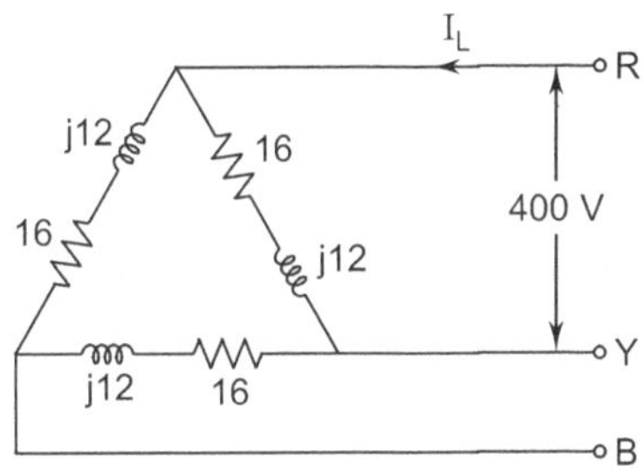

Fig. 15.34

Given $$V_L = 400 \text{ V}$$

$$V_{ph} = V_L = 400 \text{ V}$$

$$Z_{ph} = 16 + j\,12 = 20 \angle 36.87°$$

(*a*) Phase current, $$I_{ph} = \frac{V_{ph}}{Z_{ph}} = \frac{400}{20 \angle 36.87°}$$

or $$\mathbf{I_{ph} = 20 \angle -36.87° \text{ A}}$$ **Ans.**

(*b*) Line current, $$I_L = \sqrt{3}\, I_{ph} \text{ (for delta connection)}$$

$$= \sqrt{3} \times 20 \angle -36.87°$$

or $$\mathbf{I_L = 34.64 \angle -36.87° \text{ A}}$$ **Ans.**

(*c*) Power factor, $$\text{p.f.} = \cos\phi = \cos 36.87$$

$$\text{p.f.} = \mathbf{0.799 \text{ (Lagging)}}$$ **Ans.**

(*d*) Power (Active), $P = \sqrt{3}\, V_L\, I_L \cos\phi$

$= \sqrt{3} \times 400 \times 34.64 \cos 36.87°$

$= 19{,}199.41$ W

or $P_{act} = $ **19.19 kW** **Ans.**

(*e*) Power (Reactive), $P = \sqrt{3}\, V_L\, I_L \sin\phi$

$= \sqrt{3} \times 400 \times 34.64 \sin 36.87°$

$= 14399.61$ VAR

or $P_{rec} = $ **14.39 kVAR** **Ans.**

(*f*) Power (Apparent), $P = \sqrt{3}\, V_L\, I_L$

$= \sqrt{3} \times 400 \times 34.64$

$= 23999.295$ VA

or $P_{app} = $ **23.99 kVA** **Ans.**

Problem 15.19. *Three identical resistors of 20 Ω each are connected in star to a 415 V, 50 Hz, three-phase supply. Calculate:*

(*a*) *The total power consumed*

(*b*) *The total power consumed, if they are connected in delta*

(*c*) *The power consumed, if one of the resistors is opened.*

Solution.

(*a*) When the resistors are star-connected, as shown in Fig. 15.35(*a*),

$$\text{Total power consumed, } P = 3 \times \frac{V_{ph}^2}{R} = 3 \times \frac{(V_L/\sqrt{3})^2}{R} = 3 \times \frac{415 \times 415}{3 \times 20}$$

$= 8612$ W $=$ **8.612 kW** **Ans.**

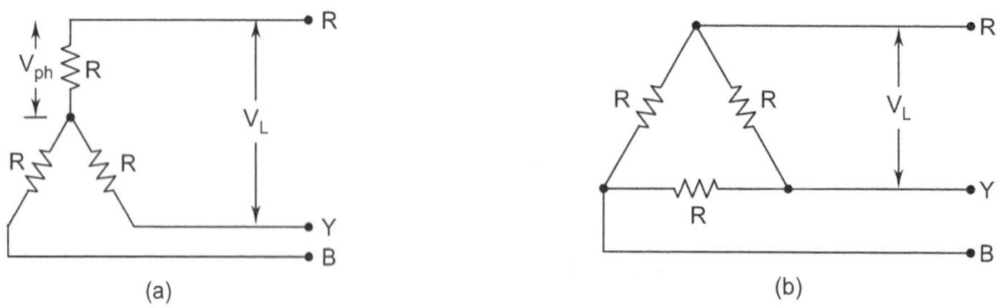

Fig. 15.35

(*b*) When the resistors are delta-connected, as shown in Fig. 15.35(*b*)

$$\text{Total power consumed, } P = 3 \times \frac{V_{ph}^2}{R} = 3 \times \frac{V_{ph}^2}{R} = 3 \times \frac{(415)^2}{20}$$

$= 25836$ W $=$ **25.836 kW** **Ans.**

(*c*) When one of the resistor is open circuited

(*i*) In case of star-connections, as shown in Fig. 15.36(*a*)

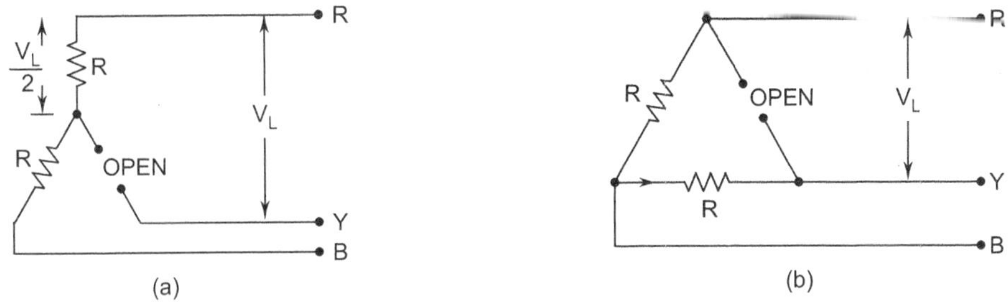

Fig. 15.36

$$\text{Total power consumed, P} = 2 \times \frac{V_{ph}^2}{R} = 2 \times \frac{(V_L/2)^2}{R} = 2 \times \frac{(415)^2}{2 \times 2 \times 20}$$

or $$P = 4306 \text{ W} = \mathbf{4.306 \text{ kW}}$$ **Ans.**

(*ii*) In case of delta-connections, as shown in Fig. 15.36(*b*)

$$\text{Total power consumed, P} = 2 \times \frac{V_L^2}{R} = 2 \times \frac{(415)^2}{20}$$

or $$P = 17224 \text{ W} = \mathbf{17.224 \text{ kW}}$$ **Ans.**

Problem 15.20. *The line current is I if balanced three-phase impedances are star-connected. Show that the line current will be 3 I if the same impedances are delta-connected.*

Solution. When the three impedances each of Z ohm are connected in star, as shown in Fig. 15.37(*a*),

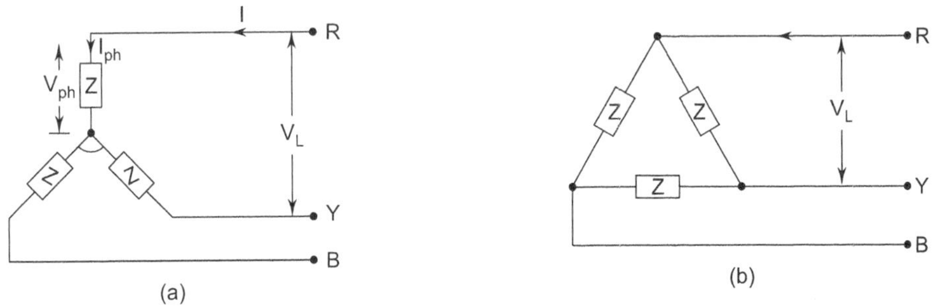

Fig. 15.37

Line current = phase current = I,

∴ Phase voltage = IZ

Line voltage = $\sqrt{3}$ phase voltage = $\sqrt{3}$ IZ

When the same impedances, each of Z ohm, are connected in delta, as shown in Fig. 15.37 (*b*)

Phase voltage = line voltage = $\sqrt{3}$ IZ

$$\text{Phase current} = \sqrt{3}\ \text{IZIZ} = \sqrt{3}\,\text{I}$$

$$\text{Line current} = \sqrt{3}\ \text{phase current} = \sqrt{3} \times \sqrt{3}\,\text{I} = \mathbf{3I}$$ **Ans.**

Problem 15.21. *Three arms of a three phase load each comprise an inductor of resistance 25 Ω and of inductance 0.15 H in the series with 120 μF capacitor. The supply is 415 V, 50 Hz. If load is delta connected, calculate:*

(*a*) *Line current* (*b*) *Line voltage*

(*c*) *Power* (*d*) *Power factor.*

Solution.

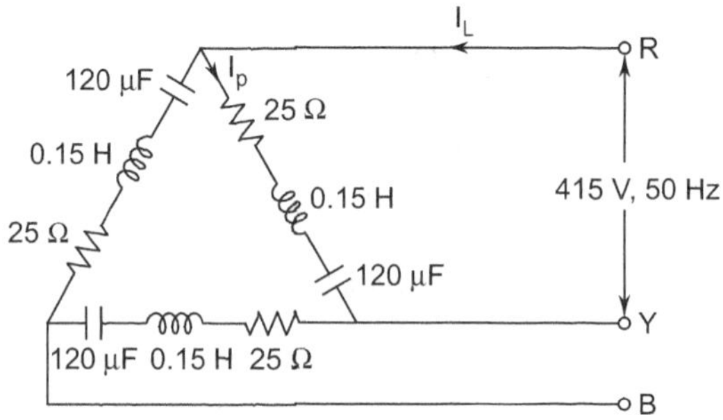

Fig. 15.38

$$X_L = \omega L = 2\pi f L = 2\pi \times 50 \times 0.15 = 47.124\ \Omega$$

$$X_C = \frac{1}{\omega C} = \frac{1}{2\pi C} = \frac{10^6}{2\pi \times 50 \times 120} = 2.653\ \Omega$$

Impedance per phase, $Z_{ph} = R + j(X_L - X_C) = 25 + j(47.124 - 2.653)$

or $Z_{ph} = 25 + j\,44.47 = 51.02\ \angle\ 60.66°\ \Omega$

In case of delta connected load,

Line voltage, $V_L = V_{ph} = 415$ V

Phase current, $I_{ph} = \frac{V_{ph}}{Z_{ph}} = \frac{415}{51.02\ \angle\ 60.66°} = 8.134\ \angle\ 60.66°$ A

(*a*) Line current, $I_L = \sqrt{3}\,I_{ph} = \sqrt{3} \times 8.134 = \mathbf{14.08\ A}$ **Ans.**

(*b*) Line voltage, $V_L = V_{ph} = \mathbf{415\ V}$ **Ans.**

(*c*) Power, $P = 3V_{ph}I_{ph}\cos\phi = 3 \times 415 \times 8.134 \times \cos(60.66)$

or $P = \mathbf{4.94\ kW}$ **Ans.**

(*d*) Power factor, $\cos\phi = \frac{R_{ph}}{Z_{ph}} = \frac{25}{51.02} = \mathbf{0.49}$ **(lagging)** **Ans.**

EXERCISE

1. Prove that the power in a three-phase balanced circuit can be deduced from the reading of two-wattmeters with the help of suitable vector diagrams. Discuss the nature of power factor when
   (*i*) two-wattmeters read equal and positive values.
   (*ii*) one of the wattmeter indicates zero value.
2. Each phase of a star connected load consists of a non-reactive resistance of 100 Ω in parallel with a capacitance of 3.18 μF connected to a 400 V, 50 Hz supply. Calculate:
   (*i*) the line current
   (*ii*) the power absorbed
   (*iii*) the total kVA
   (*iv*) the power factor.
3. Derive the relationship between line and phasor current for a delta connected 3-phase balanced system.
4. A star connected 3 phase load has a resistance of 6 Ω and inductive reactance of 8 Ω in each branch. Three-phase supply voltage is 220 V. Write the phasor expression for voltage across each branch, line voltages and line currents.
5. Derive the relationship between line and phasor current for a delta connected 3-phase balanced system.
6. In the two-wattmeter method of power measurement in a 3-phase circuit, the readings of watt-meters are 1000 W and 550 W. What is the power factor of the load?
7. A 3-phase balanced load connected across a 3-phase 400 V AC supply draws a line current of 10 A. Two-wattmeters are used to measure input power. The ratio of two-wattmeters is 2 : 1. Find the readings of two-wattmeters.
8. Three identical resistors of 20 Ω each are connected in star to a 415 V, 50 Hz, 3-phase supply. Calculate:
   (*i*) total power consumed.
   (*ii*) the total power consumed if the load is connected in delta.
   (*iii*) the power consumed if one of the resistor is opened.
9. A balanced star connected load of (8 + *j*6) Ω per phase is connected to a 3-phase 400 V supply. Find the line current, power factor and three-phase volt-amperes.
10. Derive the relation between line and phase voltage for a delta connected 3-phase balanced system.
11. A balanced delta connected load of (8 + *j*6) Ω per phase is connected to a 3-phase 430 V supply. Find the line current.
12. Show that the power in 3-phase balanced system is constant at every instant and is given by 3 V_P I_P cos ϕ, where V_P, I_P and cos ϕ are phase value of voltage and current and power factor.
13. Derive the relation between line and phase voltage for a star connected 3-phase balanced system.

❑ ❑ ❑

16

INTRODUCTION TO POWER SYSTEM

16.1 INTRODUCTION

Power system is the nomenclature given to the whole life span of electricity from the birth (*i.e.*, generation) to the end (*i.e.*, consumption). Electrical power has become almost as much of a necessity as the basic elements in our everyday life. It is used everywhere and every time.

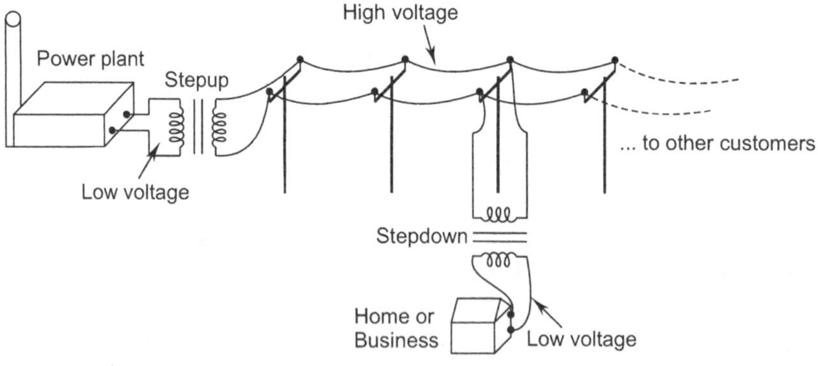

Fig. 16.1

The installed generating capacity in India on 1915 was 105 MW and increase by 90,000 MW in 1997. Now, India generates nearly 1,50,000 MW. The growth rate of installed capacity is 7%. Usually, high capacity generators give an electrical power output at 11 kV to 17 kV. Thus, it is require to increase the voltage level to 132 kV or 220 kV through transformers as per requirement. Than with the help of transmission line the power is at transmitted at very long distance.

Design of transmission line is based on the factors like transmission line voltage levels, constant like resistance, reactance of the line, interference with the neighbouring circuit, etc. Its mechanical feature are strength of the support, sag calculation, tension etc.

Transmission of power by the overhead lines is very much cheaper because repair carries out is comparatively very easy.

Than finally to bring the voltage at lower used step down transformer are used rated as 220 kV/ 33 kV or 33 kV/6.6 kV or 6.6 kV/440 V or 220 V. The transfer of power from receiving station to the substation is done with the help of conductor called feeders. This is called secondary transmission.

16.2 POWER SYSTEM COMPONENTS

In power system there are three important branches namely,

- Generation
- Transmission
- Distribution

All the branches need to be given equal importance.

16.2.1 Generation

At the generating plants the energy is produced at a relatively low voltage of up to 30 kV, then stepped up by the power station transformer to a higher voltage (115 kV to 765 kV AC, ± 250 – 500 kV AC, varying by country) for transmission over long distances to grid exit points (substations). There are mainly three types of power generating plants for large scale power generation. These are:

- Hydroelectric Power Plant
- Thermal Power Plant
- Nuclear Power Plant

Beside these three power generating plants, a lot of other renewable energy sources are employed for power generation such as wind, biomass, solar, ocean, geothermal, etc.

16.2.2 Transmission

Electric power transmission, a process in the delivery of electricity to consumers, is the bulk transfer of electrical power. Typically, power transmission is between the power plant and a substation near a populated area. Electric power transmission allows distant energy sources (such as hydroelectric power plants) to be connected to consumers in population centers, and may allow exploitation of low-grade fuel resources that would otherwise be too costly to transport to generating facilities. Due to the large amount of power involved, transmission normally takes place at high voltage (110 kV or above). Electricity is usually transmitted over long distance through overhead power transmission lines. Underground power transmission is used only in densely populated areas due to its high cost of installation and maintenance, and because the high reactive power produces large charging currents and difficulties in voltage management.

A power transmission system is sometimes referred to as a "Grid"; however, for reasons of economy, the network is not a mathematical grid. Redundant paths and lines are provided so that power can be routed from any power plant to any load center, through a variety of routes, based on the economics of the transmission path and the cost of power.

The transmission of electric power is divided broadly into two parts:

(*i*) ***Primary Transmission :*** In this the voltage level is increase to 132 kV or 220 kV or more with the help of step up transformer.

Therefore, it is called as High Voltage Transmission Line (HVTL).

This primary transmission uses 3-phase 3-wire system.

(*ii*) ***Secondary Transmission :*** At the receiving station the voltage level is reduced to 22 kV or 33 kV using the step down transformer. There can be more than one receiving stations. Than at reduced voltage level of 22 kV or 33 kV the power is transmitted to various stations using overhead 3-phase 3-wire System.

This is secondary transmission. The conductor, used for the secondary transmission are called feeders.

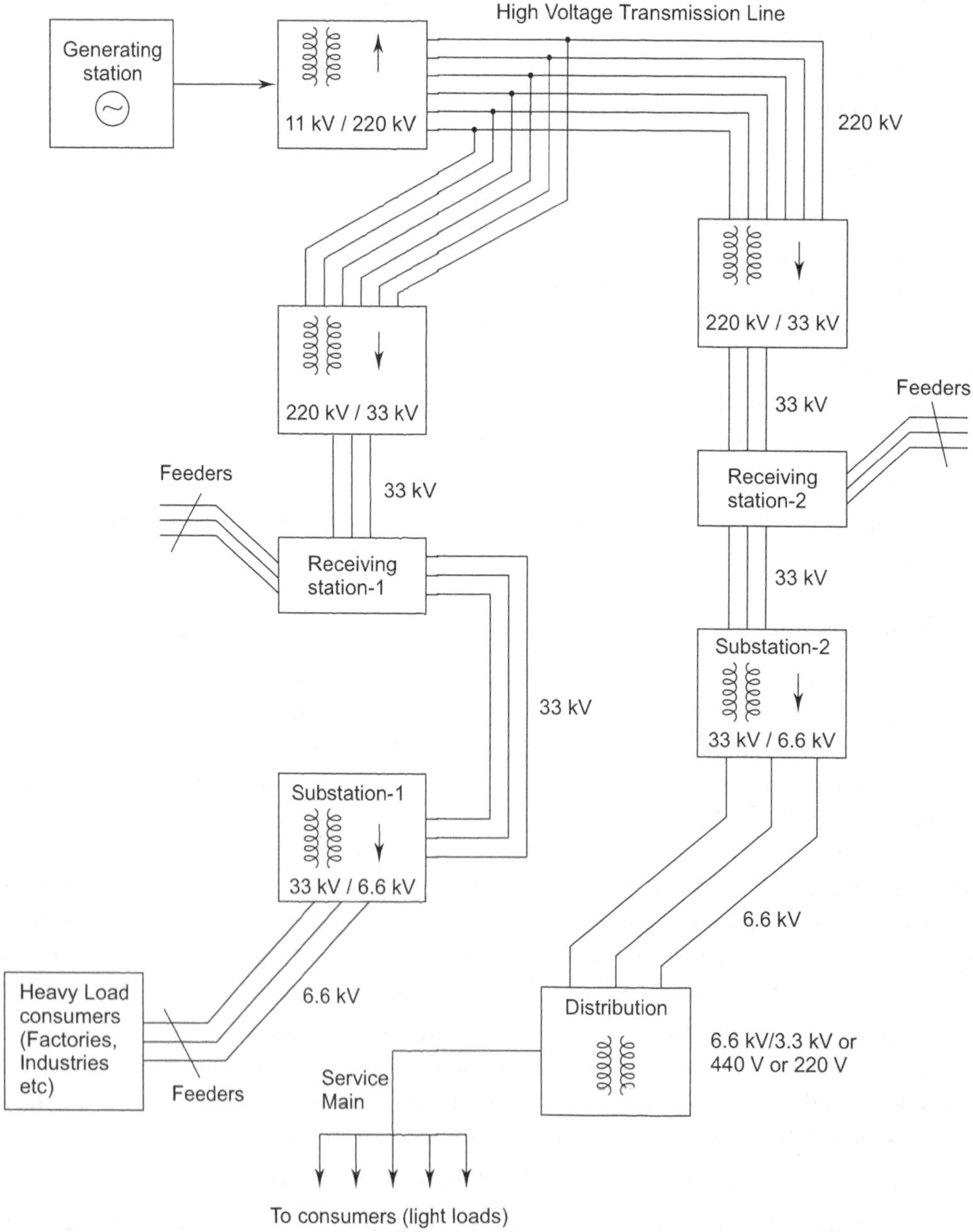

Fig. 16.2 *Schematic Diagram of Typical Transmission Distribution Scheme*

16.2.3 Distribution

At the substations, transformers are again used to step the voltage down to a lower voltage for distribution to commercial and residential users. This distribution is accomplished with a combination of sub-transmission (33 kV to 115 kV, varying by country and customer requirements) and distribution (3.3 kV to 25 kV). Finally, at the point of use, the energy is transformed to low voltage with the help of step down transformers (100 V to 600 V, varying by country and customer requirements). Similar to the transmission, the distribution of electric power is also divided broadly into two parts:

(*i*) ***Primary Distribution :*** At the substations, the voltage level in reduced to 6.6 kV, 3.3 kV and 1.1 kV with the help of step down transformer. It uses 3-phase 3-wire underground system. And the power is further transmitted to the local distribution centres. This primary distribution is also called as High Voltage (HV) Distribution. For larger consumer like factories and industries, the power is directly transmitted to such loads from a substation. Such big loads have their own substations.

(*ii*) ***Secondary Distribution :*** At local distribution centers, the voltage level of 6.6 kV, 1.1 kV is further reduced to 440 V using distribution transformers. The power is then transmitted to user or consumers; this is called a secondary distribution. The 1-phase lightening loads are supplied using a line and neutral wire. Loads like three phase motors are supplied using three phase line.

16.3 SINGLE LINE DIAGRAM OF THE TYPICAL TRANSMISSION SCHEME

A single line diagram of a power system represents the most essential components in a simplified manner. All these essential components are represented in such a way that their sequence becomes a continuous single line.

- Initially the generators (G) at power stations or power plants are considered.
- Two or more generators are connected to the bus bar (B.B.) with the circuit breakers (C.B.) in between them.
- The three phase supply from all the generators are synchronized on bus bar.
- After that this supply is transmitted to long distance with the help of transmission line (T.L.) or directly given to the near by load centers with the help of transformers (T_1, T_2 etc.).
- Of course the circuit breakers are used at appropriate locations for safety purpose and send signals to relays for isolation (if fault occur).

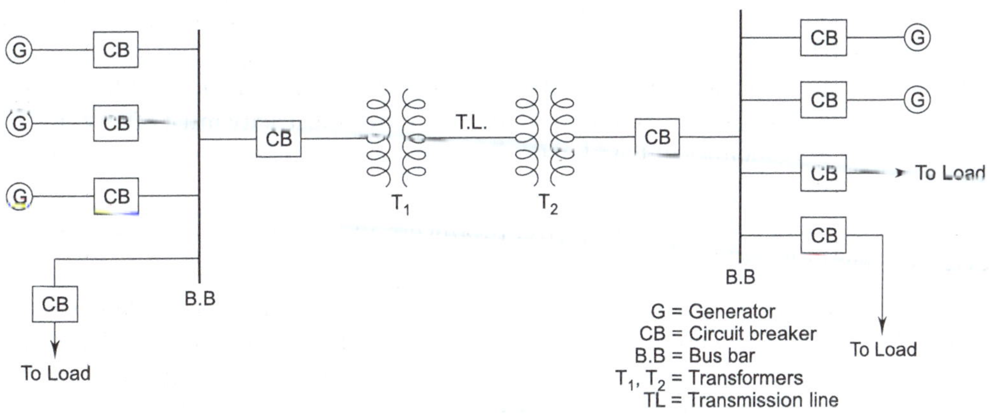

Fig. 16.3

16.4 GRID

According to the power generation, our India is divided into five regions:

1. Northern
2. Eastern
3. Southern
4. Western
5. North-Eastern.

The interconnection of generating stations in any one of the above mention particular region is known as grid.

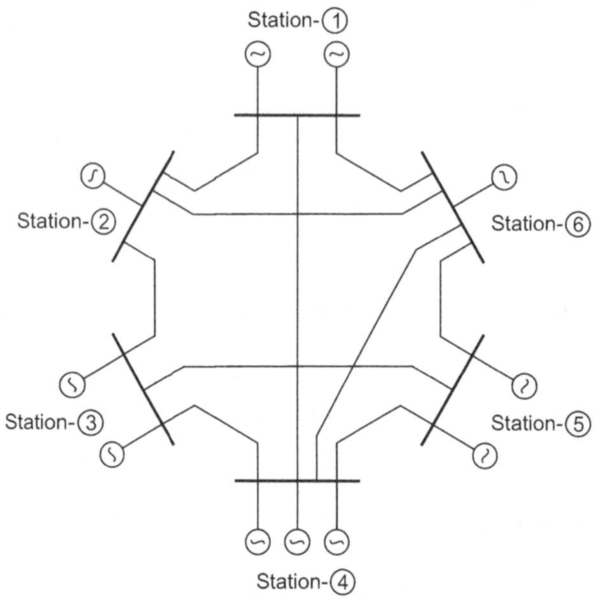

Fig. 16.4

A transmission grid is a network of power stations, transmission circuits, and substations. Energy is usually transmitted within the grid with three-phase AC. DC systems require relatively costly conversion equipment which may be economically justified for particular projects. Single phase AC is used only for distribution to end users since it is not usable for large poly-phase induction motors. The capital cost of electric power stations is so high, and electric demand is so variable, that it is often cheaper to import some portion of the variable load than to generate it locally. Because nearby loads are often correlated, imported electricity must often come from far away. Because of the economics of load balancing, transmission grids now span across countries and even large portions of continents.

The web of interconnections between power producers and consumers ensures that power can flow even if a few links are inoperative.

The unvarying (or slowly varying over many hours) portion of the electric demand is known as the "base load", and is generally served best by large facilities (and therefore efficient due to economies of scale) with low variable costs for fuel and operations, *i.e.*, nuclear, coal, hydro. Renewable sources of power generation such as solar, wind, ocean/tidal, etc., are not considered "base load" but can still add power to the grid. Smaller and higher cost sources such as combined cycle or combustion turbine plants that run on natural gas are then added as needed.

Long-distance transmission of electricity (thousands of miles) is cheap and efficient. Thus distant suppliers can be cheaper than local sources (*e.g.*, New Delhi City buys a lot of electricity from near by states such as Himachal Pradesh, Punjab etc). Multiple local sources (even if more expensive and infrequently used) can make the transmission grid more fault tollerant to weather and other disasters that can disconnect distant suppliers.

16.5 IMPORTANT TERMS AND DEFINITIONS

1. ***Substation:*** Transmission lines bring the power up to the substation at the voltage level of 22 kV or 33 kV. At the substation the level is reduced to 3.3 kV or 6.6 kV. Then using feeders the power is given to local distribution centres.
2. ***Feeder:*** The feeders are connected to substation to the area where power is to be finally distributed to the consumers. The feeders' current always remains constant.
3. ***Distributor:*** These are the conductors used to transfer power from distribution centre to the consumers. The tappings are taken for the supply to the consumers.
4. ***Service Mains:*** These are the small cables between distributors and the actual consumer premises.
5. ***Local Distribution Station:*** It consists of distribution transformers which steps down the voltage level from 3.3 kV or 6.6 kV to 440 V or 220 V. Then it is distributed to distribution and is called as distribution substation.

EXERCISE

1. Discuss the voltage structure of the electric power system.
2. Give the concept of Grid.
3. Define the following terms:
   (*i*) Substation
   (*ii*) Feeders
   (*iii*) Distributors
   (*iv*) Service mains
   (*v*) Local distribution stations.
4. Draw the single line diagram of the typical transmission scheme.
5. Draw the schematic diagram of typical transmission distribution scheme.
6. List out main components of power supply system with a brief description.

□ □ □

17

INTRODUCTION TO EARTHING AND ELECTRICAL SAFETY

17.1 INTRODUCTION

Electricity enters your home through a service head from a series of outdoor power lines or an underground connection. A typical service head consists of two 220 volt wires and one neutral wire that deliver power to lights and appliances around the home. The electric meter is mounted outdoors where electricity enters your home. This device is used to measure the amount of electricity that is consumed in your home. The meter is monitored by your electric utility company and is protected by law.

The service panel is the central distribution point for delivering electricity to switches, outlets and appliances throughout the house. Located near the electric meter, the service panel is equipped with a breaker that shuts off power to the circuits if an electrical system failure occurs.

The process of connecting metallic bodies of all the electrical apparatus and equipment to huge mass of earth by a wire having negligible resistance is called earthing. Also known as grounding is the method used to connect an electrical system to the earth with a wire. Grounding adds critical protection against electric shock and electrocution by using a grounding rod to provide a third path for conducting electricity in the event of a short-circuit or an overload. Grounding will help protect the person working on the system, the system itself, and any appliances and equipment that are connected to the system.

Grounded outlets are required to be installed in all modern homes today. If your home does not have grounded outlets, then your electrical system is likely missing critical safety features.

It doesn't take much current to kill a human being, especially if it passes through the heart. Currents above about 75 mA can cause a condition called ventricular fibrillation. If the heart goes into fibrillation, it beats very rapidly, but it doesn't pump any blood, because it's not beating in its normal rhythm. If body blood can't carry oxygen to the brain, you'll experience brain death in 3 to 4 minutes. The way to get you back involves another electric shock, from a defibrillator. If your skin is wet and you get your body across 220 volts of electricity, it's very likely that you'll have a current of 100 mA or more flowing through your heart. Currents above 10 mA can cause muscle paralysis. You may not be able to let go of energized tools or equipment. Shocks that are longer in duration are more severe.

Electrical systems must be wired with either fuses or circuit breakers. These devices are known as over-current protection and they are rated in amps. Most common household circuits are wired for 15 amps or 20 amps. Over-current protection devices protect wiring and equipment from overheating and fires.

17.2 OBJECTIVES OF THE EARTHING

Good Earthing must have low impedance enough to ensure that sufficient current can flow through the safety device so that it disconnects the supply (<0.4 sec). Fault current is much more than the full load current of the circuit which melts the fuse. Hence, the appliance is disconnected automatically from the supply mains. The earthing is provided with the following objectives:

- Provide an alternative path for the fault current to flow so that it will not endanger the user
- Ensure that all exposed conductive parts do not reach a dangerous potential
- Maintain the voltage at any part of an electrical system at a known value so as to prevent over current or excessive voltage on the appliances or equipment.
- To save human life from danger of electrical shock or death by blowing a fuse *i.e.*, to provide an alternative path for the fault current to flow so that it will not endanger the user
- To protect buildings, machinery and appliances under fault conditions *i.e.*, to ensure that all exposed conductive parts do not reach a dangerous potential.
- To provide safe path to dissipate lightning and short-circuit currents.
- To provide stable platform for operation of sensitive electronic equipments *i.e.*, to maintain the voltage at any part of an electrical system at a known value so as to prevent over current or excessive voltage on the appliances or equipment.
- To provide protection against static electricity from friction.

17.3 METHODS OF EARTHING

The important methods of earthing are the plate earthing and the pipe earthing. The earth resistance for copper wire is 1 Ω and that of GI wire less than 3 Ω. The earth resistance should be kept as low as possible so that the neutral of any electrical system, which is earthed, is maintained almost at the earth potential. The typical value of the earth resistance at powerhouse is 0.5 Ω and that at substation is 1 Ω.

17.3.1 Plate Earthing

In this method a copper plate of 60 cm × 60 cm × 3.18 cm or a GI plate of the size 60 cm × 60 cm × 6.35 cm is used for earthing (Fig. 17.1). The plate is placed vertically down inside the ground at a depth of 3 m and is embedded in alternate layers of coal and salt for a thickness of 15 cm. In addition, water is poured for keeping the earth electrode resistance value well below a maximum of 5 ohms. The earth wire is securely bolted to the earth plate. A cement masonry chamber is built with a cast iron cover for easy regular maintenance.

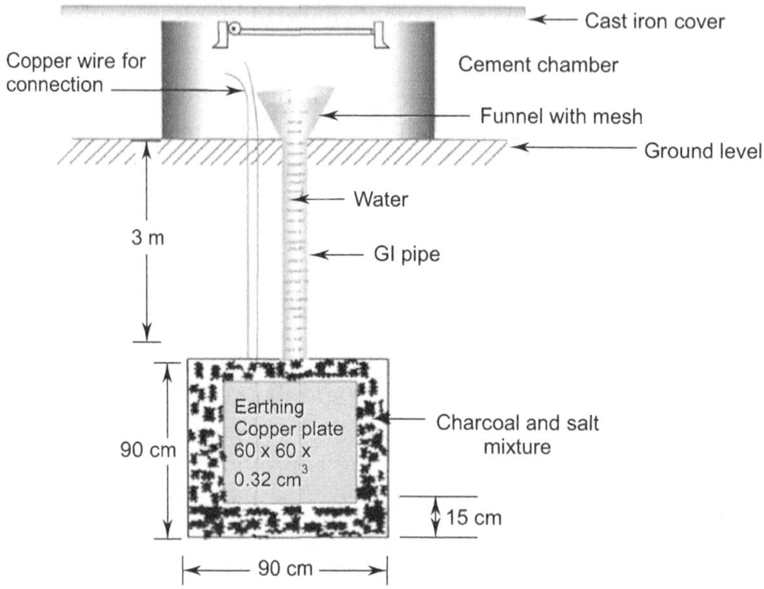

Fig. 17.1

17.3.2 Pipe Earthing

Earth electrode made of a GI (galvanized) iron pipe of 38 mm in diameter and length of 2 m (depending on the current) with 12 mm holes on the surface is placed upright at a depth of 4.75 m in a permanently wet ground. To keep the value of the earth resistance at the desired level, the area (15 cms) surrounding the GI pipe is filled with a mixture of salt and coal. The efficiency of the earthing system is improved by pouring water through the funnel periodically. The GI earth wires of sufficient cross-sectional area are run through a 12.7 mm diameter pipe (at 60 cms below) from the 19 mm diameter pipe and secured tightly at the top as shown in the following Fig. 17.2.

When compared to the plate earth system the pipe earth system can carry larger leakage currents as a much larger surface area is in contact with the soil for a given electrode size. The system also enables easy maintenance as the earth wire connection is housed at the ground level.

17.4 PROTECTIVE DEVICES FOR WIRING APPLIANCES AND INSTALLATIONS

For electrical safety, mainly there are three means employed for over current or over voltage protection:

1. Fuses
2. MCB (Miniature circuit breaker)
3. Lighting arresters (protection from surge light).

17.5 FUSES

Fuse is a primary protection component of each electronic/electric equipment or system. A switching device, which by the fusion of one or more of its specially designed and proportioned components open the circuit, in which it is inserted and breaks the current, when the same exceeds a given value for a sufficient time. The term fuse covers all the parts forming the complete switching device.

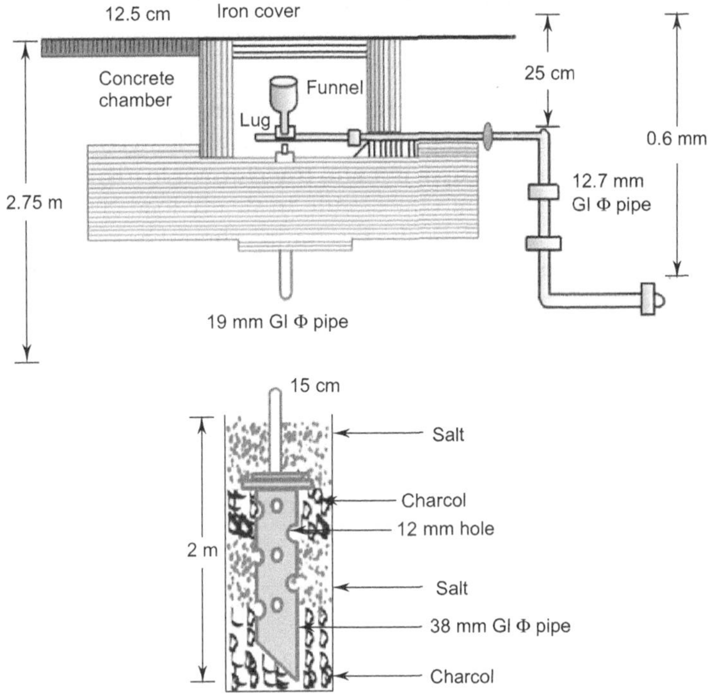

Fig. 17.2

Fuse is a safety device and the weakest part of the electrical system. It consists of a small link of soft metal, which melts when excessive current passes through it. The fuse wires are usually bare wires and made of lead and tin. Lead alloy for small current and tinned copper wire issued for large current. Up to 3 amperes load, lead fuse wires are used, up to 15 A alloy of lead and tin is used and above 15 A, tinned copper wires are used.

If a short-circuit occurs anywhere in the wiring system, an excessive current flow through the wires and fuse and if fuse operates properly, it will melt, cutting of the current before any harm can be done. If the fuse fails to operate, the wire may become hot enough to ignite the insulation, possibly causing a fire and it may harm the appliances and fittings connected in the circuit.

Fuses have a second function. Besides protecting in the case of short-circuit, they protect in the case of overload. If too many appliances are connected to one circuit, more current will flow through the supply wires than the wires were meant to carry and will cause the wires to be burnt and appliances, accessories connected to the circuits will also be damaged. In such cases, the fuse, if properly selected, will melt or blow, thus protecting the wires, appliances, accessories etc.

17.5.1 Principle of Operation

The operation of fuse depends upon the heating effect of an electric current. When the electric current is increased in a circuit due to short-circuits or over loads, it increases the rate of heat generation, which will increase the temperature of the fuse wire and thereby the rate of heat dissipation increases from the exposed surface of the fuse wire. The final temperature reaches to such stage that the heat generation will be equal to the rate dissipation. If this temperature happens to be above the melting point of the material for fuse wire, fuse must have operated.

17.5.2 Selection of Fuse Wire

To select the proper fuse wire to be inserted in a circuit two factors viz. (*a*) maximum current rating of the circuit, (*b*) current rating of the smallest size of wire of accessories is to be seen, the fuse wire inserted should be of size so that when the current with reference to factors (*a*) and (*b*) is increased, it should blow out.

The type of wire to be selected to use as a fuse wire depends upon the type of load connected to the circuit *i.e.*, steady load and fluctuating loads. The steady load covers the heating loads and the fluctuating load consists of motor, capacitor and transformer loads, all of which take transient over current when they are switched into the circuit. In steady load circuits, the fuse forms only the protection part of the circuit. Therefore, the fuse rating should be equal to or next greater than the ratings of the smallest cable used in the circuit. But if a number of fuse wires are run in parallel to augment the rating off use, total rating should not be equal to the product of rating of and strand and number of strands used. In fluctuating load circuits, fuse should allow the short time over current to flow without blowing. For this, it is necessary to select fuses of rated current greater than that of the cable of the circuit.

The fuse wire should always be inserted in live wires or phase of the circuit, or otherwise even if the fuse blows out, the faulty circuit will be fade and would cause considerable damage in case of leakage fault. Some of the metals those can be selected as fuse wire are Silver, Copper, Lead-tin alloy, etc.

17.5.3 Types of Fuses

Fuses are of various forms but the most common types are described below:

(*i*) ***Ordinary fuse:*** The ordinary or single link fuse is simply a strip of metal; usually lead (Fig. 17.3). The strip of metal, called the fuse link, is the part of the fuse through which current flows. This metal melts and breaks the connection *i.e.*, stops the flow of electricity when current more than its rating flows.

(*ii*) ***The screw plug type fuse***: It is used on electric machinery circuits of 220 volts (Fig. 17.4). This type of fuse can be had in sizes from 3 to 30 amperes and is constructed with either a porcelain or glass body, which encloses the fuse link.

(*iii*) ***A knife blade cartridge fuse***: This type of fuse is used mostly on heavy power lines of 60 to 600 amperes and 250 volts and more (Fig. 17.5). They are constructed with a heavy fibre body, fitted with brass ferrules at each end and a heavy copper blade making contact with the circuit. Removing the end ferrules and inserting the new link can easily renew the fuse link.

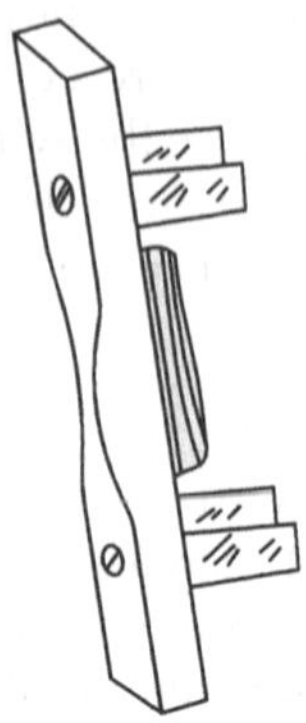

Fig. 17.3

Fig. 17.4

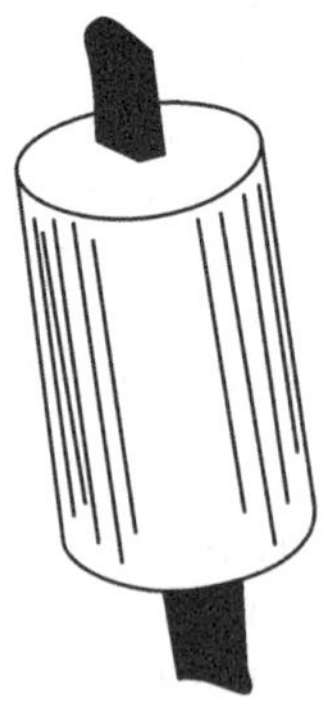

Fig. 17.5

(*iv*) ***The ferrule type cartridge fuse*:** This type of fuse is of two types–The renewal or non-renewal types (Fig 17.6). The renewal type has a small screw plug in each end that can be renewed and the new link is inserted. The body part of this fuse is made of fibre with brass ferrules fastened to each end and can be had in sizes up to 100 amperes and 250 volts.

(i) Renewal type (ii) Non-renewal type

Fig. 17.6

(*v*) ***High rupturing capacity (HRC) cartridge fuse*:** These are used where the high power is supplied. These have a definite non-breaking capacity and a high value. HRC fuse link is a unit in which the fuse element is enclosed. The fuse link is replaced when it blows.

The outer cover is usually of steatite, a ceramic material having good mechanical strength. Epoxy resins have been recently introduced and are replacing the ceramic material. The fuse elements are fitted inside the body. The ends of the fuse elements are connected to the metal end caps. The metal caps are screwed to the ceramic body by means of special forged screws to withstand the pressure developed under short-circuit condition. End contacts are welded to the metal end caps. These contacts are bolted on the stationery contacts on the panel.

An indicator pin is provided, which indicates when the fuse blows out. The fuse body is filled with powdered pure quartz. The fuse element is of silver or copper with a special shape. Normally, the element has two or more sections joined by means of a tin joint. The element consists of several identical strips, which are interconnected such that the arc spreads instantly to all the strips.

Normally the fuse elements are in parts which are connected in the middle by tin bridge. The melting point of the bridge is precise and about 230°C. The bridge does not melt at temperatures below the melting point. Since the melting point is higher than ambient temperature, the ambient temperature does not affect the melting.

The current passing through fuse element produces heat. With a certain current, the temperature rises and the tin bridges melts producing a break in the circuit. Thereby an arc is immediately spread over the neighbouring elements and they too melt. The metal vapour diffuses with quartz powder and the product of chemical reaction produces a substance of high resistance, which becomes an insulator. Thereby the space between the caps is filled with a material of high dielectric strength, as the current is interrupted.

(*vi*) ***Time delay fuse:*** It is a fuse, which has the ability to carry overload currents of short duration without melting (Fig. 17.7). The heavier overload, the less time required for fuse to "Blow". In duration, this type of fuse need not have as high rating as an ordinary fuse to permit the instrument to be started. Like the common fuse the time delay fuse is also made in plug and cartridge types.

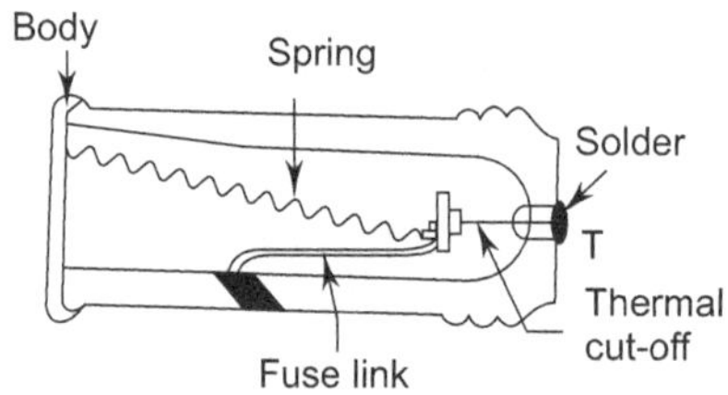

Fig. 17.7

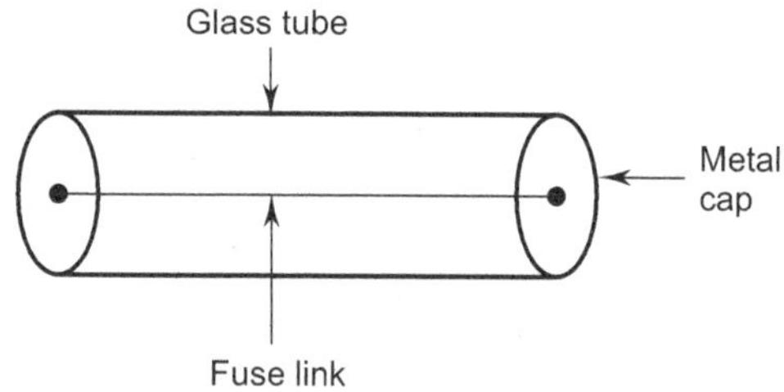

Fig. 17.8

(*vii*) ***Glass tube fuse unit:*** In this type of fuse unit the fuse link is closed in glass vacuum tubes which have shielded metal caps at both ends (Fig. 17.8). It is inserted between two stable terminals of the circuit. This fuse unit is designed for low current. On blowing of fuse, fuse link cannot be replace *i.e.*, complete fuse unit has to be replaced.

17.5.4 Comparatively Advantage and Disadvantages

A. Disadvantages of ordinary fuses: Re-wireable/ordinary fuses suffer from the following disadvantages:

(*a*) *Unreliable operation due to*: (*i*) Oxidation of fuse wire and consequent thinning of wire section with lapse of time, (*ii*) Loose connection causes the local heating, (*iii*) Heat radiating devices used in the circuit.

(*b*) *Lack of discrimination*: On account of unreliable operation, discrimination cannot be ensured always.

(*c*) *Small time lag*: On account of small time lag, these types of fuses can blow will large transient current when three phase motors, transformer, capacitors and fluorescent lights etc are used in the circuits.

(*d*) *Misuse*: Sometimes the proper rated wire for the fuse element is not available and any other piece of wire is used for the fuse element to solve the purpose, which is dangerous.

(*e*) *Low rupturing capacity*: The use of this fuse is limited to 4 kA in faulty circuits.

B. Advantages of re-wireable fuse:

(*a*) It is cheapest form of operation.

(*b*) It affords current limiting effect under short-circuits conditions due to cut-off.

(*c*) It requires minimum time to replace.

(*d*) It requires no maintenance.

(*e*) It has current limiting effect.

C. Advantages of HRC fuses:

(*a*) They are simple and have easy installation.

(*b*) They are cheaper as compared to other types of circuit interrupter of the same breaking capacity. They have high breaking capacity.

(*c*) They do not require any maintenance.

(*d*) Their operation is quick and sure.

(*e*) They have inverse time current characteristics.

(*f*) They do not deteriorate with time.

(*g*) They are quite reliable and can be selected for proper discrimination.

(*h*) They are capable of clearing high as well as low currents.

(*i*) They have closer control on sustained overloads due to low fusing factor and the time lag feature.

(*j*) They have current limitation by 'cut-off' property.

D. Disadvantages of HRC fuses:

(*a*) After each operation, its replacement is required.

(*b*) Interlocking is not possible in these types of fuses.

(*c*) They lack relays in complete discrimination.

17.6 CIRCUIT BREAKER

A circuit breaker is defined as, "A mechanical switching device, capable of making, carrying and breaking currents under normal circuit conditions. Also capable of making and carrying for a specified time and breaking currents under specified abnormal circuit conditions, such as those of a short-circuit."

Another definition of circuit breaker as given by The National Electric Code (NEC) is, "a circuit breaker is a device designed to open and close a circuit by non-automatic means, and to open the circuit automatically on a predetermined over-current without damage to itself when properly applied within its rating."

Circuit breakers provide a manual means of energizing and de-energizing a circuit. In addition, circuit breakers provide automatic over current protection of a circuit. A circuit breaker allows a circuit to be reactivated quickly after a short-circuit or overload is cleared. Unlike fuses which must be replaced when they open, a simple flip of the breaker's operating handle restores the circuit. All circuit breakers perform the following functions:

- SENSE when an over-current occurs
- MEASURE the amount of over-current
- ACT by tripping the circuit breaker in a time frame necessary to prevent damage to itself and the associated load cables.

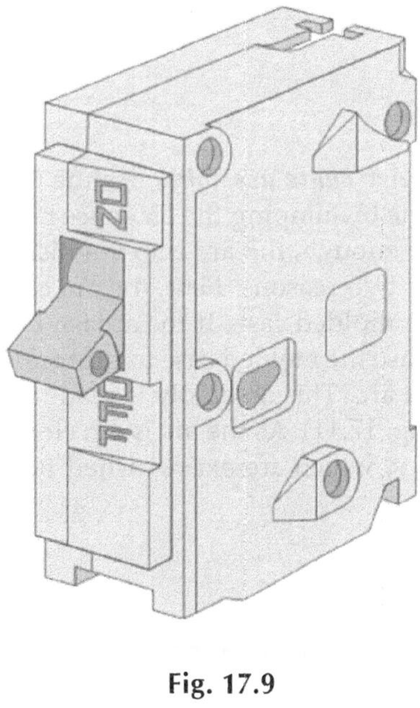

Fig. 17.9

17.6.1 Circuit Breaker Design

The following section presents some basics of circuit breaker design. Circuit breakers are constructed in five major components:

- Frame (molded case)
- Contacts
- Arc chute assembly
- Operating mechanism
- Trip unit

Frame: The frame provides an insulated housing to mount the circuit breaker components. The construction material is usually a thermal set plastic such as glass-polymer.

Contacts: The current flowing in a circuit controlled by a circuit breaker flows through the circuit breaker's contacts (Fig. 17.10). When a circuit breaker is turned off or is tripped by a fault current, the circuit breaker interrupts the flow of current by separating its contacts.

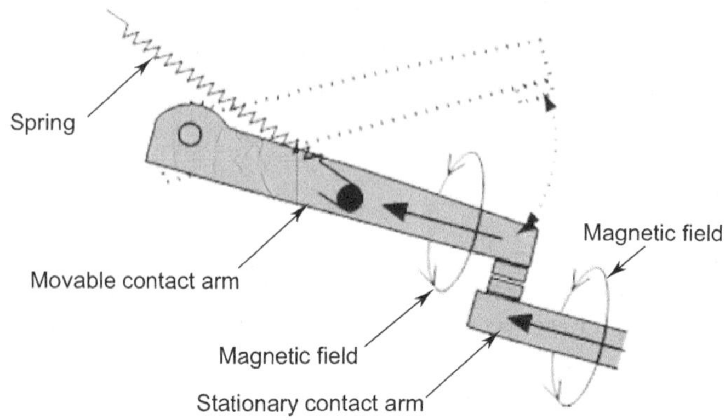

Fig. 17.10

Arc chute assembly: As the contacts open a live circuit, current continues to flow for a short time by jumping the air space between the contacts in the form of an arc. When the contacts open far enough, the arc is extinguished and the current flow stops. Minimizing the arc is important for two reasons. First, the arc can damage the contacts. In addition, the arc ionizes gases inside the molded case. If the arc isn't extinguished quickly the pressure from the ionized gases could cause the molded case to rupture. Circuit breakers commonly use an arc chute assembly to quench the arc. This assembly is made up of several "U" shaped steel plates that surround the contacts (Fig. 17.11). As the arc is developed, it is drawn into the arc chute where it is divided into smaller arcs, which are extinguished faster.

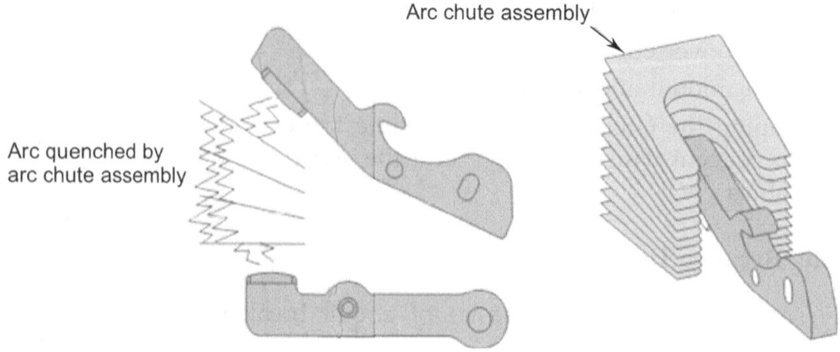

Fig. 17.11

Operating mechanism: As previously stated, a circuit breaker must provide a manual means for energizing and de-energizing a circuit and must be capable of being reset after a fault condition has been cleared. These capabilities are typically provided through use of an operating handle (Fig. 17.12). The operating handle is connected to the moveable contact arm through an operating mechanism.

Trip unit: In addition to providing a means to open and close its contacts manually, a circuit breaker must automatically open its contacts when an over-current condition is sensed. The trip unit (Fig. 17.13) is the part of the circuit breaker that determines when the contacts will open automatically.

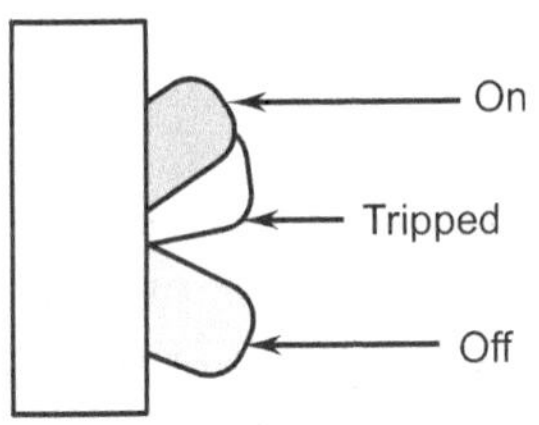

Fig. 17.12

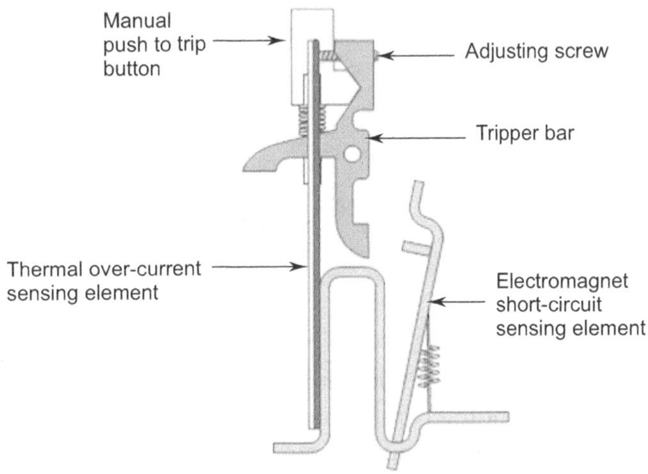

Fig. 17.13

17.7 PROTECTION OF BUILDING AGAINST LIGHTNING

A lighting strike can average 100 million volts of electricity, having current of up to 100,000 amperes and can generate heat of 54,000°F. Lightning strikes somewhere on the earth every second.

A lightning protection system is designed to protect a structure from damage due to lightning strikes by intercepting such strikes and safely passing their extremely high voltage currents to 'ground'. A lightning protection system includes a network of air terminals, bonding conductors, and ground electrodes designed to provide a low impedance path to ground for potential strikes.

Because of the high energy and current levels associated with lightning (currents can be in excess of 150,000 amps), and the very rapid rise time of a lightning strike, no lightning protection system can guarantee absolute safety from lightning. Lightning current will divide to follow every conductive path to ground, and even the divided current can cause damage. Secondary 'side-flashes' can be enough to ignite a fire, blow apart brick, stone, or concrete, or injure occupants within structure or building. However, the benefits of basic lightning protection systems have been evident for well over a century.

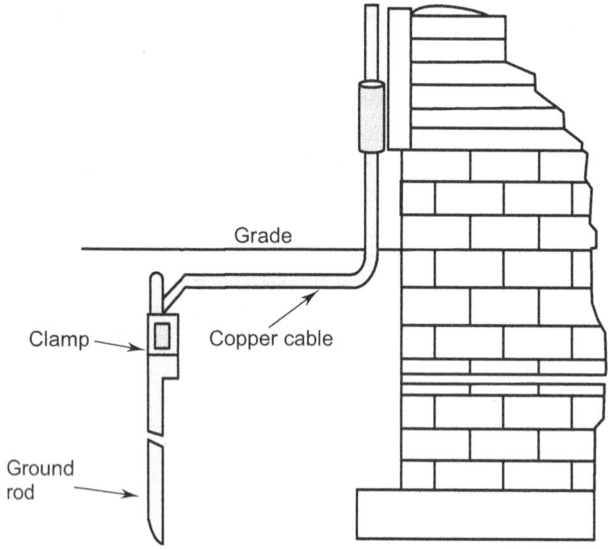

Fig. 17.14

The parts of a lightning protection system are air terminals (lightning rods or strike termination devices), bonding conductors, ground terminals (ground or earthing rods, plates, or mesh), and all of the connectors and supports to complete the system. The air terminals are typically arranged at or along the upper points of a roof structure, and are electrically bonded together by bonding conductors (called down conductors or down-leads), which are connected by the most direct route to one or more grounding or earthing terminals. Connections to the earth electrodes must not only have low resistance, but must have low self-inductance.

A lightning rod or lightning conductor is a metal rod or metallic object mounted on top of a building, electrically bonded using a wire or electrical conductor to interface with ground or 'earth' through an electrode, engineered to protect the building in the event of lightning strike. If lightning targets the building it will preferentially strike the rod and be conducted to ground through the wire, instead of passing through the building, where it could start a fire or cause electrocution. Lightning rods come in many different forms, including hollow, solid, pointed, rounded, flat strips of copper or aluminum. The main attribute of all lightning rods is that they are conductive.

EXERCISE

1. What do you understand by earthing?
2. What are the objectives of the earthing?
3. What are the various methods of earthing? Explain any one in detail.
4. What is plate earthing? Explain in detail with neat sketch diagram.
5. What is pipe earthing? Explain in detail with neat sketch diagram.

6. What are the various protective devices for wiring appliances and installations?
7. What is fuse? What is its principle of operation?
8. What is the criterion for selecting a fuse wire?
9. What are the various types of fuses?
10. What are the disadvantages of ordinary fuses?
11. What are the advantages of re-wireable fuse?
12. What are the advantages of HRC fuses?
13. What are the disadvantages of HRC fuses?
14. What are circuit breakers? Explain its design in detail.
15. Explain how to protect a building against lightning.

APPENDICES

OBJECTIVE TYPE QUESTIONS

1. In the circuit shown in Fig. 1 the current 'I' marked is given by

(*a*) 1/3 (*b*) 1/5 (*c*) 1 (*d*) 5/3

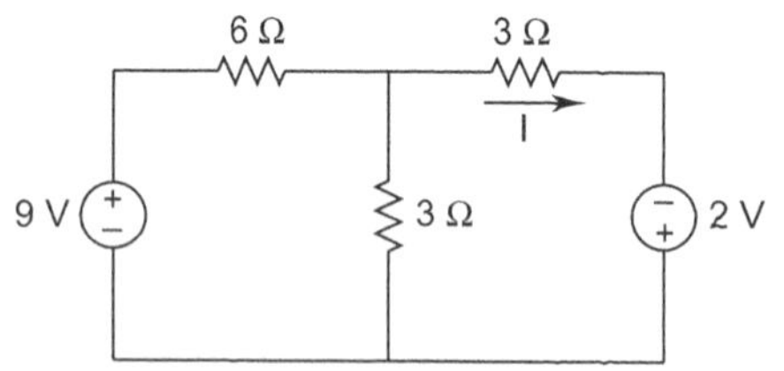

Fig. 1

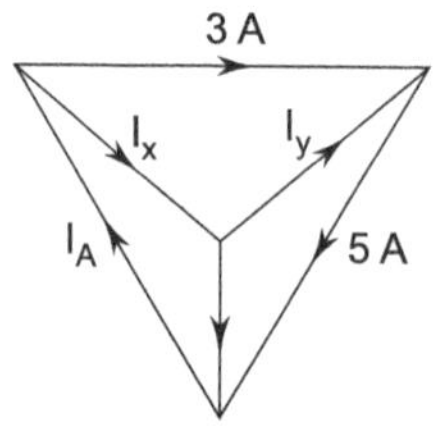

Fig. 2

2. In Fig. 2 currents in some branches are given. The value of I_x and I_y respectively are

(*a*) – 2, 2 (*b*) 2, – 2 (*c*) 2, 2 (*d*) – 2, – 2

3. The voltage labeled V in Fig. 3 is given by

(*a*) 2 (*b*) 0 (*c*) 9 (*d*) 3

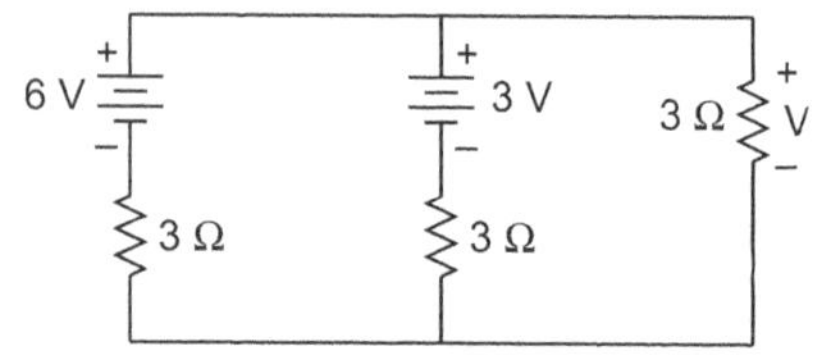

Fig. 3

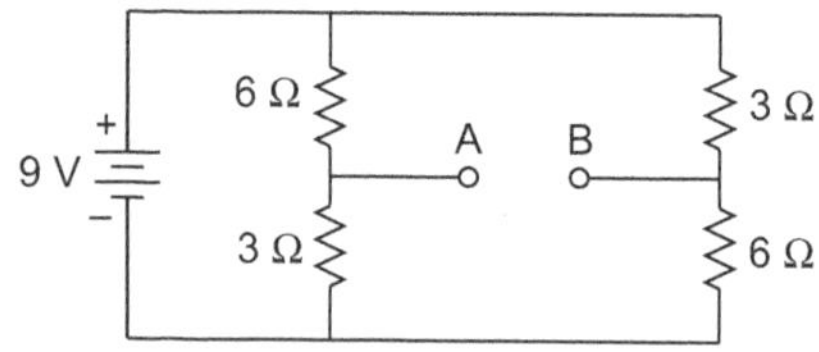

Fig. 4

4. The Thevenin's equivalent across point AB in Fig. 4 is given by V_{th} and R_{th}. Their values are

(*a*) $V_{th} = 3$ $R_{th} = 4$ (*b*) $V_{th} = -3$ $R_{th} = 4$

(*c*) $V_{th} = -3$ $R_{th} = 4.5$ (*d*) $V_{th} = 6$ $R_{th} = 4$

5. The Norton's equivalent across AB in Fig. 5 is given by

(*a*) $I_{sc} = V_s/3$ $R_{th} = 3/2$ (*b*) $I_{sc} = V_s/6$ $R_{th} = 3/2$

(*c*) $I_{sc} = V_s/4$ $R_{th} = 6$ (*d*) $I_{sc} = V_s/2$ $R_{th} = 2/3$

6. $V = \sqrt{2} \times 200 \cos 500t$, $P_{av} = 250$ Watt, power factor = 0.7 lagging. The reactive power is given by

(*a*) 250 VAR (*b*) 0 VAR

(*c*) 500 VAR (*d*) 175 VAR

7. The dimension of L/CR is

(*a*) Volt/Amp (*b*) Amp/ Volt sec^2

(*c*) Seconds (*d*) None of these

8. The Thevenin's equivalent across AB in Fig. 6 is given by

(*a*) – 16, 10/3 (*b*) – 12, 10

(*c*) 4, 16/5 (*d*) 1, 10/3

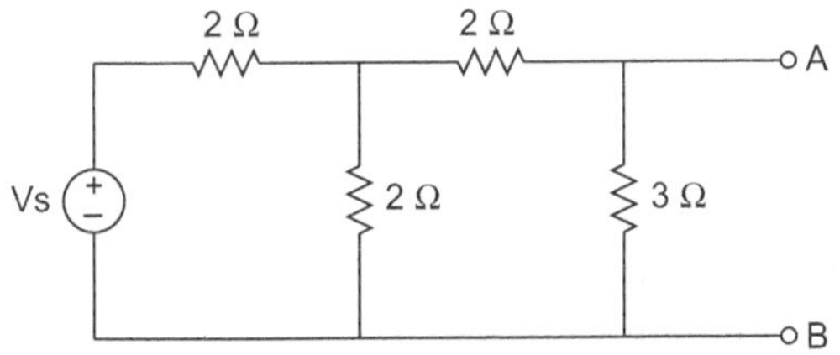

Fig. 5

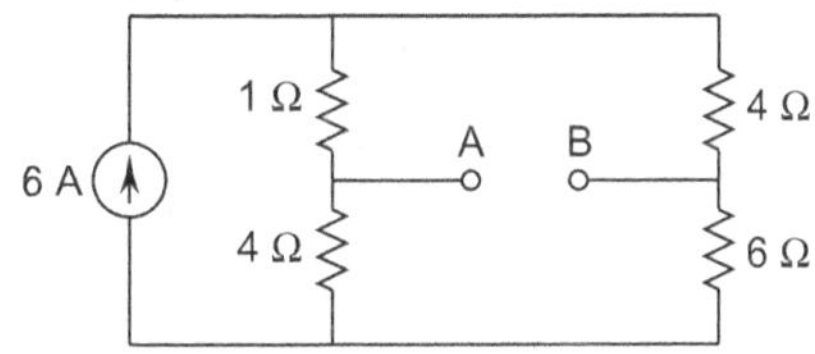

Fig. 6

9. In electronics the term IC denotes

(*a*) Industrial control (*b*) Integrated circuits

(*c*) Internal combustion (*d*) Indian calculations

10. The forbidden energy gap in semiconductors

(*a*) Lies just below the valance band

(*b*) Is the same as the valance band

(*c*) Lies just above the conduction band

(*d*) Lies between the valance band and the conduction band

11. A zener diode has a

(*a*) High forward voltage rating

(*b*) Negative resistance

(*c*) High amplification

(*d*) Sharp breakdown voltage at low reverse voltage

12. When a reverse bias is applied across a crystal diode, it

(*a*) Raises of potential barrio

(*b*) Lowers the potential barrio

(*c*) Increases the majority-carrier current greatly

(*d*) None of these

13. Under normal operating voltage, the reverse current in a silicon diode is about

(*a*) 10 mA (*b*) 1 μA (*c*) 1000 μA (*d*) None of these

14. In a single phase AC motor, condenser is uses

(*a*) For splitting the phase (*b*) Minimizing the radio interference

(*c*) Minimizing the losses (*d*) Minimizing the current

15. What is called the electromotive force (e.m.f) of a voltage source?

(*a*) Terminal voltage when load is applied

(*b*) Internal voltage when no load is applied

(*c*) Product of internal resistance and load current

(*d*) Electric pressure provided to the load

16. The condition in ohm's law is that
 (*a*) The temperature should remain constant
 (*b*) The ratio of V/I should be constant
 (*c*) The temperature should vary
 (*d*) Current should be proportional to voltage

17. Conductance is analogous to
 (*a*) Flux (*b*) Reluctance (*c*) Permeance (*d*) None of these

18. With the increase of frequency, the capacitive reactance or the circuit
 (*a*) Decreases (*b*) Increases (*c*) Remain same (*d*) None of these

19. The form factor is
 (*a*) The ratio of rms value to average value
 (*b*) The ratio of peak value to rms value
 (*c*) The ratio of average value to rms value
 (*d*) None of these

20. The maximum and minimum value of power factor can be
 (*a*) 1 and 0 (*b*) +1 and –1 (*c*) +1 and –5 (*d*) +5 and –5

21. A series RLC circuit is fed from 230 volts variable frequency supply. If the frequency is adjusted to create resonance, the voltage across the resistance is
 (*a*) 230 V (*b*) Less than 230 V (*c*) 0 V (*d*) Greater than 230 V

22. A series circuit has R = 10 Ω, L = 0.01 H and C = 10 μF. The Q-factor is
 (*a*) 100 (*b*) 10 (*c*) 115 (*d*) 10.1

23. In an RLC circuit, the current at resonance is
 (*a*) Maximum in series circuit and minimum in parallel circuit
 (*b*) Maximum in parallel circuit and minimum in series circuit
 (*c*) Maximum in both the circuits
 (*d*) Minimum in both the circuits

24. The star transformation of the following circuit shown in Fig. 7 is
 (*a*) 1, 2, 3 (*b*) 1/2, 1/6, 1/3 (*c*) 3/11, 1/11, 2/11 (*d*) 1, 1/3, 1/2

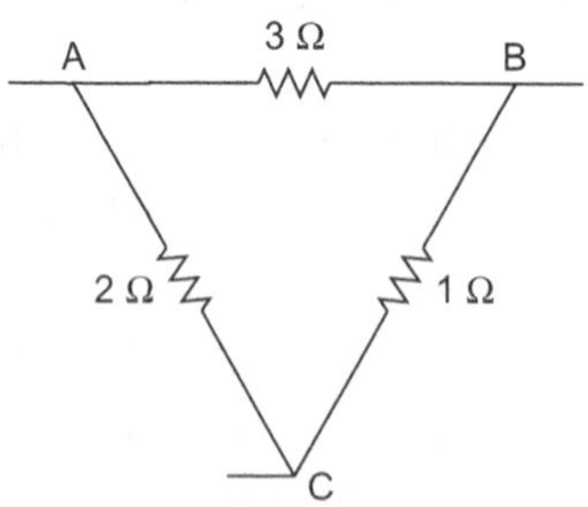

Fig. 7

25. If the speed of the DC generator is increased, the generated e.m.f. will
 (*a*) Increase (*b*) Decrease
 (*c*) Remain constant (*d*) Decreases and then increases

26. In a DC motor the iron losses occur in
 (*a*) The yoke (*b*) The armature (*c*) The field (*d*) None of these

27. If the current in the armature of a DC series motor is reduced by 5%, the torque of the motor will become
    (*a*) 50% of the previous value (*b*) 25% of the previous value
    (*c*) 150% of the previous value (*d*) 125% of the previous value
28. The armature of DC motor is laminated
    (*a*) To reduce the hysteresis losses
    (*b*) To reduce the eddy current losses
    (*c*) To reduce the inductivity of the armature
    (*d*) To reduce the mass of the armature
29. What happens if the field winding of a running shunt motor suddenly breaks open?
    (*a*) Its speed slow down (*b*) Its speed becomes dangerously high
    (*c*) It gives out spark (*d*) It stops at once
30. A synchronous motor can run at a
    (*a*) Leading power factor
    (*b*) Unity power factor
    (*c*) Lagging or leading or unity power factor
    (*d*) Zero power factor
31. The power factor of the input power to a synchronous motor is adjusted by varying
    (*a*) Number of poles (*b*) Magnitude of excitation
    (*c*) Magnitude of armature reaction (*d*) None of these
32. If a synchronous motor is switched on to 3-phase supply with its rotor winding short-circuited, it will
    (*a*) Start
    (*b*) Not start
    (*c*) Start and continue to run as induction motor
    (*d*) Start and continue to run as synchronous motor
33. Alternator generates
    (*a*) DC (*b*) AC (*c*) DC and AC both (*d*) Pulsating DC
34. The rotor of the alternator requires
    (*a*) DC (*b*) AC (*c*) Pulsating DC (*d*) None of the above
35. The rotor of the alternator has
    (*a*) 4 slip rings (*b*) 3 slip rings (*c*) 2 slip rings (*d*) No slip rings.
36. The different type of rotor of an alternator are
    (*a*) Salient pole type
    (*b*) Cylindrical pole type
    (*c*) Both salient pole and cylindrical pole type
    (*d*) None of these
37. Cylindrical pole type rotors are generally used with prime movers of
    (*a*) High speed (*b*) Low speed (*c*) Medium speed (*d*) Low and high speed
38. An alternator running at 3000 rpm generated voltage at 50 Hz. The no. of poles of the alternator will be
    (*a*) 8 poles (*b*) 6 poles (*c*) 4 poles (*d*) 2 poles

39. The frequency of a 4 pole alternator running at 1500 rpm will be

(*a*) 150 Hz (*b*) 100 Hz (*c*) 50 Hz (*d*) 25 Hz

40. The e.m.f generated in an alternator depends upon

(*a*) Coil span factor (*b*) Frequency (*c*) Flux per pole (*d*) All the above three

41. If the speed of the alternator is changed from 3000 rpm to 1500 rpm the e.m.f generated will become

(*a*) Twice (*b*) Half (*c*) One forth (*d*) Four times

42. The condition for running two alternator in parallel are

(*a*) Terminal voltage should be same (*b*) Frequency should be same

(*c*) Phase sequence should be same (*d*) All the above three

43. The frequency per revolution in an alternator is equal to

(*a*) Number of poles (*b*) Number of armature conductors

(*c*) Number of pair of poles (*d*) None of these

44. What is typical use of an autotransformer?

(*a*) Isolating transformer (*b*) Control transformer

(*c*) Toy transformer (*d*) Variable transformer.

45. The full load copper loss and iron loss of a transformer are 6400 watts and 5000 watts respectively. The above copper loss and iron-loss at half load will be

(*a*) 3200 W, 2500 W respectively (*b*) 3200 W, 5000 W respectively

(*c*) 1600 W, 1250 W respectively (*d*) 1600 W, 5000 W respectively

46. The efficiency in transformer is maximum at

(*a*) Full load (*b*) Half the full load

(*c*) One third the full load (*d*) No load

47. The primary reason why open-circuit test is performed on the low-voltage winding of the transformer is that

(*a*) It draws sufficiently large no load current for convenient reading

(*b*) It requires least voltage to perform the test

(*c*) It needs minimum power input

(*d*) It has become a universal custom

48. During short-circuit test, the iron-loss of a transformer is negligible because

(*a*) The entire input is just sufficient to meet copper losses only

(*b*) Voltage applied across the HV side is a small fraction of the rated voltage and so is the flux

(*c*) Iron core becomes fully saturated

(*d*) Supply frequency is held constant

49. The main purpose of using core is a transformer is to

(*a*) Decrease iron loss

(*b*) Prevent eddy current loss

(*c*) Eliminate magnetic hysteresis

(*d*) Decrease reluctance of the common magnetic flux path

50. A transformer has a percentage resistance of 1% and percentage reactance of 4%. Its regulation of power factor 0.8 lagging and 0.8 leading are respectively

(*a*) 3.2% and – 1.6% (*b*) 6% and – 4% (*c*) 4.8% and – 3.2% (*d*) None of these.

51. Eddy current loss in a transformer depends upon

(*a*) Both voltage and frequency (*b*) Voltage alone

(*c*) Load current alone (*d*) Thickness of core

52. Hysteresis loss in a transformer depends upon

(*a*) Frequency (*b*) Supply voltage

(*c*) Square of frequency alone (*d*) Square of the voltage alone

53. The iron losses in a 10 kVA, 240/100 V single phase transformer are 100 watts. The maximum efficiency occurs at full load. When the secondary currents is 50 A, the copper losses in the transformer will be

(*a*) 100 W (*b*) 50 W (*c*) 25 W (*d*) 10 W

54. During a short-circuit test, the iron losses are negligible because.

(*a*) The mutual flux is small (*b*) The power factor is low

(*c*) The current is high (*d*) None of these.

55. If DC supply is given to a transformer, it will

(*a*) Work

(*b*) Not work

(*c*) Gives lower voltage than the rated voltage on secondary side

(*d*) Burn the winding

56. A synchronous can run at

(*a*) A leading power factor

(*b*) Unity power factor

(*c*) Lagging or leading or unity power factor

(*d*) Zero power factor

57. An induction motor is coupled to a synchronous motor. In order to bring up the synchronous motor to synchronous speed at starting, the pair of pole of induction motor should be

(*a*) Less than that of synchronous motor

(*b*) Equal to that of synchronous motor

(*c*) Greater than that of synchronous motor

(*d*) None of these

58. A 10 pole, 25 Hz alternator is directly coupled to, and is driven by a 60 Hz synchronous motor than the number of poles in a synchronous motor are

(*a*) 48 poles (*b*) 12 poles (*c*) 24 poles (*d*) None of these

59. In a synchronous generator

(*a*) The field m.m.f lags the air-gap flux and the air-gap flux lags the armature m.m.f.

(*b*) The field m.m.f leads the air-gap flux and the air-gap flux leads the armature m.m.f.

(*c*) The fields m.m.f leads the air-gap flux lags the armature m.m.f.

(*d*) The field m.m.f lags the air-gap flux and the air-gap flux leads the armature m.m.f.

60. The synchronous motor run at

(*a*) Less than synchronous speed (*b*) Synchronous speed
(*c*) More than synchronous speed (*d*) None of the above

61. The construction of synchronous motor is similar to

(*a*) DC compound motor (*b*) Slip ring induction motor
(*c*) DC shunt generator (*d*) Alternator

62. The synchronous motor runs on

(*a*) 3-phase AC supply (*b*) 3-phase AC and DC supply
(*c*) DC supply only (*d*) 3-phase AC and single phase AC.

63. If a synchronous motor is switch on 3-phase supply with its rotor winding short-circuited, it will

(*a*) Start
(*b*) Not start
(*c*) Start and continue to run as induction motor
(*d*) Start and continue to run as synchronous motor

64. The magnitude of the e.m.f induced in the stator due to revolving flux will depends upon the

(*a*) Speed of the motor (*b*) DC excitation current
(*c*) Load on the motor (*d*) Speed and rotor flux

65. Under no load running condition, the angle between the induced voltage and supply voltage will be

(*a*) 180 degree (*b*) 90 degree
(*c*) Between 90 degree and 180 degree (*d*) Zero

66. Under running condition on load, the angle between the induced voltage and supply voltage will be

(*a*) Zero (*b*) 180 degree (*c*) 90 degree (*d*) None of these

67. If the field of synchronous motor under excited, the power factor will be

(*a*) More than unity (*b*) Unity (*c*) Lagging (*d*) Leading

68. When the synchronous motor is started, the field winding is initially

(*a*) Short circuited (*b*) Open circuited
(*c*) Excited by a DC source (*d*) None of these

69. V-curve of a synchronous motor show the relation between

(*a*) Armature current and field current (*b*) Applied voltage and field current
(*c*) Applied voltage and armature current (*d*) None of these

70. For a given load, the armature current of a synchronous motor is minimum when the power factor is

(*a*) Unity (*b*) Lightly less than unity and lagging
(*c*) Slightly less than unity and leading (*d*) None of these

71. Synchronous motor generally have

(*a*) Cylindrical rotor (*b*) Salient pole rotor
(*c*) Both (*a*) and (*b*) are correct (*d*) None of these

72. Synchronous motor are usually operated at

(*a*) Unity power factor (*b*) Leading power factor

(*c*) Lagging power factor (*d*) None of these

73. High speed alternator usually have

(*a*) Salient pole rotors (*b*) Cylindrical rotors

(*c*) Both (*a*) and (*b*) (*d*) None of these

74. Cylindrical rotor alternators have

(*a*) Large length to diameter ratio (*b*) Small length to diameter ratio

(*c*) Vertical configuration (*d*) None of these

75. Salient pole machine have

(*a*) Large number of poles (*b*) Small number of poles

(*c*) Small diameters (*d*) Longs cores.

76. Armature winding is one in which working

(*a*) Flux is produced by field current (*b*) Flux is produced by the working current

(*c*) E.m.f. is produced by the working flux (*d*) E.m.f. is produced by the leakage flux

77. A coil consists of

(*a*) Two conductors (*b*) Two coil-sides

(*c*) Two turns (*d*) Four turns.

78. When a substance is placed in a magnetic field, its ability to get magnetized depends upon its

(*a*) Permittivity (*b*) Permeability

(*c*) Susceptibility (*d*) None of these

79. For induction motors, normally

(*a*) The rotor winding is connecting to supply and the stator winding is short-circuited

(*b*) The stator winding is connected to supply and the rotor winding is short-circuited

(*c*) Both the stator and the rotor windings are connected to supply

(*d*) The stator winding is connected to supply and the rotor winding is open circuited.

80. If the speed of a induction motor be synchronous speed, then the relative speed between the rotating flux and the rotor will be

(*a*) Maximum and hence torque will be maximum

(*b*) Maximum and hence torque will be zero

(*c*) Zero and hence torque will be maximum

(*d*) Zero and hence torque will be zero

81. If the rotor winding of a wound rotor 3-phase induction motor is connected to balance 3-phase supply and the stator winding is short-circuited, then the rotor will

(*a*) Not run at all

(*b*) Run at sub-synchronous speed in the direction of rotating flux

(*c*) Run at sub-synchronous speed against the direction of rotating flux

(*d*) Run synchronously with rotating flux

82. In terms of air-gap power, p_g the rotor copper loss and the mechanical power developed are given by

(*a*) sP_g and $(1 - s)P_g$ (*b*) $(1 - s)P_g$ and sP_g
(*c*) P_g and P_g/s (*d*) P_g/s and $(1 - s)P_g$

83. In a 3-phase induction motor, internal developed torque, T_e in terms of supply voltage, V_1 is proportional to
(*a*) V_1 (*b*) $\sqrt{V_1}$ (*c*) V_1^2 (*d*) None of these

84. In a 3-phase induction motor, slip for maximum torque, in terms of rotor resistance R_2 is
(*a*) Independent of R_2 (*b*) Directly proportional to R_2
(*c*) Inversely proportional to R_2 (*d*) None of these

85. In a 3-phase induction motor, the current is produced in the rotor conductors by
(*a*) Giving AC supply (*b*) Giving DC supply
(*c*) Induction effect (*d*) Pulsating DC supply

86. The direction of rotation of 3-phase revolving field can changed by interchanging
(*a*) R and Y phase only (*b*) B and Y phase only
(*c*) R and B phase only (*d*) Any two phases.

87. Different types of 3-phase induction motors are
(*a*) Squirrel cage (*b*) Slip ring
(*c*) Commutator (*d*) All of these

88. In a 3-phase squirrel cage induction motor, the
(*a*) Rotor conductors are kept open
(*b*) Rotor conductors are shorts-circuited through end rings
(*c*) Ends of the rotor conductors are connected to slip ring
(*d*) Ends of the rotor conductors are short-circuited through slipped

89. The frequency of the induced e.m.f in the rotor circuit is
(*a*) Maximum at standstill
(*b*) Zero at standstill
(*c*) Maximum at synchronous speed
(*d*) None of these

90. Generally, a 3-phase slip ring induction motor is started with
(*a*) DOL starter (*b*) Star-delta starter
(*c*) Auto-transformer starter (*d*) Stator-rotor starter

91. The meter used for measuring electrical quantities are
(*a*) Tachometer (*b*) Micrometer
(*c*) Measuring instrument (*d*) Spherometer

92. The meter used for electrical power measurement is called
(*a*) kWh meter (*b*) Voltmeter (*c*) Ammeter (*d*) Wattmeter

93. The meter used for measuring electrical energy of consumer is called
(*a*) Wattmeter (*b*) Ampere hour meter
(*c*) kWh meter (*d*) Avometer

94. The deflecting torque can be produced by
(*a*) Gravity control (*b*) Spring control (*c*) Air fraction (*d*) Magnetically

95. The damping torque can be produced by

(*a*) Eddy currents (*b*) Gravity control (*c*) Electro statically (*d*) Thermally

96. Moving iron instrument are

(*a*) Attraction type (*b*) Repulsion type
(*c*) Attraction and repulsion type (*d*) Dynamometer type

97. Moving coil instrument are

(*a*) Permanent magnet type (*b*) Dynameters type
(*c*) Induction type (*d*) Permanent magnet and dynamometer type

98. The current in the pressure coil is proportional to

(*a*) Load current (*b*) Line current
(*c*) Supply voltage (*d*) Supply voltage and line current both

99. Megger is used for measuring

(*a*) Low resistance (*b*) High resistance
(*c*) Medium resistance (*d*) Very low resistance

100. Sensitivity of a voltmeter is expressed as

(*a*) Volt/Ω (*b*) Ω/volt (*c*) Ω volt (*d*) 1/Ω volt

101. The number of slip rings in a 3-phase wound rotor induction motor is

(*a*) 3 (*b*) 4 (*c*) 6 (*d*) 12

102. In a 50 Hz, 3-phase induction motor, frequency of rotor current is

(*a*) 50 (*b*) about 2 Hz (*c*) about 10 Hz (*d*) 0 Hz

103. A 3-phase induction motor with wound rotor has a delta-connected stator winding. For proper operation, the rotor winding

(*a*) Must be delta connected (*b*) Must be star connected
(*c*) May be connected in delta or star (*d*) None of these

104. At slip = ∞ the torque of a 3-phase induction motor is

(*a*) 0 (*b*) small (*c*) very high (*d*) high

105. The ratio of no load current to full load current of a 3-phase induction motor is about

(*a*) 0.05 (*b*) 0.2 (*c*) 0.5 (*d*) 0.8

106. What are the advantages of the slip ring induction motor over the squirrel cage induction motor?

(*a*) It is suitable for higher speeds
(*b*) Its efficiency is higher
(*c*) Its power factor is higher
(*d*) It can be started with the help of rotor resistances.

107. What are the disadvantages of starting an induction motor with a star-delta starter?

(*a*) The starting torque increases and the motor runs with jerks
(*b*) The starting torque is one third of the torque in case of the delta connection
(*c*) By changing over from star to delta connection a voltage peak is produced which endangers the insulation of the motor
(*d*) During starting, high losses are reduced

108. An induction motor is started on the 220/380 V supply mains with the helps of a star-delta starter. Which types of induction motor is this?

(*a*) 220/380 V squirrel cage induction motor

(*b*) 220–380 V slip ring induction motor

(*c*) 380–660 V squirrel cage induction motor

(*d*) 380/660 V slip ring induction motor

109. What are the advantages of starting a slip ring induction motor with the help of rotor resistances as compared to other methods?

(*a*) The starting current is reduced

(*b*) The starting torque increase due to the rotor resistances

(*c*) The starter has to be designed for only a very low current

(*d*) The starter can be built directly in to the motor

110. While starting an induction motor with star-delta starter at the stator position

(*a*) The torque developed is 3 times as high as at the run position

(*b*) The torque developed is $\sqrt{3}$ times as high as at the run position

(*c*) The voltage per phase is $1/\sqrt{3}$ of the supply voltage

(*d*) The current is $1/\sqrt{3}$ of the current at direct on line starting

111. Synchronous speed is defined as

(*a*) The speed of the rotor of an induction motor

(*b*) The speed of the synchronous motor

(*c*) The speed of an induction motor at no load

(*d*) The natural speed at which the magnetic field rotates

112. Which of the following is not a part of a squirrel cage induction motor?

(*a*) Rotor (*b*) stator (*c*) carbon brushes (*d*) shaft

113. Which of the following formula is used to calculate the synchronous speed of an induction motor?

(*a*) $N_s = P \times f$ (*b*) $N_s = 120 \times P/f$ (*c*) $N_s = 120 \times f/p$ (*d*) $N_s = f \times p/60$

114. An induction motor has a rated speed of 715 rpm. How many poles have its rotating magnetic field?

(*a*) 2 poles (*b*) 4 poles (*c*) 6 poles (*d*) None of these

115. The rated speed of an induction motor is 1410 rpm. What is meant by the statement the 'slip $s = 1$'?

(*a*) The speed of the motor is higher than the synchronous speed

(*b*) The motor runs at a synchronous speed

(*c*) The motor runs at its rated speed

(*d*) The rotor is stationary

116. A machine is driven by an induction motor running at nominal speed, what happens if the counter-torque of the machine becomes larger than the maximum torque of the motor?

(*a*) The induction motor heats up to an inadmissible extent

(*b*) The induction motor is stops

(*c*) The speed of the induction motor is reduced to less than half of its normal speed

(*d*) The winding of the motor can be damaged because of high current strength

117. A 3-phase, 400 V, 50 Hz, 4 poles induction motor is fed from a 3-phase, 400 V supply and runs at 1425 rpm the frequency of the rotor e.m.f is

(*a*) 2.5 Hz (*b*) 50 Hz (*c*) 48 Hz (*d*) zero

118. A change of 5% in supply voltage to a 3-phase induction motor will produced the approximate change in the torque of

(*a*) 5% (*b*) 7.5% (*c*) 10% (*d*) 25%

119. The stator of an unloaded 3-phase, 6-pole wound rotor induction motor is connected to a 60 Hz source and the rotor is connected to a 25 Hz source. The rotor speed at which steady state motor action will result is given by

(*a*) 1200 rpm (*b*) 700 rpm (*c*) 500 rpm (*d*) None of these

120. A 50 Hz, 10 pole, 3-phase induction motor is used to drive a 60 Hz, 3-phase alternator, the pole number of alternator should be

(*a*) 10 (*b*) 6 (*c*) 12 (*d*) None of these

121. Frequency is 50 Hz and number of poles are 4 if rotor speed is 95%, the actual speed of motor is

(*a*) 1425 rpm (*b*) 1300 rpm (*c*) 1500 rpm (*d*) None of these

122. The mechanical load in an induction motor can be represented by

(*a*) $\left(\frac{R_2}{s} - R_2\right)$ (*b*) $\frac{R_2 - s}{R_2}$ (*c*) $\frac{1}{s(R_2 - 1)}$ (*d*) $\frac{R_2}{(s+1)}$

123. Maximum power is delivered in an induction motor when the load resistance R_L and internal impedance Z are

(*a*) Equal (*b*) In 2 : 1 ratio (*c*) Not equal (*d*) In 1 : 2 ratio

124. Which of the following statement is true for an induction motor?

(*a*) If $(1 - s)$ is the power input to the rotor, the power lost in the rotor is $s(1 - s)$

(*b*) Maximum torque of an induction motor is independent of the rotor resistance.

(*c*) An induction motor can supply the load stable from $s = 0$ to $s = R_2/X_2$

(*d*) The efficiency of the induction motor is maximum when $R_2 = X_2$.

125. In an induction motor, the rotor input is 600 W and slip is 4%, the rotor copper loss is

(*a*) 700 W (*b*) 0.6 kW (*c*) 650 W (*d*) 24 W

126. Which of the following is most economical method for starting single-phase motor?

(*a*) Capacitor start method (*b*) Split phase method

(*c*) Induction start method (*d*) Resistance start method

127. What are the advantages of the slip ring induction motor over the squirrel cage induction motor?

(*a*) It is suitable for higher speeds

(*b*) Its efficiency is higher

(*c*) Its power factor is higher

(*d*) It can be started with the help of rotor resistance

128. The starting torque of the slip ring induction motor can be increased by

(*a*) Adding resistance to the rotor (*b*) Adding resistance to stator

(*c*) Adding resistance to both (*d*) None of these

129. A 400 kW, 3-phase, 440 volts and 50 Hz induction motor has a speed of 950 rpm on full load. The machine has 6 poles. The slip of machine will be

(*a*) 0.06 (*b*) 0.05 (*c*) 0.04 (*d*) 0.02

130. If the speed is 460 rpm than the number of poles of a 50 Hz induction motor are

(*a*) 4 (*b*) 8 (*c*) 12 (*d*) 18

131. If a 3-phase, 60 Hz, 10 poles induction motor has a full load slip of 0.075, than the speed of the rotor relative to revolving field will be

(*a*) 54 rpm (*b*) 1000 rpm (*c*) Zero rpm (*d*) 720 rpm

132. The purpose of skewing of rotor slots in induction motor is

(*a*) to reduced the magnetic hum of motor

(*b*) to increase the distribution factor

(*c*) to reduced the locking tendency of rotor

(*d*) to increase the breath factor

133. In the induction motor the ratio of starting torque to maximum torque is 80%. The speed at maximum torque, if the synchronous speed is 1500 rpm, is

(*a*) 1290 rpm (*b*) 1360 rpm (*c*) 1440 rpm (*d*) None of these

134. In an induction motor the rotor resistance is equal to standstill reactance than the maximum torque is

(*a*) Less than starting torque (*b*) Equal to starting torque

(*c*) More than starting torque (*d*) None of these

135. In an induction motor the ratio of starting current to full load current is 12.65 and the ratio of starting torque to full load torques is 1.6, then percentage slip at full load is

(*a*) 0.01% (*b*) 0.5% (*c*) 1% (*d*) None of these

136. A three phase induction motor connected in star takes 5 A when connected in delta, it will takes

(*a*) $5/\sqrt{3}$ A (*b*) $5\sqrt{3}$ A (*c*) 15 A (*d*) 5 A

137. Whenever any poly-phase induction motor is loaded, then

(*a*) Induced e.m.f in the rotor increases and its frequency falls

(*b*) Induced e.m.f in rotor increases and its frequency rises

(*c*) Induced e.m.f in the rotor remains constant

(*d*) Induced e.m.f decreases and frequency increases

138. What is typical use of an auto-transformer?

(*a*) Isolating transformer (*b*) Control transformer

(*c*) Toy transformer (*d*) Variable transformer

139. The full load copper loss and iron-loss of a transformer are 6400 watts and 5000 watts respectively. The above copper loss and iron-loss at half load will be

(*a*) 3200 W, 2500 W respectively (*b*) 3200 W, 5000 W respectively

(*c*) 1600 W, 1250 W respectively (*d*) 1600 W, 5000 W respectively

140. The efficiency in transformer is maximum at

(*a*) Full load (*b*) Half the full load

(*c*) One third the full load (*d*) No load

141. The primary reason why open-circuit test is performed on the low-voltage winding of the transformer is that it

(*a*) Draws sufficiently large no load current for convenient reading

(*b*) Requires least voltage to perform the test

(*c*) Needs minimum power input

(*d*) Has become a universal custom

142. During short-circuit test, the iron-loss of a transformer is negligible because

(*a*) The entire input is just sufficient to meet copper losses only

(*b*) Voltage applied across the HV side is a small fraction of the rated voltage and so is the flux

(*c*) Iron core becomes fully saturated

(*d*) Supply frequency is held constant

143. The main purpose of using core in a transformer is to

(*a*) Decrease iron loss

(*b*) Prevent eddy current loss

(*c*) Eliminate magnetic hysteresis

(*d*) Decrease reluctance of the common magnetic flux path

144. A transformer has a percentage resistance of 1% and percentage reactance of 4%. Its regulation of power factor 0.8 lagging and 0.8 leading are respectively

(*a*) 3.2% and – 1.6% (*b*) 6% and – 4% (*c*) 4.8% and – 3.2% (*d*) None of these.

145. Eddy current loss in a transformer depends upon

(*a*) Both voltage and frequency (*b*) Voltage alone

(*c*) Load current alone (*d*) Thickness of core

146. Hysteresis loss in a transformer depends upon

(*a*) Frequency (*b*) Supply voltage

(*c*) Square of frequency alone (*d*) Square of the voltage alone

147. The iron losses in a 10 kVA, 240/100 V single phase transformer are 100 watts. The maximum efficiency occurs at full load. When the secondary currents is 50 A, the copper losses in the transformer will be

(*a*) 100 W (*b*) 50 W (*c*) 25 W (*d*) 10 W

148. The no-load current in a transformer flow

(*a*) In primary winding only (*b*) In secondary winding only

(*c*) In primary and secondary both (*d*) In core only

149. How are DC motors classified?

(*a*) Self excited and separately excited (*b*) Slip ring and squirrel cage type

(*c*) Self excited and squirrel cage type (*d*) Separately excited and slip ring type

150. How is induction motors classified?

(*a*) Self excited and separately excited (*b*) Slip ring and squirrel cage type

(*c*) Self excited and squirrel cage type (*d*) Separately excited and slip ring type

151. An induction motor runs at

(*a*) Synchronous speed (*b*) More than the synchronous speed

(*c*) Less than the synchronous speed (*d*) None of these.

152. What kind of supply is given to synchronous machine?

(*a*) AC only (*b*) DC only

(*c*) Both AC and DC only (*d*) None of these

153. An auto-transformer has

(*a*) Two winding (*b*) One winding (*c*) Three windings (*d*) None of these

154. On varying the load, the speed of a DC shunt motor

(*a*) Remains constant

(*b*) It varies considerably on load

(*c*) Decreases in a small amount

(*d*) The speed of DC shunt motor is almost constant with the increase of load, but their is small decrease in speed with increase load.

155. For the circuit shown in Fig. 8, the value of R such that the same amount of power is supplied to the 10 Ω resistance by the current and by the voltage source will be

(*a*) 5 Ω (*b*) 10 Ω (*c*) 15 Ω (*d*) 20 Ω

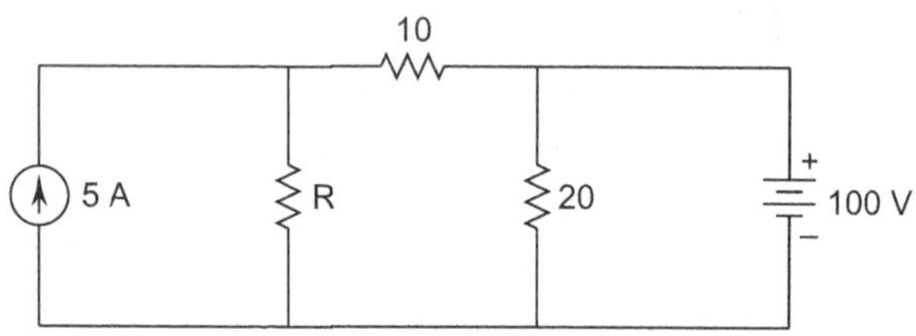

Fig. 8

156. Maximum transmission voltage in India is

(*a*) 220 kV (*b*) 400 kV (*c*) 765 kV (*d*) 1200 kV

157. A voltage source of 100 V has internal impedance 2 Ω and supplies a load having that same impedance the power absorbed by the load is

(*a*) 5000 W (*b*) 1250 W (*c*) 1500 W (*d*) 2500 W

158. A moving coil ammeter has full scale deflection of 50 μA and coil resistance of 100 Ω. The value of the shunt resistance required for the instrument to be converted to read a full-scale reading of A will be

(*a*) 5 Ω (*b*) 0.5 Ω (*c*) 0.05 Ω (*d*) 0.005 Ω

159. If W_1, W_2 and W_3 are the readings of three wattmeter used to measure the power in 3-phase, 4-wire circuit, the total power of load circuit will be

(*a*) $W_1 + W_3$ (*b*) $W_2 + W_3$ (*c*) $W_1 + W_2$ (*d*) $W_1 + W_2 + W_3$

160. Eddy current loss can be reduced by using

(*a*) Electric fan (*b*) Small size core

(*c*) Large size core (*d*) Thin laminated plates

161. A single phase transformer working at maximum efficiency. The copper losses are 100 W, the iron losses would be

(*a*) 50 W (*b*) 200 W (*c*) 100 W (*d*) 2500 W

162. The current drawn by a 120 V DC motor of armature resistance 0.4 Ω and back e.m.f 112 V is

(*a*) 50 A (*b*) 20 A (*c*) 10 A (*d*) 25 A

163. The rotor speed of a six pole 50 Hz induction motor is 960 rpm the percentage slip is

(*a*) 3% (*b*) 4% (*c*) 5% (*d*) 2%

164. The following circuit shows in Fig. 9 will resonates at

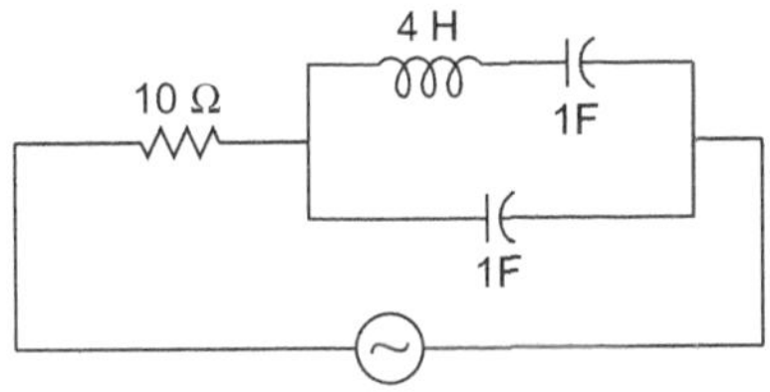

Fig. 9

(*a*) All frequencies (*b*) 0.5 rad/sec (*c*) 5 rad/sec (*d*) None of these

165. While performing the short-circuit test on a single phase transformer, the applied current is

(*a*) 3–10% of full-load current (*b*) 30–40% of full-load current

(*c*) 65–70% of full-load current (*d*) 15–20% of full-load current

166. A series RLC circuit has the following parameters: R = 10 Ω, L = 0.01 H and C = 100 mF. The quality factor of the circuit at resonance is

(*a*) 0.32 (*b*) 0.032 (*c*) 0.0032 (*d*) 32.032

167. Two coils having self inductance of L_1 and L_2 henry are differentially coupled. The mutual inductance between the coils is M. The expression for total inductance will be

(*a*) $L_1 + L_2 + M$ (*b*) $L_1 + L_2 + 2M$ (*c*) $2L_1 + 2L_2 + M$ (*d*) $2L_1 + 2L_2 + 2M$

168. In a normal 3 phase slip ring induction motor, the developed starting torque will

(*a*) Increase with the increase in the value of the rotor resistance

(*b*) Decrease with the increase in the value of the rotor resistance

(*c*) Not depend upon the value of the rotor resistance

(*d*) Increase with the increase in the value of the rotor resistance up to a certain value and then decrease with further increase in the value of the rotor resistance

169. A circuit with resistor, inductor and capacitor in series is resonant at f_o Hz. If all the components values are now doubled, the new resonant frequency will be

(*a*) Double (*b*) Half (*c*) Remain same (*d*) Become four times

170. In a transformer zero voltage regulation at full-load is

(*a*) Not possible

(*b*) Possible at unity power factor load

(*c*) Possible at leading power factor load

(*d*) Possible at lagging power factor load

171. In Fig. 10 the Thevenin's equivalent voltage source and resistance as seen from the terminals A and B will be

(*a*) R_{th} = 10 Ω and V_{th} = 40 Volts (*b*) R_{th} = 15 Ω and V_{th} = 45 Volts

(*c*) R_{th} = 10 Ω and V_{th} = 50 Volts (*d*) R_{th} = 15 Ω and V_{th} = 95 Volts

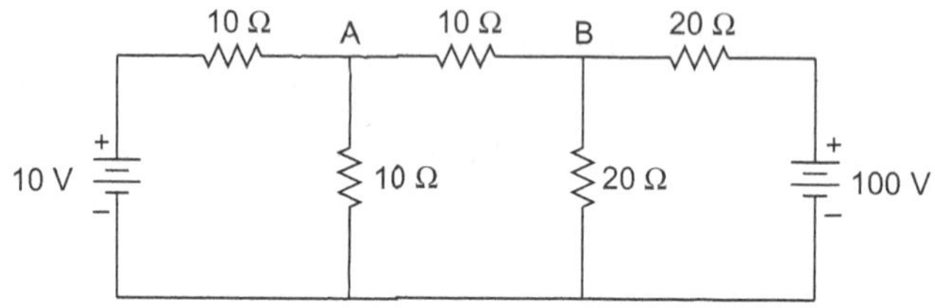

Fig. 10

172. The minimum number of wattmeter(*s*) required to measured 3-phase 3-wire balanced or unbalanced power is

(*a*) One (*b*) Two (*c*) Three (*d*) Four

173. A 4 pole DC generator is running at 1500 rpm. The frequency of current in the armature is

(*a*) 25 Hz (*b*) 50 Hz (*c*) 0 Hz (*d*) 100 Hz

174. In series RLC circuit the power factor during resonance is

(*a*) 0.8 lagging (*b*) Zero (*c*) Unity (*d*) 0.8 leading

175. Distributed winding in three phase alternator is preferred over concentrated winding as it

(*a*) Reduces noise (*b*) Reduces machine size

(*c*) Reduces amount of copper required (*d*) Reduces the harmonics in the generated e.m.f

176. The efficiency of electrical machines is maximum (where x is the fraction of load)

(*a*) Constant losses = variable losses (*b*) Constant losses = x (variable losses)

(*c*) Constant losses =$1/x$ (variable losses) (*d*) Constant losses = 2 (variable losses)

177. Single-phase induction requires starting winding because it has no

(*a*) Starting current (*b*) Current is high (*c*) Starting torque (*d*) Maximum torque

178. Single phase induction motors are used in

(*a*) Paper industry (*b*) Constant speed drive

(*c*) Electric trains (*d*) Electric fans

179. DC series motors are used in

(*a*) Welding (*b*) Paper industry

(*c*) Constant speed drive (*d*) Electric trains

(*e*) Electric fans

180. Three phase induction motor are used in

(*a*) Welding (*b*) Paper industry

(*c*) Constant speed drive (*d*) Electric trains

(*e*) Electric fans

181. Differentially compounded generator are used in

(*a*) Welding (*b*) Paper industry

(*c*) Constant speed drive (*d*) Electric trains

(*e*) Electric fans

182. Synchronous motor are used in

(*a*) Welding (*b*) Paper industry

(*c*) Constant speed drive (*d*) Electric trains

(*e*) Electric fans

183. Which of the elements in the following is not bilateral?

(*a*) Resistor (*b*) Inductor (*c*) Capacitor (*d*) Transistor

184. Three resistances of 3 ohm each are connected in delta. The value of the resistances in the equivalent star is

(*a*) 27 Ω (*b*) 9 Ω (*c*) 1.5 Ω (*d*) 1 Ω

185. In the magnetic circuit

(*a*) Flux is analogous to current

(*b*) Reluctance is analogous to resistance

(*c*) Magneto motive force is analogous to potential difference

(*d*) All of these

186. Which of the following constitute the basic components of a single phase transformer?

(*a*) Tank (*b*) Windings (*c*) Magnetic core (*d*) All of these

187. At which of the following voltage level ranges is the power generated in a power station

(*a*) 1.1–6.6 kV (*b*) 11–30 kV (*c*) 33–66 kV (*d*) None of those

188. A 100 μA ammeter has an internal resistance of 100 Ω. For extending its range to measure 500 μA, the shunt require is of resistance

(*a*) 20 Ω (*b*) 22.22 Ω (*c*) 25 Ω (*d*) 50 Ω

189. The value of resistance R shown in Fig. 11 is

(*a*) 3.5 Ω (*b*) 2.5 Ω (*c*) 1.5 Ω (*d*) 4.5 Ω

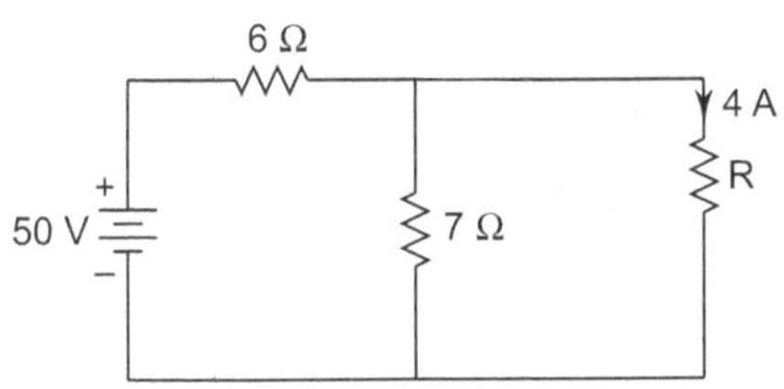

Fig. 11

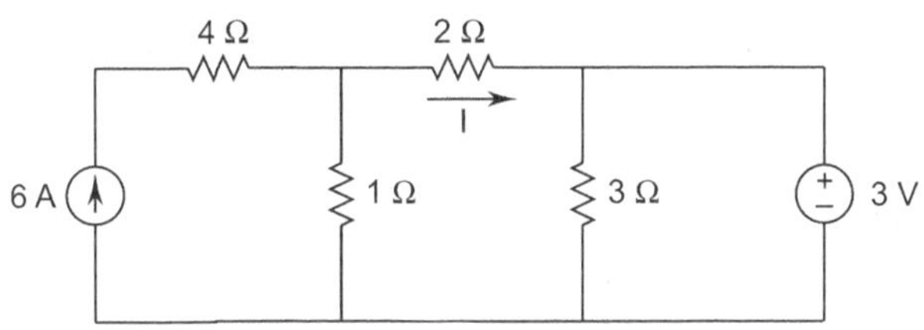

Fig. 12

190. For the circuit shown in Fig. 12 the current I is given by

(*a*) 3 A (*b*) 2 A (*c*) 1 A (*d*) 0 A

191. The maximum power that can be supplied to the load in the following circuit is

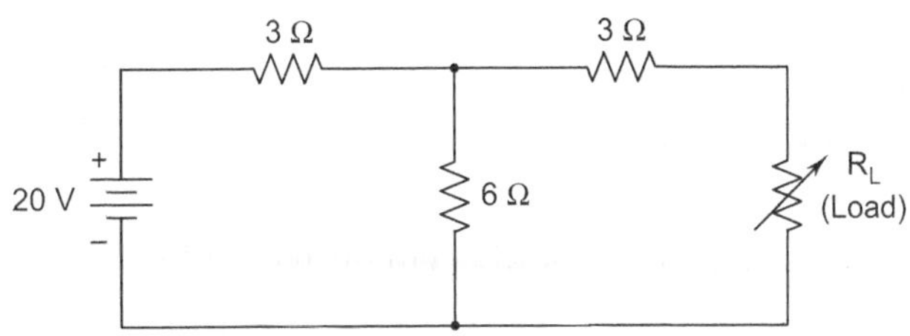

Fig. 13

(*a*) 7.26 (*b*) 8.89 (*c*) 10.08 (*d*) 40.00

192. The coupling between two magnetically coupled coils is said to be the ideal, if the coefficient of coupling is

(*a*) 0 (*b*) 0.50 (*c*) 0.70 (*d*) 1.00

193. Which of the following conditions is common to both series and parallel resonance?

(a) Current is maximum (b) Power is low
(c) Impedance is minimum (d) Power factor is unity

194. Which of the following formulae is used to express active power in a balanced 3-phase circuit?

(a) $V_L I_L \cos \phi$ (b) $\sqrt{3} V_L I_L \cos \phi$
(c) $V_{ph}, I_{ph} \cos \phi$ (d) $\sqrt{3} V_{ph} \cdot I_{ph} \cos \phi$

195. A 3-phase induction motor connected from a 3-phase, 50 Hz AC supply runs at 720 rpm and has 4% slip. The number of poles in the motor are

(a) 4 (b) 6 (c) 8 (d) 16

196. Among the following which conductor has highest conductivity?

(a) Cu (b) Al (c) Ag (d) Mg

197. A pure AC inductive circuit

(a) Consumes some power on average
(b) Does not consumers power
(c) Takes power from the line during some part of the cycle and then returns back during other part of the cycle
(d) None of these

198. Auto-transformer can do the following

(a) Stepup voltage (b) Stepdown voltage
(c) Both (a) and (b) (d) None of these

199. Power factor of the following circuit will be zero

(a) Resistive (b) Inductive
(c) Capacitive (d) Both (b) and (c)

200. An ideal voltage source should have

(a) Large value of e.m.f. (b) Small value of e.m.f.
(c) Zero source resistance (d) Infinite source resistance

201. The power measurement in balanced 3-phase circuit can be done by

(a) One wattmeter method only
(b) Two wattmeter method only
(c) Three wattmeter method only
(d) Only of the above

202. At resonance the power factor series RLC circuit would be

(a) 0 (b) 1 (c) – 1 (d) 1.1

203. A transformer transfers

(a) Voltage (b) Current
(c) Electrical power (d) Frequency

204. A transformer can be connected be DC

(a) True (b) False (c) Can't say (d) None of these

205. Slip rings are made of

(*a*) Aluminum (*b*) Iron (*c*) Cast steel (*d*) Carbon

206. The form factor of a sinusoidal alternating current is

(*a*) 1 (*b*) 0 (*c*) 1.11 (*d*) 1.15

207. The efficiency of three phase induction motor, as compared to other three phase motors, is

(*a*) Low (*b*) High (*c*) Equal (*d*) None of these

208. The torque developed in an three phase induction motor is nearly proportional to

(*a*) 1/V (*b*) V (*c*) V^2 (*d*) None of these

209. The motor which can be used as power factor improving devices is

(*a*) Induction motor (*b*) Synchronous motor

(*c*) Stepper motor (*d*) Universal motor

210. Ceiling fan is

(*a*) Three phase induction motor

(*b*) Single phase induction motor

(*c*) Single phase synchronous motor

(*d*) Three phase synchronous motor

211. In the circuit shown in Fig. 14 the value of current in 20 Ω resistance will be

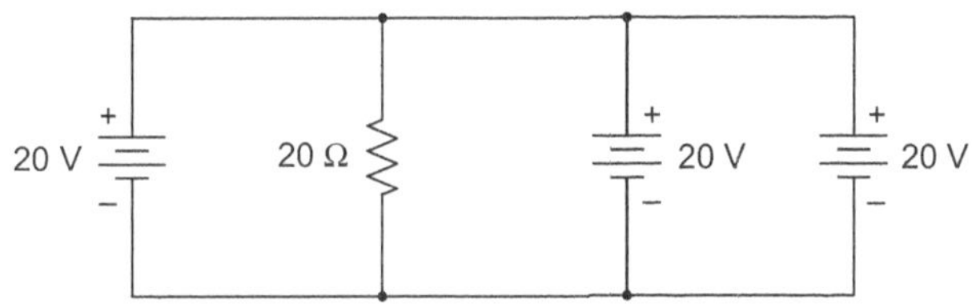

Fig. 14

(*a*) 0 (*b*) 1/3 (*c*) 1 (*d*) 3

212. An AC current is given by $i = 200 \sin \omega t$. It will attain a value of 100 A after

(*a*) 1.67 sec (*b*) 1.67 min (*c*) 1.67 m sec (*d*) 1.67 μ sec

213. In a three phase balanced star connected load, neutral current is equal to

(*a*) Zero (*b*) Phase current

(*c*) Line current (*d*) Unpredictable

214. In the measurement of three phase power by two-wattmeter method, if the readings of two wattmeter are equal the power factor of the circuit is

(*a*) 1 (*b*) 0.5 (*c*) 0 (*d*) 0.866

215. The low voltage winding of the transformer is placed near to the core to

(*a*) Increase efficiency (*b*) Minimize losses

(*c*) Minimize insulation (*d*) All of the above

216. A shunt generator supplied 50 A at 500 V. If the armature and shunt field resistances are 0.2 Ω and 12.5 Ω respectively, then the armature current will be

(*a*) 4 A (*b*) 50 A (*c*) 54 A (*d*) 58 A

217. The starting torque of a three phase induction motor depends upon

(*a*) Rotor resistance (*b*) Supply voltage

(*c*) Both (*a*) and (*b*) (*d*) None of the above

218. By which of the following system electric power may be transmitted ?

(*a*) Over head system (*b*) Underground system

(*c*) Both (*a*) and (*b*) (*d*) None of the above

219. A synchronous motor

(*a*) Always rotates at synchronous speed

(*b*) May run below synchronous speed

(*c*) Always run below synchronous speed

(*d*) All of the above

ANSWERS TO OBJECTIVE TYPE QUESTIONS

1. (*c*)	**2.** (*a*)	**3.** (*d*)	**4.** (*b*)	**5.** (*b*)
6. (*a*)	**7.** (*a*)	**8.** (*d*)	**9.** (*b*)	**10.** (*d*)
11. (*d*)	**12.** (*a*)	**13.** (*b*)	**14.** (*a*)	**15.** (*b*)
16. (*a*)	**17.** (*c*)	**18.** (*a*)	**19.** (*a*)	**20.** (*a*)
21. (*a*)	**22.** (*b*)	**23.** (*a*)	**24.** (*d*)	**25.** (*a*)
26. (*a*)	**27.** (*b*)	**28.** (*b*)	**29.** (*b*)	**30.** (*c*)
31. (*b*)	**32.** (*c*)	**33.** (*b*)	**34.** (*a*)	**35.** (*c*)
36. (*c*)	**37.** (*a*)	**38.** (*d*)	**39.** (*c*)	**40.** (*d*)
41. (*b*)	**42.** (*d*)	**43.** (*c*)	**44.** (*d*)	**45.** (*d*)
46. (*a*)	**47.** (*a*)	**48.** (*b*)	**49.** (*d*)	**50.** (*a*)
51. (*d*)	**52.** (*a*)	**53.** (*c*)	**54.** (*a*)	**55.** (*d*)
56. (*c*)	**57.** (*a*)	**58.** (*c*)	**59.** (*b*)	**60.** (*b*)
61. (*d*)	**62.** (*b*)	**63.** (*c*)	**64.** (*b*)	**65.** (*d*)
66. (*d*)	**67.** (*c*)	**68.** (*a*)	**69.** (*a*)	**70.** (*a*)
71. (*b*)	**72.** (*b*)	**73.** (*b*)	**74.** (*a*)	**75.** (*a*)
76. (*c*)	**77.** (*b*)	**78.** (*c*)	**79.** (*b*)	**80.** (*d*)
81. (*c*)	**82.** (*a*)	**83.** (*c*)	**84.** (*b*)	**85.** (*c*)
86. (*d*)	**87.** (*d*)	**88.** (*b*)	**89.** (*a*)	**90.** (*d*)
91. (*c*)	**92.** (*d*)	**93.** (*c*)	**94.** (*d*)	**95.** (*a*)
96. (*c*)	**97.** (*d*)	**98.** (*c*)	**99.** (*b*)	**100.** (*b*)
101. (*a*)	**102.** (*b*)	**103.** (*c*)	**104.** (*a*)	**105.** (*c*)
106. (*d*)	**107.** (*b*)	**108.** (*c*)	**109.** (*b*)	**110.** (*b*)
111. (*d*)	**112.** (*c*)	**113.** (*c*)	**114.** (*d*)	**115.** (*d*)
116. (*b*)	**117.** (*a*)	**118.** (*c*)	**119.** (*b*)	**120.** (*c*)
121. (*a*)	**122.** (*a*)	**123.** (*a*)	**124.** (*d*)	**125.** (*d*)
126. (*a*)	**127.** (*d*)	**128.** (*a*)	**129.** (*b*)	**130.** (*c*)
131. (*a*)	**132.** (*c*)	**133.** (*c*)	**134.** (*b*)	**135.** (*c*)
136. (*c*)	**137.** (*a*)	**138.** (*d*)	**139.** (*d*)	**140.** (*a*)
141. (*a*)	**142.** (*b*)	**143.** (*d*)	**144.** (*a*)	**145.** (*d*)

146. (*a*)	**147.** (*c*)	**148.** (*a*)	**149.** (*a*)	**150.** (*b*)
151. (*c*)	**152.** (*c*)	**153.** (*b*)	**154.** (*d*)	**155.** (*b*)
156. (*a*)	**157.** (*b*)	**158.** (*d*)	**159.** (*d*)	**160.** (*d*)
161. (*c*)	**162.** (*b*)	**163.** (*b*)	**164.** (*b*)	**165.** (*a*)
166. (*b*)	**167.** (*b*)	**168.** (*d*)	**169.** (*b*)	**170.** (*a*)
171. (*b*)	**172.** (*b*)	**173.** (*b*)	**174.** (*c*)	**175.** (*d*)
176. (*b*)	**177.** (*c*)	**178.** (*d*)	**179.** (*d*)	**180.** (*b*)
181. (*a*)	**182.** (*c*)	**183.** (*d*)	**184.** (*d*)	**185.** (*d*)
186. (*d*)	**187.** (*a*)	**188.** (*c*)	**189.** (*a*)	**190.** (*c*)
191. (*b*)	**192.** (*d*)	**193.** (*d*)	**194.** (*b*)	**195.** (*c*)
196. (*c*)	**197.** (*c*)	**198.** (*c*)	**199.** (*d*)	**200.** (*c*)
201. (*d*)	**202.** (*b*)	**203.** (*c*)	**204.** (*b*)	**205.** (*c*)
206. (*c*)	**207.** (*b*)	**208.** (*c*)	**209.** (*b*)	**210.** (*b*)
211. (*a*)	**212.** (*c*)	**213.** (*a*)	**214.** (*a*)	**215.** (*d*)
216. (*c*)	**217.** (*c*)	**218.** (*c*)	**219.** (*a*)	

□ □ □

FILL IN THE BLANKS

1. The factors determine torque of a DC machine are ....
2. Iron losses in transformer consist of .... and ....
3. DC motor starter is used for ....
4. The e.m.f equation of transformer is ....
5. The resistance of field winding of a DC shunt motor is .... than the resistance of the armature winding.
6. The slip of an induction motor at synchronous speed will be ....
7. The condition for maximum torque in a 3-phase induction motor at standstill/starting condition is ....
8. The condition for maximum torque in a 3-phase induction motor at running condition is ....
9. When the load on a synchronous motor increase, its speed ....
10. Synchronous motor can be made to work on leading power factor by increasing its ....
11. The direction of rotation of a single phase induction motor can be reversed by interchanging ....
12. An electrical motor achieves energy conversion due to the presence of .... e.m.f induced in its armature conductors.
13. Prior to magnetic saturation, the armature torque of a series motor is .... proportional to its armature current.
14. In a balanced 3-phase system, the line voltages are .... degrees apart.
15. Two-wattmeter method can be used to find the .... of a balanced/unbalanced 3-phase load.
16. Whatever the load condition, the net flux passing through the core of a transformer is nearly the .... as at no load?
17. Short-circuit test of a transformer helps in finding its .... loss at full-load and at any desired load.
18. In an induction motor, the resulting flux revolves synchronously around the inside surface of the .... core.
19. For low values of slip, the torque-slip curve of an induction motor is ....
20. Vacuum cleaners usually use .... motors.
21. A synchronous motor is electrically identical with ....
22. As the load on a synchronous motor is increased, its rotor falls back in ....
23. The function of a commutator in a DC motor is to make its armature torque ....
24. Whereas shunt motor is called an approximately constant speed motor, a series motor is called a .... speed motor.
25. As controller resistance of a DC motor is decreased, its speed is ....
26. In a 3-phase, 4-wire system supplying balance load, current in the neutral is ....
27. Sum of the two wattmeter reading gives the .... power consumption in the 3-phase balanced/unbalanced load.
28. A transformer works on the principle of .... induction.

29. Within a load of leading power factor, the full-load secondary voltage of a transformer is .... than it's no load voltage.
30. The rotor on an induction motor always revolves in the .... direction as the stator flux.
31. The slip of induction motor also represents the ratio of rotor Cu loss to the .... input.
32. An over excited synchronous motor acts as a ....
33. Stator Core is laminated to reduce .... losses.
34. Synchronous motor is .... starting motor.
35. Synchronous speed is given by the equation ....
36. In a DC shunt motor shunt field copper losses are practically ....
37. Commutator of a DC machine is made of carbon (true or false)
38. DC .... motors are best suited for electric traction.
39. Transformer core is made of silicon steel to reduce ....
40. Iron losses in a transformer can be determined by .... test.
41. Transformer can be used on DC (true or false)
42. The relationship between phase current and line current in a 3-phase star connections is ....
43. A three phase star connected system has 220 V between its phases. The line voltage is ....
44. The minimum number of wattmeter required to measure power in balanced three phase system is/are ....
45. A four pole three phase induction motor is supplied with 400 V, 50 Hz supply. The slip if it runs at 1440 rpm will be ....
46. The factors on which the starting torque of an induction motor depends are ....
47. The expression of back e.m.f induced in a DC motor is ....
48. Transformation ratio of a transformer is defined as ....
49. The transformer is rated in kVA because ....
50. The speed of a DC shunt motor may be increases by .... its shunt field flux.
51. Two-wattmeter method can be used to find the .... factor of a balanced 3-phase load.
52. A sinusoidal current having rms value of $8 \angle 0°$ A is added to another sinusoidal current of rms value $6 \angle 90°$ A. The rms value of the resultant current is ....
53. A moving coil instrument gives full scale deflection with 20 mA. The resistance of the coil is 4 Ω. The value of series resistance needed for the instrument to read upto 30 V is ....
54. A 100 kVA single phase transformer operating at 0.9 power factor has 90% maximum efficiency. The iron losses will be ....
55. A 4 pole lap wound DC generator generates 200 V at 1000 rpm. If this generator is now wave wound and runs at 500 rpm, the generated voltage will be ....
56. .... motor has self loaded properties.
57. .... motor will be preferred for elevators.
58. If the back emf of DC motor vanishes, then ....
59. The mass of electron is ....
60. While using a shunt resistance, for increasing the range of an ammeter, it should have .... resistance.

61. The shunt resistance used for increasing the range of an ammeter is usually made up of .... strips.
62. The resistance used for increasing the range of an ammeter is connected in .... to the meter.
63. The resistance used for increasing the range of an voltmeter is connected in .... to the meter.
64. The series resistance used for increasing the range of an voltmeter is connected in .... resistance.

ANSWERS TO FILL IN THE BLANKS

1. (*a*) Flux (*b*) No. of conductors (*c*) Armature current (*d*) Speed (*e*) Number of poles.
2. Hysteresis and eddy current.
3. Reducing the starting current
4. $E_1 = 4.44 f N_1 \phi_m$ volts
5. Larger/Greater
6. Zero
7. $R_2 = X_2$
8. $R_2 = sX_2$
9. Remain constant
10. Excitation
11. The negative field
12. Dynamically
13. Directly
14. 120
15. Power
16. Constant
17. Copper
18. Stator
19. Linear
20. 1-phase induction motor
21. Alternator
22. Opposite direction
23. Unidirectional
24. Variable
25. Increases
26. Zero
27. Total
28. Faraday's Law of electromagnetic.
29. Greater
30. Same

31. Rotor
32. Synchronous condenser
33. Iron/core
34. Not self
35. $N_s = \frac{120f}{P}$
36. Constant
37. False
38. Series
39. Hysteresis
40. Open circuit test
41. False
42. $I_L = I_{ph}$
43. $V_L = \sqrt{3}\,V_{ph}$, $V_L = \sqrt{3} \times 220 = 381$ V
44. two
45. $N_s = \frac{120f}{P} = \frac{120 \times 50}{4} = 1500\,\text{rpm},\ s = \frac{N_s - N_r}{N_s} = \frac{1500 - 1440}{1500} = 0.04$
46. $T_{st} = \frac{3}{2\pi N_s}\frac{E_2^2 R_2}{R_2^2 + X_2^2}$
47. $E_b = \frac{\phi ZNP}{60\,A}$
48. Transformation ratio of a transformer is defined as the ratio of secondary number of turns to the primary number of turns. It is denoted by K.
49. Transformers are rated in kVA, because losses in transformer depend upon the voltage and current. Iron losses depend on voltage and copper losses depend upon current. Hence they are rated in kVA. Actual power in kW depends upon the load power factor.
50. Decreasing
51. Power factor
52. $10 \angle 36.87°$ A
53. 1496 Ω
54. 5000 W or 5 kW
55. 200 V
56. DC shunt
57. DC series
58. Armature current becomes very high which is dangerous for the machine.
59. 9.109382×10^{-31} kg
60. Low
61. Magnanin

62. Parallel or shunt
63. Series
64. High

MATCH THE COLUMNS

1. In case of DC circuits,

(a) Linear circuit	(i) Has no source of e.m.f
(b) Bilateral circuit	(ii) Constant parameters
(c) Passive circuit	(iii) Same properties in both direction

2. In case of DC circuits,

(a) Non-linear circuit	(i) Parameters changes with voltage or current
(b) Unilateral circuit	(ii) Contains one or more than one e.m.f. source
(c) Active circuit	(iii) Properties changes with direction of operation

3. In case of DC circuits,

(a) Lumped parameters	(i) Which follow ohm's law
(b) Distributed parameters	(ii) Can be separate out from a given circuit
(c) Linear parameters	(iii) Cannot be separate out from a given circuit
(d) Non-linear parameters	(iv) Which no not follow ohms law

4. In case of DC circuits,

(a) Node	(i) Lies between two junctions
(b) Branch	(ii) Closed path with no recitation
(c) Loop	(iii) Where current divides in different branches
(d) Mesh	(iv) Contains no other loop.

5. In case of DC circuits,

(a) Voltage source	(i) Infinite internal resistance
(b) Current source	(ii) Zero internal resistance
(c) Ideal current source	(iii) Very low internal resistance
(d) Ideal voltage source	(iv) Very high internal resistance

6. In case of DC network theorems,

(a) Norton's theorem	(i) Open circuit voltage
(b) Thevenin's theorem	(ii) Algebric sum of all the currents
(c) Super position theorem	(iii) Short-circuit current
(d) Maximum power transfer theorem	(iv) Load resistance is equal to internal resistance

7. In case of single phase AC circuits,

(a) Average value (I_{av})	(i) 1.11
(b) r.m.s. value (I_{rms})	(ii) 0.637 Im
(c) Form factor (K_f)	(iii) 1.414
(d) Peak factor (K_p)	(iv) 0.707 Im

8. In case of single phase AC circuits, match various powers with their units,

(*a*) Apperent power	(*i*) kW
(*b*) Active power	(*ii*) kVA
(*c*) Reactive power	(*iii*) kVAR

9. In case of single phase AC circuits,

(*a*) RL AC circuit	(*i*) Voltage lags behind the current
(*b*) RC AC circuit	(*ii*) Current lags behind the voltage
(*c*) Pure resistive AC circuit	(*iii*) Current and voltage are in same phase
(*d*) Pure capactive AC circuit	(*iv*) Current leads the voltage by 90°

10. The three powers in AC circuit are defined as,

(*a*) Active power	(*i*) Total power
(*b*) Reactive power	(*ii*) Working power which performs usefull work
(*c*) Apperent power	(*iii*) Which produces magnetizing flux or non-working power

11. In case of tesonance in AC circuits,

(*a*) Series resonance	(*i*) Current becomes maximum
	(*ii*) Current becomes minimum
	(*iii*) Power factor becomes unity
(*b*) Parallel resonance	(*iv*) Impedance becomes minimum
	(*v*) Impedance becomes maximum
	(*vi*) Also known as voltage resonance
	(*vii*) Also known as current resonance

12. Match the circuit elements with their respective type of graph in series RLC circuit at resonance,

(*a*) Resistance	(*i*) Bell shape curve
(*b*) Inductive reactance	(*ii*) Straight line parallel to X-axis
(*c*) Capacitive reactance	(*iii*) Inverted bell shape curve
(*d*) Current	(*iv*) Rectangular hyperbola
(*e*) Impedance	(*v*) Straight line passing through origin

13. Match the following w.r.t. series RLC circuit,

(*a*) Bandwidth	(*i*) Frequency at which current becomes 70.7% of maximum value for first time.
(*b*) Lower cut-off frequency	(*ii*) Range of frequencies over which current is equal or greater than 70.7% of maximum value.
(*c*) Upper cut-off frequency	(*iii*) Voltage magnification across L or C
(*d*) Quality factor	(*iv*) Frequency at which current becomes 70.7% of maximum value for second time

14. Match the circuit elements with their respective type of graph in parallel RLC circuit at resonance,

(*a*) Inductive susceptance	(*i*) Straight line passing through origin
(*b*) Capacitance susceptance	(*ii*) Inverted bell shape curve
(*c*) Current	(*iii*) Bell shape curve

(*d*) Circuit impedance	(*iv*) Straight line parallel to x-axis
(*e*) Conductance	(*v*) Rectangular hyperbola

15. In case of magnetic circuits,

(*a*) Permeability	(*i*) Any physical cause that produces magnetic flux.
(*b*) Magnetic flux density	(*ii*) Degree of magnetization
(*c*) Magnetic field intensity	(*iii*) Amount of magnetic flux per unit area
(*d*) Magnetomotive force	(*iv*) Amount of magnetization force

16. In case of magnetic circuits,

(*a*) Reluctance	(*i*) Ability of a magnetic circuit to conduct magnetic flux.
(*b*) Permeance	(*ii*) Bulging out of flux in air gap.
(*c*) Fringing	(*iii*) Property which offers opposition to the flow of magnetic flux.

17. While doing comparison between electrical circuits and magnetic circuits, match the following:

(*a*) Magnetic circuit	(*i*) Reluctance
	(*ii*) Resistance'
	(*iii*) Flux density
(*b*) Electric circuit	(*iv*) Current density
	(*v*) Permeance
	(*vi*) Conductance
	(*vii*) m.m.f.
	(*viii*) e.m.f.

18. In case of measuring instruments,

(*a*) Deflecting torque	(*i*) Spring control
(*b*) Controlling torque	(*ii*) Air friction
(*c*) Damping torque	(*iii*) Magnetic effect
	(*iv*) Gravity control
	(*v*) Electrodynamic effect
	(*vi*) Eddy current
	(*vii*) Hydrolic

19. In case of measuring instruments,

(*a*) Moving coil instruments	(*i*) Attraction type
	(*ii*) Permanent magnet type
(*b*) Moving iron instruments	(*iii*) Dynamometer type
	(*iv*) Repulsion type

20. In case of electrical machines,

(*a*) Iron loss	(*i*) I^2R loss
	(*ii*) Friction loss
(*b*) Copper loss	(*iii*) Hysteresis loss
	(*iv*) Eddy current loss.
(*c*) Mechanical losses	

21. In case of single phase transformer, match the following:

(*a*) Open circuit test	(*i*) R_0
	(*ii*) R_{01}
(*b*) Short circuit test	(*iii*) X_0
	(*iv*) X_{01}

22. In case of DC machine, match the following:

(*a*) Stator	(*i*) Yoke
	(*ii*) Armature
(*b*) Rotor	(*iii*) Field poles
	(*iv*) Commutator
	(*v*) Interpoles

23. In case of DC machines,

(*a*) Lap winding	(*i*) No. of parallel paths-2
(*b*) Wave winding	(*ii*) Flux left over inside the field
(*c*) Armature reaction	(*iii*) A = No. of poles
(*d*) Residual magnetism	(*iv*) De-magnetize and cross-magnetize the main flux.

24. In case of DC motors,

(*a*) DC shunt motor	(*i*) Sewing machine
(*b*) DC series motor	(*ii*) Lifts
(*c*) DC compound motor	(*iii*) Wood working machines
	(*iv*) Vacuum cleaner
	(*v*) Printing press
	(*vi*) Rolling machines

25. In case of torque-slip characteristics of 3-phase induction motor, match the following:

(*a*) Motoring mode	(*i*) $1 < s < 2$
(*b*) Generating mode	(*ii*) $0.5 < s < 0$
(*c*) Breaking mode	(*iii*) $1 < s < 0$
(*d*) Operating point	(*iv*) $0 < s < -1$

26. In case of starting methods of 3-phase induction motor, match the following:

(*a*) Direct online starting	(*i*) Starting torque is reduced by x^2 times as compared to DOL method
(*b*) Auto-transformer starting	(*ii*) Usefull for small motors only
(*c*) Star-delta starting	(*iii*) Applied voltage can be reduced by *x* fraction.
(*d*) Stator resistance starting	(*iv*) Starting torque is reduced by $\frac{1}{3}$rd as compared to DOL method

27. Match the motors with their applications,

(*a*) Single-phase induction motor	(*i*) Water pumps
(*b*) Three-phase induction motor	(*ii*) Washing machine

(*c*) Synchronous motor	(*iii*) Fan
	(*iv*) Power factor correction
	(*v*) Drilling machine
	(*vi*) Crushers

28. In case of power measurement by two-wattmeter methods, match the following:

Power factor	**Wattmeter readings**
(*a*) Unity	(*i*) One negative and one positive
(*b*) Equal to 0.5	(*ii*) Equal but opposite
(*c*) Less than 0.5	(*iii*) Same and positive
(*d*) Zero	(*iv*) One positive and other zero

29. In case of three-phase AC circuits,

(*a*) Line voltage (Δ connection)	(*i*) $\sqrt{3}$ Phase voltage
(*b*) Line current (Δ connection)	(*ii*) Phase voltage
(*c*) Line voltage (Y connection)	(*iii*) Phase current
(*d*) Line current (Y connection)	(*iv*) $\sqrt{3}$ Phase current

ANSWERS TO MATCH THE COLUMNS

1. (*a*) (*ii*) (*b*) (*iii*) (*c*) (*i*)

2. (*a*) (*i*) (*b*) (*iii*) (*c*) (*ii*)

3. (*a*) (*ii*) (*b*) (*iii*) (*c*) (*i*) (*d*) (*iv*)

4. (*a*) (*iii*) (*b*) (*i*) (*c*) (*ii*) (*d*) (*iv*)

5. (*a*) (*iii*) (*b*) (*iv*) (*c*) (*i*) (*d*) (*ii*)

6. (*a*) (*iii*) (*b*) (*i*) (*c*) (*ii*) (*d*) (*iv*)

7. (*a*) (*iii*) (*b*) (*i*) (*c*) (*ii*) (*d*) (*iv*)

8. (*a*) (*ii*) (*b*) (*i*) (*c*) (*iii*)

9. (*a*) (*ii*) (*b*) (*i*) (*c*) (*iii*) (*d*) (*iv*)

10. (*a*) (*ii*) (*b*) (*iii*) (*c*) (*i*)

11. (*a*) (*i*) (*iii*) (*iv*) (*vi*) (*b*) (*ii*) (*iii*) (*v*) (*vii*)

12. (*a*) (*ii*) (*b*) (*v*) (*c*) (*iv*) (*d*) (*i*) (*e*) (*iii*)

13. (*a*) (*ii*) (*b*) (*i*) (*c*) (*iv*) (*d*) (*iii*)

14. (*a*) (*v*) (*b*) (*i*) (*c*) (*ii*) (*d*) (*iii*) (*e*) (*iv*)

15. (*a*) (*ii*) (*b*) (*iii*) (*c*) (*iv*) (*d*) (*i*)

16. (*a*) (*iii*) (*b*) (*i*) (*c*) (*ii*)

17. (*a*) (*i*) (*iii*) (*v*) (*vii*) (*b*) (*ii*) (*iv*) (*vi*) (*viii*)

18. (*a*) (*iii*) (*v*) (*b*) (*i*) (*iv*) (*c*) (*ii*) (*vi*) (*vii*)

19. (*a*) (*ii*) (*iii*) (*b*) (*i*) (*iv*)

20. (*a*) (*iii*) (*iv*) (*b*) (*i*) (*c*) (*ii*)

21. (*a*) (*i*) (*iii*)
(*b*) (*ii*) (*iv*)

22. (*a*) (*i*) (*iii*) (*v*)
(*b*) (*ii*) (*iv*)

23. (*a*) (*iii*)
(*b*) (*i*)
(*c*) (*iv*)
(*d*) (*ii*)

24. (*a*) (*i*)
(*b*) (*ii*)
(*c*) (*vi*)

25. (*a*) (*iii*)
(*b*) (*iv*)
(*c*) (*i*)
(*d*) (*ii*)

26. (*a*) (*ii*)
(*b*) (*i*)
(*c*) (*iv*)
(*d*) (*iii*)

27. (*a*) (*ii*) (*iii*) (*v*)
(*b*) (*i*) (*vi*)
(*c*) (*iv*)

28. (*a*) (*iii*)
(*b*) (*iv*)
(*c*) (*i*)
(*d*) (*ii*)

29. (*a*) (*ii*)
(*b*) (*iv*)
(*c*) (*i*)
(*d*) (*iii*)

References

1. McPherson G, and Laramore R D, *"Introduction to Electrical Machines and Transformers"*, john Wiley & Sons, 1990.
2. Bhattacharya SK, *"Electrical Machines"*, Tata McGraw Hill, New Delhi.
3. Nagrath I J and Kothari D P, *"Electrical Machines"*, Tata McGraw Hill, New Delhi.
4. Gupta S B, *"Electrical Machines"*, S K Kataria & Sons, New Delhi
5. Fitzerald A E, C Kingsley and S D Umans, *"Electric Machinery"*, 4th edition, McGraw Hill, New York.
6. Draper A, *"Electric Machines"*, 2nd edition, McGraw Hill, New York, 1967.
7. Siskind C S, *"Electric Machines, Direct and Alternating Current"*, 2nd edition, McGraw Hill, New York, 1950.
8. Gupta B R & Singhal Vandana, *"Fundamental of Electrical Machines"*, New Age International Publication,
9. Ghosh Smarajit, *"Electrical Machines"*, Pearson Education, Delhi, 2005.
10. Sharma Sanjeev, *"Basic Electrical Engineering"*, I K International Publication, 3rd edition, 2008.
11. McPherson G and Laramore R D, *"Introduction to Electrical Machines and Transformers"*, John Wiley & Sons, 1990.
12. A.E. Fitzgerald, Charles Kingsley, Jr. and Stephen D. Umans, *"Electric Machinery"* - McGraw Hill series in Electrical Engineering, Sixth edition, 2005.
13. D P Kothari and I J Nagrath, *"Basic Electrical Engineering"*, Tata McGraw Hill Publishing Company, Second Edition, 2008.
14. J B Gupta, *"A Textbook of Electrical Engineering"*, S K Kataria & Sons, Sixth Edition, 2008.
15. Ashfaq Husain, *"Fundamentals of Electrical Engineering"*, Dhanpat Rai & Co. Ltd., Third Edition, 2002.
16. S K Sahdev, *"A Text Book of Electrical Engineering"*, Uneek Publications, First Edition, 2008.
17. B L Theraja and A K Theraja, *"A Textbook of Electrical Technology Vol-I"*, S Chand & Company, First Multicolour Edition, 2005.
18. B L Theraja and A K Theraja, *"A Textbook of Electrical Technology Vol-II"*, S Chand & Company, First Multicolour Edition, 2005.
19. Jimmie J Cathey, Syed A Nasar and Prabhat Kumar, *"Basic Electrical Engineering"*, Tata McGraw Hill Publishing Company, Second Edition, 2006.
20. V N Mittle and Arvind Mittal, *"Basic Electrical Engineering"*, Tata McGraw Hill Publishing Company, Second Edition, 2006.
21. Vincent Del Toro, *"Electrical Engineering Fundamentals"*, Prentice-Hall of India, Second Edition, 2006.
22. http://powerelectrical.blogsspot.com.
23. http://en.wikipedia.org.

24. http://www.allaboutcircuits.com.
25. http://www.magnet.fsu.edu.
26. http://www.physics.uoguelph.ca.
27. http://www.pumped101.com.
28. http://lancet.mit.edu.
29. http://science.nasa.gov.
30. www.nrel.gov.
31. www.dur.ac.uk.
32. www.epsrc.ac.uk.
33. www.pvnet.org.uk.
34. http://ec.europa.eu.
35. http://www.electronicsaera.com.
36. http://www.3phasepower.org.
37. http://www.vocw.edu,vn.
38. http://www.elec-toolbog.com.
39. http://www.federalpacific.com.
40. http://www.axeonpower.com.
41. http://www.books.google.co.in.
42. http://www.toolingu.com.
43. http://en.wikipedia.org.
44. M N O Sadiku (2007). *Elements of Electromagnetics*, Fourth Edition, NY/Oxford UK: Oxford University Press.
45. Tai L. Chow (2006). *Electromagnetic Theory*. Sudbury MA: Jones and Bartlett, ISBN 0-7637-3827-1.
46. Griffiths *Introduction to Electrodynamics*, or Feynman Lectures on Physics, C.
47. Roger F Harrington (2003). *Introduction to electromagnetic engineering*. Mineola, NY: Dover Publications, ISBN 0486432416.
48. Peter Alan Davidson (2001). *An Introduction to Magnetohydrodynamics*. Cambridge UK: Cambridge University Press, ISBN 0521794870.
49. Richard Phillips Feynman, Leighton R B & Sands M L (2006). The Feynman Lectures on Physics. San Francisco: Pearson/Addison-Wesley, Vol. II, ISBN 0805390499.
50. Griffiths, David J. (1999). *Introduction to Electrodynamics*, Third Edition, Upper Saddle River NJ: Prentice Hall, pp.301-3. ISBN 0-13-805326-X.
51. BB Laud (1987). *Electromagnetics*. New Delhi: New Age International, ISBN 0852264992.
52. L. Pearce Williams (1966). *The Origins of Field Theory*. Random House,
53. Ulaby, Fawwaz (2007). *Fundamentals of applied electromagnetics*, 5th Edition, Pearson: Prentice Hall, ISBN 0-13-241326-4.
54. James Clerk Maxwell (1881). *A treatise on electricity and magnetism v. 2*. Oxford UK: Clarendon Press, ISBN 0486606376.
55. Griffiths, David J. (1999). *Introduction to Electrodynamics*, Third Edition, Upper Saddle River NJ: Prentice Hall, ISBN 0-13-805326-X.

□ □ □

INDEX

S

T

U

V

W

Y